EARTH SCIENCE

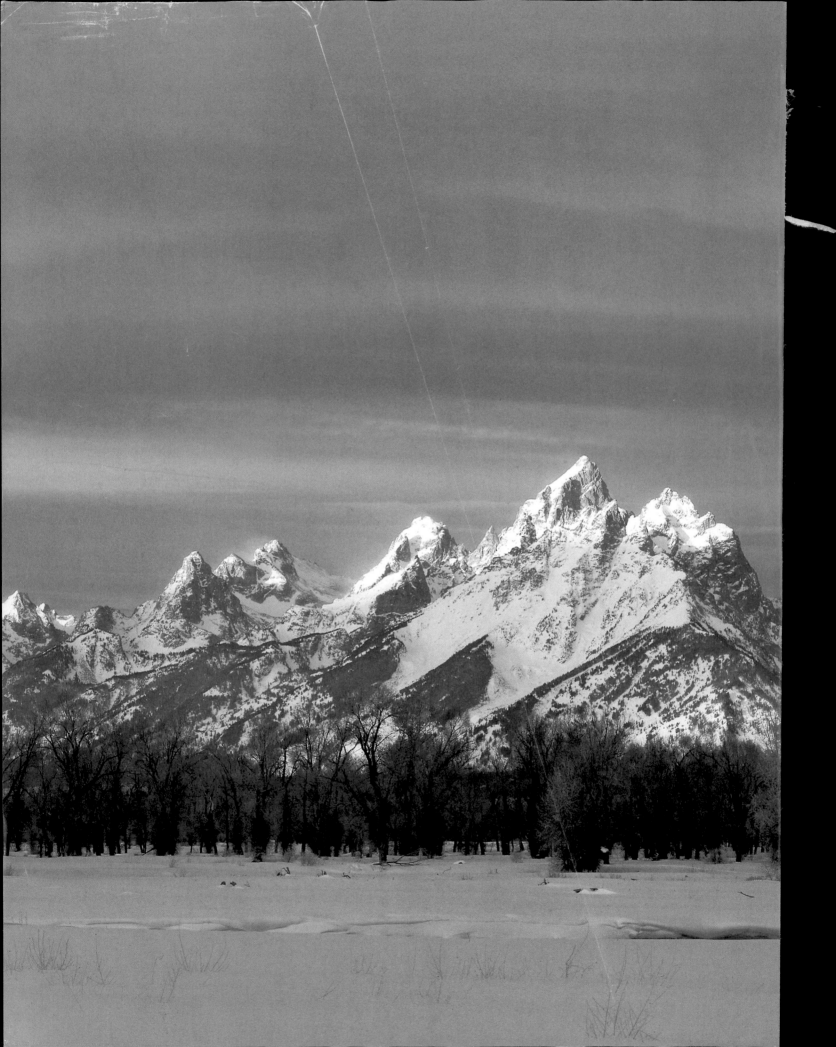

EARTH SCIENCE

TENTH EDITION

Edward J. Tarbuck
Frederick K. Lutgens

Both Illinois Central College, Emeritus

Illustrated by
Dennis Tasa

PEARSON EDUCATION, INC.
Upper Saddle River, New Jersey 07458

Library of Congress Cataloging-in-Publication Data

Tarbuck, Edward J.
   Earth science / Edward J. Tarbuck, Frederick K. Lutgens; illustrated by Dennis Tasa.—10th ed.
     p. cm.
   Includes index.
   ISBN 0-13-035390-6
   1. Earth sciences.   I. Lutgens, Frederick K. II. Title.
QE26.2 .T38 2003
550—dc21

2002070008

Senior Editor: Patrick Lynch
Editor-in-Chief, Life and Geosciences: Sheri L. Snavely
Vice President of Production and Manufacturing: David W. Riccardi
Executive Managing Editor: Kathleen Schiaparelli
Assistant Managing Editor: Beth Sweeten
Production Editor: Edward Thomas
Assistant Managing Editor, Science Media: Nicole Bush
Media Editor: Chris Rapp
Assistant Editor: Melanie Van Benthuysen
Senior Marketing Manager: Christine Henry
Manufacturing Manager: Trudy Pisciotti
Assistant Manufacturing Manager: Michael Bell
Art Editor: Adam Velthaus
Director of Creative Services: Paul Belfanti
Director of Design: Carole Anson
Art Director: Maureen Eide
Assistant to Art Director: Geoffrey Cassar
Cover Design: Daniel Conte
Interior Design: Thomas Nery/Geoffrey Cassar
Managing Editor, Audio/Video Assets: Patty Burns
Editorial Assistant: Sean Hale
Director, Image Resource Center: Melinda Reo
Manager, Rights and Permissions: Zina Arabia
Interior Image Specialist: Beth Boyd-Brenzel
Cover Image Specialist: Karen Sanatar
Photo Researcher: Yvonne Gerin
Copy Editor: Barbara Booth
Image Permission Coordinator: Debbie Hewitson
Color Scanning Supervisor: Joseph Conti
Production Assistant: Nancy Bauer
Composition: Pine Tree Composition
Cover Photo: Hoodoo at Bryce Canyon National Park (photo by Carr Clifton).
Title Page Photo: Grand Teton National Park (photo by Carr Clifton).

ISBN 0-13-035390-6 (college edition)
ISBN 0-13-048456-3 (school edition)

Pearson Education Ltd., *London*
Pearson Education Australia Pty., Limited, *Sydney*
Pearson Education Singapore, Pte. Ltd
Pearson Education North Asia Ltd., *Hong Kong*
Pearson Education Canada, Ltd., *Toronto*
Pearson Educacion de Mexico, S.A. de C.V.
Pearson Education—Japan, *Tokyo*
Pearson Education Malaysia, Pte. Ltd.

BRIEF CONTENTS

GEODe: EARTH SCIENCE

A copy of the *GEODe: Earth Science* CD-ROM is packaged with each copy of *Earth Science,* Tenth Edition. This dynamic learning aid reinforces key Earth science concepts by using tutorials, animations and interactive exercises.

 This *GEODe: Earth Science* icon appears throughout the book wherever a text discussion has a corresponding activity on the CD-ROM.

CONTENTS

UNIT ONE
Earth Materials   19

1   Minerals: Building Blocks of Rocks   19

2   Rocks: Materials of the Solid Earth   39

UNIT TWO
Sculpturing Earth's Surface   67

3   Weathering, Soil, and Mass Wasting   67

4   Running Water and Groundwater 97

5   Glaciers, Deserts, and Wind 129

UNIT THREE
Forces Within 163

6   Earthquakes and Earth's Interior 163

7 Plate Tectonics 191

8 Volcanoes and Other Igneous Activity 225

9 Mountain Building 259

UNIT SEVEN
Earth's Place in the Universe 549

21 Touring Our Solar System   575

22 Light, Astronomical Observations, and the Sun   603

23 Beyond Our Solar System   623

PREFACE

*E*arth Science Tenth Edition, like its predecessors, is a college-level text designed for an introductory course that often has the same name as this text. It consists of seven units that emphasize broad and up-to-date coverage of basic topics and principles in geology, oceanography, meteorology, and astronomy. The book is intended to be a meaningful, nontechnical survey for undergraduate students with little background in science. Usually these students are taking an Earth science class to meet a portion of their college's or university's general requirements.

In addition to being informative and up-to-date, a major goal of *Earth Science* is to meet the need of beginning students for a readable and user-friendly text, a book that is a highly usable "tool" for learning basic Earth science principles and concepts.

Distinguishing Features

Readability

The language of this book is straightforward and *written to be understood*. Clear, readable discussions with a minimum of technical language are the rule. The frequent headings and subheadings help students follow discussions and identify the important ideas presented in each chapter. In the Tenth Edition, improved readability was achieved by examining chapter organization and flow, and writing in a more personal style. Large portions of the text were substantially rewritten in an effort to make the material more understandable.

Focus on Learning

When a chapter has been completed, three useful devices help students review. First, the Chapter Summary recaps all of the major points. Next is a checklist of Key Terms with page references. Learning the language of Earth science helps students learn the material. This is followed by Review Questions that help students examine their knowledge of significant facts and ideas. Each chapter closes with a reminder to visit the Website for *Earth Science*, Tenth Edition (http://www.prenhall.com/tarbuck). It contains many excellent opportunities for review and exploration.

Illustrations and Photographs

The Earth sciences are highly visual. Therefore, photographs and artwork are a very important part of an introductory book. *Earth Science*, Tenth Edition, contains dozens of new high-quality photographs that were carefully selected to aid understanding, add realism, and heighten the interest of the reader.

The illustrations in each new edition of *Earth Science* keep getting better and better. In the Tenth Edition more than 100 pieces of line art are new or redesigned. The new art illustrates ideas and concepts more clearly and realistically than ever before. The art program was carried out by Dennis Tasa, a gifted artist and respected Earth science illustrator.

Focus on Basic Principles and Instructor Flexibility

Although many topical issues are treated in *Earth Science*, Tenth Edition, it should be emphasized that the main focus of this new edition remains the same as its predecessors—to foster student understanding of basic Earth science principles. Whereas student use of the text is a primary concern, the book's adaptability to the needs and desires of the instructor is equally important. Realizing the broad diversity of Earth science courses in both content and approach, we have continued to use a relatively nonintegrated format to allow maximum flexibility for the instructor. Each of the major units stands alone; hence, they can be taught in any order. A unit can be omitted entirely without appreciable loss of continuity, and portions of some chapters may be interchanged or excluded at the instructor's discretion.

Three Important Themes

The newly revised and expanded "Introduction to Earth Science" that opens the text acquaints students with three important themes that recur throughout the book—*Earth as a System, People and the Environment,* and *Understanding Earth.*

Earth as a System

 An important occurrence in modern science has been the realization that Earth is a giant multi-dimensional system. Our planet consists of many separate but interacting parts. A change in any one part can produce changes in any or all of the other parts—often in ways that are neither obvious nor immediately apparent. Although it is not possible to study the entire system at once, it is possible to develop an awareness and appreciation for the concept and for many of the system's important interrelationships. Therefore, starting with the new section on "Earth System Science" in the Introduction, the theme of "Earth as a System" keeps recurring through all major units of the book. It is a thread that "weaves" through the chapters and helps tie them together.

Several new and revised special interest boxes relate to *Earth as a system*. To remind the reader of this important theme, the small icon you see at the beginning of this section is used to mark these boxes. Finally, each chapter concludes with a section on *Examining the Earth*

System. The questions and problems found here are intended to develop an awareness and appreciation for some of the Earth system's many interrelationships.

People and the Environment

 Because knowledge about our planet and how it works is necessary to our survival and well being, the treatment of environmental issues has always been an important part of *Earth Science.* Such discussions serve to illustrate the relevance and application of Earth science knowledge. With each new edition this focus has been given greater emphasis. This is certainly the case with the Tenth Edition. The text integrates a great deal of information about the relationship between people and the natural environment and explores the application of the Earth sciences to understanding and solving problems that arise from these interactions.

In addition to many basic text discussions, 22 of the text's special interest boxes involve the "People and the Environment" theme and are quickly recognized by the distinctive icon you see at the beginning of this section.

Understanding Earth

 As members of a modern society, we are constantly reminded of the benefits derived from science. But what exactly is the nature of scientific inquiry? Developing an understanding of how science is done and how scientists work is a third important theme that appears throughout this book beginning with the section on "The Nature of Scientific Inquiry" in the Introduction. Students will examine some of the difficulties encountered by scientists as they attempt to acquire reliable data about our planet and some of the ingenious methods that have been developed to overcome these difficulties. Students will also explore many examples of how hypotheses are formulated and tested as well as learn about the evolution and development of some major scientific theories. Many basic text discussions as well as a number of the special interest boxes on "Understanding Earth" provide the reader with a sense of the observational techniques and reasoning processes involved in developing scientific knowledge. The emphasis is not just on what scientists know, but how they figured it out.

Highlights of the Tenth Edition

The Tenth Edition of *Earth Science* represent a thorough revision. *Every* part of the book was examined carefully with the dual goals of keeping topics current and improving the clarity of text discussions. People familiar with preceding editions will see much that is new in the Tenth Edition. Among the most obvious changes are those highlighted below.

Expanded Oceanography Coverage

In order to provide more balanced coverage, the oceanography unit now consists of three chapters instead of two: Chapter 12, "The Ocean Floor"; Chapter 13, "Ocean Water and Ocean Life"; and Chapter 14, "The Dynamic Ocean." This change does not represent a 50 percent expansion over the previous edition. Rather it involves the addition of some new material and a modest rearrangement of topics. In addition to broader coverage, this expanded unit also has a greater emphasis on how scientists learn about the ocean. Professor Alan Trujillo of Palomar College is largely responsible for revising Chapters 12 and 14 and for preparing Chapter 13, which has a high percentage of new material.

New CD-ROM

 Each copy of *Earth Science,* Tenth Edition, comes with a new interactive CD-ROM: *GEODe: Earth Science.* In addition to a revised treatment of topics in geology, the new CD-ROM now includes broad coverage of the oceans, basic meteorology, and the solar system. This greatly expanded coverage includes many new tutorials, interactive exercises, animations, and video clips. Every unit in the text now has a corresponding unit in *GEODe: Earth Science.* A special *GEODe: Earth Science* icon appears throughout the book wherever a text discussion has a corresponding *GEODe: Earth Science* activity.

Other Highlights

- A new feature, "Students Sometimes Ask . . ." is now part of every chapter. A few questions and answers are distributed at appropriate places in each chapter and identified by a large question mark. They are the kind of questions inquisitive students might ask during class sessions and are intended to add interest and relevance to text discussions. We are indebted to Professor Alan Trujillo at Palomar College for this excellent idea and for providing many of the questions and answers we used.

- Twenty-seven of the special interest boxes are new. Most are intended to illustrate and reinforce the three themes of "Earth as a System," "People and the Environment," and "Understanding Earth." The greatest number of new boxes (16) are devoted to the last of these.

- *Earth Science,* Tenth Edition now consists of seven units rather than the four units of previous editions. Instead of having all of geology together in one unit, the material on geology is now divided into four parts: "Earth Materials," "Sculpturing Earth's Surface," "Forces Within," and "Deciphering Earth's History." This change *does not* represent a reorganization or significant expansion of the geology coverage. Rather this change was done so that the units in the text and in the new CD-ROM would be compatible.

The Teaching and Learning Package

The authors and publisher have been pleased to work with a number of talented people who produced an excellent supplements package. This package includes the traditional supplements that students and professors have come to expect from authors and publishers, as well as some new kinds of supplements that involve electronic media.

For the Student

GEODe: Earth Science CD-ROM. Each copy of *Earth Science* Tenth Edition comes with *GEODe: Earth Science* by Ed Tarbuck, Fred Lutgens, and Dennis Tasa of Tasa Graphic Arts, Inc. *GEODe: Earth Science* is a dynamic program that reinforces key concepts by using animations, tutorials, and interactive exercises. This *new product* includes much broader coverage than its predecessor. Each unit in *Earth Science,* Tenth Edition now has a corresponding unit in *GEODe: Earth Science.* A special *GEODe: Earth Science* icon appears throughout the book wherever a text discussion has a corresponding *GEODe: Earth Science* activity. This special offering gives the students two valuable products (*GEODe: Earth Science* and the textbook) for the price of one.

An Explorer's Guide to the Earth System. (ISBN 0-13-093335-X) Written by Ellen Metzger of San Jose State University, this is a new project designed for Earth science courses emphasizing an "Earth as a System" approach. This innovative workbook contains readings, questions, and exercises that will cultivate—through active learning and inquiry—a greater appreciation for the planet Earth and its inhabitants, and demonstrate how relevant Earth science is to our lives and communities. Five recurring themes—interactions of spheres, scale, cycles, energy, humans and the Earth system—are woven throughout the five modules: Introduction to the Earth System, Geosphere, Hydrosphere, Atmosphere, and Cosmosphere. *An Explorer's Guide* is written to accompany *Earth Science,* Tenth Edition, and the close integration makes it easy to incorporate into existing courses.

Companion Website. This on-line study guide, specific to the text, contains numerous review exercises (from which students get immediate feedback), exercises to expand one's understanding of Earth science, and resources for further exploration. This Website provides an excellent platform from which to start using the Internet for the study of Earth science. Please visit the site at **http://www.prenhall.com/tarbuck.**

Study Guide. (ISBN 0-13-035443-0) Written by experienced educator Kenneth Pinzke, the *Study Guide* helps students identify the important points from the text, and then provides them with review exercises, study questions, self-check exercises, and vocabulary review.

For the Professor

A wealth of valuable teaching tools are available.

Instructor's Manual. (ISBN 0-13-035435-X) Written by Kenneth Pinzke and Stanley Hatfield, both of Southwestern Illinois College, the *Instructor's Manual* is intended as a resource for both new and experienced instructors. It includes a variety of lecture outlines, additional source materials, teaching tips, advice about how to integrate visual supplements (including the Web-based resources and *GEODe: Earth Science*), and various other ideas for the classroom. Also included in every chapter are a wide variety of test questions.

Test Item File. Included in the Instructor's Manual, this resource averages more than 100 multiple-choice, true/false, and short-answer questions per chapter.

Transparencies. (ISBN 0-13-035428-7) Includes nearly 400 transparencies, including every Dennis Tasa illustration from the text.

Slides. (ISBN 0-13-035420-1) Includes 200 photos, illustrating many of the natural phenomena discussed in the text.

Digital Image Gallery for Interactive Teaching (DIGIT) CD-ROM. (ISBN 0-13-035431-7) All illustrations from the text, the slide-set images, animations from *GEODe: Earth Science,* and many text photos in high-resolution JPEG files that can be easily imported into PowerPoint®.

PowerPoint® lecture presentations. Ken Pinzke and Stan Hatfield of Southwestern Illinois College have created compelling, heavily illustrated *PowerPoint Lecture Presentations* for each chapter of this text. The presentations are fully customizable or suitable for use as provided. They can be found on the Digital Image Gallery (DIGIT) CD-ROM, in the appropriate chapters.

TestGen EQ CD-ROM. (ISBN 0-13-035432-5) This test generator is fully customizable and compatible with both Mac and Windows.

New York Times *Themes of the Times—Changing Earth.* This unique newspaper-format supplement features recent articles about Earth science from the pages of *The New York Times.* This supplement, available at no extra charge from your local Prentice Hall representative, encourages students to make connections between the classroom and the world around them.

For the Laboratory

Applications and Investigations in Earth Science, **Fourth Edition.** (ISBN 0-13-046095-8) Written by Ed

Tarbuck, Fred Lutgens, and Ken Pinzke, this full-color laboratory manual contains 22 exercises that provide students with hands-on experiences in geology, oceanography, meteorology, astronomy, and Earth science skills.

Acknowledgments

Writing a college textbook requires the talents and co-operation of many individuals. We appreciate the contributions of Professor Alan Trujillo at Palomar College. Alan not only revised and expanded the oceanography unit, he also contributed the idea and much of the material for the new "Students Sometimes Ask . . ." feature. Working with Dennis Tasa, who is responsible for all of the outstanding illustrations, is always special for us. We not only value his outstanding artistic talents and imagination but his friendship. We are grateful to Professor Ken Pinzke at Southwestern Illinois College. In addition to his many helpful suggestions regarding the manuscript, Ken is responsible for the text's Website, student study guide, and laboratory manual. Ken is an important part of our team and a valued friend as well.

Special thanks goes to those colleagues who prepared in-depth reviews. Their critical comments and thoughtful input helped guide our work and clearly strengthened the text. We wish to thank:

Theodore Chamberlain, Colorado State University

Frank Florence, Jefferson Community College

M. E. Folkoff, Salisbury State University

Roberto Garza, University of Houston

Christopher Hooker, Waubonsee Community College

Barry Metz, Delaware County Community College

Tina Gayle Osborn, Palm Beach Community College

We also want to acknowledge the team of professionals at Prentice Hall. We sincerely appreciate the company's continuing strong support for excellence and innovation. Special thanks to our senior editor Patrick Lynch. We value his leadership and appreciate his attention to detail, excellent communication skills, and easy-going style. We also want to acknowledge our marketing manager Christine Henry for her helpful input, enthusiasm, hard work, and friendship. The production team, led by Ed Thomas, has once again done an outstanding job. The strong visual impact of *Earth Science,* Tenth Edition benefited greatly from the work of photo researcher Yvonne Gerin and image permission coordinator Debbie Hewitson. Thanks also to Barbara Booth for her excellent copyediting skills and to Michelle Gardner at Pine Tree Composition who once again did an excellent job of putting everything together. All are true professionals with whom we are very fortunate to be associated.

Edward J. Tarbuck
Frederick K. Lutgens

Figure I.1 British Columbia's Mount Robson is the highest point in the Canadian Rockies. Understanding the forces that created this majestic range is just one of the many areas of Earth science study. (Photo by Art Wolfe)

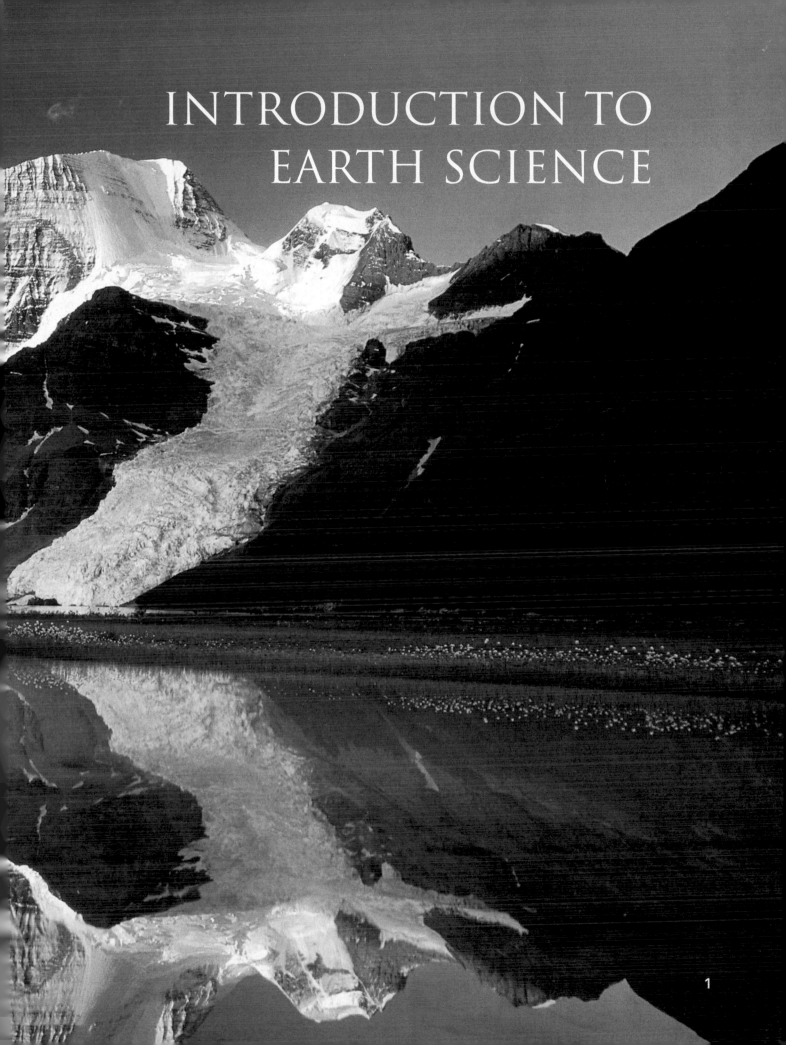

INTRODUCTION TO
EARTH SCIENCE

The spectacular eruption of a volcano, the magnificent scenery of a rocky coast, and the destruction created by a hurricane are all subjects for the Earth scientist. The study of Earth science deals with many fascinating and practical questions about our environment (Figure I.1, opening photo). What forces produce mountains? Why is our daily weather so variable? Is climate really changing? How old is Earth, and how is it related to the other planets in the solar system? What causes ocean tides? What was the Ice Age like? Will there be another? Can a successful well be located at this site?

The subject of this text is **Earth science.** To understand Earth is not an easy task because our planet is not a static and unchanging mass. Rather, it is a dynamic body with many interacting parts and a long and complex history.

What Is Earth Science?

Earth science is the name for all the sciences that collectively seek to understand Earth and its neighbors in space. It includes geology, oceanography, meteorology, and astronomy.

In this book, Units One through Four focus on the science of **geology,** a word that literally means "study of Earth." Geology is traditionally divided into two broad areas—physical and historical.

Physical geology examines the materials composing Earth and seeks to understand the many processes that operate beneath and upon its surface. Earth is a dynamic, ever changing planet. Internal forces create earthquakes, build mountains, and produce volcanic structures. At the surface, external processes break rock apart and sculpture a broad array of landforms. The erosional effects of water, wind, and ice result in a great diversity of landscapes. Because rocks and minerals form in response to Earth's internal and external processes, their interpretation is basic to an understanding of our planet.

In contrast to physical geology, the aim of *historical geology* is to understand the origin of Earth and the development of the planet through its 4.5-billion-year history. It strives to establish an orderly chronological arrangement of the multitude of physical and biological changes that have occurred in the geologic past. The study of physical geology logically precedes the study of Earth history because we must first understand how Earth works before we attempt to unravel its past.

Unit Five, *The Global Ocean,* is devoted to **oceanography.** Oceanography is actually not a separate and distinct science. Rather, it involves the application of all sciences in a comprehensive and interrelated study of the oceans in all their aspects and relationships. Oceanography integrates chemistry, physics, geology, and biology. It includes the study of the composition and movements of seawater, as well as coastal processes, seafloor topography, and marine life.

Unit Six, *Earth's Dynamic Atmosphere*, examines the mixture of gases that is held to the planet by gravity and thins rapidly with altitude. Acted on by the combined effects of Earth's motions and energy from the Sun, the formless and invisible atmosphere reacts by producing an infinite variety of weather, which in turn creates the basic pattern of global climates. **Meteorology** is the study of the atmosphere and the processes that produce weather and climate. Like oceanography, meteorology involves the application of other sciences in an integrated study of the thin layer of air that surrounds Earth.

Unit Seven, *Earth's Place in the Universe,* demonstrates that an understanding of Earth requires that we relate our planet to the larger universe. Because Earth is related to all of the other objects in space, the science of **astronomy**—the study of the universe—is very useful in probing the origins of our own environment. As we are so closely acquainted with the planet on which we live, it is easy to forget that Earth is just a tiny object in a vast universe. Indeed, Earth is subject to the same physical laws that govern the many other objects populating the great expanses of space. Thus, to understand explanations of our planet's origin, it is useful to learn something about the other members of our solar system. Moreover, it is helpful to view the solar system as a part of the great assemblage of stars that comprise our galaxy, which in turn is but one of many galaxies (see Box I.1)

To understand Earth science is challenging because our planet is a dynamic body with many interacting parts and a complex history. Throughout its long existence, Earth has been changing. In fact, it is changing as you read this page and will continue to do so into the foreseeable future. Sometimes the changes are rapid and violent, as when severe storms, landslides, or volcanic eruptions occur. Just as often, change takes place so gradually that it goes unnoticed during a lifetime. Scales of size and space also vary greatly among the phenomena studied in Earth science. Sometimes the focus is on phenomena that are submicroscopic, and at other times it is on features and processes that are continental or global in scale.

Early Evolution of Earth

Earth is one of nine planets that along with several dozen moons and numerous smaller bodies revolve around the Sun. The orderly nature of our solar system leads most researchers to conclude that Earth and the other planets formed during essentially the same time span and from the same primordial material as the Sun. The **nebular hypothesis** suggests that the bodies of our solar system evolved from an enormous rotating cloud called the *solar nebula*, composed mostly of hydrogen and helium, with a small percentage of the heavier elements (Figure I.2A).

About 5 billion years ago this huge cloud of gases and minute particles began to gravitationally contract.

As this slowly spiraling nebula contracted, it rotated faster and faster for the same reason ice skaters do when they draw their arms toward their bodies. Eventually the inward pull of gravity came into balance with the outward force caused by the rotational motion of the nebula (Figure I.2B). By this time the once vast cloud

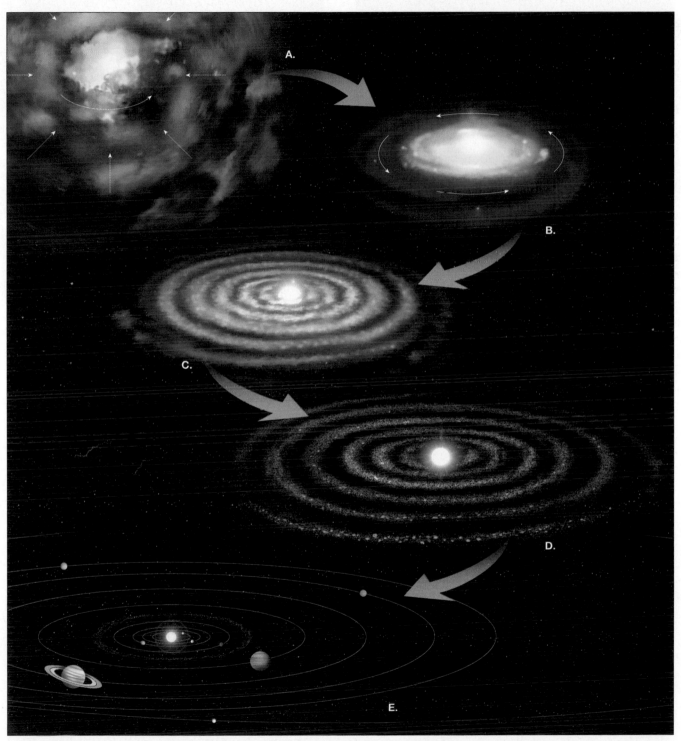

Figure I.2   Formation of the solar system according to the nebular hypothesis. **A.** The birth of our solar system began as dust and gases (nebula) started to gravitationally collapse. **B.** The nebula contracted into a rotating disk that was heated by the conversion of gravitational energy into thermal energy. **C.** Cooling of the nebular cloud caused rocky and metallic material to condense into tiny solid particles. **D.** Repeated collisions caused the dust-sized particles to gradually coalesce into asteroid-sized bodies. **E.** Within a few million years these bodies acreted into the planets.

A.   B.

Figure I.3 **A.** View that greeted the *Apollo 8* astronauts as their spacecraft emerged from behind the Moon. (NASA Headquarters) **B.** Africa and Arabia are prominent in this image of Earth taken from *Apollo 17*. The tan cloud-free zones over the land coincide with major desert regions. The band of clouds across central Africa is associated with a much wetter climate that in places sustains tropical rain forests. The dark blue of the oceans and the swirling cloud patterns remind us of the importance of the oceans and the atmosphere. Antarctica, a continent covered by glacial ice, is visible at the South Pole. (NASA/Science Source/Photo Researchers, Inc.)

had assumed a flat disk shape with a large concentration of material at its center called the *protosun* (pre-Sun). During the collapse, gravitational energy was converted to thermal energy (heat), causing the temperature of the inner portion of the nebula to dramatically rise.

The formation of the Sun marked the end of the period of contraction and thus the end of gravitational heating. Temperatures in the region where the inner planets now reside began to decline. The decrease in temperature caused those substances with high melting points to coalesce (join together). Materials such as iron and nickel and the elements of which the rock-forming minerals are composed—silicon, calcium, sodium, and so forth—formed metallic and rocky clumps that orbited the Sun (Figure I.2C) Repeated collisions caused these fragments to coalesce into larger asteroid-size bodies (Figure I.2D), which in a few tens of millions of years accreted into the four inner planets we call Mercury, Venus, Earth, and Mars (Figure I.2E). Not all of these clumps of matter were incorporated into the *protoplanets* (pre-planets). Those that remained in orbit are called *meteorites* when they survive an impact with Earth.

As more and more material was swept up by the protoplanets, the high-velocity impact of nebular debris caused the temperature of these bodies to rise. Because of their relatively high temperatures and weak gravitational fields, the inner planets were unable to accumulate much of the lighter components of the nebular cloud. The lightest of these, hydrogen and helium, were eventually whisked from the inner solar system by the solar winds. However, Earth, Mars, and Venus were able to retain various amounts of the heavier gases, including water vapor and carbon dioxide. Mercury, the smallest of the inner planets, and Earth's moon, were unable to retain even the heaviest of these gases.

At the same time that the inner planets were forming, the larger, outer planets (Jupiter, Saturn, Uranus, and Neptune), along with their extensive satellite systems, were also developing. Because of the frigid temperatures existing far from the Sun, the fragments from which these planets formed contained a high percentage of ices—water, carbon dioxide, ammonia, and methane. The two most massive planets, Jupiter and Saturn, had a surface gravity sufficient to attract and hold even the lightest elements—hydrogen and helium.

Shortly after Earth formed, the decay of radioactive elements, coupled with heat released by colliding particles, produced at least some melting of the interior. Melting allowed the denser elements, principally iron and nickel, to sink to Earth's center, while the lighter, rocky components floated outward, toward the surface. The sorting of material by density, which began early in Earth's history, is believed to still be occurring, but on a much smaller scale. As a result of this differentiation, Earth's interior is not homogeneous. Rather, it consists of shells or spheres of materials that have different properties.

An important consequence of this period of differentiation is that gaseous materials were allowed to escape from Earth's interior, just as gases are emitted today during volcanic eruptions. By this process an atmosphere gradually evolved, composed chiefly of gases expelled from within the planet.

A View of Earth

A view such as the one in Figure I.3A provided the *Apollo 8* astronauts as well as the rest of humanity with a unique perspective of our home. Seen from space, Earth is breathtaking in its beauty and startling in its solitude. Such an image reminds us that our home is, after all, a planet—small, self-contained, and in some ways even fragile.

As we look closely at our planet from space, it becomes apparent that Earth is much more than rock and soil. In fact, the most conspicuous features in Figure I.3A are not continents but swirling clouds suspended above the surface of the vast global ocean. These features emphasize the importance of water on our planet.

The closer view of Earth from space shown in Figure I.3B helps us appreciate why the physical environment is traditionally divided into three major spheres: the water portion of our planet, the hydrosphere; Earth's gaseous envelope, the atmosphere; and, of course, the solid Earth.

It should be emphasized that our environment is highly integrated and is not dominated by rock, water, or air alone. It is instead characterized by continuous interactions as air comes in contact with rock, rock with water, and water with air. Moreover, the biosphere, the totality of life forms on our planet, extends into each of the three physical realms and is an equally integral part of the planet. Thus, Earth can be thought of as consisting of four major spheres: the hydrosphere, atmosphere, solid Earth, and biosphere.

The interactions among the spheres of Earth's environment are uncountable. Figure I.4 provides us with one easy-to-visualize example. The shoreline is an obvious meeting place for rock, water, and air. In this scene, ocean waves that were created by the drag of air moving across the water are breaking against the rocky shore. The force of the water can be powerful, and the erosional work that is accomplished can be great.

Figure I.4 The shoreline is one obvious meeting place for rock, water, and air. In this scene along the Oregon coast, ocean waves that were created by the force of moving air break against the rocky shore. The force of the water can be powerful, and the erosional work that is accomplished can be great. *(Photo by David Muench)*

BOX I.1

EARTH AS A SYSTEM:
Earth's Place in the Cosmos*

For centuries, those who have gazed at the night sky have wondered about the nature of the universe, Earth's place within it, and whether or not we are "alone." Today, many exciting discoveries in astronomy are beginning to provide some answers concerning the origin of the universe, the formation and evolution of stars, and how Earth and its materials came into existence.

The realization that the universe is immense and orderly began in the early 1900s when Edwin Hubble and other dedicated scientists demonstrated that the Milky Way Galaxy is one of hundreds of billions of galaxies, each of which contains billions of stars. Today we understand that Earth, its materials, and indeed all living things are the consequence of a sequence of events, governed by natural laws, that occurred within the universe during the past 10–15 billion years.

Astronomical evidence supports the theory, called the Big Bang, that the universe began perhaps as long as 15 billion years ago when a dense, hot, supermassive concentration of material exploded with cataclysmic force (Fig-

*This box was prepared by Kenneth Pinzke, Southwestern Illinois College.

ure I.A). Within about one second, the temperature of the expanding universe cooled to approximately 10 billion degrees, and fundamental atomic particles called protons and neutrons began to appear. After a few minutes, atoms of the least complex elements—hydrogen and helium—had formed, and the initial conversion of energy to matter in the young universe was completed.

During the first billion years or so, matter (essentially hydrogen and to a lesser extent helium) in the expanding universe became clumpy and fragmented into enormous clouds and groups of clouds that eventually collapsed to become galaxies and clusters of galaxies. Inside these collapsing clouds, smaller concentrations of matter formed into stars. One of the billions of galaxies to form was the one we call the Milky Way.

The Milky Way is a collection of several hundred billion stars, the oldest of which is about 10 billion years. It is one of a cluster of approximately 28 galaxies, called the Local Group, that exist in our region of the universe. Initially, the oldest stars in the Milky Way formed from nearly pure hydrogen. Later, succeeding generations of younger stars, including our Sun, would have heavier,

more complex atoms available for their formation.

After forming, all stars go through a series of changes before finally dying out. During the life of most stars, energy is produced as hydrogen nuclei (protons) fuse with other hydrogen nuclei to form helium. During this process, called *nuclear fusion*, matter is converted to energy. Stars begin to die when their nuclear fuel becomes exhausted. Massive stars often have indescribably explosive deaths. During these violent events, called supernovas, nuclear fusion advances beyond helium and produces heavier, more complex atoms such as oxygen, carbon, and iron. In turn, these products of stellar death may become the materials that make up future generations of stars. It was from the debris scattered during the death of a preexisting star, or stars, that our Sun, as well as the rest of the solar system, formed.

Our star, the Sun, is at the very least a second-generation star. It, along with the planets and other members of the solar system, began forming nearly 5 billion years ago, some 10 billion years after the Big Bang, from a large, interstellar cloud, called a *nebula*, which consisted of dust particles and gases

Hydrosphere

Earth is sometimes called the *blue* planet. Water more than anything else makes Earth unique. The **hydrosphere** is a dynamic mass of liquid that is continually on the move, evaporating from the oceans to the atmosphere, precipitating to the land, and running back to the ocean again. The global ocean is certainly the most prominent feature of the hydrosphere, blanketing nearly 71 percent of Earth's surface to an average depth of about 3800 meters (12,500 feet). It accounts for about 97 percent of Earth's water. However, the hydrosphere also includes the fresh water found in streams, lakes, and glaciers, as well as that found underground.

Although these latter sources constitute just a tiny fraction of the total, they are much more important than their meager percentage indicates. In addition to providing the fresh water that is so vital to life on the land, streams, glaciers, and groundwater are responsible for sculpturing and creating many of our planet's varied landforms.

Atmosphere

Earth is surrounded by a life-giving gaseous envelope called the **atmosphere.** Compared with the solid Earth, the atmosphere is thin and tenuous. One half lies below an altitude of 5.6 kilometers (3.5 miles), and 90 percent occurs within just 16 kilometers (10 miles) of Earth's surface. By comparison, the radius of the solid Earth (distance from the surface to the center) is about 6400 kilometers (nearly 4000 miles)! Despite its modest dimensions, this thin blanket of air is an integral part of the planet. It not only provides the air that we breathe but also acts to protect us from the Sun's intense heat and dangerous radiation. The energy exchanges that continually occur between the atmosphere and Earth's surface and between the atmosphere and space produce the effects we call weather and climate.

If, like the Moon, Earth had no atmosphere, our planet would not only be lifeless but also many of the processes and interactions that make the surface such a dynamic place could not operate. Without weathering

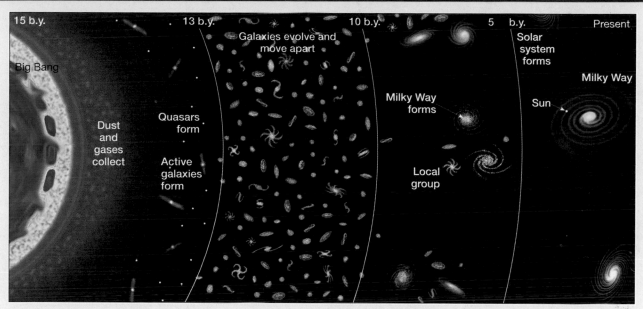

Figure I.A About 15 billion years ago, an incomprehensibly large explosion sent all the matter of the universe flying outward at incredible speeds. After a few billion years, the material cooled and condensed into the first stars and galaxies. Because the universe is expanding, the evolving galaxies continue to move away from one another. About 5 billion years ago, our solar system began forming in one of these galaxies, the Milky Way.

enriched in heavy elements from supernova explosions. Gravitational energy caused the nebula to contract and, in so doing, it began to rotate and flatten. Inside, smaller concentrations of matter began condensing to form the planets. At the center of the nebula there was sufficient pressure and heat to initiate hydrogen nuclear fusion, and our Sun was born.

It has been said that all life on Earth is related to the stars. This is true because the very atoms in our bodies, as well as the atoms that make up every other thing on Earth, owe their origin to a supernova event that occurred billions of years ago, trillions of kilometers away.

and erosion, the face of our planet might more closely resemble the lunar surface, which has not changed appreciably in nearly 3 billion years.

Biosphere

The **biosphere** includes all life on Earth. It is concentrated near the surface in a zone that extends from the ocean floor upward for several kilometers into the atmosphere. Plants and animals depend on the physical environment for the basics of life. However, organisms do more than just respond to their physical environment. Through countless interactions, life forms help maintain and alter their physical environment. Without life, the makeup and nature of the solid Earth, hydrosphere, and atmosphere would be very different.

Solid Earth

Lying beneath the atmosphere and the ocean is the solid Earth. Because the solid Earth is not uniform, it is divided into various regions or units. Based on compositional differences, three principal regions are identified: a dense inner sphere called the **core;** the less dense **mantle;** and the **crust,** which is the light and very thin outer skin of Earth (Figure I.5A). The crust is not a layer of uniform thickness. It is thinnest beneath the oceans and thickest where continents exist. Although the crust may seem insignificant when compared with the much thicker units of solid Earth, it was created by the same general processes that formed Earth's present structure. Thus, the crust is important in understanding the history and nature of our planet.

The outer portion of the solid Earth is also divided based on how materials behave when various forces and stresses are applied. The term **lithosphere** refers to the rigid outer layer that includes the crust and uppermost mantle (Figure I.5B). Beneath the rigid rocks that compose the lithosphere, the rocks become weak and are able to slowly flow in response to the uneven distribution of heat deep within Earth.

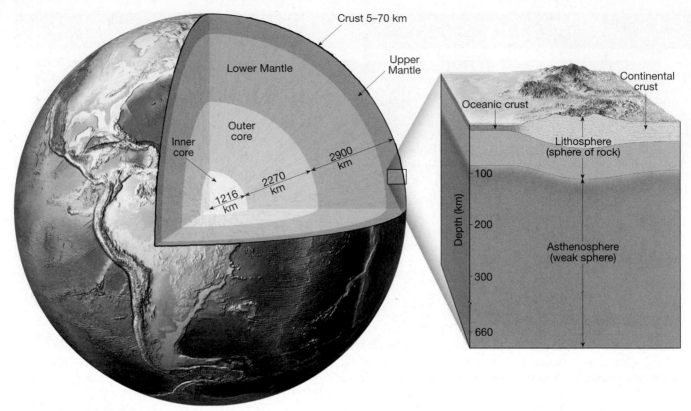

Figure I.5 View of Earth's layered structure. **A.** The inner core, outer core, and mantle are drawn to scale, but the thickness of the crust is exaggerated by about five times. **B.** A blowup of Earth's outer shell. It shows the two types of crust (oceanic, continental), the rigid lithosphere, and weak asthenosphere.

The two principal divisions of Earth's surface are the continents and the ocean basins. The most obvious difference between these two diverse provinces is their relative levels. The average elevation of the continents above sea level is about 840 meters (2750 feet), whereas the average depth of the oceans is about 3800 meters (12,500 feet). Thus, the continents stand, on average, 4640 meters (about 4.6 kilometers or nearly 3 miles) above the level of the ocean floor.

? STUDENTS SOMETIMES ASK...
How do we know about the internal structure of Earth?

You might suspect that the internal structure of Earth has been sampled directly. However, humans have never penetrated beneath the crust!

The internal structure of Earth is determined by using indirect observations. Every time there is an earthquake, waves of energy (called *seismic waves*) penetrate Earth's interior. Seismic waves change their speed and are bent and reflected as they move through zones having different properties. An extensive series of monitoring stations around the world detects and records this energy. The data are analyzed and used to work out the structure of Earth's interior. For more about this technique, see Chapter 6, "Earthquakes and Earth's Interior."

Earth System Science

As we study Earth, it becomes clear that our planet is a dynamic body with many separate but highly interactive parts or *spheres*. The hydrosphere, atmosphere, biosphere, and solid Earth and all of their components can be studied separately. However, the parts are not isolated; each is connected in some way to the others. A simple example of this interdependence occurs every winter as moisture evaporates from the Pacific Ocean and subsequently falls as rain in the hills of southern California, triggering destructive landslides. The processes that move water from the hydrosphere to the atmosphere and then to the solid Earth have a profound impact on the plants and animals (including humans) that inhabit the affected regions.

Scientists have recognized that in order to more fully understand our planet, they must learn how its individual components (land, water, air, and life forms) are interconnected. This endeavor, called **Earth system science,** aims to study Earth as a *system* composed of numerous interacting parts, or *subsystems.* Rather than looking through the limited lens of only one of the traditional sciences—geology, atmospheric science, chemistry, biology, etc.—Earth system science attempts to integrate the knowledge of several academic fields. Using this interdisciplinary approach, we hope to achieve the level of understanding necessary to comprehend and solve many of our global environmental problems.

What Is a System?

Most of us hear and use the term *system* frequently. We may service our car's cooling *system,* make use of the city's transportation *system,* and be a participant in the political *system.* A news report might inform us of an approaching weather *system.* Further, we know that Earth is just a small part of a larger system known as the *solar system* and that it is a *subsystem* of the even larger system called the Milky Way Galaxy.

Loosely defined, a **system** can be any size group of interacting parts that form a complex whole. Most natural systems are driven by sources of energy that move matter and/or energy from one place to another. A simple analogy is a car's cooling system, which contains a liquid (usually water and antifreeze) that is driven from the engine to the radiator and back again. The role of this system is to transfer heat generated by combustion in the engine to the radiator, where moving air removes it from the system. Hence, the term cooling system.

Systems like a car's cooling system are self-contained with regard to matter and are called **closed systems.** Although energy moves freely in and out of a closed system, no matter (liquid in the case of our auto's cooling systems) enters or leaves the system. (This assumes you don't get a leak in your radiator.) By contrast, most natural systems are **open systems** and are far more complicated than the foregoing example. In an open system both energy and matter flow into and out of the system. In a river system, for example, the amount of water flowing in the channel can fluctuate a great deal. So, too, can the amount of sediment in the system. At different times and places the river is accumulating and depositing varying quantities of silt, clay, and sand.

Feedback Mechanisms

Most natural systems have mechanisms that tend to enhance change, as well as other mechanisms that tend to resist change and thus stabilize the system. For example, when we get too hot, we perspire to cool down. This cooling phenomenon works to stabilize our body temperature and is referred to as a **negative feedback mechanism.** Negative feedback mechanisms work to maintain the system as it is, or in other words to maintain the status quo. By contrast, mechanisms that enhance or drive change are called **positive feedback mechanisms.**

Most of Earth's systems, particularly the climate system, contain a wide variety of negative and positive feedback mechanisms. For example, substantial scientific evidence indicates that Earth has entered a period of global warming. One consequence of global warming is that some of the world's glaciers and ice caps have begun to melt. Highly reflective snow- and ice-covered surfaces are gradually being replaced by brown soils, green trees, or blue oceans, all of which are darker, so they absorb more sunlight. Therefore, as Earth warms and some snow and ice melts, our planet absorbs more sunlight. The result is a positive feedback that contributes to the warming.

On the other hand, an increase in global temperature also causes greater evaporation of water from Earth's land-sea surface. One result of having more water vapor in the air is an increase in cloud cover. Because cloud tops are white and highly reflective, more sunlight is reflected back to space, which diminishes the amount of sunshine reaching Earth's surface and thus reduces global temperatures. Further, warmer temperatures tend to promote the growth of vegetation. Plants in turn remove carbon dioxide (CO_2) from the air. Since carbon dioxide is one of the atmosphere's *greenhouse gases,* its removal has a negative impact on global warming.*

In addition to natural processes, we must also consider the human element. Extensive cutting and clearing of the tropical rainforests and the burning of fossil fuels (oil, natural gas, and coal) result in an increase in atmospheric CO_2. Such activity appears to have contributed to the increase in global temperature that our planet is experiencing. One of the daunting tasks of Earth system scientists is to predict what the climate will be like in the future by taking into account many variables, including technological changes, population trends, and the overall impact of the numerous competing positive and negative feedback mechanisms.

Earth as a System

The Earth system has a nearly endless array of subsystems in which matter is recycled over and over again. One familiar subsystem briefly considered earlier is the *hydrologic cycle* (Figure I.6). It represents the unending circulation of Earth's water among the hydrosphere, atmosphere, biosphere, and solid Earth. Water enters the atmosphere by evaporation from Earth's surface and by transpiration from plants. Water vapor condenses in the atmosphere to form clouds, which in turn produce precipitation that falls back to Earth's surface. Some of the rain that falls onto the land sinks in to be taken up by plants or become groundwater, and some flows across the surface toward the ocean.

The Earth system is powered by energy from two sources. The Sun drives external processes that occur in the atmosphere, hydrosphere, and at Earth's surface.

---

*Greenhouse gases absorb heat energy emitted by Earth and thus help keep the atmosphere warm.

Figure I.6 The hydrologic cycle is just one of Earth's many subsystems. Our planet's water is constantly cycled among all four of Earth's major spheres.

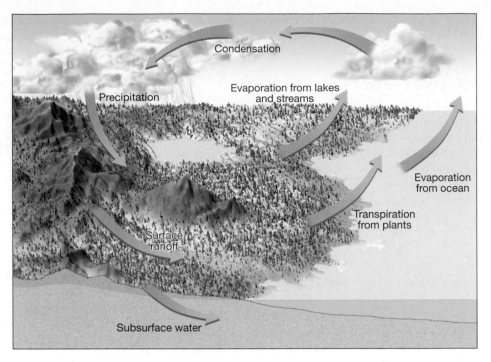

Weather and climate, ocean circulation, and erosional processes are driven by energy from the Sun. Earth's interior is the second source of energy. Heat remaining from when our planet formed, and heat that is continuously generated by decay of radioactive elements powers the internal processes that produce volcanoes, earthquakes, and mountains.

The parts of the Earth system are linked so that a change in one part can produce changes in any or all of the other parts. For example, when a volcano erupts, lava from Earth's interior may flow out at the surface and block a nearby valley. This new obstruction influences the region's drainage system by creating a lake or causing streams to change course. The large quantities of volcanic ash and gases that can be emitted during an eruption might be blown high into the atmosphere and influence the amount of solar energy that can reach Earth's surface. The result could be a drop in air temperatures over the entire hemisphere.

Where the surface is covered by lava flows or a thick layer of volcanic ash, existing soils are buried. This causes the soil-forming processes to begin anew to transform the new surface material into soil (Figure I.7). The soil that eventually forms will reflect the interactions among many parts of the Earth system—the volcanic parent material, the type and rate of weathering, and the impact of biological activity. Of course, there would also be significant changes in the biosphere. Some organisms and their habitats would be eliminated by the lava and ash, whereas new settings for life, such as the lake, would be created. The potential climate change could also impact sensitive life forms.

The Earth system is characterized by processes that vary on spatial scales from fractions of millimeters to thousands of kilometers. Time scales for Earth's processes range from milliseconds to billions of years. As we learn about Earth, it becomes increasingly clear that despite significant separations in distance or time, many processes are connected, and a change in one component can influence the entire system.

Humans are *part of* the Earth system, a system in which the living and nonliving components are entwined and interconnected. Therefore, our actions produce changes in all of the other parts. When we burn

Figure I.7 When Mount St. Helens erupted in May 1980, the area shown here was buried by a volcanic mudflow. Now, plants are reestablished and new soil is forming. (Photo by Jack Dykinga)

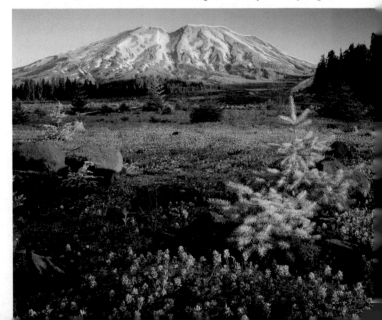

gasoline and coal, build breakwaters along the shoreline, dispose of our wastes, and clear the land, we cause other parts of the system to respond, often in unforeseen ways. Throughout this book you will learn about many of Earth's subsystems: the hydrologic system, the tectonic (mountain-building) system, and the climate system, to name a few. Remember that these components *and we humans* are all part of the complex interacting whole we call the Earth system.

The organization of this text involves traditional groupings of chapters that focus on closely related topics. Nevertheless, the theme of *Earth as a system* keeps recurring through *all* major units of *Earth Science.* It is a thread that "weaves" through the chapters and helps tie them together. At the end of each chapter there is a section entitled *Examining the Earth System.* The questions and problems found there are intended to help you develop an awareness and appreciation for some of Earth system's important interrelationships.

People and the Environment

Environment refers to everything that surrounds and influences an organism. Some of these things are biological and social, but others are nonliving. The factors in this latter category are collectively called our **physical environment.** The physical environment encompasses water, air, soil, and rock, as well as conditions such as temperature, humidity, and sunlight. The phenomena and processes studied by the Earth sciences are basic to an understanding of the physical environment. In this sense, most of Earth science can be characterized as environmental science.

However, when the term *environmental* is applied to Earth science today, it is usually reserved for those aspects that focus on the relationships between people and the natural environment.

We can dramatically influence natural processes. For example, river flooding is natural, but the magnitude and frequency of flooding can be changed significantly by human activities such as clearing forests, building cities, and constructing dams. Unfortunately, natural systems do not always adjust to artificial changes in ways we can anticipate. Thus, an alteration to the environment that was intended to benefit society may have the opposite effect.

Interactions between people and the natural environment is a second important theme that recurs throughout this text and helps to tie its units together. The primary focus of this book is to develop an understanding of basic Earth science principles, but along the way, we will explore numerous important relationships between people and the natural environment.

Resources

Resources are an important focus of the Earth sciences that is of great value to people. They include water and soil, a great variety of metallic and nonmetallic minerals, and energy. Together they form the very foundation of modern civilization. The Earth sciences deal not only with the formation and occurrence of these vital resources but also with maintaining supplies and the environmental impact of their extraction and use.

Few people who live in highly industrialized nations realize the quantity of resources needed to maintain their present standard of living. For example, the annual per capita consumption of metallic and nonmetallic mineral resources for the United States is nearly 10,000 kilograms (11 tons). This is each person's prorated share of the materials required by industry to provide the vast array of products modern society demands. Figures for other highly industrialized countries are comparable.

Resources are commonly divided into two broad categories. Some are classified as **renewable,** which means that they can be replenished over relatively short time spans. Common examples are plants and animals for food, natural fibers for clothing, and forest products for lumber and paper. Energy from flowing water, wind, and the Sun are also considered renewable.

By contrast, many other basic resources are classified as **nonrenewable.** Important metals such as iron, aluminum, and copper fall into this category, as do our most important fuels: oil, natural gas, and coal. Although these and other resources continue to form, the processes that create them are so slow that significant deposits take millions of years to accumulate. In essence, Earth contains fixed quantities of these substances. When the present supplies are mined or pumped from the ground, there will be no more. Although some nonrenewable resources, such as aluminum, can be used over and over again, others, such as oil, cannot be recycled.

Population

A glance at Figure I.8 shows that the population of our planet is growing rapidly. Although it took until the beginning of the nineteenth century for the number to reach 1 billion, just 130 years were needed for the population to double to 2 billion. Between 1930 and 1975 the figure doubled again to 4 billion, and by about 2010 as many as 7 billion people may inhabit the planet. Clearly, as population grows, the demand for resources expands as well. However, the rate of mineral and energy resource usage has climbed more rapidly than the overall growth of population. This fact results from an increasing standard of living. This is certainly true in the United States, where about 6 percent of the world's population uses approximately 30 percent of the annual production of mineral and energy resources.

How long will the remaining supplies of basic resources last? How long can we sustain the rising standard of living in today's industrialized countries and

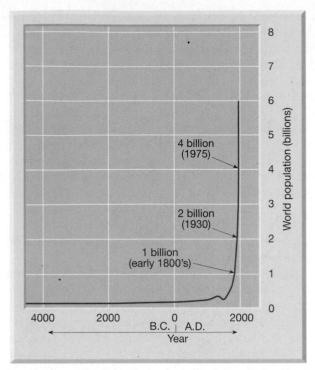

Figure I.8 Growth of world population. It took until 1800 for the number to reach 1 billion. By the year 2010, nearly 7 billion people may inhabit the planet. The demand for basic resources is growing faster than the rate of population increase. (Data from the Population Reference Bureau)

Environmental Problems

In addition to the quest for adequate mineral and energy resources, the Earth sciences must also deal with a broad array of other environmental problems. Some are local, some are regional, and still others are global in extent. Developed and developing nations alike face serious difficulties. Urban air pollution, acid rain, ozone depletion, and global warming are just a few concerns that pose significant threats (Figure I.9). Other problems involve the loss of fertile soils to erosion, the disposal of toxic wastes, and the contamination and depletion of water resources. The list continues to grow.

In addition to human-induced and human-accentuated problems, people must also cope with the many natural hazards posed by the physical environment (Figure I.10). Earthquakes, landslides, floods, and hurricanes are just four of the many risks. Others, such as drought, although not as spectacular, are nevertheless equally important environmental concerns. Of course, environmental hazards are simply *natural* processes. They become hazards only when people try to live where these processes occur (Figure I.11). In many cases, the threat of natural hazards is aggravated by increases in population as more people crowd into places where an impending danger exists.

It is clear that as world population continues its rapid growth, pressures on the environment will increase as well. Therefore, an understanding of Earth is not only essential for the location and recovery of basic resources but also for dealing with the human impact on the environment and minimizing the effects of natural hazards. Knowledge about our planet and how it works is necessary to our survival and well-being. Earth is the only suitable habitat we have, and its resources are limited.

still provide for the growing needs of developing regions? How much environmental deterioration are we willing to accept in pursuit of basic resources? Can alternatives be found? If we are to cope with an increasing per capita demand and a growing world population, it is important that we have some understanding of our present and potential resources.

Figure I.9 An air pollution episode in the Chinese city of Guangzhou. Air-quality problems affect many cities. Fuel combustion by motor vehicles and power plants provide a high proportion of the pollutants. Meteorological factors determine whether pollutants remain "trapped" in the city or are dispersed. (Photo by Mike Yamashita)

Figure I.10 Two geologic hazards are represented in this image. On January 13, 2001, a magnitude 7.6 earthquake caused considerable damage in El Salvador. The damage pictured here was caused by a landslide that was triggered by the earthquake. As many as 1000 people were buried under 8 meters (26 feet) of landslide debris. (Photo by Reuters/STR/Archive Photos)

The Nature of Scientific Inquiry

As members of a modern society, we are constantly reminded of the benefits derived from science. But what exactly is the nature of scientific inquiry? Developing an understanding of how science is done and how scientists work is a third important theme that appears throughout this book. You will explore the difficulties in gathering data and some of the ingenious methods that have been developed to overcome these difficulties. You will also see many examples of how hypotheses are formulated and tested, as well as learn about the evolution and development of some major scientific theories.

All science is based on the assumption that the natural world behaves in a consistent and predictable manner that is comprehensible through careful, systematic study. The overall goal of science is to discover the un-

Figure I.11 This is an image of Italy's Mt. Vesuvius in September 2000. This major volcano is surrounded by the city of Naples and the Bay of Naples. In 79 A.D. Vesuvius explosively erupted, burying the towns of Pompeii and Herculanaeum in volcanic ash. Will it happen again? Such hazards are *natural* processes. They only become hazards when people try to live where these processes occur. (Image courtesy of NASA)

derlying patterns in nature and then to use this knowledge to make predictions about what should or should not be expected, given certain facts or circumstances. For example, by knowing how oil deposits form, geologists are able to predict the most favorable sites for exploration and, perhaps as important, how to avoid regions having little or no potential.

The development of new scientific knowledge involves some basic logical processes that are universally accepted. To determine what is occurring in the natural world, scientists collect scientific "*facts*" through observation and measurement. Because some error is inevitable, the accuracy of a particular measurement or observation is always open to question. Nevertheless, these data are essential to science and serve as the springboard for the development of scientific theories (see Box I.2).

Hypothesis

Once facts have been gathered and principles have been formulated to describe a natural phenomenon, investigators try to explain how or why things happen in the manner observed. They often do this by constructing a tentative (or untested) explanation, which is called a scientific **hypothesis** or **model.** (The term *model,* although often used synonymously with hypothesis, is a less precise term because it is sometimes used to describe a scientific theory as well.) It is best if an investigator can formulate more than one hypothesis to explain a given set of observations. If an individual scientist is unable to devise multiple models, others in the scientific community will almost always develop alternative explanations. A spirited debate frequently ensues. As a result, extensive research is conducted by proponents of opposing models, and the results are made available to the wider scientific community in scientific journals.

Before a hypothesis can become an accepted part of scientific knowledge, it must pass objective testing and analysis. (If a hypothesis cannot be tested, it is not scientifically useful, no matter how interesting it might seem.) The verification process requires that *predictions* be made based on the model being considered and that the predictions be tested by comparing them against objective observations of nature. Put another

---

BOX I.2

UNDERSTANDING EARTH
Studying Earth from Space

Scientific facts are gathered in many ways, including laboratory studies and field observations and measurements. Satellite images like the one in Figure 1.B are another useful source of data. Such images provide perspectives that are difficult to gain from more traditional sources. Moreover, the high-tech instruments aboard many satellites enable scientists to gather information from remote regions where data are otherwise scarce.

The image in Figure 1.B makes use of the Advanced Spaceborne Thermal Emission and Reflection Radiometer (ASTER). Because different materials reflect and emit energy in different ways, ASTER can provide detailed information about the composition of Earth's surface. Figure 1.B is a three-dimensional view looking north over Death Valley, California. The data have been computer enhanced to exaggerate the color variations that highlight differences in types of surface materials.

Salt deposits on the floor of Death Valley appear in shades of yellow, green, purple, and pink, indicating the presence of carbonate, sulfate, and chloride minerals. The Panamint Mountains to the west (left) and the Black Mountains to the east are made up of sedimentary limestones, sandstones, shales, and metamorphic rocks. The bright red areas are dominated by the mineral quartz, found in sandstone; green areas are limestone. In the lower center of the image is Badwater, the lowest point in North America.

Figure I.B This satellite image shows detailed information about the composition of surface materials in Death Valley, California. It was produced by superimposing nighttime thermal infrared data, acquired on April 7, 2000, over topographic data from the U.S. Geological Survey. (Image courtesy of NASA)

way, hypotheses must fit observations other than those used to formulate them in the first place. Those hypotheses that fail rigorous testing are ultimately discarded. The history of science is littered with discarded hypotheses. One of the best known is the Earth-centered model of the universe—a proposal that was supported by the apparent daily motion of the Sun, Moon, and stars around Earth. As the mathematician Jacob Bronowski so ably stated, "Science is a great many things, but in the end they all return to this: Science is the acceptance of what works and the rejection of what does not."

Theory

When a hypothesis has survived extensive scrutiny and when competing models have been eliminated, a hypothesis may be elevated to the status of a scientific **theory.** In everyday language we may say "that's only a theory." But a scientific theory is a well-tested and widely accepted view that the scientific community agrees best explains certain observable facts.

Theories that are extensively documented are held with a very high degree of confidence. Theories of this stature that are comprehensive in scope have a special status. They are called **paradigms** because they explain a large number of interrelated aspects of the natural world. For example, the theory of plate tectonics is a paradigm of the geological sciences that provides the framework for understanding the origin of mountains, earthquakes, and volcanic activity. In addition, plate tectonics explains the evolution of the continents and the ocean basins through time—a topic we will consider later in the book.

Scientific Methods

The process just described, in which researchers gather facts through observations and formulate scientific hypotheses and theories, is called the *scientific method.* Contrary to popular belief, the scientific method is not a standard recipe that scientists apply in a routine manner to unravel the secrets of our natural world. Rather, it is an endeavor that involves creativity and insight. Rutherford and Ahlgren put it this way: "Inventing hypotheses or theories to imagine how the world works and then figuring out how they can be put to the test of reality is as creative as writing poetry, composing music, or designing skyscrapers."*

There is not a fixed path that scientists always follow that leads unerringly to scientific knowledge. Nevertheless, many scientific investigations involve the following steps: (1) the collection of scientific facts

*F. James Rutherford and Andrew Ahlgren, *Science for All Americans* (New York: Oxford University Press, 1990), p. 7.

Figure I.12 An automated weather station. The meteorologist is downloading data into a laptop computer. (Photo by David Parker/Science Photo Library/Photo Researchers, Inc.)

through observation and measurement (Figure I.12); (2) the development of one or more working hypotheses or models to explain these facts; (3) development of observations and experiments to test the hypotheses; and (4) the acceptance, modification, or rejection of the model based on extensive testing.

Other scientific discoveries may result from purely theoretical ideas, which stand up to extensive examination. Some researchers use high-speed computers to simulate what is happening in the "real" world. These models are useful when dealing with natural processes that occur on very long time scales or take place in extreme or inaccessible locations. Still other scientific advancements are made when a totally unexpected happening occurs during an experiment. These serendipitous discoveries are more than pure luck, for as Louis Pasteur said, "In the field of observation, chance favors only the prepared mind."

Scientific knowledge is acquired through several avenues, so it might be best to describe the nature of scientific inquiry as the methods of science rather than the scientific method. In addition, it should always be remembered that even the most compelling scientific theories are still simplified explanations of the natural world.

In this book you will discover the results of centuries of scientific work. You will see the end product of millions of observations, thousands of hypotheses, and hundreds of theories. We have distilled all of this to give you a "briefing" on Earth science.

But realize that our knowledge of Earth is changing daily, as thousands of scientists worldwide make satellite observations, analyze drill cores from the seafloor, measure earthquakes, develop computer models to predict climate, examine the genetic codes of organisms, and discover new facts about our planet's long history. This new knowledge often updates hypotheses and theories. Expect to see many new discoveries and changes in scientific thinking in your lifetime.

STUDENTS SOMETIMES ASK ...

In class you compared a hypothesis to a theory. How is each one different from a scientific law?

A scientific *law* is a basic principle that describes a particular behavior of nature that is generally narrow in scope and can be stated briefly—often as a simple mathematical equation. Because scientific laws have been shown time and time again to be consistent with observations and measurements, they are rarely discarded. Laws may, however, require modifications to fit new findings. For example, Newton's laws of motion are still useful for everyday applications (NASA uses them to calculate satellite trajectories), but they do not work at velocities approaching the speed of light. For these circumstances, they have been supplanted by Einstein's theory of relativity.

Chapter Summary

• *Earth science* is the name for all the sciences that collectively seek to understand Earth and its neighbors in space. It includes *geology, oceanography, meteorology,* and *astronomy.* Geology is traditionally divided into two broad areas—*physical* and *historical.*

• The *nebular hypothesis* describes the formation of the solar system. The planets and Sun began forming about 5 billion years ago from a large cloud of dust and gases. As the cloud contracted, it began to rotate and assume a disk shape. Material that was gravitationally pulled toward the center became the *protosun.* Within the rotating disk, small centers, called *protoplanets,* swept up more and more of the cloud's debris. Because of their high temperatures and weak gravitational fields, the inner planets were unable to accumulate and retain many of the lighter components. Because of the very cold temperatures existing far from the Sun, the large outer planets consist of huge amounts of lighter materials. These gaseous substances account for the comparatively large sizes and low densities of the outer planets.

• Earth's physical environment is traditionally divided into three major parts: the solid Earth; the water portion of our planet, the *hydrosphere;* and Earth's gaseous envelope, the *atmosphere.* In addition, the *biosphere,* the totality of life on Earth, interacts with each of the three physical realms and is an equally integral part of Earth.

• Although each of Earth's four spheres can be studied separately, they are all related in a complex and continuously interacting whole that we call the *Earth system. Earth system science* uses an interdisciplinary approach to integrate the knowledge of several academic fields in the study of our planet and its global environmental problems.

• A *system* is a group of interacting parts that form a complex whole. *Closed systems* are those in which energy moves freely in and out, but matter does not enter or leave the system. In an *open system,* both energy and matter flow into and out of the system.

• Most natural systems have mechanisms that tend to enhance change, called *positive feedback mechanisms,* and other mechanisms, called *negative feedback mechanisms,* that tend to resist change and thus stabilize the system.

• The *two sources of energy that power the Earth system are* (1) *the Sun,* which drives the external processes that occur in the atmosphere, hydrosphere, and at Earth's surface, and (2) *heat from Earth's interior* that powers the internal processes that produce volcanoes, earthquakes, and mountains.

• *Environment* refers to everything that surrounds and influences an organism. These influences can be biological, social, or physical. When applied to Earth science today, the term *environmental* is usually reserved for those aspects that focus on the relationships between people and the natural environment.

• *Resources* are an important environmental concern. The two broad categories of resources are (1) *renewable,* which means that they can be replenished over relatively short time spans, and 2) *nonrenewable.* As population grows, the demand for resources expands as well.

• Environmental problems can be local, regional, or global. Human-induced problems include urban air pollution, acid rain, ozone depletion, and global warming. Natural hazards include earthquakes, landslides, floods, and hurricanes. As world population grows, pressures on the environment also increase.

• All science is based on the assumption that the natural world behaves in a consistent and predictable manner. The process by which scientists gather facts through observation and careful measurement, and formulate scientific *hypotheses* and *theories* is called the *scientific method.* To determine what is occurring in the natural world, scientists often (1) collect facts, (2) develop a scientific hypothesis, (3) construct experiments to validate the hypothesis, and (4) accept, modify, or reject the hypothesis on the basis of extensive testing. Other discoveries represent purely theoretical ideas that have stood up to extensive examination. Still other scientific advancements have been made when a totally unexpected happening occurred during an experiment.

Key Terms

astronomy (p. 2)
atmosphere (p. 6)
biosphere (p. 7)
closed system (p. 9)
core (p. 7)
crust (p. 7)
Earth science (p. 2)
Earth system science (p. 8)
environment (p. 11)

geology (p. 2)
hydrosphere (p. 6)
hypothesis (p. 14)
lithosphere (p. 7)
mantle (p. 7)
meteorology (p. 2)
model (p. 14)
nebular hypothesis (p. 2)
negative feedback mechanism (p. 9)

nonrenewable resource (p. 11)
oceanography (p. 2)
open system (p. 9)
paradigm (p. 15)
physical environment (p. 11)
positive feedback mechanism (p. 9)
renewable resource (p. 11)
system (p. 9)
theory (p. 15)

Review Questions

1. Name the specific Earth science described by each of the following statements.
   (a) The science that deals with the dynamics of the oceans.
   (b) This word literally means "the study of Earth."
   (c) An understanding of the atmosphere is the primary focus of this science.
   (d) This science helps us understand Earth's place in the universe.

2. Briefly describe the events that led to the formation of the solar system.

3. List and briefly define the four "spheres" that constitute our environment.

4. Name the three principal divisions of the solid Earth based on compositional differences. Which one is thinnest?

5. The oceans cover nearly _____ percent of Earth's surface and contain about _____ percent of the planet's total water supply.

6. How is an open system different from a closed system?

7. Contrast positive feedback mechanisms and negative feedback mechanisms.

8. What are the two sources of energy for the Earth system?

9. Contrast renewable and nonrenewable resources. Give one or more examples of each.

10. World population may reach nearly _____ billion by the year 2010. How does this compare to the world population near the beginning of the nineteenth century?

11. List at least four phenomena that could be regarded as natural hazards.

12. How is a scientific hypothesis different from a scientific theory?

Examining the Earth System

1. Examine Figure I.1 closely. Selecting among Earth's four major spheres (atmosphere, solid Earth, hydrosphere, and biosphere), indicate which sphere is represented by each of the following features: (a) mountain, (b) lake, (c) tree, (d) cloud, (e) ice and snow.

2. Humans are a part of the Earth system. List at least three examples of how you in particular influence one or more of Earth's major spheres.

Web Resources

The *Earth Science* Website uses the resources and flexibility of the Internet to aid in your study of the topics in this chapter. Written and developed by Earth science instructors, this site will help improve your understanding of Earth science. Visit **http://www.prenhall.com/tarbuck** and click on the cover of *Earth Science* 10e to find:

- **Online review quizzes.**
- **Web-based critical thinking and writing exercises.**
- **Links to chapter-specific Web resources.**
- **Internet-wide key term searches.**

http://www.prenhall.com/tarbuck

Wulfenite crystals from the Red Cloud Mine, Arizona. (Photo by Jeff Scovill)

CHAPTER **1**

MINERALS: BUILDING BLOCKS OF ROCKS

The study of geology appropriately begins with the study of rocks and minerals, because a knowledge of the materials that make up Earth is basic to our understanding of much of the rest of geology. In this chapter we will examine atoms and elements, the building blocks of minerals, and then investigate the major mineral groups. Our inquiry will attempt to answer some important questions: How are minerals identified? How are rocks and minerals similar, and how are they different? What are the major rock-forming minerals? What defines a mineral resource?

Earth's crust is the source of a wide variety of minerals, many of which are useful and essential to people (Figure 1.1). In fact, practically every manufactured product contains materials obtained from minerals. Most people are familiar with the common uses of basic metals, including aluminum in beverage cans, copper in electrical wiring, gold in jewelry, and silicon in computer chips. But fewer are aware that pencil "lead" does not contain lead metal but is really made of the soft black mineral called *graphite*. Baby powder ("talcum powder") is ground-up rock made of the mineral *talc*, with perfume added. Drill bits impregnated with bits of *diamond* (a mineral) are used by dentists to drill through tooth enamel. The common mineral *quartz* is the main ingredient in ordinary glass and is the source of silicon for computer chips. As the material demands of modern society grow, the need to locate additional supplies of useful minerals also grows, and becomes more challenging as the easily mined sources become depleted.

In addition to the economic uses of rocks and minerals, every process studied by geologists depends in some way on the properties of these basic Earth materials. Events such as volcanic eruptions, mountain building, weathering and erosion, and even earthquakes involve rocks and minerals. Consequently, a basic knowledge of Earth materials is essential to the understanding of many Earth science phenomena.

Figure 1.1   Mineral samples. **A.** Quartz; **B.** Olivine (variety Fosterite); **C.** Fluorite; **D.** Realgar; **E.** Beryl (variety aquamarine); **F.** Bornite and Chalcopyrite; **G.** Native copper; **H.** Gold nugget; **I.** Cut diamond. (Photos A–F by Dennis Tasa; G by E. J. Tarbuck; H and I by Dane Pendland, courtesy of Smithsonian Institution)

Minerals: The Building Blocks of Rocks

 Earth Materials
↳ Minerals

We begin our discussion of Earth materials with an overview of **mineralogy** (*mineral* = mineral, *ology* = the study of), because minerals are the building blocks of rocks. Geologists define **minerals** as any naturally occurring inorganic solids that possess an orderly internal structure and a definite chemical composition. Thus, for any Earth material to be considered a mineral, it must exhibit the following characteristics:

1. It must occur naturally.
2. It must be inorganic (was never alive).
3. It must be a solid.
4. It must possess an orderly internal structure—that is, its atoms must be arranged in a definite pattern.
5. It must have a definite chemical composition that may vary within specified limits.

When the term *mineral* is used by geologists, only those substances that meet these criteria are considered minerals. Consequently, synthetic diamonds, and a wide variety of other useful materials produced by chemists, are not considered minerals. Further, the gemstone *opal* is classified as a *mineraloid* rather than a mineral, because it lacks an orderly internal structure.

Rocks, on the other hand, are more loosely defined. Simply, a **rock** is any solid mass of mineral, or mineral-like, matter that occurs naturally as part of our planet. A few rocks are composed almost entirely of one mineral. A common example is the sedimentary rock *limestone*, which consists of impure masses of the mineral calcite. However, most rocks, like the common rock granite (shown in Figure 1.2), occur as aggregates of several kinds of minerals. Here, the term *aggregate* implies that the minerals are joined in such a way that the properties of each mineral are retained. Note that you can easily identify the mineral constituents of the sample of granite shown in Figure 1.2.

A few rocks are composed of nonmineral matter. These include the volcanic rocks *obsidian* and *pumice*, which are noncrystalline glassy substances, and *coal*, which consists of solid organic debris.

Granite
(Rock)

Quartz
(Mineral)

Hornblende
(Mineral)

Feldspar
(Mineral)

Figure 1.2   Rocks are aggregates of one or more minerals.

Figure 1.3 Periodic Table of the Elements.

Although this chapter deals primarily with the nature of minerals, keep in mind that most rocks are simply aggregates of minerals. Because the properties of rocks are determined largely by the chemical composition and internal structure of those minerals contained within them, we will first consider these Earth materials. In the next chapter we will consider the major rock types.

Composition and Structure of Minerals

Each of Earth's nearly 4000 minerals is a unique substance that is defined by its chemical composition and internal structure. What elements make up a particular mineral? In what pattern are its elements bonded together? And what important properties does this give each mineral?

At this point, let us briefly review the basic building blocks of minerals, the chemical **elements.** At present, 112 elements are known. Of these, only 92 are naturally occurring (Figure 1.3). Some minerals, such as gold and sulfur, are made entirely from one element. But most elements are not stable, and thus most minerals are a combination of two or more elements, joined to form a chemically stable compound.

To understand how elements combine to form minerals, we must first consider the atom. The **atom** (*a* = not, *tomos* = cut) is the smallest particle of mat-

A.

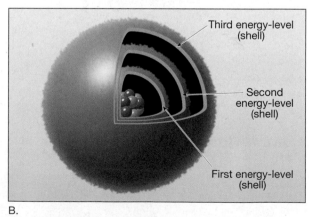

B.

Figure 1.4 Two models of the atom. **A.** A very simplified view of the atom, which consists of a central nucleus, consisting of protons and neutrons, encircled by high-speed electrons. **B.** Another model of the atom showing spherically shaped electron clouds (energy-level shells). Note that these models are not drawn to scale. Electrons are minuscule in size compared to protons and neutrons, and the relative space between the nucleus and electron shells is much greater than illustrated.

ter that retains the essential characteristics of an element. It is this extremely small particle that does the combining.

How Atoms Are Constructed

Two simplified models of an atom are shown in Figure 1.4. Each atom has a central region, called the **nucleus,** which contains very dense **protons** (particles with positive electrical charges) and equally dense **neutrons** (particles with neutral electrical charges). Orbiting the nucleus are **electrons,** which have negative electrical charges. For convenience, we often diagram atoms to show the electrons orbiting the nucleus, like the orderly orbiting of the planets around the Sun (Figure 1.4A). However, electrons move so rapidly that they create a sphere-shaped negative zone around the nucleus. Hence, a more realistic picture of the positions of electrons can be obtained by envisioning a cloud of negatively charged electrons surrounding the nucleus (Figure 1.4B).

It is also known that individual electrons are located at given distances from the nucleus in regions called **energy levels,** or **shells.** Three such shells are shown in Figure 1.4. As you will see, an important fact about these shells is that each can have only a specific number of electrons.

The number of protons in an atom's nucleus determines its **atomic number** and name of the element. For example, all atoms with six protons are carbon atoms, all those with eight protons are oxygen atoms, and so forth. "Free" atoms (those not combined with other atoms) have the same number of electrons as protons, so the atomic number also equals the number of electrons surrounding the nucleus. Therefore, carbon has six electrons to match its six protons, and oxygen has eight electrons to match its eight protons. Neutrons have no charge, so the positive charge of the protons is exactly balanced by the negative charge of the electrons. Consequently, uncombined atoms are neutral electrically and have no overall electrical charge. Thus, the *element* is a large collection of electrically neutral atoms, all having the same atomic number.

How Atoms Bond

Elements combine with each other to form a wide variety of more complex compounds. A **compound** consists of two or more elements bonded together in definite proportions. When the elements separate, the bonds are broken and the compound no longer exists; only its component elements do. Through experimentation, scientists have learned that the forces bonding atoms together are electrical. Further, it is known that chemical bonding results in a change in the arrangement of electrons in the shells of the bonded atoms.

When an atom combines chemically, it either gains, loses, or shares electrons with another atom. An atom that gains electrons becomes negatively charged because it has more electrons than protons. Atoms that lose electrons become positively charged. Atoms that have an electrical charge because of a gain or loss of electrons are called **ions.** Simply stated, oppositely charged ions attract one another to produce a neutral chemical compound.

An example of chemical bonding involves sodium (Na) atoms and chlorine (Cl) atoms to produce sodium chloride molecules (NaCl). Sodium chloride is common table salt (Figure 1.5). When the sodium atom loses one

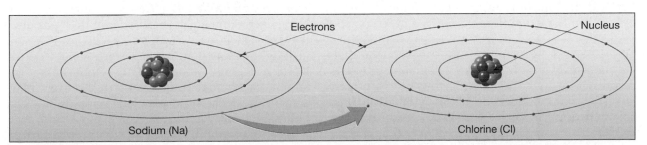

Figure 1.5   Chemical bonding of sodium and chlorine atoms to produce sodium chloride. Through the transfer of one electron in the outer shell of a sodium atom to the outer shell of a chlorine atom, the sodium becomes a positive ion and chlorine a negative ion.

BOX 1.1

PEOPLE AND THE ENVIRONMENT:
Asbestos—What Are the Risks?

Once considered safe enough to use in toothpaste, asbestos may have become the most feared contaminant on Earth. Although health concerns arose two decades earlier, the asbestos panic began in 1986 when the Environmental Protection Agency (EPA) instituted the Asbestos Hazard Emergency Response Act. It required inspection of all public and private schools for asbestos. This brought asbestos to public attention and raised parental fears that their children could contract asbestos-related cancers because of high levels of airborne fibers in schools. Since that time, billions of dollars have been spent on asbestos testing and removal.

Figure 1.A Chrysotile asbestos. This sample is a fibrous form of the mineral serpentine. (Photo by E. J. Tarbuck)

What Is Asbestos?

Asbestos is not a single material. Rather, asbestos is a general term for a group of silicate minerals that *readily separate into thin, strong fibers* (Figure 1.A). Because these fibers are flexible, heat resistant, and relatively chemically inert, they have many uses. Asbestos has been widely used to strengthen concrete, make fireproof fabrics, and insulate boilers and pipes that carry hot fluids. It is a component in floor tiles and the major ingredient in automobile brake linings. Wall coatings rich in asbestos fibers were used

extensively during the U.S. building boom of the 1950s and early 1960s.

Most asbestos comes from three minerals. *Chrysotile* ("white asbestos") is a fibrous form of the mineral serpentine. It is the only type of asbestos mined in North America and once made up 95 percent of world production. *Crocidolite* ("blue asbestos") and *amosite* ("brown asbestos") are currently mined in South Africa and make up about 5 percent of world production.

Exposure and Risk

There is no question that prolonged exposure to air laden with certain types of asbestos dust in an unregulated workplace can be dangerous. When the thin, rodlike fibers are inhaled into the lungs, they are neither easily broken down nor easily expelled, but can remain *for life*. Three lung diseases can result: (1) *asbestosis*, a scarring of tissue, that decreases the lung's ability to absorb oxygen; (2) *mesothelioma*, a rare

---

electron, it becomes a positive ion, and when the chlorine atom gains one electron, it becomes a negative ion. These opposite charges act as an "electrical glue" that bonds the atoms together.

The properties of a chemical compound are dramatically different from the properties of the elements composing it. For example, chlorine is a green poisonous gas that is so toxic it was used as a weapon during World War I. Sodium is a soft, silvery metal that reacts vigorously with water and, if held in your hand, could burn it severely. Together, however, these atoms produce the compound sodium chloride (table salt), the familiar clear, crystalline solid that is essential for human life.

This example also illustrates an important difference between a rock and a mineral. When elements join

to form a mineral, the constituent elements lose their physical properties. In a *rock,* on the other hand, which is a *mixture* of minerals, each mineral retains its own distinctive physical properties.

Isotopes and Radioactive Decay. Subatomic particles are so incredibly small that a special unit was devised to express their mass. A proton or a neutron has a mass just slightly more than one *atomic mass unit*, whereas an electron is only about one two-thousandth of an atomic mass unit. Thus, although electrons play an active role in chemical reactions, they do not contribute significantly to the mass of an atom.

The **mass number** of an atom is simply the total of its neutrons and protons in the nucleus. Atoms of the same element always have the same number of protons, tumor that develops in the chest or gut; and (3) *lung cancer*.

Evidence that incriminates blue asbestos and brown asbestos comes from health studies at mines in South Africa and western Australia. Miners and mill-workers showed an extremely high incidence of mesothelioma, sometimes after less than a year of exposure.

However, the U.S. Geological Survey concludes that the risks from the most widely used form of asbestos (chrysotile or white asbestos) are minimal to nonexistent. They cite studies of miners of white asbestos in Canada and in northern Italy, where mortality rates from mesothelioma and lung cancer differ very little from the general public.

Another study was conducted on miner's wives in the area of Thetford Mines, Quebec, once the largest chrysotile mine in the world. For many years there were no dust controls on mining and milling operations, so these women were exposed to extremely high levels of airborne asbestos. Nevertheless, they exhibited *below-normal* levels of the diseases thought to be associated with asbestos exposure.

The various types of asbestos fibers differ in their chemical composition, shape, and durability. The thin, rodlike "blue asbestos" and "brown asbestos" fibers, which can easily penetrate the lining of the lungs, are certainly pathogenic. But chrysotile fibers are curly and can be expelled more easily than can the rodlike fibers. Further, if inhaled, chrysotile fibers break down within a year. This is not the case for the other forms of asbestos, or for fiberglass, which is frequently used as a substitute for asbestos. These differences are thought to explain the fact that the mortality rates for chrysotile workers differ very little from the mortality rates for the general population. Although the EPA has lifted its ban on asbestos, very little of this once exalted mineral is presently used in the United States.

Risk in U.S. Schools?

Does asbestos present a risk to the nation's students? Available data indicate that the levels of airborne asbestos in schools are approximately 0.01 of the permissible exposure levels for the U.S. workplace. In addition, the indoor concentrations of the most biologically active fibers are comparable to outdoor levels. With few exceptions, the type of asbestos found in schools is chrysotile, which in low concentrations has been shown to be relatively harmless.

A comparison of the risk from asbestos exposure in schools to other risks in society is shown in Table 1.A. Playing high school football poses over 100 times the risk as exposure to airborne asbestos. These data clearly demonstrate that the asbestos panic was and is unwarranted. Misguided public policies, no matter how well intentioned they are, are expensive.

Table 1.A Estimates of Risk from Asbestos Exposure in Schools in Comparison with Other Risks

Cause	Annual Death Rate (per million)
Asbestos exposure in schools	0.005–0.093
Whooping cough vaccination	1–6
Aircraft accidents	6
High school football	10
Drowning (ages 5–14)	27
Motor vehicle accident, Pedestrian (ages 5–14)	32
Home accidents (ages 1–14)	60
Long-term smoking	1200

but commonly have varying numbers of neutrons. This means that an element can have more than one mass number. These variants of the same element are called **isotopes** of that element.

For example, carbon has two well-known isotopes, one having a mass number of 12 (carbon-12), the other a mass number of 14 (carbon-14). All atoms of the same element must have the same number of protons, and carbon always has six (atomic number = 6). Hence, carbon-12 must have six protons *plus six neutrons* to give it a mass number of 12, whereas carbon-14 must have six protons *plus eight neutrons* to give it a mass number of 14.

The *average* atomic mass of any random sample of carbon is much closer to 12 than 14, because carbon-12 is the more common isotope. This average is called *atomic weight*.*

In chemical behavior, all isotopes of the same element are nearly identical. To distinguish among them is like trying to differentiate identical twins, with one slightly heavier. Because isotopes of an element react the same chemically, different isotopes can become parts of the same mineral. For example, when calcite is formed from calcium, carbon, and oxygen, some of its carbon atoms are carbon-12 and some are carbon-14.

---

*The term *weight* is a misnomer that has resulted from long use. The correct term is *atomic mass*.

Although the nuclei of most atoms are stable, many elements do have isotopes in which the nuclei are unstable. "Unstable" means that the isotopes disintegrate through a process called **radioactive decay.** It is so called because a decaying atom actively radiates energy and particles. Radioactive decay occurs when the forces that bind the nucleus are not strong enough to keep it together.

The rate at which the unstable nuclei decay is steady and measurable. This makes such isotopes useful "clocks" for dating the events of Earth history. A discussion of radioactive decay and its applications in dating past geologic events is found in Chapter 10.

Properties of Minerals

Earth Materials
↳ Minerals

Minerals occur in a fascinating palette of colors, shapes, and lusters. They vary in hardness and may even have a distinctive taste or scent. We will now look at these properties, which are useful when identifying minerals.

Crystal Form

Crystal form is the external expression of a mineral's internal orderly arrangement of atoms. Generally, when a mineral forms without space restrictions, it will develop individual crystals with well-formed crystal faces. Figure 1.6A illustrates this for pyrite, an

iron-and-sulfur mineral with cubic crystals. Figure 1.6B shows the distinctive hexagonal crystals of quartz that form when space and time permit. However, most of the time, crystal growth is severely constrained. It is stunted because of competition for space, resulting in an intergrown mass of small, jammed crystals, none of which exhibits its crystal form. This is what happened to the minerals in the granite in Figure 1.2. Thus, most inorganic solid objects are composed of crystals, but they are not clearly visible to the unaided eye.

Luster

Luster is the appearance or quality of light reflected from the surface of a mineral. Minerals that have the appearance of metals, regardless of color, are said to have a *metallic luster*, like the pyrite crystals in Figure 1.6A. Minerals with a *nonmetallic luster* are described by various adjectives. These include vitreous (glassy, like the quartz crystals in Figure 1.6B), pearly, silky, resinous, and earthy (dull). Some minerals appear somewhat metallic in luster and are said to be *submetallic*.

Color

Although **color** is an obvious feature of a mineral, it is often an unreliable diagnostic property. Slight impurities in the common mineral quartz, for example, give it a variety of colors, including pink, purple (amethyst), milky white, and even black.

A.

B.

Figure 1.6 Crystal form is the external expression of a mineral's orderly internal structure. **A.** Pyrite, commonly known as "fool's gold," often forms cubic crystals that may contain parallel lines called striations. (Photo by GeoScience/PH) **B.** Quartz sample that exhibits well-developed hexagonal (six-sided) crystals with pyramidal-shaped ends. (Photo B. by Breck P. Kent)

Streak

Streak is the color of a mineral in its powdered form, which is a much more reliable indication of color. Streak is obtained by rubbing the mineral across a piece of hard, unglazed porcelain, termed a *streak plate*. Whereas the color of a mineral may vary from sample to sample, the streak usually does not and is therefore the more consistent property. Streak can also help to distinguish minerals with metallic lusters from those having nonmetallic lusters. Metallic minerals generally have a dense, dark streak, whereas minerals with nonmetallic lusters do not.

Hardness

One of the most useful diagnostic properties is **hardness,** a measure of the resistance of a mineral to abrasion or scratching. This property is determined by rubbing the mineral to be identified against another mineral of known hardness. One will scratch the other (unless they have the same hardness).

Geologists use a standard hardness scale, called the **Mohs scale.** It consists of 10 minerals arranged in order from 10 (hardest) to 1 (softest), as shown in Table 1.1. Any mineral of unknown hardness can be rubbed against these to determine its hardness. In the field, other handy objects work, too. For example, your fingernail has a hardness of 2.5, a copper penny 3.5, and a piece of glass 5.5. The mineral gypsum, which has a hardness of 2, can be easily scratched by your fingernail. Conversely, the similar-looking mineral calcite, which has a hardness of 3, cannot be scratched by your fingernail. Calcite cannot scratch glass, because its hardness is less than 5.5. Quartz, the hardest of the common minerals at 7, will scratch a glass plate. Diamonds, hardest of all, scratch anything. You can see that by rubbing an unknown mineral against objects of known hardness, you can quickly narrow its hardness range.

Figure 1.7　Sheet-type cleavage common to the micas. (Photo by Chip Clark)

Cleavage

In the crystal structure of a mineral, some bonds are weaker than others. These bonds are where a mineral will break when it is stressed. **Cleavage** (*kleiben* = carve) is the tendency of a mineral to cleave, or break, along planes of weak bonding. Not all minerals have definite planes of weak bonding, but those that possess cleavage can be identified by the distinctive smooth surfaces that are produced when the mineral is broken.

The simplest type of cleavage is exhibited by minerals called micas (Figure 1.7). Because the micas have weak bonds in one direction, they cleave to form thin, flat sheets. Some minerals have several cleavage planes that produce smooth surfaces when broken, whereas others exhibit poor cleavage, and still others have no cleavage at all. When minerals break evenly in more than one direction, cleavage is described by the *number of planes* exhibited and the *angles at which they meet* (Figure 1.8).

Do not confuse *cleavage* with *crystal form!* When a mineral exhibits cleavage, it will break into pieces *that have the same geometry as each other.* By contrast, the quartz crystals shown in Figure 1.6B do not have cleavage. If broken, they fracture into pieces that do not resemble each other or the original crystals.

Fracture

Minerals that do not exhibit cleavage when broken, such as quartz, are said to **fracture.** Even fracturing has variety: Minerals that break into smooth curved surfaces like those seen in broken glass have a *conchoidal fracture* (Figure 1.9). Others break into splinters or fibers, like asbestos, but most minerals fracture irregularly (see Box 1.1).

Table 1.1　Mohs scale of mineral hardness.		
Relative Scale	**Mineral**	**Hardness of Some Common Objects**
Hardest 10	Diamond	
9	Corundum	
8	Topaz	
7	Quartz	
6	Potassium Feldspar	
5	Apatite	5.5 Glass, Pocketknife
4	Fluorite	
3	Calcite	3.5 Copper Penny
2	Gypsum	2.5 Fingernail
Softest 1	Talc	

Figure 1.8 Smooth surfaces produced when a mineral with cleavage is broken. The sample on the left (fluorite) exhibits four planes of cleavage (eight sides), whereas the other two samples exhibit three planes of cleavage (six sides). Also notice that the mineral in the center (halite) has cleavage planes that meet at 90-degree angles, whereas the mineral on the right (calcite) has cleavage planes that meet at 75-degree angles.

Specific Gravity

Specific gravity compares the weight of a mineral to the weight of an equal volume of water. For example, if a cubic centimeter of a mineral weighs three times as much as a cubic centimeter of water, its specific gravity is 3. With a little practice, you can estimate the specific gravity of minerals by hefting them in your hand. For instance, if a mineral feels as heavy as the common rocks you have handled, its specific gravity will probably be somewhere between 2.5 and 3. Some metallic minerals have a specific gravity noticeably greater than that of common rock-forming minerals. Galena, which is an ore of lead, has a specific gravity of roughly 7.5 (Figure 1.10). The specific gravity of pure 24-karat gold is a very hefty 20.

STUDENTS SOMETIMES ASK...
Are there any artificial materials harder than diamonds?

Yes, but you won't be seeing them anytime soon. A hard form of carbon nitride (C_3N_4), described in 1989 and synthesized in a laboratory shortly thereafter may be harder than diamond but hasn't been produced in large enough amounts for a proper test. In 1999, researchers discovered that a form of carbon made from fused spheres of 20 and 28 carbon atoms—relatives of the famous "buckyballs"—also could be as hard as a diamond. These materials are expensive to produce, so diamonds continue to be used as abrasives and in certain kinds of cutting tools. Synthetic diamonds, produced since 1955, are now widely used in these industrial applications.

Figure 1.9 Conchoidal fracture. The smooth curved surfaces result when minerals break in a glasslike manner. (Photo by E. J. Tarbuck)

Figure 1.10 Galena is lead sulfide. Like other metallic ores, it has a relatively high specific gravity and therefore feels heavy when hefted.

Other Properties of Minerals

In addition, some minerals can be recognized by other distinctive properties. For example, halite is ordinary salt, so it is quickly identified with your tongue. Thin sheets of mica will bend and elastically snap back. Gold is malleable, which means it can be easily hammered or shaped. Talc and graphite both have distinctive feels. Talc feels soapy. Graphite feels greasy (it is a principal ingredient in dry lubricants).

A few minerals, such as magnetite, have a high iron content and can be picked up with a magnet. Some varieties of magnetite (lodestone) are natural magnets and will pick up small iron-based objects such as pins and paper clips. Some minerals exhibit special optical properties. For example, when a transparent piece of calcite is placed over printed material, the letters appear doubled. This optical property is known as *double refraction.* In addition, the streak of many sulfur-bearing minerals smells like rotten eggs.

One very simple chemical test involves placing a drop of dilute hydrochloric acid from a dropper bottle on a freshly broken mineral surface. Certain minerals, called carbonates, will effervesce (fizz) with hydrochloric acid. This test is useful in identifying the mineral calcite, which is a common carbonate mineral.

In summary, a number of special physical and chemical properties are useful in identifying particular minerals. These include taste, smell, elasticity, malleability, feel, magnetism, double refraction, and chemical reaction to hydrochloric acid. Remember that every one of these properties depends on the composition (elements) of a mineral and its structure (how the atoms are arranged).

Mineral Groups

 Earth Materials
↳ Minerals

Nearly 4000 minerals have been named, and several new ones are identified each year. Fortunately, for students who are beginning to study minerals, no more than a few dozen are abundant! Collectively, these few make up most of the rocks of Earth's crust and, as such, are classified as the **rock-forming minerals.** It is also interesting to note that *only eight elements* make up the bulk of these minerals and represent over 98 percent (by weight) of the continental crust (Table 1.2). These elements, in order of abundance, are oxygen (O), silicon (Si), aluminum (Al), iron (Fe), calcium (Ca), sodium (Na), potassium (K), and magnesium (Mg).

The two most abundant elements are silicon and oxygen, comprising nearly three-fourths of Earth's continental crust. Silicon and oxygen combine to form the framework of the most common mineral group, the **silicates.** Perhaps the next most common mineral group is the carbonates (carbon plus oxygen plus other elements), of which calcite is the most prominent member.

Table 1.2 Relative abundance of the most common elements in Earth's continental crust.	
Element	**Approximate Percentage by Weight**
Oxygen (O)	46.6
Silicon (Si)	27.7
Aluminum (Al)	8.1
Iron (Fe)	5.0
Calcium (Ca)	3.6
Sodium (Na)	2.8
Potassium (K)	2.6
Magnesium (Mg)	2.1
All others	1.7
Total	100

Source: Data from Brian Mason.

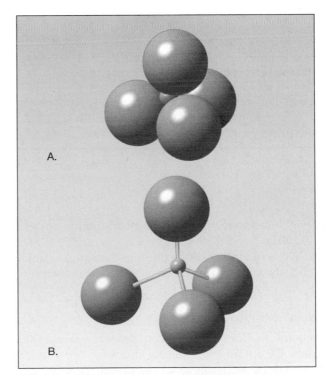

Figure 1.11 Two representations of the silicon-oxygen tetrahedron. **A.** The four large spheres represent oxygen atoms and the blue sphere represents a silicon atom. The spheres are drawn in proportion to the radii of the atoms. **B.** A model of the tetrahedron using rods to depict the bonds that connect the atoms.

BOX 1.2

UNDERSTANDING EARTH:
Gemstones

Precious stones have been prized by people since antiquity. This tradition survives today. Gems are not just prized keepsakes of the rich but are possessed by people of modest means as well. Misinformation abounds about the nature of gems and the minerals of which they are composed. Part of the misinformation stems from the ancient practice of grouping precious stones by color rather than mineral makeup. For example, the more common red spinels were often passed off to royalty as rubies, which are more valuable gems. Even today, when modern techniques of mineral identification are commonplace, yellow quartz is frequently sold as topaz.

Naming Gemstones

Compounding the confusion is the fact that many gems have names that are different from their mineral names. For example, diamond is composed of the mineral of the same name, whereas sapphire is a form of corundum, an aluminum oxide-rich mineral. Although pure aluminum oxide is colorless, a minute amount of a foreign element can produce a vividly colored gemstone. Hence, depending on the impurity, sapphires of nearly every color are known to exist. Pure aluminum oxide with trace amounts of ti-

Table 1.B Important gemstones.

Gem	Mineral Name	Prized Hues
Precious		
Diamond	Diamond	Colorless, yellows
Emerald	Beryl	Greens
Opal	Opal	Brilliant hues
Ruby	Corundum	Reds
Sapphire	Corundum	Blues
Semiprecious		
Alexandrite	Chrysoberyl	Variable
Amethyst	Quartz	Purples
Cat's-eye	Chrysoberyl	Yellows
Chalcedony	Quartz (agate)	Banded
Citrine	Quartz	Yellows
Garnet	Garnet	Reds, greens
Jade	Jadeite or nephrite	Greens
Moonstone	Feldspar	Transparent blues
Peridot	Olivine	Olive greens
Smoky quartz	Quartz	Browns
Spinel	Spinel	Reds
Topaz	Topaz	Purples, reds
Tourmaline	Tourmaline	Reds, blue-greens
Turquoise	Turquoise	Blues
Zircon	Zircon	Reds

tanium and iron produce the most prized blue sapphires. When the mineral corundum contains sufficient quantities of chromium, it exhibits a brilliant red color, and the gem is called *ruby*. Large gem-quality rubies are much rarer than diamonds and thus command a very high price.

To summarize, when a gem-quality sample of corundum exhibits a red hue, it is called *ruby*, but if it exhibits any other color, the gem is called *sapphire*. If

---

Other common rock-forming minerals include gypsum and halite.

We will first discuss the most common mineral group, the silicates, and then consider other prominent mineral groups.

Rock-Forming Silicates

Each of the silicate minerals contains oxygen and silicon atoms. Except for a few silicate minerals such as quartz, most silicate minerals also contain one or more additional elements in their crystalline structure. These elements give rise to the great variety of silicate minerals and their varied properties.

All silicates have the same fundamental building block, the **silicon-oxygen tetrahedron** (*tetra* = four, *hedra* = a base). This structure consists of four oxygen atoms surrounding a much smaller silicon atom, as shown in Figure 1.11. Thus, a typical hand-size silicate mineral specimen contains millions of these silicon-oxygen tetrahedra, joined together in a variety of ways.

In some minerals, the tetrahedra are joined into chains, sheets, or three-dimensional networks by sharing oxygen atoms (Figure 1.12). These larger silicate structures are then connected to one another by other elements. The primary elements that join silicate structures are iron (Fe), magnesium (Mg), potassium (K), sodium (Na), and calcium (Ca).

Major groups of silicate minerals and common examples are given in Figure 1.12. The **feldspars** are by far the most plentiful group, comprising over 50 per-the specimen is not suitable as a gem, it simply goes by the mineral name *corundum*. Although common corundum is not a gemstone, it does have value as an abrasive material. Whereas two gems—rubies and sapphires—are composed of the mineral corundum, quartz is the parent mineral of more than a dozen gems. Table 1.B lists some well-known gemstones and their mineral names.

What Constitutes a Gemstone?

In essence, certain mineral specimens when cut and polished possess beauty of such quality that they can command a price that makes the process of producing the gem profitable. Gemstones can be divided into two categories: precious and semiprecious. A *precious* gem has beauty, durability, size, and rarity, whereas a *semiprecious* gem generally has only one or two of these qualities. The gems that have traditionally enjoyed the highest esteem are diamonds, rubies, sapphires, emeralds, and some varieties of opal (Table 1.B). All other gemstones are classified as semiprecious. It should be noted, however, that large, high-quality specimens of the so-called semiprecious stones can often command a very high price.

Obviously, beauty is the most important quality that a gem can possess. Today we prefer translucent stones

Figure 1.B   Australian sapphires depicting variations in cuts and colors. (Photo by Fred Ward, Black Star)

with evenly tinted colors. The most favored hues appear to be red, blue, green, purple, rose, and yellow. The most prized stones are pigeon-blood rubies, blue sapphires, grass-green emeralds, and canary-yellow diamonds. Colorless gems are generally less than desirable except in the case of diamonds that display "flashes of color" known as *brilliance*. Note that gemstones in the "rough" are dull and would be passed over by most laypersons as "just another rock." Gemstones must be cut and polished by experienced artisans before their true beauty can be displayed (Figure 1.B).

The durability of a gem depends on its hardness—that is, its resistance to abrasion by objects normally encountered in everyday living. For good durability, gems should be as hard or harder than quartz, as defined by the Mohs scale of hardness, which consists of 10 minerals arranged in order from 1 (softest) to 10 (hardest). One notable exception is opal, which is comparatively soft (hardness 5 to 6.5) and brittle. Opal's esteem comes from its fire, which is a display of a variety of brilliant colors including greens, blues, and reds.

It seems to be human nature to treasure that which is rare. In the case of gemstones, large, high-quality specimens are much rarer than smaller stones. Thus, large rubies, diamonds, and emeralds, which are rare in addition to being beautiful and durable, command the very highest prices.

---

cent of Earth's crust. **Quartz,** the second most abundant mineral in the continental crust, is the only common mineral made completely of silicon and oxygen.

Notice in Figure 1.12 that each mineral *group* has a particular silicate *structure.* A relationship exists between this internal structure of a mineral and the *cleavage* it exhibits. Because the silicon-oxygen bonds are strong, silicate minerals tend to cleave between the silicon-oxygen structures rather than across them. For example, the micas have a sheet structure and thus tend to cleave into flat plates (see muscovite in Figure 1.13). Quartz, which has equally strong silicon-oxygen bonds in all directions, has no cleavage but fractures instead.

How do silicate minerals form? Most crystallize from molten rock as it cools. This cooling can occur at or near Earth's surface (low temperature and pressure) or at great depths (high temperature and pressure). The *environment* during crystallization and the *chemical composition of the molten rock* mainly determine which minerals are produced. For example, the silicate mineral olivine crystallizes at high temperatures (about 1200°C), whereas quartz crystallizes at much lower temperatures (about 700°C).

In addition, some silicate minerals form at Earth's surface from the weathered products of other silicate minerals. Clay minerals are an example. Still other silicate minerals are formed under the extreme pressures associated with mountain building. Each silicate mineral, therefore, has a structure and a chemical composition that *indicate the conditions under which it formed.* Thus, by carefully examining the mineral

Mineral	Idealized Formula	Cleavage	Silicate Structure
Olivine	$(Mg, Fe)_2SiO_4$	None	Single tetrahedron
Pyroxene group (Augite)	$(Mg,Fe)SiO_3$	Two planes at right angles	Single chains
Amphibole group (Hornblende)	$Ca_2(Fe,Mg)_5Si_8O_{22}(OH)_2$	Two planes at 60° and 120°	Double chains
Micas — Biotite	$K(Mg,Fe)_3AlSi_3O_{10}(OH)_2$	One plane	Sheets
Micas — Muscovite	$KAl_2(AlSi_3O_{10})(OH)_2$	One plane	Sheets
Feldspars — Orthoclase (Potassium feldspar)	$KAlSi_3O_8$	Two planes at 90°	Three-dimensional networks
Feldspars — Plagioclase	$(Ca,Na)AlSi_3O_8$	Two planes at 90°	Three-dimensional networks
Quartz	SiO_2	None	Three-dimensional networks

Figure 1.12   Common silicate minerals. Note that the complexity of the silicate structure increases down the chart.

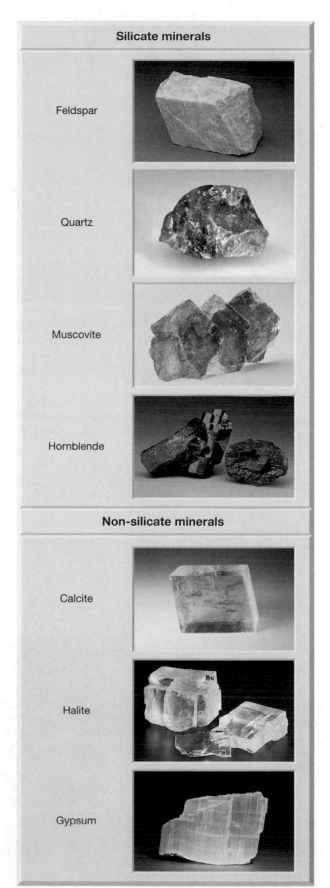

Figure 1.13 Some common rock-forming minerals. Silicate minerals are common constituents of igneous rocks, while the nonsilicates along with quartz and clay are prominent minerals in many sedimentary rocks.

makeup of rocks, geologists can often determine the circumstances under which the rocks formed.

? STUDENTS SOMETIMES ASK...

Are these silicates the same materials used in silicon computer chips and silicone breast implants?

Not really, but all three contain the element silicon (Si). Further, the source of silicon for numerous products, including computer chips and breast implants, comes from silicate minerals. Pure silicon (without the oxygen that silicates have) is used to make computer chips, giving rise to the term "Silicon Valley" for the high-tech region of San Francisco, California's south bay area, where many of these devices are designed. Manufacturers of computer chips engrave silicon wafers with incredibly narrow conductive lines, squeezing millions of circuits into every fingernail-size chip.

Silicone—the material used in breast implants—is a silicon-oxygen polymer gel that feels rubbery and is water repellent, chemically inert, and stable at extreme temperatures. Concern about the long-term safety of these implants limited their use after 1992.

Important Nonsilicate Minerals

Nonsilicate minerals make up only about one-fourth of the continental crust, so they can be considered scarce compared to the silicates. However, many are important economically. Table 1.3 lists examples of nonsilicate minerals of economic value: oxides, sulfides, sulfates, halides, and native elements.

A major rock-forming group worldwide is the **carbonates**, which includes the mineral calcite ($CaCO_3$). This mineral is the major constituent in two well-known rocks: limestone and marble. Limestone is used commercially for road paving, building stone, and as the main ingredient in portland cement. Marble is used decoratively.

Two other nonsilicate minerals frequently found in sedimentary rocks are *halite* and *gypsum*. Both minerals are commonly found in thick layers, which are the last vestiges of ancient seas that long ago evaporated, leaving behind their salts (Figure 1.14). Like limestone, halite and gypsum both are important nonmetallic resources. Halite is the mineral name for common table salt, sodium chloride. Gypsum is the mineral from which plaster and other similar building materials are composed.

In addition, a number of other minerals have economic value (Table 1.3). Ores of metals include hematite (iron), sphalerite (zinc), and galena (lead). Native (free-occurring, not in compounds) elements include gold, silver, copper, and carbon (diamonds and graphite). Other important minerals are fluorite, corundum, and sylvite (see Box 1.2).

Mineral Group	Name	Chemical Formula	Economic Use
Oxides	Hematite	Fe_2O_3	Ore of iron, pigment
	Magnetite	Fe_3O_4	Ore of iron
	Corundum	Al_2O_3	Gemstone, abrasive
	Ice	H_2O	Solid form of water
Sulfides	Galena	PbS	Ore of lead
	Sphalerite	ZnS	Ore of zinc
	Pyrite	FeS_2	Sulfuric acid production
	Chalcopyrite	$CuFeS_2$	Ore of copper
	Cinnabar	HgS	Ore of mercury
Sulfates	Gypsum	$CaSO_4 \cdot 2H_2O$	Plaster
	Anhydrite	$CaSO_4$	Plaster
	Barite	$BaSO_4$	Drilling mud
Native elements	Gold	Au	Trade, jewelry
	Copper	Cu	Electrical conductor
	Diamond	C	Gemstone, abrasive
	Sulfur	S	Sulfa drugs, chemicals
	Graphite	C	Pencil lead, dry lubricant
	Silver	Ag	Jewelry, photography
	Platinum	Pt	Catalyst
Halides	Halite	NaCl	Common salt
	Fluorite	CaF_2	Used in steel making
	Sylvite	KCl	Fertilizer
Carbonates	Calcite	$CaCO_3$	Portland cement, lime
	Dolomite	$CaMg(CO_3)_2$	Portland cement, lime

Table 1.3 Common nonsilicate mineral groups.

Figure 1.14 Thick bed of halite (salt) at an underground mine near Lansing, New York. (Photo by Michael J. Okoniewski/Liaison Agency, Inc.)

Mineral Resources

Mineral resources are Earth's storehouse of useful minerals that can be recovered for use. Resources include already identified deposits from which minerals can be extracted profitably, called **reserves,** as well as known deposits that are not yet recoverable under present economic conditions or technology. Deposits inferred to exist but not yet discovered are also considered mineral resources.

The term **ore** denotes useful metallic minerals that can be mined at a profit. In common usage, the term *ore* is also applied to some nonmetallic minerals, such as fluorite and sulfur. However, materials used for such purposes as building stone, road paving, abrasives, ceramics, and fertilizers are not usually called ores; rather, they are classified as *industrial rocks* and *minerals.*

Recall that more than 98 percent of Earth's crust is composed of only eight elements. Except for oxygen and silicon, all other elements make up a relatively small fraction (about one-fourth) of common crustal rocks (see Table 1.2). Indeed, the natural concentrations of many elements are exceedingly small. A deposit containing only the average crustal percentage of a valuable element like gold is worthless because the cost of extracting it greatly exceeds the value of the material recovered.

To be considered of value, an element must be concentrated above the level of its average crustal abundance. For example, copper makes up about 0.0135 percent of the crust. However, for a material to be considered as copper ore, it must contain a concentration that is about 100 times this amount. Aluminum, in contrast, represents 8.13 percent of the crust and must be concentrated to only about four times its average crustal percentage before it can be extracted profitably.

It is important to realize that economic changes may make a deposit profitable to extract, or lose its profitability. If demand for a metal increases and prices rise sufficiently, the status of a previously unprofitable deposit changes, and it becomes an ore. The status of unprofitable deposits may also change if a technological advance allows the ore to be extracted at a lower cost than before.

Conversely, changing economic factors can turn a once profitable ore deposit into an unprofitable deposit that can no longer be called an ore. This situation is illustrated by a copper mine at Bingham Canyon, Utah, one of the largest open-pit mines on Earth (Figure 1.15). Mining was halted there in 1985 because outmoded equipment had driven the cost of extracting the copper beyond the current selling price. The owners responded by replacing an antiquated 1000-car railroad with conveyor belts and pipelines for transporting the ore and waste. These devices achieved a cost reduction of nearly 30 percent and returned this mining operation to profitability.

Figure 1.15 Aerial view of Bingham Canyon copper mine near Salt Lake City, Utah. This huge open-pit mine is about 4 kilometers across and 900 meters deep. Although the amount of copper in the rock is less than 1 percent, the huge volumes of material removed and processed each day (about 200,000 tons) yield significant quantities of metal. (Photo by Michael Collier)

Chapter Summary

• A *mineral* is a naturally occurring inorganic solid possessing a definite chemical structure that gives it a unique set of physical properties. Most *rocks* are aggregates composed of two or more minerals.

• The building blocks of minerals are *elements*. An *atom* is the smallest particle of matter that still retains the characteristics of an element. Each atom has a *nucleus* containing *protons* and *neutrons*. Orbiting the nucleus of an atom are *electrons*. The number of protons in an atom's nucleus determines its *atomic number* and the name of the element. Atoms bond together to form a *compound* by either gaining, losing, or sharing electrons with another atom.

• *Isotopes* are variants of the same element but with a different *mass number* (the total number of neutrons plus protons found in an atom's nucleus). Some isotopes are unstable and disintegrate naturally through a process called *radioactive decay*.

• The properties of minerals include *crystal form, luster, color, streak, hardness, cleavage, fracture,* and *specific gravity*. In addition, a number of special physical and chemical properties (*taste, smell, elasticity, malleability, feel, magnetism, double refraction,* and *chemical reaction to hydrochloric acid*) are useful in identifying certain minerals. Each mineral has a unique set of properties that can be used for identification.

• The eight most abundant elements found in Earth's continental crust (oxygen, silicon, aluminum, iron, calcium, sodium, potassium, and magnesium) also make up the majority of minerals.

• The most common mineral group is the *silicates*. All silicate minerals have the *silicon-oxygen tetrahedron* as their fundamental building block. In some silicate minerals the tetrahedra are joined in chains; in others the tetrahedra are arranged into sheets, or three-dimensional networks. Each silicate mineral has a structure and a chemical composition that indicates the conditions under which it was formed.

• The *nonsilicate* mineral groups include the *oxides* (e.g., magnetite, mined for iron), *sulfides* (e.g., sphalerite, mined for zinc), *sulfates* (e.g., gypsum, used in plaster and frequently found in sedimentary rocks), *native elements* (e.g., graphite, a dry lubricant), *halides* (e.g., halite, common salt and frequently found in sedimentary rocks), and *carbonates* (e.g., calcite, used in portland cement and a major constituent in two well-known rocks: limestone and marble).

• The term *ore* is used to denote useful metallic minerals, like hematite (mined for iron) and galena (mined for lead), that can be mined for a profit, as well as some nonmetallic minerals, such as fluorite and sulfur, that contain useful substances.

Key Terms

atom (p. 22)
atomic number (p. 23)
carbonates (p. 33)
cleavage (p. 27)
color (p. 26)
compound (p. 23)
crystal form (p. 26)
electron (p. 23)
energy levels (shells) (p. 23)
element (p. 22)
feldspar (p. 30)
fracture (p. 27)

hardness (p. 27)
ions (p. 23)
isotope (p. 25)
luster (p. 26)
mass number (p. 24)
mineral (p. 21)
mineralogy (p. 21)
mineral resource (p. 35)
Mohs hardness scale (p. 27)
neutron (p. 23)
nucleus (p. 23)

ore (p. 35)
proton (p. 23)
quartz (p. 31)
radioactive decay (p. 26)
reserve (p. 35)
rock (p. 21)
rock-forming minerals (p. 29)
silicate (p. 29)
silicon-oxygen tetrahedron (p. 30)
specific gravity (p. 28)
streak (p. 27)

Review Questions

1. List five characteristics an Earth material must have in order to be considered a mineral.

2. Define the term *rock*.

3. List the three main particles of an atom and explain how they differ from one another.

4. If the number of electrons in an atom is 35 and its mass number is 80, calculate the following:
   (a) The number of protons.
   (b) The atomic number.
   (c) The number of neutrons.

5. What occurs in an atom to produce an ion?

6. What is an isotope?

7. Although all minerals have an orderly internal arrangement of atoms (crystalline structure), most mineral samples do not visibly demonstrate their crystal form. Why?

8. Why might it be difficult to identify a mineral by its color?

9. If you found a glassy-appearing mineral while rock hunting and had hopes that it was a diamond, what simple test might help you make a determination?

10. Table 1.3 (p. 34) lists a use for corundum as an abrasive. Explain why it makes a good abrasive in terms of the Mohs hardness scale.

11. Gold has a specific gravity of almost 20. A five-gallon pail of water weighs about 40 pounds, how much would a five-gallon pail of gold weigh?

12. What are the two most common elements in Earth's crust? What is the term used to describe the basic building block of all silicate minerals?

13. What are the two most common silicate minerals?

14. List three nonsilicate minerals that are commonly found in rocks.

15. Contrast a mineral *resource* and a mineral *reserve*.

16. What might cause a mineral deposit that had not been considered an ore to become reclassified as an ore?

Examining the Earth System

1. Perhaps one of the most significant interrelationships between humans and the Earth system involves the extraction, refinement, and distribution of the planet's mineral wealth. To help you understand these associations, begin by thoroughly researching a mineral commodity that is mined in your local region or state. (You might find the information at these United States Geological Survey (USGS) Websites, **http://minerals.er.usgs.gov/minerals/pubs/state/mo.html** and **http://minerals.er.usgs.gov:80/minerals/pubs/mcs/,** useful.) What products are made from this mineral? Do you use any of these products? Describe the mining and refining of the mineral and the local impact these processes have on each of Earth's spheres (atmosphere, hydrosphere, solid Earth, and biosphere). Are any of the effects negative? If so, what, if anything, is being done to end or minimize the damage?

2. Referring to the mineral you described above, in your opinion does the environmental impact of extracting this mineral outweigh the benefits derived from the products produced from this mineral?

Web Resources

The *Earth Science* Website uses the resources and flexibility of the Internet to aid in your study of the topics in this chapter. Written and developed by Earth science instructors, this site will help improve your understanding of Earth science. Visit **http://www.prenhall.com/tarbuck** and click on the cover of *Earth Science, 10e* to find:

- **Online review quizzes.**
- **Web-based critical thinking and writing exercises.**
- **Links to chapter-specific Web resources.**
- **Internet-wide key term searches.**

http://www.prenhall.com/tarbuck

Climbers ascending the colorful sedimentary rocks in Arizona's Oak Creek Canyon (Photo by Mark Muench)

CHAPTER 2

ROCKS: MATERIALS OF THE SOLID EARTH

Why study rocks? You have already learned that rocks and minerals have great economic value. Furthermore, all Earth processes in some way depend on the properties of these basic materials. Events such as volcanic eruptions, mountain building, weathering, erosion, and even earthquakes involve rocks and minerals. Consequently, a basic knowledge of Earth materials is essential in understanding Earth phenomena.

Every rock contains clues about the environments in which it formed (Figure 2.1). For example, some rocks are composed entirely of small shell fragments. This tells Earth scientists that the particles composing the rock originated in a shallow marine environment. Other rocks contain clues that indicate they formed from a volcanic eruption or deep in Earth during mountain building. Thus, rocks contain a wealth of information about events that have occurred over Earth's long history.

We divide rocks into three groups based on their mode of origin. The groups are igneous, sedimentary, and metamorphic. Before examining each group, we will view the *rock cycle*, which depicts the interrelationships among these rock groups.

Earth as a System: The Rock Cycle

GEODe Earth Materials
↳ The Rock Cycle

Earth is a system. This means that our planet consists of many interacting parts that form a complex whole. Nowhere is this idea better illustrated than when we examine the rock cycle (Figure 2.2). The **rock cycle** allows us to view many of the interrelationships among different parts of the Earth system. It helps us understand the origin of igneous, sedimentary, and metamorphic rocks and to see that each type is linked to the others by the processes that act upon and within the planet. Learn the rock cycle well; you will be examining its interrelationships in greater detail throughout this chapter and many other chapters as well.

Figure 2.1 Rocks contain information about the processes that produce them. This massive granitic monolith (El Capitan) located in Yosemite National Park, California, was once a molten mass found deep within Earth. (Photo by Jeff Gness/The Stock Market)

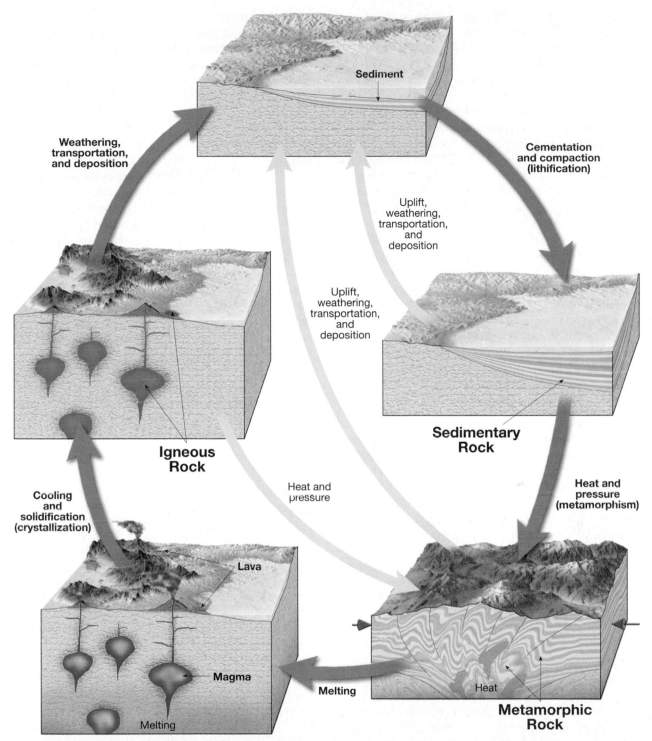

Figure 2.2   Viewed over long spans, rocks are constantly forming, changing, and reforming. The rock cycle helps us understand the origin of the three basic rock groups. Arrows represent processes that link each group to the others.

The Basic Cycle

We begin at the bottom of Figure 2.2. *Magma* is molten material that forms inside Earth. Eventually magma cools and solidifies. This process, called *crystallization*, may occur either beneath the surface or, following a volcanic eruption, at the surface. In either situation, the resulting rocks are called **igneous rocks** (*ignis* = fire).

If igneous rocks are exposed at the surface, they will undergo *weathering*, in which the day-in and day-out influences of the atmosphere slowly disintegrate and decompose rocks. The materials that result are often moved downslope by gravity before being picked up and transported by any of a number of erosional agents, such as running water, glaciers, wind, or waves. Eventually these particles and dissolved substances, called *sediment*, are deposited. Although most sediment ultimately comes to rest in the ocean, other sites of deposition include river floodplains, desert basins, swamps, and dunes.

Next the sediments undergo *lithification*, a term meaning "conversion into rock." Sediment is usually lithified into *sedimentary rock* when compacted by the weight of overlying layers or when cemented as percolating groundwater fills the pores with mineral matter.

If the resulting sedimentary rock is buried deep within Earth and involved in the dynamics of mountain building or intruded by a mass of magma, it will be subjected to great pressures and/or intense heat. The sedimentary rock will react to the changing environment and turn into the third rock type, **metamorphic rock**. When metamorphic rock is subjected to additional pressure changes or to still higher temperatures, it will melt, creating magma, which will eventually crystallize into igneous rock.

Processes driven by heat from Earth's interior are responsible for creating igneous and metamorphic rocks. Weathering and erosion, external processes powered by energy from the Sun, produce the sediment from which sedimentary rocks form.

Alternative Paths

The paths shown in the basic cycle are not the only ones that are possible. To the contrary, other paths are just as likely to be followed as those described in the preceding section. These alternatives are indicated by the blue arrows in Figure 2.2.

Igneous rocks, rather than being exposed to weathering and erosion at Earth's surface, may remain deeply buried. Eventually these masses may be subjected to the strong compressional forces and high temperatures associated with mountain building. When this occurs, they are transformed directly into metamorphic rocks.

Metamorphic and sedimentary rocks, as well as sediment, do not always remain buried. Rather, overlying layers may be stripped away, exposing the once buried rock. When this happens, the material is attacked by weathering processes and turned into new raw materials for sedimentary rocks.

Although rocks may seem to be unchanging masses, the rock cycle shows that they are not. The changes, however, take time—great amounts of time.

Igneous Rocks: "Formed by Fire"

 GEODe
Earth materials
↳ Igneous Rocks

In our discussion of the rock cycle, we pointed out that igneous rocks form as **magma** cools and crystallizes. This molten rock, which originates at depths as great

as 200 kilometers within Earth, consists primarily of the elements found in silicate minerals (silicon and oxygen, plus aluminum, iron, calcium, sodium, potassium, magnesium, and others). Magma also contains some gases, particularly water vapor, which are confined within the magma by the surrounding rocks. Because the magma body is less dense than the surrounding rocks, it works its way toward the surface over time spans of thousands to millions of years. On occasion, the magma breaks through the surface, producing a volcanic eruption.

The spectacular explosions that sometimes accompany an eruption are produced by the gases that escape as the confining pressure lessens near the surface. Sometimes blockage of the volcanic vent, coupled with surface-water seepage into the magma chamber, can produce catastrophic explosions.

Along with ejected rock fragments, a volcanic eruption often generates extensive lava flows. **Lava** is similar to magma, except that most of the gaseous component has escaped (Figure 2.3). The rocks that result when lava solidifies are classified as **volcanic** or **extrusive**, because they are *extruded* onto the surface. The magma not able to reach the surface eventually crystallizes at depth. Igneous rocks produced in this manner are termed **plutonic**, or **intrusive**, because they *intrude* existing rocks. These deep rocks would never be observed were it not for erosion stripping away the overlying rock.

Magma Crystallizes to Form Igneous Rocks

Magma is basically a very hot, thick fluid, but it also contains solids and gases. The solids are mineral crystals. The liquid portion of a magma body, termed the *melt*, is composed of ions that move about freely. However, as magma cools, the random movements of the ions slow, and the ions begin to arrange themselves into orderly patterns. This process is called **crystallization**. Usually all of the molten material does not solidify at the same time. Rather, as it cools, numerous small crystals develop. In a systematic fashion, ions are added to these centers of crystal growth. When the crystals grow large enough for their edges to meet, their growth ceases for lack of space, and crystallization continues elsewhere. Eventually, all of the liquid is transformed into a solid mass of interlocking crystals.

The rate of cooling strongly influences crystal size. If magma cools very slowly, relatively few centers of crystal growth develop. Slow cooling also allows ions to migrate over relatively great distances. Consequently, *slow cooling results in the formation of large crystals*. Conversely, if cooling occurs quite rapidly, the ions lose their motion and quickly combine. This results in a large number of tiny crystals that all compete for the available ions. Therefore, the outcome of rapid cooling is the formation of a solid mass of *small intergrown crystals*.

Figure 2.3 Fluid basaltic lava moving down the slopes of Hawaii's Kilauea Volcano Inset photo is a sample of the igneous rock basalt, which forms as lava solidifies. (Photo by G. Brad Lewis/Liaison Agency, Inc. Inset photo by E. J. Tarbuck.)

Thus, if a geologist discovers rock containing large, finger-size crystals, it means the rock cooled very slowly. But if the crystals can be seen only with a microscope, the geologist knows that the magma cooled very quickly.

If the molten material is quenched almost instantly, there is not sufficient time for the ions to arrange themselves into a crystalline network. Therefore, solids produced in this manner consist of randomly distributed ions. Such rocks are called *glass* and are quite similar to ordinary manufactured glass. Instant quenching occurs during violent volcanic eruptions that produce tiny shards of glass termed volcanic ash.

? STUDENTS SOMETIMES ASK...
Are lava and magma the same thing?

No, but their *composition* might be similar. Both are terms that describe molten or liquid rock: Magma exists beneath Earth's surface, and lava is molten rock that has reached the surface. That's the reason why they can be similar in composition: Lava is produced from magma, but it generally has lost materials that escape as a gas, such as water vapor.

Classifying Igneous Rocks

In addition to the rate of cooling, the composition of a magma and the amount of dissolved gases influence crystallization. Because magmas differ in each of these aspects, the physical appearance and mineral composition of igneous rocks vary widely. Nevertheless, it is possible to classify **igneous rocks** based on their *texture* and *mineral constituents*. We will now look at both features.

Igneous Textures. **Texture** describes the overall appearance of an igneous rock based on the *size*, *shape*, and *arrangement* of its interlocking crystals. Texture is a very important characteristic, because it reveals a great deal about the environment in which the rock formed. You learned that rapid cooling produces small crystals, whereas very slow cooling produces much larger crystals. As you might expect, the rate of cooling is quite slow in magma chambers lying deep within the crust, whereas a thin layer of lava extruded upon Earth's surface may chill rock solid in a matter of hours. Small molten blobs ejected into the air during a violent eruption can solidify almost instantly.

Igneous rocks that form rapidly at the surface or as small masses within the upper crust have a **fine-grained texture**, with the individual crystals too small to be seen with the unaided eye (Figure 2.4A). Common in many

A.

B.

Figure 2.4   **A**. Rhyolite exhibits a fine-grained texture. **B**. Granite is a common igneous rock that has a coarse-grained texture. (Photos by E. J. Tarbuck)

fine-grained igneous rocks are voids, called *vesicles,* left by gas bubbles that formed as the lava solidified (Figure 2.5).

When large masses of magma solidify far below the surface, they form igneous rocks that exhibit a **coarse-grained texture**. These coarse-grained rocks have the appearance of a mass of intergrown crystals, which are roughly equal in size and large enough that the individual minerals can be identified with the unaided eye. Granite is a classic example (Figure 2.4B).

A large mass of magma located at depth may require tens of thousands of years to solidify. Because all materials within a magma do not crystallize at the same rate or at the same time during cooling, it is possible for some crystals to become quite large before others even

start to form. If magma that already contains some large crystals suddenly erupts at the surface, the remaining molten portion of the lava would cool quickly. The resulting rock, which has large crystals embedded in a matrix of smaller crystals, is said to have a **porphyritic texture** (Figure 2.6).

During some volcanic eruptions, molten rock is ejected into the atmosphere, where it is quenched very quickly. Rapid cooling of this type may generate rock with a **glassy texture**. As was indicated, glass results when the ions do not have sufficient time to unite into an orderly crystalline structure. *Obsidian*, a common type of natural glass, is similar in appearance to a dark chunk of manufactured glass (Figure 2.7A). Another volcanic rock that exhibits a glassy texture is *pumice*.

Figure 2.5   Scoria is a volcanic rock that exhibits vesicles, which form as gas bubbles escape near the top of a lava flow. (Photo courtesy of E. J. Tarbuck)

Figure 2.6   Andesite porphyry. Notice the two distinctively different sizes of crystals. (Photo by E. J. Tarbuck)

A   B

Figure 2.7   Igneous rocks that exhibit a glassy texture. **A.** Obsidian, a glassy volcanic rock. **B.** Pumice, a glassy rock containing numerous tiny voids. (Photos courtesy of E. J. Tarbuck)

Usually found with obsidian, pumice forms when large amounts of gas escape from a melt to generate a gray, frothy mass (Figure 2.7B). In some samples, the vesicles are quite noticeable; in others, the pumice resembles fine shards of intertwined glass. Because of the large volume of air-filled voids, many samples of pumice will float in water.

Mineral Composition.   The mineral makeup of an igneous rock depends on the chemical composition of the magma from which it crystallizes. Because a large variety of igneous rocks exists, you might assume an equally large variety of magmas must also exist. However, various eruptive stages of the same volcano often extrude lavas exhibiting somewhat different mineral compositions, particularly if long periods separate the eruptions. Such evidence led geologists to examine the possibility that a single magma might produce rocks of varying mineral content.

N. L. Bowen studied this idea in the first quarter of the twentieth century. Bowen discovered that as magma cools in the laboratory, certain minerals crystallize first, at very high temperatures (top of Figure 2.8). At successively lower temperatures, other minerals crystallize.

Bowen also demonstrated that if a mineral remains in the molten solution after crystallization, it will react with the remaining liquid to produce the next mineral in the sequence shown in Figure 2.8. For this reason, this arrangement of minerals became known as **Bowen's reaction series**.

On the upper left branch of this reaction series, olivine, the first mineral to form, begins to react with the remaining melt to produce the mineral pyroxene. Pyroxene, in turn, may react with the melt to produce the mineral amphibole, and so forth. The right branch of the reaction series is a continuum in which the earliest-formed calcium-rich feldspar crystals react with the sodium ions contained in the melt to become progressively more sodium-rich. Rarely are these reactions complete, so that various amounts of several of these minerals typically exist at any given time.

During the last stage of crystallization, after most of the magma is solidified, the minerals muscovite and potassium feldspar may form. Finally, quartz crystalizes from any remaining liquid.

A consequence of this crystallization scheme is that minerals that form at about the same temperature are

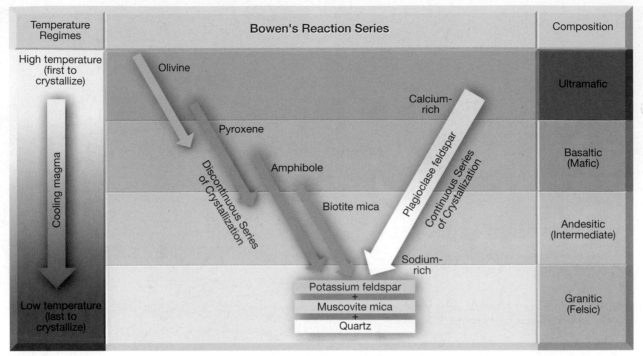

Figure 2.8   Bowen's reaction series shows the sequence in which minerals crystallize from a magma. Compare this figure to the mineral composition of the rock groups in Figure 2.10. Note that each rock group consists of minerals that crystallize in the same temperature range.

found together in the same igneous rock. For example, notice that the minerals potassium feldspar, muscovite mica, and quartz are found in the same region of the diagram, and they are the major constituents of the igneous rock granite.

How does Bowen's reaction series account for the great diversity of igneous rocks? During crystallization, a separation of the solid and liquid components of a magma can occur. This happens if the earlier formed minerals are denser (heavier) than the liquid portion and settle to the bottom, as shown in Figure 2.9. This process, called **crystal settling**, is thought to occur with dense silicate minerals, such as olivine and pyroxene. When the remaining melt solidifies—either in place or in another location if it migrates into fractures in the surrounding rocks—it will form a rock with a chemical composition much different from the parent magma. Consequently, the separation of the solid and liquid components at various stages of crystallization can produce a wide variety of igneous rocks.

Naming Igneous Rocks

Igneous rocks are classified by their texture and mineral composition. Various igneous textures result from different cooling histories, while the mineral compositions are a consequence of the chemical makeup of the parent magma and the environment of crystallization.

Despite their great compositional diversity, igneous rocks can be divided into broad groups according to

their proportions of light and dark minerals. A general classification scheme based on texture and mineral composition is provided in Figure 2.10.

Granitic Rocks.   Near one end of the continuum are rocks composed almost entirely of light-colored silicates—quartz and feldspar. Igneous rocks in which these are the dominant minerals have a **granitic composition**. Geologists also refer to granitic rocks as being **felsic**, a term derived from *fe*ldspar and *si*lica (quartz). In addition to quartz and feldspar, most granitic rocks contain about 10 percent dark silicate minerals, usually biotite mica and amphibole. Granitic rocks are rich in silica (about 70 percent) and are major constituents of the continental crust.

Granite is a coarse-grained igneous rock that forms where large masses of magma slowly solidify at depth. During episodes of mountain building, granite and related crystalline rocks may be uplifted, whereupon the processes of weathering and erosion strip away the overlying crust. Pikes Peak in the Rockies, Mount Rushmore in the Black Hills, Stone Mountain in Georgia, and Yosemite National Park in the Sierra Nevada are all areas where large quantities of granite are exposed at the surface.

Granite is perhaps the best-known igneous rock (see Figure 2.4B). This is partly because of its natural beauty, which is enhanced when polished, and partly because of its abundance. Slabs of polished granite are commonly used for tombstones, monuments, and as building stones.

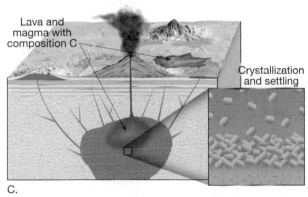

Figure 2.9   Illustration of how a magma evolves as the earliest-formed minerals (those richer in iron, magnesium, and calcium) crystallize and settle to the bottom of the magma chamber, leaving the remaining melt richer in sodium, potassium, and silica (SiO_2). **A.** Emplacement of a magma body and associated igneous activity generates rocks having a composition similar to that of the initial magma. **B.** After a period of time, crystallization and settling changes the composition of the melt, while generating rocks having a composition quite different than the original magma. **C.** Further magmatic differentiation results in another more highly evolved melt with its associated rock types.

Rhyolite is the extrusive equivalent of granite and, like granite, is composed essentially of the light-colored silicates (see Figure 2.4A). This fact accounts for its color, which is usually buff to pink or light gray. Rhyolite is fine-grained and frequently contains glass fragments and voids, indicating rapid cooling in a surface environment. In contrast to granite, which is widely dis-

tributed as large plutonic masses, rhyolite deposits are less common and generally less voluminous. Yellowstone Park is one well-known exception. Here rhyolite lava flows and thick ash deposits of similar composition are extensive.

Basaltic Rocks.   Rocks that contain substantial dark silicate minerals and calcium-rich plagioclase feldspar (but no quartz) are said to have a **basaltic composition** (Figure 2.10). Because basaltic rocks contain a high percentage of dark silicate minerals, geologists also refer to them as **mafic** (from *ma*gnesium and *f*errum, the Latin name for iron). Because of their iron content, basaltic rocks are typically darker and denser than granitic rocks.

Basalt is a very dark green to black fine-grained volcanic rock composed primarily of pyroxene and plagioclase feldspar (see Figure 2.3). Basalt is the most common extrusive igneous rock. Many volcanic islands, such as the Hawaiian Islands and Iceland, are composed mainly of basalt. Further, the upper layers of the oceanic crust consist of basalt. In the United States, large portions of central Oregon and Washington were the sites of extensive basaltic outpourings (see Figure 8.19).

Other Compositional Groups.   As you can see in Figure 2.10, rocks with a composition between granitic and basaltic rocks are said to have an **andesitic** or **intermediate composition** after the common volcanic rock *andesite*. Andesitic rocks contain at least 25 percent dark silicate minerals, mainly amphibole, pyroxene, and biotite mica, with the other dominant mineral being plagioclase feldspar. This important category of igneous rocks is associated with volcanic activity that is typically confined to the margins of continents.

Another important igneous rock, *peridotite*, contains mostly olivine and pyroxene and thus falls on the opposite side of the compositional spectrum from granitic rocks (Figure 2.10). Because peridotite is composed almost entirely of dark silicate minerals, its chemical composition is referred to as **ultramafic**. Although ultramafic rocks are rare at Earth's surface, peridotite is believed to be the main constituent of the upper mantle.

Sedimentary Rocks: Compacted and Cemented Sediment

   Earth Materials
↳ Sedimentary Rocks

Recall the rock cycle, which shows the origin of sedimentary rocks. Weathering begins the process. Next, gravity and erosional agents (running water, wind,

iron, magnesium

Chemical Composition	Granitic (Felsic)	Andesitic (Intermediate)	Basaltic (Mafic)	Ultramafic
Dominant Minerals	Quartz Potassium feldspar Sodium-rich plagioclase feldspar	Amphibole Sodium- and calcium-rich plagioclase feldspar	Pyroxene Calcium-rich plagioclase feldspar	Olivine Pyroxene
Phaneritic (coarse-grained) *Plutonic*	**Granite**	**Diorite**	**Gabbro**	**Peridotite**
Aphanitic (fine-grained) *volcanic*	**Rhyolite**	**Andesite**	**Basalt**	**Komatiite** (rare)
Porphyritic	"Porphyritic" precedes any of the above names whenever there are appreciable phenocrysts			Uncommon
Glassy	**Obsidian** (compact glass) **Pumice** (frothy glass)			
Rock Color (based on % of dark minerals)	0% to 25%	25% to 45%	45% to 85%	85% to 100%

(left column label: T E X T U R E)

Figure 2.10 Classification of the major groups of igneous rocks based on their mineral composition and texture. Coarse-grained rocks are plutonic, solidifying deep underground. Fine-grained rocks are volcanic, or solidify as shallow, thin plutons. Ultramafic rocks are dark, dense rocks, composed almost entirely of minerals containing iron and magnesium. Although relatively rare on Earth's surface, these rocks are believed to be major constituents of the upper mantle.

waves, and glacial ice) remove the products of weathering and carry them to a new location where they are deposited. Usually the particles are broken down further during this transport phase. Following deposition, this **sediment** may become lithified, or turned to rock. Commonly, *compaction* and *cementation* transform the sediment into solid sedimentary rock.

The word *sedimentary* indicates the nature of these rocks, for it is derived from the Latin *sedimentum*, which means settling, a reference to a solid material settling out of a fluid. Most sediment is deposited in this fashion. Weathered debris is constantly being swept from bedrock and carried away by water, ice, or wind. Eventually the material is deposited in lakes, river valleys, seas, and countless other places. The particles in a desert sand dune, the mud on the floor of a swamp, the gravels in a stream bed, and even household dust are examples of sediment produced by this never-ending process.

The weathering of bedrock and the transport and deposition of the weathering products are continuous. As piles of sediment accumulate, the materials near the bottom are compacted by the weight of the overlying layers. Over long periods, these sediments are cemented together by mineral matter deposited from water in the spaces between particles. This forms solid sedimentary rock.

Geologists estimate that sedimentary rocks account for only about 5 percent (by volume) of Earth's outer 16 kilometers (10 miles). However, the importance of this group of rocks is far greater than this percentage

implies. If you sampled the rocks exposed at Earth's surface, you would find that the great majority are sedimentary (Figure 2.11). Indeed, about 75 percent of all rock outcrops on the continents are sedimentary. Therefore, we can think of sedimentary rocks as comprising a relatively thin and somewhat discontinuous layer in the uppermost portion of the crust. This makes sense because sediment accumulates at the surface.

It is from sedimentary rocks that geologists reconstruct many details of Earth's history. Because sediments are deposited in a variety of different settings at the surface, the rock layers that they eventually form hold many clues to past surface environments. They may also exhibit characteristics that allow geologists to decipher information about the method and distance of sediment transport. Furthermore, it is sedimentary rocks that contain fossils which are vital evidence in the study of the geologic past.

Finally, many sedimentary rocks are very important economically. Coal, for example, is a sedimentary rock that presently is the source for over half the electric power generated in the United States. Other major energy resources (petroleum and natural gas) occur in pores within sedimentary rocks. Other sedimentary rocks are major sources of iron, aluminum, manganese, fertilizer, and sand and gravel.

Classifying Sedimentary Rocks

Materials accumulating as sediment have two principal sources. First, sediments may originate as solid particles from weathered rocks, such as the igneous rocks

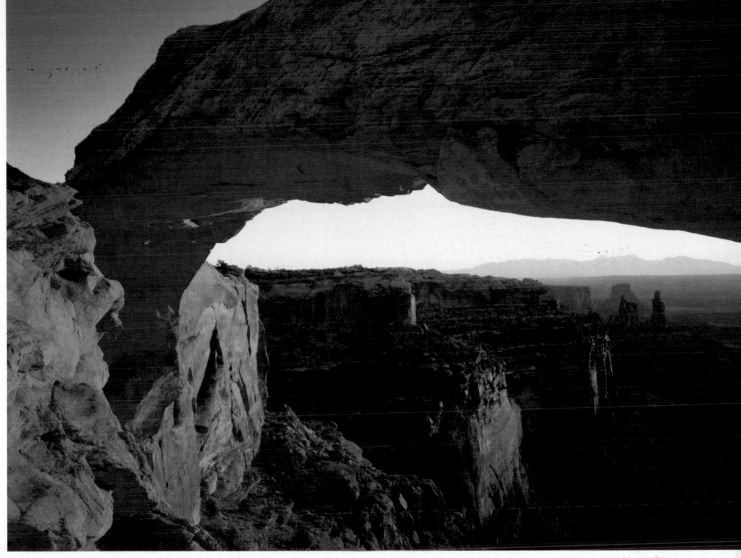

Figure 2.11 Sedimentary rocks exposed in Canyonlands National Park, Utah. Sedimentary rocks occur in layers called strata. About 75 percent of all rock outcrops on the continents are sedimentary rocks. (Photo by Jeff Gnass)

earlier described. These particles are called *detritus*, and the sedimentary rocks that they form are called **detrital sedimentary rocks** (Figure 2.12).

The second major source of sediment is soluble material produced largely by chemical weathering. When these dissolved substances are precipitated back as solids, they are called chemical sediment, and they form **chemical sedimentary rocks**. We will now look at detrital and chemical sedimentary rocks.

Detrital Sedimentary Rocks. Though a wide variety of minerals and rock fragments may be found in detrital rocks, clay minerals and quartz dominate. Clay minerals are the most abundant product of the chemical weathering of silicate minerals, especially the feldspars. Quartz, in contrast, is abundant because it is extremely durable and very resistant to chemical weathering. Thus, when igneous rocks such as granite are weathered, individual quartz grains are set free.

Geologists use particle size to distinguish among detrital sedimentary rocks. Figure 2.12 presents the four size categories for particles making up detrital rocks. When gravel-size particles predominate, the rock is

called *conglomerate* if the sediment is rounded (Figure 2.13A) and *breccia* if the pieces are angular (Figure 2.13B). Angular fragments indicate that the particles were not transported very far from their source prior to deposition and so have not had corners and rough edges abraded. *Sandstone* is the name given rocks when sand-size grains prevail (Figure 2.13C). *Shale*, the most common sedimentary rock, is made of very fine-grained sediment (Figure 2.13D). *Siltstone*, another rather fine-grained rock, is sometimes difficult to differentiate from rocks such as shale that are composed of even smaller clay-size sediment.

Particle size is not only a convenient method of dividing detrital rocks; the sizes of the component grains also provide useful information about the environment in which the sediment was deposited. Currents of water or air sort the particles by size. The stronger the current, the larger the particle size carried. Gravels, for example, are moved by swiftly flowing rivers, rockslides, and glaciers. Less energy is required to transport sand; thus, it is common in windblown dunes, river deposits, and beaches. Because silts and clays settle very slowly, accumulations of these materials are generally associated

49

Detrital Sedimentary Rocks		
Texture (grain size)	Sediment Name	Rock Name
Coarse (over 2 mm)	Gravel (Rounded fragments)	Conglomerate
Coarse (over 2 mm)	Gravel (Angular fragments)	Breccia
Medium (1/16 to 2 mm)	Sand (If abundant feldspar is present the rock is called **Arkose**)	Sandstone
Fine (1/16 to 1/256 mm)	Mud	Siltstone
Very fine (less than 1/256 mm)	Mud	Shale

Chemical Sedimentary Rocks			
Composition	Texture (grain size)	Rock Name	
Calcite, $CaCO_3$	Fine to coarse crystalline	Crystalline Limestone	
Calcite, $CaCO_3$	Fine to coarse crystalline	Travertine	
Calcite, $CaCO_3$	Visible shells and shell fragments loosely cemented	Coquina	Biochemical Limestone
Calcite, $CaCO_3$	Various size shells and shell fragments cemented with calcite cement	Fossiliferous Limestone	Biochemical Limestone
Calcite, $CaCO_3$	Microscopic shells and clay	Chalk	Biochemical Limestone
Quartz, SiO_2	Very fine crystalline	Chert (light colored) Flint (dark colored)	
Gypsum $CaSO_4 \cdot 2H_2O$	Fine to coarse crystalline	Rock Gypsum	
Halite, NaCl	Fine to coarse crystalline	Rock Salt	
Altered plant fragments	Fine-grained organic matter	Bituminous Coal	

Figure 2.12 Identification of sedimentary rocks. Sedimentary rocks are divided into two major groups, detrital and chemical, based on their source of sediment. The main criterion for naming detrital rocks is particle size, whereas the primary basis for distinguishing among chemical rocks is their mineral composition.

with the quiet waters of a lake, lagoon, swamp, or marine environment.

Although detrital sedimentary rocks are classified by particle size, in certain cases the mineral composition is also part of naming a rock. For example, most sandstones are predominantly quartz-rich and are often referred to as quartz sandstone. In addition, rocks consisting of detrital sediments are rarely composed of grains of just one size. Consequently, a rock containing quantities of both sand and silt can be correctly classified as sandy siltstone or silty sandstone, depending on which particle size dominates.

Chemical Sedimentary Rocks. In contrast to detrital rocks, which form from the solid products of weathering, chemical sediments are derived from material that is carried in solution to lakes and seas. This material does not remain dissolved in the water indefinitely. When conditions are right, it precipitates to form chemical sediments. This precipitation may occur directly as the result of physical processes, or indirectly through life processes of water-dwelling organisms. Sediment formed in this second way has a *biochemical* origin (see Box 2.1).

An example of a deposit resulting from physical processes is the salt left behind as a body of salt water evaporates. In contrast, many water-dwelling animals and plants extract dissolved mineral matter to form shells and other hard parts. After the organisms die, their skeletons may accumulate on the floor of a lake or ocean.

Limestone is the most abundant chemical sedimentary rock. It is composed chiefly of the mineral calcite ($CaCO_3$). Ninety percent of limestone is biochemical sediment. The rest precipitates directly from water.

One easily identified biochemical limestone is *coquina*, a coarse rock composed of loosely cemented shells and shell fragments (Figure 2.14). Another less obvious but familiar example is *chalk*, a soft, porous rock made up almost entirely of the hard parts of microscopic organisms (Figure 2.15).

Inorganic limestones form when chemical changes or high water temperatures increase the concentration of calcium carbonate to the point that it precipitates. *Travertine*, the type of limestone that decorates caverns, is one example. Groundwater is the source of travertine that is deposited in caves. As water drops reach the air in a cavern, some of the carbon dioxide dissolved in the water escapes, causing calcium carbonate to precipitate.

Conglomerate

Breccia

sandstone

Shale

← 5 cm →

Figure 2.13   Common detrital sedimentary rocks. **A**. Conglomerate (rounded particles). **B**. Breccia (angular particles). **C**. Sandstone **D**. Shale with plant fossil. (Photos by E. J. Tarbuck)

Dissolved silica (SiO_2) precipitates to form varieties of microcrystalline quartz including chert (light color), flint (dark), jasper (red), and agate (banded). These chemical sedimentary rocks may have either an inorganic or biochemical origin, but the mode of origin is usually difficult to determine.

Very often, evaporation causes minerals to precipitate from water. Such minerals include halite, the chief component of *rock salt*, and gypsum, the main ingredient of *rock gypsum*. Both materials have significant commercial importance. Halite is familiar to everyone as the common salt used in cooking and seasoning foods. Of course, it has many other uses and has been considered important enough that people have sought, traded, and fought over it for much of human history. Gypsum is the basic ingredient of plaster of Paris. This material is used most extensively in the construction industry for drywall and plaster.

In the geologic past, many areas that are now dry land were covered by shallow arms of the sea that had only narrow connections to the open ocean. Under these conditions, water continually moved into the bay to replace water lost by evaporation. Eventually the waters of the bay became saturated and salt deposition began. Today, these arms of the sea are gone, and the remaining deposits are called **evaporites**.

On a smaller scale, evaporite deposits can be seen in such places as Death Valley, California. Here, following rains or periods of snowmelt in the mountains, streams flow from surrounding mountains into an enclosed basin. As the water evaporates, *salt flats* form from dissolved materials left behind as a white crust on the ground (Figure 2.16).

Coal is quite different from other chemical sedimentary rocks. Unlike other rocks in this category, which are calcite- or silica-rich, coal is made mostly of organic matter. Close examination of a piece of coal under a microscope or magnifying glass often reveals plant structures such as leaves, bark, and wood that have been chemically altered but are still identifiable. This supports the conclusion that coal is the end product of the burial of large amounts of plant material over extended periods.

The initial stage in coal formation is the accumulation of large quantities of plant remains. However, special conditions are required for such accumulations,

BOX 2.1

EARTH AS A SYSTEM
The Carbon Cycle and Sedimentary Rocks

To illustrate the movement of material and energy in the Earth system, let us take a brief look at the *carbon cycle* (Figure 2.A). Pure carbon is relatively rare in nature. It is found predominantly in two minerals: diamond and graphite. Most carbon is bonded chemically to other elements to form compounds such as carbon dioxide, calcium carbonate, and the hydrocarbons found in coal and petroleum. Carbon is also the basic building block of life as it readily combines with hydrogen and oxygen to form the fundamental organic compounds that compose living things.

In the atmosphere, carbon is found mainly as carbon dioxide (CO_2). Atmospheric carbon dioxide is significant because it is a greenhouse gas, which means it is an efficient absorber of energy emitted by Earth and thus influences the heating of the atmosphere. Because many of the processes that operate on Earth involve carbon dioxide, this gas is constantly moving into and out of the atmosphere. For example, through the process of photosynthesis, plants absorb carbon dioxide from the atmosphere to produce the essential organic compounds needed for growth. Animals that consume these plants (or consume other animals that eat plants) use these organic compounds as a source of energy and, through the process of respiration, return carbon dioxide to the atmosphere. (Plants also return some CO_2 to the atmosphere via respiration.) Further, when plants die and decay or are burned, this biomass is oxidized, and carbon dioxide is returned to the atmosphere.

Not all dead plant material decays immediately back to carbon dioxide. A small percentage is deposited as sediment. Over long spans of geologic time, considerable biomass is buried with sediment. Under the right conditions, some of these carbon-rich deposits are converted to fossil fuels—coal, petroleum, or natural gas. Eventually some of the fuels are recovered (mined or

Figure 2.A Simplified diagram of the carbon cycle, with emphasis on the flow of carbon between the atmosphere and the hydrosphere, lithosphere and biosphere. The colored arrows show whether the flow of carbon is into or out of the atmosphere.

pumped from a well) and burned to run factories and fuel our transportation system. One result of fossil-fuel combustion is the release of huge quantities of CO_2 into the atmosphere. Certainly one of the most active parts of the carbon cycle is the movement of CO_2 from the atmosphere to the biosphere and back again.

Carbon also moves from the lithosphere and hydrosphere to the atmosphere and back again. For example, volcanic activity early in Earth's history is thought to be the source of much of the carbon dioxide found in the atmosphere. One way that carbon dioxide makes its way back to the hydrosphere and then to the solid Earth is by first combining with water to form carbonic acid (H_2CO_3), which then attacks the rocks that compose the lithosphere. One product of this chemical weathering of solid rock is the soluble bicarbonate ion ($2HCO_3^-$), which is carried by groundwater and streams to the ocean. Here water-dwelling organisms extract this

dissolved material to produce hard parts of calcium carbonate ($CaCO_3$). When the organisms die, these skeletal remains settle to the ocean floor as biochemical sediment and become sedimentary rock. In fact, the lithosphere is by far Earth's largest depository of carbon, where it is a constituent of a variety of rocks, the most abundant being limestone. Eventually the limestone may be exposed at Earth's surface, where chemical weathering will cause the carbon stored in the rock to be released to the atmosphere as CO_2.

In summary, carbon moves among all four of Earth's major spheres. It is essential to every living thing in the biosphere. In the atmosphere carbon dioxide is an important greenhouse gas. In the hydrosphere, carbon dioxide is dissolved in lakes, rivers, and the ocean. In the lithosphere carbon is contained in carbonate sediments and sedimentary rocks and is stored as organic matter dispersed through sedimentary rocks and as deposits of coal and petroleum.

Close up

Figure 2.14   This rock, called coquina, consists of shell fragments; therefore, it has a biochemical origin. (Photo by F. J. Tarbuck)

because dead plants normally decompose when exposed to the atmosphere. An ideal environment that allows for the buildup of plant material is a swamp. Because stagnant swamp water is oxygen-deficient, complete decay (oxidation) of the plant material is not possible. At various times during Earth history, such environments have been common. Coal undergoes successive stages of formation. With each successive stage,

higher temperatures and pressures drive off impurities and volatiles, as shown in Figure 2.17.

Lignite and bituminous coals are sedimentary rocks, but anthracite is a metamorphic rock. Anthracite forms when sedimentary layers are subjected to the folding and deformation associated with mountain building.

In summary, we divide sedimentary rocks into two major groups: detrital and chemical. The main criterion for classifying detrital rocks is particle size, whereas chemical rocks are distinguished by their mineral composition. The categories presented here are more rigid than is the actual state of nature. Many detrital sedimentary rocks are a mixture of more than one particle size. Furthermore, many sedimentary rocks classified as chemical also contain at least small quantities of detrital sediment, and practically all detrital rocks are cemented with material that was originally dissolved in water.

? STUDENTS SOMETIMES ASK...

You mentioned that a body of seawater can evaporate. Has a large area of the sea ever evaporated?

Yes—in fact, several have. The evidence for this includes thick and extensive deposits of evaporite minerals. A large-scale example of evaporite deposition occurred in the Mediterranean Sea about 6 million years ago. Evidence suggests that the Mediterranean Sea was cut off from the Atlantic Ocean at the narrow Strait of Gibraltar—in effect, creating a dam. Because of the warm and dry climate in this region, the Mediterranean Sea nearly evaporated in a few thousand years, leaving a thick deposit of evaporites on the hot, dry basin floor. A half million years later a huge waterfall spilled into the Mediterranean from the Atlantic and filled the basin with seawater once again.

Figure 2.15   The White Chalk Cliffs, Sussex, England. (Photo by Art Wolfe)

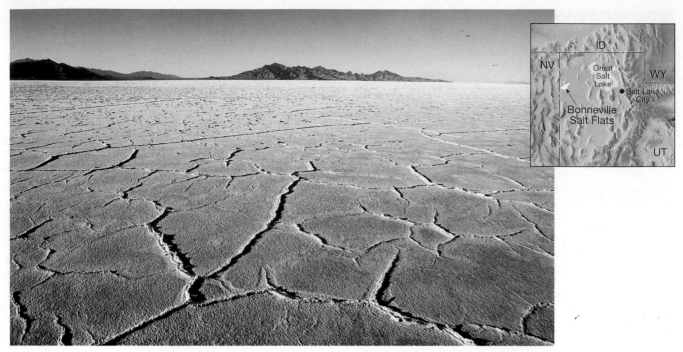

Figure 2.16 The Bonneville salt flats in Utah are a well-known example of evaporite deposits. (Photo by Tom & Susan Bean, Inc.)

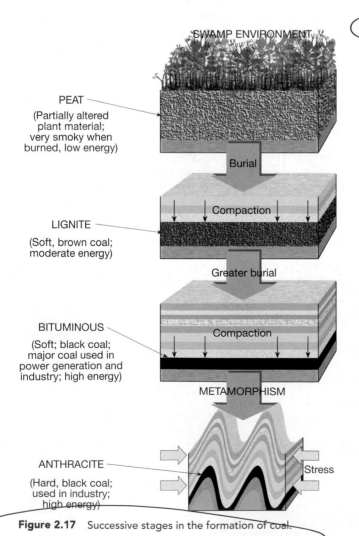

Figure 2.17 Successive stages in the formation of coal.

Lithification of Sediment

Lithification refers to the processes by which sediments are transformed into solid sedimentary rocks. One of the most common processes is compaction. As sediments accumulate through time, the weight of overlying material compresses the deeper sediments. As the grains are pressed closer and closer, pore space is greatly reduced. For example, when clays are buried beneath several thousand meters of material, the volume of the clay might be reduced as much as 40 percent. Compaction is most significant in fine-grained sedimentary rocks such as shale, because sand and other coarse sediments compress little.

Cementation is another important means by which sediments are converted to sedimentary rock. The cementing materials are carried in solution by water percolating through the pore spaces between particles. Through time, the cement precipitates onto the sediment grains, fills the open spaces, and joins the particles. Calcite, silica, and iron oxide are the most common cements. Identification of the cementing material is simple. Calcite cement will effervesce (fizz) with dilute hydrochloric acid. Silica is the hardest cement and thus produces the hardest sedimentary rocks. When a sedimentary rock has an orange or red color, this usually means iron oxide is present.

Features of Sedimentary Rocks

Sedimentary rocks are particularly important evidence of Earth's long history. These rocks form at Earth's surface, and as layer upon layer of sediment accumulates, each records the nature of the environment at the time the sediment was deposited. These layers, called **strata**, or **beds**, are the *single most characteristic feature of sedimentary rocks* (see Figure 2.11).

The thickness of beds range from microscopically thin to dozens of meters thick. Separating the strata are *bedding planes*, flat surfaces along which rocks tend to separate or break. Generally, each bedding plane marks the end of one episode of sedimentation and the beginning of another.

Sedimentary rocks provide geologists with evidence for deciphering past environments. A conglomerate, for example, indicates a high-energy environment, such as a rushing stream, where only the coarse materials can settle out. By contrast, black shale and coal are associated with a low-energy, organic-rich environment such as a swamp or lagoon. Other features found in some sedimentary rocks also give clues to past environments (Figure 2.18).

Fossils, the traces or remains of prehistoric life, are perhaps the most important inclusions found in some sedimentary rock. Knowing the nature of the life forms that existed at a particular time may help to answer many questions about the environment. Was it land or ocean? A lake or swamp? Was the climate hot or cold, rainy or dry? Was the ocean water shallow or deep, turbid or clear? Furthermore, fossils are important time indicators and play a key role in matching up rocks from different places that are the same age. Fossils are important tools used in interpreting the geologic past and will be examined in some detail in Chapter 10.

Metamorphic Rocks: New Rock from Old

Earth Materials
↳ Metamorphic Rocks

Recall from the section on the rock cycle that metamorphism is the transformation of one rock type into another. Metamorphic rocks are produced from preexisting igneous, sedimentary, or even from other metamorphic rocks. Thus, every metamorphic rock has a *parent rock*—the rock from which it was formed.

Metamorphism is a very appropriate name for this process because it literally means to "change form." Metamorphism takes place *where preexisting rock is subjected to temperatures and pressures unlike those in which it formed*. In response to these new conditions, the rock gradually changes until a state of equilibrium with the new environment is reached. Most metamorphic changes occur at the elevated temperatures and pressures that exist in the zone beginning a few kilometers below Earth's surface and extending into the upper mantle.

A

B

Figure 2.18   A. Ripple marks preserved in sedimentary rocks may indicate a beach or stream channel environment. (Photo by Stephen Trimble) **B.** Mud cracks form when wet mud or clay dries and shrinks, perhaps signifying a tidal flat or desert basin. (Photo by Gary Yeowell/Tony Stone Images)

Metamorphism often progresses incrementally, from slight changes (*low-grade metamorphism*) to substantial changes (*high-grade metamorphism*). For example, under low-grade metamorphism, the common sedimentary rock *shale* becomes the more compact metamorphic rock called *slate*. Hand samples of these rocks are sometimes difficult to distinguish, illustrating that the stage at which a sedimentary rock becomes a metamorphic rock is somewhat arbitrary.

In more extreme environments, metamorphism causes a transformation so complete that the identity of the parent rock cannot be determined. In high-grade metamorphism, such features as bedding planes, fossils, and vesicles that may have existed in the parent rock are obliterated. Further, when rocks at depth (where temperatures are high) are subjected to directed pressure, they slowly deform by flowing to generate intricate folds (Figure 2.19). In the most extreme metamorphic environments, the temperatures approach those at which rocks melt. However, *during metamorphism the rock must remain essentially solid*, for if complete melting occurs, we have entered the realm of igneous activity.

Most metamorphism occurs in one of two settings:

1. When rock is intruded by a magma body, **contact** or **thermal metamorphism** may take place. Here, change is driven by a rise in temperature within the host rock surrounding a molten igneous body.

2. During mountain building, great quantities of rock are subjected to directed pressures and high temperatures associated with large-scale deformation called **regional metamorphism**.

Extensive areas of metamorphic rocks are exposed on every continent. Metamorphic rocks are an important component of many mountain belts, where they make up a large portion of a mountain's crystalline core (Figure 2.19). Even the stable continental interiors, which are generally covered by sedimentary rocks, are underlain by metamorphic basement rocks. In all of these settings, the metamorphic rocks are usually highly deformed and intruded by igneous masses. Indeed, significant parts of Earth's continental crust are composed of metamorphic and associated igneous rocks.

Agents of Metamorphism

The agents of metamorphism include *heat, pressure (stress), and chemically active fluids*. During metamorphism, rocks are usually subjected to all three metamorphic agents simultaneously. However, the degree of metamorphism and the contribution of each agent vary greatly from one environment to another.

Heat as a Metamorphic Agent.   The most important agent of metamorphism is heat because it provides the energy to drive chemical reactions that result in the recrystallization of existing minerals and/or the formation of new minerals. The heat to metamorphose rocks comes mainly from two sources. First, rocks experience a rise in temperature when they are intruded by magma rising from below. This is called *contact metamorphism*. Here the adjacent host rock is "baked" by the emplaced magma.

Second, rocks that formed at Earth's surface will experience a gradual increase in temperature as they are transported to greater depths. In the upper crust, this increase in temperature averages between 20°C and 30°C

Figure 2.19 Deformed strata; north end of Cottonwood Mountains, Death Valley, California. (Photo by Michael Collier)

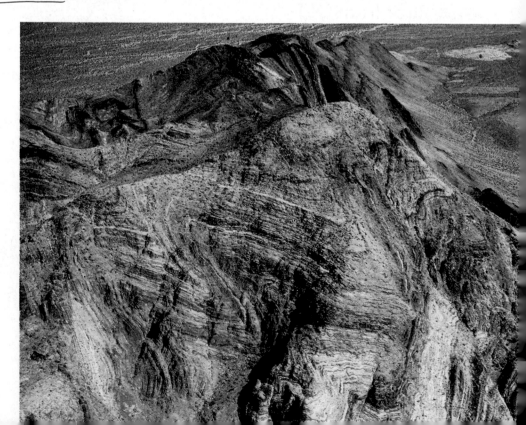

per kilometer. When buried to a depth of about 8 kilometers (5 miles), where temperatures are between 150°C and 200°C, clay minerals tend to become unstable and begin to recrystallize into minerals, such as chlorite and muscovite, that are stable in this environment. (Chlorite is a micalike mineral formed by the metamorphism of iron- and magnesian-rich silicates.) However, many silicate minerals, particularly those found in crystalline igneous rocks—quartz, for example—remain stable at these temperatures. Thus, metamorphic changes in these minerals occur at much higher temperatures.

Pressure (Stress) as a Metamorphic Agent. Pressure, like temperature, also increases with depth as the thickness of the overlying rock increases. Buried rocks are subjected to *confining pressure*, which is analogous to water pressure, where the forces are applied equally in all directions (Figure 2.20A). The deeper you go in the ocean, the greater the confining pressure. The same is true for rock that is buried. Confining pressure causes the spaces between mineral grains to close,

producing a more compact rock having a greater density. Further, at great depths, confining pressure may cause minerals to recrystallize into new minerals that display a more compact crystalline form.

During episodes of mountain building, large rock bodies become highly crumpled and metamorphosed (Figure 2.19). The forces that generate mountains are unequal in different directions and are called *differential stress*. Unlike confining pressure, which "squeezes" the rock equally from all directions, differential stress acts mainly along one plane. As shown in Figure 2.20B, rocks subjected to differential stress are shortened in the direction pressure is applied and elongated, or lengthened, in the direction perpendicular to that force. The deformation caused by differential stresses plays a major role in developing metamorphic textures.

In surface environments where temperatures are comparatively low, rocks are *brittle* and tend to fracture when subjected to differential stress. Continued deformation grinds and pulverizes the mineral grains into small fragments. By contrast, in high-temperature environments rocks are *ductile*. When rocks exhibit ductile behavior, their mineral grains tend to flatten and elongate when subjected to differential stress. This accounts for their ability to deform by flowing (rather than fracturing) to generate intricate folds (Figure 2.21).

Chemically Active Fluids. Fluids composed mainly of water and other volatiles (materials that readily change to a gas at surface conditions), including carbon dioxide, are believed to play an important role in some types of metamorphism. Fluids that surround mineral grains act as catalysts to promote recrystallization by enhancing ion migration. Further, in progressively hotter environments these ion-rich fluids become correspondingly more reactive. For example, when two mineral grains are squeezed together, the part of their crystalline structures that touch are the most highly stressed. Ions located at these sites are readily dissolved by the hot fluids and migrate along the

Figure 2.20 Pressure (stress) as a metamorphic agent. **A.** In a depositional environment, as confining pressure increases, rocks deform by decreasing in volume. **B.** During mountain building, rocks subjected to differential stress are shortened in the direction that pressure is applied, and lengthened in the direction perpendicular to that force.

Figure 2.21 Deformed metamorphic rocks exposed in a road cut in the Eastern Highland of Connecticut. Imagine the tremendous force required to fold rock in this manner. (Photo by Phil Dombrowski)

surface of the grain to locations where the two grains are not in contact. Thus, hot fluids aid in the recrystallization of mineral grains by dissolving material from regions of high stress and then precipitating (depositing) this material in areas of low stress. As a result, *minerals tend to recrystallize and grow longer in a direction perpendicular to compressional stresses.*

Where hot fluids circulate freely through rocks, ionic exchange may occur between two adjacent rock layers, or ions may migrate great distances before they are finally deposited. When ions are exchanged between two or more rock units, a change in the overall composition may result.

STUDENTS SOMETIMES ASK...
How hot is it deep in the crust?

The increase in temperature with depth, based on the geothermal gradient, can be expressed as *the deeper one goes, the hotter it gets.* This relationship has been observed by miners in deep mines and in deeply drilled wells. In the deepest mine in the world (the Western Deep Levels mine in South Africa, which is 4 kilometers or 2.5 miles deep), the temperature of the surrounding rock is so hot that it can scorch human skin! In fact, miners often work in groups of two: one to mine the rock, and the other to operate a large fan that keeps the other worker cool.

The temperature is even higher at the bottom of the deepest well in the world, which was completed in the Kola Peninsula of Russia in 1992 and goes down a record 12.3 kilometers (7.7 miles). At this depth it is 245°C (473°F), much greater than the boiling point of water. What keeps water from boiling is the high confining pressure at depth.

Metamorphic Textures

The degree of metamorphism is reflected in the rock's texture and mineralogy. (Recall that the term *texture* is used to describe the size, shape, and arrangement of grains within a rock.) When rocks are subjected to low-grade metamorphism, they become more compact and thus more dense. A common example is the metamorphic rock slate, which forms when shale is subjected to temperatures and pressures only slightly greater than those associated with the compaction that lithifies sediment. In this case, differential stress causes the microscopic clay minerals in shale to align into the more compact arrangement found in slate.

Under more extreme conditions, stress causes certain minerals to recrystallize. In general, recrystallization encourages the growth of larger crystals. Consequently, many metamorphic rocks consist of visible crystals, much like coarse-grained igneous rocks.

The crystals of some minerals will recrystallize with a preferred orientation, essentially perpendicular to the direction of the compressional force. The resulting mineral alignment usually gives the rock a layered or band-

Figure 2.22 Under directed pressure, planar minerals, such as the micas, become reoriented or recrystallized so that their surfaces are aligned at right angles to the stress. The resulting planar orientation of mineral grains gives the rock a foliated texture. If the coarse-grained igneous rock (granite) on the left underwent intense metamorphism, it could end up closely resembling the metamorphic rock on the right (gneiss). (Photos by E. J. Tarbuck)

A. Before metamorphism (Uniform stress)

B. After metamorphism (Differential stress)

Rock Name	Texture		Grain Size	Comments	Parent Rock
Slate	Increasing Metamorphism	Foliated	Very fine	Excellent rock cleavage, smooth dull surfaces	Shale, mudstone, or siltstone
Phyllite		Foliated	Fine	Breaks along wavey surfaces, glossy sheen	Slate
Schist		Foliated	Medium to Coarse	Micaceous minerals dominate, scaly foliation	Phyllite
Gneiss		Foliated	Medium to Coarse	Compositional banding due to segregation of minerals	Schist, granite, or volcanic rocks
Marble		Nonfoliated	Medium to coarse	Interlocking calcite or dolomite grains	Limestone, dolostone
Quartzite		Nonfoliated	Medium to coarse	Fused quartz grains, massive, very hard	Quartz sandstone
Anthracite		Nonfoliated	Fine	Shiny black organic rock that may exhibit conchoidal fracture	Bituminous coal

Figure 2.23   Classification of common metamorphic rocks.

ed appearance termed **foliated texture** (Figure 2.22). Simply, foliation results whenever the minerals of a rock are brought into parallel alignment.

Not all metamorphic rocks have a foliated texture. Such rocks are said to exhibit a **nonfoliated texture**. Metamorphic rocks composed of only one mineral that forms equidimensional crystals are, as a rule, not visibly foliated. For example, limestone, if pure, is composed of only a single mineral, calcite. When a fine-grained limestone is metamorphosed, the small calcite crystals combine to form larger interlocking crystals. The resulting rock resembles a coarse-grained igneous rock. This nonfoliated metamorphic equivalent of limestone is called *marble*.

Figure 2.24   Because slate breaks into flat slabs, it has a number of uses. Here, it is used to roof this house in Switzerland. (Photo by E. J. Tarbuck)

Common Metamorphic Rocks

To review, metamorphic processes cause many changes in existing rocks, including increased density, growth of larger crystals, foliation (reorientation of the mineral grains into a layered or banded appearance), and the transformation of low-temperature minerals into high-temperature minerals (Figure 2.23). Further, the introduction of ions generates new minerals, some of which are economically important.

Here is a brief look at common rocks produced by metamorphic processes.

Foliated Rocks. Slate is a very fine-grained foliated rock composed of minute mica flakes. The most noteworthy characteristic of slate is its excellent rock cleavage, meaning that it splits easily into flat slabs. This property has made slate a most useful rock for roof and floor tile, chalkboards, and billiard tables (Figure 2.24). Slate is most often generated by the low-grade metamorphism of shale, although less frequently it forms from the metamorphism of volcanic ash. Slate can be almost any color, depending on its mineral constituents. Black slate contains organic material; red slate gets its color from iron oxide; and green slate is usually composed of chlorite, a micalike mineral.

Schists are strongly foliated rocks, formed by regional metamorphism. They are platy and can be readily split into thin flakes or slabs. Like slate, the parent material from which many schists originate is shale, but in the case of schist, the metamorphism is more intense.

The term *schist* describes the *texture* of a rock regardless of composition. For example, schists composed primarily of muscovite and biotite are called *mica schists*. Gneiss (pronounced "nice") is the term applied to banded metamorphic rocks that contain mostly elongated and granular, as opposed to platy, minerals (see Figure 2.22B). The most common minerals in gneisses are quartz and feldspar, with lesser amounts of muscovite, biotite, and hornblende. Gneisses exhibit strong segregation of light and dark silicates, giving them a characteristic banded texture. While in a plastic state, these banded gneisses can be deformed into intricate folds.

Nonfoliated Rocks. Marble is a coarse, crystalline rock whose parent rock is limestone. Marble is composed of large interlocking calcite crystals, which form from the recrystallization of smaller grains in the parent rock.

Because of its color and relative softness (hardness of only 3 on the Mohs scale), marble is a popular building stone. White marble is particularly prized as a stone from which to carve monuments and statues, such as the famous statue of *David* by Michelangelo (Figure 2.25). Often the limestone from which marble forms contains impurities that color the marble. Thus, marble can be pink, gray, green, or even black.

Quartzite is a very hard metamorphic rock most often formed from quartz sandstone. Under moderate- to high-grade metamorphism, the quartz grains in sandstone fuse. Pure quartzite is white, but iron oxide may produce reddish or pinkish stains, and dark minerals may impart a gray color.

Figure 2.25 Replica of the statue of David by Michelangelo. Like the original, this sculpture was created from a large block of white marble. (Courtesy Getty images/Photo Disk)

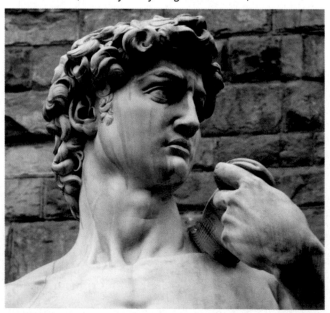

Resources from Rocks and Minerals

The outer layer of Earth, which we call the crust, is only as thick when compared to the remainder of the Earth as a peach skin is to a peach, yet it is of supreme importance to us. We depend on it for fossil fuels and as a source of such diverse minerals as the talc for baby powder, salt to flavor food, and gold for world trade. In fact, on occasion, the availability or absence of certain Earth materials has altered the course of history. As the material requirements of modern society grow, the need to locate additional supplies of useful minerals also grows and becomes more challenging (see Box 2.2).

Metallic Mineral Resources

Some of the most important accumulations of metals, such as gold, silver, copper, mercury, lead, platinum, and nickel, are produced by igneous and metamorphic processes (Table 2.1). These mineral resources, like most others, result from processes that concentrate desirable materials to the extent that extraction is economically feasible.

The igneous processes that generate some metal deposits are quite straightforward. For example, as a large magma body cools, the heavy minerals that crystallize

Table 2.1 Ore minerals of important metals.

Metal	Ore Mineral
Aluminum	Bauxite
Chromium	Chromite
Copper	Chalcopyrite
	Bornite
	Chalcocite
Gold	Native gold
Iron	Hematite
	Magnetite
	Limonite
Lead	Galena
Magnesium	Magnesite
	Dolomite
Manganese	Pyrolusite
Mercury	Cinnabar
Molybdenum	Molybdenite
Nickel	Pentlandite
Platinum	Native platinum
Silver	Native silver
	Argentite
Tin	Cassiterite
Titanium	Ilmenite
	Rutile
Tungsten	Wolframite
	Scheelite
Uranium	Uraninite (pitchblende)
Zinc	Sphalerite

early tend to settle to the lower portion of the magma chamber. This type of magmatic segregation is particularly active in large basaltic magmas where chromite (ore of chromium), magnetite, and platinum are occasionally generated. Layers of chromite, interbedded with other heavy minerals, are mined from such deposits in the Bushveld Complex in South Africa, which contains over 70 percent of the world's known reserves of platinum.

Among the best-known and most important ore deposits are those generated from **hydrothermal** (*hydra* = water, *therm* = heat) **solutions**. Included in this group are the gold deposits of the Homestake mine in South Dakota; the lead, zinc, and silver ores near Coeur d'Alene, Idaho; the silver deposits of the Comstock Lode in Nevada; and the copper ores of the Keweenaw Peninsula in Michigan.

The majority of hydrothermal deposits are thought to originate from hot, metal-rich fluids that are associated with cooling magma bodies. During solidification, liquids plus various metallic ions accumulate near the top of the magma chamber. Because these hot fluids are very mobile, they can migrate great distances through the surrounding rock before they are eventually deposited. Some of this fluid moves along fractures or bedding planes, where it cools and precipitates the metallic ions to produce **vein deposits** (Figure 2.26). Many of the most productive deposits of gold, silver, and mercury occur as hydrothermal vein deposits.

Another important type of accumulation generated by hydrothermal activity is called a **disseminated deposit**. Rather than being concentrated in narrow veins and dikes, these ores are distributed as minute masses throughout the entire rock mass. Much of the world's copper is extracted from disseminated deposits, including the huge Bingham Canyon copper mine in Utah. Because these accumulations contain only 0.4 to 0.8 percent copper, between 125 and 250 metric tons of ore must be mined for every ton of metal recovered. The environmental impact of these large excavations, including the problems of waste disposal, is significant.

Table 2.2 Uses of nonmetallic minerals.

Mineral	Uses
Apatite	Phosphorous fertilizers
Asbestos (chrysotile)	Incombustible fibers
Calcite	Aggregate; steelmaking; soil conditioning; chemicals; cement; building stone
Clay minerals (kaolinite)	Ceramics; china
Corundum	Gemstones; abrasives
Diamond	Gemstones; abrasives
Fluorite	Steelmaking; aluminum refining; glass; chemicals
Garnet	Abrasives; gemstones
Graphite	Pencil lead; lubricant; refractories
Gypsum	Plaster of Paris
Halite	Table salt; chemicals; ice control
Muscovite	Insulator in electrical applications
Quartz	Primary ingredient in glass
Sulfur	Chemicals; fertilizer manufacture
Sylvite	Potassium fertilizers
Talc	Powder used in paints, cosmetics, etc.

Figure 2.26 Light-colored vein deposits emplaced along a series of fractures in dark-colored igneous rock. (Photo by James E. Patterson)

Figure 2.27 Rock of Ages granite quarry, Barre, Vermont. Granite is used as a building stone as well as for making monuments. (Photo by Paul Rocheleau/Index Stock Imagery)

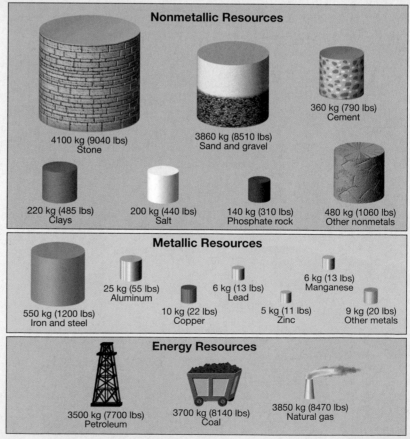

BOX 2.2

PEOPLE AND THE ENVIRONMENT:
United States Per Capita Use of Mineral and Energy Resources

Mineral and energy resources from Earth's crust are the raw materials from which the products used by our modern industrial economy are made. Like most people who live in highly industrialized nations, you may not realize the quantity of resources needed to maintain your present standard of living. Figure 2.B shows the approximate annual per capita consumption of several important metallic and nonmetallic mineral resources for the United States. Per capita use of energy resources is also shown. This is each person's proportional share of the materials required by industry to provide the vast array of homes, cars, electronics, cosmetics, packaging, and so on that modern society demands. It represents the energy we need to drive cars, fly airplanes, run machinery, and heat buildings.

Every year about 9000 kilograms (20,000 pounds) of stone, sand, and gravel is mined for each person in the United States. Further, the per capita use of oil, coal, and natural gas exceeds 11,000 kilograms (24,000 pounds). Figures for countries such as Canada, Japan, the United Kingdom, as well as for most other highly industrialized countries are comparable.

Nonmetallic Resources

4100 kg (9040 lbs)
Stone

3860 kg (8510 lbs)
Sand and gravel

360 kg (790 lbs)
Cement

220 kg (485 lbs)
Clays

200 kg (440 lbs)
Salt

140 kg (310 lbs)
Phosphate rock

480 kg (1060 lbs)
Other nonmetals

Metallic Resources

550 kg (1200 lbs)
Iron and steel

25 kg (55 lbs)
Aluminum

10 kg (22 lbs)
Copper

6 kg (13 lbs)
Lead

5 kg (11 lbs)
Zinc

6 kg (13 lbs)
Manganese

9 kg (20 lbs)
Other metals

Energy Resources

3500 kg (7700 lbs)
Petroleum

3700 kg (8140 lbs)
Coal

3850 kg (8470 lbs)
Natural gas

Figure 2.B The annual per capita consumption of nonmetallic and metallic mineral resources for the United States is nearly 10,000 kilograms! About 94 percent of the materials used are nonmetallic. The per capita use of oil, coal, and natural gas exceeds 11,000 kilograms.

Nonmetallic Mineral Resources

Mineral resources not used as fuels or processed for the metals they contain are referred to as *nonmetallic mineral resources*. These materials are extracted and processed either to make use of the nonmetallic elements they contain or for the physical and chemical properties they possess (Table 2.2). Nonmetallic mineral resources are commonly divided into two broad groups: *building materials* and *industrial minerals* (Figure 2.27). As some substances have many different uses, they are found in both categories. Limestone, perhaps the most versatile and widely used rock of all, is the best example. As a building material, it is used not only as crushed rock and building stone but also in the making of cement. Moreover, as an industrial mineral, it is an ingredient in the manufacture of steel and is used in agriculture to neutralize acidic soils.

Besides aggregate (sand, gravel, and crushed rock) and cut stone, the other important building materials include gypsum for plaster and wallboard, clay for tile and bricks, and cement, which is made from limestone and shale. Cement and aggregate go into the making of concrete, a material that is essential to practically all construction.

Many and various nonmetallic resources are classified as industrial minerals. People often do not realize the importance of industrial minerals because they see only the products that resulted from their use and not the minerals themselves. That is, many nonmetallics are used up in the process of creating other products. Examples include fluorite and limestone, which are part of the steelmaking process; corundum and garnet, which are used as abrasives to make machinery parts; and sylvite, which is employed in the production of the fertilizers used to grow a food crop.

Chapter Summary

• *Igneous rock* forms from *magma* that cools and solidifies in a process called *crystallization. Sedimentary rock* forms from the *lithification of sediment. Metamorphic rock* forms from rock that has been subjected to great pressure and heat in a process called *metamorphism.*

• The rate of cooling of magma greatly influences the size of mineral crystals in igneous rock. The four basic igneous rock textures are (1) *fine-grained*, (2) *coarse-grained*, (3) *porphyritic*, and (4) *glassy.*

• The mineral makeup of an igneous rock is ultimately determined by the chemical composition of the magma from which it crystallized. N. L. Bowen showed that as magma cools, minerals crystallize in an orderly fashion. *Crystal settling* can change the composition of magma and cause more than one rock type to form from a common parent magma.

• Igneous rocks are classified by their *texture* and *mineral composition.* Igneous rocks are divided into broad compositional groups based on the percentage of dark and light silicate minerals they contain. *Felsic rocks* (e.g., granite and rhyolite) are composed mostly of the light-colored silicate minerals potassium feldspar and quartz. Rocks of *intermediate* composition (e.g., andesite) contain plagioclase feldspar and amphibole. *Mafic rocks* (e.g., basalt) contain abundant olivine, pyroxene, and calcium feldspar.

• *Detrital sediments* are materials that originate and are transported as solid particles derived from weathering. *Chemical sediments* are soluble materials produced largely by chemical weathering that are precipitated by either inorganic or organic processes. *Detrital sedimentary rocks*, which are classified by particle size, contain a variety of mineral and rock fragments, with clay minerals and quartz the chief constituents. *Chemical sedimentary rocks* often contain the products of biological processes such as shells or mineral crystals that form as water evaporates and minerals precipitate. *Lithification* refers to the processes by which sediments are transformed into solid sedimentary rocks.

• Common detrital sedimentary rocks include *shale* (the most common sedimentary rock), *sandstone*, and *conglomerate.* The most abundant chemical sedimentary rock is *limestone*, composed chiefly of the mineral calcite. *Rock gypsum* and *rock salt* are chemical rocks that form as water evaporates and triggers the deposition of chemical precipates.

• Some of the features of sedimentary rocks that are often used in the interpretation of Earth history and past environments include *strata*, or *beds* (the single most characteristic feature), *fossils, ripple marks*, and *mud cracks.*

• Two types of metamorphism are (1) *regional metamorphism* and (2) *contact metamorphism.* The agents of metamorphism include *heat, pressure* (stress), and *chemically active fluids.* Heat is perhaps the most important because it provides the energy to drive the reactions that result in the *recrystallization* of minerals. Metamorphic processes cause many changes in rocks, including *increased density*, growth of *larger mineral crystals, reorientation of the mineral grains* into a layered or banded appearance known as *foliation*, and the formation of *new minerals.*

• Some common metamorphic rocks with a *foliated texture* include *slate, schist*, and *gneiss.* Metamorphic rocks with a *nonfoliated texture* include *marble* and *quartzite.*

• Some of the most important accumulations of *metallic mineral resources* are produced by igneous and metamorphic processes. *Vein deposits* (deposits in fractures or bedding planes) and *disseminated deposits* (deposits distributed throughout the entire rock mass) are produced from *hydrothermal solutions*—hot, metal-rich fluids associated with cooling magma bodies.

• *Nonmetallic mineral resources* are mined for the nonmetallic elements they contain or for the physical and chemical properties they possess. The two groups of nonmetallic mineral resources are (1) *building materials* (e.g., limestone and gypsum) and (2) *industrial minerals* (e.g., fluorite and corundum).

Key Terms

andesitic composition (p. 47)
basaltic composition (p. 47)
Bowen's reaction series (p. 45)
chemical sedimentary rock (p. 49)
coarse-grained texture (p. 44)
contact metamorphism (p. 56)
crystallization (p. 42)
crystal settling (p. 46)
detrital sedimentary rock (p. 49)
disseminated deposit (p. 62)
evaporite (p. 51)
extrusive (volcanic) (p. 42)
felsic (p. 46)

fine-grained texture (p. 43)
fossils (p. 55)
foliated texture (p. 59)
glassy texture (p. 44)
granitic composition (p. 46)
hydrothermal solution (p. 61)
igneous rock (p. 43)
intermediate composition (p. 47)
intrusive (plutonic) (p. 42)
lava (p. 42)
lithification (p. 54)
mafic (p. 47)
magma (p. 42)

metamorphic rock (p. 42)
metamorphism (p. 55)
nonfoliated texture (p. 59)
porphyritic texture (p. 44)
regional metamorphism (p. 56)
rock cycle (p. 40)
sediment (p. 48)
sedimentary rock (p. 49)
strata (beds) (p. 55)
texture (p. 43)
thermal metamorphism (p. 56)
ultramafic composition (p. 47)
vein deposit (p. 62)

Review Questions

1. Using the rock cycle, explain the statement "One rock is the raw material for another."

2. If a lava flow at Earth's surface had a basaltic composition, what rock type would the flow likely be (see Figure 2.10)? What igneous rock would form from the same magma if it did not reach the surface but instead crystallized at great depth?

3. What does a porphyritic texture indicate about the history of an igneous rock?

4. How are granite and rhyolite different? The same (see Figure 2.10)?

5. Relate the classification of igneous rocks to Bowen's reaction series.

6. What minerals are most common in detrital sedimentary rocks? Why are these minerals so abundant?

7. What is the primary basis for distinguishing among various detrital sedimentary rocks?

8. Distinguish between the two categories of chemical sedimentary rocks.

9. What are evaporite deposits? Name a rock that is an evaporite.

10. Compaction is an important lithification process with which sediment size?

11. What is probably the single most characteristic feature of sedimentary rocks?

12. What is metamorphism?

13. List the three agents of metamorphism and describe the role of each.

14. Distinguish between regional and contact metamorphism.

15. What feature would easily distinguish schist and gneiss from quartzite and marble?

16. In what ways do metamorphic rocks differ from the igneous and sedimentary rocks from which they formed?

17. List two general types of hydrothermal deposits.

18. Nonmetallic resources are commonly divided into two broad groups. List the two groups and some examples of materials that belong to each.

Examining the Earth System

1. The sedimentary rocks coquina and shale have each formed in response to interactions among two or more of Earth's spheres. List the spheres associated with the formation of each of the rocks and write a short explanation for each of your choices.

2. Of the two main sources of energy that drive the rock cycle—Earth's internal heat and solar energy—which is primarily responsible for each of the three groups of rocks found on and within Earth? Explain your reasoning.

3. Every year about 20,000 pounds of stone, sand, and gravel are mined for each person in the United States. Calculate how many pounds of stone, sand, and gravel will be needed for an individual during a 75-year lifespan. Since a cubic yard of rock weighs roughly 1700 pounds, how big a hole must be dug to supply an individual with 75 years worth of stone, sand, and gravel? (Note that a typical pickup truck will carry about a half of a cubic yard of rock.)

Web Resources

 The *Earth Science* Website uses the resources and flexibility of the Internet to aid in your study of the topics in this chapter. Written and developed by Earth science instructors, this site will help improve your understanding of Earth science. Visit **http://www.prenhall.com/tarbuck** and click on the cover of *Earth Science 10e* to find:

- **On-line review quizzes.**
- **Web-based critical thinking and writing exercises.**
- **Links to chapter-specific Web resources.**
- **Internet-wide key term searches.**

http://www.prenhall.com/tarbuck

Differential weathering is responsible for much of the spectacular scenery in Arizona's Monument Valley. (Photo by David Muench)

CHAPTER 3

WEATHERING, SOIL, AND MASS WASTING

Earth's surface is constantly changing. Rock is disintegrated and decomposed, moved to lower elevations by gravity, and carried away by water, wind, or ice. In this manner Earth's physical landscape is sculptured. This chapter focuses on the first two steps of this never-ending process—weathering and mass wasting. What causes solid rock to crumble, and why does the type and rate of weathering vary from place to place? What mechanisms act to move weathered debris downslope? Soil, an important product of the weathering process and a vital resource, is also examined.

Earth's External Processes

 Sculpturing Earth's Surface
↳ External vs Internal Processes

Weathering, mass wasting, and erosion are called **external processes** because they occur at or near Earth's surface and are powered by energy from the Sun. External processes are a basic part of the rock cycle because they are responsible for transforming solid rock into sediment.

To the casual observer, the face of Earth may appear to be without change, unaffected by time. In fact, 200 years ago most people believed that mountains, lakes, and deserts were permanent features of an Earth that was thought to be no more than a few thousand years old. Today we know that Earth is 4.5 billion years old and that mountains eventually succumb to weathering and erosion, lakes fill with sediment or are drained by streams, and deserts come and go with changes in climate.

Earth is a dynamic body. Some parts of Earth's surface are gradually elevated by mountain building and volcanic activity. These **internal processes** derive their energy from Earth's interior. Meanwhile, opposing external processes are continually breaking rock apart and moving the debris to lower elevations (Figure 3.1). The latter processes include:

1. **Weathering**—the physical breakdown (disintegration) and chemical alteration (decomposition) of rocks at or near Earth's surface.
2. **Mass wasting**—the transfer of rock and soil downslope under the influence of gravity.

Figure 3.1 Slopes are places where materials are continually moving from higher to lower elevations. Weathering begins the process by attacking the solid rock exposed at the surface. Next, gravity moves the weathered debris downslope. This step, termed mass wasting, may range from a slow and gradual creep to a thundering landslide. Eventually, the material that was once high up the slope reaches the stream at the bottom. The moving water then transports the debris away. (Photo © Carr Clifton)

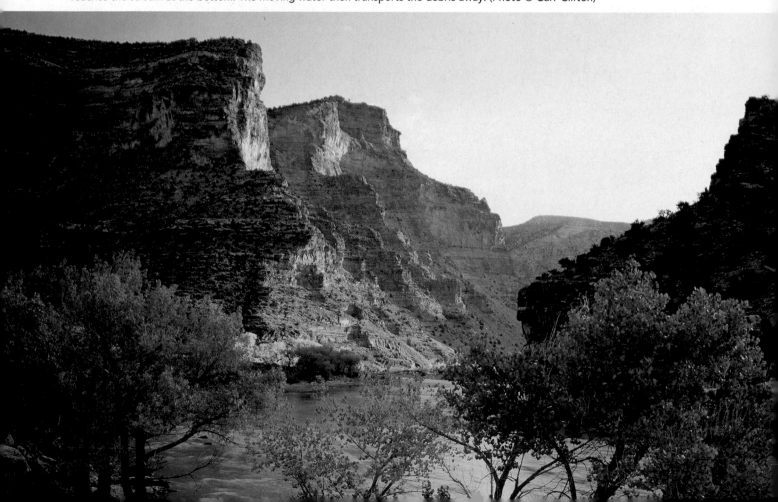

3. **Erosion**—the physical removal of material by mobile agents such as water, wind, or ice.

We will first turn our attention to the process of weathering and the products generated by this activity. However, weathering cannot be easily separated from the other two processes because, as weathering breaks rocks apart, it facilitates the movement of rock debris by mass wasting and erosion. Conversely, the transport of material by mass wasting and erosion further disintegrates and decomposes the rock.

Weathering

Weathering goes on all around us, but it seems like such a slow and subtle process that it is easy to underestimate its importance. However, it is worth remembering that weathering is a basic part of the rock cycle and thus a key process in the Earth system.

All materials are susceptible to weathering. Consider, for example, the synthetic rock we call concrete, which closely resembles the sedimentary rock called conglomerate. A newly poured concrete sidewalk has a smooth, fresh look. However, not many years later, the same sidewalk will appear chipped, cracked, and rough, with pebbles exposed at the surface. If a tree is nearby, its roots may grow under the concrete, heaving and buckling it. The same natural processes that eventually break apart a concrete sidewalk also act to disintegrate rocks, regardless of their type or strength.

Weathering occurs when rock is mechanically fragmented (disintegrated) and/or chemically altered (decomposed). **Mechanical weathering** is accomplished by physical forces that break rock into smaller and smaller pieces without changing the rock's mineral composition. **Chemical weathering** involves a chemical transformation of rock into one or more new compounds. These two concepts can be illustrated with a piece of paper. The paper can be disintegrated by tearing it into smaller and smaller pieces, whereas decomposition occurs when the paper is set afire and burned.

In the following sections we will discuss the various modes of mechanical and chemical weathering. Although we will consider these two processes separately, keep in mind that they usually work simultaneously in nature.

Mechanical Weathering

 Earth Materials
↳ Sedimentary Rocks

When a rock undergoes *mechanical weathering*, it is broken into smaller and smaller pieces, each retaining the characteristics of the original material. The end result is many small pieces from a single large one. Figure 3.2 shows that breaking a rock into smaller pieces increases the surface area available for chemical attack. Hence, by breaking rocks into smaller pieces, mechanical weathering increases the amount of surface area available for chemical weathering.

In nature, three physical processes are especially important in breaking rocks into smaller fragments: frost wedging, expansion resulting from unloading, and biological activity.

Frost Wedging

Repeated cycles of freezing and thawing of water represent an important process of mechanical weathering. Liquid water has the unique property of expanding about 9 percent when it freezes. This increase in volume occurs because, as ice forms, the water molecules arrange themselves into a very open crystalline structure. As a result, when water freezes, it expands and exerts a tremendous outward force. This can be verified by completely filling a container with water and freezing it. The formation of ice will rupture the container.

In nature, water works its way into every crack or void in rock and, upon freezing, expands and enlarges these openings. After many freeze-thaw cycles, the rock is broken into pieces. This process is appropriately called frost wedging (Figure 3.3). Frost wedging is most pronounced in mountainous regions in the middle latitudes where a daily freeze-thaw cycle often exists.

Figure 3.2 Chemical weathering can occur only to those portions of a rock that are exposed to the elements. Mechanical weathering breaks rock into smaller and smaller pieces, thereby increasing the surface area available for chemical attack.

4 square units ×
6 sides ×
1 cube =
_____
24 square units

1 square unit ×
6 sides ×
8 cubes =
_____
48 square units

.25 square unit ×
6 sides ×
64 cubes =
_____
96 square units

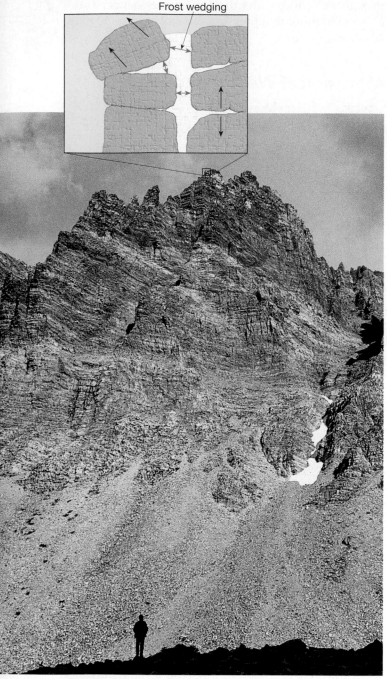

Frost wedging

Figure 3.3 Frost wedging. As water freezes, it expands, exerting a force great enough to break rock. When frost wedging occurs in a setting such as this, the broken rock fragments fall to the base of the cliff and create a cone-shaped accumulation known as talus. (Photo by Tom & Susan Bean, Inc.)

Here, sections of rock are wedged loose and may tumble into large piles called **talus slopes**, which often form at the base of steep rocky cliffs (Figure 3.3).

Unloading

When large masses of igneous rock, particularly those composed of granite, are exposed by erosion, slabs begin to break loose like the layers of an onion. The process is called **sheeting** and probably takes place in response to the great reduction in pressure that occurs when the overlying rock is eroded away. Accompanying this *unloading*, the outer layers expand more than the rock below and thus separate from the rock body (Figure 3.4A,B). Continued weathering eventually causes the slabs to separate and spall off, creating **exfoliation domes** (*ex* = off, *folium* = leaf). Excellent examples of exfoliation domes include Stone Mountain, Georgia, and Half Dome (Figure 3.4C) and Liberty Cap in Yosemite National Park.

Deep underground mining provides us with another example of how rocks behave once the confining pressure is removed. Large rock slabs sometimes explode off the walls of newly cut mine tunnels because of the abruptly reduced pressure. Evidence of this type, plus the fact that fracturing occurs parallel to the floor of a quarry when large blocks of rock are removed, strongly supports the process of unloading as the cause of sheeting.

Biological Activity

Weathering is also accomplished by the activities of organisms, including plants, burrowing animals, and humans. Plant roots in search of minerals and water grow into fractures, and as the roots grow, they wedge the rock apart (Figure 3.5). Burrowing animals further break down the rock by moving fresh material to the surface, where physical and chemical processes can more effectively attack it. Decaying organisms also produce acids, which contribute to chemical weathering. Where rock has been blasted in search of minerals or for road construction, the impact of humans is particularly noticeable.

Chemical Weathering

GEODe   Earth Materials
↳ Sedimentary Rocks
H_2O

Chemical weathering involves the complex processes that alter the internal structures of minerals by removing and/or adding elements. During this transformation, the original rock decomposes into substances that are stable in the surface environment. Consequently, the products of chemical weathering will remain essentially unchanged as long as they remain in an environment similar to the one in which they formed.

Water and Carbonic Acid

Water is by far the most important agent of chemical weathering. Although pure water is nonreactive, a small amount of dissolved material is generally all that is needed to activate it. Oxygen dissolved in water will oxidize some materials. For example, when an iron nail is found in moist soil, it will have a coating of rust (iron oxide), and if the time of exposure has been long, the

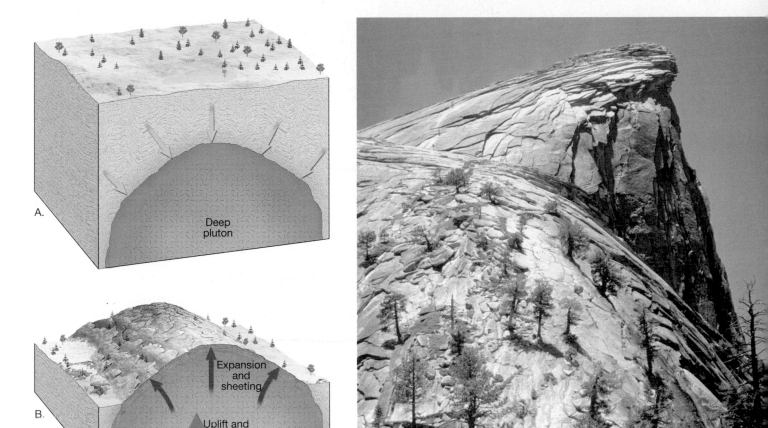

Figure 3.4   Sheeting is caused by the expansion of crystalline rock as erosion removes the overlying material. When the deeply buried pluton in **A** is exposed at the surface following uplift and erosion in **B**, the igneous mass fractures into thin slabs. The photo in **C** is of the summit of Half Dome in Yosemite National Park, California. It is an exfoliation dome and illustrates the onionlike layers created by sheeting. (Photo by Breck P. Kent)

Figure 3.5   Root wedging widens fractures in rock and aids the process of mechanical weathering. (Photo by Tom and Susan Bean, Inc).

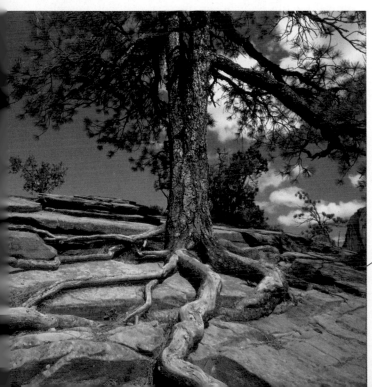

nail will be so weak that it can be broken as easily as a toothpick. When rocks containing iron-rich minerals oxidize, a yellow to reddish-brown rust will appear on the surface (Figure 3.6).

Carbon dioxide (CO_2) dissolved in water (H_2O) forms carbonic acid (H_2CO_3), the same weak acid produced when soft drinks are carbonated. Rain dissolves some carbon dioxide as it falls through the atmosphere, and additional amounts released by decaying organic matter are acquired as the water percolates through the soil. Carbonic acid ionizes to form the very reactive hydrogen ion (H^+) and the bicarbonate ion (HCO_3^-).

How Granite Weathers

To illustrate how rock chemically weathers when attacked by carbonic acid, we will consider the weathering of granite, an abundant continental rock. Recall that granite consists mainly of quartz and potassium feldspar. The weathering of the potassium feldspar component of granite takes place as follows:

$$2KAlSi_3O_8 + 2(H^+ + HCO_3^-) + H_2O \rightarrow$$

potassium feldspar   carbonic acid   water

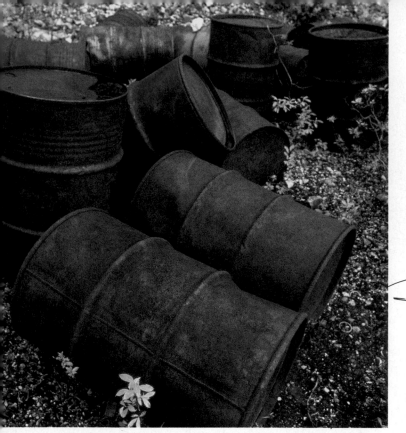

Figure 3.6 Iron reacts with oxygen to form iron oxide as seen on these rusted barrels. (Photo by Stephen J. Krasemann/DRK Photo)

$$Al_2Si_2O_5(OH)_4 + 2K^+ + 2HCO_3^- + 4SiO_2$$

clay mineral $\underbrace{\text{potassium ion} \quad \text{bicarbonite ion} \quad \text{silica}}_{\text{in solution}}$

In this reaction, the hydrogen ions (H^+) attack and replace potassium ions (K^+) in the feldspar structure, thereby disrupting the crystalline network. Once removed, the potassium is available as a nutrient for plants or becomes the soluble salt potassium bicarbonate ($KHCO_3$), which may be incorporated into other minerals or carried to the ocean in dissolved form by streams.

The most abundant products of the chemical breakdown of feldspar are residual clay minerals. Clay minerals are the end product of weathering and are very stable under surface conditions. Consequently, clay minerals make up a high percentage of the inorganic material in soils. Moreover, the most abundant sedimentary rock, shale, contains a high proportion of clay minerals.

In addition to the formation of clay minerals during this reaction, some silica is removed from the feldspar structure and is carried away by groundwater (water beneath Earth's surface). This dissolved silica will eventually precipitate to produce nodules of chert or flint, fill in the pore spaces between sediment grains, or be carried to the ocean where microscopic animals will remove it to build hard silica shells.

To summarize, the weathering of potassium feldspar generates a residual clay mineral, a soluble salt

(potassium bicarbonate), and some silica that enters into solution.

Quartz, the other main component of granite, is *very resistant* to chemical weathering; it remains substantially unaltered when attacked by weakly acidic solutions. As a result, when granite weathers, the feldspar crystals dull and slowly turn to clay, releasing the once interlocked quartz grains, which still retain their fresh, glassy appearance. Although some quartz remains in the soil, much is transported to the sea or to other sites of deposition, where it becomes the main constituent of such features as sandy beaches and sand dunes. In time it may become lithified to form the sedimentary rock *sandstone.*

Weathering of Silicate Minerals

Table 3.1 lists the weathered products of some of the most common silicate minerals. Remember that silicate minerals make up most of Earth's crust and that these minerals are composed essentially of only eight elements. When chemically weathered, silicate minerals yield sodium, calcium, potassium, and magnesium ions, which form soluble products that may be removed by groundwater. The element iron combines with oxygen, producing relatively insoluble iron oxides, which give soil a reddish-brown or yellowish color. Under most conditions the three remaining elements—aluminum, silicon, and oxygen—join with water to produce residual clay minerals. However, even the highly insoluble clay minerals are very slowly removed by subsurface water.

Spheroidal Weathering

In addition to altering the internal structure of minerals, chemical weathering causes physical changes as well. For instance, when angular rock masses are attacked by water that enters along joints, the rocks tend to take on a spherical shape. Gradually the corners and edges of the angular blocks become more rounded. The corners are attacked most readily because of their greater surface area, as compared to the edges and faces. This

Table 3.1 Products of weathering.

Mineral	Residual Products	Material in Solution
Quartz	Quartz grains	Silica
Feldspars	Clay minerals	Silica
		K^+, Na^+, Ca^{2+}
Amphibole (hornblende)	Clay minerals	Silica
	Limonite	Ca^{2+}, Mg^{2+}
	Hematite	
Olivine	Limonite	Silica
	Hematite	Mg^{2+}

A.   B.

Figure 3.7 **A.** Spheroidal weathering is evident in this exposure of granite in California's Joshua Tree National Monument. Because the rocks are attacked more vigorously on the corners and edges, they take on a spherical shape. The lines visible in the rock are called *joints*. Joints are important rock structures that allow water to penetrate and start the weathering process long before the rock is exposed. (Photo by E. J. Tarbuck) **B.** Sometimes successive shells are loosened as the weathering process continues to penetrate ever deeper into the rock. (Photo by Martin Schmidt, Jr.)

process, called **spheroidal weathering**, gives the weathered rock a more rounded or spherical shape (Figure 3.7).

Sometimes during the formation of spheroidal boulders, successive shells separate from the rock's main body. Eventually the outer shells spall off, allowing the chemical weathering activity to penetrate deeper into the boulder. This spherical scaling results because, as the minerals in the rock weather to clay, they increase in size through the addition of water to their structure. This increased bulk exerts an outward force that causes concentric layers of rock to break loose and fall off. Hence, chemical weathering does produce forces great enough to cause mechanical weathering.

This type of spheroidal weathering, in which shells spall off, should not be confused with the phenomenon of sheeting discussed earlier. In sheeting, the fracturing occurs as a result of unloading, and the rock layers that separate from the main body are largely unaltered at the time of separation.

Rates of Weathering

Several factors influence the type and rate of rock weathering. We have already seen how mechanical weathering affects the rate of weathering. By breaking rock into smaller pieces, the amount of surface area exposed to chemical weathering is increased. Other im-

portant factors examined here include rock characteristics and climate.

Rock Characteristics

Rock characteristics encompass all of the chemical traits of rocks, including mineral composition and solubility. In addition, any physical features, such as joints (cracks), can be important because they influence the ability of water to penetrate rock.

The variations in weathering rates, attributable to the mineral constituents, can be demonstrated by comparing old headstones made from different rock types. Headstones of granite, which is composed of silicate minerals, are relatively resistant to chemical weathering. We can see this by examining the inscriptions on the headstones shown in Figure 3.8. This is not true of the marble headstone, which shows signs of extensive chemical alteration over a relatively short period. Marble is composed of calcite (calcium carbonate), which readily dissolves even in a weakly acidic solution.

The silicates, the most abundant mineral group, weather in essentially the same sequence as their order of crystallization. By examining Bowen's reaction series (see Figure 2.8, p. 46), we see that olivine crystallizes first and is therefore the least resistant to chemical weathering, whereas quartz, which crystallizes last, is the most resistant.

Figure 3.8 An examination of headstones reveals the rate of chemical weathering on diverse rock types. The granite headstone (left) was erected four years before the marble headstone (right). The inscription date of 1872 on the marble monument is nearly illegible. (Photos by E. J. Tarbuck)

Climate

Climatic factors, particularly temperature and moisture, are crucial to the rate of rock weathering. One important example from mechanical weathering is that the frequency of freeze-thaw cycles greatly affects the amount of frost wedging. Temperature and moisture also exert a strong influence on the rates of chemical weathering and on the kind and amount of vegetation present. Regions with lush vegetation generally have a thick mantle of soil rich in decayed organic matter from which chemically active fluids such as carbonic and humic acids are derived.

The optimum environment for chemical weathering is a combination of warm temperatures and abundant moisture. In polar regions chemical weathering is ineffective because frigid temperatures keep the available moisture locked up as ice, whereas in arid regions there is insufficient moisture to foster rapid chemical weathering.

Human activities can influence the composition of the atmosphere, which in turn can impact the rate of chemical weathering. One well-known example is acid rain (Figure 3.9).

Differential Weathering

Masses of rock do not weather uniformly. Take a moment to look at the photo of Shiprock, New Mexico, in Figure 8.21 (p. 243). The durable volcanic neck protrudes high above the surrounding terrain. A glance at the chapter-opening photo shows an additional example of this phenomenon, called **differential weathering**. The results vary in scale from the rough, uneven surface of the marble headstone in Figure 3.8 to the boldly sculpted exposures in Bryce Canyon (Figure 3.10).

Many factors influence the rate of rock weathering. Among the most important are variations in the composition of the rock. More resistant rock protrudes as ridges or pinnacles, or as steeper cliffs on a canyon wall (see Figure 10.2 p. 289). The number and spacing of joints can also be a significant factor (see Figure 3.7A and Figure 9.16, p. 272). Differential weathering and subsequent erosion are responsible for creating many unusual and sometimes spectacular rock formations and landforms.

Soil

Soil covers most land surfaces. Along with air and water, it is one of our most indispensable resources (Figure 3.11). Also like air and water, soil is taken for granted by many of us. The following quote helps put this vital layer in perspective.

> Science, in recent years, has focused more and more on the Earth as a planet, one that for all we know is unique—where a thin blanket of air, a thinner film of

Figure 3.9 As a consequence of burning large quantities of coal and petroleum, more than 40 million tons of sulfur and nitrogen oxides are released into the atmosphere each year in the United States. Through a series of complex chemical reactions, some of these pollutants are converted into acids that then fall to Earth's surface as rain or snow. Among its many environmental effects, acid rain accelerates the chemical weathering of stone monuments and structures, including this building façade in Leipzig, Germany. (Photo by Doug Plummer/Photo Researchers, Inc.)

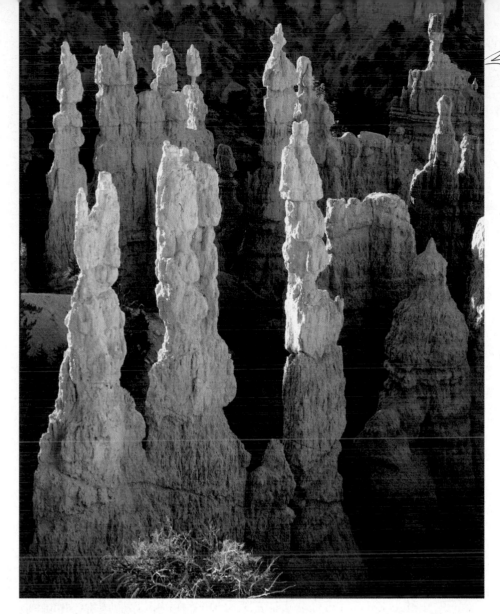

Figure 3.10 Differential weathering is illustrated by these sculpted rock pinnacles in Bryce National Park, Utah. (Photo by Art Wolfe)

Figure 3.11 Soil is an essential resource that we often take for granted. Soil is not a living entity, but it contains a great deal of life. Moreover, this complex medium supports nearly all plant life, which in turn supports animal life. (Photo by Tom Bean/Corbis/Stock Market)

water, and the thinnest veneer of soil combine to support a web of life of wondrous diversity in continuous change.*

Soil has accurately been called "the bridge between life and the inanimate world." All life—the entire biosphere—owes its existence to a dozen or so elements that must ultimately come from Earth's crust. Once weathering and other processes create soil, plants carry out the intermediary role of assimilating the necessary elements and making them available to animals, including humans.

An Interface in the Earth System

When Earth is viewed as a system, soil is referred to as an *interface*—a common boundary where different parts of a system interact. This is an appropriate designation because soil forms where the solid Earth, the atmosphere, the hydrosphere, and the biosphere meet. Soil is a material that develops in response to complex envi-

---

*Jack Eddy. "A Fragile Seam of Dark Blue Light," in *Proceedings of the Global Change Research Forum*. U.S. Geological Survey Circular 1086, 1993, p. 15.

ronmental interactions among different parts of the Earth system. Over time, soil gradually evolves to a state of equilibrium or balance with the environment. Soil is dynamic and sensitive to almost every aspect of its surroundings. Thus, when environmental changes occur, such as climate, vegetative cover, and animal (including human) activity, the soil responds. Any such change produces a gradual alteration of soil characteristics until a new balance is reached. Although thinly distributed over the land surface, soil functions as a fundamental interface, providing an excellent example of the integration among many parts of the Earth system.

? STUDENTS SOMETIMES ASK...

I've seen photos of footprints left on the lunar surface by astronauts. Does this mean the Moon has soil?

Not exactly. The Moon has no atmosphere, no water, and lacks biological activity. Therefore, the chemical weathering, frost wedging, and other weathering processes we are familiar with on Earth are lacking on the Moon. However, all lunar terrains are mantled with a soil-like layer of gray debris, called *lunar regolith*, derived from a few billion years of bombardment by meteorites. The rate of surface change is so slow that the footprints left by the *Apollo* astronauts (see Figure 21.8 on p. 583) will likely remain fresh-looking for millions of years to come.

What Is Soil?

With few exceptions, Earth's land surface is covered by **regolith** (*rhegos* = blanket, *lithos* = stone), the layer of rock and mineral fragments produced by weathering. Some would call this material soil, but soil is more than an accumulation of weathered debris. **Soil** is a combination of mineral and organic matter, water, and air—that portion of the regolith that supports the growth of plants. Although the proportions of the major components in soil vary, the same four components always are present to some extent (Figure 3.12). About one half of the total volume of a good-quality surface soil is a mixture of disintegrated and decomposed rock (mineral matter) and *humus*, the decayed remains of animal and plant life (organic matter). The remaining half consists of pore spaces among the solid particles where air and water circulate.

Although the mineral portion of the soil is usually much greater than the organic portion, humus is an essential component. In addition to being an important source of plant nutrients, humus enhances the soil's ability to retain water. Because plants require air and water to live and grow, the portion of the soil consisting of pore spaces that allow for the circulation of these fluids is as vital as the solid soil constituents.

25% air

25% water

45% mineral matter

5% organic matter

Figure 3.12 Composition (by volume) of a soil in good condition for plant growth. Although the percentages vary, each soil is composed of mineral and organic matter, water, and air.

Soil water is far from "pure" water; instead, it is a complex solution containing many soluble nutrients. Soil water not only provides the necessary moisture for the chemical reactions that sustain life; it also supplies plants with nutrients in a form they can use. The pore spaces not filled with water contain air. This air is the source of necessary oxygen and carbon dioxide for most microorganisms and plants that live in the soil.

Soil Texture and Structure

Most soils are far from uniform and contain particles of different sizes. **Soil texture** refers to the proportions of different particle sizes. Texture is a very basic soil property because it strongly influences the soil's ability to retain and transmit water and air, both of which are essential to plant growth. Sandy soils may drain too rapidly and dry out quickly. At the opposite extreme, the pore spaces of clay-rich soils may be so small that they inhibit drainage, and long-lasting puddles result. Moreover, when the clay and silt content is very high, plant roots may have difficulty penetrating the soil.

Because soils rarely consist of particles of only one size, *textural categories* have been established based upon the varying proportions of clay, silt, and sand. The standard system of classes used by the U.S. Department of Agriculture is shown in Figure 3.13. For example, point *A* on this triangular diagram (left center) represents a soil composed of 10 percent silt, 40 percent clay, and 50 percent sand. Such a soil is called a *sandy clay*. The soils called *loam*, which occupy the central portion of the diagram, are those in which no single particle size predominates over the other two. Loam soils are best suited to support plant life because they generally have better moisture characteristics and nutrient storage ability than do soils composed predominantly of clay or coarse sand.

Figure 3.13   The texture of any soil can be represented by a point on this soil-texture diagram. Soil texture is one of the most significant factors used to estimate agricultural potential and engineering characteristics. (After U.S. Department of Agriculture)

Soil particles are seldom completely independent of one another. Rather, they usually form clumps called *peds* that give soils a particular structure. Four basic soil structures are recognized: platy, prismatic, blocky, and spheroidal. Soil structure is important because it influences the ease of a soil's cultivation as well as the susceptibility of a soil to erosion. In addition, soil structure affects the porosity and permeability of soil (that is, the ease with which water can penetrate). This in turn influences the movement of nutrients to plant roots. Prismatic and blocky peds usually allow for moderate water infiltration, whereas platy and spheroidal structures are characterized by slower infiltration rates.

Controls of Soil Formation

Soil is the product of the complex interplay of several factors. The most important of these are parent material, time, climate, plants, animals, and slope. Although all of these factors are interdependent, their roles will be examined separately.

Parent Material

The source of the weathered mineral matter from which soils develop is called the **parent material**, and it is a major factor influencing a newly forming soil. Gradually it undergoes physical and chemical changes as the processes of soil formation progress. Parent material may be the underlying bedrock, or it can be a layer of unconsolidated deposits, as in a stream valley. When the parent material is bedrock, the soils are termed *residual soils*. By contrast, those developed on unconsolidated sediment are called *transported soils* (Figure 3.14).

Note that transported soils form *in place* on parent materials that have been carried from elsewhere and deposited by gravity, water, wind, or ice.

The nature of the parent material influences soils in two ways. First, the type of parent material affects the rate of weathering and thus the rate of soil formation. (Consider the weathering rates of granite versus limestone.) Also, because unconsolidated deposits are already partly weathered and provide more surface area for chemical weathering, soil development on such material usually progresses more rapidly. Second, the chemical makeup of the parent material affects the soil's fertility. This influences the character of the natural vegetation the soil can support.

At one time the parent material was believed to be the primary factor causing differences among soils. Today soil scientists realize that other factors, especially climate, are more important. In fact, it has been found that similar soils are often produced from different parent materials and that dissimilar soils have developed from the same parent material. Such discoveries reinforce the importance of the other soil-forming factors.

Time

Time is an important component of every geological process, and soil formation is no exception. The nature of soil is strongly influenced by the length of time processes have been operating. If weathering has been going on for a comparatively short time, the parent material determines to a large extent the characteristics of the soil. As weathering processes continue, the influence of parent material on soil is overshadowed by the other soil-forming factors, especially climate. The amount of time required for various soils to evolve cannot be specified because the soil-forming processes act at varying rates under different circumstances. However, as a rule, the longer a soil has been forming, the thicker it becomes and the less it resembles the parent material.

Climate

Climate is the most influential control of soil formation. Just as temperature and precipitation are the climatic elements that influence people the most, so too are they the elements that exert the strongest impact on soil formation. Variations in temperature and precipitation determine whether chemical or mechanical weathering predominates. They also greatly influence the rate and depth of weathering. For instance, a hot, wet climate may produce a thick layer of chemically weathered soil in the same amount of time that a cold, dry climate produces a thin mantle of mechanically weathered debris. Also, the amount of precipitation influences the degree to which various materials are removed (leached) from the soil, thereby affecting soil fertility. Finally, climatic

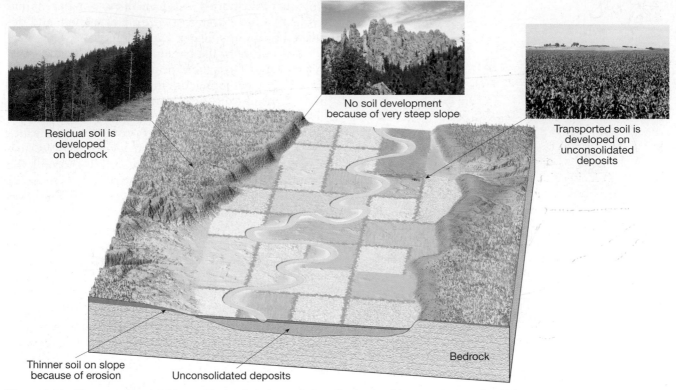

Residual soil is developed on bedrock

No soil development because of very steep slope

Transported soil is developed on unconsolidated deposits

Thinner soil on slope because of erosion

Unconsolidated deposits

Bedrock

Figure 3.14 The parent material for residual soils is the underlying bedrock, whereas transported soils form on unconsolidated deposits. Also note that as slopes become steeper, soil becomes thinner.

conditions are important factors controlling the type of plant and animal life present.

Plants and Animals

The biosphere plays a vital role in soil formation. The types and abundance of organisms present have a strong influence on the physical and chemical properties of a soil. In fact, for well-developed soils in many regions, the significance of natural vegetation is frequently implied in the description used by soil scientists. Such phrases as *prairie soil*, *forest soil*, and *tundra soil* are common.

The chief function of plants and animals is to furnish organic matter to the soil. Certain bog soils are composed almost entirely of organic matter, whereas desert soils may contain only a tiny percentage. Although the quantity of organic matter varies substantially among soils, it is the rare soil that completely lacks it.

The primary source of organic matter is plants, although animals and the uncountable microorganisms also contribute. When organic matter decomposes, important nutrients are supplied to plants, as well as to animals and microorganisms living in the soil. Consequently, soil fertility depends in part on the amount of organic matter present. Furthermore, the decay of plant and animal remains causes the formation of various organic acids. These complex acids hasten the weathering process. Organic matter also has a high water-holding ability and thus aids water retention in a soil.

Microorganisms, including fungi, bacteria, and single-celled protozoa, play an active role in the decay of plant and animal remains. The end product is *humus*, a material that no longer resembles the plants and animals from which it is formed. In addition, certain microorganisms aid soil fertility because they have the ability to *fix* (alter) atmospheric nitrogen gas into soil nitrogen compounds.

Earthworms and other burrowing animals act to mix the mineral and organic portions of a soil. Earthworms, for example, feed on organic matter and thoroughly mix soils in which they live, often moving and enriching many tons per acre each year. Burrows and holes also aid the passage of water and air through the soil.

Slope

The slope of the land can vary greatly over short distances. Such variations, in turn, can lead to widely varied local soil types. Many of the differences exist because slope has a significant impact on the amount of erosion and the water content of soil.

On steep slopes, soils are often poorly developed. In such situations little water can soak in, and as a result, soil moisture may be insufficient for vigorous plant growth. Further, because of accelerated erosion on steep slopes, the soils are thin or nonexistent (Figure 3.14).

In contrast, poorly drained and waterlogged soils in bottomlands have a much different character. Such soils are usually thick and dark. The dark color results from generous organic matter that accumulates because saturated conditions retard the decay of vegetation. The optimum terrain for soil development is a flat-to-undulating upland surface. Here we find good drainage, minimum erosion, and sufficient infiltration of water into the soil.

Slope orientation, the direction the slope is facing, also is significant. In the midlatitudes of the Northern Hemisphere, a south-facing slope receives a great deal more sunlight than does a north-facing slope. In fact, a steep north-facing slope may receive no direct sunlight at all. The difference in the amount of solar radiation received causes substantial differences in soil temperature and moisture, which in turn may influence the nature of the vegetation and the character of the soil.

Although we have dealt separately with each of the soil-forming factors, remember that *all work together* to form soil. No single factor is responsible for a soil being as it is. Rather, it is the combined influence of parent material, time, climate, plants, animals, and slope that determines a soil's character.

The Soil Profile

It is important to realize that soil-forming processes *operate from the surface downward*. Thus, variations in composition, texture, structure, and color gradually evolve at varying depths. These vertical differences, which usually become more pronounced as time passes, divide the soil into zones or layers known as **horizons**. If you were to dig a trench in soil, you would see that its walls are layered. Such a vertical section through all of the soil horizons constitutes the **soil profile** (Figure 3.15).

Figure 3.16 presents an idealized view of a well-developed soil profile in which five horizons are identified. From the surface downward, they are designated as *O, A, E, B,* and *C*, respectively. These five horizons are common to soils in temperate regions. The characteristics and extent of development of horizons vary in different environments. Thus, different localities exhibit soil profiles that can contrast greatly with one another.

The *O* horizon consists largely of organic material, unlike the layers beneath it that consist mainly of mineral matter. The upper portion of the *O* horizon is primarily plant litter such as loose leaves and other organic debris that are still recognizable. By contrast, the lower portion of the *O* horizon is made up of partly decomposed organic matter (humus) in which plant structures can no longer be identified. In addition to plants, the *O* horizon is teeming with microscopic life, including bacteria, fungi, algae, and insects. All of these organisms contribute oxygen, carbon dioxide, and organic acids to the developing soil.

Underlying the organic-rich *O* horizon is the *A* horizon. This zone is largely mineral matter, yet biological activity is high and humus is generally present—up to

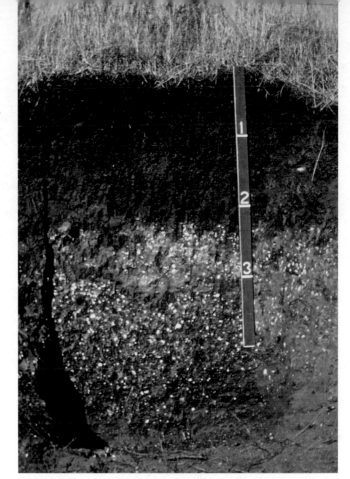

Figure 3.15 A soil profile is a vertical cross-section from the surface down to the parent material. Well-developed soils show distinct layers called horizons. (Photo by E. J. Tarbuck)

30 percent in some instances. Together the *O* and *A* horizons make up what is commonly called *topsoil*. Below the *A* horizon, the *E* horizon is a light-colored layer that contains little organic material. As water percolates downward through this zone, finer particles are carried away. This washing out of the fine soil components is termed **eluviation** (*elu* = get away from, *via* = a way). Water percolating downward also dissolves soluble inorganic soil components and carries them to deeper zones. This depletion of soluble materials from the upper soil is termed **leaching**.

Immediately below the *E* horizon is the *B* horizon, or *subsoil*. Much of the material removed from the *E* horizon by eluviation is deposited in the *B* horizon, which is often referred to as the *zone of accumulation*. The accumulation of the fine clay particles enhances water retention in the subsoil. However, in extreme cases clay accumulation can form a very compact and impermeable layer called *hardpan*. The *O, A, E,* and *B* horizons together constitute the **solum**, or "true soil." It is in the solum that the soil-forming processes are active and that living roots and other plant and animal life are largely confined.

Below the solum and above the unaltered parent material is the *C* horizon, a layer characterized by partially altered parent material. Whereas the *O, A, E,* and *B* horizons bear little resemblance to the parent mater-

Figure 3.16   Idealized soil profile from a humid climate in the middle latitudes.

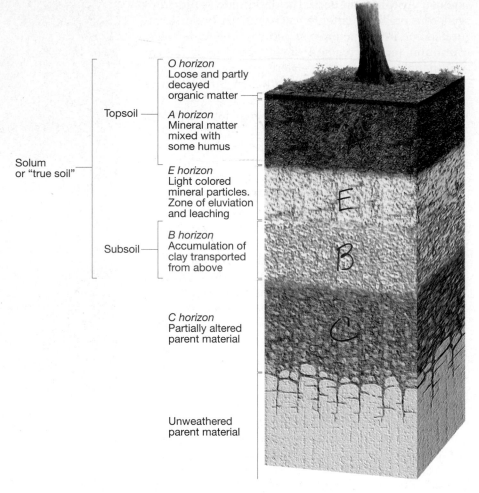

Solum or "true soil"

Topsoil

O horizon
Loose and partly decayed organic matter

A horizon
Mineral matter mixed with some humus

Subsoil

E horizon
Light colored mineral particles. Zone of eluviation and leaching

B horizon
Accumulation of clay transported from above

C horizon
Partially altered parent material

Unweathered parent material

ial, it is easily identifiable in the *C* horizon. Although this material is undergoing changes that will eventually transform it into soil, it has not yet crossed the threshold that separates regolith from soil.

Soil characteristics and development can vary greatly in different environments. The boundaries between soil horizons can be very distinct, as in Figures 3.15 and 3.16, or the horizons may blend gradually from one to another. A well-developed soil profile indicates that environmental conditions have been relatively stable over an extended time span.

Soil Types

Hundreds of soil types and subtypes exist worldwide, far too many to consider here. Instead, we shall look at just three very generic types—pedalfers, pedocals, and laterites. As you read, notice that the characteristics of each soil primarily reflect the prevailing climatic conditions. A summary of the characteristics is provided in Table 3.2. (For more information on different soil groups and their distribution, see Appendix C.)

In cold or dry climates soils are generally thin and poorly developed. The reasons for this are fairly obvious. Chemical weathering progresses very slowly in such climates, and the scant plant life yields very little organic matter.

Pedalfer

The term **pedalfer** is derived from the Greek **ped**on, meaning "soil," and the chemical symbols **Al** (aluminum) and **Fe** (iron). Pedalfers are characterized by an accumulation of iron oxides and aluminum-rich clays in the *B* horizon. In midlatitude areas where the annual rainfall exceeds 63 centimeters (25 inches), most of the soluble materials, such as calcium carbonate, are leached from the soil and carried away by underground water. The less-soluble iron oxides and clays are carried from the *E* horizon and deposited in the *B* horizon, giving it a brown to red-brown color. These soils are best developed under forest vegetation, where large quantities of decomposing organic matter provide the acid conditions necessary for leaching. In the United States pedalfers are found east of a line from northwestern Minnesota to south-central Texas.

Table 3.2 Summary of soil types.				
Climate	Temperate humid (> 63 cm rainfall)	Temperate dry (< 63 cm rainfall)	Tropical (heavy rainfall)	Extreme arctic or desert
Vegetation	Forest	Grass and brush	Grass and trees	Almost none, so no humus develops
Typical Area	Eastern U.S.	Western U.S.		
Soil Type	Pedalfer	Pedocal	Laterite	
Topsoil	Sandy, light-colored; acid	Commonly enriched in calcite; whitish color	*Zones not developed* — Enriched in iron, (and aluminum); brick red color	No real soil forms because there is no organic material. Chemical weathering is very slow.
Subsoil	Enriched in aluminum, iron, and clay; brown color	Enriched in calcite; whitish color	All other elements removed by leaching	
Remarks	Extreme development in conifer forests, because abundant humus makes groundwater very acidic. Produces light gray soil because of removal of iron	*Caliche* is name applied to the accumulation of calcite	Apparently bacteria destroy humus, so no acid is available to remove iron	

Pedocal

The term **pedocal** is also derived from the Greek **ped**on, meaning "soil," and the first three letters of **cal**cite (calcium carbonate). As the name implies, pedocals are characterized by an accumulation of calcium carbonate. This soil type is found in the drier western United States in association with grassland and brush vegetation. Because chemical weathering is less intense in drier areas, pedocals generally contain a smaller percentage of clay materials than pedalfers.

In the arid and semiarid western states, a calcite-enriched layer called *caliche* may be present in the soils. In these areas, little of the rain that falls penetrates to great depths. Rather, it is held by the soil particles near the surface until it evaporates. As a result, the soluble materials, chiefly calcium carbonate, are removed from the uppermost layer and redeposited below, forming the caliche layer.

Laterite

In hot, wet tropical climates, soils called **laterites** may develop. Chemical weathering is intense under such climate conditions, so these soils are usually deeper than soils developing over a similar period in the midlatitudes. Not only does leaching remove the soluble materials such as calcite but the great quantities of percolating water also remove much of the silica, with the result that oxides of iron and aluminum become concentrated in the soil. The iron gives the soil a distinctive orange to red color.

Because bacterial activity is great in the tropics, laterites contain practically no humus. This fact, coupled with the highly leached and bricklike nature of these soils, makes laterites poor for growing crops. Their infertility has been demonstrated repeatedly in tropical countries where cultivation has been attempted (see Box 3.1).

Soil Erosion

Soils are just a tiny fraction of all Earth materials, yet they are a vital resource. Because soils are necessary for the growth of rooted plants, they are the very foundation of the human life-support system. Just as human ingenuity can increase the agricultural productivity of soils through fertilization and irrigation, soils can be damaged or destroyed by carelessness. Despite their basic role in providing food, fiber, and other basic materials, soils are among our most abused resources.

Perhaps this neglect and indifference has occurred because a substantial amount of soil seems to remain even where soil erosion is serious. Nevertheless, although the loss of fertile topsoil may not be obvious to the untrained eye, it is a growing problem as human activities expand and disturb more and more of Earth's surface.

How Soil Is Eroded

Soil erosion is a natural process; it is part of the constant recycling of Earth materials that we call the *rock cycle*. Once soil forms, erosional forces, especially water and

BOX 3.1

PEOPLE AND THE ENVIRONMENT:
Laterites and the Clearing of the Rain Forest

Laterites are thick red soils that form in the wet tropics and subtropics and are the end product of extreme chemical weathering. Because lush tropical rain forests have lateritic soils, it might seem as though they would have great potential for agriculture. However, this is not the case. In fact, just the opposite is true: Laterites are among the poorest soils for farming. Why is this the case?

Because laterites develop under conditions of high temperatures and heavy rainfall, they are severely leached. Leaching leads to low fertility because plant nutrients are removed by the large volume of downward-percolating water. Therefore, even though the vegetation may be dense and luxuriant, the soil itself contains few available nutrients. Most of the nutrients that support the rain forest are locked up in the trees themselves. The forest is maintained by the decay of plant material. That is, as vegetation dies and decomposes, the roots of the rain forest trees quickly absorb the nutrients before they are leached from the soil. The nutrients are continuously recycled as trees die and decompose.

Therefore, when forests are cleared to provide land for farming or to harvest the timber, most of the nutrients are removed as well (Figure 3.A). What remains is a soil that contains little to nourish crop growth.

The clearing of rain forests not only removes the supply of plant nutrients but also leads to accelerated erosion. When vegetation is present, its roots anchor the soil while its leaves and branches provide a canopy that protects the ground from the full force of the frequent heavy rains. Removal of the vegetation also exposes the ground to strong direct sunlight. When baked by the Sun, laterites can harden to a bricklike consistency and become practically impenetrable to water and crop roots. In only a few years, lateritic soils in a freshly cleared area may no longer be cultivable.

The term *laterite* is derived from the Latin word *latere*, meaning "brick," and was first applied to the use of this material for brick making in India and Cambodia. Laborers simply excavated

Figure 3.A   Clearing the tropical rain forest in West Kalimantan (Borneo), Indonesia. The thick lateric soil is highly leached. (Photo by Wayne Lawler/Photo Researchers, Inc.)

the soil, shaped it, and allowed it to harden in the Sun. Ancient but still well-preserved structures built of laterite remain standing today in the wet tropics (Figure 3.B). Such structures have withstood centuries of weathering because all of the original soluble materials were already removed from the soil by chemical weathering. Laterites are therefore virtually insoluble and thus very stable.

In summary, laterites are highly leached soils that are the products of extreme chemical weathering in the warm, wet tropics. Although they may be associated with lush tropical rain forests, these soils are unproductive when vegetation is removed. Moreover, when cleared of plants, laterites are subject to accelerated erosion and can be baked to bricklike hardness by the Sun.

Figure 3.B   This ancient temple at Angor Wat, Cambodia, was built of bricks made of laterite. (Photo by R. Ian Lloyd/The Stock Market)

Figure 3.17 When it is raining, millions of water drops are falling at velocities approaching 10 meters per second (35 kilometers per hour). When water drops strike an exposed surface, soil particles may splash as high as 1 meter into the air and land more than a meter away from the point of raindrop impact. Soil dislodged by splash erosion is more easily moved by sheet erosion. (Photo courtesy of U.S. Department of Agriculture)

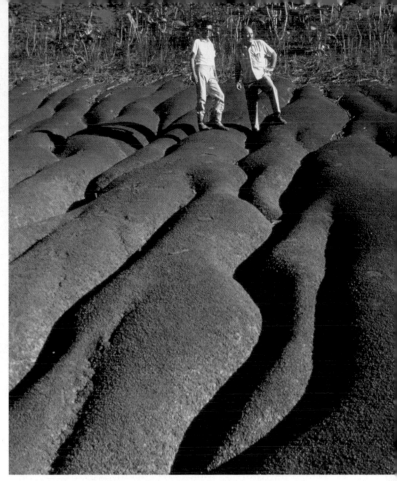

Figure 3.18 Gully erosion in poorly protected soil, southern Colombia. (Photo by Carl Purcell/Photo Researchers, Inc.)

wind, move soil components from one place to another. Every time it rains, raindrops strike the land with surprising force (Figure 3.17). Each drop acts like a tiny bomb, blasting movable soil particles out of their positions in the soil mass. Then, water flowing across the surface carries away the dislodged soil particles. Because the soil is moved by thin sheets of water, this process is termed *sheet erosion.*

After flowing as a thin, unconfined sheet for a relatively short distance, threads of current typically develop and tiny channels called *rills* begin to form. Still deeper cuts in the soil, known as *gullies*, are created as rills enlarge (Figure 3.18). When normal farm cultivation cannot eliminate the channels, we know the rills have grown large enough to be called gullies. Although most dislodged soil particles move only a short distance during each rainfall, substantial quantities eventually leave the fields and make their way downslope to a stream. Once in the stream channel, these soil particles, which can now be called *sediment*, are transported downstream and eventually deposited.

Rates of Erosion

We know that soil erosion is the ultimate fate of practically all soils. In the past, erosion occurred at slower rates than it does today because more of the land surface was covered and protected by trees, shrubs, grasses, and other plants. However, human activities such as farming, logging, and construction, which remove or disrupt the natural vegetation, have greatly accelerated the rate of soil erosion. Without the stabilizing effect of plants, the soil is more easily swept away by the wind or carried downslope by sheet wash.

Natural rates of soil erosion vary greatly from one place to another and depend on soil characteristics as well as such factors as climate, slope, and type of vegetation. Over a broad area, erosion caused by surface runoff may be estimated by determining the sediment loads of the streams that drain the region. When studies of this kind were made on a global scale, they indicated that prior to the appearance of humans, sediment transport by rivers to the oceans amounted to just over 9 billion metric tons per year (1 metric ton = 1000 kilograms). By contrast, the amount of material currently transported to the sea by rivers is about 24 billion metric tons per year, or more than two and a half times the earlier rate.

It is more difficult to measure the loss of soil due to wind erosion. However, the removal of soil by wind is generally much less significant than erosion by flowing water except during periods of prolonged drought. When dry conditions prevail, strong winds can remove large quantities of soil from unprotected fields. Such was the case in the 1930s in the portions of the Great Plains that came to be called the *Dust Bowl* (see Box 5.3, p. 155).

In many regions the rate of soil erosion is significantly greater than the rate of soil formation. This means that a renewable resource has become nonrenewable in these places. At present, it is estimated that topsoil is eroding faster than it forms on more than one-third of the world's croplands. The result is lower productivity, poorer crop quality, reduced agricultural income, and an ominous future.

Sedimentation and Chemical Pollution

Another problem related to excessive soil erosion involves the deposition of sediment. Each year in the United States hundreds of millions of tons of eroded soil are deposited in lakes, reservoirs, and streams. The detrimental impact of this process can be significant. For example, as more and more sediment is deposited in a reservoir, the capacity of the reservoir is reduced, limiting its usefulness for flood control, water supply, and/or hydroelectric power generation. In addition, sedimentation in streams and other waterways can restrict navigation and lead to costly dredging operations.

In some cases soil particles are contaminated with pesticides used in farming. When these chemicals are introduced into a lake or reservoir, the quality of the water supply is threatened and aquatic organisms can be endangered. In addition to pesticides, nutrients found naturally in soils as well as those added by agricultural fertilizers make their way into streams and lakes, where they stimulate the growth of plants. Over a period of time, excessive nutrients accelerate the process by which plant growth leads to the depletion of oxygen and an early death of the lake.

The availability of good soils is critical if the world's rapidly growing population is to be fed. On every continent, unnecessary soil loss is occurring because appropriate conservation measures are not being used. Although it is a recognized fact that soil erosion can never be completely eliminated, soil conservation programs can substantially reduce the loss of this basic resource. Windbreaks (rows of trees), terracing, and plowing along the contours of hills are some of the effective measures, as are special tillage practices and crop rotation.

Weathering Creates Ore Deposits

Weathering creates many important mineral deposits by concentrating minor amounts of metals that are scattered through unweathered rock into economically valuable concentrations. Such a transformation is often termed **secondary enrichment** and takes place in one of two ways. In one situation, chemical weathering coupled with downward-percolating water removes undesired materials from decomposing rock, leaving the desired elements enriched in the upper zones of the soil. The second way is basically the reverse of the first. That is, the desirable elements that are found in low concentrations near the surface are removed and carried to

Figure 3.19   Bauxite is the ore of aluminum and forms as a result of weathering processes under tropical conditions. Its color varies from red or brown to nearly white. (Photo by E. J. Tarbuck)

lower zones, where they are redeposited and become more concentrated.

Bauxite

The formation of *bauxite*, the principal ore of aluminum, is one important example of an ore created as a result of enrichment by weathering processes (Figure 3.19). Although aluminum is the third most abundant element in Earth's crust, economically valuable concentrations of this important metal are not common, because most aluminum is tied up in silicate minerals from which it is extremely difficult to extract.

Bauxite forms in rainy tropical climates in association with laterites. (In fact, bauxite is sometimes called aluminum laterite.) When aluminum-rich source rocks are subjected to the intense and prolonged chemical weathering of the tropics, most of the common elements, including calcium, sodium, and silicon, are removed by leaching. Because aluminum is extremely insoluble, it becomes concentrated in the soil (as bauxite, a hydrated aluminum oxide). Thus, the formation of bauxite depends on climatic conditions in which chemical weathering and leaching are pronounced, plus, of course, the presence of aluminum-rich source rock. Important deposits of nickel and cobalt are also found in laterite soils that develop from igneous rocks rich in other silicate minerals.

Other Deposits

Many copper and silver deposits result when weathering processes concentrate metals that are dispersed through a low-grade primary ore. Usually such enrichment occurs in deposits containing pyrite (FeS_2), the most common and widespread sulfide mineral. Pyrite is important because when it chemically weathers, sulfuric acid forms, which enables percolating waters to dissolve the ore metals. Once dissolved, the metals gradually migrate downward through the primary ore body until they are precipitated. Deposition takes place because of changes that occur in the chemistry of the solution when it reaches the groundwater zone (the zone beneath the surface where all pore spaces are filled with water). In this manner, the small percentage of dispersed metal can be removed from a large volume of rock and redeposited as a higher-grade ore in a smaller volume of rock.

Mass Wasting: The Work of Gravity

Earth's surface is never perfectly flat but instead consists of slopes. Some are steep and precipitous; others are moderate or gentle. Some are long and gradual; others are short and abrupt. Some slopes are mantled with soil and covered by vegetation; others consist of barren rock and rubble. Their form and variety are great. Although most slopes appear to be stable and unchanging, they are not static features because the force of gravity causes material to move downslope. At one extreme the movement may be gradual and practically imperceptible. At the other extreme it may consist of a roaring debris flow or a thundering rock avalanche. Landslides are a worldwide natural hazard. When these natural processes lead to loss of life and property, they become natural disasters.

Most mass wasting, whether spectacular or subtle, is the result of circumstances that are completely independent of human activities. Very few landslides occur where they cannot be anticipated. In places where mass wasting is a recognized threat, steps can often be taken to control downslope movements or limit the damages that such movements can cause. If the potential for mass wasting goes unrecognized or is ignored, the results can be costly and dangerous. We should also note that although most downslope movements occur whether people are present or not, many occurrences are aggravated or even triggered by human actions.

Mass Wasting and Landform Development

Landslides are spectacular examples of a basic geologic process called *mass wasting*. Mass wasting refers to the downslope movement of rock, regolith, and soil under the direct influence of gravity. It is distinct from the erosional processes that are examined in subsequent chapters because mass wasting does not require a transporting medium such as water, wind, or glacial ice.

In the evolution of most landforms, mass wasting is the step that follows weathering. By itself, weathering does not produce significant landforms. Rather, landforms develop as the products of weathering are removed from the places where they originate. Once weathering weakens and breaks rock apart, mass wasting transfers the debris downslope, where a stream, acting as a conveyor belt, usually carries it away (see Figure 3.1). Although there may be many intermediate stops along the way, the sediment is eventually transported to its ultimate destination: the sea.

The combined effects of mass wasting and running water produce stream valleys, which are the most common and conspicuous of Earth's landforms. If streams alone were responsible for creating the valleys in which they flow, the valleys would be very narrow features. However, the fact that most river valleys are much wider than they are deep is a strong indication of the significance of mass-wasting processes in supplying material to streams. The walls of a canyon extend far from the river because of the transfer of weathered debris downslope to the river and its tributaries by mass-wasting processes. In this manner, streams and mass wasting combine to modify and sculpture the surface. Of course, glaciers, groundwater, waves, and wind are also important agents in shaping landforms and developing landscapes.

Controls and Triggers of Mass Wasting

Gravity is the controlling force of mass wasting, but several factors play an important role in overcoming inertia and creating downslope movements. Long before a landslide occurs, various processes work to weaken slope material, gradually making it more and more susceptible to the pull of gravity. During this span, the slope remains stable but gets closer and closer to being unstable. Eventually, the strength of the slope is weakened to the point that something causes it to cross the threshold from stability to instability. Such an event that initiates downslope movement is called a *trigger*. Remember that the trigger is not the sole cause of the mass-wasting event, but just the last of many causes. Among the common factors that trigger mass-wasting processes are saturation of material with water, oversteepening of slopes, removal of anchoring vegetation, and ground vibrations from earthquakes.

The Role of Water

Mass wasting is sometimes triggered when heavy rains or periods of snowmelt saturate surface materials. This was the case in October 1998 when torrential downpours associated with Hurricane Mitch triggered devastating mudflows in Central America (Figure 3.20).

When the pores in sediment become filled with water, the cohesion among particles is destroyed, allowing them to slide past one another with relative ease. For example, when sand is slightly moist, it sticks together quite well. However, if enough water is added to fill the openings between the grains, the sand will ooze

Figure 3.20 Heavy rains from Hurricane Mitch in November 1998 created devastating mudflows in Central America. (Photo by Noel Quidu/Liaison Agency, Inc.)

out in all directions. Thus, saturation reduces the internal resistance of materials, which are then easily set in motion by the force of gravity. When clay is wetted, it becomes very slick—another example of the "lubricating" effect of water. Water also adds considerable weight to a mass of material. The added weight in itself may be enough to cause the material to slide or flow downslope.

Oversteepened Slopes

Oversteepening of slopes is another trigger of many mass movements. There are many situations in nature where this takes place. A stream undercutting a valley wall and waves pounding against the base of a cliff are two familiar examples. Furthermore, through their activities, people often create oversteepened and unstable slopes that become prime sites for mass wasting.

Unconsolidated, granular (sand-size or coarser) particles assume a stable slope called the **angle of repose** (*reposen* = to be at rest), the steepest angle at which material remains stable. Depending on the size and shape of the particles, the angle varies from 25 to 40 degrees. The larger, more angular particles maintain the steepest slopes. If the angle is increased, the rock debris will adjust by moving downslope.

Oversteepening is important not only because it triggers movements of unconsolidated granular materials, but it also produces unstable slopes and mass movements in cohesive soils, regolith, and bedrock. The response will not be immediate, as with loose, granular

material, but sooner or later one or more mass-wasting processes will eliminate the oversteepening and restore stability to the slope.

Removal of Vegetation

Plants protect against erosion and contribute to the stability of slopes because their root systems bind soil and regolith together. Where plants are lacking, mass wasting is enhanced, especially if slopes are steep and water is plentiful. When anchoring vegetation is removed by forest fires or by people (for timber, farming, or development), surface materials frequently move downslope.

An unusual example illustrating the anchoring effect of plants occurred several decades ago on steep slopes near Menton, France. Farmers replaced olive trees, which have deep roots, with a more profitable but shallow-rooted crop: carnations. When the less stable slope failed, the landslide took 11 lives.

Earthquakes as Triggers

Conditions favoring mass wasting can exist in an area for a long time without movement occurring. An additional factor is sometimes necessary to trigger the movement. Among the more important and dramatic triggers are earthquakes. An earthquake and its aftershocks can dislodge enormous volumes of rock and unconsolidated material. The mass-wasting events shown in Figures 3.21 and 3.24 were both triggered by earthquakes. In many areas that are jolted by earthquakes, it is not ground vibrations directly but landslides and ground

Figure 3.21 This home in Pacific Palisades, California, was destroyed by a landslide triggered by the January 1994 Northridge earthquake. In some cases, damages from earthquake-induced mass wasting are greater than damages caused directly by an earthquake's ground vibrations. Also see Figure I.10, p.13. (Photo by ChromoSohm/Sohm/The Stock Market)

subsidence triggered by the vibrations that cause the greatest damage.

Classifying Mass-Wasting Processes

Geologists include several processes under the name of mass wasting. Four are illustrated in Figure 3.22. Generally, each process is defined by the type of material involved, the kind of motion, and the velocity of the movement.

If soil and regolith dominate, terms such as *debris*, *mud*, or *earth* are used. In contrast, when a mass of bedrock breaks loose and moves downslope, the term *rock* may be part of the description. Generally, the kind of motion is described as either a fall, a slide, or a flow.

Type of Motion

When the movement involves the freefall of detached individual pieces of any size, it is termed a **fall.** Falls are common on slopes that are too steep for loose material to remain on the surface (Figure 3.23). Many falls result when freeze and thaw cycles or the action of plant roots loosen rock to the point that gravity takes over. Rockfall is the primary way in which talus slopes are built and maintained (see Figure 3.3). Sometimes falls may trigger other forms of downslope movement.

Many mass-wasting processes are **slides,** which occur whenever material remains fairly coherent and moves along a well-defined surface. Sometimes the surface is a joint, a fault, or a bedding plane that is approximately parallel to the slope. However, in the movement called *slump*, the descending material moves en masse along a curved surface of rupture.

Sometimes the word *slide* is used as a synonym for the word *landslide*. It should be pointed out that although many people, including geologists, use the term, the word *landslide* has no specific definition in geology. Rather, it should be considered as a popular nontechnical term to describe all relatively rapid forms of mass wasting, including those in which sliding does not occur.

A third type of movement common to mass wasting is termed flow. Flow occurs when material moves downslope as a viscous fluid. Most flows are saturated with water and typically move as lobes or tongues.

Rate of Movement

When mass-wasting events make the news, a large quantity of material has in all likelihood moved rapidly downslope and has had a disastrous effect upon people and property. Indeed, during events called *rock avalanches*, rock and debris can hurtle downslope at speeds exceeding 200 kilometers (125 miles) per hour. Many researchers believe that rock avalanches, such as the one that produced the scene in Figure 3.24, must literally "float on air" as they move downslope. That is, high velocities result when air becomes trapped and compressed beneath the falling mass of debris, allowing it to move as a buoyant, flexible sheet across the surface.

Most mass movements, however, do not move with the speed of a rock avalanche. In fact, a great deal of mass wasting is imperceptibly slow. One process that we will examine later, termed *creep*, results in particle movements that are usually measured in millimeters or centimeters per year. Thus, as you can see, rates of movement can be spectacularly sudden or exceptionally gradual. Even the velocity of a single process at a particular site can vary considerably from one time to another.

? STUDENTS SOMETIMES ASK...
Are snow avalanches considered a type of mass wasting?

Sure. Sometimes these thundering downslope movements of snow and ice move large quantities of rock, soil, and trees. Of course, snow avalanches are very dangerous, especially to skiers on high mountain slopes and to buildings and roads at the bottom of slopes in avalanche-prone regions.

About 10,000 snow avalanches occur each year in the mountainous western United States. In an average year they claim between 15 and 25 lives in the U.S. and Canada. They are a growing problem as more people become involved in winter sports and recreation.

A. Slump

B. Rockside

C. Debris flow

D. Earthflow

Figure 3.22   The four processes illustrated here are all considered to be relatively rapid forms of mass wasting. Because material in slumps **A.** and rockslides **B.** move along well-defined surfaces, they are said to move by sliding. By contrast, when material moves downslope as a viscous fluid, the movement is described as a flow. Debris flow **C.** and earthflow **D.** advance downslope in this manner.

Slump

Slump refers to the downward sliding of a mass of rock or unconsolidated material moving as a unit along a *curved* surface (see Figure 3.22A). Usually the slumped material does not travel spectacularly fast nor very far. This is a common form of mass wasting, especially in

Figure 3.23   Rockfall blocking the Karakoram Highway in Pakistan's Hunza Valley. (Photo by Robert Holmes/CORBIS)

thick accumulations of cohesive materials such as clay. As the movement occurs, a crescent-shaped scarp (cliff) is created at the head, and the block's upper surface is sometimes tilted backward.

Slump commonly occurs because a slope has been oversteepened. The material on the upper portion of a slope is held in place by the material at the bottom of the slope. As this anchoring material at the base is removed, the material above is made unstable and reacts to the pull of gravity. A common example is a valley wall that becomes oversteepened by a meandering river. Another is a coastal area that has been undercut by wave activity at its base.

Rockslide

Rockslides frequently occur in high mountain areas such as the Andes, Alps, and Canadian Rockies. They are sudden and rapid movements that happen when detached segments of bedrock break loose and slide downslope (Figure 3.22B). As the moving mass thunders along the surface, it breaks into many smaller pieces. Such events are among the fastest and most destructive mass movements.

Rockslides usually take place where there is an inclined surface of weakness. Such surfaces tend to form

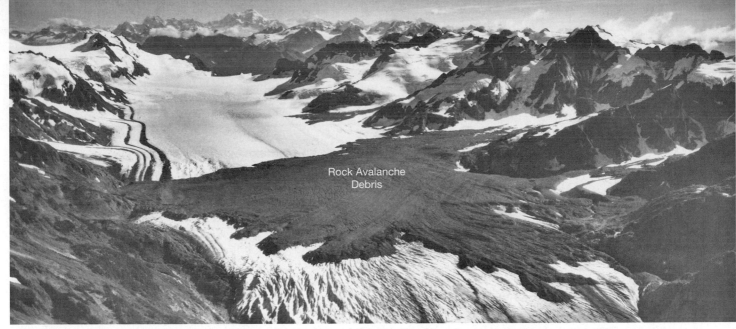

Figure 3.24 This 4-kilometer-long tongue of rubble was deposited atop Alaska's Sherman Glacier by a rock avalanche. The event was triggered by a tremendous earthquake in March 1964. (Photo by Austin Post, U.S. Geological Survey)

where strata are tilted or where joints and fractures exist parallel to the slope. When rock in such a setting is undercut at the base of the slope, it loses support and eventually gives way. Sometimes an earthquake is the trigger. On other occasions the rockslide is triggered when rain or melting snow lubricates the underlying surface to the point that friction is no longer sufficient to hold the rock unit in place. As a result, rockslides tend to be more common during the spring, when heavy rains and melting snow are most prevalent. The massive Gros Ventre slide shown in Figure 3.25 is a classic example.

A.

B.

Figure 3.25 On June 23, 1925, a massive rockslide took place in the valley of the Gros Ventre River in northwestern Wyoming following heavy spring rains and snowmelt. The volume of debris, estimated at 38 million cubic meters, created a 70-meter-high dam. Later, the lake created by the debris dam overflowed, resulting in a devastating flood downstream. **A.** Cross-sectional view. The slide occurred when the tilted and undercut sandstone bed could no longer maintain its position atop the saturated bed of clay. **B.** Even though the Gros Ventre rockslide occurred in 1925, the scar left on the side of Sheep Mountain is still a prominent feature. (Part A after W. C. Alden, "Landslide and Flood at Gros Ventre, Wyoming," *Transactions* (AIME) 76 (1928); 348. Part B photo by Stephen Trimble)

Debris Flow

Debris flow is a relatively rapid type of mass wasting that involves a flow of soil and regolith containing a large amount of water (Figure 3.22C). Debris flows, which are also called *mudflows*, are most characteristic of semiarid mountainous regions and are also common on the slopes of some volcanoes. Because of their fluid properties, debris flows follow canyons and stream channels. As Figure 3.26 illustrates, in populated areas debris flows can pose a significant hazard to life and property (see Box 3.2).

Debris Flows in Semiarid Regions

When a cloudburst or rapidly melting mountain snows create a sudden flood in a semiarid region, large quantities of soil and regolith are washed into nearby stream channels because there is usually little vegetation to anchor the surface material. The end product is a flowing tongue of well-mixed mud, soil, rock, and water. Its consistency may range from that of wet concrete to a soupy mixture not much thicker than muddy water. The rate of flow therefore depends not only on the slope but also on the water content. When dense, debris flows are capable of carrying or pushing large boulders, trees, and even houses with relative ease.

Debris flows pose a serious hazard to development in dry mountainous areas such as southern California. The construction of homes on canyon hillsides and the removal of anchoring vegetation by brush fires and other means have increased the frequency of these destructive events.

Lahars

Debris flows composed mostly of volcanic materials on the flanks of volcanoes are called *lahars.* The word originated in Indonesia, a volcanic region that has experienced many of these often destructive events. His-

torically, lahars have been one of the deadliest volcano hazards. They can occur either during an eruption or when a volcano is quiet. They take place when highly unstable layers of ash and debris become saturated with water and flow down steep volcanic slopes, generally following existing stream channels. Heavy rainfalls often trigger these flows. Others are triggered when large volumes of ice and snow are suddenly melted by heat flowing to the surface from within the volcano or by the hot gases and near-molten debris emitted during a violent eruption.

In November 1985 lahars were produced when Nevado del Ruiz, a 5300-meter (17,400-foot) volcano in the Andes Mountains of Colombia, erupted. The eruption melted much of the snow and ice that capped the uppermost 600 meters (2000 feet) of the peak, producing torrents of hot, thick mud, ash, and debris. The lahars moved outward from the volcano, following the valleys of three rain-swollen rivers that radiate from the peak. The flow that moved down the valley of the Lagunilla River was the most destructive, devastating the town of Armero, 48 kilometers (30 miles) from the mountain. Most of the more than 25,000 deaths caused by the event occurred in this once-thriving agricultural community.

Earthflow

We have seen that debris flows are frequently confined to channels in semiarid regions. In contrast, **earthflows** most often form on hillsides in humid areas during times of heavy precipitation or snowmelt (see Figure 3.22D). When water saturates the soil and regolith on a hillside, the material may break away, leaving a scar on the slope and forming a tongue- or teardrop-shaped mass that flows downslope (Figure 3.27). The materials most commonly involved are rich in clay and silt and contain only small proportions of sand and coars-

Figure 3.26 On January 25, 1997, a debris flow literally buried this one-story home in Mill Valley, California. Heavy rains from a powerful Pacific storm triggered the event. (Justin Sullivan/AP/Wide World Photos)

BOX 3.2

PEOPLE AND THE ENVIRONMENT
Real-Time Monitoring of Active Landslides*

Landslides threaten lives and property in all 50 states. To reduce the risk from active landslides, the U.S. Geological Survey (USGS) develops and uses real-time landslide monitoring systems. Monitoring can detect early indications of rapid catastrophic movement.

During heavy rains in January 1997, thousands of tons of rock and debris gave way in the Sierra Nevada, destroying three homes, blocking a major northern California highway (U.S. 50), and briefly damming the adjacent American River (Figure 3.C). Waters dammed by the landslide quickly flooded two vehicles traveling on the highway, leading to dramatic rescues. Reopening the highway cost $4.5 million, and indirect economic losses from the highway closure were estimated to exceed $50 million.

Several large, active landslides in the same area continue to threaten U.S.

*Based on material prepared by the U.S. Geological Survey.

Highway 50. Although these slides often move slowly, some have the potential to collapse catastrophically. To help reduce the risk posed by these rapid movements, the USGS, in cooperation with other agencies, established continuous real-time monitoring of the landslide activity using systems developed for monitoring active volcanoes in remote areas. Data from a variety of sensors installed at the landslides are transmitted by radio to USGS computers (Figure 3.D).

Why is it necessary to collect real-time landslide data remotely? The immediate detection of landslide activity provided by real-time systems can be crucial in saving human lives and protecting property. Monitoring can determine the speed of landslide movement and can also detect early indications of catastrophic movement. Rapid movement at some landslides, for example, is preceded by gradual acceleration. By using real-time data from such slides, geologists can anticipate possible catastrophic movement. USGS landslide monitoring focus-

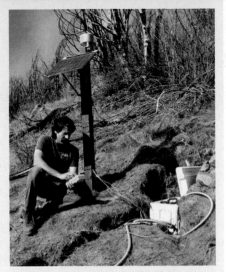

Figure 3.D   Testing a solar-powered radio telemetry system for remote transmission of real-time landslide data. At remotely monitored landslide areas, ground movement and groundwater pressures are measured every second. The amount of downslope movement is recorded by extensometers that can detect stretching or shortening of the ground. Ground vibrations are monitored by geophones buried within the slides. Groundwater conditions within slides are monitored by porewater pressure sensors, and on-site rain gauges record precipitation. (Photo by Mark Reid, U.S. Geological Survey)

es on detecting (1) precipitation and groundwater conditions that could destabilize a hillslope, (2) acceleration of slide movement, and (3) ground vibrations associated with movement.

Currently, five landslides along U.S. Highway 50 are monitored in real time using 58 instruments. This network, operated in cooperation with the California Department of Transportation, provides early notification of landslide activity and information useful in the design of remedial measures to halt these slides. The USGS operates other remote real-time landslide monitoring sites elsewhere in California and in Washington, New Mexico, and Colorado.

Figure 3.C   Home destroyed by a sudden catastrophic landslide in 1997 along U.S. Highway 50, about 25 miles east of Placerville, California. (Photo by Mark Reid U.S. Geological Survey)

Figure 3.27 This small, tongue-shaped earthflow occurred on a newly formed slope along a recently constructed highway. It formed in clay-rich material following a period of heavy rain. Notice the small slump at the head of the earthflow. (Photo by E. J. Tarbuck)

er particles. Earthflows range in size from bodies a few meters long, a few meters wide, and less than 1 meter deep to masses more than 1 kilometer long, several hundred meters wide, and more than 10 meters deep.

Because earthflows are quite viscous, they generally move at slower rates than the more fluid debris flows. They move slowly and persistently for periods ranging from days to years. Depending on slope steepness and the material's consistency, velocities range from less than 1 millimeter per day up to several meters per day. Movement is typically faster during wet periods. In addition to occurring as isolated hillside phenomena, earthflows commonly take place in association with large slumps. In this situation, they may be seen as tonguelike flows at the base of the slump.

A special type of earthflow, known as *liquefaction,* sometimes occurs in association with earthquakes. Porous, clay- to sand-size sediments that are saturated with water are most vulnerable. When shaken suddenly, the grains lose cohesion and the ground flows. Liquefaction can cause buildings to sink or tip on their sides and underground storage tanks and sewer lines to float upward. To say the least, damage can be substantial. You will learn more about this in Chapter 6.

Slow Movements

Movements such as rockslides, rock avalanches, and lahars are certainly the most spectacular and catastrophic forms of mass wasting. These dangerous events deserve intensive study to enable more effective prediction, timely warnings, and better controls to save lives. However, because of their large size and spectacular nature, they give us a false impression of their importance as a mass-wasting process. Indeed, sudden movements transport less material than the slow, subtle action of creep. Whereas rapid types of mass wasting are characteristic of mountains and steep hillsides, creep takes place on both steep and gentle slopes and is thus much more widespread.

Creep *most widespread form of mass wasting*

Creep is a type of mass wasting that involves the gradual downhill movement of soil and regolith. One factor that contributes to creep is the alternate expansion and contraction of surface material caused by freezing and thawing or wetting and drying. As shown in Figure 3.28, freezing or wetting lifts particles at right angles to the slope, and thawing or drying allows the particles to fall back to a slightly lower level. Each cycle therefore moves the material a short distance downhill.

Creep is aided by anything that disturbs the soil. For example, raindrop impact and disturbance by plant roots and burrowing animals may contribute. Creep is also promoted when the ground becomes saturated with water. Following a heavy rain or snowmelt, a waterlogged soil may lose its internal cohesion, allowing gravity to pull the material downslope. Because creep is imperceptibly slow, the process cannot be observed in action. What can be observed, however, are the effects of creep. Creep causes fences and utility poles to tilt and retaining walls to be displaced.

? STUDENTS SOMETIMES ASK...
How much damage is caused by mass wasting?

The annual cost of damage by all types of mass wasting in the U.S. is about $4 billion. However, this figure is probably conservative in that slow movements like creep do a substantial amount of damage that goes unreported. The best way to avoid damage from mass wasting is to avoid building on slopes (or just downhill from them). Also, mass-wasting events tend to occur where they have occurred before, so good site analysis done by qualified geologists is a critical step toward avoiding problems caused by mass-wasting events.

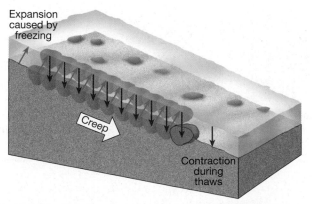

Figure 3.28 The repeated expansion and contraction of the surface material causes a net downslope migration of rock particles—a process called *creep.*

Solifluction

When soil is saturated with water, the soggy mass may flow downslope at a rate of a few millimeters or a few centimeters per day or per year. Such a process is called **solifluction** (literally, "soil flow"). It is a type of mass wasting that is common wherever water cannot escape from the saturated surface layer by infiltrating to deeper levels. A dense clay hardpan in soil or an impermeable bedrock layer can promote solifluction.

Solifluction is also common in regions underlain by **permafrost.** Permafrost refers to the permanently frozen ground that occurs in Earth's harsh tundra and ice-cap climates. It occurs in a zone above the permafrost called the *active layer*, which thaws in summer and refreezes in winter. During summer, water is unable to percolate into the impervious permafrost layer below. As a result, the active layer becomes saturated and slowly flows. The process can occur on slopes as gentle as 2 to 3 degrees. Where there is a well-developed mat of vegetation, a solifluction sheet may move in a series of well-defined lobes or overriding folds (Figure 3.29).

Figure 3.29   Solifluction lobes northeast of Fairbanks, Alaska. Solifluction occurs in permafrost regions when the active layer thaws in summer. (Photo by James E. Patterson)

Chapter Summary

- External processes include (1) *weathering*—the disintegration and decomposition of rock at or near the surface, (2) *mass wasting*—the transfer of rock material downslope under the influence of gravity, and (3) *erosion*—the incorporation and transportation of material by a mobile agent, usually water, wind, or ice. They are called *external processes* because they occur at or near Earth's surface and are powered by energy from the Sun. By contrast, *internal processes*, such as volcanism and mountain building, derive their energy from Earth's interior.

- *Mechanical weathering* is the physical breaking up of rock into smaller pieces. *Chemical weathering* alters a rock's chemistry, changing it into different substances. Rocks can be broken into smaller fragments by *frost wedging, unloading*, and *biological activity*. Water is by far the most important agent of chemical weathering. Oxygen in water can *oxidize* some materials, while carbon dioxide (CO_2) dissolved in water forms *carbonic acid*. The chemical weathering of silicate minerals frequently produces (1) soluble products containing sodium, calcium, potassium, and magnesium, (2) insoluble iron oxides, and (3) clay minerals.

- The rate at which rock weathers depends on such factors as (1) *particle size*—small pieces generally weather faster than large pieces; (2) *mineral makeup*—calcite readily dissolves in mildly acidic solutions, and silicate minerals that form first from magma are least resistant to chemical weathering; and (3) *climatic factors*, particularly temperature and moisture. Frequently, rocks exposed at Earth's surface do not weather at the same rate. This *differential weathering* of rocks is influenced by such factors as mineral makeup and degree of jointing.

- *Soil* is a combination of mineral and organic matter, water, and air—that portion of the *regolith* (the layer of rock and mineral fragments produced by weathering) that supports the growth of plants. *Soil texture* refers to the proportions of different particle sizes (clay, silt, and sand) found in soil. The most important *factors that control soil formation are parent material, time, climate, plants, animals, and slope.*

- Soil-forming processes operate from the surface downward and produce zones or layers in the soil called *horizons*. From the surface downward the horizons are designated as O, A, E, B, and C, respectively. Although there are hundreds of soil types and subtypes worldwide, the three very generic types are (1) *pedalfer*—characterized by an accumulation of iron oxides and aluminum-rich clays in the B horizon, (2) *pedocal*—characterized by an accumulation of calcium carbonate, and (3) *laterite*—deep soils that develop in the hot, wet tropics that are poor for growing crops because they are highly leached and bricklike.

- Soil erosion by wind and water is the ultimate fate of practically all soils. *Rates of soil erosion vary* from one place to another and depend on the soil's characteristics as well as such factors as climate, slope, and type of vegetation.

- Weathering creates mineral deposits by concentrating metals into economically valuable deposits. The process, called *secondary enrichment*, is accomplished by either (1) removing undesirable materials and leaving the desired elements enriched in the upper zones of the soil or (2) removing and carrying the desirable elements to lower soil zones where they are redeposited and thus become more concentrated. *Bauxite*, the principal ore of aluminum, is one important ore created by secondary enrichment.

- In the evolution of most landforms, mass wasting is the step that follows weathering. The combined effect of mass wasting and erosion by running water produce stream valleys. *Gravity is the controlling force of mass wasting.* Other factors that influence or trigger downslope movements are saturation of the material with water, oversteepening of slopes beyond the *angle of repose*, removal of anchoring vegetation, and ground vibrations from earthquakes.

• The various processes included under the name of mass wasting are classified and described on the basis of (1) the type of material involved (debris, mud, earth, or rock), (2) the kind of motion (fall, slide, or flow), and (3) the rate of movement (fast, slow). The various kinds of mass wasting include the more rapid forms called *slump, rockslide, debris flow,* and *earthflow,* as well as the slow movements referred to as *creep* and *solifluction.*

Key Terms

angle of repose (p. 86)
chemical weathering (p. 69)
creep (p. 92)
debris flow (p. 90)
differential weathering (p. 74)
earthflow (p. 90)
eluviation (p. 79)
erosion (p. 69)
exfoliation dome (p. 70)
external processes (p. 68)
fall (p. 87)
flow (p. 87)
frost wedging (p. 69)

horizon (p. 79)
internal processes (p. 68)
lahar (p. 90)
laterite (p. 81)
leaching (p. 79)
mass wasting (p. 68)
mechanical weathering (p. 69)
parent material (p. 77)
pedalfer (p. 80)
pedocal (p. 81)
permafrost (p. 93)
regolith (p. 76)
rockslide (p. 88)

secondary enrichment (p. 84)
sheeting (p. 70)
slide (p. 87)
slump (p. 88)
soil (p. 76)
soil profile (p. 79)
soil texture (p. 76)
solifluction (p. 93)
solum (p. 79)
spheroidal weathering (p. 73)
talus slope (p. 70)
weathering (p. 68)

Review Questions

1. Describe the role of external processes in the rock cycle.

2. If two identical rocks were weathered—one mechanically and the other chemically—how would the products of weathering for the two rocks differ?

3. How does mechanical weathering add to the effectiveness of chemical weathering?

4. Describe the formation of an exfoliation dome. Give an example of such a feature.

5. Granite and basalt are exposed at the surface in a hot, wet region.
   (a) Which type of weathering (mechanical or chemical) will predominate?
   (b) Which of the rocks will weather most rapidly? Why?

6. Heat speeds up a chemical reaction. Why then does chemical weathering proceed slowly in a hot desert?

7. How is carbonic acid (H_2CO_3) formed in nature? What results when this acid reacts with potassium feldspar?

8. Using the soil-texture diagram (Figure 3.13), name the soil that consists of 60 percent sand, 30 percent silt, and 10 percent clay.

9. What factors might cause different soils to develop from the same parent material, or similar soils to form from different parent materials?

10. Which of the controls of soil formation is most important? Explain.

11. How can slope affect the development of soil? What is meant by the term *slope orientation?*

12. List the characteristics associated with each of the horizons in a well-developed soil profile. Which of the horizons constitute the solum? Under what circumstances do soils lack horizons?

13. Distinguish between pedalfers and pedocals.

14. What soil type is associated with tropical rain forests? Since this soil supports the growth of lush natural vegetation, is it also excellent for growing crops? Briefly explain (see Box 3.1.).

15. Is soil erosion a natural process or primarily the result of inappropriate land use by people? Explain.

16. List three detrimental effects of soil erosion other than the loss of topsoil from croplands.

17. Name the primary ore of aluminum and describe its formation.

18. Describe how mass-wasting processes contribute to the development of stream valleys.

19. What is the controlling force of mass wasting? What other factors can influence or trigger mass wasting?

20. Distinguish between fall, slide, and flow.

21. List the criteria that are used to classify mass-wasting processes.

22. What factors led to the massive rockslide at Gros Ventre, Wyoming?

23. What type of mass-wasting event killed thousands of people in the vicinity of the Colombian volcano Nevado del Ruiz in 1985?

24. Describe the mechanisms or factors that contribute to the slow downslope movement called creep.

25. During what season does solifluction occur in permafrost regions?

Examining the Earth System

1. Which of Earth's spheres are associated with the production of carbonic acid?

2. The level of carbon dioxide (CO_2) in the atmosphere has been increasing for more than a century. Should this increase tend to accelerate or slow down the rate of chemical weathering of Earth's surface rocks? Explain how you arrived at your conclusion.

3. Many gases are emitted into the atmosphere as a result of human activities. List some gases other than carbon dioxide that create acids and therefore accelerate the rate of chemical weathering. Do these acids have other effects on the Earth system? You will find the information about acid rain at these Web sites helpful: **http://www.ec.gc.ca/acidrain/acidfact.html**

   http://epa.gov/airmarkets/acidrain/

4. Discuss the interaction of the atmosphere, solid Earth, biosphere, and hydrosphere in the formation of soil.

Web Resources

The *Earth Science* Website uses the resources and flexibility of the Internet to aid in your study of the topics in this chapter. Written and developed by Earth science instructors, this site will help improve your understanding of Earth science. Visit **http://www.prenhall.com/tarbuck** and click on the cover of *Earth Science 10e* to find:

- **On-line review quizzes.**
- **Web-based critical thinking and writing exercises.**
- **Links to chapter-specific Web resources.**
- **Internet-wide key term searches.**

http://www.prenhall.com/tarbuck

Lost Yak Rapids on Chile's Rio Bio Bio. (Photo © by Carr Clifton)

RUNNING WATER AND GROUNDWATER

All the rivers run into the sea; yet the sea is not full; unto the place from whence the rivers come, thither they return again. (Ecclesiastes 1:7)

As the perceptive writer of Ecclesiastes indicated, water is continually on the move, from the ocean to the land and back again in an endless cycle. This chapter deals with that part of the hydrologic cycle that returns water to the sea. Some water travels quickly via a rushing stream, and some moves more slowly below the surface. When viewed as part of the Earth system, streams and groundwater represent basic links in the constant cycling of the planet's water. In Chapter 4, we will examine the factors that influence the distribution and movement of water, as well as look at how water sculptures the landscape. To a great extent, the Grand Canyon, Niagara Falls, Old Faithful, and Mammoth Cave all owe their existence to the action of water on its way to the sea.

Earth as a System: The Hydrologic Cycle

 Sculpturing Earth's Surface
↳ Hydrologic Cycle

Water is everywhere on Earth—in the oceans, glaciers, rivers, lakes, the air, soil, and in living tissue. All of these reservoirs constitute Earth's hydrosphere. In all, the water content of the hydrosphere is about 1.36 billion cubic kilometers (326 million cubic miles). The vast bulk of it, about 97.2 percent, is stored in the global oceans (Figure 4.1). Ice sheets and glaciers account for another 2.15 percent, leaving only 0.65 percent to be divided among lakes, streams, groundwater, and the atmos-

phere (Figure 4.1). Although the percentages of Earth's total water found in each of the latter sources is but a small fraction of the total inventory, the absolute quantities are great.

The water found in each of the reservoirs depicted in Figure 4.1 does not remain in these places indefinitely. Water can readily change from one state of matter (solid, liquid, or gas) to another at the temperatures and pressures occurring at Earth's surface. Therefore, water is constantly moving among the oceans, the atmosphere, the solid Earth, and the biosphere. This unending circulation of Earth's water supply is called the hydrologic cycle. The cycle shows us many critical interrelationships among different parts of the Earth system.

The hydrologic cycle is a gigantic worldwide system powered by energy from the Sun in which the atmosphere provides the vital link between the oceans and continents (Figure 4.2). Water evaporates into the atmosphere from the ocean and to a much lesser extent from the continents. Winds transport this moisture-laden air, often great distances, until conditions cause the moisture to condense into clouds, and precipitation to fall. The precipitation that falls into the ocean has completed its cycle and is ready to begin another. The water that falls on the continents, however, must make its way back to the ocean.

What happens to precipitation once it has fallen on land? A portion of the water soaks into the ground (called **infiltration**), slowly moving downward, then laterally, finally seeping into lakes, streams, or directly into the ocean. When the rate of rainfall exceeds Earth's ability to absorb it, the surplus water flows over the surface into lakes and streams, a process called **runoff.** Much of the water that infiltrates or runs off eventually returns to the atmosphere because of evaporation from the soil, lakes, and streams. Also, some of the water that infiltrates the ground surface is absorbed by plants, which then release it into the atmosphere. This process is called **transpiration** (*trans* = across, *spiro* = to breathe). Each year a field of crops may transpire the equivalent of a water layer 60 centimeters (2 feet) deep over the entire field. The same area of trees may pump twice this amount into the atmosphere.

When precipitation falls in very cold areas—at high elevations or high latitudes—the water may not immediately soak in, run off, or evaporate. Instead, it may become part of a snowfield or a glacier. In this way, glaciers store large quantities of water on land. If present-day glaciers were to melt and release all their water, sea level would rise by several dozen meters. This would submerge many heavily populated coastal areas. As you will see in Chapter 5, over the past 2 million years, huge ice sheets have formed and melted on several occasions, each time changing the balance of the hydrologic cycle.

Freshwater lakes
0.009%
Saline lakes and
inland seas
0.008%
Soil moisture
0.005%
Stream channels
0.0001%
Atmosphere
0.001%

Oceans
97.2%

2.8%

Hydrosphere

Glaciers
2.15%

Groundwater
0.62%

Nonocean Component
(% of total hydrosphere)

Figure 4.1   Distribution of Earth's water.

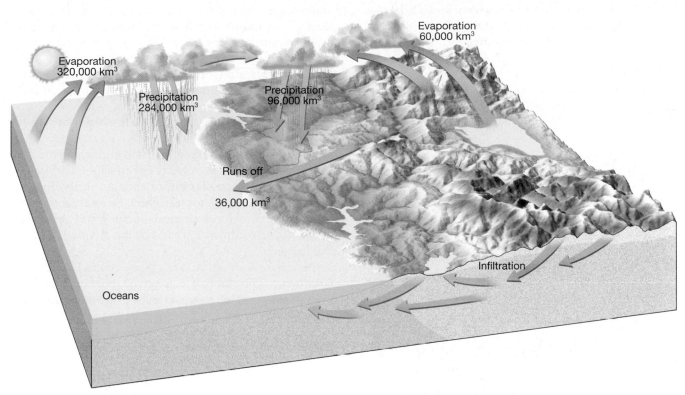

Figure 4.2   Earth's water balance. About 320,000 cubic kilometers of water are evaporated each year from the oceans, while evaporation from the land (including lakes and streams) contributes 60,000 cubic kilometers of water. Of this total of 380,000 cubic kilometers of water, about 284,000 cubic kilometers fall back to the ocean, and the remaining 96,000 cubic kilometers fall on Earth's land surface. Since 60,000 cubic kilometers of water evaporate from the land, 36,000 cubic kilometers of water remain to erode the land during the journey back to the ocean.

Figure 4.2 also shows Earth's overall *water balance,* or the volume of water that passes through each part of the cycle annually. The amount of water vapor in the air at any one time is just a tiny fraction of Earth's total water supply. But the *absolute* quantities that are cycled through the atmosphere over a one-year period are immense—some 380,000 cubic kilometers—enough to cover Earth's entire surface to a depth of about 1 meter (39 inches). Estimates show that over North America almost six times more water is carried by moving currents of air than is transported by all the continents' rivers.

It is important to know that the hydrologic cycle is *balanced.* Because the total amount of water vapor in the atmosphere remains about the same, the average annual precipitation over Earth must be equal to the quantity of water evaporated. However, for all of the continents taken together, precipitation exceeds evaporation. Conversely, over the oceans, evaporation exceeds precipitation. Because the level of the world ocean is not dropping, the system must be in balance. In Figure 4.2, the 36,000 cubic kilometers of water that annually runs off from the land to the ocean causes enormous erosion. In fact, this immense volume of moving water is the *single most important agent sculpturing Earth's land surface.*

To summarize, the hydrologic cycle is the continuous movement of water from the oceans to the atmosphere, from the atmosphere to the land, and from the land back to the sea. The land-back-to-the-sea step is the primary action that wears down Earth's land surface. In this chapter, we will first observe the work of water running over the surface, including floods, erosion, and the formation of valleys. Then we will look underground at the slow labors of groundwater as it forms springs and caverns and provides drinking water on its long migration to the sea.

Running Water

   Sculpturing Earth's Surface
↳ Running Water

Running water is of great importance to people. We depend on rivers for energy, transportation, and irrigation. Their fertile floodplains have been favored sites for agriculture and industry since the dawn of civilization. As the dominant agent of erosion, running water has shaped much of our physical environment.

Although we have always depended on running water, its source eluded us for centuries. Not until the

1500s did we realize that streams were supplied by surface runoff and underground water, which ultimately had their sources as rain and snow.

Streamflow

Water makes its way to the sea under the influence of gravity. The time required for the journey depends on the velocity of the stream. Velocity is the distance that water travels in a unit of time. Some sluggish streams flow at less than 1 kilometer per hour, whereas a few rapid ones may exceed 30 kilometers per hour. Velocities are measured at gaging stations (Figure 4.3A). Along straight stretches, the highest velocities are near the center of the channel just below the surface, where friction is lowest (Figure 4.3B). But when a stream curves, its zone of maximum speed shifts toward its outer bank (Figure 4.3C).

The ability of a stream to erode and transport materials depends on its velocity. Even slight changes in velocity can lead to significant changes in the load of sediment that water can transport. Several factors determine the velocity of a stream, including (1) gradient; (2) shape, size, and roughness of the channel; and (3) discharge.

Gradient.   The slope of a stream channel expressed as the vertical drop of a stream over a specified distance is **gradient**. Portions of the lower Mississippi River, for example, have very low gradients of 10 centimeters per kilometer or less. By contrast, some mountain stream channels decrease in elevation at a rate of more than 40 meters per kilometer, or a gradient 400 times steeper than the lower Mississippi (Figure 4.4). Gradient varies not only between streams but over a stream's length. The steeper the gradient, the more energy available for streamflow. If two streams were identical in every respect except gradient, the stream with the higher gradient would obviously have the greater velocity.

Channel Characteristics.   A stream's channel is a conduit that guides the flow of water, but the water encounters friction as it flows. The shape, size, and roughness of the channel affect the amount of friction. Larger channels have more efficient flow because a

A.

B.

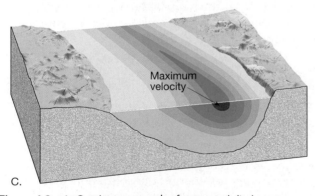

C.

Figure 4.3   **A.** Continuous records of stage and discharge are collected by the U.S. Geological Survey at more than 7000 gaging stations in the United States. Average velocities are determined by using measurements from several spots across the stream. This station is on the Rio Grande south of Taos, New Mexico. (Photo by E. J. Tarbuck) **B.** Along straight stretches, stream velocity is highest at the center of the channel. **C.** When a stream curves, its zone of maximum speed shifts toward the outer bank.

Figure 4.4   Rapids can form where a portion of a stream channel has a steep gradient. These rapids are on the Rio Futaleufu in Chile. (Photo © by Carr Clifton)

| Rank | River | Country | Drainage Area | | Average Discharge | |
			Square kilometers	Square miles	Cubic meters	Cubic feet
1	Amazon	Brazil	5,778,000	2,231,000	212,400	7,500,000
2	Congo	Zaire	4,014,500	1,550,000	39,650	1,400,000
3	Yangtze	China	1,942,500	750,000	21,800	770,000
4	Brahmaputra	Bangladesh	935,000	361,000	19,800	700,000
5	Ganges	India	1,059,300	409,000	18,700	660,000
6	Yenisei	Russia	2,590,000	1,000,000	17,400	614,000
7	Mississippi	United States	3,222,000	1,244,000	17,300	611,000
8	Orinoco	Venezuela	880,600	340,000	17,000	600,000
9	Lena	Russia	2,424,000	936,000	15,500	547,000
10	Parana	Argentina	2,305,000	890,000	14,900	526,000

Table 4.1 World's largest rivers ranked by discharge.

smaller proportion of water is in contact with the channel. A smooth channel promotes a more uniform flow, whereas an irregular channel filled with boulders creates enough turbulence to slow the stream significantly.

Discharge.   The **discharge** of a stream is the volume of water flowing past a certain point in a given unit of time. This is usually measured in cubic meters per second or cubic feet per second. Table 4.1 lists the world's largest rivers in terms of discharge. The largest river in North America, the Mississippi, discharges an average of 17,300 cubic meters (611,000 cubic feet) per second. Although this is a huge quantity of water, it is nevertheless dwarfed by the mighty Amazon in South America, the world's largest river. Draining a rainy region that is nearly three-fourths the size of the conterminous United States, the Amazon discharges 12 times more water than the Mississippi.

The discharges of most rivers are far from constant (see Box 4.1). This is true because of such variables as rainfall and snowmelt. When discharge changes, the factors noted earlier must also change. Indeed, measurements show that when the amount of water in a stream increases, the width, depth, and velocity all increase. To handle the additional water, the stream will increase the size of its channel by widening and deepening it. The larger the channel, the less friction, and the more swiftly the water will flow.

Changes from Upstream to Downstream

One useful way of studying a stream is to examine its *profile*. A profile is simply a cross-sectional view of a stream from its source area (called the *head* or *headwaters*) to its *mouth*, the point downstream where the river empties into another water body. By examining Figure 4.5, you can see that the most obvious feature of a typical profile is a constantly decreasing gradient from the head to the mouth. Although many local irregularities

may exist, the overall profile is a smooth concave-upward curve.

The profile shows that the gradient decreases downstream. To see how other factors change in a downstream direction, observations and measurements must be made. When data are collected from several gaging stations along a river, they show that discharge increases from the head toward the mouth. This should come as no surprise because, as we move downstream, more and more tributaries contribute water to the main channel (Figure 4.5). Furthermore, in most humid regions, additional water is added from the groundwater supply. Thus, as you move downstream, the stream's width, depth, and velocity change in response to the increased volume of water carried by the stream.

The observed increase in average velocity that occurs downstream contradicts our intuitive impressions about wild, turbulent mountain streams and wide, placid rivers in the lowlands. Although a mountain stream may have the appearance of a raging torrent, its average velocity is often less than that of the river near its mouth. The difference is primarily attributable to the greater efficiency of the larger channel in a downstream direction.

In the headwaters region where the gradient is steep, the water must flow in a relatively small and often boulder-strewn channel. The small channel and rough bed create great friction and inhibit movement by scattering water in all directions, with almost as much backward motion as forward motion. However, downstream the material on the bed of the stream becomes much smaller, offering less resistance to flow, and the width and depth of the channel increase to accommodate the greater discharge. These factors, especially the wider and deeper channel, permit the water to flow more freely and hence more rapidly.

In summary, you have seen an inverse relationship between gradient and discharge. Where gradient is steepest, discharge is smallest, and where discharge is greatest, gradient is least. Stated another way, a stream

PEOPLE AND THE ENVIRONMENT:
The Effect of Urbanization on Discharge

When rains occur, stream discharge increases. If the rains are sufficiently heavy, the ability of the channel to contain the discharge is exceeded, and water spills over the banks as a flood. Floods are natural events that should be expected. However, when cities are built, the magnitude and frequency of flooding increases. The top portion of Figure 4.A is a hypothetical hydrograph that shows the time relationship between a rainstorm and the occurrence of flooding. Notice that the water level in the stream does not rise at the onset of precipitation because time is needed for water to move from the place where it fell to the stream. This time difference is called the *lag time*.

When an area changes from being predominantly rural to largely urban, streamflow is affected. The effect of urbanization on streamflow is illustrated by the bottom hydrograph in Figure 4.A. Notice that after urbanization the peak discharge during a flood is greater and that the lag time between precipitation and flood peak is shorter than before urbanization. The explanation for this effect is relatively simple. The construction of streets, parking lots, and buildings covers over the ground that once soaked up water. Thus, less water infiltrates the ground, and the rate and amount of runoff increase. Further, because much less water soaks into the ground, the low-water (dry-season) flow in urban streams, which is maintained by the seepage of groundwater into the channel, is greatly reduced. As one might expect, the magnitude of these effects is a function of the per-

Typical lag time between rainfall and runoff

Lag time between rainfall and runoff after urbanization

Figure 4.A When an area changes from rural to urban, the lag time between rainfall and flood peak is shortened. The flood peak is also higher following urbanization. (After L B. Leopold, U.S. Geological Survey)

centage of land that is covered by impermeable surfaces.

Urbanization is just one example of human interference with streams. There are many other ways that land use inadvertently influences the flow of streams and the work they carry out. Moreover, there are also many ways by which people intentionally attempt to manipulate and control streams. Some of these are discussed in the section on Floods and Flood Control (p. 109).

can maintain a higher velocity near its mouth even though it has a lower gradient than upstream because of the greater discharge, larger channel, and smoother bed.

Base Level

GEODe Sculpturing Earth's Surface
↳ Running Water

Streams cannot endlessly erode their channels deeper and deeper. There is a lower limit to how deep a

stream can erode, and that limit is the stream's base level. **Base level** is the lowest point to which a stream can erode its channel. Essentially this is the level at which the mouth of a stream enters the ocean, a lake, or another stream.

Two general types of base level are recognized. Sea level is considered the *ultimate base level*, because it is the lowest level to which stream erosion could lower the land. *Temporary*, or *local, base levels* include lakes, resistant layers of rock, and main streams that act as base level for their tributaries. For example, when a stream enters a lake, its velocity quickly approaches zero and

Figure 4.5   A longitudinal profile is a cross-section along the length of a stream. Note the concave-upward curve of the profile, with a steeper gradient upstream and a gentler gradient downstream.

its ability to erode ceases. Thus, the lake prevents the stream from eroding below its level at any point upstream from the lake. However, because the outlet of the lake can cut downward and drain the lake, the lake is only a temporary hindrance to the stream's ability to downcut its channel. In a similar manner, the layer of resistant rock at the lip of the waterfall in Figure 4.6 acts as a temporary base level. Until the ledge of hard rock is eliminated, it will limit the amount of downcutting upstream.

Any change in base level will cause a corresponding readjustment of stream activities. When a dam is built along a stream, the reservoir that forms behind it raises the base level of the stream (Figure 4.7). Upstream from the dam the stream gradient is reduced, lowering its velocity and, hence, its sediment-transporting ability. The stream, now having too little energy to transport all of its load, will deposit sediment. This builds up its channel. This process continues until the stream again has a gradient sufficient to carry its load.

Figure 4.6   A resistant layer of rock can act as a local (temporary) base level. Because the durable layer is eroded more slowly, it limits the amount of downcutting upstream.

Figure 4.7 When a dam is built and a reservoir forms, the stream's base level is raised. This reduces the stream's velocity and leads to deposition and a reduction of the gradient upstream from the reservoir.

The opposite situation is illustrated in Figure 4.8. A stream in a broad, flat-floored valley that is near base level often develops a course with many bends and turns called *meanders* (Figure 4.8A). What would happen to such a stream if base level dropped or if the land were uplifted? The river, now considerably above base level, would have excess energy and would downcut its channel to establish a balance with its new base level. The result could be *incised meanders*—that is, a winding river in a steep, narrow valley (Figure 4.8B).

The Work of Streams

Streams are Earth's most important erosional agent. Not only do they have the ability to downcut their channels but streams also have the capacity to transport the enormous quantities of sediment produced by weathering. Although erosion by running water in a stream's channel contributes significant amounts of material for transport, by far the greatest quantity of sediment carried by a stream is derived from the products of weathering. Weathering produces tremendous amounts of material that are delivered to the stream by sheet flow, mass wasting, and groundwater. Eventually much of this material is dropped by the water to create a variety of depositional features.

Erosion

Streams erode their channels by lifting loose particles, by abrasion (grinding), and by dissolving soluble materials. The last of these is by far the least significant.

When the flow of water is sufficiently turbulent, it can dislodge particles from the channel and lift them into the moving water. In this manner, the force of running water swiftly erodes poorly consolidated materials on the bed and banks of a stream. The stronger the current, the more effectively the water will pick up particles.

A stream is capable of eroding solid rock in a manner similar to sandpapering. Just as the particles of grit on sandpaper can wear away a piece of wood, so too can the sand and gravel carried by a stream abrade a bedrock channel. Moreover, pebbles caught in swirling eddies can act like cutting tools and bore circular "potholes" into the channel floor.

Transportation

Streams transport their load of sediment in three ways: (1) in solution (**dissolved load**), (2) in suspension (**suspended load**), and (3) scooting or rolling along the bottom (**bed load**).

Dissolved Load. Most of the *dissolved load* is brought to the stream by groundwater, and to a lesser degree it is acquired by dissolving rock along the stream's course. The quantity of material carried in solution is highly variable and depends on climate and the geologic setting. Usually the dissolved load is expressed as parts of dissolved material per million parts of water (parts per million, or ppm). Although some rivers may have a dissolved load of 1000 ppm or more, the average figure for the world's rivers is estimated at 115 to 120 ppm. Almost 4 billion metric tons of dissolved mineral matter are supplied to the oceans each year by streams.

Suspended Load. Most streams (but not all) carry the largest part of their load in *suspension*. Indeed, the

A.

B.

Figure 4.8   **A.** A river in a broad, flat-floored valley near base level often exhibits a meandering channel. **B.** Incised meanders of the Colorado River in Canyonlands National Park, Utah. Here, as the Colorado Plateau was gradually uplifted, the meandering river adjusts to being higher above base level by downcutting. (Photos by Michael Collier)

visible cloud of sediment suspended in the water is the most obvious portion of a stream's load. Usually only sand, silt, and clay can be carried this way, but during a flood larger particles are transported as well. Also, during a flood, the total quantity of material carried in suspension increases dramatically, as can be verified by anyone whose home has been a site for the deposition of this material.

Bed Load.   A portion of a stream's load of solid material consists of sediment that is too large to be carried in suspension. These coarser particles move along the bottom (bed) of the stream channel and constitute the *bed load*. Unlike the suspended and dissolved loads, which are constantly in motion, the bed load is in motion only intermittently, when the force of the water is sufficient to move the larger particles. In terms of the erosional work accomplished by a downcutting stream, the grinding action of the bed load is of great importance.

Each year, the Mississippi River transports about 750 million tons of material to the Gulf of Mexico. Of this total, it is estimated that approximately 500 million tons are carried in suspension, 200 million tons in solution, and the remaining 50 million tons as bed load; however, such proportions vary widely from stream to stream.

Competence and Capacity.   Streams vary in their ability to carry a load. Their ability is determined by two criteria. First, the **competence** of a stream measures the maximum size of particles it is capable of transporting. The stream's velocity determines its competence. If the velocity of a stream doubles, its competence increases four times; if the velocity triples, its competence increases nine times; and so forth. This explains how large boulders that seem immovable can be transported during a flood, which greatly increases a stream's velocity.

Second, the **capacity** of a stream is the maximum load it can carry. The capacity of a stream is directly related to its discharge. The greater the volume of water flowing in a stream, the greater is its capacity for hauling sediment.

By now it should be clear why the greatest erosion and transportation of sediment occur during floods (Figure 4.9). The increase in discharge results in a greater capacity, and the increase in velocity results in greater competence. With rising velocity the water becomes more turbulent, and larger and larger particles are set in motion. In just a few days or perhaps a few hours a stream in flood stage can erode and transport more sediment than it does during months of normal flow.

Deposition

Whenever a stream slows down, the situation reverses. As its velocity decreases, its competence is reduced and sediment begins to drop out, largest particles first. Each

Figure 4.9 The suspended load is clearly visible because it gives this flooding river a "muddy" appearance. During floods both capacity and competency increase. Therefore the greatest erosion and sediment transport occur during these high-water periods. The flooding Ohio River at Cincinnati, Ohio, March 1997. This regional flood caused more than 50 deaths and approximately $500 million in damages. (Photo by Mark Lyons/Liaison Agency, Inc.)

particle size has a *critical settling velocity*. As streamflow drops below the critical settling velocity of a certain particle size, sediment in that category begins to settle out. Thus, stream transport provides a mechanism by which solid particles of various sizes are separated. This process, called **sorting**, explains why particles of similar size are deposited together.

The well-sorted material typically deposited by a stream is called **alluvium**, the general term for any stream-deposited sediment. Many different depositional features are composed of alluvium. Some occur within stream channels, some occur on the valley floor adjacent to the channel, and some exist at the mouth of the stream.

Deltas. When a stream enters the relatively still waters of an ocean or lake, its velocity drops abruptly, and the resulting deposits form a **delta** (Figure 4.10). As the delta grows outward, the stream's gradient continually lessens. This circumstance eventually causes the channel to become choked with sediment from the slowing water. As a consequence, the river seeks a shorter, higher-gradient route to base level, as illustrated in Figure 4.10B. This illustration shows the main channel dividing into several smaller ones called **distributaries**. Most deltas are characterized by these shifting channels that act in an opposite way to that of tributaries.

Rather than carrying water into the main channel, distributaries carry water away from the main channel. After numerous shifts of the channel, a delta may grow into a rough triangular shape like the Greek letter delta (Δ), for which it is named. Note, however, that many deltas do not exhibit the idealized shape. Differences in the configurations of shorelines and variations in the nature and strength of wave activity result in many shapes.

Many large rivers have deltas extending over thousands of square kilometers. The delta of the Mississippi River is one example. It resulted from the accumulation of huge quantities of sediment derived from the vast region drained by the river and its tributaries. Today, New Orleans rests where there was ocean less than 5000 years ago. Figure 4.11 shows that portion of the Mississippi delta that has been built over the past 5000 to 6000 years. As shown, the delta is actually a series of seven coalescing subdeltas. Each formed when the river left its existing channel in favor of a shorter, more direct path to the Gulf of Mexico. The individual subdeltas interfinger and partially cover one another to produce a very complex structure. The present subdelta, called a *bird-foot* delta because of the configuration of its distributaries, has been built by the Mississippi in the last 500 years.

Natural Levees. Some rivers occupy valleys with broad, flat floors and build **natural levees** that parallel their channels on both banks (Figure 4.12). Natural levees are built by successive floods over many years. When a stream overflows its banks, its velocity immediately diminishes, leaving coarse sediment deposited in strips bordering the channel. As the water spreads out over the valley, a lesser amount of fine sediment is deposited over the valley floor. This uneven distribution of material produces the very gentle slope of the natural levee.

The natural levees of the lower Mississippi rise 6 meters (20 feet) above the valley floor. The area behind the levee is characteristically poorly drained for the obvious reason that water cannot flow up the levee and into the river. Marshes called **backswamps** result. A tributary stream that cannot enter a river because levees block the way often has to flow parallel to the river until it can breach the levee. Such streams are called **yazoo**

Figure 4.10   **A.** Structure of a simple delta that forms in the relatively quiet waters of a lake. **B.** Growth of a simple delta. As a stream extends its channel, the gradient is reduced. Frequently, during flood stage the river is diverted to a higher-gradient route, forming a new distributary. Old abandoned distributaries are gradually invaded by aquatic vegetation and fill with sediment. (After Ward's Natural Science Establishment, Inc., Rochester, N.Y.)

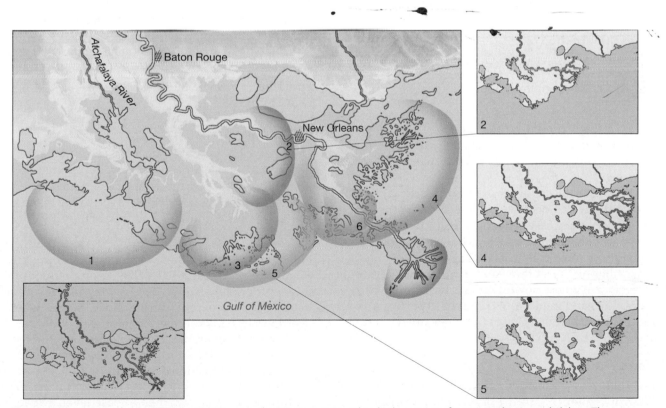

Figure 4.11   During the past 5000 to 6000 years, the Mississippi River has built a series of seven coalescing subdeltas. The numbers indicate the order in which the subdeltas were deposited. The present bird-foot delta (number 7) represents the activity of the past 500 years. Without ongoing human efforts, the present course will shift and follow the path of the Atchafalaya River. The inset on left shows the point where the Mississippi may someday break through (arrow) and the shorter path it would take to the Gulf of Mexico. (After C. R. Kolb and J. R. Van Lopik)

Figure 4.12 Natural levees are gently sloping deposits that are created by repeated floods. Because the ground next to the stream channel is higher than the adjacent floodplain, back swamps and yazoo tributaries may develop.

tributaries after the Yazoo River, which parallels the Mississippi for over 300 kilometers (about 190 miles).

Stream Valleys

GEODe
EARTH SCIENCE

Sculpturing Earth's Surface
↳ Running Water

Stream valleys can be divided into two general types. Narrow V-shaped valleys and wide valleys with flat floors exist as the ideal forms, with many gradations between.

Most stream valleys are much broader at the top than is the width of their channel at the bottom. This would not be the case if the only agent responsible for eroding valleys were the streams flowing through them. The sides of most valleys are shaped by a combination of weathering, sheet flow, and mass wasting. In some arid regions, where downcutting is rapid and weathering is slow, and in places where rock is particularly resistant, narrow valleys may not be V-shaped but rather may have nearly vertical walls.

Narrow Valleys

The Yellowstone River provides an excellent example of a narrow valley (Figure 4.13). A narrow V-shaped valley indicates that the primary work of the stream has been downcutting toward base level. The most prominent features of a narrow valley are *rapids* and *waterfalls*. Both occur where the stream profile drops rapidly,

a situation usually caused by variations in the erodibility of the bedrock into which a stream channel is cutting. Resistant beds create rapids by acting as a temporary base level upstream while allowing downcutting to continue downstream. Once erosion has eliminated the resistant rock, the stream profile smooths out again. Waterfalls are places where the stream profile makes an abrupt vertical drop.

Wide Valleys

Once a stream has cut its channel closer to base level, downward erosion becomes less dominant. At this point more of the stream's energy is directed from side to side. The result is a widening of the valley as the river cuts away first at one bank and then at the other (Figure 4.14).

Floodplains.   The side-to-side cutting of a stream eventually produces a flat valley floor, or **floodplain**. It is appropriately named because the river is confined to its channel, except during flood stage, when it overflows its banks and inundates the floodplain.

When a river erodes laterally, creating a floodplain as just described, it is called an *erosional floodplain*. Floodplains can be depositional in nature as well. *Depositional floodplains* are produced by a major fluctuation in conditions, such as a change in base level. The floodplain in California's Yosemite Valley is one such feature and was produced when a glacier gouged the former stream valley deeper by about 300 meters (1000 feet). After the glacial ice melted, the stream readjusted

Figure 4.14 Stream eroding its floodplain.

Figure 4.13 V-shaped valley of the Yellowstone River. The rapids and waterfalls indicate that the river is vigorously downcutting. (Photo by Art Wolfe)

to its former base level by refilling the valley with alluvium.

Meanders. Streams that flow on floodplains move in sweeping bends called **meanders** (Figure 4.14). Once a bend in a channel begins to form, it grows larger. Erosion occurs on the outside of the meander where velocity and turbulence are greatest. Most often, the outside bank is undermined, especially during periods of high water. Because the outside of a meander is a zone of active erosion, it is often referred to as the *cut bank* (Figure 4.15). Much of the debris acquired by the stream at the cut bank moves downstream and is soon deposited as *point bars* in zones of decreased velocity on the insides of meanders. In this manner, meanders migrate laterally by eroding the outside of the bends and depositing on the inside.

Because of the slope of the channel, erosion is more effective on the downstream side of a meander. Therefore, in addition to migrating laterally, the bends also gradually migrate down the valley. Sometimes the downstream migration of a meander is slowed when it reaches a more resistant portion of the floodplain. This allows the next meander upstream to "catch up" and overtake it. Gradually the neck of land between the meanders is narrowed. Eventually the river may erode through the narrow neck of land to the next loop (Figure 4.16). The new, shorter channel segment is called a **cutoff** and, because of its shape, the abandoned bend is called an **oxbow lake**. Such a situation is shown in the bottom of portion Figure 4.8A.

Floods and Flood Control

When the discharge of a stream becomes so great that it exceeds the capacity of its channel, it overflows its banks as a **flood**. Floods are the most common and most destructive of all geologic hazards. They are, nevertheless, simply part of the *natural* behavior of streams.

Figure 4.15 When a stream meanders, its zone of maximum speed shifts toward the outer bank. A point bar is deposited when the water on the inside of a meander slows. By eroding its outer bank and depositing material on the inside of the bend, a stream is able to shift its channel.

Figure 4.16 Formation of a cutoff and oxbow lake.

Causes of Floods

Rivers flood because of the weather. Rapid melting of snow in the spring and/or major storms that bring heavy rains over a large region cause most floods. The extensive 1997 flood along the Red River of the North is a recent example of an event triggered by rapid snowmelt. Exceptional rains caused the devastating floods in the upper Mississippi River Valley during the summer of 1993 (Figure 4.17). Heavy rains were also responsible for the Ohio River Flood shown in Figure 4.9.

Unlike the extensive regional floods just mentioned, *flash floods* are more limited in extent. Flash floods occur with little warning and can be deadly because they produce a rapid rise in water levels and can have a devastating flow velocity (see Box 4.2). Several factors influence flash flooding. Among them are rainfall intensity and duration, surface conditions, and topography. Urban areas are susceptible to flash floods because a high percentage of the surface area is composed of impervious roofs, streets, and parking lots where runoff is very rapid (see Box 4.1). Mountainous areas are susceptible because steep slopes can funnel runoff into narrow canyons.

Human interference with the stream system can worsen or even cause floods. A prime example is the failure of a dam or an artificial levee. These structures are built for flood protection. They are designed to contain floods of a certain magnitude. If a larger flood occurs, the dam or levee is overtopped. If the dam or levee fails or is washed out, the water behind it is released to become a flash flood. The bursting of a dam in 1889 on the Little Conemaugh River caused the devastating Johnstown, Pennsylvania, flood that took some 3000 lives. A second dam failure occurred there again in 1977 and caused 77 fatalities.

Flood Control

Several strategies have been devised to eliminate or lessen the catastrophic effects of floods. Engineering efforts include the construction of artificial levees, the building of flood-control dams, and river channelization.

Many flood designations are reevaluated and changed over time as more data are collected or when a river basin is altered in a way that affects the flow of water. Dams and urban development are examples of some human influences in a basin that affect floods.

Figure 4.17   Satellite views of the Missouri River flowing into the Mississippi River. St. Louis is just south of their confluence. The upper image shows the rivers during a drought that occurred in summer 1988. The lower image depicts the peak of the record-breaking 1993 flood. Exceptional rains produced the wettest spring and early summer of the twentieth century in the upper Mississippi River basin. In all, nearly 14 million acres were inundated, displacing at least 50,000 people. (Photos by courtesy Spaceimaging.com)

? STUDENTS SOMETIMES ASK...

Sometimes when there is a major flood, it is described as a 100-year flood. What does that mean?

The phrase "100-year flood" is misleading because it leads people to believe that such an event happens only once every 100 years. The truth is that an uncommonly big flood can happen *any* year. The phrase "100-year flood" is really a statistical designation, indicating that there is a *1-in-100 chance* that a flood this size will happen during any year. Perhaps a better term would be the "1-in-100-chance flood."

Artificial Levees.   *Artificial levees* are earthen mounds built on the banks of a river to increase the volume of water the channel can hold. These most common of stream-containment structures have been used since ancient times and continue to be used today.

Artificial levees are usually easy to distinguish from natural levees because their slopes are much steeper. When a river is confined by levees during periods of high water, it frequently deposits material in its channel as the discharge diminishes. This is sediment that otherwise would have been dropped on the floodplain. Thus, each time there is a high flow, deposits are left on the river bed, and the bottom of the channel is built up. With the buildup of the bed, less water is required to overflow the original levee. As a result, the height of the levee may have to be raised periodically to protect the floodplain. Moreover, many artificial levees are not built to withstand periods of extreme flooding. For example, levee failures were numerous in the Midwest during the summer of 1993, when the upper Mississippi and many of its tributaries experienced record floods (Figure 4.18).

Flood-Control Dams.   *Flood-control dams* are built to store floodwater and then let it out slowly. This lowers the flood crest by spreading it out over a longer time span. Since the 1920s, thousands of dams have been built on nearly every major river in the United States. Many dams have significant nonflood-related functions, such as providing water for irrigated agriculture and for hydroelectric power generation. Many reservoirs are also major regional recreational facilities.

Although dams may reduce flooding and provide other benefits, building these structures also has significant costs and consequences. For example, reservoirs created by dams may cover fertile farmland, useful forests, historic sites, and scenic valleys. Of course, dams trap sediment. Therefore, deltas and floodplains downstream erode because they are no longer replenished with silt during floods. Large dams can also cause significant ecological damage to river environments that took thousands of years to establish.

Building a dam is not a permanent solution to flooding. Sedimentation behind a dam means that the volume of its reservoir will gradually diminish, reducing the effectiveness of this flood-control measure.

Channelization.   *Channelization* involves altering a stream channel in order to speed the flow of water to

BOX 4.2

PEOPLE AND THE ENVIRONMENT:
Flash Floods

Tornadoes and hurricanes are nature's most awesome storms. Yet surprisingly, these dreaded events are not responsible for the greatest number of storm-related deaths. That distinction is reserved for flash floods. For a recent 30-year period, the number of storm-related deaths in the United States from flooding averaged 135 per year. By contrast, tornado fatalities averaged 73 annually and hurricanes 25.

Flash floods are local floods of great volume and short duration. The rapidly rising surge of water usually occurs with little advance warning and can destroy roads, bridges, homes, and other substantial structures (Figure 4.B). Discharges quickly reach a maximum and diminish almost as rapidly. Flood flows often contain large quantities of sediment and debris as they sweep channels clean.

Frequently flash floods result from the torrential rains associated with a slow-moving severe thunderstorm or take place when a series of thunderstorms repeatedly pass over the same location. Occasionally floating debris or ice can accumulate at a natural or artificial obstruction and restrict the flow of water. When such temporary dams fail, torrents of water can be released as a flash flood.

Flash floods can take place in almost any area of the country. They are particularly common in mountainous terrain, where steep slopes can quickly channel runoff into narrow valleys. The

Figure 4.B   The disastrous nature of flash floods is illustrated by the Big Thompson River flood of July 31, 1976, in Colorado. During a four-hour period more than 30 centimeters (12 inches) of rain fell on portions of the small area drained by the river. This amounted to nearly three-quarters of the average yearly total. The flash flood in the narrow canyon lasted only a few hours but cost 139 people their lives. Damages were estimated at $39 million. (U.S. Geological Survey, Denver)

hazard is most acute when the soil is already nearly saturated from earlier rains or consists of impermeable materials.

Why do so many people perish in flash floods? Aside from the factor of surprise (many are caught sleeping), people do not appreciate the power of moving water. A glance at Figure 4.B helps illustrate the force of a flood wave. Just 15 centimeters (6 inches) of fast-moving floodwater can knock a person down. Most automobiles will float and be swept away in only 0.6

meter (2 feet) of water. *More than half of all U.S. flash-flood fatalities are auto related!* Clearly, people should never attempt to drive over a flooded road. The depth of water is not always obvious. Also, the road bed may have been washed out under water. Present-day flash floods are calamities with potential for very high death tolls and huge property losses. Although efforts are being made to improve observations and warnings, flash floods remain elusive natural killers.

---

prevent it from reaching flood height. This may simply involve clearing a channel of obstructions or dredging a channel to make it wider and deeper.

A more radical alteration involves straightening a channel by creating *artificial cutoffs.* The idea is that by shortening the stream, the gradient and hence the velocity are both increased. By increasing velocity, the larger discharge associated with flooding can be dispersed more rapidly.

Since the early 1930s, the U.S. Army Corps of Engineers has created many artificial cutoffs on the Mississippi for the purpose of increasing the efficiency of the channel and reducing the threat of flooding. In all, the river has been shortened more than 240 kilometers (150 miles). The program has been somewhat successful in reducing the height of the river in flood stage.

However, because the river's tendency toward meandering still exists, preventing the river from returning to its previous course has been difficult.

A Nonstructural Approach.   All of the flood-control measures described so far have involved structural solutions aimed at "controlling" a river. These solutions are expensive and often give people residing on the floodplain a false sense of security.

Today many scientists and engineers advocate a nonstructural approach to flood control. They suggest that an alternative to artificial levees, dams, and channelization is sound floodplain management. By identifying high-risk areas, appropriate zoning regulations can be implemented to minimize development and promote more appropriate land use.

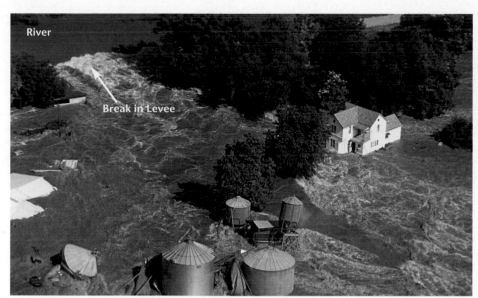

Figure 4.18 Water rushes through a break in an artificial levee in Monroe County, Illinois. During the record-breaking 1993 Midwest floods, many artificial levees could not withstand the force of the floodwaters. Sections of many weakened structures were overtopped or simply collapsed. (Photo by James A. Finley/AP/Wide World Photos)

Drainage Basins and Patterns

GEODe Sculpturing Earth's Surface
↳ Running Water

Every stream, no matter how large or small, has a **drainage basin**. This is the land area that contributes water to the stream. The drainage basin of one stream is separated from the drainage basin of another by an imaginary line called a **divide**. Divides range in scale from a ridge separating two small gullies on a hillside to a *continental divide*, which splits continents into enormous drainage basins. The Mississippi River has the largest drainage basin in North America (Figure 4.19). Extending between the Rocky Mountains in the West and the Appalachian Mountains in the East, the Mississippi River and its tributaries collect water from more than 3.2 million square kilometers (1.2 million square miles) of the continent.

Drainage systems are networks of streams that together form distinctive patterns. The nature of a drainage pattern can vary greatly from one type of terrain to another, primarily in response to the kinds of rock on which the streams developed or the structural pattern of faults and folds.

The most commonly encountered drainage pattern is the **dendritic pattern** (Figure 4.20A). This pattern of irregularly branching tributary streams resembles the branching pattern of a deciduous tree. In fact, the word *dendritic* means "treelike." The dendritic pattern forms where the underlying material is relatively uniform. Because the surface material is essentially uniform in its resistance to erosion, it does not control the pattern of streamflow. Rather, the pattern is determined chiefly by the direction of slope of the land.

When streams diverge from a central area like spokes from the hub of a wheel, the pattern is said to be radial (Figure 4.20B). This pattern typically develops on isolated volcanic cones and domal uplifts.

Figure 4.20C illustrates a rectangular pattern, in which many right-angle bends can be seen. This pattern develops when the bedrock is crisscrossed by a series of joints and/or faults. Because these structures are eroded more easily than unbroken rock, their geometric pattern guides the directions of valleys.

Figure 4.20D illustrates a **trellis** drainage pattern, a rectangular pattern in which tributary streams are nearly parallel to one another and have the appearance of a garden trellis. This pattern forms in areas underlain by alternating bands of resistant and less-resistant rock.

Water Beneath the Surface

GEODe Sculpturing Earth's Surface
↳ Groundwater

Groundwater is one of our most important and widely available resources. Yet people's perceptions of groundwater are often unclear and incorrect. The reason is that groundwater is hidden from view except in caves and mines, and the impressions people gain from these subsurface openings are often misleading. Observations on the land surface give an impression that Earth is solid. This view is not changed very much when we enter a cave and see water flowing in a channel that appears to have been cut into solid rock. Because of such observations, many people believe that groundwater occurs only in underground "rivers." But actual rivers underground are extremely rare.

In reality, most of the subsurface environment is not solid at all. It includes countless tiny *pore spaces* between grains of soil and sediment, plus narrow joints

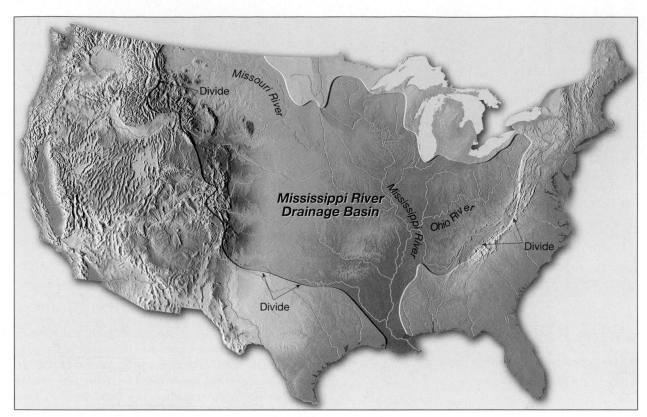

Figure 4.19 A drainage basin is the land area drained by a stream and its tributaries. The drainage basin of the Mississippi River, North America's largest river, covers about 3 million square kilometers. Divides are the boundaries that separate drainage basins from each other. Drainage basins and divides exist for all streams regardless of size.

and fractures in bedrock. Together these spaces add up to an immense volume. It is in these tiny openings that groundwater collects and moves.

The Importance of Groundwater

Considering the entire hydrosphere, or all of Earth's water, only about six-tenths of 1 percent occurs underground. Nevertheless, this small percentage, stored in the rocks and sediments beneath Earth's surface, is a vast quantity. When the oceans are excluded and only sources of freshwater are considered, the significance of groundwater becomes more apparent. Figure 4.21 contains estimates of the distribution of freshwater in the hydrosphere. Clearly, the largest volume occurs as glacial ice. Second in rank is groundwater, with slightly more than 14 percent of the total. However, when ice is excluded and just liquid water is considered, more than 94 percent is groundwater. Without question, *groundwater represents the largest reservoir of freshwater that is readily available to humans.* Its value in terms of economics and human well-being is incalculable.

Worldwide, wells and springs provide water for cities, crops, livestock, and industry. In the United States, groundwater is the source of about 40 percent of the water used for all purposes (except hydroelectric

power generation and power-plant cooling). Groundwater is the drinking water for more than 50 percent of the population, is 40 percent of the water for irrigation, and provides more than 25 percent of industry's needs. In some areas, however, overuse of this basic resource has caused serious problems, including streamflow depletion, land subsidence, and increased pumping costs. In addition, groundwater contamination due to human activities is a real and growing threat in many places.

Groundwater's Geological Roles

Geologically, groundwater is important as an erosional agent. The dissolving action of groundwater slowly removes soluble rock, allowing surface depressions known as sinkholes to form as well as creating subterranean caverns (Figure 4.22). Groundwater is also an equalizer of streamflow. Much of the water that flows in rivers is not direct runoff from rain and snowmelt. Rather, a large percentage of precipitation soaks in and then moves slowly underground to stream channels. Groundwater is thus a form of storage that sustains streams during periods when rain does not fall. When we observe water flowing in a river during a dry period, it is water from rain that fell at some earlier time and was stored underground.

Figure 4.20   Drainage patterns. **A.** Dendritic. **B.** Radial. **C.** Rectangular. **D.** Trellis.

Distribution and Movement of Groundwater

GEODe   Sculpturing Earth's Surface
EARTH SCIENCE   ↳ Groundwater

When rain falls, some of the water runs off, some evaporates, and the remainder soaks into the ground. This

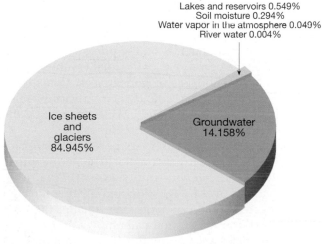

Figure 4.21   Estimated distribution of freshwater in the hydrosphere according to the U.S. Geological Survey.

last path is the primary source of practically all subsurface water. The amount of water that takes each of these paths, however, varies considerably both in time and space. Influential factors include the steepness of the slopes, the nature of the surface materials, the intensity of the rainfall, and the type and amount of vegetation. Heavy rains falling on steep slopes underlain by impervious materials will obviously result in a high percentage of the water running off. Conversely, if rain falls steadily and gently on more gradual slopes composed of materials that are easily penetrated by water, a much larger percentage of the water soaks into the ground.

Distribution

Some of the water that soaks in does not travel far, because it is held by molecular attraction as a surface film on soil particles. This near-surface zone is called the *belt of soil moisture.* It is crisscrossed by roots, voids left by decayed roots, and animal and worm burrows that enhance the infiltration of rainwater into the soil. Soil water is used by plants in life functions and transpiration. Some water also evaporates directly back into the atmosphere.

Water that is not held as soil moisture percolates downward until it reaches a zone where all of the open spaces in sediment and rock are completely filled with water. This is the **zone of saturation**. Water within it is called **groundwater**. The upper limit of this zone is

Figure 4.22 The dissolving action of groundwater creates caves. The dripstone features found in caves are of many types, including stalactites, stalagmites, and columns. Three Fingers Cave, Lincoln County, New Mexico. (Photo by Harris Photographic/Tom Stack and Associates)

known as the **water table**. The area above the water table where the soil, sediment, and rock are not saturated is called the **zone of aeration** (Figure 4.23). Although a considerable amount of water can be present in the zone of aeration, this water cannot be pumped by wells, because it clings too tightly to rock and soil particles. By contrast, below the water table, the water pressure is great enough to allow water to enter wells, thus permitting groundwater to be withdrawn for use. We will examine wells more closely later in the chapter.

The water table is rarely level as we might expect a table to be. Instead, its shape is usually a subdued replica of the surface, reaching its highest elevations beneath hills and decreasing in height toward valleys (Figure 4.23). When you see a wetland (swamp), it indicates that the water table is right at the surface. Lakes and streams generally occupy areas low enough that the water table is above the land surface.

Several factors contribute to the irregular surface of the water table. One important influence is the fact that groundwater moves very slowly. Because of this, water tends to "pile up" beneath high areas between stream valleys. If rainfall were to cease completely, these water "hills" would slowly subside and gradually approach the level of the valleys. However, new supplies of rain-

water are usually added often enough to prevent this. Nevertheless, in times of extended drought, the water table may drop enough to dry up shallow wells. Other causes for the uneven water table are variations in rainfall and permeability from place to place.

Movement

Depending on the nature of the subsurface material, the flow of groundwater and the amount of water that can be stored vary greatly. The quantity of groundwater that can be stored depends on the **porosity** of the material; that is, the percentage of the total volume of rock or sediment that consists of pore spaces. Voids most often are spaces between sedimentary particles, but also common are joints, faults, and cavities formed by the dissolving of soluble rocks such as limestone.

Rock or sediment may be very porous and still prohibit water from moving through it. The **permeability** of a material indicates its ability to *transmit a fluid*. Groundwater moves by twisting and turning through interconnected small openings. The smaller the pore spaces, the slower the groundwater moves. If the spaces between particles are too small, water cannot move at all. For example, clay's ability to store water can be great, owing to its high porosity, but its pore spaces are

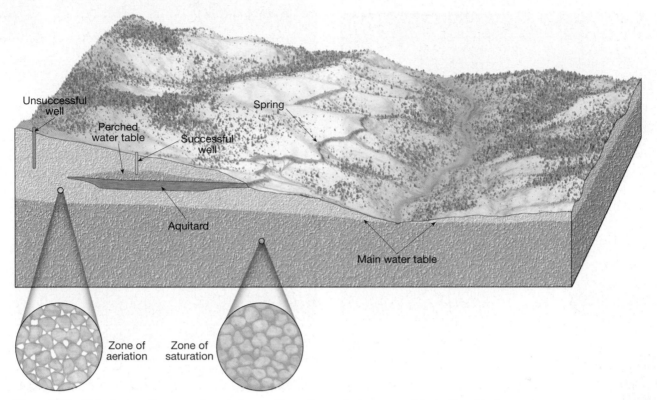

Unsuccessful
well

Spring

Perched
water table

Successful
well

Aquitard

Main water table

Zone of
aeration

Zone of
saturation

Figure 4.23   This diagram illustrates the relative positions of many features associated with subsurface water.

so small that water is unable to move through it. Thus, we say that clay is *impermeable.*

Impermeable layers such as clay that hinder or prevent water movement are termed **aquitards.** In contrast, larger particles, such as sand or gravel, have larger pore spaces. Therefore, the water moves with relative ease. Permeable rock strata or sediments that transmit groundwater freely are called **aquifers** ("water carriers"). Aquifers are important because they are the water-bearing layers sought after by well drillers.

Springs

Sculpturing Earth's Surface
↳ Groundwater

Springs have aroused the curiosity and wonder of people for thousands of years. The fact that springs were (and to some people still are) rather mysterious phenomena is not difficult to understand, for here is water flowing freely from the ground in all kinds of weather in seemingly inexhaustible supply but with no obvious source. Today we know that the source of springs is water from the zone of saturation and that the ultimate source of this water is precipitation.

Whenever the water table intersects the ground surface, a natural flow of groundwater results, which we call a **spring.** Springs such as the one in Figure 4.24 form when an aquitard blocks the downward movement of groundwater and forces it to move laterally. When the

permeable bed (aquifer) outcrops in a valley, a spring or series of springs results.

Another situation that can produce a spring is illustrated in Figure 4.23. Here an aquitard is situated above the main water table. As water percolates downward, a portion accumulates above the aquitard to create a localized zone of saturation and a *perched water table.* Springs, however, are not confined to places where a perched water table creates a flow at the surface. Many geological situations lead to the formation of springs because subsurface conditions vary greatly from place to place.

Hot Springs

By definition, the water in **hot springs** is 6–9°C (10–15°F) warmer than the mean annual air temperature for the localities where they occur. In the United States alone, there are well over 1000 such springs.

Temperatures in deep mines and oil wells usually rise with an increase in depth averaging about 2°C per 100 meters (1°F per 100 feet). Therefore, when groundwater circulates at great depths, it becomes heated, and if it rises to the surface, the water may emerge as a hot spring. The water of some hot springs in the United States, particularly in the East, is heated in this manner. However, the great majority (over 95 percent) of the hot springs (and geysers) in the United States are found in the West. The reason for such a distribution is that the source of heat for most hot springs is cooling igneous rock, and it is in the West that igneous activity has been most recent.

Figure 4.24   Spring flowing from a valley wall into a stream near Guilin, China. (Photo by W. Kenneth Hamblin)

Figure 4.25   A wintertime erruption of Old Faithful, one of the world's most famous geysers. It emits as much as 45,000 liters (almost 12,000 gallons) of hot water and steam about once each hour. (Photo by Marc Muench/David Muench Photography, Inc.)

Geyser

Geysers are intermittent hot springs or fountains in which columns of water are ejected with great force at various intervals, often rising 30 to 60 meters (100 to 200 feet; Figure 4.25). After the jet of water ceases, a column of steam rushes out, usually with a thundering roar. Perhaps the most famous geyser in the world is Old Faithful in Yellowstone National Park, which erupts about once each hour. Geysers are also found in other parts of the world, including New Zealand and Iceland. In fact, the Icelandic word *geysa*—to gush—gives us the name *geyser*.

Geysers occur where extensive underground chambers exist within hot igneous rocks. How they operate is shown in Figure 4.26. As relatively cool groundwater enters the chambers, it is heated by the surrounding rock. At the bottom of the chamber, the water is under great pressure because of the weight of the overlying water. This great pressure prevents the water from boiling at the normal surface temperature of 100°C (212°F). For example, at the bottom of a 300-meter (1000-foot) water-filled chamber, water must attain a temperature of nearly 230°C (450°F) before it will boil. The heating causes the water to expand, with the result that some is forced out at the surface. This loss of water reduces the pressure on the remaining water in the chamber, which lowers the boiling point. A portion of the water deep within the chamber quickly turns to steam and causes the geyser to erupt. Following the eruption, cool groundwater again seeps into the chamber, and the cycle begins anew.

Wells

 Sculpturing Earth's Surface
↳ Groundwater

The most common method for removing groundwater is the **well**, a hole bored into the zone of saturation (see Figure 4.23). Wells serve as small reservoirs into which groundwater moves and from which it can be pumped to the surface. The use of wells dates back many centuries and continues to be an important method of obtaining water today. By far the single greatest use of this water in the United States is irrigation for agriculture. More than 65 percent of the groundwater used each year is for this purpose. Industrial uses rank a distant second, followed by the amount used by homes in cities and rural areas.

The level of the water table may fluctuate considerably during the course of a year, dropping during dry seasons and rising following periods of rain. Therefore, to ensure a continuous supply of water, a well must penetrate far below the water table. Whenever a substantial amount of water is withdrawn from a well, the water table around the well is lowered. This effect, termed **drawdown,** decreases with increasing distance from the well. The result is a depression in the water table, roughly conical in shape, known as a **cone of depression** (Figure 4.27). For most small domestic wells, the cone of depression is negligible. However, when wells are used

Figure 4.26   Idealized diagrams illustrating the stages in the erruption cycle of a geyser. **A.** Groundwater enters underground caverns and fractures in hot igneous rock where it is heated to near its boiling point. **B.** Heating causes the water to expand, with some being forced out at the surface. The loss of water reduces the pressure on the remaining water, thus reducing its boiling temperature. Some of the water flashes to steam. **C.** The rapidly expanding steam forces the hot water out of the chambers to produce a geyser. The empty chambers fill again, and the cycle starts anew.

for irrigation or for industrial purposes, the withdrawal of water can be great enough to create a very wide and steep cone of depression that may substantially lower the water table in an area and cause nearby shallow wells to become dry. Figure 4.26 illustrates this situation.

STUDENTS SOMETIMES ASK...

I have heard people say that supplies of groundwater can be located using a forked stick. Can this actually be done?

What you describe is a practice called "water dowsing." In the classic method, a person holding a forked stick walks back and forth over an area. When water is detected, the bottom of the "Y" is supposed to be attracted downward.

Geologists and engineers are dubious, to say the least. Case histories and demonstrations may seem convincing, but when dowsing is exposed to scientific scrutiny, it fails. Most "successful" examples of water dowsing occur in places where water would be hard to miss. In a region of adequate rainfall and favorable geology, it is difficult to drill and *not* find water!

Artesian Wells

GEODe   External Processes
EARTH SCIENCE   ↳ Groundwater

In most wells, water does not rise on its own. If water is first encountered at 30 meters of depth, it remains at that level, fluctuating perhaps a meter or two with seasonal wet and dry periods. However, in some wells, water rises, sometimes overflowing at the surface.

The term **artesian** is applied to any situation in which groundwater rises in a well above the level where it was initially encountered. For such a situation to occur, two conditions must exist (Figure 4.28): (1) water must be confined to an aquifer that is inclined so that one end is exposed at the surface, where it can receive water; and (2) aquitards both above and below the aquifer must be present to prevent the water from escaping. When such a layer is tapped, the pressure created by the weight of the water above will force the water to rise. If there were no friction, the water in the well would rise to the level of the water at the top of the aquifer. However, friction reduces the height of this pressure surface. The greater the distance from the recharge area (area where water enters the inclined aquifer), the greater the friction and the less the rise of water.

In Figure 4.28, Well 1 is a *nonflowing artesian well*, because at this location the pressure surface is below ground level. When the pressure surface is above the ground and a well is drilled into the aquifer, a *flowing artesian well* is created (Well 2, Figure 4.28).

Artesian systems act as conduits, transmitting water from remote areas of recharge great distances to the points of discharge. In this manner, water that fell in cen-

Figure 4.27 A cone of depression in the water table often forms around a pumping well. If heavy pumping lowers the water table, some wells may be left dry.

Before heavy pumping

After heavy pumping

tral Wisconsin years ago is now taken from the ground and used by communities many kilometers to the south in Illinois. In South Dakota, such a system brings water eastward across the state from the Black Hills in the west.

On a different scale, city water systems may be considered as examples of artificial artesian systems. The water tower, into which water is pumped, may be considered the area of recharge, the pipes the confined aquifer, and the faucets in homes the flowing artesian wells.

Environmental Problems Associated with Groundwater

As with many of our valuable natural resources, groundwater is being exploited at an increasing rate. In some areas, overuse threatens the groundwater supply. In other places, groundwater withdrawal has caused the ground and everything resting upon it to sink. Still other localities are concerned with the possible contamination of their groundwater supply.

Treating Groundwater as a Nonrenewable Resource

For many, groundwater appears to be an endlessly renewable resource, for it is continually replenished by rainfall and melting snow. But in some regions, groundwater has been and continues to be treated as a *nonrenewable* resource. Where this occurs, the amount of water available to recharge the aquifer is significantly less than the amount being withdrawn.

The High Plains provides one example. Here an extensive agricultural economy is largely dependent on irrigation (Figure 4.29). In some parts of the region, where intense irrigation has been practiced for an extended period, depletion of groundwater has been severe. Under these circumstances, it can be said that the groundwater is literally being "mined." Even if pumping were to cease immediately, it could take hundreds or thousands of years for the groundwater to be fully replenished. Box 4.3 explores this issue more fully.

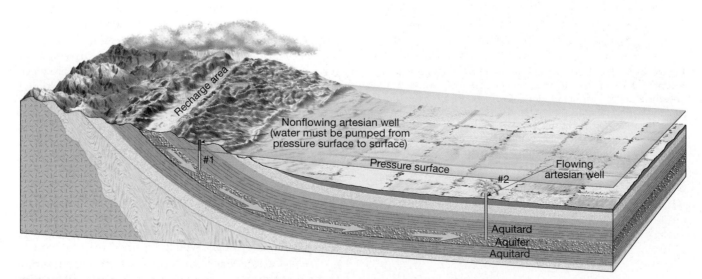

Figure 4.28 Artesian systems occur when an inclined aquifer is surrounded by impermeable beds.

Figure 4.29   In some agricultural regions water is pumped from the ground faster than it is replenished. In such instances, groundwater is being treated as a nonrenewable resource. (Photo by Getty Images, Inc./PhotoDisc, Inc.)

Land Subsidence Caused by Groundwater Withdrawal

As you will see later in this chapter, surface subsidence can result from natural processes related to groundwater. However, the ground may also sink when water is pumped from wells faster than natural recharge processes can replace it. This effect is particularly pronounced in areas underlain by thick layers of loose sediments. As water is withdrawn, the ground subsides because the weight of the overburden packs the sediment grains more tightly together.

Many areas can be used to illustrate land subsidence caused by excessive pumping of groundwater from relatively loose sediment. A classic example in the United States occurred in the San Joaquin Valley of California (Figure 4.30). This important agricultural region relies heavily on irrigation. Land subsidence due to groundwater withdrawal began in the valley in the mid-1920s and locally exceeded 8 meters (28 feet) by 1970. Then, because of the importation of surface water and a decrease in groundwater pumping, water levels in the aquifer recovered and subsidence ceased. However, during a drought in 1976 and 1977, heavy groundwater pumping led to renewed subsidence. This time, water levels dropped at a much faster rate than during the previous period, because of the reduced storage capacity caused by earlier compaction of material in the aquifer. In all, more than 13,400 square kilometers (5200 square miles) of irrigable land—one half the entire valley—were affected by subsidence. Damage to structures, including highways, bridges, water lines, and wells, was extensive. Many other examples of land subsidence due to groundwater pumping occur in the United States and elsewhere in the world.

Groundwater Contamination

The pollution of groundwater is a serious matter, particularly in areas where aquifers provide a large part of the water supply. One common source of groundwater pollution is sewage. Its sources include an ever increasing number of septic tanks, as well as farm wastes and inadequate or broken sewer systems.

If sewage water that is contaminated with bacteria enters the groundwater system, it may become purified through natural processes. The harmful bacteria can be mechanically filtered by the sediment through which the water percolates, destroyed by chemical oxidation, and/or assimilated by other organisms. For purification to occur, however, the aquifer must be of the correct composition.

For example, extremely permeable aquifers (such as highly fractured crystalline rock, coarse gravel, or cavernous limestone) have such large openings that contaminated groundwater may travel long distances without being cleansed. In this case, the water flows too rapidly and is not in contact with the surrounding material long enough for purification to occur. This is the problem at Well 1 in Figure 4.31A.

Conversely, when the aquifer is composed of sand or permeable sandstone, the water can sometimes be purified after traveling only a few dozen meters through it. The openings between sand grains are large enough to permit water movement, yet the movement of the water is slow enough to allow ample time for its purification (Well 2, Figure 4.31B).

Other sources and types of contamination also threaten groundwater supplies (Figure 4.32). These include widely used substances such as highway salt, fertilizers that are spread across the land surface, and pesticides. In addition, a wide array of chemicals and industrial materials may leak from pipelines, storage tanks, landfills, and holding ponds. Some of these pollutants are classified as *hazardous*, meaning that they are either flammable, corrosive, explosive, or toxic. As rainwater oozes through the refuse, it may dissolve a variety of potential contaminants. If the leached material reaches the water table, it will mix with the groundwater and contaminate the supply.

Because groundwater movement is usually slow, polluted water might go undetected for a long time. In fact, most contamination is discovered only after drinking water has been affected and people become ill. By this time, the volume of polluted water might be very large, and even if the source of contamination is removed immediately, the problem is not solved. Although the sources of groundwater contamination are numerous, the solutions are relatively few.

Once the source of the problem has been identified and eliminated, the most common practice is simply to abandon the water supply and allow the pollutants to be flushed away gradually. This is the least costly and easiest solution, but the aquifer must remain unused

BOX 4.3

PEOPLE AND THE ENVIRONMENT:
The Ogallala Aquifer—
How Long Will the Water Last?

The High Plains extend from the western Dakotas south to Texas. Despite being a land of little rain, this is one of the most important agricultural regions in the United States. The reason is a vast endowment of groundwater that makes irrigation possible throughout most of the region's 450,000 square kilometers (175,000 square miles). The source of most of this water is the Ogallala Formation, the largest aquifer in the United States (Figure 4.C).

Geologically, the Ogallala Formation consists of a number of sandy and gravelly rock layers. The sediments were derived from the erosion of the Rocky Mountains and carried eastward by sluggish streams. Erosion has removed much of the formation from eastern Colorado, severing the Ogallala's connection to the Rockies.

The Ogallala Formation averages 60 meters (200 feet) thick, and in places it is as much as 180 meters (600 feet) thick. Groundwater in the aquifer originally came downslope from the Rockies as well as from surface precipitation that soaked into the ground over thousands of years. Because of its high porosity and great size, the Ogallala accumulated huge amounts of groundwater—enough to fill Lake Huron! Today, with the connection between the aquifer and the Rockies severed, all of the Ogallala's recharge must come from the region's meager rainfall.

Figure 4.C   The Ogallala Formation underlies about 450,000 square kilometers (175,000 square miles) of the High Plains, making it the largest aquifer in the United States.

The Ogallala aquifer was first used for irrigation in the late 1800s, but its use was limited by the capacity of the pumps available at the time. In the 1920s, with the development of large-capacity irrigation pumps, High Plains farmers, particularly in Texas, began tapping the Ogallala for irrigation. Then in the 1950s, improved technology brought large-scale exploitation of the aquifer. Today, nearly 170,000 wells are being used to irrigate more than 65,000 square kilometers (16 million acres) of land.

With the increase in irrigation has come a drastic drop in the Ogallala's water table, especially in the southern High Plains. Declines in the water table of 3 to 15 meters (10 to 50 feet) are common, and in places the water table is now 60 meters (200 feet) below its level prior to irrigation.

Since the 1980s, irrigated acreage has declined in the High Plains. One important reason has been higher energy costs. As water levels have dropped, the costs of pumping groundwater to the surface have increased.

Although the water-table decline has slowed in parts of the southern High Plains, substantial pumping continues, often in excess of recharge. The future of irrigated farming in this region is clearly in jeopardy.

The southern High Plains represents an area of the United States that will return sooner or later to dry-land farming. The transition will come sooner and with fewer ecological and economic crises if the agricultural industry is weaned gradually from its dependence on groundwater irrigation. If nothing is done until all the accessible water in the Ogallala aquifer has been removed, the transition will be ecologically dangerous and economically dreadful.*

*National Research Council. *Solid-Earth Sciences and Society.* Washington, DC: National Academy Press, 1993, p. 148.

---

for many years. To accelerate this process, polluted water is sometimes pumped out and treated. Following removal of the tainted water, the aquifer is allowed to recharge naturally, or in some cases, the treated water or other freshwater is pumped back in. This process is costly, time consuming, and it may be risky because there is no way to be certain that all of the contamination has been removed. Clearly, the most effective solution to groundwater contamination is prevention.

The Geologic Work of Groundwater

Groundwater dissolves rock. This fact is key to understanding how caverns and sinkholes form. Because soluble rocks, especially limestone, underlie millions of square kilometers of Earth's surface, it is here that groundwater carries on its important role as an erosional agent. Limestone is nearly insoluble in pure water but is quite easily dissolved by water containing small quantities of carbonic acid. Most natural water contains this weak acid because rainwater readily dissolves carbon dioxide from the air and from decaying plants. Therefore, when groundwater comes in contact with limestone, the carbonic acid reacts with calcite in the rocks to form calcium bicarbonate, a soluble material that is then carried away in solution.

Caverns

The most spectacular results of groundwater's erosional handiwork are limestone **caverns**. In the United States alone about 17,000 caves have been discovered. Although most are relatively small, some have spectacular dimensions. Carlsbad Caverns in southeastern New

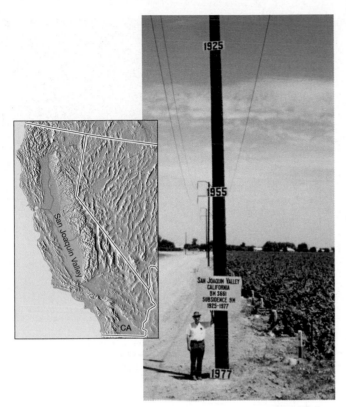

Figure 4.30 The shaded area on the map shows California's San Joaquin Valley. The marks on the utility pole in the photo indicate the level of the surrounding land in preceding years. Between 1925 and 1975 this part of the San Joaquin Valley subsided almost 9 meters because of the withdrawal of groundwater and the resulting compaction of sediments. (Photo courtesy of U.S. Geological Survey)

Mexico and Mammoth Cave in Kentucky are famous examples. One chamber in Carlsbad Caverns has an area equivalent to 14 football fields and enough height to accommodate the U.S. Capitol Building. At Mammoth Cave, the total length of interconnected caverns extends for more than 540 kilometers (340 miles).

Most caverns are created at or below the water table in the zone of saturation. Here acidic groundwater follows lines of weakness in the rock, such as joints and bedding planes. As time passes, the dissolving process slowly creates cavities and gradually enlarges them into caverns. Material that is dissolved by the groundwater is eventually discharged into streams and carried to the ocean.

Certainly the features that arouse the greatest curiosity for most cavern visitors are the stone formations that give some caverns a wonderland appearance (see Figure 4.22). These are not erosional features, like the caverns in which they reside, but depositional features. They are created by the seemingly endless dripping of water over great spans of time. The calcium carbonate that is left behind produces the limestone we call *travertine*. These cave deposits, however, are also commonly called *dripstone*, an obvious reference to their mode of origin.

Although the formation of caverns takes place in the zone of saturation, the deposition of dripstone is not possible until the caverns are above the water table in the zone of aeration. This commonly occurs as nearby streams cut their valleys deeper, lowering the water table as the elevation of the rivers drops. As soon as the chamber is filled with air, the conditions are right for the decoration phase of cavern building to begin.

Figure 4.31 **A.** Although the contaminated water has traveled more than 100 meters before reaching Well 1, the water moves too rapidly through the cavernous limestone to be purified. **B.** As the discharge from the septic tank percolates through the permeable sandstone, it is purified in a relatively short distance.

A.

B.

Figure 4.32 Sometimes agricultural chemicals **A.** and materials leached from landfills **B.** find their way into the groundwater. These are two of the potential sources of groundwater contamination. (Photo A by Roy Morsch/The Stock Market; Photo B by F. Rossotto/The Stock Market)

Of the various dripstone features found in caverns, perhaps the most familiar are **stalactites**. These icicle-like pendants hang from the ceiling of the cavern and form where water seeps through cracks above. When water reaches air in the cave, some of the dissolved carbon dioxide escapes from the drop and calcite begins to precipitate. Deposition occurs as a ring around the edge of the water drop. As drop after drop follows, each leaves an infinitesimal trace of calcite behind, and a hollow limestone tube is created. Water then moves through the tube, remains suspended momentarily at the end, contributes a tiny ring of calcite, and falls to the cavern floor. The stalactite just described is appropriately called a *soda straw* (Figure 4.33). Often the hollow tube of the soda straw becomes plugged or its supply of water increases. In either case, the water is forced to flow and deposit along the outside of the tube. As deposition continues, the stalactite takes on the more common conical shape.

Formations that develop on the floor of a cavern and reach upward toward the ceiling are called **stalagmites.** The water supplying the calcite for stalagmite growth falls from the ceiling and splatters over the surface. As a result, stalagmites do not have a central tube and are usually more massive in appearance and more rounded on their upper ends than stalactites. Given enough time, a downward-growing stalactite and an upward-growing stalagmite may join to form a *column*.

Karst Topography

Many areas of the world have landscapes that to a large extent have been shaped by the dissolving power of groundwater. Such areas are said to exhibit **karst topography**, named for the *Krs* region in Slovenia (formerly a

Figure 4.33 "Live" solitary soda-straw stalactites. Lehman Caves. Great Basin National Park, Nevada. (Photo by Tom Bean)

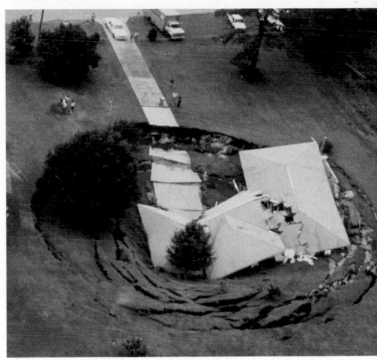

A.   **B.**

Figure 4.34 **A.** This high-altitude infrared image shows an area of karst topography in central Florida. The numerous lakes occupy sinkholes. (Courtesy of USDA-ASCS) **B.** This small sinkhole formed suddenly in 1991 when the roof of a cavern collapsed, destroying this home in Frostproof, Florida. (Photo by *St. Petersburg Times*/Liaison Agency, Inc.)

part of Yugoslavia) where such topography is strikingly developed. In the United States, karst landscapes occur in many areas that are underlain by limestone, including portions of Kentucky, Tennessee, Alabama, southern Indiana, and central and northern Florida. Generally, arid and semiarid areas do not develop karst topography because there is insufficient groundwater. When solution features exist in such regions, they are likely to be remnants of a time when rainier conditions prevailed.

Karst areas typically have irregular terrain punctuated with many depressions called **sinkholes** or, simply, **sinks**. In the limestone areas of Florida, Kentucky, and southern Indiana, there are literally tens of thousands of these depressions varying in depth from just a meter or two to a maximum of more than 50 meters (Figure 4.34A).

Sinkholes commonly form in one of two ways. Some develop gradually over many years without any physical disturbance to the rock. In these situations, the limestone immediately below the soil is dissolved by downward-seeping rainwater that is freshly charged with carbon dioxide. These depressions are usually not deep and are characterized by relatively gentle slopes. By contrast, sinkholes can also form suddenly and without warning when the roof of a cavern collapses under its own weight. Typically, the depressions created in this manner are steep-sided and deep. When they form in populous areas, they may represent a serious geologic hazard. Such a situation is clearly the case in Figure 4.34B.

In addition to a surface pockmarked by sinkholes, karst regions characteristically show a striking lack of surface drainage (streams). Following a rainfall, runoff is quickly funneled below ground through sinks. It then flows through caverns until it finally reaches the water table. Where streams do exist at the surface, their paths are usually short. The names of such streams often give a clue to their fate. In the Mammoth Cave area of Kentucky, for example, there is Sinking Creek, Little Sinking Creek, and Sinking Branch. Some sinkholes become plugged with clay and debris, creating small lakes or ponds.

Chapter Summary

• The *hydrologic cycle* describes the continuous interchange of water among the oceans, atmosphere, and continents. Powered by energy from the Sun, it is a global system in which the atmosphere provides the link between the oceans and continents. The processes involved in the hydrologic cycle include *precipitation, evaporation, infiltration* (the move-

ment of water into rocks or soil through cracks and pore spaces), *runoff* (water that flows over the land rather than infiltrating into the ground), and *transpiration* (the release of water vapor to the atmosphere by plants). *Running water is the single most important agent sculpturing Earth's land surface.*

- The factors that determine a stream's *velocity* are *gradient* (slope of the stream channel), *shape, size,* and *roughness* of the channel, and the stream's *discharge* (amount of water passing a given point per unit of time, frequently measured in cubic feet per second). Most often, the gradient and roughness of a stream decrease downstream, while width, depth, discharge, and velocity increase.

- The two general types of *base level* (the lowest point to which a stream may erode its channel) are (1) *ultimate base level* and (2) *temporary,* or *local base level.* Any change in base level will cause a stream to adjust and establish a new balance. Lowering base level will cause a stream to downcut, whereas raising base level results in deposition of material in the channel.

- The work of a stream includes *erosion* (the incorporation of material), *transportation* (as *dissolved load, suspended load,* and *bed load*), and *deposition,* whenever a stream's velocity decreases.

- Although many gradations exist, the two general types of stream valleys are (1) *narrow V-shaped valleys* and (2) *wide valleys with flat floors.* Because the dominant activity is downcutting toward base level, narrow valleys often contain *waterfalls* and *rapids.* When a stream has cut its channel closer to base level, its energy is directed from side to side, and erosion produces a flat valley floor, or *floodplain.* Streams that flow upon floodplains often move in sweeping bends called *meanders.* Widespread meandering may result in shorter channel segments, called *cutoffs,* and/or abandoned bends, called *oxbow lakes.*

- *Floods* are triggered by heavy rains and/or snowmelt. Sometimes human interference can worsen or even cause floods. Flood-control measures include the building of *artificial levees* and dams, as well as *channelization,* which could involve creating *artificial cutoffs.* Many scientists and engineers advocate a nonstructural approach to flood control that involves more appropriate land use.

- The land area that contributes water to a stream is called a *drainage basin.* Drainage basins are separated by an imaginary line called a *divide.* Common *drainage patterns* produced by streams include (1) *dendritic,* (2) *radial,* (3) *rectangular,* and (4) *trellis.*

- As a resource, *groundwater* represents the largest reservoir of freshwater that is readily available to humans. Geologically, the dissolving action of groundwater produces *caves* and *sinkholes.* Groundwater is also an equalizer of streamflow.

- Groundwater is water that occupies the pore spaces in sediment and rock in a zone beneath the surface called the *zone of saturation.* The upper limit of this zone is the *water table.* The *zone of aeration* is above the water table where the soil, sediment, and rock are not saturated.

- The quantity of water that can be stored depends on the *porosity* (the volume of open spaces) of the material. The *permeability* (the ability to transmit a fluid through interconnected pore spaces) of a material is a very important factor controlling the movement of groundwater.

- Materials with very small pore spaces (such as clay) hinder or prevent groundwater movement and are called *aquitards.* *Aquifers* consist of materials with larger pore spaces (such as sand) that are permeable and transmit groundwater freely.

- *Springs* occur whenever the water table intersects the land surface and a natural flow of groundwater results. *Wells,* openings drilled into the zone of saturation, withdraw groundwater and create roughly conical depressions in the water table known as *cones of depression.* Artesian wells occur when water rises above the level at which it was initially encountered.

- When groundwater circulates at great depths, it becomes heated. If it rises, the water may emerge as a *hot spring.* Geysers occur when groundwater is heated in underground chambers, expands, and some water quickly changes to steam, causing the geyser to erupt. The source of heat for most hot springs and geysers is hot igneous rock.

- Some of the current environmental problems involving groundwater include (1) *overuse* by intense irrigation, (2) *land subsidence* caused by groundwater withdrawal, and (3) *contamination* by pollutants.

- Most *caverns* form in limestone at or below the water table when acidic groundwater dissolves rock along lines of weakness, such as joints and bedding planes. *Karst topography* exhibits an irregular terrain punctuated with many depressions, called *sinkholes.*

Key Terms

Review Questions

1. Once precipitation has fallen on land, what paths are available to it?

2. A stream starts out 2000 meters above sea level and travels 250 kilometers to the ocean. What is its average gradient in meters per kilometer?

3. Suppose that the stream mentioned in Question 2 developed extensive meanders so that its course was lengthened to 500 kilometers. Calculate its new gradient. How does meandering affect gradient?

4. When the discharge of a stream increases, what happens to the stream's velocity?

5. Define base level. Name the main river in your area. For what streams does it act as base level? What is the base level for the Mississippi River? The Missouri River?

6. When an area changes from being predominantly rural to largely urban, how is streamflow affected (see Box 4.1)?

7. In what three ways does a stream transport its load?

8. If you collect a jar of water from a stream, what part of its load will settle to the bottom of the jar? What portion will remain in the water? What part of a stream's load would probably *not* be present in your sample?

9. Differentiate between competency and capacity.

10. Briefly describe the formation of a natural levee. How is this feature related to backswamps and yazoo tributaries?

11. List and briefly describe three basic flood-control strategies. What are some drawbacks of each?

12. Each of the following statements refers to a particular drainage pattern. Identify the pattern. (a) Streams diverge from a central high area such as a volcano. (b) Streams form a branching, treelike pattern. (c) A pattern that develops when bedrock is crisscrossed by joints and faults.

13. What percentage of freshwater is groundwater (see Table 4.2)? If glacial ice is excluded and only liquid freshwater is considered, about what percentage is groundwater?

14. Geologically, groundwater is important as an erosional agent. Name another significant geological role of groundwater.

15. Define groundwater and relate it to the water table.

16. How do porosity and permeability differ?

17. What is the source of heat for most hot springs and geysers? How is this reflected in the distribution of these features?

18. What is meant by the term *artesian*? In order for artesian wells to exist, two conditions must be present. List these conditions.

19. What problem is associated with the pumping of groundwater for irrigation in the southern part of the High Plains (see Box 4.3)?

20. Briefly explain what happened in the San Joaquin Valley of California as the result of excessive groundwater withdrawal.

21. Which would be most effective in purifying polluted groundwater: an aquifer composed mainly of coarse gravel, sand, or cavernous limestone?

22. Differentiate between stalactites and stalagmites. How do these features form?

23. Describe two ways in which sinkholes are created.

Examining the Earth System

1. List the process(es) involved in moving water through the hydrologic cycle from the (a) hydrosphere to the atmosphere, (b) atmosphere to the solid Earth, (c) biosphere to the atmosphere, and (d) hydrosphere (land) to the hydrosphere (ocean).

2. Over the oceans, evaporation exceeds precipitation, yet sea level does not drop. Why?

3. List at least three specific examples of interactions between humans and the hydrologic cycle (for example, the construction of a dam). Briefly describe the consequence(s) of each of these interactions.

4. Describe the role of the atmosphere, solid Earth, and biosphere in determining the quantity and quality of groundwater available for human consumption.

Web Resources

The *Earth Science* Website uses the resources and flexibility of the Internet to aid in your study of the topics in this chapter. Written and developed by Earth science instructors, this site will help improve your understanding of Earth science. Visit **http://www.prenhall.com/tarbuck** and click on the cover of *Earth Science* 10e to find:

- **On-line review quizzes.**
- **Web-based critical thinking and writing exercises.**
- **Links to chapter-specific Web resources.**
- **Internet-wide key term searches.**
http://www.prenhall.com/tarbuck

These sand dunes and salt-encrusted surface are in Sassusvlei National Park, Namibia. (Photo by Art Wolfe)

CHAPTER 5

GLACIERS, DESERTS, AND WIND

Climate has a strong influence on the nature and intensity of Earth's external processes. This fact is dramatically illustrated in this chapter. The existence and extent of glaciers is largely controlled by Earth's changing climate. Another excellent example of the strong link between climate and geology is seen when we examine the development of arid landscapes.

Today glaciers cover nearly 10 percent of Earth's land surface; however, in the recent geologic past, ice sheets were three times more extensive, covering vast areas with ice thousands of meters thick. Many regions still bear the mark of these glaciers. The first part of this chapter examines glaciers and the erosional and depositional features they create. The second part is devoted to dry lands and the geologic work of wind. Because desert and near-desert conditions prevail over an area as large as that affected by the massive glaciers of the Ice Age, the nature of such landscapes is indeed worth investigating.

Glaciers: A Part of Two Basic Cycles in the Earth System

 GEODe Sculpturing Earth's Surface
↳ Glaciers

Many present-day landscapes were modified by the widespread glaciers of the most recent Ice Age and still strongly reflect the handiwork of ice. The basic character of such diverse places as the Alps, Cape Cod, and Yosemite Valley was fashioned by now vanished masses of glacial ice. Moreover, Long Island, the Great Lakes, and the fiords of Norway and Alaska all owe their existence to glaciers. Glaciers, of course, are not just a phenomenon of the geologic past. As you will see, they are still sculpting and depositing debris in many regions today.

Glaciers are a part of two fundamental cycles in the Earth system—the hydrologic cycle and the rock cycle. Earlier you learned that the water of the hydrosphere is constantly cycled through the atmosphere, biosphere, and solid Earth. Time and time again the same water is evaporated from the oceans into the atmosphere, precipitated upon the land, and carried by rivers and underground flow back to the sea. However, when precipitation falls at high elevations or high latitudes, the water may not immediately make its way toward the sea. Instead, it may become part of a glacier. Although the ice will eventually melt, allowing the water to continue its path to the sea, water can be stored as glacial ice for many tens, hundreds, or even thousands of years.

A **glacier** is a thick ice mass that forms over hundreds or thousands of years. It originates on land from the accumulation, compaction, and recrystallization of snow. A glacier appears to be motionless, but it is not—glaciers move very slowly. Like running water, groundwater, wind, and waves, glaciers are dynamic erosional agents that accumulate, transport, and deposit sediment. As such, glaciers are among the processes that perform a very basic function in the rock cycle. Although glaciers are found in many parts of the world, most are located in remote areas, either near Earth's poles or in high mountains.

Valley (Alpine) Glaciers

Literally thousands of relatively small glaciers exist in lofty mountain areas, where they usually follow valleys originally occupied by streams. Unlike the rivers that previously flowed in these valleys, the glaciers advance slowly, perhaps only a few centimeters each day. Because of their setting, these moving ice masses are termed **valley glaciers** or **alpine glaciers** (Figure 5.1). Each glacier is a stream of ice, bounded by precipitous rock walls, that flows downvalley from a snow accumulation center near its head. Like rivers, valley glaciers can be long or short, wide or narrow, single or with branching tributaries. Generally the widths of alpine glaciers are small compared to their lengths; some extend for just a fraction of a kilometer, whereas others go on for many dozens of kilometers. The west branch of the Hubbard Glacier, for example, runs through 112 kilometers (nearly 70 miles) of mountainous terrain in Alaska and the Yukon Territory.

Ice Sheets

In contrast to valley glaciers, **ice sheets** exist on a *much* larger scale. In fact, they are often referred to as *continental ice sheets*. These enormous masses flow out in all directions from one or more centers and completely obscure all but the highest areas of underlying terrain. Although many ice sheets have existed in the past, just two achieve this status at present (Figure 5.2). Nevertheless, their combined areas represent almost 10 percent of Earth's land area. In the Northern Hemisphere, Greenland is covered by an imposing ice sheet that occupies 1.7 million square kilometers (0.7 million square miles), or about 80 percent of this large island. Averaging nearly 1500 meters (5000 feet) thick, in places the ice extends 3000 meters (10,000 feet) above the island's bedrock floor.

In the Southern Hemisphere, the huge Antarctic Ice Sheet attains a maximum thickness of nearly 4300 meters (14,000 feet) and covers an area of more than 13.9 million square kilometers (5.4 million square miles). The Antarctic Ice Sheet represents 80 percent of the world's ice and nearly two-thirds of Earth's fresh water. If this ice melted, sea level would rise an estimated 60 to 70 meters (200 to 230 feet) and the ocean would inundate many densely populated coastal areas (Figure 5.3).

Figure 5.1 Barry Glacier, in Alaska's Chugach Mountains, is an example of a valley glacier. (Photo © by Carr Clifton)

Other Types of Glaciers

In addition to valley glaciers and ice sheets, other types of glaciers are also identified. Covering some uplands and plateaus are masses of glacial ice called **ice caps.** Like ice sheets, ice caps completely bury the underlying landscape but are much smaller than the continental-scale features. Ice caps occur in many places, including Iceland and several of the large islands in the Arctic Ocean. Another type, known as **piedmont glaciers,** occupy broad lowlands at the bases of steep mountains and form when one or more valley glaciers emerge from the confining walls of mountain valleys. Here the advancing ice spreads out to form a broad sheet. The size of individual piedmont glaciers varies greatly. Among the largest is the broad Malaspina Glacier along the coast of southern Alaska. It covers more than 5000 square kilometers (2000 square miles) of the flat coastal plain at the foot of the lofty St. Elias range.

STUDENTS SOMETIMES ASK...

*Can glaciers form
in tropical regions?*

Certainly. Glaciers form wherever there are low temperatures and adequate supplies of snow. Because temperatures drop with an increase in altitude, glaciers can occur in the tropics at high elevations. Even near the equator, glaciers may form at elevations above about 5000 meters (16,400 feet). Examples of equatorial glaciers include those atop Mt. Kenya and Mt. Kilimanjaro in East Africa and those in the tall mountains of New Guinea.

How Glaciers Move

Sculpturing Earth's Surface
↳ Glaciers

The movement of glacial ice is generally referred to as *flow.* The fact that glacial movement is described in this way seems paradoxical—how can a solid flow? The way in which ice flows is complex and is of two basic types. The first of these, *plastic flow,* involves movement *within* the ice. Ice behaves as a brittle solid until the pressure upon it is equivalent to the weight of about 50 meters (165 feet) of ice. Once that load is surpassed, ice behaves as a plastic material and flow begins. A second and often equally important mechanism of glacial movement consists of the entire ice mass slipping along the ground. The lowest por-

Figure 5.2 The only present-day continental ice sheets are those covering Greenland and Antarctica. Their combined areas represent almost 10 percent of Earth's land area. Greenland's ice sheet occupies 1.7 million square kilometers, or about 80 percent of the island. The area of the Antarctic Ice Sheet is almost 14 million square kilometers. Along portions of the Antarctic coast, glacial ice flows into bays, creating features called *ice shelves*. These large masses of floating ice remain attached to the land on one or more sides. Ice shelves occupy an additional 1.4 million square kilometers adjacent to the Antarctic Ice Sheet.

tions of most glaciers are thought to move by this sliding process.

The upper 50 meters or so of a glacier is not under sufficient pressure to exhibit plastic flow. Rather, the ice in this uppermost zone is brittle and is appropriately referred to as the *zone of fracture.* The ice in this zone is carried along "piggyback" style by the ice below. When the glacier moves over irregular terrain, the zone of fracture is subjected to tension, resulting in cracks called **crevasses** (Figure 5.4). These gaping cracks can make travel across glaciers dangerous and can extend to depths of 50 meters (165 feet). Beyond this depth, plastic flow seals them off.

Rates of Glacial Movement

Unlike streamflow, glacial movement is not obvious (see Box 5.1). How rapidly does glacial ice move? Average rates vary considerably from one glacier to another. Some move so slowly that trees and other vegetation may become well established in the debris that accumulates on the glacier's surface. Others advance up to several meters per day. The movement of some glaciers is characterized by periods of extremely rapid advance followed by periods during which movement is practically nonexistent.

Budget of a Glacier

Snow is the raw material from which glacial ice originates. Therefore, glaciers form in areas where more snow falls in winter than can melt during the summer. Glaciers are constantly gaining and losing ice. Snow accumulation and ice formation occur in the **zone of accumulation** (Figure 5.5). Here the addition of snow thickens the glacier and promotes movement. Beyond this area of ice formation is the **zone of wastage.** Here there is a net loss to the glacier as all of the snow from the previous winter melts, as does some of the glacial ice (Figure 5.5).

In addition to melting, glaciers also waste as large pieces of ice break off the front of a glacier in a process called *calving.* Where glaciers reach the sea, calving creates *icebergs* (Figure 5.6). Because icebergs are just slightly less dense than seawater, they float very low in the water, with more than 80 percent of their mass submerged. The margins of the Greenland Ice Sheet produce thousands of icebergs each year. Many drift southward and find their way into the North Atlantic, where they are a hazard to navigation.

Whether the margin of a glacier is advancing, retreating, or remaining stationary depends on the *budget* of the glacier. The glacial budget is the balance or lack of balance between accumulation at the upper end of a glacier and loss at the lower end. If ice accumulation exceeds wastage, the glacial front advances until the two factors balance. At this point, the terminus of the glacier becomes stationary.

If a warming trend increases wastage and/or if a drop in snowfall decreases accumulation, the ice front will retreat. As the terminus of the glacier retreats, the extent of the zone of wastage diminishes. Therefore, in time a new balance will be reached between accumulation and wastage, and the ice front will again become stationary.

Whether the margin of a glacier is advancing, retreating, or stationary, the ice within the glacier contin-

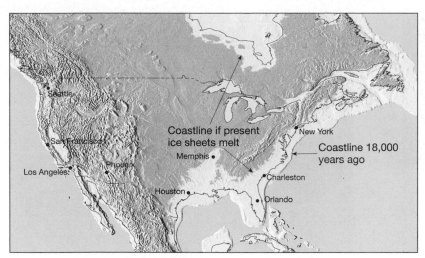

Figure 5.3   This map of a portion of North America shows the present-day coastline compared to the coastline that existed during the last ice-age maximum (18,000 years ago) and the coastline that would exist if present ice sheets in Greenland and Antarctica melted. (After R. H. Dott, Jr. and R. L. Battan, *Evolution of the Earth*, New York: McGraw-Hill, 1971.)

ues to flow forward. In the case of a receding glacier, the ice still flows forward, but not rapidly enough to offset wastage. This point is illustrated in Figure 5.A in Box 5.1. As the line of stakes within the Rhone Glacier continued to move downvalley, the terminus of the glacier slowly retreated upvalley.

Glacial Erosion

Sculpturing Earth's Surface
↳ Glaciers

Glaciers are capable of great erosion. For anyone who has observed the terminus of an alpine glacier, the evidence of its erosive force is clear. You can witness firsthand the release of rock fragments of various sizes from the ice as it melts. All signs lead to the conclusion that the ice has scraped, scoured, and torn rock debris from the floor and walls of the valley and carried it downvalley. Indeed, as a transporter of sediment, ice has no equal.

Once rock debris is acquired by the glacier, it cannot settle out as does the load carried by a stream or by the wind. Consequently, glaciers can carry huge blocks that no other erosional agent could possibly budge. Although today's glaciers are of limited importance as erosional agents, many landscapes that were modified by the widespread glaciers of the recent Ice Age still reflect to a high degree the work of ice.

How Glaciers Erode

Glaciers erode land primarily in two ways—plucking and abrasion. First, as a glacier flows over a fractured bedrock surface, it loosens and lifts blocks of rock and incorporates them into the ice. This process, known as **plucking,** occurs when meltwater penetrates the cracks and joints along the rock floor of the glacier and freezes. As the water expands, it exerts tremendous leverage

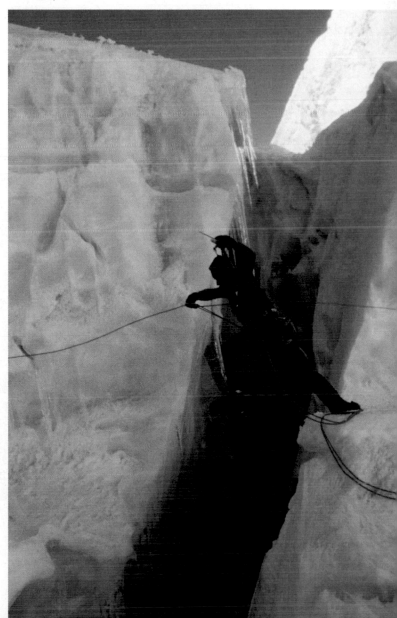

Figure 5.4   Crevasses form in the brittle ice of the zone of fracture. They can extend to depths of 50 meters and can obviously make travel across glaciers dangerous. Gasherbrum II expedition, Pakistan. (Photo by Galen Rowell/Mountain Light Photography, Inc.)

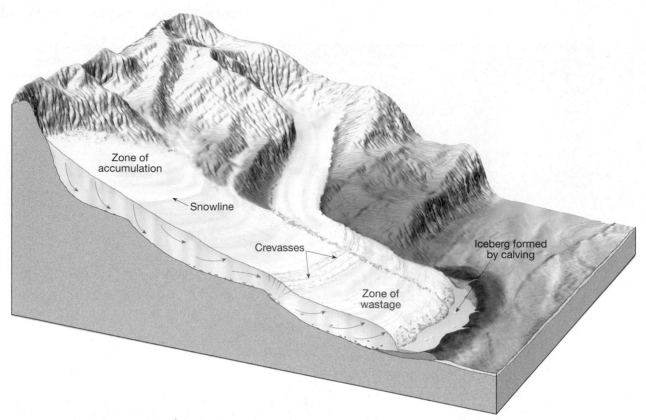

Figure 5.5 The snowline separates the zone of accumulation and the zone of wastage. Above the snowline, more snow falls each winter than melts each summer. Below the snowline, the snow from the previous winter completely melts, as does some of the underlying ice. Whether the margin of a glacier advances, retreats, or remains stationary depends on the balance or lack of balance between accumulation and wastage. When a glacier moves across irregular terrain, *crevasses* form in the brittle portion.

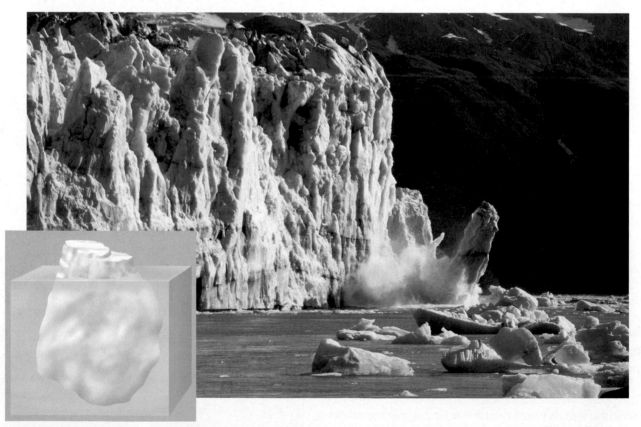

Figure 5.6 Icebergs are created when large pieces calve from the front of a glacier after it reaches a water body. Here ice is calving from the terminus of Alaska's Hubbard Glacier in Wrangell–St. Elias National Park, Alaska. Only about 20 percent of an iceberg protrudes above the water line. (Photo by Tom & Susan Bean, Inc.)

BOX 5.1

UNDERSTANDING EARTH:
Do Glaciers Move? An Application of the Scientific Method

The study of glaciers provides an early application of the scientific method. High in the Alps of Switzerland and France, small glaciers exist in the upper portions of some valleys. In the late eighteenth and early nineteenth centuries, people who farmed and herded animals in these valleys suggested that glaciers in the upper reaches of the valleys had previously been much larger and had occupied downvalley areas. They based their belief on the fact that the valley floors were littered with angular boulders and other rock debris that seemed identical to the materials that they could see in and near the glaciers at the heads of the valleys.

Although the explanation of these observations seemed logical, others did not accept the notion that masses of ice hundreds of meters thick were capable of movement. The disagreement was settled after a simple experiment was designed and carried out to test the hypothesis that glacial ice can move.

Markers were placed in a straight line completely across an alpine glacier. The position of the line was marked on the valley walls so that if the ice moved, the change in position could be detected. After a year or two the results were clear: The markers on the glacier had advanced down the valley, proving that glacial ice indeed moves. In addition, the experiment demonstrated that ice within a glacier does not move at a uniform rate, be-

cause the markers in the center advanced farther than did those along the margins. Although most glaciers move too slowly for direct visual detection, the experiment succeeded in demonstrating that movement nevertheless occurs. In the years that followed, this experiment was repeated many times with greater accuracy, using more modern surveying techniques. Each time, the basic relationships established by earlier attempts were verified.

The experiment illustrated in Figure 5.A was carried out at Switzerland's Rhone Glacier later in the nineteenth century. It not only traced the movement of markers within the ice but also mapped the position of the glacier's terminus. Notice that even though the ice within the glacier was advancing, the ice front was retreating. As often occurs in science, experiments and observations designed to test one hypothesis yield new information that requires further analysis and explanation.

Figure 5.A   Ice movement and changes in the terminus at Rhone Glacier, Switzerland. In this classic study of a valley glacier, the movement of stakes clearly showed that ice along the sides of the glacier moves slowest. Also notice that even though the ice front was retreating, the ice within the glacier was advancing.

that pries the rock loose. In this manner, sediment of all sizes becomes part of the glacier's load.

The second major erosional process is **abrasion** (Figure 5.7). As the ice and its load of rock fragments slide over bedrock, they function like sandpaper to smooth and polish the surface below. The pulverized rock produced by the glacial gristmill is appropriately called **rock flour.** So much rock flour may be produced that meltwater streams leaving a glacier often have the grayish appearance of skim milk—visible evidence of the grinding power of the ice.

When the ice at the bottom of a glacier contains large fragments, long scratches and grooves called **glacial striations** may be gouged into the bedrock (Figure 5.7). These linear scratches on the bedrock surface

provide clues to the direction of glacial movement. By mapping the striations over large areas, glacial flow patterns can often be reconstructed. Conversely, not all abrasive action produces striations. The rock surface over which the glacier moves may also become highly polished by the ice and its load of finer particles.

As is the case with other agents of erosion, the rate of glacial erosion is highly variable. This differential erosion by ice is largely controlled by four factors: (1) rate of glacial movement; (2) thickness of the ice; (3) shape, abundance, and hardness of the rock fragments contained in the ice at the base of the glacier; and (4) the erodibility of the surface beneath the glacier. Variations in any or all of these factors from time to time and/or from place to place mean that the features, ef-

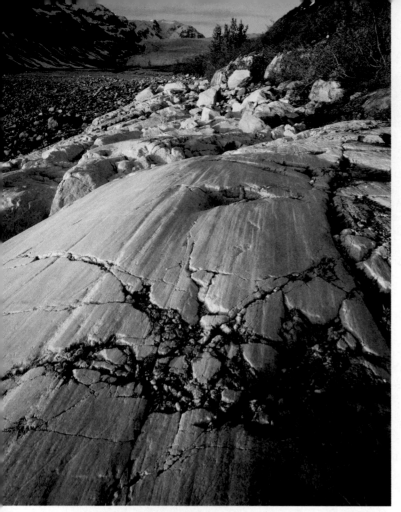

Figure 5.7 Glacial abrasion created the scratches and grooves in this bedrock. Glacier Bay National Park, Alaska. (Photo © by Carr Clifton)

fects, and degree of landscape modification in glaciated regions can vary greatly.

Landforms Created by Glacial Erosion

Although the erosional accomplishments of ice sheets can be tremendous, landforms carved by these huge ice masses usually do not inspire the same awe as do the erosional features created by valley glaciers. In regions where the erosional effects of ice sheets are significant, glacially scoured surfaces and subdued terrain are the rule. By contrast, in mountainous areas, erosion by valley glaciers produces many truly spectacular features. Much of the rugged mountain scenery so celebrated for its majestic beauty is the product of erosion by valley glaciers.

Take a moment to study Figure 5.8, which shows a mountain setting before, during, and after glaciation. You will refer to this often in the following discussion.

Glaciated Valleys. Prior to glaciation, alpine valleys are characteristically V-shaped because streams are well above base level and are therefore downcutting (Figure 5.8A). However, in mountainous regions that

have been glaciated, the valleys are no longer narrow. As a glacier moves down a valley once occupied by a stream, the ice modifies it in three ways: The glacier widens, deepens, and straightens the valley, so that what was once a narrow V-shaped valley is transformed into a U-shaped **glacial trough** (Figures 5.8C and 5.9).

The amount of glacial erosion depends in part on the thickness of the ice. Consequently, main glaciers, also called *trunk glaciers,* cut their valleys deeper than do their smaller tributary glaciers. Thus, after the ice has receded, the valleys of tributary glaciers are left standing above the main glacial trough and are termed **hanging valleys** (Figure 5.8C). Rivers flowing through hanging valleys can produce spectacular waterfalls, such as those in Yosemite National Park, California (Figure 5.10).

Cirques. At the head of a glacial valley is a characteristic and often imposing feature associated with an alpine glacier—a **cirque.** As Figure 5.11 illustrates, these bowl-shaped depressions have precipitous walls on three sides but are open on the downvalley side. The cirque is the focal point of the glacier's growth because it is the area of snow accumulation and ice formation. Cirques begin as irregularities in the mountainside that are subsequently enlarged by frost wedging and plucking along the sides and bottom of the glacier. The glacier in turn acts as a conveyor belt that carries away the debris. After the glacier has melted away, the cirque basin is sometimes occupied by a small lake.

Arêtes and Horns. The Alps, Northern Rockies, and many other mountain landscapes carved by valley glaciers reveal more than glacial troughs and cirques. In addition, sinuous, sharp-edged ridges called **arêtes** and sharp, pyramid-like peaks termed **horns** project above the surroundings (Figure 5.8C). Both features can originate from the same basic process the enlargement of cirques produced by plucking and frost action. Several cirques around a single high mountain create the spires of rock called horns. As the cirques enlarge and converge, an isolated horn is produced. A famous example is the Matterhorn in the Swiss Alps (Figure 5.12).

Arêtes can form in a similar manner except that the cirques are not clustered around a point but rather exist on opposite sides of a divide. As the cirques grow, the divide separating them is reduced to a very narrow, knifelike partition. An arête can also be created in another way. When two glaciers occupy parallel valleys, an arête can form when the land separating the moving tongues of ice is progressively narrowed as the glaciers scour and widen their valleys.

Fiords. **Fiords** are deep, often spectacular, steep-sided inlets of the sea that exist in many high-latitude

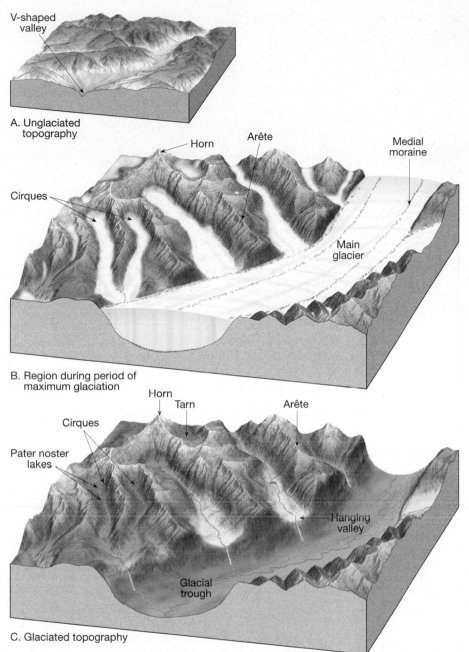

A. Unglaciated topography

B. Region during period of maximum glaciation

C. Glaciated topography

Figure 5.8 Erosional landforms created by alpine glaciers. The unglaciated landscape in part **A** is modified by valley glaciers in part **B**. After the ice recedes, in part **C**, the terrain looks very different than before glaciation.

areas of the world where mountains are adjacent to the ocean (Figure 5.13). Norway, British Columbia, Greenland, New Zealand, Chile, and Alaska all have coastlines characterized by fiords. They are glacial troughs that became submerged as the ice left the valley and sea level rose following the Ice Age.

The depths of fiords can exceed 1000 meters (3300 feet). However, the great depths of these flooded troughs is only partly explained by the post–Ice Age rise in sea level. Unlike the situation governing the downward erosional work of rivers, sea level does not act as base level for glaciers. As a consequence, glaciers are capable of eroding their beds far below the surface of the sea. For example, a valley glacier 300 meters (1000

feet) thick can carve its valley floor more than 250 meters (800 feet) below sea level before downward erosion ceases and the ice begins to float.

Glacial Deposits

 Sculpturing Earth's Surface
  ↳ Glaciers

Glaciers are capable of picking up and transporting a huge load of debris as they slowly advance across the land. Ultimately these materials are deposited when the ice melts. In regions where glacial sediment is deposit-

Figure 5.9 Prior to glaciation, a mountain valley is typically narrow and V-shaped. During glaciation, an alpine glacier widens, deepens, and straightens the valley, creating the U-shaped glacial trough seen here. The string of lakes is called pater noster lakes. This valley is in Glacier National Park, Montana. (Photo by John Montagne)

Figure 5.10 Bridalveil Falls in Yosemite National Park cascades from a hanging valley into the glacial trough below. (Photo by Marc Muench)

Figure 5.11 Glacier flowing from a cirque in Kluane National Park, Yukon Territory, Canada. These bowl-shaped depressions are found at the heads of glacial valleys and represent a focal point of ice formation. (Photo by Martin G. Miller)

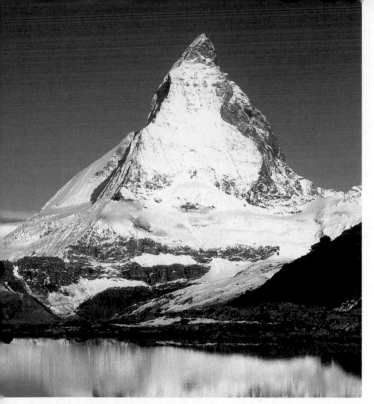

Figure 5.12   Horns are sharp, pyramid-like peaks that are fashioned by alpine glaciers. This example is the famous Matterhorn in the Swiss Alps. (Photo by Gavriel Jecan)

ed, it can play a truly significant role in forming the physical landscape. For example, in many areas once covered by the ice sheets of the recent Ice Age, the bedrock is rarely exposed, because glacial deposits that are dozens or even hundreds of meters thick completely mantle the terrain. The general effect of these deposits is to reduce the local relief and thus level the topography. Indeed, rural country scenes that are familiar to many of us—rocky pastures in New England, wheat

fields in the Dakotas, rolling farmland in the Midwest—result directly from glacial deposition.

Types of Glacial Drift

Long before the theory of an extensive Ice Age was proposed, much of the soil and rock debris covering portions of Europe was recognized as coming from elsewhere. At the time, these foreign materials were believed to have been "drifted" into their present positions by floating ice during an ancient flood. As a consequence, the term *drift* was applied to this sediment. Although rooted in a concept that was not correct, this term was so well established by the time the true glacial origin of the debris became widely recognized that it remained in the glacial vocabulary. Today, **glacial drift** is an all-embracing term for sediments of glacial origin, no matter how, where, or in what form they were deposited.

Glacial drift is divided into two distinct types: (1) materials deposited directly by the glacier, which are known as *till,* and (2) sediments laid down by glacial meltwater, called *stratified drift*. Here is the difference: **Till** is deposited as glacial ice melts and drops its load of rock debris. Unlike moving water and wind, ice cannot sort the sediment it carries; therefore, deposits of till are characteristically unsorted mixtures of many particle sizes (Figure 5.14). **Stratified drift** is sorted according to the size and weight of the fragments. As ice is not capable of such sorting activity, these sediments are not deposited directly by the glacier. Rather, they reflect the sorting action of glacial meltwater.

Some deposits of stratified drift are made by streams issuing directly from the glacier. Other strat-

Figure 5.13   Like other fiords, this one at Tracy Arm, Alaska is a drowned glacial trough. (Photo by Tom & Susan Bean, Inc.)

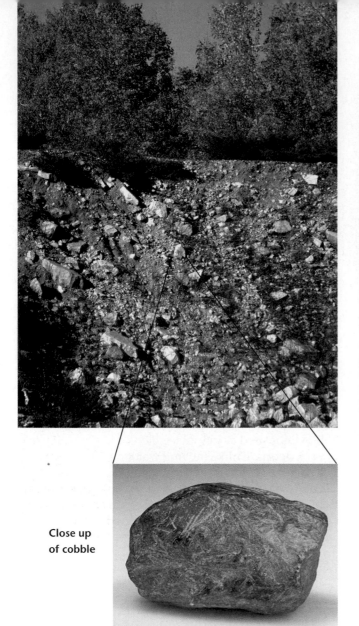

Close up of cobble

Figure 5.14 Glacial till is an unsorted mixture of many different sediment sizes. A close examination often reveals cobbles that have been scratched as they were dragged along by the glacier. (Photos by E. J. Tarbuck)

Figure 5.15 A large, glacial erratic, central Wyoming. (Photo by Yva Momatiuk and John Eastcott/Photo Researchers, Inc.)

ified deposits involve sediment that was originally laid down as till and later picked up, transported, and redeposited by meltwater beyond the margin of the ice. Accumulations of stratified drift often consist largely of sand and gravel, because the meltwater is not capable of moving larger material and because the finer rock flour remains suspended and is commonly carried far from the glacier. An indication that stratified drift consists primarily of sand and gravel can be seen in many areas where these deposits are actively mined as aggregate for road work and other construction projects.

When boulders are found in the till or lying free on the surface, they are called **glacial erratics** if they are different from the bedrock below (Figure 5.15). Of course, this means that they must have been derived from a source outside the area where they are found. Although the locality of origin for most erratics is unknown, the origin of some can be determined. Therefore, by studying glacial erratics as well as the mineral composition of the till, geologists can sometimes trace the path of a lobe of ice. In portions of New England as well as other areas, erratics can be seen dotting pastures and farm fields. In some places, these rocks were cleared from fields and piled to make fences and walls.

Moraines, Outwash Plains, and Kettles

Perhaps the most widespread features created by glacial deposition are *moraines,* which are simply layers or ridges of till. Several types of moraines are identified; some are common only to mountain valleys, and others are associated with areas affected by either ice sheets or valley glaciers. Lateral and medial moraines fall in the first category, whereas end moraines and ground moraines are in the second.

Lateral and Medial Moraines. The sides of a valley glacier accumulate large quantities of debris from the valley walls. When the glacier wastes away, these materials are left as ridges, called **lateral moraines,** along the sides of the valley (Figure 5.16). **Medial moraines** are formed when two valley glaciers coalesce to form a single ice stream. The till that was

Figure 5.16 Lateral moraines form from the accumulation of debris along the sides of a valley glacier. Medial moraines form when the lateral moraines of merging valley glaciers join. Medial moraines could not form if the ice did not advance downvalley. Therefore, these dark stripes are proof that glacial ice moves. Saint Elias National Park, Alaska. (Photo by Tom Bean)

once carried along the edges of each glacier joins to form a single dark stripe of debris within the newly enlarged glacier. Creation of these dark stripes within the ice stream is one obvious proof that glacial ice moves, because the medial moraine could not form if the ice did not flow downvalley (Figure 5.16). It is common to see several medial moraines within a large alpine glacier, because a streak will form whenever a tributary glacier joins the main valley.

End Moraines and Ground Moraines. An **end moraine** is a ridge of till that forms at the terminus of a glacier. These relatively common landforms are deposited when a state of equilibrium is attained between wastage and ice accumulation. That is, the end moraine forms when the ice is melting near the end of the glacier at a rate equal to the forward advance of the glacier from its region of nourishment. Although the terminus of the glacier is stationary, the ice continues to flow forward, delivering a continuous supply of sediment in the same manner a conveyor belt delivers goods to the end of a production line. As the ice melts, the till is dropped and the end moraine grows. The longer the ice front remains stable, the larger the ridge of till will become.

Eventually the time comes when wastage exceeds nourishment. At this point, the front of the glacier begins to recede in the direction from which it originally advanced. However, as the ice front retreats, the conveyor-belt action of the glacier continues to provide fresh supplies of sediment to the terminus. In this manner a large quantity of till is deposited as the ice melts away, creating a rock-strewn, undulating plain. This gently rolling layer of till deposited as the ice front re-

cedes is termed **ground moraine.** Ground moraine has a leveling effect, filling in low spots and clogging old stream channels, often leading to a derangement of the existing drainage system. In areas where this layer of till is still relatively fresh, such as the northern Great Lakes region, poorly drained swampy lands are quite common.

Periodically, a glacier will retreat to a point where wastage and nourishment once again balance. When this happens, the ice front stabilizes and a new end moraine forms.

The pattern of end moraine formation and ground moraine deposition may be repeated many times before the glacier has completely vanished. The very first end moraine to form marks the farthest advance of the glacier and is called the *terminal end moraine.* Those end moraines that form as the ice front occasionally stabilized during retreat are termed *recessional end moraines.* Terminal and recessional moraines are essentially alike; the only difference between them is their relative positions.

End moraines deposited by the most recent major stage of Ice Age glaciation are prominent features in many parts of the Midwest and Northeast. In Wisconsin, the wooded, hilly terrain of the Kettle Moraine near Milwaukee is a particularly picturesque example. A well-known example in the Northeast is Long Island. This linear strip of glacial sediment that extends northeastward from New York City is part of an end moraine complex that stretches from eastern Pennsylvania to Cape Cod, Massachusetts (Figure 5.17).

Figure 5.18 represents a hypothetical area during and following glaciation. It shows the end moraines that were just described as well as the depositional features that are discussed in the sections that follow. This fig-

Figure 5.17 End moraines make up substantial parts of Long Island, Cape Cod, Martha's Vineyard, and Nantucket. Although portions are submerged, the Ronkonkoma moraine (a terminal end moraine) extends through central Long Island, Martha's Vineyard, and Nantucket. It was deposited about 20,000 years ago. The recessional Harbor Hill moraine, which formed about 14,000 years ago, extends along the north shore of Long Island, through southern Rhode Island and Cape Cod.

ure depicts landscape features similar to what might be encountered if you were traveling in the upper Midwest or New England. As you read about other glacial deposits, you will be referred to this figure again.

Outwash Plains and Valley Trains. At the same time that an end moraine is forming, meltwater emerges from the ice in rapidly moving streams. Often they are choked with suspended material and carry a substantial bed load. As the water leaves the glacier, it rapidly loses velocity and much of its bed load is dropped. In this way a broad, ramplike accumulation of stratified drift is built adjacent to the downstream edge of most end moraines. When the feature is formed in association with an ice sheet, it is termed an **outwash plain,** and when it is confined to a mountain valley, it is usually referred to as a **valley train** (Figure 5.18).

Kettles. Often end moraines, outwash plains, and valley trains are pockmarked with basins or depressions known as **kettles** (Figure 5.18). Kettles form when blocks of stagnant ice become buried in drift and eventually melt, leaving pits in the glacial sediment. Most kettles do not exceed 2 kilometers in diameter, and the typical depth of most kettles is less than 10 meters (33 feet). Water often fills the depression and forms a pond or lake. One well-known example is Walden Pond near Concord, Massachusetts. It is here that Henry David Thoreau lived alone for two years in the 1840s and about which he wrote *Walden,* his classic of American literature.

Drumlins, Eskers, and Kames

Moraines are not the only landforms deposited by glaciers. Some landscapes are characterized by numerous elongate parallel hills made of till. Other areas exhibit conical hills and relatively narrow winding ridges composed mainly of stratified drift.

Drumlins. **Drumlins** are streamlined asymmetrical hills composed of till (Figure 5.18). They range in height from 15 to 60 meters (50 to 200 feet) and average 0.4 to 0.8 kilometer (0.25–0.50 mile) in length. The steep side of the hill faces the direction from

which the ice advanced, whereas the gentler slope points in the direction the ice moved. Drumlins are not found singly but rather occur in clusters, called *drumlin fields.* One such cluster, east of Rochester, New York, is estimated to contain about 10,000 drumlins. Their streamlined shape indicates that they were molded in the zone of flow within an active glacier. It is thought that drumlins originate when glaciers advance over previously deposited drift and reshape the material.

Eskers and Kames. In some areas that were once occupied by glaciers, sinuous ridges composed largely of sand and gravel might be found. These ridges, called **eskers,** are deposits made by streams flowing in tunnels beneath the ice, near the terminus of a glacier (Figure 5.18). They may be several meters high and extend for many kilometers. In some areas they are mined for sand and gravel, and for this reason, eskers are disappearing in some localities.

Kames are steep-sided hills that, like eskers, are composed of sand and gravel (Figure 5.18). Kames originate when glacial meltwater washes sediment into openings and depressions in the stagnant wasting terminus of a glacier. When the ice eventually melts away, the stratified drift is left behind as mounds or hills.

Glaciers of the Ice Age

At various points in the preceding pages, we mentioned the Ice Age, a time when ice sheets and alpine glaciers were far more extensive than they are today. There was a time when the most popular explanation for what we now know to be glacial deposits was that the material had been drifted in by means of icebergs or perhaps simply swept across the landscape by a catastrophic flood. However, during the nineteenth century, field investigations by many scientists provided convincing evidence that an extensive Ice Age was responsible for these deposits and for many other features.

By the beginning of the twentieth century, geologists had largely determined the extent of Ice Age glaciation. Further, they discovered that many glaciated regions had not one layer of drift but several. Close examination of these older deposits showed well-developed zones of chemical weathering and soil formation as well as the remains of plants that require warm temperatures. The evidence was clear: There had not been just one glacial advance but many, each separated by extended periods when climates were as warm or warmer than at present. The Ice Age had not simply been a time when the ice advanced over the land, lingered for a while, and then receded. Rather, the period was a very complex event characterized by a number of advances and withdrawals of glacial ice.

The glacial record on land is punctuated by many erosional gaps. This makes it difficult to reconstruct the episodes of the Ice Age. But sediment on the ocean floor provides an uninterrupted record of climate cycles for this period. Studies of cores drilled from these seafloor sediments show that glacial/interglacial cycles have occurred about every 100,000 years. Approximately 20 such cycles of cooling and warming were identified for the span we call the Ice Age.

During the glacial age, ice left its imprint on almost 30 percent of Earth's land area, including about 10 million square kilometers of North America, 5 million square kilometers of Europe, and 4 million square kilometers of Siberia (Figure 5.19). The amount of glacial ice in the Northern Hemisphere was roughly twice that of the Southern Hemisphere. The primary reason is that the Southern Hemisphere has little land in the middle latitudes, and therefore the southern polar ice could not spread far beyond the margins of Antarctica. By contrast, North America and Eurasia provided great expanses of land for the spread of ice sheets.

Today we know that the Ice Age began between 2 million and 3 million years ago. This means that most of the major glacial episodes occurred during a division of the geologic time scale called the **Pleistocene epoch.** Although the Pleistocene is commonly used as a synonym for the Ice Age, this epoch does not encompass it all. The Antarctic Ice Sheet, for example, formed at least 14 million years ago and, in fact, might be much older.

Some Indirect Effects of Ice Age Glaciers

In addition to the massive erosional and depositional work carried on by Pleistocene glaciers, the ice sheets had other, sometimes profound, effects on the landscape. For example, as the ice advanced and retreated, animals and plants were forced to migrate. This led to stresses that some organisms could not tolerate. Furthermore, many present-day stream courses bear little resemblance to their preglacial routes. The Missouri River once flowed northward toward Hudson Bay in Canada. The Mississippi River followed a path through central Illinois, and the head of the Ohio River reached

only as far as Indiana (Figure 5.20). Other rivers that today carry only a trickle of water but nevertheless occupy broad channels are a testimony to the fact that they once carried torrents of glacial meltwater.

In areas that were centers of ice accumulation, such as Scandinavia and northern Canada, the land has been slowly rising for the past several thousand years. The land had downwarped under the tremendous weight of 3-kilometer-thick masses of ice. Following the removal of this immense load, the crust has been adjusting by gradually rebounding upward ever since.

A far-reaching effect of the Ice Age was the worldwide change in sea level that accompanied each advance and retreat of the ice sheets. The snow that nourishes glaciers ultimately comes from moisture evaporated from the oceans. Therefore, when the ice sheets increased in size, sea level fell and the shoreline shifted seaward. Estimates suggest that sea level was as much as 100 meters (330 feet) lower than today. Thus, land that is presently flooded by the oceans was dry. The Atlantic Coast of the United States lay more than 100 kilometers (60 miles) to the east of New York City (see Figure 5.3). Moreover, France and Britain were joined where the English Channel is today; Alaska and Siberia were connected across the Bering Strait; and Southeast Asia was tied by dry land to the islands of Indonesia.

The formation and growth of ice sheets was an obvious response to significant changes in climate. But the existence of the glaciers themselves triggered climatic changes in the regions beyond their margins. In arid and semiarid areas on all continents, temperatures were lower, which meant evaporation rates were also lower. At the same time, precipitation was moderate. This cooler, wetter climate resulted in the formation of many lakes called **pluvial lakes** (from the Latin term *pluvia*, meaning rain). In North America, pluvial lakes were concentrated in the vast Basin and Range region of Nevada and Utah (Figure 5.21). Although most are now gone, a few remnants remain, the largest being Utah's Great Salt Lake.

Causes of Glaciation

A great deal is known about glaciers and glaciation. Much has been learned about glacier formation and movement, the extent of glaciers past and present, and the features created by glaciers, both erosional and depositional. However, a widely accepted theory for the causes of glacial ages has not yet been established.

Although widespread glaciation has been rare in Earth's history, the Pleistocene Ice Age is not the only glacial period for which a record exists. Two Precambrian glacial episodes have been identified in the geologic record, the first approximately 2 billion years ago and the second about 600 million years ago. Further, a well-documented record of an earlier glacial age is

Labels: Esker, Retreating glacier, Kame, Ground moraine, Bedrock

Figure 5.18 This hypothetical area illustrates many common depositional landforms. The outermost end moraine marks the limit of glacial advance and is called the *terminal end moraine*. End moraines that form as the ice front occasionally becomes stationary during retreat are called *recessional end moraines*. (Drumlin photo courtesy by Ward's Natural Science Establishment; Kame, Esker, and Kettle photos by Richard P. Jacobs/JLM Visuals)

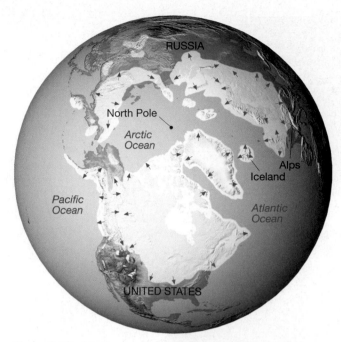

Figure 5.19 Maximum extent of ice sheets in the Northern Hemisphere during the Ice Age.

A.

B.

Figure 5.20 **A.** This map shows the Great Lakes and the familiar present-day pattern of rivers in the central United States. Pleistocene ice sheets played a major role in creating this pattern. **B.** Reconstruction of drainage systems in the central United States prior to the Ice Age. The pattern was very different from today, and there were no Great Lakes.

found in late Paleozoic rocks that are about 250 million years old and which exist on several landmasses.*

Any theory that attempts to explain the causes of glacial ages must successfully answer two basic questions: (1) *What causes the onset of glacial conditions?* For continental ice sheets to have formed, average temperatures must have been somewhat lower than at present and perhaps substantially lower than throughout much of geologic time. Thus, a successful theory would have to account for the cooling that finally leads to glacial conditions. (2) *What caused the alternation of glacial and interglacial stages that have been documented for the Pleistocene epoch?* The first question deals with long-term trends in temperature on a scale of millions of years, but this second question relates to much shorter-term changes.

Although the literature of science contains many hypotheses relating to the possible causes of glacial periods, we will discuss only a few major ideas to summarize current thought.

Plate Tectonics

Probably the most attractive proposal for explaining the fact that extensive glaciations have occurred only a few times in the geologic past comes from the theory of plate tectonics.[†] Because glaciers can form only on land, we know that landmasses must exist somewhere in the

higher latitudes before an ice age can commence. Many scientists suggest that ice ages have occurred only when Earth's shifting crustal plates have carried the continents from tropical latitudes to more poleward positions.

Glacial features in present-day Africa, Australia, South America, and India indicate that these regions, which are now tropical or subtropical, experienced an Ice Age near the end of the Paleozoic era, about 250 million years ago. However, there is no evidence that ice sheets existed during this same period in what are today the higher latitudes of North America and Eurasia. For many years this puzzled scientists. Was the climate in these relatively tropical latitudes once like it is today in Greenland and Antarctica? Why did glaciers not form in North America and Eurasia? Until the plate tectonics theory was formulated, there had been no reasonable explanation.

---

*The terms *Precambrian* and *Paleozoic* refer to time spans on the geologic time scale of Earth history. For more on the geologic time scale, see Chapter 10 and Figure 10.17.

[†]A complete discussion of plate tectonics is presented in Chapter 7.

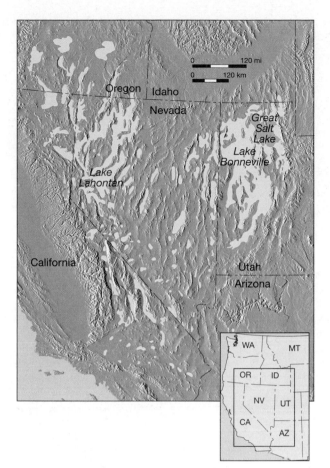

Figure 5.21   Pluvial lakes of the Western United States. By far the largest pluvial lake in the vast Basin and Range region of Nevada and Utah was Lake Bonneville. With maximum depths exceeding 300 meters and an area of 50,000 square kilometers, Lake Bonneville was nearly the same size as present-day Lake Michigan. The Great Salt Lake is a remnant of this huge pluvial lake. (After R. F. Flint, *Glacial and Quaternary Geology*, New York: John Wiley & Sons.)

Today scientists realize that the areas containing these ancient glacial features were joined together as a single supercontinent (Pangaea) located at latitudes far to the south of their present positions. Later this landmass broke apart, and its pieces, each moving on a different plate, migrated toward their present locations (Figure 5.22). It is now understood that during the geologic past, plate movements accounted for many dramatic climatic changes as landmasses shifted in relation to one another and moved to different latitudinal positions. Changes in oceanic circulation also must have occurred, altering the transport of heat and moisture and consequently the climate as well. Because the rate of plate movement is very slow—a few centimeters per year—appreciable changes in the positions of the continents occur only over great spans of geologic time. Thus, climatic changes brought about by shifting plates are extremely gradual and occur on a scale of millions of years.

Variations in Earth's Orbit

Because climatic changes brought about by moving plates are extremely gradual, the plate tectonics theory cannot be used to explain the alternation between glacial and interglacial climates that occurred during the Pleistocene epoch. Therefore, we must look to some other triggering mechanism that may cause climatic change on a scale of thousands rather than millions of years. Today many scientists strongly suspect that the climatic oscillations that characterized the Pleistocene may be linked to variations in Earth's orbit. This hypothesis was first developed and strongly advocated by the Yugoslavian scientist Milutin Milankovitch and is based on the premise that variations in incoming solar radiation are a principal factor in controlling Earth's climate.

Milankovitch formulated a comprehensive mathematical model based on the following elements (Figure 5.23):

1. Variations in the shape (*eccentricity*) of Earth's orbit about the Sun;
2. Changes in *obliquity*—that is, changes in the angle that the axis makes with the plane of Earth's orbit; and
3. The wobbling of Earth's axis, called *precession*.

Using these factors, Milankovitch calculated variations in the receipt of solar energy and the corresponding surface temperature of Earth back into time in an attempt to correlate these changes with the climatic fluctuations of the Pleistocene. In explaining climatic changes that result from these three variables, note that they cause little or no variation in the total solar energy reaching the ground. Instead, their impact is felt because they change the degree of contrast between the seasons. Somewhat milder winters in the middle to high latitudes means greater snowfall totals, whereas cooler summers would bring a reduction in snowmelt.

Among the studies that have added credibility to this astronomical hypothesis is one in which deep-sea sediments containing certain climatically sensitive microorganisms were analyzed to establish a chronology of temperature changes going back nearly 500,000 years.[*] This time scale of climatic change was then compared to astronomical calculations of eccentricity, obliquity, and precession to determine if a correlation did indeed exist. Although the study was very involved and mathematically complex, the conclusions were straightforward. The researchers found that major variations in climate over the past several hundred thou-

---

[*]J. D. Hays, John Imbrie, and N. J. Shackelton, Variations in the Earth's Orbit: Pacemaker of the Ice Ages, *Science* 194 (1976): 1121–32.

Figure 5.22 **A.** The supercontinent Pangaea showing the area covered by glacial ice 300 million years ago. **B.** The continents as they are today. The white areas indicate where evidence of the old ice sheets exists.

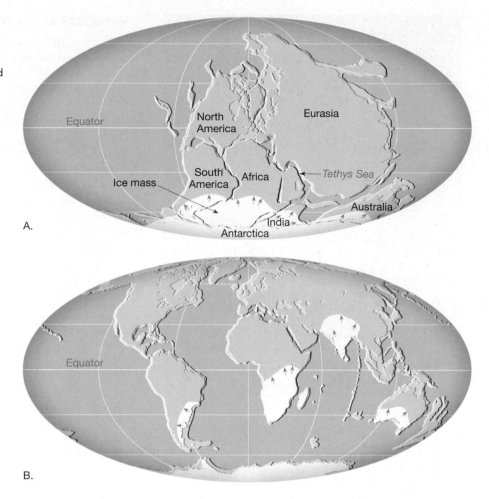

A.

B.

sand years were closely associated with changes in the geometry of Earth's orbit; that is, cycles of climatic change were shown to correspond closely with the periods of obliquity, precession, and orbital eccentricity. More specifically, the researchers stated: "It is concluded that changes in the earth's orbital geometry are the fundamental cause of the succession of Quaternary ice ages."*

Let us briefly summarize the ideas that were just described. The theory of plate tectonics provides an explanation for the widely spaced and nonperiodic onset of glacial conditions at various times in the geologic past, whereas the astronomical model proposed by Milankovitch and supported by the work of J. D. Hays and his colleagues furnishes an explanation for the alternating glacial and interglacial episodes of the Pleistocene.

In conclusion, we emphasize that the ideas just discussed do not represent the only possible explanations for glacial ages. Although interesting and attractive, these proposals are certainly not without critics, nor are they the only possibilities currently under study. Other factors may be, and probably are, involved.

---

*J. D. Hays et al., p. 1131. The term *Quaternary* refers to the period on the geologic time scale that encompasses the last 1.6 million years.

Deserts

 GEODe Sculpturing Earth's Surface
↳ Deserts

The dry regions of the world encompass about 42 million square kilometers—a surprising 30 percent of Earth's land surface. No other climatic group covers so large a land area.† The word *desert* literally means deserted or unoccupied. For many dry regions this is a very appropriate description. Yet where water is available in deserts, plants and animals thrive. Nevertheless, the world's dry regions are among the least familiar land areas on Earth outside of the polar realm.

Desert landscapes frequently appear stark. Their profiles are not softened by a carpet of soil and abundant plant life. Instead, barren rocky outcrops with steep, angular slopes are common. At some places the rocks are tinted orange and red. At others they are gray and brown and streaked with black. For many visitors desert scenery exhibits a striking beauty; to others the terrain seems bleak. No matter which feeling is elicited, it is clear that deserts are very different from the more humid places where most people live.

---

†An examination of dry climates is found in Chapter 18.

C.

Figure 5.23 Orbital variations. **A.** The shape of Earth's orbit changes during a cycle that spans about 100,000 years. It gradually changes from nearly circular to one that is more elliptical and then back again. This diagram greatly exaggerates the amount of change. **B.** Today the axis of rotation is tilted about 23.5° to the plane of Earth's orbit. During a cycle of 41,000 years, this angle varies from 22° to 24.5°. **C.** Precession. Earth's axis wobbles like that of a spinning top. Consequently, the axis points to different spots in the sky during a cycle of about 26,000 years.

As you will see, arid regions are not dominated by a single geologic process. Rather, the effects of tectonic forces, running water, and wind are all apparent. Because these processes combine in different ways from place to place, the appearance of desert landscapes varies a great deal as well (Figure 5.24).

STUDENTS SOMETIMES ASK...

I heard somewhere that deserts are expanding. Is that actually occurring?

Yes. The problem is called *desertification,* and it refers to the alteration of land to desertlike conditions as the result of human activities. It commonly takes place on the margins of deserts and results primarily from inappropriate land use. It is triggered when the modest natural vegetation in marginal areas is removed by plowing or grazing. When drought occurs, as it inevitably does in these regions, and the vegetative cover has been destroyed beyond the minimum to hold the soil against erosion, the destruction becomes irreversible. Desertification is occurring in many places but is particularly serious in the region south of the Sahara Desert known as the Sahel.

Geologic Processes in Arid Climates

GEODe   Sculpturing Earth's Surface
↳ Deserts

The angular rock exposures, the sheer canyon walls, and the pebble- or sand-covered surface of the desert contrast sharply with the rounded hills and curving slopes of more humid places. To a visitor from a humid region, a desert landscape may seem to have been shaped by forces different from those operating in wetter areas. However, although the contrasts might be striking, they do not reflect different processes. They merely disclose the differing effects of the same processes that operate under contrasting climatic conditions.

Weathering

In humid regions, relatively well-developed soils support an almost continuous cover of vegetation. Here the slopes and rock edges are rounded. Such a landscape reflects the strong influence of chemical weathering in a humid climate. By contrast, much of the weathered debris in deserts consists of unaltered rock and mineral fragments—the result of mechanical weathering processes. In dry lands rock weathering of any type is greatly reduced because of the lack of moisture and the scarcity of organic acids from decaying plants. Chemi-

Figure 5.24 A scene in California's Death Valley. The appearance of desert landscapes varies a great deal from place to place. (Photo by David Muench.)

cal weathering, however, is not completely lacking in deserts. Over long spans of time, clays and thin soils do form, and many iron-bearing silicate minerals oxidize, producing the rust-colored stain found tinting some desert landscapes.

The Role of Water

Permanent streams are normal in humid regions, but practically all desert streams are dry most of the time (Figure 5.25A). Desert streams are said to be **ephemeral,** which means that they carry water only in response to specific episodes of rainfall. A typical ephemeral stream might flow only a few days or perhaps just a few hours during the year. In some years the channel may carry no water at all.

This fact is obvious even to the casual observer who, while traveling in a dry region, notices the number of bridges with no streams beneath them or the number of dips in the road where dry channels cross. However, when the rare heavy showers do occur, so much rain falls in such a short time that all of it cannot soak in. Because the vegetative cover is sparse, runoff is largely unhindered and consequently rapid, often creating flash floods along valley floors (Figure 5.25B). Such floods, however, are quite unlike floods in humid regions. A flood on a river such as the Mississippi may take many days to reach its crest and then subside. But desert floods arrive suddenly and subside quickly. Because much of the surface material is not anchored by vege-

tation, the amount of erosional work that occurs during a single short-lived rain event is impressive.

In the dry western United States a number of different names are used for ephemeral streams. Two of the most common are *wash* and *arroyo*. In other parts of the world, a dry desert stream may be called a *wadi* (Arabia and North Africa), a *donga* (South America), or a *nullah* (India).

Humid regions are notable for their integrated drainage systems. But in arid regions streams usually lack an extensive system of tributaries. In fact, a basic characteristic of desert streams are that they are small and die out before reaching the sea. Because the water table is usually far below the surface, few desert streams can draw upon it as streams do in humid regions. Without a steady supply of water, the combination of evaporation and infiltration soon depletes the stream.

The few permanent streams that do cross arid regions, such as the Colorado and Nile rivers, originate *outside* the desert, often in well-watered mountains. Here the water supply must be great to compensate for the losses occurring as the stream crosses the desert (see Box 5.2). For example, after the Nile leaves the lakes and mountains of central Africa that are its source, it traverses almost 3000 kilometers (nearly 1900 miles) of the Sahara *without a single tributary*. By contrast, in humid regions the discharge of a river increases as it flows downstream because tributaries and groundwater contribute additional water along the way.

A.

B.

Figure 5.25   **A.** Most of the time, desert stream channels are dry. **B.** An ephemeral stream shortly after a heavy shower. Although such floods are short-lived, large amounts of erosion occur. (Photos by E. J. Tarbuck)

It should be emphasized that *running water, although infrequent, nevertheless does most of the erosional work in deserts.* This is contrary to a common belief that wind is the most important erosional agent sculpturing desert landscapes. Although wind erosion is indeed more significant in dry areas than elsewhere, most desert landforms are nevertheless carved by running water. As you will see shortly, the main role of wind is in the transportation and deposition of sediment, which creates and shapes the ridges and mounds we call dunes.

Basin and Range: The Evolution of a Desert Landscape

Because arid regions typically lack permanent streams, they are characterized as having **interior drainage.** This means that they have a discontinuous pattern of intermittent streams that do not flow out of the desert to the ocean. In the United States, the dry Basin and Range region provides an excellent example. The region includes southern Oregon, all of Nevada, western Utah, southeastern California, southern Arizona, and southern New Mexico. The name Basin and Range is an apt description for this almost 800,000-square-kilometer (more than 300,000-square-mile) region, because it is characterized by more than 200 relatively small mountain ranges that rise 900 to 1500 meters (3000 to 5000 feet) above the basins that separate them.

In this region, as in others like it around the world, most erosion occurs without reference to the ocean (ul-

timate base level), because the interior drainage never reaches the sea. Even where permanent streams flow to the ocean, few tributaries exist, and thus only a narrow strip of land adjacent to the stream has sea level as its ultimate level of land reduction.

The block models in Figure 5.26 depict how the landscape has evolved in the Basin and Range region. During and following uplift of the mountains, running water begins carving the elevated mass and depositing large quantities of debris in the basin. In this early stage, relief is greatest, and as erosion lowers the mountains and sediment fills the basins, elevation differences diminish.

When the occasional torrents of water produced by sporadic rains move down the mountain canyons, they are heavily loaded with sediment. Emerging from the confines of the canyon, the runoff spreads over the gentler slopes at the base of the mountains and quickly loses velocity. Consequently, most of its load is dumped within a short distance. The result is a cone of debris known as an **alluvial fan** at the mouth of a canyon (Figure 5.27). Over the years, a fan enlarges, eventually coalescing with fans from adjacent canyons to produce an apron of sediment (*bajada*) along the mountain front.

On the rare occasions of abundant rainfall, or snowmelt in the mountains, streams may flow across the alluvial fans to the center of the basin, converting the basin floor into a shallow **playa lake.** Playa lakes last only a few days or weeks, before evaporation and infiltration remove the water. The dry, flat lake bed that remains is termed a *playa.*

BOX 5.2

PEOPLE AND THE ENVIRONMENT:
The Shrinking Aral Sea

The Aral Sea lies on the border between Uzbekistan and Kazakhstan in central Asia (Figure 5.B). The setting is the Turkestan desert, a middle-latitude desert in the rainshadow of Afghanistan's high mountains. In this region of interior drainage, two large rivers, the Amu Darya and the Syr Darya, carry water from the mountains of northern Afghanistan across the desert to the Aral Sea. Water leaves the sea by evaporation. Thus, the size of the water body depends upon the balance between river inflow and evaporation.

In 1960 the Aral Sea was one of the world's largest inland water bodies, with an area of about 67,000 square kilometers (26,000 square miles). Only the Caspian Sea, Lake Superior, and Lake Victoria were larger. By the year 2000 the area of the Aral Sea was less than 50 percent of its 1960 size, and its volume was reduced by 80 percent. The shrinking of this water body is depicted in Figure 5.C. By about 2010 all that will remain will be three shallow remnants.

What caused the Aral Sea to dry up over the past 40 years? The answer is that the flow of water from the mountains that supplied the sea was significantly reduced and then all but eliminated. The waters of the Amu Darya and Syr Darya were diverted to

Figure 5.C The shrinking Aral Sea. By the year 2010 all that will remain are three small remnants.

Figure 5.B The Aral Sea lies east of the Caspian Sea in the Turkestan Desert. Two rivers, the Amu Darya and Syr Darya, bring water from the mountains to the south.

supply a major expansion of irrigated agriculture in this dry realm.

The intensive irrigation greatly increased agricultural productivity, but not without a significant cost. Some of the effects are listed below.

• The once thriving fishing industry is dead, and the 24 species of fish that once lived in the Aral Sea are no longer there. The shoreline is now tens of kilometers from the towns that were once fishing centers (Figure 5.D).

• The shrinking sea has exposed millions of acres of former seabed to Sun and wind. The former seabed is the source of an increasing number of dust storms that blow salts and toxic herbicides and pesticides into nearby towns with detrimental health effects.

Figure 5.D In the town of Jamboul, Kazakhstan, boats now lie in the sand because the Aral Sea has dried up. (Photo by Ergun Cagatay/Liaison Agency, Inc.)

• The deltas of the two major rivers have lost their wetlands, and wildlife has disappeared.

• The region has lost a moderating climatic influence. Without the large water body, there are greater extremes of temperature and reduced local precipitation.

If enough fresh water were to once again flow into the Aral Sea, could the situation be reversed? Prospects appear grim. Experts estimate that restoring the Aral Sea to about twice its present size would require stopping all irrigation from the two major rivers for 50 years. This could not be done without ruining the economies of the countries that rely on that water.*

The decline of the Aral Sea is a major environmental disaster that sadly is of human making.

*For more on this, see "Coming to Grips with the Aral Sea's Grim Legacy," in *Science*, vol. 284, 2 April 1999 pp. 30–31.

Figure 5.26 Stages of landscape evolution in a mountainous desert such as the Basin and Range region of the West. As erosion of the mountains and deposition in the basins continue, relief diminishes. **A.** Early stage. **B.** Middle stage. **C.** Late stage.

Playas occasionally become encrusted with salts left behind by evaporation. These precipitated salts may be unusual. A case in point is the sodium borate (better known as borax) mined from ancient playa lake deposits in Death Valley, California.

With the ongoing erosion of the mountain mass and the accompanying sedimentation, the local relief continues to diminish. Eventually nearly the entire mountain mass is gone. Thus, by the late stages of erosion, the mountain areas are reduced to a few large bedrock knobs (called *inselbergs*) projecting above the sediment-filled basin.

Each of the stages of landscape evolution in an arid climate depicted in Figure 5.26 can be observed in the Basin and Range region. Recently uplifted mountains in an early stage of erosion are found in southern Oregon and northern Nevada. Death Valley, California, and southern Nevada fit into the more advanced middle stage, whereas the late stage, with its inselbergs, can be seen in southern Arizona.

Wind Erosion

 Sculpturing Earth's Surface
↳ Deserts

Compared to running water and glaciers, wind is a relatively insignificant erosional agent. Recall that even in deserts, most erosion is performed by intermittent running water, not by the wind. Wind erosion is more effective in arid lands than in humid areas because in humid places moisture binds particles together and vegetation anchors the soil. For wind to be an effective erosional force, dryness and scanty vegetation are essential. When such circumstances exist, wind may pick up, transport, and deposit great quantities of fine sediment. During the 1930s, parts of the Great Plains experienced vast dust storms. The plowing under of the natural vegetative cover for farming, followed by severe drought, exposed the land to wind

Figure 5.27 Alluvial fans develop where the gradient of a stream changes abruptly from steep to flat. Such a situation exists in Death Valley, California, where streams emerge from the mountains into a flat basin. As a result, Death Valley has many large alluvial fans. (Photo by Michael Collier)

erosion and led to the area being labeled the Dust Bowl (see Box 5.3).

One way that wind erodes is by **deflation** (*de* = out, *flat* = blow), the lifting and removal of loose material. Because the competence (ability to transport different-size particles) of moving air is low, it can suspend only fine sediment, such as clay and silt. Larger grains of sand are rolled or skipped along the surface (a process called *saltation*) and comprise the bed load. Particles larger than sand are usually not transported by wind. Deflation sometimes is difficult to notice because the entire surface is being lowered at the same time, but it can be significant. In portions of the 1930s Dust Bowl, the land was lowered by a meter or more in only a few years (Figure 5.28).

The most noticeable results of deflation in some places are shallow depressions called **blowouts**. In the Great Plains region, from Texas north to Montana, thousands of blowouts can be seen. They range from small dimples less than 1 meter deep and 3 meters wide to depressions that are over 45 meters deep and several kilometers across.

In portions of many deserts, the surface is characterized by a layer of coarse pebbles and cobbles that are too large to be moved by the wind. This stony veneer, called **desert pavement,** is created as deflation lowers the surface by removing sand and silt until eventually only a continuous cover of coarse particles remains (Figure 5.29). Once desert pavement becomes established, a process that might take hundreds of years, the surface is effectively protected from further deflation if left

undisturbed. However, as the layer is only one or two stones thick, the passage of vehicles or animals can dislodge the pavement and expose the fine-grained material below. If this happens, the surface is once again subject to deflation.

Like glaciers and streams, wind erodes in part by *abrasion.* In dry regions as well as along some beaches, windblown sand will cut and polish exposed rock surfaces. Abrasion is often credited for accomplishments beyond its actual capabilities. Such features as balanced rocks that stand high atop narrow pedestals and intricate detailing on tall pinnacles are not the results of abrasion. Sand seldom travels more than a meter above the surface, so the wind's sandblasting effect is obvi-

Figure 5.28 This photo was taken north of Granville, North Dakota, in July 1936, during a prolonged drought. Strong winds removed the soil that was not anchored by vegetation. The mounds are 1.2 meters (4 feet) high and show the level of the land prior to deflation. (Photo courtesy of the State Historical Society of North Dakota, 0278-01)

PEOPLE AND THE ENVIRONMENT:
Dust Bowl—Soil Erosion in the Great Plains

During a span of dry years in the 1930s, large dust storms plagued the Great Plains. Because of the size and severity of these storms, the region came to be called the Dust Bowl, and the time period the Dirty Thirties. The heart of the Dust Bowl consisted of nearly 100 million acres in the panhandles of Texas and Oklahoma, as well as adjacent parts of Colorado, New Mexico, and Kansas. To a lesser extent, dust storms were also a problem over much of the Great Plains, from North Dakota to west-central Texas.

At times dust storms were so severe that they were called "black blizzards" and "black rollers" because visibility was reduced to only a few feet. Numerous storms lasted for hours and stripped huge volumes of topsoil from the land (Figure 5.E).

In the spring of 1934, a windstorm that lasted for a day and a half created a dust cloud that extended for 2000 kilometers (1200 miles). As the sediment moved east, "muddy rains" were experienced in New York, and "black snows" in Vermont. Less than a year later, another storm carried dust more than 3 kilometers (2 miles) into the atmosphere and transported it 3000 kilometers from its source in Colorado to create twilight conditions in the middle of the day in parts of New England and New York.

Figure 5.E   Dust blackens the sky on May 21, 1937, near Elkhart, Kansas. It was because of storms like this that portions of the Great Plains were called the Dust Bowl in the 1930s. (Photo reproduced from the collection of the Library of Congress)

What caused the Dust Bowl? Clearly, the fact that portions of the Great Plains experience some of North America's strongest winds is important. However, it was the expansion of agriculture that set the stage for the disastrous period of soil erosion. Mechanization allowed the rapid transformation of the grass-covered prairies of this semiarid region into farms. Between the 1870s and 1930, the area of cultivation in the region expanded nearly tenfold, from about 10 million acres to more than 100 million acres. As long as precipitation was adequate, the soil remained in place. How-ever, when a prolonged drought struck in the 1930s, the unprotected fields were vulnerable to the wind. The results were severe soil loss, crop failures, and economic hardship.

Beginning in 1939, a return to rainier conditions brought relief. Moreover, farming practices that were designed to reduce soil loss by wind had also been instituted. Although dust storms are less numerous and not as severe as in the Dirty Thirties, soil erosion by strong winds still occurs periodically whenever the combination of drought and unprotected soil exists.

Figure 5.29   A. Formation of desert pavement. As these cross-sections illustrate, coarse particles gradually become concentrated into a tightly packed layer as deflation lowers the surface by removing sand and silt. B. If left undisturbed, desert pavement such as this in Arizona's Sonoran Desert will protect the surface from further deflation. (Photo by David Muench)

Figure 5.30 This vertical loess bluff near the Mississippi River in southern Illinois is about 3 meters high. (Photo by James E. Patterson)

ously limited in vertical extent. However, in areas prone to such activity, telephone poles have actually been cut through near their bases. For this reason, collars are often fitted on the poles to protect them from being sawed down.

Wind Deposits

Sculpturing Earth's Surface
↳ Deserts

Although wind is relatively unimportant in producing *erosional* landforms, significant *depositional* landforms are created by the wind in some regions. Accumulations of windblown sediment are particularly conspicuous in the world's dry lands and along many sandy coasts. Wind deposits are of two distinctive types: (1) extensive blankets of silt, called *loess,* which once were carried in suspension, and (2) mounds and ridges of sand from the wind's bed load, which we call *dunes.*

Loess

In some parts of the world the surface topography is mantled with deposits of windblown silt termed **loess.** Over thousands of years dust storms deposited this material. When loess is breached by streams or road cuts, it tends to maintain vertical cliffs and lacks any visible layers, as you can see in Figure 5.30.

The distribution of loess worldwide indicates that there are two primary sources for this sediment: deserts and glacial deposits of stratified drift. The thickest and most extensive deposits of loess on Earth occur in western and northern China. They were blown there from the extensive desert basins of central Asia. Accumulations of 30 meters are not uncommon, and thicknesses of more than 100 meters have been measured. It is this fine, buff-colored sediment that gives the Yellow River (Hwang Ho) its name.

In the United States, deposits of loess are significant in many areas, including South Dakota, Nebraska, Iowa, Missouri, and Illinois, as well as portions of the Columbia Plateau in the Pacific Northwest. Unlike the deposits in China, the loess in the United States, as well as in Europe, is an indirect product of glaciation. Its source is deposits of stratified drift. During the retreat of the ice sheets, many river valleys were choked with sediment deposited by meltwater. Strong westerly winds sweeping across the barren floodplains picked up the finer sediment and dropped it as a blanket on the eastern sides of the valleys.

Sand Dunes

Like running water, wind releases its load of sediment when its velocity falls and the energy available for transport diminishes. Thus, sand begins to accumulate wherever an obstruction across the path of the wind

Figure 5.31 Sand sliding down the steep slip face of a dune in White Sands National Monument, New Mexico. (Photo by Michael Collier)

slows its movement. Unlike deposits of loess, which form blanketlike layers over broad areas, winds commonly deposit sand in mounds or ridges called *dunes* (Figure 5.31).

As moving air encounters an object, such as a clump of vegetation or a rock, the wind sweeps around and over it, leaving a shadow of more slowly moving air behind the obstacle as well as a smaller zone of quieter air just in front of the obstacle. Some of the saltating sand grains moving with the wind come to rest in these wind shadows. As the accumulation of sand continues, it forms an increasingly efficient wind barrier to trap even more sand. If there is a sufficient supply of sand and the wind blows steadily long enough, the mound of sand grows into a dune.

Many dunes have an asymmetrical profile, with the leeward (sheltered) slope being steep and the windward slope more gently inclined (Figure 5.30). Sand moves up the gentler slope on the windward side by saltation. Just beyond the crest of the dune, where wind velocity is reduced, the sand accumulates. As more sand collects, the slope steepens, and eventually some of it slides or slumps under the pull of gravity (Figure 5.31). In this way, the leeward slope of the dune, called the **slip face**, maintains an angle of about 34 degrees. Continued sand accumulation, coupled with periodic slides down the slip face, results in the slow migration of the dune in the direction of air movement.

As sand is deposited on the slip face, it forms layers inclined in the direction the wind is blowing. These sloping layers are called **cross beds** (Figure 5.32). When the dunes are eventually buried under other layers of sediment and become part of the sedimentary rock record, their asymmetrical shape is destroyed, but the cross beds remain as a testimony to their origin. Nowhere is cross-bedding more prominent than in the sandstone walls of Zion Canyon in Utah (Figure 5.32D).

? STUDENTS SOMETIMES ASK...

Aren't deserts mostly covered with sand dunes?

A common misconception about deserts is that they consist of mile after mile of drifting sand dunes. It is true that sand accumulations do exist in some areas and may be striking features. But, perhaps surprisingly, sand accumulations worldwide represent only a small percentage of the total desert area. For example, in the Sahara—the world's largest desert—accumulations of sand cover only *one-tenth* of its area. The sandiest of all deserts is the Arabian, one-third of which consists of sand.

D.

Figure 5.32   As parts **A** and **B** illustrate, dunes commonly have an asymmetrical shape. The steeper leeward side is called the *slip face*. Sand grains deposited on the slip face create the cross-bedding of the dunes. **C.** A complex pattern develops in response to changes in wind direction. Also notice that when dunes are buried and become part of the sedimentary record, the cross-bedded structure is preserved. **D.** Cross beds are an obvious characteristic of the Navajo Sandstone in Zion National Park, Utah. (Photo by David Muench)

Types of Sand Dunes

Dunes are not just random heaps of windblown sediment. Rather, they are accumulations that usually assume patterns that are surprisingly consistent (Figure 5.33). A broad assortment of dune forms exist, generally simplified to a few major types for discussion. Of

Figure 5.33 Sand dune types. **A.** Barchan dunes. **B.** Transverse dunes. **C.** Barchanoid dunes. **D.** Longitudinal dunes. **E.** Parabolic dunes. **F.** Star dunes.

course, gradations exist among different forms as well as irregularly shaped dunes that do not fit easily into any category. Several factors influence the form and size that dunes ultimately assume. These include wind direction and velocity, availability of sand, and the amount of vegetation present. Six basic dune types are shown in Figure 5.33, with arrows indicating wind directions.

Barchan Dunes. Solitary sand dunes shaped like crescents and with their tips pointing downwind are called **barchan dunes** (Figure 5.33A). These dunes form where supplies of sand are limited and the surface is relatively flat, hard, and lacking vegetation. They migrate slowly with the wind at a rate of up to 15 meters annually. Their size is usually modest, with the largest barchans reaching heights of about 30

meters while the maximum spread between their horns approaches 300 meters. When the wind direction is nearly constant, the crescent form of these dunes is nearly symmetrical. However, when the wind direction is not perfectly fixed, one tip becomes larger than the other.

Transverse Dunes. In regions where the prevailing winds are steady, sand is plentiful, and vegetation is sparse or absent, the dunes form a series of long ridges that are separated by troughs and oriented at right angles to the prevailing wind. Because of this orientation, they are termed **transverse dunes** (Figure 5.33B). Typically, many coastal dunes are of this type. In addition, transverse dunes are common in many arid regions where the extensive surface of wavy sand is sometimes called a *sand sea.* In some parts of the Sahara and Arabian deserts, transverse dunes reach heights of 200 meters, are 1 to 3 kilometers across, and can extend for distances of 100 kilometers or more.

There is a relatively common dune form that is intermediate between isolated barchans and extensive waves of transverse dunes. Such dunes, called **barchanoid dunes,** form scalloped rows of sand oriented at right angles to the wind (Figure 5.33C). The rows resemble a series of barchans that have been positioned side by side. Visitors exploring the gypsum dunes at White Sands National Monument, New Mexico, will recognize this form.

Longitudinal Dunes. **Longitudinal dunes** are long ridges of sand that form more or less parallel to the prevailing wind and where sand supplies are moderate (Figure 5.33D). Apparently the prevailing wind direction must vary somewhat but still remain in the same quadrant of the compass. Although the smaller types are only 3 or 4 meters high and several tens of meters long, in some large deserts longitudinal dunes can reach great size. For example, in portions of North Africa, Arabia, and central Australia, these dunes may approach a height of 100 meters and extend for distances of more than 100 kilometers (62 miles).

Parabolic Dunes. Unlike the other dunes that have been described thus far, **parabolic dunes** form where vegetation partially covers the sand. The shape of these dunes resembles the shape of barchans except that their tips point into the wind rather than downwind (Figure 5.33E). Parabolic dunes often form along coasts where there are strong onshore winds and abundant sand. If the sand's sparse vegetative cover is disturbed at some spot, deflation creates a blowout. Sand is then transported out of the depression and deposited as a curved rim that grows higher as deflation enlarges the blowout.

Star Dunes. Confined largely to parts of the Sahara and Arabian deserts, **star dunes** are isolated hills of sand that exhibit a complex form (Figure 5.33F). Their name is derived from the fact that the bases of these dunes resemble multipointed stars. Usually three or four sharp-crested ridges diverge from a central high point that in some cases may approach a height of 90 meters. As their form suggests, star dunes develop where wind directions are variable.

Chapter Summary

• A *glacier* is a thick mass of ice originating on land from the compaction and recrystallization of snow, and it shows evidence of past or present flow. Today, *valley* or *alpine glaciers* are found in mountain areas where they usually follow valleys that were originally occupied by streams. *Ice sheets* exist on a much larger scale, covering most of Greenland and Antarctica.

• Near the surface of a glacier, in the *zone of fracture,* ice is brittle. However, below about 50 meters, pressure is great, causing ice to *flow* like a *plastic material.* A second important mechanism of glacial movement consists of the whole ice mass *slipping* along the ground.

• Glaciers form in areas where more snow falls in winter than melts during summer. Snow accumulation and ice formation occur in the *zone of accumulation.* Beyond this area is the *zone of wastage,* where there is a net loss to the glacier. The *glacial budget* is the balance, or lack of balance, between accumulation at the upper end of the glacier, and loss at the lower end.

• Glaciers erode land by *plucking* (lifting pieces of bedrock out of place) and *abrasion* (grinding and scraping of a rock surface). Erosional features produced by valley glaciers include *glacial troughs, hanging valleys, cirques, arêtes, horns,* and *fiords.*

• Any sediment of glacial origin is called *drift.* The two distinct types of glacial drift are (1) *till,* which is unsorted sediment deposited directly by the ice; and (2) *stratified drift,* which is relatively well-sorted sediment laid down by glacial meltwater.

• The most widespread features created by glacial deposition are layers or ridges of till, called *moraines.* Associated with valley glaciers are *lateral moraines,* formed along the sides of the valley, and *medial moraines,* formed between two valley glaciers that have joined. *End moraines,* which mark the former position of the front of a glacier, and *ground moraine,* an undulating layer of till deposited as the ice front retreats, are common to both valley glaciers and ice sheets.

• Perhaps the most convincing evidence for the occurrence of several glacial advances during the *Ice Age* is the widespread existence of *multiple layers of drift* and an uninterrupted record of climate cycles preserved in *seafloor sediments.* In addition to massive erosional and depositional work, other effects of Ice Age glaciers included the *migration of organisms, changes in stream courses, adjustment of the crust* by rebounding after the removal of the immense load of ice, and *climate changes* caused by the existence of the glaciers themselves. In the sea, the most far-reaching effect of the Ice Age was the *worldwide change in sea level* that accompanied each advance and retreat of the ice sheets.

• Any theory that attempts to explain the causes of glacial ages must answer the two basic questions: (1) What causes the onset of glacial conditions? and (2) What caused the alternating glacial and interglacial stages that have been documented for the Pleistocene epoch? Two of the many hypotheses for the cause of glacial ages involve (1) plate tectonics and (2) variations in Earth's orbit.

• Practically all desert streams are dry most of the time and are said to be *ephemeral.* Nevertheless, *running water is responsible for most of the erosional work in a desert.* Although wind erosion is more significant in dry areas than elsewhere, the main role of wind in a desert is in the transportation and deposition of sediment.

• Many of the landscapes of the Basin and Range region of the western and southwestern United States are the result of streams eroding uplifted mountain blocks and depositing the sediment in interior basins. *Alluvial fans, playas,* and *playa lakes* are features often associated with these landscapes.

• For wind erosion to be effective, dryness and scant vegetation are essential. *Deflation,* the lifting and removal of loose material, often produces shallow depressions called *blowouts* and can also lower the surface by removing sand and silt, leaving behind a stony veneer called *desert pavement. Abrasion,* the sandblasting effect of wind, is often given too much credit for producing desert features. However, abrasion does cut and polish rock near the surface.

• Wind deposits are of two distinct types: (1) extensive *blankets of silt,* called *loess,* carried by wind in *suspension,* and (2) *mounds and ridges of sand,* called *dunes,* which are formed from sediment that is carried as part of the wind's *bed load.*

Key Terms

abrasion (p. 134)
alluvial fan (p. 151)
alpine glacier (p. 130)
arête (p. 136)
barchan dune (p. 158)
barchanoid dune (p. 159)
blowout (p. 154)
cirque (p. 136)
crevasse (p. 132)
cross beds (p. 157)
deflation (p. 154)
desert pavement (p. 154)
drumlin (p. 142)
end moraine (p. 141)
ephemeral (p. 150)
esker (p. 142)
fiord (p. 136)

glacial drift (p. 139)
glacial erratic (p. 140)
glacial striations (p. 134)
glacial trough (p. 136)
glacier (p. 130)
ground moraine (p. 141)
hanging valley (p. 136)
horn (p. 136)
ice cap (p. 131)
ice sheet (p. 130)
interior drainage (p. 151)
kame (p. 142)
kettle (p. 142)
lateral moraine (p. 140)
loess (p. 156)
longitudinal dune (p. 159)
medial moraine (p. 140)

outwash plain (p. 142)
parabolic dune (p. 159)
piedmont glacier (p. 131)
playa lake (p. 151)
Pleistocene epoch (p. 143)
plucking (p. 133)
pluvial lake (p. 143)
rock flour (p. 134)
slip face (p. 157)
star dune (p. 159)
stratified drift (p. 139)
till (p. 139)
transverse dune (p. 158)
valley glacier (p. 130)
valley train (p. 142)
zone of accumulation (p. 132)
zone of wastage (p. 132)

Review Questions

1. What is a glacier? What percentage of Earth's land area do glaciers cover? Where are glaciers found today?

2. Each of the following statements refers to a particular type of glacier. Name the type of glacier.
   (a) The term *continental* is often used to describe this type of glacier.
   (b) This type of glacier is also called an *alpine glacier.*
   (c) This is a glacier formed when one or more valley glaciers spreads out at the base of a steep mountain front.
   (d) Greenland is the only example in the Northern Hemisphere.

3. Describe how glaciers fit into the hydrologic cycle. What role do they play in the rock cycle?

4. Describe the two components of glacial flow. At what rates do glaciers move? In a valley glacier does all of the ice move at the same rate? Explain.

5. Why do crevasses form in the upper portion of a glacier but not below a depth of about 50 meters?

6. Under what circumstances will the front of a glacier advance? Retreat? Remain stationary?

7. Describe two basic processes of glacial erosion.

8. How does the shape of a glaciated mountain valley differ from a mountain valley that was not glaciated?

9. List and describe the erosional features you might expect to see in an area where valley glaciers exist or have recently existed.

10. What is glacial drift? What is the difference between till and stratified drift? What general effect do glacial deposits have on the landscape?

11. List and briefly describe the four basic moraine types. What do all moraines have in common?

12. List and briefly describe four depositional features other than moraines.

13. About what percentage of Earth's land surface was covered at some time by Pleistocene glaciers? How does this compare to the area presently covered by ice sheets and glaciers? (Check your answer with Question 1.)

14. List three indirect effects of Ice Age glaciers.

15. How might plate tectonics help us understand the cause of ice ages? Can plate tectonics explain the alternation between glacial and interglacial climates during the Pleistocene?

16. How extensive are Earth's dry regions?

17. What is the most important erosional process in deserts?

18. Describe the characteristics associated with each of the stages in the evolution of a mountainous desert region such as the Basin and Range region in the western United States.

19. Why is wind erosion relatively more important in dry regions than in humid areas?

20. Although sand dunes are the best-known wind deposits, accumulations of loess are very significant in some parts of the world. What is loess? Where are such deposits found? What are the origins of this sediment?

21. How do sand dunes migrate?

22. Identify each of the dunes described in the following statements.

(a) Long ridges of sand oriented parallel to the prevailing wind.

(b) Solitary, crescent-shaped dunes oriented with their tips pointing downwind.

(c) Ridges of sand oriented at right angles to the prevailing wind.

(d) Dunes whose tips point into the wind.

(e) Scalloped rows of sand oriented at right angles to the wind.

(f) Isolated dunes consisting of three or four sharp-crested ridges diverging from a central high point.

Examining the Earth System

1. Glaciers are solid, but they are a basic part of the hydrologic cycle. Should glaciers be considered a part of the sphere we call the solid Earth, or do they belong to the hydrosphere?

2. Some scientists think that ice should be a separate sphere of the Earth system, called the *cryosphere*. Does such an idea have merit?

3. Wind erosion occurs at the interface of the atmosphere, solid Earth, and biosphere and is influenced by the hydrosphere and human activity. With this in mind, describe how human activity contributed to the Dust Bowl, the period of intense wind erosion in the Great Plains in the 1930s. You might find it helpful to research Dust Bowl on the Internet using a search engine such as HotBot (**http://www.hotbot.lycos.com/**) or Yahoo (**http://www. yahoo.com/**). You may also find the Wind Erosion Research Unit site to be informative: (**http://www.weru. ksu.edu/**).

Web Resources

The *Earth Science* Website uses the resources and flexibility of the Internet to aid in your study of the topics in this chapter. Written and developed by Earth science instructors, this site will help improve your understanding of Earth science. Visit **http://www.prenhall.com/tarbuck** and click on the cover of *Earth Science* 10e to find:

• **Online review quizzes.**

• **Web-based critical thinking and writing exercises.**

• **Links to chapter-specific Web resources.**

• **Internet-wide key term searches.**

http://www.prenhall.com/tarbuck

Destruction caused by a major earthquake that struck northwestern Turkey on August 17, 1999. More than 17,000 people perished. (Photo © by Yann Arthus-Bertrand "Earth From Above"/Peter Arnold, Inc.)

CHAPTER

6

EARTHQUAKES AND EARTH'S INTERIOR

What is an earthquake? How does a seismograph record the location of a quake? Can earthquakes ever be predicted? If we could view the Earth's interior, what would it look like? In this chapter we shall answer these and other questions. The study of earthquakes is important, not only because of the devastating effect that some earthquakes have on us but also because they furnish clues about the structure of Earth's interior.

On October 17, 1989, at 5:04 p.m. Pacific daylight time, millions of television viewers around the world were settling in to watch the third game of the World Series. Instead, they saw their television sets go black as tremors hit San Francisco's Candlestick Park. Although the earthquake was centered in a remote section of the Santa Cruz Mountains, 100 kilometers to the south, major damage occurred in the Marina District of San Francisco.

The most tragic result of the violent shaking was the collapse of some double-decked sections of Interstate 880, also known as the Nimitz Freeway. The ground motions caused the upper deck to sway, shattering the concrete support columns along a mile-long section of the freeway. The upper deck then collapsed onto the lower roadway, flattening cars as if they were aluminum beverage cans. This earthquake, named the Loma Prieta quake for its point of origin, claimed 67 lives.

In mid-January 1994, less than five years after the Loma Prieta earthquake devastated portions of the San Francisco Bay Area, a major earthquake struck the Northridge area of Los Angeles. Although not the fabled Big One, this moderate 6.7-magnitude earthquake left 57 dead, over 5000 injured, and tens of thousands of households without water and electricity. The damage exceeded $40 billion and was attributed to an apparently unknown fault that ruptured 18 kilometers (11 miles) beneath Northridge

The Northridge earthquake began at 4:31 a.m. and lasted roughly 40 seconds. During this brief period, the quake terrorized the entire Los Angeles area. In the three-story Northridge Meadows apartment complex, 16 people died when sections of the upper floors collapsed onto the first-floor units. Nearly 300 schools were seriously damaged, and a dozen major roadways buckled. Among these were two of California's major arteries—the Golden State Freeway (Interstate 5), where an overpass collapsed completely and blocked the roadway, and the Santa Monica Freeway. Fortunately, these roadways had practically no traffic at this early morning hour.

In nearby Granada Hills, broken gas lines were set ablaze while the streets flooded from broken water mains. Seventy homes burned in the Sylmar area. A 64-car freight train derailed, including some cars carrying hazardous cargo. But it is remarkable that the destruc-

Figure 6.1 Damage in San Francisco's Marina District following the 1989 Loma Prieta earthquake. (Photo by David Weintraub/Photo Researchers, Inc.)

Figure 6.2 San Francisco in flames after the 1906 earthquake. (Reproduced from the collection of the Library of Congress)

tion was not greater. Unquestionably, the upgrading of structures to meet the requirements of building codes developed for this earthquake-prone area helped minimize what could have been a much greater human tragedy.

Over 30,000 earthquakes that are strong enough to be felt occur worldwide annually (Figure 6.1). Fortunately, most are minor tremors and do very little damage. Generally, only about 75 significant earthquakes take place each year, and many of these occur in remote regions. However, occasionally a large earthquake occurs near a large population center. Under these conditions, an earthquake is among the most destructive natural forces on Earth (see Box 6.1).

The shaking of the ground, coupled with the liquefaction of some soils, wreaks havoc on buildings and other structures. In addition, when a quake occurs in a populated area, power and gas lines are often ruptured, causing numerous fires. In the famous 1906 San Francisco earthquake, much of the damage was caused by fires (Figure 6.2), which quickly became uncontrollable when broken water mains left firefighters with only trickles of water.

? STUDENTS SOMETIMES ASK...
How often do earthquakes occur?

All the time—in fact, there are literally thousands of earthquakes daily! Fortunately, the majority of them are too small to be felt by people (see Table 6.2, p. 175), and many of them occur in remote regions. Their existence is known only because of sensitive seismographs.

What Is an Earthquake?

Forces Within
↳ Earthquakes

An **earthquake** is the vibration of Earth produced by the rapid release of energy. Most often earthquakes are caused by slippage along a fault in Earth's crust. The energy released radiates in all directions from its source, the **focus** (*foci* = a point), in the form of waves. These waves are analogous to those produced when a stone is dropped into a calm pond (Figure 6.3). Just as the impact of the stone sets water waves in motion, an earthquake generates seismic waves that radiate throughout the Earth. Even though the energy dissipates rapidly with increasing distance from the focus, sensitive instruments located throughout the world record the event.

Earthquakes and Faults

The tremendous energy released by atomic explosions or by volcanic eruptions can produce an earthquake, but these events are relatively weak and infrequent. What mechanism produces a destructive earthquake? Ample evidence exists that Earth is not a static planet. We know that Earth's crust has been uplifted at times, because we have found numerous ancient wave-cut benches many meters above the level of the highest tides. Other regions exhibit evidence of extensive subsidence. In addition to these vertical displacements, offsets in fence lines, roads, and other structures indicate that horizontal movement is common (Figure 6.4). These movements are usually associated with large fractures in Earth's crust called **faults**.

165

Most of the motion along faults can be satisfactorily explained by the plate tectonics theory. This theory states that large slabs of Earth's crust are in continual slow motion. These mobile plates interact with neighboring plates, straining and deforming the rocks at their edges. In fact, it is along faults associated with plate boundaries that most earthquakes occur. Furthermore, earthquakes are repetitive—as soon as one is over, the continuous motion of the plates resumes, adding strain to the rocks until they fail again.

Discovering the Cause of Earthquakes: Applying the Scientific Method

The actual mechanism of earthquake generation eluded geologists until H. F. Reid of Johns Hopkins University conducted a study following the great 1906 San Francisco earthquake. The earthquake was accompanied by horizontal surface displacements of several meters along the northern portion of the San Andreas Fault. This 1300-kilometer (780-mile) fracture extends north-south through southern California. It is a large fault zone that separates two great sections of Earth's crust—the North American plate and the Pacific plate. Field investigations following the 1906 quake determined that during this single event, the land on the western side of the San Andreas Fault lurched as much as 4.7 meters (15 feet) northward relative to the land on the eastern side of the fault.

Based on these measurements and related field investigations, Reid formulated a hypothesis to explain what had been observed. Figure 6.5 illustrates Reid's hypothesis. Part A shows an existing fault. In part B, tec-

tonic forces ever so slowly deform the crustal rocks on both sides of the fault, as demonstrated by the bent features. Under these conditions, rocks are bending and storing elastic energy, much like a wooden stick does if bent. Eventually, the frictional resistance holding the rocks together is overcome. As slippage occurs at the weakest point (the focus), displacement will exert stress farther along the fault, where additional slippage will occur until most of the built-up strain is released (Figure 6.5C). This slippage allows the deformed rock to snap back. The vibrations we know as an earthquake occur as the rock elastically returns to its original shape. The springing back of the rock was termed **elastic rebound** by Reid, because the rock behaves elastically, much like a stretched rubber band does when it is released.

To test his hypothesis, Reid examined land surveys that had been made *prior* to the great 1906 earthquake. What Reid discovered by analyzing the surveys was that features on opposite sides of the San Andreas Fault had experienced a relative displacement of about 3 meters (10 feet) during a 50-year span prior to the quake. Just as Reid's hypothesis had predicted, there had been a gradual bending of crustal rocks on either side of the fault before the earthquake occurred. Since Reid's initial test, many subsequent studies have validated his hypothesis, that for shallow earthquakes, at least, elastic rebound of crustal rocks produces the destructive ground shaking associated with earthquakes.

In summary, most earthquakes are produced by the rapid release of elastic energy stored in rock that has

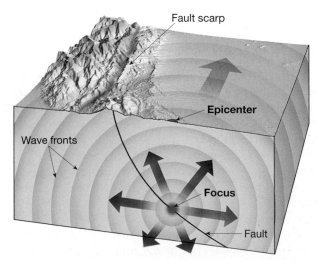

Figure 6.3 The focus of all earthquakes is located at depth. The surface location directly above it is called the epicenter.

Figure 6.4 Slippage along a fault produced an offset in this orange grove east of Calexico, California. (Photo by John S. Shelton)

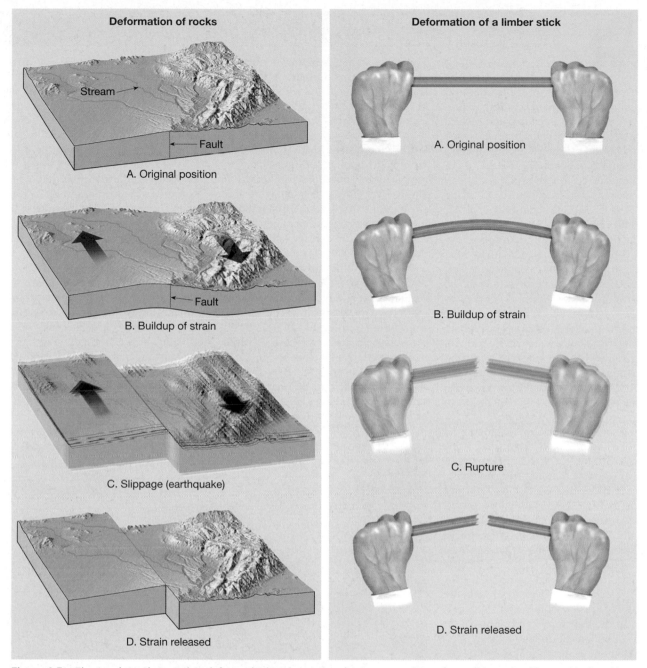

Figure 6.5 Elastic rebound. As rock is deformed it bends, storing elastic energy. Once the rock is strained beyond its breaking point it ruptures, releasing the stored-up energy in the form of earthquake waves.

been subjected to great stress. Once the strength of the rock is exceeded, it suddenly ruptures, causing the vibrations of an earthquake. Earthquakes most often occur along existing faults whenever the frictional forces on the fault surfaces are overcome.

Foreshocks and Aftershocks

The intense vibrations of the 1906 San Francisco earthquake lasted about 40 seconds. Although most of the displacement along the fault occurred in this rather

short period, additional movements along this and other nearby faults continued for several days following the main quake. The adjustments that follow a major earthquake often generate smaller earthquakes called **aftershocks.** Although these aftershocks are usually much weaker than the main earthquake, they can sometimes destroy already badly weakened structures. This occurred, for example, during a 1988 earthquake in Armenia. A large aftershock of magnitude 5.8 collapsed many structures that had been weakened by the main tremor.

When you think earthquake, you probably think of California and Japan. However, six major earthquakes have occurred in the central and eastern United States since colonial times. Three of these had estimated Richter magnitudes of 7.5, 7.3, and 7.8, and they were centered near the Mississippi River Valley in southeastern Missouri. Occurring on December 16, 1811, January 23, 1812, and February 7, 1812, these earthquakes, plus numerous smaller tremors, destroyed the town of New Madrid, Missouri, triggered massive landslides, and caused damage over a six-state area. The course of the Mississippi River was altered, and Tennessee's Reelfoot Lake was enlarged. The distances over which these earthquakes were felt are truly remarkable. Chimneys were reported downed in Cincinnati, Ohio, and Richmond, Virginia, while Boston residents, located 1770 kilometers (1100 miles) to the northeast, felt the tremor.

Despite the history of the New Madrid earthquake, Memphis, Tennessee, the largest population center in the area today, does not have adequate earthquake provisions in its building code. Further, because Memphis is located on unconsolidated floodplain deposits, buildings are more susceptible to damage than similar structures built on bedrock. It has been estimated that if an earthquake the size of the 1811–1812 New Madrid event were to strike in the next decade, it would result in casualties in the thousands and damages in tens of billions of dollars.

Damaging earthquakes that occurred in Aurora, Illinois (1909), and Valentine, Texas (1931), remind us that other areas in the central United States are vulnerable.

The greatest historical earthquake in the eastern states occurred August 31, 1886, in Charleston, South Carolina.

The event, which spanned 1 minute, caused 60 deaths, numerous injuries, and great economic loss within a radius of 200 kilometers (120 miles) of Charleston. Within 8 minutes, effects were felt as far away as Chicago and St. Louis, where strong vibrations shook the upper floors of buildings, causing people to rush outdoors. In Charleston alone, over 100 buildings were destroyed, and 90 percent of the remaining structures were damaged. It was difficult to find a chimney still standing (Figure 6.A).

Numerous other strong earthquakes have been recorded in the eastern United States. New England and adjacent areas have experienced sizable shocks since colonial times. The first reported earthquake in the Northeast took place in Plymouth, Massachusetts, in 1683, and was followed in 1755 by the destructive Cambridge, Massachusetts, earthquake. Moreover, ever since records have been kept, New York State alone has experienced over 300 earthquakes large enough to be felt.

Earthquakes in the central and eastern United States occur far less frequently than in California. Yet history indicates that the East is vulnerable. Further, these shocks east of the Rockies have generally produced structural damage over a larger area than counterparts of similar magnitude in California. The reason is that the underlying bedrock in the central and eastern United States is older and more rigid. As a result, seismic waves are able to travel greater distances with less weakening than in the western United States. It is estimated that for earthquakes of similar magnitude, the region of maximum ground motion in the East may be up to 10 times larger than in the West. Consequently, the higher rate of earthquake occurrence in the western United States is balanced somewhat by the fact that central and eastern U.S. quakes can damage larger areas.

Figure 6.A Damage to Charleston, South Carolina, caused by the August 31, 1886 earthquake. Damage ranged from toppled chimneys and broken plaster to total collapse. (Photo courtesy of U.S. Geological Survey)

In addition, small earthquakes called **foreshocks** often precede a major earthquake by days or, in some cases, by as much as several years. Monitoring of these foreshocks has been used as a means of predicting forthcoming major earthquakes, with mixed success. We will consider the topic of earthquake prediction in a later section of this chapter.

San Andreas Fault: An Active Earthquake Zone

The San Andreas is undoubtedly the most studied fault system in the world. Over the years, investigations have shown that displacement occurs along discrete segments that are 100 to 200 kilometers long.

Further, each fault segment behaves somewhat differently from the others. Some portions of the San Andreas exhibit a slow, gradual displacement known as **fault creep,** which occurs relatively smoothly and therefore with little noticable seismic activity. Other segments regularly slip, producing small earthquakes.

Still other segments remain locked and store elastic energy for hundreds of years before rupturing in great earthquakes. The latter process is described as *stick-slip* motion, because the fault exhibits alternating periods of locked behavior followed by sudden slippage. It is estimated that great earthquakes should occur about every 50 to 200 years along those sections of the San Andreas Fault that exhibit stick-slip motion. This knowledge is useful when assigning a potential earthquake risk to a given segment of the fault zone.

STUDENTS SOMETIMES ASK...

*Do moderate earthquakes
decrease the chances of a major
quake in the same region?*

No. This is due to the vast increase in release of energy associated with higher-magnitude earthquakes. For instance, an earthquake with a magnitude of 8.5 releases millions of times more energy than the smallest earthquakes felt by humans. Similarly, thousands of moderate tremors would be needed to release the huge amount of energy equal to one "great" earthquake.

The tectonic forces along the San Andreas Fault zone that were responsible for the 1906 San Francisco earthquake are still active. Currently, laser beams are used to measure the relative motion between the opposite sides of this fault. These measurements reveal a displacement of 2 to 5 centimeters (1 to 2 inches) per year. Although this seems slow, it produces substantial movement over millions of years.

To illustrate, in 30 million years this rate of displacement would slide the western portion of California northward so that Los Angeles, on the Pacific plate, would be adjacent to San Francisco on the North American plate! More important in the short term, a displacement of just 2 centimeters each year produces 2 meters of offset every 100 years. Consequently, the 4 meters of displacement produced during the 1906 San Francisco earthquake should occur at least every 200 years along this segment of the fault zone. This fact lies behind California's concern for making buildings earthquake-resistant in anticipation of the inevitable big one.

Seismology: The Study of Earthquake Waves

GEODe   Forces Within
          ↳ Earthquakes

The study of earthquake waves, **seismology** (*seismos* = shake, *ology* = the study of), dates back to attempts by the Chinese almost 2000 years ago to determine the direction of the source of each earthquake. Modern **seismographs** (*seismos* = shake, *graph* = write) are instruments that record earthquake waves. Their principle is simple: A weight is freely suspended from a support that is attached to bedrock (Figure 6.6). When waves from an earthquake reach the instrument, the inertia of the weight keeps it stationary, while Earth and the support vibrate. The movement of Earth in relation to the stationary weight is recorded on a rotating drum. (*Inertia* is the tendency of a stationary object to remain still, or a moving object to stay in motion.)

Modern seismographs amplify and record ground motion, producing a trace as shown in Figure 6.7. These records, called **seismograms** (*seismos* = shake, *gramma* = what is written) reveal that seismic waves are elastic energy. This energy radiates outward in all directions from the focus, as you saw in Figure 6.3. The transmission of this energy can be compared to the shaking of gelatin in a bowl that is jarred. Seismograms reveal that two main types of seismic waves are generated by the slippage of a rock mass. Some travel along Earth's outer layer and are called **surface waves.** Others travel through Earth's interior and are called **body waves.** Body waves are further divided into **primary waves (P waves)** and **secondary waves (S waves).**

Body waves are divided into P and S waves by their mode of travel through intervening materials. P waves are push-pull waves—they push (compress) and pull (expand) rocks in the direction the wave is traveling (Figure 6.8A). Imagine holding someone by the shoulders and shaking the person. This push-pull movement is how P waves move through the Earth. This wave motion is analogous to that generated by human vocal cords as they move air to create sound. Solids, liquids, and gases resist a change in volume when compressed and will elastically spring back once the force is removed (Figure 6.8B). Therefore, P waves, which are compressional waves, can travel through all these materials.

In contrast, S waves shake the particles at right angles to their direction of travel. This can be illustrated by fastening one end of a rope and shaking the other end, as shown in Figure 6.8C. Unlike P waves, which temporarily change the *volume* of the intervening material by alternately compressing and expanding it, S waves

Figure 6.6 Principle of the seismograph. The inertia of the suspended mass tends to keep it motionless, while the recording drum, which is anchored to bedrock, vibrates in response to seismic waves. Thus, the stationary mass provides a reference point from which to measure the amount of displacement occurring as the seismic wave passes through the ground below.

temporarily change the *shape* of the material that transmits them (Figure 6.8D). Because fluids (gases and liquids) do not respond elastically to changes in shape, they will not transmit S waves.

The motion of surface waves is somewhat more complex. As surface waves travel along the ground, they cause the ground and anything resting upon it to move, much like ocean swells toss a ship (Figure 6.8F). In addition to their up-and-down motion, surface waves have a side-to-side motion similar to an S wave oriented in a horizontal plane (Figure 6.8E). This latter motion is particularly damaging to the foundations of structures.

By observing a typical seismic record, as shown in Figure 6.7, you can see a major difference among these seismic waves: P waves arrive at the recording station first, then S waves, and then surface waves. This is a consequence of their speeds. To illustrate, the velocity of P waves through granite within the crust is about 6 kilometers per second. Under the same conditions, S waves travel at 3.5 kilometers per second. Differences in density and elastic properties of the rock greatly influence the velocities of these waves. Generally, in any solid material, P waves travel about 1.7 times faster than S waves, and surface waves can be expected to travel at 90 percent of the velocity of the S waves.

As you shall see, seismic waves allow us to determine the location and magnitude of earthquakes. In addition, seismic waves provide us with a tool for probing Earth's interior.

Locating an Earthquake

GEODe Forces Within
        ↳ Earthquakes

Recall that the *focus* is the place within Earth where earthquake waves originate. The **epicenter** (*epi* = upon, *center* = a point) is the location on the surface directly above the focus (see Figure 6.3).

The difference in velocities of P and S waves provides a method for locating the epicenter. The principle used is analogous to a race between two autos, one

Figure 6.7 Typical seismic record. Note the time interval (about 5 minutes) between the arrival of the first P waves and the arrival of the first S waves.

A. P waves generated using a slinky

B. P waves traveling along the surface

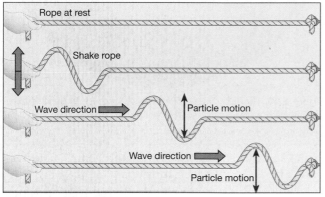

C. S waves generated using a rope

D. S waves traveling along the surface

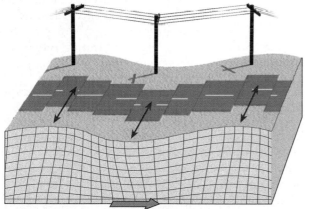

E. Surface waves that are like S waves

F. Surface waves that are like ocean waves

Figure 6.8 Types of seismic waves and their characteristic motion. (Note that during a strong earthquake, ground shaking consists of a combination of various kinds of seismic waves.) **A.** As illustrated by a slinky, P waves are compressional waves that alternately compress and expand the material through which they pass. **B.** The back-and-forth motion produced as compressional waves travel along the surface can cause the ground to buckle and fracture, and may cause power lines to break. **C.** S waves cause material to oscillate at right angles to the direction of wave motion. **D.** Because S waves can travel in any plane, they produce up-and-down and sideways shaking of the ground. **E.** One type of surface wave is essentially the same as that of an S wave that exhibits only horizontal motion. This kind of surface wave moves the ground from side to side and can be particularly damaging to the foundations of buildings. **F.** Another type of surface wave travels along Earth's surface much like rolling ocean waves. The arrows show the elliptical movement of rock as the wave passes.

faster than the other. The P wave always wins the race, arriving ahead of the S wave. But, the greater the length of the race, the greater will be the difference in the arrival times at the finish line (the seismic station). Therefore, the greater the interval measured on a seismogram between the arrival of the first P wave and the first S wave, the greater the distance to the earthquake source.

A system for locating earthquake epicenters was developed by using seismograms from earthquakes whose epicenters could be easily pinpointed from physical evidence. From these seismograms, travel-time graphs were constructed (Figure 6.9). The first travel-time graphs were greatly improved when seismograms became available from nuclear explosions, because the precise location and time of detonation were known.

Using the sample seismogram in Figure 6.7 and the travel-time curves in Figure 6.9, we can determine the distance separating the recording station from the earthquake in two steps: (1) Determine the time interval between the arrival of the first P wave and the first S wave, and (2) find on the travel-time graph the equivalent time spread between the P and S wave curves. From this information, we can determine that this earthquake occurred 3800 kilometers (2350 miles) from the recording instrument.

Now we know the *distance,* but what *direction*? The epicenter could be in any direction from the seismic station. As shown in Figure 6.10, the precise location can be found when the distance is known from three or more different seismic stations. On a globe, we draw a circle around each seismic station. Each circle represents the epicenter distance for each station. The point where the three circles intersect is the epicenter of the quake. This method is called triangulation.

About 95 percent of the energy released by earthquakes originates in a few relatively narrow zones (Figure 6.11). The greatest energy is released along a path around the outer edge of the Pacific Ocean known as the *circum-Pacific belt.* Included in this zone are regions of great seismic activity, such as Japan, the Philippines,

Figure 6.10 Earthquake epicenter is located using the distance obtained from three seismic stations.

Chile, and numerous volcanic island chains, as exemplified by Alaska's Aleutian Islands.

Figure 6.11 reveals another continuous belt that extends for thousands of kilometers through the world's oceans. This zone coincides with the oceanic ridge system, an area of frequent but low-intensity seismic activity. By comparing this figure with Figure 7.8, you can see a close correlation between the location of earthquake epicenters and plate boundaries.

Measuring the Size of Earthquakes

Historically, seismologists have employed a variety of methods to obtain two fundamentally different measures that describe the size of an earthquake—intensity and magnitude. The first of these to be used was **intensity**—a measure of the degree of earthquake shaking at a given locale based on the amount of damage. With the development of seismographs, it became clear that a quantitative measure of an earthquake based on seismic records rather than uncertain personal estimates of damage was desirable. The measurement that was developed, called **magnitude,** relies on calculations that use data provided by seismic records (and other techniques) to estimate the amount of energy released at the source of the earthquake.

Intensity Scales

Until a little over a century ago, historical records provided the only accounts of the severity of earthquake shaking and destruction. Using these descriptions—which were compiled without any established stan-

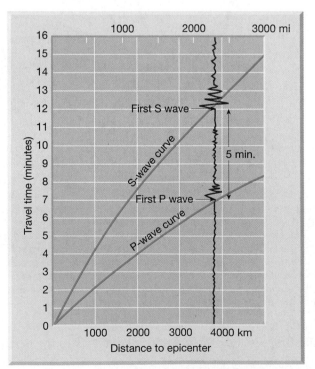

Figure 6.9 A travel-time graph is used to determine the distance to the epicenter. The difference in arrival times of the first P wave and the first S wave in the example is 5 minutes. Thus, the epicenter is roughly 3800 kilometers (2350 miles) away.

Figure 6.11 Distribution of the 14,229 earthquakes with magnitudes equal to or greater than 5 for the years 1980 to 1990. (Data from National Geophysical Data Center/NOAA)

dards for reporting—made accurate comparisons of earthquake sizes difficult, at best.

In order to standardize the study of earthquake severity, workers developed various intensity scales that considered damage done to buildings, as well as individual descriptions of the event, and secondary effects such as landslides and the extent of ground rupture. By 1902, Guiseppe Mercalli had developed a relatively reliable intensity scale, which in a modified form is still used today. The **Modified Mercalli Intensity Scale** shown in Table 6.1 was developed using California buildings as its standard, but it is appropriate for use throughout most of the United States and Canada to estimate the strength of an earthquake. For example, if some well-built wood structures and most masonry buildings are destroyed by an earthquake, a region would be assigned an intensity of X on the Mercalli scale (Table 6.1).

Despite their usefulness in providing seismologists with a tool to compare earthquake severity, particularly in regions where there are no seismographs, intensity scales have severe drawbacks. In particular, intensity scales are based on effects (largely destruction) of earthquakes that depend not only on the severity of ground shaking but also on factors such as population density, building design, and the nature of surface materials.

Magnitude Scales

In order to compare earthquakes across the globe, a measure was needed that does not rely on parameters that vary considerably from one part of the world to an-

other, such as construction practices. As a consequence, a number of magnitude scales were developed.

Richter Magnitude. In 1935 Charles Richter of the California Institute of Technology developed the first magnitude scale using seismic records to estimate the relative sizes of earthquakes. As shown in Figure 6.12 (top), the **Richter scale** is based on the amplitude of the largest seismic wave (P, S, or surface wave) recorded on a seismogram. Because seismic waves weaken as the distance between the earthquake focus and the seismograph increases (in a manner similar to light), Richter developed a method that accounted for the decrease in wave amplitude with increased distance. Theoretically, as long as the same, or equivalent, instruments were used, monitoring stations at various locations would obtain the same Richter magnitude for every recorded earthquake. (Richter selected the Wood-Anderson seismograph as the standard recording device.)

Although the Richter scale has no upper limit, the largest magnitude recorded on a Wood-Anderson seismograph was 8.9. These great shocks release approximately 10^{26} ergs of energy—roughly equivalent to the detonation of 1 billion tons of TNT. Conversely, earthquakes with a Richter magnitude of less than 2.0 are not felt by humans. With the development of more sensitive instruments, tremors of a magnitude of minus 2 were recorded. Table 6.2 shows how Richter magnitudes and their effects are related.

Earthquakes vary enormously in strength, and great earthquakes produce wave amplitudes that are

Table 6.1 Modified Mercalli Intensity Scale.

I	Not felt except by a very few under especially favorable circumstances.
II	Felt only by a few persons at rest, especially on upper floors of buildings.
III	Felt quite noticeably indoors, especially on upper floors of buildings, but many people do not recognize it as an earthquake.
IV	During the day felt indoors by many, outdoors by few. Sensation like heavy truck striking building.
V	Felt by nearly everyone, many awakened. Disturbances of trees, poles, and other tall objects sometimes noticed.
VI	Felt by all; many frightened and run outdoors. Some heavy furniture moved; few instances of fallen plaster or damaged chimneys. Damage slight.
VII	Everybody runs outdoors. Damage negligible in buildings of good design and construction; slight-to-moderate in well-built ordinary structures; considerable in poorly built or badly designed structures.
VIII	Damage slight in specially designed structures; considerable in ordinary substantial buildings with partial collapse; great in poorly built structures. (Fall of chimneys, factory stacks, columns, monuments, walls.)
IX	Damage considerable in specially designed structures. Buildings shifted off foundations. Ground cracked conspicuously.
X	Some well-built wooden structures destroyed. Most masonry and frame structures destroyed. Ground badly cracked.
XI	Few, if any, (masonry) structures remain standing. Bridges destroyed. Broad fissures in ground.
XII	Damage total. Waves seen on ground surfaces. Objects thrown upward into air.

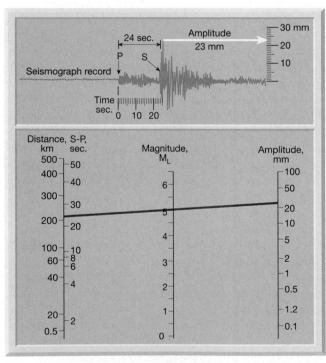

Figure 6.12 Illustration showing how the Richter magnitude of an earthquake can be determined graphically using a seismograph record from a Wood-Anderson instrument. First, measure the height (amplitude) of the largest wave on the seismogram (23 mm) and then the distance to the focus using the time interval between S and P waves (24 seconds). Next, draw a line between the distance scale (left) and the wave amplitude scale (right). By doing this, you should obtain the Richter magnitude (M_L) of 5. (Data from California Institute of Technology)

thousands of times larger than those generated by weak tremors. To accommodate this wide variation, Richter used a *logarithmic scale* to express magnitude, where a *tenfold* increase in wave amplitude corresponds to an increase of 1 on the magnitude scale. Thus, the amount of ground shaking for a 5-magnitude earthquake is 10 times greater than that produced by an earthquake having a Richter magnitude of 4.

In addition, each unit of Richter magnitude equates to roughly a *32-fold energy increase.* Thus, an earthquake with a magnitude of 6.5 releases 32 times more energy than one with a magnitude of 5.5, and roughly 1000 times more energy than a 4.5-magnitude quake. A major earthquake with a magnitude of 8.5 releases millions of times more energy than the smallest earthquakes felt by humans.

Other Magnitude Scales. Richter's original goal was modest in that he only attempted to rank the earthquakes of southern California (shallow-focus earthquakes) into groups of large, medium, and small magnitude. Hence, Richter magnitude was designed to study nearby (or local) earthquakes and is denoted by the symbol (M_L)—where M is for *magnitude* and L is for *local.*

The convenience of describing the size of an earthquake by a single number that could be calculated quickly from seismograms makes the Richter scale a powerful tool. Further, unlike intensity scales that could only be applied to populated areas of the globe, Richter magnitudes could be assigned to earthquakes in more remote regions and even to events that occurred in the ocean basins. As a result, the method devised by Richter was adapted to a number of different seismographs located throughout the world. In time, seismologists mod-

Table 6.2 Earthquake Magnitudes and Expected World Incidence.		
Richter Magnitudes	**Effects Near Epicenter**	**Estimated Number per Year**
< 2.0	Generally not felt, but recorded.	600,000
2.0–2.9	Potentially perceptible.	300,000
3.0–3.9	Felt by some.	49,000
4.0–4.9	Felt by most.	6200
5.0–5.9	Damaging shocks.	800
6.0–6.9	Destructive in populous regions.	266
7.0–7.9	Major earthquakes. Inflict serious damage.	18
8.0 and above	Great earthquakes. Destroy communities near epicenter.	1.4

SOURCE: *Earthquake Information Bulletin* and others.

ified Richter's work and developed new magnitude scales.

However, despite their usefulness, none of these "Richter-like" magnitude scales are adequate for describing very large earthquakes. For example, the 1906 San Francisco earthquake and the 1964 Alaskan earthquake have roughly the same Richter magnitudes (and surface-wave magnitudes) of about 8.3. However, the Alaskan earthquake released considerably more energy than the San Francisco quake, based on the size of the fault zone and the amount of displacement observed. Thus, the Richter scale (as well as the other related magnitude scales) are said to be *saturated* for large earthquakes because they cannot distinguish between the size of these events.

Moment Magnitude. In recent years seismologists have been employing a more precise measure called **moment magnitude** (M_W), which is derived from the amount of displacement that occurs along a fault zone rather than by measuring the ground motion at some distant point. The moment magnitude is calculated using a combination of factors that include the average amount of displacement along the fault, the area of the rupture surface, and the shear strength of the faulted rock—a measure of how much strain energy a rock can store before it suddenly slips and releases this energy in the form of an earthquake (and heat).

The moment magnitude can also be readily calculated from seismograms by examining very long period seismic waves. The values obtained have been calibrated so that small- and moderate-size earthquakes have moment magnitudes that are roughly equivalent to Richter magnitudes. However, moment magnitudes are much better for describing very large earthquakes. For example, on the moment magnitude scale, the 1906 San Francisco earthquake, which had a surface-wave magnitude of 8.3, would be demoted to 7.8 on the moment magnitude scale, whereas the 1964

Alaskan earthquake with an 8.3 Richter magnitude would be increased to 9.2. The strongest earthquake on record is the 1960 Chilean earthquake, with a moment magnitude of 9.6.

Moment magnitude has gained wide acceptance among seismologists and engineers because: (1) it is the only magnitude scale that estimates adequately the size of very large earthquakes; (2) it is a measure established from the size of the rupture surface and the amount of displacement, thus it better reflects the total energy released during an earthquake; and (3) it can be verified by two independent methods—field studies that are based on measurements of fault displacement and by seismographic methods using long-period waves.

In summary, the Richter scale was designed to differentiate the sizes of shallow-focus earthquakes in southern California occurring near a particular type of seismograph. Richter's original magnitude scale (M_L) was extended to measure earthquakes at any distance and focal depth, as well as taking into account geological variations from place to place. All of these "Richter-like" magnitude scales were designed so that the final computed magnitude is reasonably consistent with the Richter scale. Because none of these magnitude scales estimate adequately the size of very large earthquakes, the moment magnitude scale (M_W), derived from the amount of displacement that occurs along a fault zone, was developed.

Destruction from Earthquakes

The most violent earthquake to jar North America this century—the Good Friday Alaskan Earthquake—occurred in 1964. Felt throughout that state, the earthquake had a moment magnitude of 9.2 and reportedly lasted 3 to 4 minutes. This event left 131 people dead, thousands homeless, and the economy of the state badly disrupted because it occurred near major towns and seaports (Figure 6.13). Had the schools and business districts been open on this holiday, the toll surely would have been higher. Within 24 hours of the initial shock, 28 aftershocks were recorded, 10 of which exceeded a Richter magnitude of 6.

STUDENTS SOMETIMES ASK...

I've heard that the safest place to be in a house during an earthquake is in a door frame. Is that really the best place?

No! An enduring earthquake image of California is a collapsed adobe home with the door frame as the only standing part. From this came the belief that a doorway is the safest place to be during an earthquake. In modern homes, doorways are no stronger than any other part of the house and usually have doors that will swing and can injure you.

If you're inside, the best advice is to *duck, cover, and hold*. When you feel an earthquake, *duck* under a desk or sturdy table. Stay away from windows, bookcases, file cabinets, heavy mirrors, hanging plants, and other heavy objects that could fall. Stay under *cover* until the shaking stops. And, *hold* on to the desk or table: If it moves, move with it.

Destruction from Seismic Vibrations

The 1964 Alaskan earthquake provided geologists with new insights into the role of ground shaking as a destructive force. Recall that the structural damage from earthquake waves depends on several factors, including (1) the intensity, (2) duration of the vibrations, (3) the nature of the material upon which the structure rests, and (4) the design of the structure.

All multistory structures in Anchorage were damaged by the vibrations, but the more flexible wood-frame residential buildings fared best. Figure 6.14 offers a striking example of how construction variations affect earthquake damage; you can see that the steel-frame building on the left withstood the vibrations, whereas the poorly designed J.C. Penney building was badly damaged. Engineers have learned that unreinforced masonry buildings are the most serious safety threat in earthquakes.

Most large structures in Anchorage were damaged even though they were built to conform to the earthquake provisions of the Uniform Building Code. Perhaps some of that destruction can be attributed to the unusually long duration of this earthquake (estimated at 3 to 4 minutes). (The San Francisco earthquake of 1906 was felt for about 40 seconds; the strong vibrations of the 1989 Loma Prieta earthquake lasted under 15 seconds.)

The buildings in Anchorage are situated on soft, unconsolidated sediments, which amplify vibrations more than solid bedrock. Thus, damage in Anchorage was more severe than in Whittier, which rests on a firm foundation of granite. However, Whittier was damaged by a seismic sea wave (described in the next section).

The 1985 Mexican earthquake gave seismologists and engineers a vivid reminder of what had been learned from the 1964 Alaskan earthquake. In central Mexico City, nearly 400 kilometers (250 miles) from the epicenter, the vibrations intensified to five times that experienced in outlying districts, owing to the city's construction on soft sediments, remnants of an ancient lake bed (Figure 6.15).

Where unconsolidated materials are saturated with water, earthquakes can generate a phenomenon known as **liquefaction.** Under these conditions, what had been

Figure 6.13 Region most affected by the Good Friday earthquake of 1964. Note the location of the epicenter (red dot). (After U.S. Geological Survey)

Figure 6.14   Damage to the five-story JC Penney Co. building, Anchorage, Alaska. Very little structural damage was incurred by the adjacent building. (Courtesy of NOAA)

a stable soil turns into a fluid that is no longer capable of supporting buildings or other structures (Figure 6.16). As a result, underground objects such as storage tanks and sewer lines may float toward the surface of their newly liquefied environment. Buildings and other structures may settle and collapse. During the Loma Prieta earthquake, in San Francisco's Marina District, foundations failed and geysers of sand and water shot from the ground, indicating that liquefaction had occurred (Figure 6.17).

Tsunami

Most deaths associated with the 1964 Alaskan quake were caused by **seismic sea waves,** or **tsunami.** (The name *tsunami* is a Japanese word meaning "harbor wave," for Japanese harbors have suffered from them many times.) These destructive waves often are called tidal waves by the media. However, this name is inappropriate, for these waves are most often generated by earthquakes and to a lesser extent by submarine landslides, volcanic eruptions, and meteorite impacts. They are *not* produced by the tidal effect of the Moon or Sun.

Tsunami triggered by an earthquake occur where a slab of oceanic crust is displaced vertically along a fault (Figure 6.18), or where the vibration of a quake sets an underwater landslide into motion.

Once formed, a tsunami resembles the ripples created when a pebble is dropped into a pond. In contrast to ripples, tsunami advance across the ocean at speeds between 500 and 950 kilometers (300 and 600 miles) per hour. Despite this, a tsunami in the open ocean can pass undetected because its height is usually less than 1 meter and the distance between wave crests is great, ranging from 100 to 700 kilometers. However, upon entering shallower coastal water, these destructive waves are slowed and the water begins to pile up to heights that occasionally exceed 30 meters (100 feet), as shown in Figure 6.18. As the crest of a tsunami approaches shore, it appears as a rapid rise in sea level with a turbulent and chaotic surface.

Usually the first warning of an approaching tsunami is a rather rapid withdrawal of water from beaches. Coastal residents have learned to heed this warning and move to higher ground, for about 5 to 30 minutes later the retreat of water is followed by a surge capable of extending hundreds of meters inland. In a successive fashion, each surge is followed by a rapid oceanward retreat of the water. These waves are separated by intervals of between 10 and 60 minutes. They are able to traverse thousands of kilometers of the ocean before their energy is dissipated.

Figure 6.15   During the 1985 Mexican earthquake, multistory buildings swayed back and forth as much as 1 meter. (Photo by James L. Beck)

Figure 6.16 Effects of liquefaction. This tilted building rests on unconsolidated sediment that imitated quicksand during the 1985 Mexican earthquake. (Photo by James L. Beck)

Figure 6.17 These "mud volcanoes" were produced by the Loma Prieta earthquake of 1989. They formed when geysers of sand and water shot from the ground, an indication that liquefaction occurred. (Photo by Richard Hilton, courtesy of Dennis Fox)

Tsunami Damage from the 1964 Alaskan Earthquake. The tsunami generated in the 1964 Alaskan earthquake heavily damaged communities along the Gulf of Alaska and killed 107 people. By contrast, only nine people died in Anchorage as a direct result of the vibrations.

Tsunami damage following the Alaskan earthquake extended along much of the west coast of North America. Despite a one-hour warning, 12 people perished in Crescent City, California, from the fifth wave. The first wave crested about 4 meters (13 feet) above low tide and was followed by three progressively smaller waves. Believing that the tsunami had ceased, people returned to the shore, only to be met by the fifth and most devastating wave. Superimposed upon high tide, it crested about 6 meters higher than the level of low tide.

Tsunami Warning System. In 1946, a large tsunami struck the Hawaiian Islands without warning. A wave more than 15 meters (50 feet) high left several coastal villages in shambles (Figure 6.19). This destruction motivated the U.S. Coast and Geodetic Survey to establish a tsunami warning system for coastal areas of the Pacific. From seismic observatories throughout the region, large earthquakes are reported to the Tsunami Warning Center in Honolulu. Scientists at the Center use tidal gauges to determine whether a tsunami has formed. Within an hour a warning is issued. Although tsunamis travel very rapidly, there is sufficient time to evacuate all but the region nearest the epicenter. For example, a tsunami generated near the Aleutian Islands would take five hours to reach Hawaii, and one generated near the coast of Chile would travel 15 hours before reaching Hawaii (Figure 6.20).

Fortunately, most earthquakes do not generate tsunamis. On the average, only about 1.5 destructive tsunamis are generated worldwide annually. Of these, only about one every 10 years is catastrophic.

Figure 6.18 Schematic drawing of a tsunami generated by displacement of the ocean floor. The speed of a wave moving across the ocean correlates with ocean depth. As shown, waves moving in deep water advance at speeds in excess of 800 kilometers per hour. Speed gradually slows to 50 kilometers per hour at depths of 20 meters. Decreasing depth slows the movement of the wave column. As waves slow in shallow water, they grow in height until they topple and rush onto shore with tremendous force. The size and spacing of these swells are not to scale.

STUDENTS SOMETIMES ASK...

What is the largest wave triggered by an earthquake?

The largest wave ever recorded occurred in Lituya Bay, about 200 kilometers (125 miles) west of Juneau, Alaska's capital. On July 9, 1958, an earthquake triggered an enormous rockslide that dumped 90 million tons of rock into the upper part of the bay. The rockslide created a huge *splash wave* (different than a tsunami, these waves are produced when an object splashes into water) that swept over the ridge facing the rockslide and uprooted or snapped off trees 1740 feet above the bay. Even larger splash waves have occurred in prehistoric times, including an estimated 3000-foot wave that is thought to have resulted from a meteorite impact in the Gulf of Mexico about 65 million years ago.

Landslides and Ground Subsidence

In the 1964 Alaskan earthquake, the greatest damage to structures was from landslides and ground subsidence triggered by the vibrations. At Valdez and Seward, the violent shaking caused river-delta materials to experience liquefaction; the subsequent slumping carried both waterfronts away. Because the disaster could happen again, the entire town of Valdez was relocated about 7 kilometers away on more stable ground. In Valdez, 31 people on a dock died when it slid into the sea.

Most of the damage in Anchorage was attributed to landslides. Many homes were destroyed in Turnagain Heights when a layer of clay lost its strength and over 200 acres of land slid toward the ocean (Figure 6.21). A portion of this landslide has been left in its natural condition as a reminder of this destructive event.

Figure 6.19 A man stands before a wall of water about to engulf him at Hilo, Hawaii, on April 1, 1946. This tsunami, which originated in the Aleutian Islands near Alaska, was still powerful enough when it hit Hawaii to rise 9 to 16 meters (33-55 feet). The S.S. *Brigham Victory*, from which this photograph was taken, managed to survive the onslaught, but 159 people in Hawaii, including the man seen here, were killed. (Photo courtesy of Water Resources Center Archives, University of California, Berkeley)

Figure 6.20 Tsunami travel times to Honolulu, Hawaii, from selected locations throughout the Pacific. (Data from NOAA)

The site was named "Earthquake Park." Downtown Anchorage was also disrupted as sections of the main business district dropped by as much as 3 meters (10 feet).

Fire

The 1906 San Francisco earthquake reminds us of the formidable threat of fire. The central city contained mostly large, older wooden structures and brick buildings. The greatest destruction was caused by fires that started when gas and electrical lines were severed. The fires raged uncontrolled for three days and devastated over 500 city blocks (see Figure 6.2). The problem was compounded by the initial ground shaking, which broke the city's water lines into hundreds of unconnected pieces.

The fire was finally contained when buildings were dynamited along a wide boulevard to create a *fire break*,

Figure 6.21 Photo of a small portion of the Turnagain Heights slide. (Photo courtesy of U.S. Geological Survey)

the same strategy used in fighting a forest fire. Although only a few deaths were attributed to the fires, such is not always the case. A 1923 earthquake in Japan (their worst quake prior to the 1995 Kobe tremor) triggered an estimated 250 fires, which devastated the city of Yokohama and destroyed more than half the homes in Tokyo. Over 100,000 deaths were attributed to the fires, which were driven by unusually high winds.

Can Earthquakes Be Predicted?

The vibrations that shook Northridge, California, in 1994 inflicted 57 deaths and about $40 billion in damage (Figure 6.22). This was from a brief earthquake (about 40 seconds) of moderate rating (M_W 6.7). Seismologists warn that earthquakes of comparable or greater strength will occur along the San Andreas Fault, which cuts a 1300-kilometer (800-mile) path through the state. The obvious question is, Can earthquakes be predicted?

Short-Range Predictions

The goal of short-range earthquake prediction is to provide a warning of the location and magnitude of a large earthquake within a narrow time frame. Substantial efforts to achieve this objective are being put forth in Japan, the United States, China, and Russia—countries where earthquake risks are high (Table 6.3). This research has concentrated on monitoring possible *precursors*—phenomena that precede and thus provide a warning of a forthcoming earthquake. In California, for example, some seismologists are measuring uplift, subsidence, and strain in the rocks near active faults. Some Japanese scientists are studying peculiar anomalous behavior that may precede a quake.

One claim of a successful short-range prediction was made by Chinese seismologists after the February 4, 1975, earthquake in Liaoning Province. According to reports, very few people were killed, although more than 1 million lived near the epicenter, because the earthquake was predicted and the population was evacuated. Recently, some Western seismologists have questioned this claim and suggest instead that an intense swarm of foreshocks that began 24 hours before the main earthquake may have caused many people to evacuate spontaneously. Further, an official Chinese government report issued 10 years later stated that 1328 people died and 16,980 injuries resulted from this earthquake.

One year after the Liaoning earthquake at least 240,000 people died in the Tangshan, China, earthquake, which was not predicted. The Chinese have also issued false alarms. In a province near Hong Kong, people reportedly left their dwellings for over a month, but no earthquake followed. Clearly, whatever method the Chinese employ for short-range predictions, it is *not* reliable.

Figure 6.22 Damage to interstate 5 during the January 17, 1994, Northridge earthquake. (Photo by Tom McHugh/Photo Researchers, Inc.)

For a short-range prediction scheme to warrant general acceptance, it must be both accurate and reliable. Thus, *it must have a small range of uncertainty as regards to location and timing, and it must produce few failures, or false alarms.* Can you imagine the debate that would precede an order to evacuate a large city in the United States, such as Los Angeles or San Francisco? The cost of evacuating millions of people, arranging for living accommodations, and providing for their lost work time and wages would be staggering.

Long-Range Forecasts

In contrast to short-range predictions, which aim to predict earthquakes within a time frame of hours or at most days, long-range forecasts give the probability of a certain magnitude earthquake occurring on a time scale of 30 to 100 years or more. Stated another way, these forecasts give statistical estimates of the expected intensity of ground motion for a given area over a specified time frame. Although long-range forecasts may not be as informative as we might like, the data is important for updating the Uniform Building Code, which contains nationwide standards for designing earthquake-resistant structures.

Long-range forecasts are based on the premise that earthquakes are repetitive or cyclical, like the weather. In other words, as soon as one earthquake is over, the continuing motions of Earth's plates begin to build strain in the rocks again, until they fail once more. This has led seismologists to study historical records of earthquakes to see if there are any discernible patterns so that the probability of recurrence might be established.

One study conducted by the U.S. Geological Survey gives the probability of a rupture occurring along various segments of the San Andreas Fault for the 30 years between 1988 and 2018 (Figure 6.23). From this investigation, the Santa Cruz Mountains area was given a 30 percent probability of producing a 6.5-magnitude earthquake during this time period. In fact, it produced the Loma Prieta quake in 1989, of 7.1 magnitude.

The region along the San Andreas Fault given the highest probability (90 percent) of generating a quake is the Parkfield section. This area has been called the "Old Faithful" of earthquake zones because activity here has been very regular since record keeping began in 1857. (Although this section has had an earthquake on the average of once every 22 years, the last one occurred in 1966—it is overdue by more than 16 years and counting!) Another section between Parkfield and the Santa Cruz Mountains is given a very low probability of generating an earthquake. This area has experienced very little seismic activity in historical times; rather, it exhibits a slow, continual movement known as *fault creep.* Such movement is beneficial because it prevents strain from building to high levels in the rocks.

In summary, it appears that the best prospects for making useful earthquake predictions involve fore-

Table 6.3 Some Notable Earthquakes.

Year	Location	Deaths (est.)	Magnitude[†]	Comments
1556	Shensi, China	830,000		Possibly the greatest natural disaster.
1755	Lisbon, Portugal	70,000		Tsunami damage extensive.
*1811–1812	New Madrid, Missouri	Few	7.9	Three major earthquakes.
*1886	Charleston, South Carolina	60		Greatest historical earthquake in the eastern United States.
*1906	San Francisco, California	1500	7.8	Fires caused extensive damage.
1908	Messina, Italy	120,000		
1923	Tokyo, Japan	143,000	7.9	Fire caused extensive destruction.
1960	Southern Chile	5700	9.6	Possibly the largest-magnitude earthquake ever recorded.
*1964	Alaska	131	9.2	Greatest North American earthquake.
1970	Peru	66,000	7.8	Great rockslide.
*1971	San Fernando, California	65	6.5	Damage exceeded $1 billion.
1975	Liaoning Province, China	1328	7.5	First major earthquake to be predicted.
1976	Tangshan, China	240,000	7.6	Not predicted.
1985	Mexico City	9500	8.1	Major damage occurred 400 km from epicenter.
1988	Armenia	25,000	6.9	Poor construction practices.
*1989	Loma Prieta, California	62	6.9	Damages exceeded $6 billion.
1990	Iran	50,000	7.3	Landslides and poor construction practices caused great damage.
1993	Latur, India	10,000	6.4	Located in stable continental interior.
*1994	Northridge, California	57	6.7	Damages in excess of $40 billion.
1995	Kobe, Japan	5472	6.9	Damages estimated to exceed $100 billion.
1999	Izmit, Turkey	17,127	7.4	Nearly 44,000 injured and more than 250,000 displaced
1999	Chi-Chi, Taiwan	2300	7.6	Severe destruction; 8700 injuries.
2001	El Salvador	1000	7.6	Triggered many landslides.
2001	Bhuj, India	20,000[†]	7.9	1 million or more homeless.

*U.S. earthquakes.
[†]Widely differing magnitudes have been estimated for some of these earthquakes. When available, moment magnitudes are used.
SOURCE: U.S. Geological Survey

casting magnitudes and locations in time scales of years, or perhaps even decades. These forecasts are important because they provide information used to develop the Uniform Building Code and assist in land-use planning.

Earth's Layered Structure

GEODe Forces Within
  ↳ Earthquakes

Earth's interior lies just below us; however, access is very limited. The deepest well yet drilled has penetrated Earth's crust only 12 kilometers (7.5 miles), less than 0.2 percent of the distance to the planet's center. Consequently, most knowledge of our planet's interior comes from the study of earthquake waves that travel through Earth and vibrate the surface at some distant point.

If Earth were a perfectly homogeneous body, seismic waves would spread through it in a straight line at a constant speed. However, this is not the case. It so happens that the seismic waves reaching seismographs located farther from an earthquake travel at faster average speeds than those recorded at locations closer to the event. This general increase in speed with depth is a consequence of increased pressure, which enhances the elastic properties of deeply buried rock. As a result, the paths of seismic rays through Earth are refracted (bent) as they travel (Figure 6.24).

As more sensitive seismographs were developed, it became apparent that in addition to gradual changes in seismic-wave velocities, rather abrupt velocity changes also occur at particular depths. Because these changes were detected worldwide, seismologists concluded that Earth must be composed of distinct shells having varying compositions and/or mechanical properties (Figure 6.25).

Layers Defined by Composition

Compositional layering occurred early in Earth's history as meteorite impacts and decay of radioactive elements caused the temperature of our planet to steadily

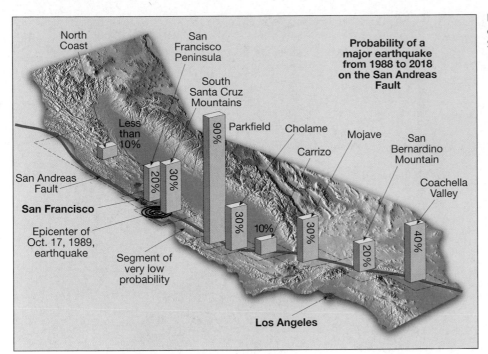

Figure 6.23 Probability of a major earthquake from 1988 to 2018 on the San Andreas Fault.

increase. Eventually Earth became hot enough so that at least some melting occurred. During this period of partial melting the heavier elements, principally iron and nickel, sank as the lighter rocky components floated upward. This segregation of material is still occurring, but at a much reduced rate. Because of this chemical differentiation, Earth's interior is not homogeneous. Rather, it consists of three major zones each defined by its chemical composition—the crust, mantle, and core (Figure 6.25).

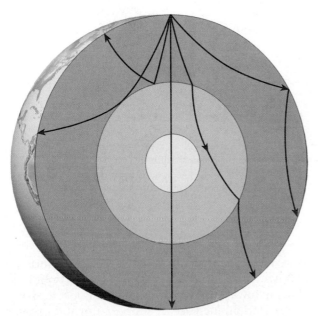

Figure 6.24 A few of the many possible paths that seismic rays take through Earth.

Crust. The **crust,** Earth's comparatively thin, rocky outer skin, is generally divided into oceanic and continental crust. The oceanic crust is roughly 7 kilometers (5 miles) thick and composed of the dark igneous rock called *basalt.* By contrast, the continental crust averages 35–40 kilometers (25 miles) thick but may exceed 70 kilometers (40 miles) in some mountainous regions. Unlike the oceanic crust, which has a relatively homogeneous chemical composition, the continental crust consists of many rock types. The upper crust has an average composition of a *granitic rock* called *granodiorite,* whereas the composition of the lowermost continental crust is more akin to basalt. Continental rocks have an average density of about 2.7 g/cm^3, and some have been discovered that are 4 billion years old. The rocks of the oceanic crust are younger (180 million years or less) and more dense (about 3.0 g/cm^3) than continental rocks.

Mantle. Over 82 percent of Earth's volume is contained in the **mantle,** a solid, rocky shell that extends to a depth of 2900 kilometers (1800 miles). The boundary between the crust and mantle represents a marked change in chemical composition. The dominant rock type in the uppermost mantle is *peridotite,* which has a density of 3.3 g/cm^3. At greater depths peridotite changes by assuming a more compact crystalline structure and hence a greater density.

Core. The **core** is a sphere composed of an iron-nickel alloy having a radius of 3486 kilometers (2161 miles). At the extreme pressures found in the core, the iron-rich material has an average density of nearly 11 g/cm^3 and approaches 14 times the density of water at Earth's center.

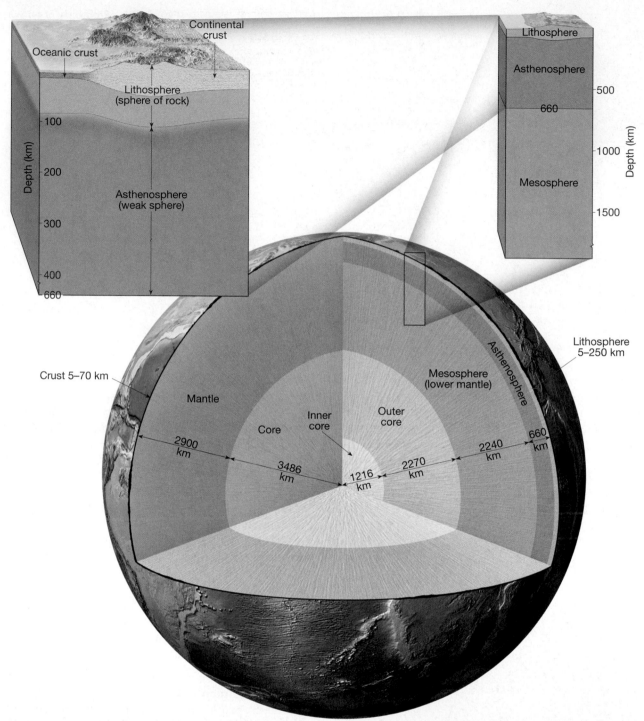

Figure 6.25 Views of Earth's layered structure. The left side of the large cross section shows that Earth's interior is divided into three different layers based on compositional differences—the crust, mantle, and core. The right side of the large cross section depicts the five main layers of Earth's interior based on physical properties and hence mechanical strength—the lithosphere, asthenosphere, mesosphere, outer core, and inner core. The block diagrams above the large cross section show enlarged views of the upper portion of Earth's interior.

Layers Defined by Physical Properties

Earth's interior is characterized by a gradual increase in temperature, pressure, and density with depth. Estimates put the temperature at a depth of 100 kilometers at between 1200°C and 1400°C, whereas the temperature at Earth's center may exceed 6700°C. Clearly, Earth's inte-

rior has retained much of the energy acquired during its formative years, despite the fact that heat is continuously flowing toward the surface, where it is lost to space.

The gradual increase in temperature and pressure with depth affects the physical properties and hence the mechanical behavior of Earth materials. When a substance is heated, its chemical bonds weaken and its mechanical strength (resistance to deformation) is reduced. If the temperature exceeds the melting point of an Earth material, the material's chemical bonds break and melting ensues. If temperature were the only factor that determined whether a substance melted, our planet would be a molten ball covered with a thin, solid outer shell. However, pressure also increases with depth and tends to increase rock strength. Furthermore, because melting is accompanied by an increase in volume, it occurs at higher temperatures at depth because of greater confining pressure. Thus, depending on the physical environment (temperature and pressure), a particular Earth material may behave like a brittle solid, deform in a puttylike manner, or even melt and become liquid.

Earth can be divided into five main layers based on physical properties and hence mechanical strength—the *lithosphere, asthenosphere, mesosphere (lower mantle), outer core,* and *inner core.*

Lithosphere and Asthenosphere. Based on physical properties, Earth's outermost layer consists of the crust and uppermost mantle and forms a relatively cool, rigid shell. Although this layer is composed of materials with markedly different chemical compositions, it tends to act as a unit that exhibits rigid behavior—mainly because it is cool and thus strong. This layer, called the **lithosphere** (*sphere of rock*), averages about 100 kilometers in thickness but may be 250 kilometers thick or more below the older portions of the continents (Figure 6.25). Within the ocean basins, the lithosphere is only a few kilometers thick along the oceanic ridges but increases to perhaps 100 kilometers in regions of older and cooler oceanic crust.

Beneath the lithosphere in the upper mantle (to a depth of about 660 kilometers) lies a soft, comparatively weak layer known as the **asthenosphere** (*weak sphere*). The top portion of the asthenosphere has a temperature/pressure regime that results in a small amount of melting. Within this very weak zone, the lithosphere is mechanically detached from the layer below. The result is that the lithosphere is able to move independently of the asthenosphere, a topic we will consider in the next chapter.

It is important to emphasize that the strength of various Earth materials is a function of both their composition and of the temperature and pressure of their environment. You should not get the idea that the entire lithosphere behaves like a brittle solid similar to rocks found on the surface. Rather, the rocks of the lithosphere get progressively hotter and weaker (more easily deformed) with increasing depth. At the depth of the uppermost asthenosphere, the rocks are close enough to their melting temperatures (some melting may actually occur) that they are very easily deformed. Thus, the uppermost asthenosphere is weak because it is near its melting point, just as hot wax is weaker than cold wax.

Mesosphere or Lower Mantle. Below the zone of weakness in the uppermost asthenosphere, increased pressure counteracts the effects of higher temperature, and the rocks gradually strengthen with depth. Between the depths of 660 kilometers and 2900 kilometers a more rigid layer called the **mesosphere** (*middle sphere*) or **lower mantle** is found (Figure 6.25). Despite their strength, the rocks of the mesosphere are still very hot and capable of very gradual flow.

Inner and Outer Core. The core, which is composed mostly of an iron-nickel alloy, is divided into two regions that exhibit very different mechanical strengths (Figure 6.25). The **outer core** is a *liquid layer* 2270 kilometers (1410 miles) thick. It is the convective flow of metallic iron within this zone that generates Earth's magnetic field. The **inner core** is a sphere having a radius of 1216 kilometers (754 miles). Despite its higher temperature, the material in the inner core is stronger (because of immense pressure) than the outer core and behaves like a *solid.*

Discovering Earth's Major Layers

The story of how seismologists discovered Earth's core and layers is interesting. In 1909, a pioneering Yugoslavian seismologist, Andrija Mohorovičić, presented the first convincing evidence for layering within Earth. By studying seismic records, he found that the velocity of seismic waves increases abruptly below about 50 kilometers of depth. This boundary separates the crust from the underlying mantle and is known as the **Mohorovičić discontinuity** in his honor. For reasons that are obvious, the name for this boundary was quickly shortened to **Moho.**

A few years later another boundary was discovered by the German seismologist Beno Gutenberg. Generally, seismic waves from even small earthquakes are strong enough to travel around the world. This is why a seismograph in Antarctica can record earthquakes in California or Italy. But Gutenberg observed that P waves diminish and eventually die out about 105 degrees around the globe from an earthquake. Then, about 140 degrees away, the P waves reappear, but about 2 minutes later than would be expected, based on the distance traveled. This belt, where direct P waves are absent, is about 35 degrees wide and has been named the **shadow zone.** Figure 6.26 illustrates how this works.

Gutenberg realized that the shadow zone could be explained if Earth contained a core composed of material unlike the overlying mantle. The core must somehow hinder the transmission of P waves in a manner similar to the light rays being blocked by an opaque object that casts a shadow. However, rather than actually stopping the P waves, the shadow zone bends them, as shown in Figure 6.26. It was further learned that S waves could not travel through the core. Therefore, geologists concluded that at least a portion of this region is liquid.

In 1936 the last major subdivision of Earth's interior was predicted by Inge Lehmann, a Danish seismologist (see Box 6.2). Lehmann discovered a new region of seismic reflection within the core. Hence, Earth has a core within a core. The size of the inner core was not accurately calculated until the early 1960s, when underground nuclear tests were conducted in Nevada. Because the precise locations and times of the explosions were known, echoes from seismic waves that bounced off the inner core provided an accurate means of determining its size.

Over the past few decades, advances in seismology have allowed for much refinement of the gross view of Earth's interior presented so far. Of major importance was the discovery of the lithosphere and asthenosphere.

STUDENTS SOMETIMES ASK...

As compared to continental crust, the ocean crust is quite thin. Has there ever been an attempt to drill through it to obtain a sample of the mantle?

Yes. Project Mohole was initiated in 1958 to retrieve a sample of material from Earth's mantle by drilling a hole through Earth's crust to the *Mohorovičić discontinuity,* or *Moho.* The plan was to drill to the Moho to gain valuable information on Earth's age, makeup, and internal processes. Despite a successful test phase, drilling was halted as control of the project was shifted from one governmental organization to another. Although Project Mohole failed in its intended purpose, it did show that deep-ocean drilling is a viable means of obtaining geological samples (see Box 12.3, p. 349).

Discovering Earth's Composition

We have examined Earth's structure, so let us now look at the composition of each layer. Composition tells us much about how our planet has developed over its estimated age of 4.5 billion years.

Earth's crust varies in thickness, exceeding 70 kilometers (40 miles) in some mountainous regions and being thinner than 3 kilometers (2 miles) in some oceanic areas (see Figure 6.25). Early seismic data indicated that the continental crust, which is mostly made of lighter, granitic rocks, is quite different in composition from denser oceanic crust. Until the late 1960s, however, scientists had only seismic evidence from which to determine the composition of oceanic crust, because it lies beneath an average of 3 kilometers of water as well as hundreds of meters of sediment. With the develop-

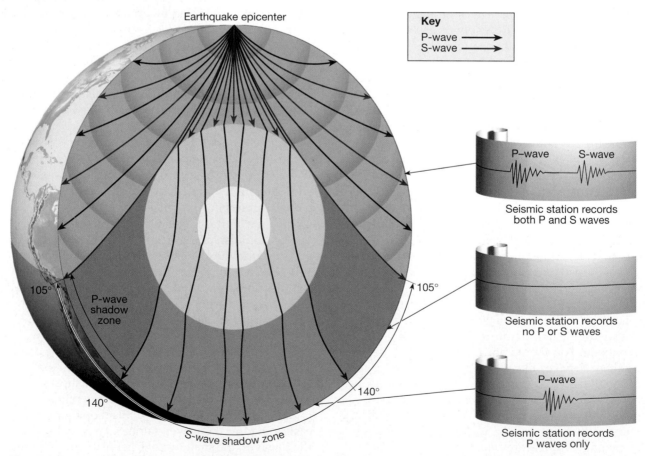

Figure 6.26 View of Earth's interior showing P and S wave paths. The abrupt change in physical properties at the mantle-core boundary causes the wave paths to bend sharply. This change in wave direction results in a shadow zone for P waves between about 105 and 140 degrees. Further, any location more than 105 degrees from an earthquake epicenter will not receive direct S waves since the outer core will not transmit them.

BOX 6.2

UNDERSTANDING EARTH:
Inge Lehmann: A Pioneering Seismologist*

Inge Lehmann was a pioneering scientist at a time when few women had careers in science and mathematics (Figure 6.B). Born in Denmark in 1888, Lehmann lived a long and productive life that included important contributions to our understanding of Earth's interior. She died in 1993 at the age of 105.

Inge Lehmann's career in seismology started in 1925 when she helped establish seismic networks in Denmark and Greenland. It was a paper that she published in 1936 that established her place in the history of geophysics. Known simply as *P'* (P-*prime*), the paper identified a new region of seismic reflection and refraction in Earth's interior, now known as the *Lehmann discontinuity* (Figure 6.C). Because of her exacting scrutiny of seismic records, Inge Lehmann had discovered the boundary that divides Earth's core into inner and outer parts.

Many honors came to Lehmann in recognition of her extraordinary accomplishments. In 1997 the Lehmann Medal was created by the American

*This box was prepared by Nancy L. Lutgens.

Figure 6.B   Inge Lehmann, 1888–1993. (Photo courtesy of Susan M. Landon)

Geophysical Union (AGU) to recognize outstanding research on the structure, composition, and dynamics of Earth's mantle and core. It was the first medal awarded by the AGU that is named for a woman and the first to be named for someone who worked outside the United States.

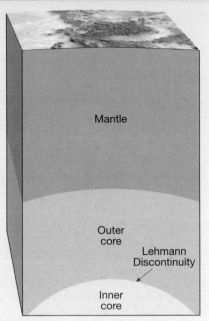

Figure 6.C   Location of Lehmann discontinuity.

---

ment of deep-sea drilling technology the recovery of ocean-floor samples was made possible. The samples were of basaltic composition very different from the rocks that compose the continents.

Our knowledge of the rocks of the mantle and core is much more speculative. However, we do have some clues. Recall that some of the lava that reaches Earth's surface originates in the partially melted asthenosphere within the mantle. In the laboratory, experiments have shown that partial melting of a rock called *peridotite* results in a melt that has a basaltic composition similar to lava that emerges during volcanic activity of oceanic islands. Denser rocks like peridotite are thought to make up the mantle and provide the lava for oceanic eruptions.

Surprisingly, meteorites—or shooting stars—that collide with Earth provide evidence of Earth's inner composition. Because meteorites are part of the solar system, they are assumed to be representative samples. Their composition ranges from metallic meteorites made of iron and nickel to stony meteorites composed of dense rock similar to peridotite.

Because Earth's crust contains a much smaller percentage of iron than do meteorites, geologists believe that the dense iron, and other dense metals, literally

sank toward Earth's center during the planet's early history. By the same token, lighter substances may have floated to the surface, creating the less-dense crust. Thus, Earth's core is thought to be mainly dense iron and nickel, similar to metallic meteorites, whereas the surrounding mantle is believed to be composed of rocks similar to stony meteorites.

The concept of a molten iron outer core is further supported by Earth's magnetic field. Our planet acts as a large magnet. The most widely accepted mechanism explaining why Earth has a magnetic field requires that the core be made of a material that conducts electricity, such as iron, and which is mobile enough to allow circulation. Both of these conditions are met by the model of Earth's core that was established on the basis of seismic data.

Not only does an iron core explain Earth's magnetic field but it also explains the high density of inner earth, about 14 times that of water at Earth's center. Even under the extreme pressure at those depths, average crustal rocks with densities 2.8 times that of water would not have the density calculated for the core. But iron, which is 3 times more dense than crustal rocks, has the required density.

In summary, although earthquakes can be very destructive, much of our knowledge about Earth's interior comes from the study of these phenomena. As our knowledge of earthquakes and their causes improve, we learn more of our planet's internal workings and possibly how to reduce the consequences of tremors. In the next chapter, you will see that most earthquakes originate at the boundaries of Earth's great lithospheric plates.

Chapter Summary

• *Earthquakes* are vibrations of Earth produced by the rapid release of energy from rocks that rupture because they have been subjected to stresses beyond their limit. This energy, which takes the form of waves, radiates in all directions from the earthquake's source, called the *focus*. The movements that produce most earthquakes occur along large fractures, called *faults*, that are associated with plate boundaries.

• Two main groups of *seismic waves* are generated during an earthquake: (1) *surface waves*, which travel along the outer layer of Earth; and (2) *body waves*, which travel through Earth's interior. Body waves are further divided into *primary*, or *P, waves*, which push (compress) and pull (expand) rocks in the direction the wave is traveling, and *secondary*, or *S, waves*, which shake the particles in rock at right angles to their direction of travel. P waves can travel through solids, liquids, and gases. Fluids (gases and liquids) will not transmit S waves. In any solid material, P waves travel about 1.7 times faster than do S waves.

• The location on Earth's surface directly above the focus of an earthquake is the *epicenter*. An epicenter is determined using the difference in velocities of P and S waves.

• *There is a close correlation between earthquake epicenters and plate boundaries.* The principal earthquake epicenter zones are along the outer margin of the Pacific Ocean, known as the *circum-Pacific belt* and through the world's oceans along the *oceanic ridge system.*

• Seismologists use two fundamentally different measures to describe the size of an earthquake—intensity and magnitude. *Intensity* is a measure of the degree of ground shaking at a given locale based on the amount of damage. The *Modified Mercalli Intensity Scale* uses damages to buildings in California to estimate the intensity of ground shaking for a local earthquake. *Magnitude* is calculated from seismic records and estimates the amount of energy released at the source of an earthquake. Using the *Richter scale*, the magnitude of an earthquake is estimated by measuring the *amplitude* (maximum displacement) of the largest seismic wave recorded. A logarithmic scale is used to express magnitude, in which a tenfold increase in ground shaking corresponds to an increase of 1 on the magnitude scale. *Moment magnitude* is currently used to estimate the size of moderate and large earthquakes. It is calculated using the average displacement of the fault, the area of the fault surface, and the sheer strength of the faulted rock.

• The most obvious factors that determine the amount of destruction accompanying an earthquake are the *magnitude* of the earthquake and the *proximity* of the quake to a populated area. *Structural damage* attributable to earthquake vibrations depends on several factors, including (1) *intensity*, (2) *duration* of the vibrations, (3) *nature of the material* upon which the structure rests, and (4) the *design* of the structure. Secondary effects of earthquakes include *tsunamis, landslides, ground subsidence,* and *fire.*

• Substantial research to predict earthquakes is under way in Japan, the United States, China, and Russia—countries where earthquake risk is high. No consistent method of short-range prediction has yet been devised. Long-range forecasts are based on the premise that earthquakes are repetitive or cyclical. Seismologists study the history of earthquakes for patterns, so their occurrences might be predicted.

• As indicated by the behavior of P and S waves as they travel through Earth, the four major zones of Earth's interior are the (1) *crust* (the very thin outer layer), (2) *mantle* (a rocky layer located below the crust with a thickness of 2885 kilometers), (3) *outer core* (a layer about 2270 kilometers thick, which exhibits the characteristics of a mobile liquid), and (4) *inner core* (a solid metallic sphere with a radius of about 1216 kilometers).

• The *continental crust* is primarily made of *granitic* rocks, while the *oceanic crust* is of *basaltic* composition. *Ultramafic rocks*, such as *peridotite*, are thought to make up the *mantle.* The *core* is composed mainly of *iron* and *nickel.*

• The crust and uppermost mantle form Earth's cool rigid outer shell called the *lithosphere*. Beneath the lithosphere lies a soft, relatively weak layer of the mantle known as the *asthenosphere.*

Key Terms

aftershock (p. 168)
asthenosphere (p. 185)
body wave (p. 169)
core (p. 183)
crust (p. 183)
earthquake (p. 164)
elastic rebound (p. 166)
epicenter (p. 170)
fault (p. 165)
fault creep (p. 169)
focus (p. 164)

foreshock (p. 168)
inner core (p. 185)
intensity (p. 172)
liquefaction (p. 176)
lithosphere (p. 185)
magnitude (p. 172)
mantle (p. 188)
Modified Mercalli Intensity Scale
  (p. 173)
Mohorovičić discontinuity (Moho)
  (p. 185)

moment magnitude (p. 175)
outer core (p. 185)
primary (P) wave (p. 169)
Richter scale (p. 173)
secondary (S) wave (p. 169)
seismic sea wave (tsunami) (p. 177)
seismogram (p. 169)
seismograph (p. 169)
seismology (p. 169)
shadow zone (p. 185)
surface wave (p. 169)

Review Questions

1. What is an earthquake? Under what circumstances do earthquakes occur?

2. How are faults, foci, and epicenters related?

3. Who was first to explain the actual mechanism by which earthquakes are generated?

4. Explain what is meant by elastic rebound.

5. Faults that are experiencing no active creep may be considered safe. Rebut or defend this statement.

6. Describe the principle of a seismograph.

7. Using Figure 6.9, determine the distance between an earthquake and a seismic station if the first S wave arrives 3 minutes after the first P wave.

8. List the major differences between P and S waves.

9. Which type of seismic wave causes the greatest destruction to buildings?

10. Most strong earthquakes occur in a zone on the globe known as the _____.

11. What factor contributed most to the extensive damage that occurred in the central portion of Mexico City during the 1985 earthquake?

12. In 1988 an earthquake in Armenia caused approximately 25,000 deaths. A year later an earthquake in the Bay Area of California, having the same Richter magnitude of 6.9, had only 62 deaths. Suggest a reason for the greater loss of life in the Armenian event.

13. An earthquake measuring 7 on the Richter scale releases about _____ times more energy than an earthquake with a magnitude of 6.

14. List three reasons why the moment magnitude scale has gained wide acceptance among seismologists.

15. List four factors that affect the amount of destruction caused by seismic vibrations.

16. In addition to the destruction created directly by seismic vibrations, list three other types of destruction associated with earthquakes.

17. Distinguish between the Mercalli scale and the Richter scale.

18. What is a tsunami? In what two ways do earthquakes generate tsunami?

19. Cite some reasons why an earthquake with a moderate magnitude might cause more extensive damage than a quake with a high magnitude.

20. What evidence do we have that Earth's outer core is molten?

21. Contrast the physical makeup of the asthenosphere and the lithosphere.

22. Why are meteorites considered important clues to the composition of Earth's interior?

23. Describe the composition of the following:
    (a) continental crust   (c) mantle
    (b) oceanic crust   (d) core

Examining the Earth System

1. What potentially disastrous phenomenon often occurs when the energy of an earthquake is tranferred from the solid Earth to the hydrosphere (ocean) at their interface on the floor of the ocean? When the energy from this event is expended along a coast, how might coastal lands and the biosphere be altered?

2. Several research projects are considering ways that humans could interact with the earthquake-generating process and prevent, or significantly reduce, the release of destructive energy. Begin your research of earthquake prevention at this University of Colorado at Boulder Website, **http://www.colorado.edu/hazards/sites/earthquakes.html.** Following your investigation, describe some of the current earthquake-prevention proposals being considered by researchers. Do you believe that the proposals are feasible? What are some other ways to lessen the devastating impact of earthquakes?

Web Resources

The *Earth Science* Website uses the resources and flexibility of the Internet to aid in your study of the topics in this chapter. Written and developed by Earth science instructors, this site will help improve your understanding of Earth science. Visit **http://www.prenhall.com/tarbuck** and click on the cover of *Earth Science, 10e* to find:

- **On-line review quizzes.**
- **Web-based critical thinking and writing exercises.**
- **Links to chapter-specific Web resources.**
- **Internet-wide key term searches.**

http://www.prenhall.com/tarbuck

Composite satellite image of Europe, North Africa, and the Arabian Penisula. (Image © by Worldsat International, Inc., 2001. www.worldsat.ca. All Rights Reserved.)

CHAPTER 7

PLATE TECTONICS

Will California eventually slide into the ocean as some predict? Have continents really drifted apart over the centuries? Answers to these questions and many others that have intrigued geologists for decades are now being provided by an exciting theory on large-scale movements taking place within Earth. This theory, called plate tectonics, represents the real frontier of the Earth sciences, and its implications are so far-reaching that it can be considered the framework from which most other geological processes should be viewed.

Early in this century, most geologists thought that the geographic positions of the ocean basins and continents were fixed. During the last few decades, however, vast amounts of new data have dramatically changed our understanding of the nature and workings of our planet. Earth scientists now realize that the continents gradually migrate across the globe. Where landmasses split apart, new ocean basins are created between the diverging blocks. Meanwhile, older portions of the seafloor are carried back into the mantle in regions where trenches occur in the deep-ocean floor. Because of these movements, blocks of continental material eventually collide and form Earth's great mountain ranges (Figure 7.1). In short, a revolutionary new model of Earth's tectonic* processes has emerged.

This profound reversal of scientific understanding has been appropriately described as a scientific revolution. Like other scientific revolutions, considerable time elapsed between the idea's inception and its general acceptance. The revolution began early in the twentieth century as a relatively straightforward proposal that the continents drift about the face of Earth. After many years of heated debate, the idea of drifting continents was rejected by the vast majority of Earth scientists. However, during the 1950s and 1960s, new evidence rekindled interest in this proposal. By 1968, these new developments led to the unfolding of a far more encompassing theory than continental drift—a theory known as *plate tectonics.*

Continental Drift: An Idea Before Its Time

The idea that continents, particularly South America and Africa, fit together like pieces of a jigsaw puzzle originated with improved world maps. However, little

---

*Tectonics refers to the deformation of Earth's crust and results in the formation of structural features such as mountains.

Figure 7.1 Waputik Range, Banff National Park, Alberta, Canada. (Photo © by Carr Clifton)

significance was given this idea until 1915, when Alfred Wegener, a German meteorologist and geophysicist, published *The Origin of Continents and Oceans.* In this book, Wegener set forth his radical hypothesis of **continental drift.***

Wegener suggested that a supercontinent he called **Pangaea** (meaning all land) once existed (Figure 7.2). He further hypothesized that about 200 million years ago this supercontinent began breaking into smaller continents, which then drifted to their present positions (see Box 7.1).

Wegener and others collected substantial evidence to support these claims. The fit of South America and Africa and the geographic distribution of fossils, rock structures, and ancient climates all seemed to support the idea that these now separate landmasses were once joined. Let us examine their evidence.

Evidence: The Continental Jigsaw Puzzle

Like a few others before him, Wegener first suspected that the continents might have been joined when he noticed the remarkable similarity between the coastlines on opposite sides of the South Atlantic. However, his use of present-day shorelines to make a fit of the continents was challenged immediately by other Earth scientists. These opponents correctly argued that shorelines are continually modified by erosional processes,

---

*Wegener's ideas were actually preceded by those of an American geologist, F. B. Taylor, who in 1910 published a paper on continental drift. Taylor's paper provided little supporting evidence for continental drift, which may have been the reason that it had a relatively small impact on the scientific community.

STUDENTS SOMETIMES ASK...

If all the continents were joined during the time of Pangaea, what did the rest of Earth look like?

When all the continents were together, there must also have been one huge ocean surrounding them. This ocean is called *Panthalassa* (*pan* = all, *thalassa* = sea). About 200 million years ago the supercontinent of Pangaea began to split apart, and the various continental masses we know today started to drift toward their present geographic positions. Today all that remains of Panthalassa is the Pacific Ocean, which has been decreasing in size since the breakup of Pangaea.

---

and even if continental displacement had taken place, a good fit today would be unlikely. Wegener appeared to be aware of this problem, and, in fact, his original jigsaw fit of the continents was only very crude.

A much better approximation of the true outer boundary of the continents is the continental shelf. Today, the seaward edge of the continental shelf lies submerged, several hundred meters below sea level. In the early 1960s, scientists produced a map that attempted to fit the edges of the continental shelves at a depth of 900 meters. The remarkable fit that was obtained is shown in Figure 7.3. Although the continents overlap in a few places, these are regions where streams have deposited large quantities of sediment, thus enlarging the continental shelves. The overall fit was even better than what the supporters of continental drift suspected it would be.

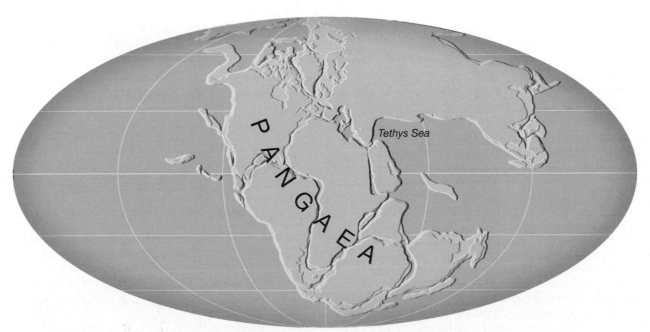

Figure 7.2   Reconstruction of Pangaea as it is thought to have appeared 200 million years ago. (After R. S. Deitz and J. C. Holden. *Journal of Geophysical Research* 75: 4943.)

Evidence: Fossils Match Across the Seas

Although Wegener was intrigued by the jigsaw fit of the continental margins that lie on opposite sides of the Atlantic, he at first thought the idea of a mobile Earth improbable. Not until he came across an article citing fossil evidence for the existence of a land bridge connecting South America and Africa did he begin to take his own idea seriously. Through a search of the literature, Wegener learned that most paleontologists were in agreement that some type of land connection was needed to explain the existence of identical fossils on the widely separated landmasses.

Mesosaurus. To add credibility to his argument for the existence of the supercontinent of Pangaea, Wegener cited documented cases of several fossil organisms that had been found on different landmasses but that could not have crossed the vast oceans presently separating the continents. The classic example is *Mesosaurus,* a presumably aquatic, snaggle-toothed reptile whose fossil remains are limited to eastern South America and southern Africa (Figure 7.4). If *Mesosaurus* had been able to swim well enough to cross the vast South Atlantic Ocean, its remains should be more widely distributed. As this is not the case, Wegener argued that South America and Africa must have been joined somehow.

How did scientists explain the discovery of identical fossil organisms separated by thousands of kilometers of open ocean? The idea of land bridges was the most widely accepted solution to the problem of migration (Figure 7.5). We know, for example, that during the most recent glacial period, the lowering of sea level allowed animals to cross the narrow Bering Strait between Asia and North America. Was it possible, then, that one or more land bridges once connected Africa and South America? We are now quite certain that land bridges of this magnitude did not exist, for their remnants should still lie below sea level. But they are nowhere to be found.

Present-Day Organisms. In his book, Wegener also cited the distribution of present-day organisms as evidence to support the concept of drifting continents. For example, modern organisms with similar ancestries clearly had to evolve in isolation during the last few tens of millions of years. Most obvious of these are the Australian marsupials, which have a direct fossil link to the marsupial opossums found in the Americas.

Evidence: Rock Types and Structures Match

Anyone who has worked a picture puzzle knows that, in addition to the pieces fitting together, the picture must also be continuous. The picture that must match in the Continental Drift Puzzle is one of rock types and mountain belts. If the continents were once together to form Pangaea, the rocks found in a particular region on one continent should closely match in age and type those in adjacent positions on the adjoining continent.

Such evidence exists in the form of several mountain belts that terminate at one coastline, only to reappear on a landmass across the ocean. For instance, the mountain belt that includes the Appalachians trends

---

BOX 7.1

UNDERSTANDING EARTH:
The Breakup of a Supercontinent

A great deal of evidence has been gathered to support the fact that Alfred Wegener's supercontinent of Pangaea began to break apart about 200 million years ago as he had originally suggested (Figure 7.A, part A). An important consequence of this continental rifting was the creation of a "new" ocean basin, the Atlantic. As you can see in Figure 7.A, part B, the rifting that eventually resulted in the Atlantic basin occurred over an extended period of time. The first rift developed between North America and Africa and generated great volumes of Jurassic-age basalts. These basalts are present as weathered flows and sills along the Eastern Seaboard of the United States but primarily buried beneath rocks that form the continental shelf. Radiometric dating of this material indicates that rifting began in various stages between 200 million and 165 million years ago. This time span can be used as the birth date for this section of the North Atlantic.

Rifting that formed the South Atlantic began about 130 million years ago near the tips of what are now Africa and South America. As this rift propagated northward, it gradually opened the South Atlantic (compare Figure 7.A, parts B and C). Continued rifting of the southern landmass of Gondwanaland sent India on a northward journey. By the early Cenozoic, about 50 million years ago, Australia began to separate from Antarctica, and the South Atlantic had emerged as a full-fledged ocean (Figure 7.A, part D).

A modern map (Figure 7.A, part E) shows that India eventually collided with Asia, an event that began about 45 million years ago and created the Himalayas as well as the Tibetan Highlands. About the same time, the separation of Greenland from Eurasia completed the breakup of the northern landmass (Laurasia). Also, notice the recent formation of the Baja Peninsula along with the Gulf of California. This event began less than 10 million years ago.

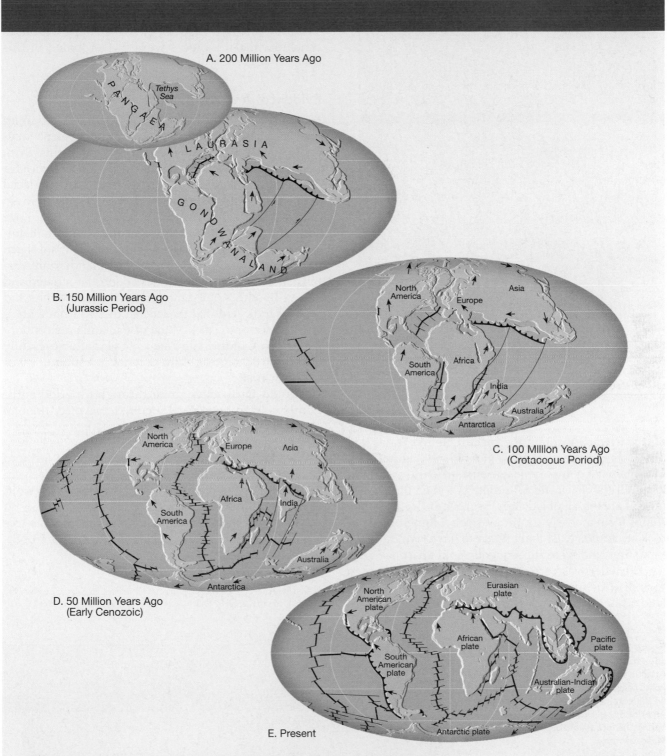

A. 200 Million Years Ago

B. 150 Million Years Ago
(Jurassic Period)

C. 100 Million Years Ago
(Cretaceous Period)

D. 50 Million Years Ago
(Early Cenozoic)

E. Present

Figure 7.A   Several views of the breakup of Pangaea over a period of 200 million years.

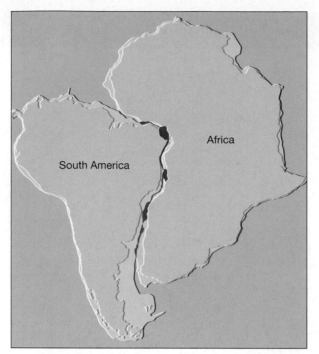

Figure 7.3 This shows the best fit of South America and Africa along the continental slope at a depth of 500 fathoms (about 900 meters). The areas where continental blocks overlap appear in brown. (After A. G. Smith, "Continental Drift," In *Understanding the Earth*, edited by I. G. Gass.)

to refit the torn pieces of a newspaper by matching their edges and then check whether the lines of print run smoothly across. If they do, there is nothing left but to conclude that the pieces were in fact joined in this way."*

Evidence: Ancient Climates

Because Alfred Wegener was a meteorologist by training, he was keenly interested in obtaining paleoclimatic (ancient climatic) data in support of continental drift. His efforts were rewarded when he found evidence for dramatic global climatic changes. For instance, glacial deposits indicate that near the end of the Paleozoic era (between 220 million and 300 million years ago), ice sheets covered extensive areas of the Southern Hemisphere. Layers of glacial till were found in southern Africa and South America, as well as in India and Australia. Below these beds of glacial debris lay striated and grooved bedrock. In some locations the striations and grooves indicated that the ice had moved from what is now the sea onto land. Much of the land area containing evidence of this late Paleozoic glaciation presently lies within 30 degrees of the equator in a subtropical or tropical climate.

Could Earth have gone through a period sufficiently cold to have generated extensive continental glaciers in what is presently a tropical region? Wegener rejected this explanation because, during the late Paleozoic, large tropical swamps existed in the Northern Hemisphere. The lush vegetation of these swamps eventually became the major coal fields of the eastern United States, Europe, and Siberia.

Fossils from these coal fields indicate that the tree ferns that produced the coal deposits had large fronds.

northeastward through the eastern United States and disappears off the coast of Newfoundland (Figure 7.6A). Mountains of comparable age and structure are found in the British Isles and Scandinavia. When these landmasses are reassembled as in Figure 7.6B, the mountain chains form a nearly continuous belt. Numerous other rock structures exist that appear to have formed at the same time and were subsequently split apart.

Wegener was very satisfied that the similarities in rock structure on both sides of the Atlantic linked these landmasses. In his own words, "It is just as if we were

---

*Alfred Wegener, *The Origin of Continents and Oceans*. Translated from the 4th revised German edition of 1929 by J. Birman (London: Methuen, 1966).

Figure 7.4 Fossils of *Mesosaurus* have been found on both sides of the South Atlantic and nowhere else in the world. Fossil remains of this and other organisms on the continents of Africa and South America appear to link these landmasses during the late Paleozoic and early Mesozoic eras.

Figure 7.5 These sketches by John Holden illustrate various explanations for the occurrence of similar species on landmasses that are presently separated by vast oceans. (Reprinted with permission of John Holden)

This indicates a tropical setting. Furthermore, unlike trees in colder climates, the tree trunks lacked growth rings. Growth rings do not form in tropical plants, because there are minimal seasonal fluctuations in temperature.

Wegener believed that a better explanation for the paleoclimatic regimes he observed is provided by fitting together the landmasses as a supercontinent, with South Africa centered over the South Pole (Figure 7.7). This would account for the conditions necessary to generate extensive expanses of glacial ice over much of the Southern Hemisphere. At the same time, this geogra-

phy would place the northern landmasses nearer the tropics and account for their vast coal deposits.

Wegener was so convinced that his explanation was correct that he wrote, "This evidence is so compelling that by comparison all other criteria must take a back seat."

How does a glacier develop in hot, arid Australia? How do land animals migrate across wide expanses of open water? As compelling as this evidence may have been, 50 years passed before most of the scientific community would accept it and the logical conclusions to which it led.

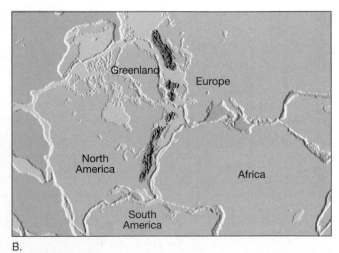

Figure 7.6 Matching mountain ranges across the North Atlantic. **A.** The Appalachian Mountains trend along the eastern flank of North America and disappear off the coast of Newfoundland. Mountains of comparable age and structure are found in the British Isles and Scandinavia. **B.** When these landmasses are placed in their predrift locations, these ancient mountain chains form a nearly continuous belt. These folded mountain belts formed roughly 300 million years ago as the landmasses collided during the formation of the supercontinent of Pangaea.

STUDENTS SOMETIMES ASK...

*Someday will the continents
come back together and form
a single landmass?*

Yes. It is very likely that the continents will come back together, but not anytime soon. Since all of the continents are on the same planetary body, there is only so far a continent can travel before it collides with other continents. Based on current plate motions, it appears that the continents may meet up again in the Pacific Ocean. Recent research suggests that the continents may form a supercontinent about once every 500 million years or so. Since it has been about 200 million years since Pangaea broke up, we have only about 300 million years before the next supercontinent is completed.

The Great Debate: Rejecting an Hypothesis

Wegener's proposal did not attract much open criticism until 1924 when his book was translated into English. From this time on, until his death in 1930, his drift hypothesis encountered a great deal of hostile criticism. To quote the respected American geologist T. C. Cham-

berlin, Wegener's hypothesis takes considerable liberty with our globe, and is less bound by restrictions or tied down by awkward, ugly facts than most of its rival theories. Its appeal seems to lie in the fact that it plays a game in which there are few restrictive rules and no sharply drawn code of conduct.

One of the main objections to Wegener's hypothesis stemmed from his inability to provide a mechanism that was capable of moving the continents across the globe. Wegener proposed that the tidal influence of the Moon was strong enough to give the continents a westward motion. However, the prominent physicist Harold Jeffreys quickly countered with the argument that tidal friction of the magnitude that's needed to displace the continents would bring Earth's rotation to a halt in a matter of a few years.

Wegener also proposed that the larger and sturdier continents broke through the oceanic crust, much like ice breakers cut through ice. However, no evidence existed to suggest that the ocean floor was weak enough to permit passage of the continents without themselves being appreciably deformed in the process.

Although most of Wegener's contemporaries opposed his views, even to the point of open ridicule, a few considered his ideas plausible. For these few geologists who continued the search for additional evidence,

Figure 7.7   **A.** The supercontinent Pangaea showing the area covered by glacial ice 300 million years ago. **B.** The continents as they are today. The shading outlines areas where evidence of the old ice sheets exists.

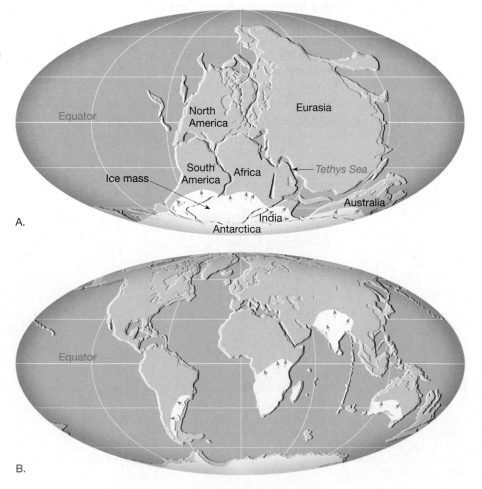

the exciting concept of continents in motion held their interest. Others viewed continental drift as a solution to previously unexplainable observations.

Plate Tectonics: The New Paradigm

GEODe   Forces Within
   ↳ Plate Tectonics

During the years that followed Wegener's proposal, major strides in technology permitted mapping of the ocean floor. Moreover, extensive data on seismic activity and Earth's magnetic field became available. By 1968, these developments led to the unfolding of a far more encompassing theory than continental drift, known as **plate tectonics.** The implications of plate tectonics are so far-reaching that this theory is today the framework within which to view most geologic processes.

Earth's Major Plates

According to the plate tectonics model, the uppermost mantle, along with the overlying crust, behaves as a strong, rigid layer, known as the **lithosphere.** This outermost shell overlies a weaker region in the mantle known as the **asthenosphere.** Further, the lithosphere is broken into numerous segments called **plates,** which are in motion and are continually changing in shape and size. As shown in Figure 7.8, seven major lithospheric plates are recognized. They are the North American, South American, Pacific, African, Eurasian, Australian-Indian, and Antarctic plates. The largest is the Pacific plate, which is located mostly within the Pacific Ocean. Notice from Figure 7.8 that several of the large plates include an entire continent plus a large area of seafloor (for example, the South American plate). This is a major departure from Wegener's continental drift hypothesis, which proposed that the continents moved through the ocean floor, not with it. Note also that none of the plates are defined entirely by the margins of a continent.

Intermediate-sized plates include the Caribbean, Nazca, Philippine, Arabian, Cocos, and Scotia plates. In addition, there are over a dozen smaller plates that have been identified but are not shown in Figure 7.8.

The lithospheric plates move relative to each other at a very slow but continuous rate that averages about 5 centimeters (2 inches) per year. This movement is ultimately driven by the unequal distribution of heat within Earth. Hot material found deep in the mantle moves slowly upward and serves as one part of our planet's internal convection system. Concurrently, cooler, denser slabs of oceanic lithosphere descend into the mantle, setting Earth's rigid outer shell into motion. Ultimately, the titanic, grinding movements of Earth's lithospheric plates generate earthquakes, create volcanoes, and deform large masses of rock into mountains.

Plate Boundaries

Lithospheric plates move as coherent units relative to all other plates. Although the interiors of plates may experience some deformation, all major interactions among individual plates (and therefore most deformation) occur along their *boundaries.* In fact, the first attempts to outline plate boundaries were made using locations of earthquakes. Later work showed that plates are bounded by three distinct types of boundaries, which are differentiated by the type of movement they exhibit. These boundaries are depicted at the bottom of Figure 7.8 and are briefly described here:

1. **Divergent plate boundaries** (*constructive margins*)—where two plates move apart, resulting in upwelling of material from the mantle to create new seafloor (Figure 7.8A).
2. **Convergent plate boundaries** (*destructive margins*)—where two plates move together, resulting in oceanic lithosphere being thrust beneath an overriding plate, eventually to be reabsorbed into the mantle (Figure 7.8B). Convergence can also result in the collision of two continental plates to create a mountain system.
3. **Transform fault boundaries** (*conservative margins*)—where two plates grind past each other without the production or destruction of lithosphere (Figure 7.8C).

Each plate is bounded by a combination of these three types of boundaries. For example, as shown in Figure 7.8, the Nazca plate has a divergent zone on the west, a convergent boundary on the east, and numerous transform faults, which offset segments of the divergent boundary. Although the total surface area of Earth does not change, individual plates may diminish or grow in area depending on the distribution of convergent and divergent boundaries. The Antarctic and African plates are almost entirely bounded by spreading centers and hence are growing larger. By contrast, the Pacific plate is being consumed into the mantle along its northern and western flanks and is therefore diminishing in size.

Furthermore, new plate boundaries can be created in response to changes in the forces acting on these rigid slabs. For example, a relatively new divergent boundary is located in Africa, in a region known as the East African Rift Valleys. If spreading continues there, the African plate will split into two plates separated by a new ocean basin. At other locations, plates carrying continental crust are presently moving toward each other. Eventually, these continents may collide and be sutured together. Thus, the boundary that once separated two plates disappears as the plates become one. The result

Figure 7.8 A mosaic of rigid plates constitutes Earth's outer shell.

A. Divergent boundary

B. Convergent boundary

C. Transform fault boundary

Figure 7.8   Continued

of such a continental collision is a majestic mountain range such as the Himalayas.

In the following sections we will briefly summarize the nature of the three types of plate boundaries.

Divergent Plate Boundaries

GEODe   Forces Within
↳ Plate Tectonics

Most divergent (*di* = apart, *vergere* = to move) plate boundaries are located along the crests of oceanic ridges and can be thought of as *constructive plate margins* since this is where new oceanic lithosphere is generated (Figure 7.9). Here, as the plates move away from the ridge axis, the fractures created are filled with molten rock that wells up from the hot mantle below. Gradually, this magma cools to produce new silvers of seafloor. In a continuous manner, spreading and upwelling of magma adds oceanic lithosphere between the diverging plates. As we shall see later, divergent boundaries can also form on the continents (Figure 7.10).

Oceanic Ridges and Seafloor Spreading

Along well-developed divergent plate boundaries, the seafloor is elevated forming the **oceanic ridge.** The interconnected ocean ridge system is the longest topographic feature on Earth's surface, exceeding 70,000 kilometers (43,000 miles) in length with crests commonly 2 to 3 kilometers higher than the adjacent ocean basins. Representing 20 percent of Earth's surface, the ocean ridge system winds through all major ocean basins in a manner similar to the seam on a baseball. The term "ridge" may be misleading, as these features are not narrow but have widths from 1000 to 4000 kilometers. Further, along the axis of some segments are deep down-faulted structures called **rift valleys.**

The mechanism that operates along the oceanic ridge system to create new seafloor is appropriately called **seafloor spreading.** Typical rates of spreading average around 5 centimeters (2 inches) per year. Comparatively slow spreading rates of 2 centimeters per year are found in the North Atlantic, whereas spreading rates exceeding 15 centimeters (6 inches) have been measured along

Figure 7.9 Most divergent plate boundaries are situated along the crests of oceanic ridges.

sections of the East Pacific Rise. Although these rates of lithospheric production are slow on a human time scale, they are nevertheless rapid enough so that all of Earth's ocean basins could have been generated within the last 200 million years. In fact, none of the ocean floor that has been dated exceeds 180 million years in age.

During seafloor spreading, the magma that is injected into newly developed fractures forms dikes that tend to cool from their outer borders inward toward their centers. Because the warm interiors of these newly formed dikes are weak, continued spreading produces new fractures that split these young rocks roughly in half. As a result, new material is added about equally to the two diverging plates. Consequently, new ocean floor grows symmetrically on each side of a centrally located ridge

crest. Indeed, the ridge systems of the Atlantic and Indian oceans are located near the middle of these water bodies and as a consequence are called mid-ocean ridges. However, the East Pacific Rise is situated far from the center of the Pacific Ocean. Despite uniform spreading along the East Pacific Rise, much of the Pacific Basin that once lay east of this spreading center has been overridden by the westward migration of the American plates.

Spreading Rates and Ridge Topography

When various segments of the oceanic ridge system were studied in detail, some topographic differences came to light. Most of these differences appear to be

Figure 7.10 **A.** Rising magma forces the crust upward, causing numerous cracks in the rigid lithosphere. **B.** As the crust is pulled apart, large slabs of rock sink, generating a rift zone. **C.** Further spreading generates a narrow sea. **D.** Eventually, an expansive ocean basin and ridge system are created.

controlled by spreading rates. At comparatively slow spreading rates of 1 to 5 centimeters per year, such as occur at the Mid-Atlantic and Mid-Indian ridges, a prominent rift valley develops along the ridge crest. This structure is usually 30 to 50 kilometers across and between 1500 to 3000 meters deep. Here, the displacement of large slabs of oceanic crust along nearly vertical faults and outpourings of pillow lavas contribute to the characteristically rugged topography of these rift valleys.

Along the Galapagos ridge and the northernmost section of the East Pacific Rise, an intermediate spreading rate of 5 to 9 centimeters per year is the norm. In these settings the rift valleys that develop are shallow, often less than 200 meters deep, and their topography is relatively smooth.

At faster spreading rates (greater than 9 centimeters per year), such as those which occur along much of the East Pacific Rise, no median rift valleys develop. Here oceanic ridges are usually narrow (roughly 10 kilometers wide) topographic highs. These elevated structures are extensively faulted and exhibit topography consisting of numerous horsts and grabens (see Chapter 9). Faulting is the primary cause of topo-

graphic variations along fast spreading centers, whereas the buildup of volcanic structures is significant at slow spreading centers.

Continental Rifts

Spreading centers can also develop within a continent, in which case the landmass may split into two or more smaller segments, as Alfred Wegener had proposed for the breakup of the Pangaea. Examples of active continental rifts include the East African rift valleys, the Baikal Rift (south central Siberia), the Rhine Valley (Northwest Europe), the Rio Grande Rift, and the Basin and Range province in the western United States. Whether any of these rifts will continue to develop and eventually split a continent is a matter of much speculation.

The most widely accepted model for continental splitting suggests that extensional forces must be acting on the lithospheric plate. These forces are thought to arise from the "pull" of cold lithospheric plates as they subduct along the margins of a continent. It appears that by themselves these extensional forces are not great enough to actually tear the lithosphere apart. Rather, the rupture of the lithosphere is initiated only in those settings where plumes of hot rock rise from the mantle. The

Figure 7.11 East African rift valleys and associated features.

effect of this hot-spot activity is weakening of the lithosphere and doming of the crust directly above the hot rising plume. Uplifting stretches and thins the crust, as shown in Figure 7.10A. Extension is accompanied by alternating episodes of faulting and volcanism that result in the development of a rift valley that resembles those found along the axes of some ridges (Figure 7.10B).

The East African rift valleys may represent the initial stage in the breakup of a continent as described above (Figure 7.11). Large volcanic mountains such as Kilimanjaro and Mount Kenya exemplify the extensive volcanic activity that accompanies continental rifting.

In those settings where the extensional forces are maintained, the rift valley will lengthen and deepen, eventually extending out to the margin of the plate and splitting it in two (Figure 7.10C). At this point the rift becomes a narrow linear sea with an outlet to the ocean, similar to the Red Sea (Figure 7.11). The Red Sea formed

Figure 7.12 Three types of convergent plate boundaries. **A.** Oceanic-continental. **B.** Oceanic-oceanic. **C.** Continental-continental.

A.

B.

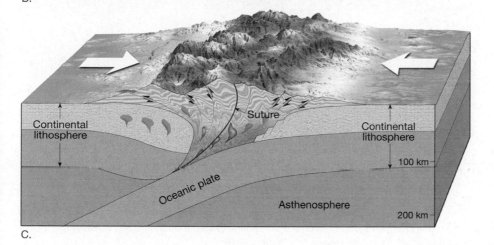

C.

when the Arabian Peninsula rifted from Africa, an event that began about 20 million years ago. Consequently, the Red Sea provides oceanographers with a view of how the Atlantic Ocean may have looked in its infancy.

Not all rift valleys develop into full-fledged spreading centers. Running through the central United States is a failed rift extending from Lake Superior to Oklahoma. This once active rift valley is filled with volcanic rock that was extruded onto the crust more than a billion years ago. Why one rift valley develops into an active spreading center while others are abandoned is not yet known.

Convergent Plate Boundaries

Forces Within
↳ Plate Tectonics

Although new lithosphere is constantly being added at the oceanic ridges, our planet is not growing larger—its total surface area remains constant. To accommodate the newly created lithosphere, older portions of oceanic plates return to the mantle along convergent (*con* = together, *vergere* = to move) plate boundaries. (Because lithosphere is "destroyed" at convergent boundaries, they are also called *destructive plate margins*.) As two plates slowly converge, the leading edge of one is bent downward, allowing it to slide beneath the other. The surface expression produced by the descending plate is an ocean **trench,** like the Peru–Chile trench (Figure 7.12, on previous page). Trenches formed in this manner may be thousands of kilometers long, 8 to 12 kilometers deep, and between 50 and 100 kilometers wide (Figure 7.13). Destructive plate margins where oceanic crust is being consumed in the mantle are called **subduction** (*sub* = under, *duct* = lead) **zones.**

The average angle at which oceanic lithosphere descends into the mantle is about 45 degrees. However, depending on its buoyancy, a plate might descend at an angle as small as a few degrees or it might plunge vertically (90 degrees) into the mantle. When a spreading center is located near a subduction zone, the litho-

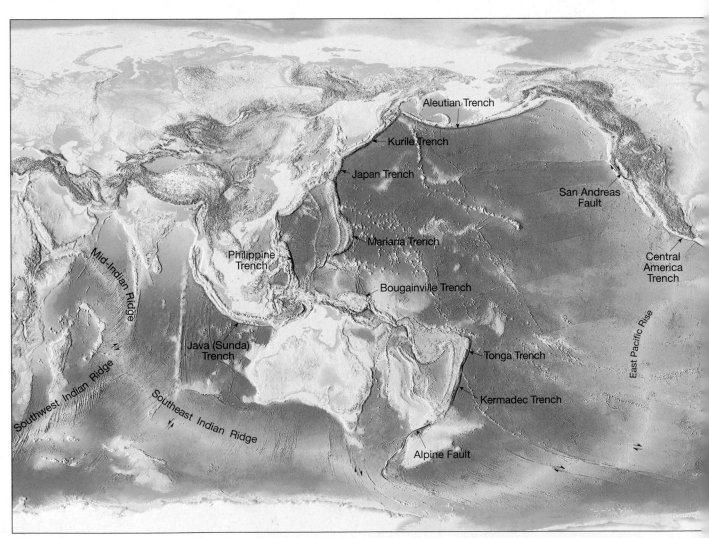

Figure 7.13 Distribution of the world's oceanic trenches, ridge system, and transform faults. Where transform faults offset ridge segments, they permit the ridge to change direction (curve) as can be seen in the Atlantic Ocean.

sphere is young and, therefore, warm and buoyant. Thus, the angle of descent is small. This is the situation along parts of the Peru–Chile trench. Low dip angles are usually associated with a strong coupling between the descending slab and the overriding plate. Consequently, these regions experience great earthquakes. By contrast, some subduction zones, such as the Mariana trench, have steep angles of descent and few strong earthquakes.

Although all convergent zones have the same basic characteristics, they are highly variable features. Each is controlled by the type of crustal material involved and the tectonic setting. Convergent boundaries can form between two oceanic plates, one oceanic and one continental plate, or two continental plates. All these situations are illustrated in Figure 7.12.

Oceanic-Continental Convergence

Whenever the leading edge of a continental plate converges with an oceanic plate, the buoyant continental plate remains floating, whereas the denser oceanic slab

sinks into the asthenosphere (Figure 7.12A). When a descending plate reaches a depth of about 100 to 150 kilometers, heat drives water and other volatile components from the subducted sediments into the overlying mantle. These substances act as a flux does at a foundry, inducing partial melting of mantle rocks at reduced temperatures. The partial melting of mantle rock generates magmas having a basaltic or occasionally an andesitic composition. The newly formed magma, being less dense than the rocks of the mantle, will buoyantly rise. Often the magma, being more dense than continental rocks, will pond beneath the overlying continental crust where it may melt some of the silica-enriched rocks. Eventually some of this silica-rich magma may migrate to the surface, where it can give rise to volcanic eruptions, some of which are explosive.

The volcanoes of the Andean arc located along the western flank of South America are the product of magma generated as the Nazca plate descends beneath the continent (see Figure 7.8). In the central section of the southern Andes the subduction angle is very shallow, which probably accounts for the lack of volcanism in that area. As the South American plate moves westward, it overruns the Nazca plate. The result is a seaward migration of the Peru–Chile trench and a reduction in the size of the Nazca plate.

Mountains such as the Andes, which are produced in part by volcanic activity associated with the subduction of oceanic lithosphere, are called **continental volcanic arcs.** Another active continental volcanic arc is located in the western United States. The Cascade Range of Washington, Oregon, and California consists of several well-known volcanic mountains, including Mount Rainier, Mount Shasta, and Mount St. Helens. As the continuing activity of Mount St. Helens testifies, the Cascade Range is still active. The magma here arises from melting triggered by the subduction of the Juan de Fuca plate—a small remaining remnant of a once large plate.

A remnant of a formerly extensive continental volcanic arc is California's Sierra Nevada, in which Yosemite National Park is located. The Sierra Nevada is much older than the Cascade Range, and has been inactive for several million years as evidenced by the absence of volcanic cones. Here erosion has stripped away most of the obvious traces of volcanic activity and left exposed the large, crystallized magma chambers that once fed lofty volcanoes.

Oceanic-Oceanic Convergence

When two oceanic slabs converge, one descends beneath the other, initiating volcanic activity in a manner similar to that which occurs at an oceanic-continental boundary. In this case, however, the volcanoes form on the ocean floor rather than on a continent (Figure 7.12B). If this activity is sustained, it will eventually build a chain of volcanic structures that emerge as islands. The volcanic islands are spaced about 80 kilometers apart and are built upon submerged ridges a few hundred kilometers wide. This newly formed land consisting of an arc-shaped chain of small volcanic islands is called a **volcanic island arc.** The Aleutian, Mariana, and Tonga islands are examples of volcanic island arcs (see Box 7.2). Island arcs such as these are gen-

erally located 200 to 300 kilometers from the trench axis. Located adjacent to the island arcs just mentioned are the Aleutian trench, the Mariana trench, and the Tonga trench (Figure 7.13).

Only two volcanic island arcs are located in the Atlantic, the Lesser Antilles arc adjacent to the Caribbean

BOX 7.2
UNDERSTANDING EARTH:
Susan DeBari—A Career in Geology

I discovered geology the summer that I worked doing trail maintenance in the North Cascades mountains of Washington State. I had just finished my freshman year in college and had never before studied Earth Science. But a coworker (now my best friend) began to describe the geological features of the mountains that we were hiking in—the classic cone shape of Mount Baker volcano, the U-shaped glacial valleys, the advance of active glaciers, and other wonders. I was hooked and went back to college that fall with a geology passion that hasn't abated. As an undergraduate, I worked as a field assistant to a graduate student and did a senior thesis project on rocks from the Aleutian island arc. From that initial spark, island arcs have remained my top research interest, on through Ph.D. research at Stanford University, postdoctoral work at the University of Hawaii, and as a faculty member at San Jose State University and Western Washington University.

Especially the deep crust of arcs, the material that lies close to the Mohorovičić discontinuity (fondly known as the Moho). What kinds of processes are occurring down there at the base of the crust in island arcs? What is the source of magmas that make their way to the surface (the mantle? the deep crust itself?). How do these magmas interact with the crust as they make their way upwards? What do these early magmas look like chemically? Are they very different from what is erupted at the surface?

Obviously, geologists cannot go down to the base of the crust (typically 20 to 40 kilometers beneath Earth's surface). So what they do is play a bit of a detective game. They must use rocks that are *now exposed at the surface* that were originally formed in the deep crust of an island arc. The rocks must have been brought to the surface rapid-

ly along fault zones to preserve their original features. Thus, I can walk on rocks of the deep crust without really leaving the Earth's surface! There are a few places around the world where these rare rocks are exposed. Some of the places that I have worked include the Chugach Mountains of Alaska, the Sierras Pampeanas of Argentina, the Karakorum range in Pakistan, Vancouver Island's west coast, and the North Cascades of Washington. Fieldwork has involved hiking most commonly, but also extensive use of mules and trucks.

I also went looking for exposed pieces of the deep crust of island arcs in a less obvious place, in one of the deepest oceanic trenches of the world, the Izu Bonin trench (Figure 7.B). Here I dove into the ocean in a submersible called the *Shinkai 6500* (pictured to my right in the background). The *Shinkai 6500* is a Japanese submersible that has the capability to dive to 6500 meters below the surface of the ocean (approx-

imately 4 miles). My plan was to take rock samples from the wall of the trench at its deepest levels using the submersible's mechanical arm. Because preliminary data suggested that vast amounts of rock were exposed for several kilometers in a vertical sense, this could be a great way to sample the deep arc basement. I dove in the submersible three times, reaching a maximum depth of 6497 meters. Each dive lasted 9 hours, spent in a space no bigger than the front seat of a Honda, shared with two of the Japanese pilots that controlled the submersible's movements. It was an exhilarating experience!

I am now on the faculty at Western Washington University where I continue to do research on the deep roots of volcanic arcs, and get students involved as well. I am also involved in Science Education training for K–12 teachers, hoping to get young people motivated to ask questions about the fascinating world that surrounds them!

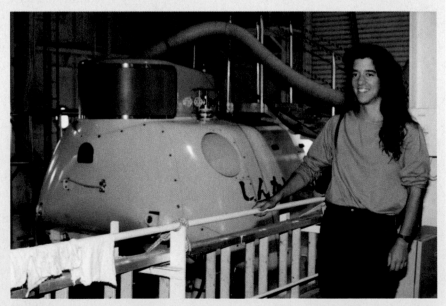

Figure 7.B Susan DeBari photographed with the Japanese submersible, *Shinkai 6500*, which she used to collect rock samples from the Izu Bonin trench. (Photo courtesy of Susan DeBari)

Sea, and the Sandwich Islands in the South Atlantic. The Lesser Antilles are a product of the subduction of the Atlantic beneath the Caribbean plate. Located within this arc is the island of Martinique where Mount Pelé erupted in 1902, destroying the town of St. Pierre and killing an estimated 28,000 people, and the island of Montserrat, where volcanic activity has occurred very recently.

Relatively young island arcs are fairly simple structures that are underlain by crust that is generally less than 20 kilometers (12 miles) thick. Examples include the arcs of Tonga, the Aleutians, and the Lesser Antilles. By contrast, older island arcs are more complex and are underlain by crust that ranges in thickness from 20 to 35 kilometers. Examples include the Japanese and Indonesian arcs, which are built upon material generated by earlier episodes of subduction. In a few places, volcanic island arcs are gradually transformed into an Andean-type volcanic chain. For example, the western section of the Aleutian arc consists of numerous volcanic islands built on oceanic crust whereas the volcanoes at the eastern end of the chain are part of the Alaskan Peninsula—a piece of continental crust.

Continental–Continental Convergence

As you saw earlier, when an oceanic plate is subducted beneath continental lithosphere, an Andean-type volcanic arc develops along the margin of the continent. However, if the subducting plate also contains conti-

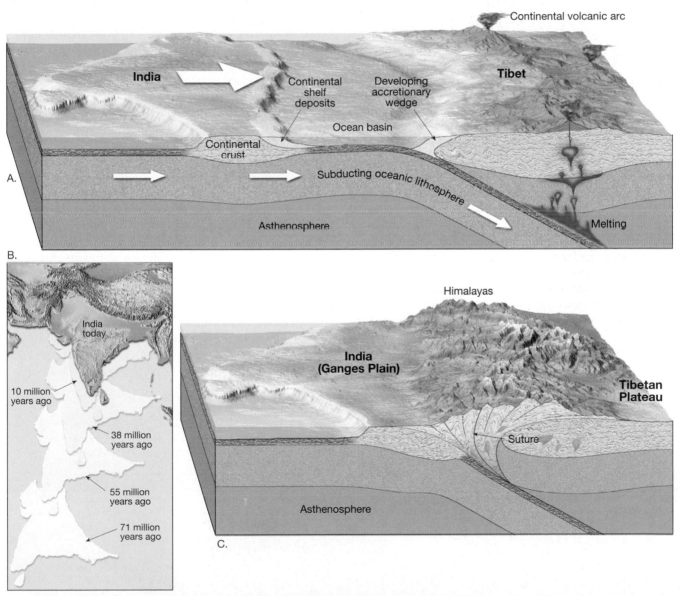

Figure 7.14   The ongoing collision of India and Asia, starting about 45 million years ago, produced the majestic Himalayas. **A.** Converging plates generated a subduction zone, while partial melting triggered by the subducting oceanic slab produced a continental volcanic arc. Sediments scraped from the subducting plate were added to the accretionary wedge. **B.** Position of India in relation to Eurasia at various times. (Modified after Peter Molnar) **C.** Eventually the two landmasses collided, deforming and elevating the accetionary wedge and continental shelf deposits. In addition, slices of the Indian crust were thrust up onto the Indian plate.

nental lithosphere, continued subduction eventually brings the two continents together (Figure 7.12C). Whereas oceanic lithosphere is relatively dense and sinks into the asthenosphere, continental lithosphere is buoyant, which prevents it from being subducted to any great depth. The result is a collision between the two continental blocks (Figure 7.14).

Such a collision occurred when the subcontinent of India rammed into Asia and produced the Himalayas—the most spectacular mountain range on Earth (Figure 7.14). During this collision, the continental crust buckled, fractured, and was generally shortened and thickened. In addition to the Himalayas, several other major mountain systems, including the Alps, Appalachians, and Urals, formed during continental collisions.

Prior to a continental collision, the landmasses involved are separated by an ocean basin. As the continental blocks converge, the intervening seafloor is subducted beneath one of the plates. Subduction initiates partial melting in the overlying mantle rocks, which in turn results in the growth of a volcanic arc. Depending on the location of the subduction zone, the volcanic arc could develop on either of the converging landmasses, or if the subduction zone developed several hundred kilometers seaward from the coast, a volcanic island arc would form. Eventually, as the intervening seafloor is consumed, these continental masses collide. This folds and deforms the accumulation of sediments along the continental margin as if they were placed in a gigantic vise. The result is the formation of a new mountain range, composed of deformed and metamorphosed sedimentary rocks, fragments of the volcanic arc, and possibly slivers of oceanic crust.

After continents collide, the subducted oceanic plate may separate from the continental block and continue its downward movement. However, because of its buoyancy, continental lithosphere cannot be carried very far into the mantle. In the case of the Himalayas, the leading edge of the Indian plate was forced partial-ly under Asia, generating an unusually great thickness of continental lithosphere. This accumulation accounts, in part, for the high elevation of the Himalayas and may help explain the elevated Tibetan Plateau to the north.

Transform Fault Boundaries

GEODe   Forces Within
↳ Plate Tectonics

The third type of plate boundary is the transform fault, where plates grind past one another without the production or destruction of lithosphere. Transform faults were first identified where they join offset segments of an ocean ridge. At first it was erroneously assumed that the ridge system originally formed a long and continuous chain that was later offset by horizontal displacement along these large faults. However, the displacement along these faults was found to be in the exact opposite direction required to produce the offset ridge segments.

The true nature of transform faults was discovered in 1965 by a Canadian researcher who suggested that these large faults connect the global active belts (convergent boundaries, divergent boundaries, and other transform faults) into a continuous network that divides Earth's outer shell into several rigid plates. Thus, he became the first to suggest that Earth was made of individual plates, while at the same time identifying the faults along which relative motion between the plates is made possible.

Most transform faults join two segments of a mid-ocean ridge (Figure 7.15). Here, they are part of prominent linear breaks in the oceanic crust known as *fracture zones* that include both the transform fault as well as their inactive extension into the plate interior. These fracture zones are present approximately every 100 kilometers along the trend of a ridge axis. As shown in Figure 7.15, active transform faults lie only *between* the two offset

Figure 7.15 Diagram illustrating a transform fault boundary offsetting segments of a divergent boundary (oceanic ridge).

ridge segments. Here seafloor produced at one ridge axis moves in the opposite direction as seafloor is produced at an opposing ridge segment. Thus, between the ridge segments these adjacent slabs of oceanic crust are grinding past each other along a transform fault. Beyond the ridge crests are the inactive zones, which are preserved as linear topographic scars. These fracture zones tend to curve such that small segments roughly parallel the direction of plate motion at the time of their formation.

In another role, transform faults provide the means by which the oceanic crust created at ridge crests can be transported to a site of destruction, the deep-ocean trenches. Figure 7.16 illustrates this situation. Notice that the Juan de Fuca plate moves in a southeasterly direction, eventually being subducted under the West Coast of the United States. The southern end of this relatively small plate is bounded by the Mendocino transform fault. This transform fault boundary connects the Juan de Fuca ridge to the Cascade subduction zone (Fig-

ure 7.16). Therefore, it facilitates the movement of the crustal material created at the ridge crest to its destination beneath the North American continent (Figure 7.16).

Another example of a *ridge-trench* transform fault is found southeast of the tip of South America. Here transform faults on the north and south margins of the Scotia plate connect the trench to a short spreading axis (see Figure 7.8).

Although most transform faults are located within the ocean basins, a few cut through continental crust. Two examples are the earthquake-prone San Andreas Fault of California and the Alpine Fault of New Zealand. Notice in Figure 7.16 that the San Andreas Fault connects a spreading center located in the Gulf of California to the Cascade subduction zone and the Mendocino transform fault located along the northwest coast of the United States. Along the San Andreas Fault, the Pacific plate is moving toward the northwest, past the North American plate. If this movement continues,

Figure 7.16   The role of transform faults. The Mendocino transform fault permits seafloor generated at the Juan de Fuca ridge to move southeastward past the Pacific plate and beneath the North American plate. Thus, this transform fault connects a divergent boundary to a subduction zone. Futhermore, the San Andreas Fault, also a transform fault, connects two spreading centers—the Juan de Fuca Ridge and a divergent zone located in the Gulf of California.

that part of California west of the fault zone, including the Baja Peninsula, will become an island off the West Coast of the United States and Canada. It could eventually reach Alaska. However, a more immediate concern is the earthquake activity triggered by movements along this fault system.

STUDENTS SOMETIMES ASK...

*If the continents move,
do other features like segments
of the oceanic ridge also move?*

That's a good observation, and yes, they do! It is interesting to note that very little is really fixed in place on Earth's surface. When we talk about movement of features on Earth, we must consider the question, "Moving relative to what?" Certainly, the oceanic ridge does move relative to the continents (which sometimes causes segments of the oceanic ridges to be subducted beneath the continents). In addition, the oceanic ridge is moving relative to a fixed location outside Earth. This means that an observer orbiting above Earth would notice, after only a few million years, that all continental and seafloor features—as well as plate boundaries—are indeed moving. The exception to this are hot spots, which seem to be relatively stationary and can be used to determine the motions of other features.

Testing the Plate Tectonics Model

With the birth of the plate tectonics model, researchers from all of the Earth sciences began testing it. Some of the evidence supporting continental drift and seafloor spreading has already been presented. Some of the evidence that was instrumental in solidifying the support for this new concept follows. Note that some of the evidence was not new; rather, it was a new interpretation of old data that swayed the tide of opinion.

Evidence: Paleomagnetism

Probably the most persuasive evidence to the geologic community for the acceptance of the plate tectonics theory comes from the study of Earth's magnetic field. Anyone who has used a compass to find direction knows that the magnetic field has a north pole and a south pole. These magnetic poles align closely, but not exactly, with the geographic poles. (The geographic poles are simply the top and bottom of the spinning sphere we live on, the points through which passes the imaginary axis of rotation.)

In many respects the magnetic field is very much like that produced by a simple bar magnet. Invisible lines of force pass through Earth and extend from one pole to the other. A compass needle, itself a small magnet free to move about, becomes aligned with these lines of force and thus points toward the magnetic poles.

The technique used to study ancient magnetic fields relies on the fact that certain rocks contain minerals that serve as fossil compasses. These iron-rich minerals, such as magnetite, are abundant in lava flows of basaltic composition. When heated above a certain temperature called the *Curie point*, these magnetic minerals lose their magnetism. However, when these iron-rich grains cool below their Curie point (about 580°C), they become magnetized in the direction parallel to the existing magnetic field. Once the minerals solidify, the magnetism they possess will remain frozen in this position. In this regard, they behave much like a compass needle inasmuch as they point toward the existing magnetic poles. Then, if the rock is moved or if the magnetic pole changes position, the rock magnetism will, in most instances, retain its original alignment. Rocks formed thousands of millions of years ago thus remember the location of the magnetic poles at the time of their formation and are said to possess fossil magnetism, or **paleomagnetism.**

Polar Wandering. A study of lava flows conducted in Europe in the 1950s led to an amazing discovery. The magnetic alignment in the iron-rich minerals in lava flows of different ages was found to vary widely. A plot of the apparent positions of the magnetic north pole revealed that during the past 500 million years, the location of the pole had gradually wandered from a spot near Hawaii northward through eastern Siberia and finally to its present site (Figure 7.17A). This was clear evidence that either the magnetic poles had migrated through time, an idea known as **polar wandering,** or that the lava flows had moved—in other words, the continents had drifted.

Although the magnetic poles are known to move, studies of the magnetic field indicated that the average positions of the magnetic poles correspond closely to the positions of the geographic poles. This is consistent with our knowledge of Earth's magnetic field, which is generated in part by the rotation of Earth about its axis. If the geographic poles do not wander appreciably, which we believe is true, neither can the magnetic poles. Therefore, a more acceptable explanation for the apparent polar wandering is provided by the plate tectonics theory. *If the magnetic poles remain stationary, their apparent movement was produced by the drifting of the continents.*

Further evidence for plate tectonics came a few years later when polar wandering curves were constructed for North America and Europe (Figure 7.17A). To nearly everyone's surprise, the curves for North America and Europe had similar paths, except that they were separated by about 24 degrees of longitude. When these rocks solidified, could there have been two magnetic north poles that migrated parallel to each other? This is very unlikely. The differences in these migration paths, however, can be reconciled if the two presently

Figure 7.17 Simplified apparent polar wandering paths as established from North American and Eurasian paleomagnetic data. **A.** The more westerly path determined from North American data is thought to have been caused by the westward drift of North America by about 24 degrees from Eurasia. **B.** The positions of the wandering paths when the landmasses are reassembled in their predrift locations.

separated continents are placed next to one another, as we now believe they were prior to the opening of the Atlantic Ocean (Figure 7.17B).

Magnetic Reversals and Seafloor Spreading.   Another discovery came when geophysicists learned that Earth's magnetic field periodically reverses polarity; that is, the north magnetic pole becomes the south magnetic pole,

and vice versa. A rock solidifying during one of the periods of reverse polarity will be magnetized with the polarity opposite that of rocks being formed today.

When rocks exhibit the same magnetism as the present magnetic field, they are said to possess **normal polarity,** whereas those rocks exhibiting the opposite magnetism are said to have **reverse polarity.** Evidence for magnetic reversals was obtained from lavas and sediments from around the world.

Once the concept of magnetic reversals was confirmed, researchers set out to establish a time scale for polarity reversals. Many areas exist where volcanic activity has occurred sporadically for periods of millions of years (Figure 7.18). The task was to measure the directions of paleomagnetism in numerous lava flows of various ages. These data were collected from several places and were used to determine the dates when the polarity of Earth's magnetic field changed. Figure 7.19 shows the time scale of the polarity reversals established for the last few million years.

A significant relationship was uncovered between the magnetic reversals and the seafloor-spreading hypothesis. Very sensitive instruments called *magnetometers* were towed by research vessels across a segment of the ocean floor located off the West Coast of the United States. Here workers from the Scripps Institute of Oceanography discovered alternating strips of high- and low-intensity magnetism that trended in roughly a north-south direction. This relatively simple pattern of magnetic variation defied explanation until 1963, when it was tied to the concept of seafloor spreading. The strips of high-intensity magnetism are regions where the paleomagnetism of the ocean crust is of the normal type. Consequently, these positively magnetized rocks *enhance* the existing magnetic field. Conversely, the low-intensity strips represent regions where the ocean crust is polarized in the reverse direction and, therefore, *weaken* the existing magnetic field. But how do parallel strips of normally and reversely magnetized rock become distributed across the ocean floor?

As new basalt is added to the ocean floor at the oceanic ridges, it becomes magnetized according to the existing magnetic field (Figure 7.20). Because new rock is added in approximately equal amounts to the trailing edges of both plates, we should expect strips of equal size and polarity to parallel both sides of the ocean ridges, as shown in Figure 7.20. This explanation of the alternating strips of normal and reverse polarity, which lay as mirror images across the ocean ridges, was the strongest evidence so far presented in support of the concept of seafloor spreading.

Now that the dates of the most recent magnetic reversals have been established, the rate at which spreading occurs at the various ridges can be determined accurately. In the Pacific Ocean, for example, the magnetic strips are much wider for corresponding time intervals than those of the Atlantic Ocean. Hence, we

Figure 7.18 Schematic illustration of paleomagnetism preserved in lava flows of various ages. Data such as these from various locales were used to establish the time scale of polarity reversals shown in Figure 7.19.

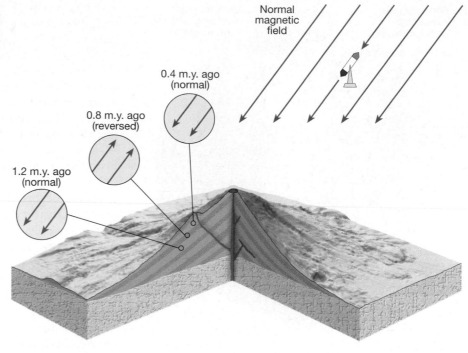

conclude that a faster spreading rate exists for the spreading center of the Pacific as compared to the Atlantic. When we apply absolute dates to these magnetic events, we find that the spreading rate for the North Atlantic Ridge is only 2 centimeters per year. The rate is somewhat faster for the South Atlantic. The spreading rates for the East Pacific Rise generally range between 6 and 12 centimeters per year, with a maximum rate exceeding 15 centimeters (6 inches) per year in one segment. Thus, we have a magnetic tape recorder that records changes in Earth's magnetic field. This recorder also permits us to determine the rate of seafloor spreading.

Evidence: Earthquake Patterns

By 1968, the basic outline of global tectonics was firmly established. In that same year, three seismologists published papers demonstrating how successfully the new plate tectonics model accounted for the global distribution of earthquakes (Figure 7.21). In particular, these scientists were able to account for the close association between deep-focus earthquakes and ocean trenches. Furthermore, the absence of deep-focus earthquakes along the oceanic ridge system was also shown to be consistent with the new theory.

The close association between plate boundaries and earthquakes can be seen by comparing the distribution of earthquakes shown in Figure 7.21 with the map of plate boundaries in Figure 7.8 and trenches in Figure 7.13. In trench regions where dense slabs of lithosphere plunge into the mantle, this association is especially striking. When the depths of earthquake foci and their locations within the trench systems are plotted, an interesting pattern emerges. Figure 7.22, which shows the distribution of earthquakes in the vicinity of the Japan trench, is an example. Here most shallow-focus earthquakes occur within, or adjacent to, the trench, whereas intermediate- and deep-focus earthquakes occur toward the mainland.

In the plate tectonics model, deep-ocean trenches are produced where cool, dense slabs of oceanic litho-

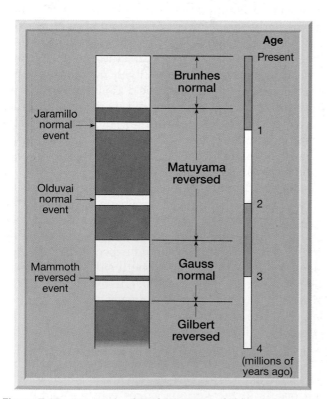

Figure 7.19 Time scale of Earth's magnetic field in the recent past. This time scale was developed by establishing the magnetic polarity for lava flows of known age. (Data from Allen Cox and G. B. Dalrymple)

A. Period of normal magnetism

Figure 7.20   As new basalt is added to the ocean floor at the oceanic ridges, it is magnetized according to Earth's existing magnetic field. Hence, it behaves much like a tape recorder as it records each reversal of the planet's magnetic field.

B. Period of reverse magnetism

C. Period of normal magnetism

sphere plunge into the mantle. Shallow-focus earthquakes are produced as the descending plate interacts with the overriding lithosphere. As the slab descends farther into the asthenosphere, deeper-focus earthquakes are generated (Figure 7.22). Because the earthquakes occur within the rigid subducting plate rather than in the plastic mantle, they provide a method for tracking the plate's descent. Very few earthquakes have

been recorded below 700 kilometers (435 miles), possibly because the slab has been heated sufficiently to lose its rigidity.

Evidence: Ocean Drilling

Some of the most convincing evidence confirming the plate tectonics theory has come from drilling directly

Key
- ● Shallow
- ○ Intermediate
- ● Deep

Figure 7.21   Distribution of shallow-, intermediate-, and deep-focus earthquakes. Note that intermediate- and deep-focus earthquakes only occur in association with subduction zones. (Data from NOAA)

Figure 7.22 Distribution of earthquake foci in the vicinity of the Japan trench. Note that intermediate- and deep-focus earthquakes occur only within the sinking slab of oceanic lithosphere. (Data from NOAA)

into ocean-floor sediment. From 1968 until 1983, the source of these important data was the Deep Sea Drilling Project, an international program sponsored by several major oceanographic institutions and the National Science Foundation. A new drilling ship was built. The *Glomar Challenger* represented a significant technological breakthrough, because this ship could lower drill pipe thousands of meters to the ocean floor and then drill hundreds of meters into the sediments and underlying basaltic crust.

Operations began in August 1968, in the South Atlantic. At several sites, holes were drilled through the entire thickness of sediments to the basaltic rock below. An important objective was to gather samples of sediment from just above the igneous crust as a means of dating the seafloor at each site. (Radiometric dates of the ocean crust itself are unreliable because seawater alters basalt.)

When the oldest sediment from each drill site was plotted against its distance from the ridge crest, it was revealed that the age of the sediment increased with increasing distance from the ridge. This finding agreed with the seafloor-spreading hypothesis, which predicted that the youngest oceanic crust would be found at the ridge crest and that the oldest oceanic crust would be at the continental margins.

The data from the Deep Sea Drilling Project also reinforced the idea that the ocean basins are geologically youthful, because no sediment with an age in excess of 180 million years was found. By comparison, some continental crust has been dated at 3.9 billion years.

During its 15 years of operation, the *Glomar Challenger* drilled 1092 holes and obtained more than 96 kilo-

meters (60 miles) of invaluable core samples. The Ocean Drilling Program has succeeded the Deep Sea Drilling Project and, like its predecessor, is a major international program. A more technologically advanced drilling ship, the *JOIDES Resolution*, now continues the work of the *Glomar Challenger* (Figure 7.23).

Evidence: Hot Spots

Mapping of seafloor volcanoes (called seamounts) in the Pacific revealed a chain of volcanic structures extending from the Hawaiian Islands to Midway Island and then continuing northward toward the Aleutian trench (Figure 7.24). Radiometric dates of volcanoes in this chain showed that the volcanoes increase in age with increasing distance from Hawaii. Suiko Seamount, which is located near the Aleutian trench, is 65 million years old, Midway Island is 27 million years old, and the island of Hawaii was built up from the seafloor less than a million years ago (Figure 7.24).

Researchers are in agreement that a rising plume of mantle material is located below the island of Hawaii. Decompression melting of this hot rock as it enters the low-pressure environment near the surface generates a volcanic area, or **hot spot.** As the Pacific plate moved over the hot spot, successive volcanic mountains have been built. The age of each volcano indicates the time when it was situated over the relatively stationary mantle plume.

This pattern is shown in Figure 7.24. Kauai is the oldest of the large islands in the Hawaiian chain. Five million years ago, when it was positioned over the hot spot, Kauai was the only Hawaiian island in existence

Figure 7.23   The *JOIDES Resolution,* the drilling ship of the Ocean Drilling Program. This modern drilling ship replaced the *Glomar Challenger* in the important work of sampling the floors of the world's oceans. JOIDES is an acronym for Joint Oceanographic Institutions for Deep Earth Sampling. (Photo courtesy of Ocean Drilling Program)

(Figure 7.24). Visible evidence of the age of Kauai can be seen by examining its extinct volcanoes, which have been eroded into jagged peaks and vast canyons. By contrast, the south slopes of the relatively youthful island of Hawaii consist of fresh lava flows, and two of Hawaii's volcanoes, Mauna Loa and Kilauea, remain active.

This evidence supports the fact that the plates do indeed move relative to Earth's interior. The hot spot tracks also trace the direction of plate motion. Notice, for example, in Figure 7.24 (lower left) that the Hawaiian Island–Emperor Seamount chain bends. This particular bend in the trace occurred about 40 million years ago when the motion of the Pacific plate changed from nearly due north to its present northwesterly path.

Measuring Plate Motion

A number of methods have been employed to establish the direction and rate of plate motion. As noted earlier, hot spot "tracks" like those of the Hawaiian Island–Emperor Seamount chain trace the movement of the Pacific Plate relative to the mantle below. Further, by measuring the length of this volcanic chain and the time interval between the formation of the oldest structure (Suiko Seamount) and youngest structure (Hawaii), an average rate of plate motion can be calculated. In this case the volcanic chain is roughly 3000 kilometers long and has formed over the past 65 million years—making the average rate of movement about 9 centimeters

Figure 7.24   The chain of islands and seamounts that extends from Hawaii to the Aleutian trench results from the movement of the Pacific plate over an apparently stationary hot spot. Radiometric dating of the Hawaiian Islands shows that the volcanic activity decreases in age toward the island of Hawaii.

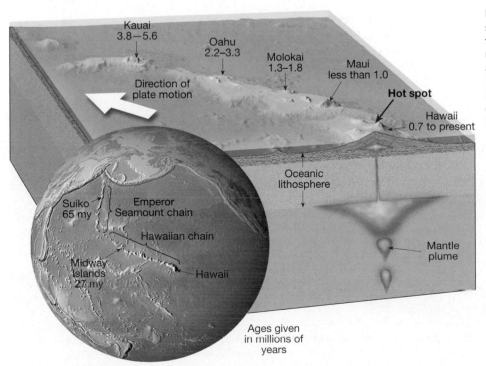

(4 inches) per year. The accuracy of this calculation hinges on the hot spot maintaining a fixed position in the mantle. Based on current evidence, this appears to be a reasonable assumption.

Recall that the magnetic stripes measured on the floor of the ocean also provide a method to measure rates of plate motion—at least as averaged over millions of years. Using paleomagnetism and other indirect techniques, researchers have been able to work out relative plate velocities as shown on the map in Figure 7.25.

It is currently possible, with the use of space-age technology, to directly measure the relative motion between plates. This is accomplished by periodically establishing the exact locations, and hence the distance between two observing stations situated on opposite sides of a plate boundary. Two of the methods used for this calculation are *Very Long Baseline Interferometry* (VLBI) and a satellite positioning technique that employs the *Global Positioning System* (GPS). The Very Long Baseline Interferometry system utilizes large radio telescopes to record signals from very distant quasars (quasi-stellar objects) (Figure 7.26). Quasars lie billions of light-years from Earth, so they act as stationary reference points. The millisecond differences in the arrival times of the same signal at different earthbound observatories provide a means of establishing the precise distance between receivers. A typical survey may take a day to perform and involves two widely spaced radio telescopes observing perhaps a dozen quasars, 5 to 10 times each. This scheme provides an estimate of the distance between these observatories, which is accurate to

about 2 centimeters. By repeating this experiment at a later date, researchers can establish the relative motion of these sites. This method has been particularly useful in establishing large-scale plate motions, such as the separation that is occurring between the United States and Europe (see Box 7.3).

You may be familiar with GPS, which uses 21 satellites to accurately locate any individual who is equipped with a handheld receiver. By using two spaced receivers, signals obtained by these instruments can be used to calculate their relative positions with considerable accuracy. Techniques using GPS receivers have been shown to be useful in establishing small-scale crustal movements such as those that occur along local faults in regions known to be tectonically active.

Confirming data obtained from these and other techniques leave little doubt that real plate motion has been detected (Figure 7.25). Calculations show that Hawaii is moving in a northwesterly direction and is approaching Japan at 8.3 centimeters per year. A site located in Maryland is retreating from one in England at a rate of about 1.7 centimeters per year—a rate that is close to the 2.3-centimeters-per-year spreading rate that was established from paleomagnetic evidence.

The Driving Mechanism

The plate tectonics theory *describes* plate motion and the effects of this motion. Therefore, acceptance of this model does not rely on the mechanism that drives plate motion. This is fortunate, because none of the driving

Figure 7.25 This map illustrates directions and rates of plate motion in centimeters per year. Seafloor-spreading velocities (as shown with black arrows and labels) are based on the spacing of dated magnetic stripes (anomalies). The colored arrows show Very Long Baseline Interferometry (VLBI) data of plate motion at selected locations. The data obtained by these methods are typically consistent. (Seafloor data from DeMets and others, VLBI data from Ryan and others.)

Figure 7.26 Radio telescopes like this one located at Green Bank, West Virginia, are used to determine accurately the distance between two distant sites. Data collected by repeated measurements have detected relative plate motions of 1 to 15 centimeters per year between various sites worldwide. (Courtesy of National Radio Astronomy Observatory)

mechanisms yet proposed can account for all major facets of plate tectonics. Nevertheless, researchers generally agree on the following:

1. Convective flow in the rocky 2900-kilometer-thick mantle—in which warm, less dense rock rises and cooler, more dense material sinks—is the basic driving force for plate movement.

2. Mantle convection and plate tectonics are part of the same system. Oceanic plates represent the cold downward-moving portion of convective flow.

3. The slow movements of the plates and mantle are driven by the unequal distribution of heat within Earth's interior. This flow is the mechanism that transports heat away from Earth's interior.

What is not known with any large degree of certainty is the precise nature of this convective flow.

Some researchers have argued that the mantle is like a giant layer cake, divided at a depth of 660 kilometers. Convection operates in both layers, but mixing between layers is minimal. At the other end of the spectrum is a model that loosely resembles a pot of just barely boiling water, churning ever so slowly from top to bottom over eons of geologic time. Neither model fits all of the available data. We will first look at some of the processes that are thought to contribute to plate motion and then examine a few of the models that have been proposed to describe plate-mantle convection.

Slab-Pull, Ridge-Push, and Mantle Plumes

Several mechanisms generate forces that contribute to plate motion. One relies on the fact that old oceanic crust, which is relatively cool and dense, sinks into the asthenosphere and "pulls" the trailing lithosphere along. This mechanism, called **slab-pull,** is thought to

be the primary downward arm of the convective flow operating in the mantle. By contrast, **ridge-push** results from the elevated position of the oceanic ridge system and causes oceanic lithosphere to gravitationally slide down the flanks of the ridge. Ridge-push, although apparently active in some spreading centers, is probably less important than slab-pull.

Most models suggest that hot, buoyant plumes of rock are the upward flowing arms in the convective mechanism at work in the mantle. These rising **mantle plumes** manifest themselves on Earth's surface as hot spots with their associated volcanic activity. Mapping of Earth's interior using seismic tomography indicates that at least some of these hot plumes extend upward from the vicinity of the mantle-core boundary. Others, however, appear to originate higher in the mantle. Further, some upward flow in the mantle is not related to plumes and occurs just below the ridge crests as a result of seafloor spreading.

Models of Plate-Mantle Convection

Any model describing mantle convection must be consistent with the observed physical and chemical properties of the mantle. In particular, these models must explain why basalts that erupt along the oceanic ridge are fairly homogeneous in composition and depleted in certain trace elements. It is assumed that these ridge basalts are derived from rocks located in the upper mantle that experienced an earlier period of chemical differentiation, in which trace elements were removed. By contrast, higher concentrations of these elements are evident in basaltic eruptions associated with hot-spot volcanism. Because basalts that erupted in different settings have different concentrations of trace elements, they are assumed to be derived from chemically distinct regions of the mantle. Basalts associated with mantle plumes are thought to come from a primitive (less differentiated) source, which more closely resembles the average chemical composition of the early mantle.

Layering at 660 Kilometers. Earlier we referred to the "layer cake" version of mantle convection. As shown in Figure 7.27A, one of these layered models has two zones of convection—a thin, convective layer above 660 kilometers and a thick one located below. This model successfully explains why the basalts that erupt along the oceanic ridges have a somewhat different composition than those that erupt in Hawaii as a result of hot-spot activity. The oceanic ridge basalts come from the upper convective layer, which is well mixed, whereas the mantle plume that feeds the Hawaiian volcanoes taps a deeper, more primitive source that resides in the lower convective layer.

Despite evidence that supports this model, recent seismic imaging has shown that subducting slabs of cold oceanic lithosphere are able to penetrate the 660-kilometer boundary. The subducting lithosphere serves to mix the upper and lower layers together. As a result, the layered mantle structure is lost.

UNDERSTANDING EARTH:
Plate Tectonics into the Future

Two geologists, Robert Dietz and John Holden, extrapolated present-day plate movements into the future. Figure 7.C illustrates where they envision Earth's landmasses will be 50 million years from now if present plate movements persist for this time span.

In North America we see that the Baja Peninsula and the portion of southern California that lies west of the San Andreas Fault will have slid past the North American plate. If this northward migration takes place, Los Angeles and San Francisco will pass each other in about 10 million years, and in about 60 million years Los Angeles will begin to descend into the Aleutian trench.

Significant changes are seen in Africa, where a new sea emerges as East Africa parts company with the mainland. In addition, Africa will have moved slowly into Europe, perhaps initiating the next major mountain-building stage on our dynamic planet. Meanwhile, the Arabian Peninsula continues to diverge from Africa, allowing the Red Sea to widen and close the Persian Gulf.

In other parts of the world, Australia is now astride the equator and, along with New Guinea, is on a collision course with Asia. Meanwhile, North and South America are beginning to separate, while the Atlantic and Indian oceans continue to grow at the expense of the Pacific Ocean.

These projections into the future, although interesting, must be viewed with caution because many assumptions must be correct for these events to unfold as just described. Nevertheless, changes in the shapes and positions of continents that are equally profound will undoubtedly occur for millions of years to come.

Figure 7.C The world as it may look 50 million years from now. (From "The Breakup of Pangaea," Robert S. Dietz and John C. Holden. Copyright 1970 by Scientific American, Inc., All rights reserved.)

Whole-Mantle Convection. Because of problems with the layered model, researchers began to favor whole-mantle convection. In a whole-mantle convection model, slabs of cold oceanic lithosphere descend into the lower mantle, providing the downward arm of convective flow (Figure 7.27B). Simultaneously, hot mantle plumes originating near the mantle-core boundary transport heat toward the surface. It was also suggested that at extremely slow rates of convection, primitive (undepleted) mantle rock would exist at depth in quantities sufficient to feed the rising mantle plumes.

Recent work, however, has shown that whole mantle mixing would cause the lower mantle to become appreciably mixed in a matter of a few hundred million years. This mixing would tend to eliminate the source of primitive magma observed in hot-spot volcanism.

Deep-Layer Model. A remaining possibility is layering deeper in the mantle. One deep-layer model has been described as analogous to a lava lamp on a low setting. As shown in Figure 7.27C, the lower perhaps one-third of the mantle is like the colored fluid in the bottom layer of a lava lamp. Like a lava lamp on low, heat from Earth's interior causes the two layers to slowly swell and shrink in complex patterns without substantial mixing. A small amount of material from the lower layer flows upward as mantle plumes to generate hot-spot volcanism at the surface.

This model provides the two chemically different mantle sources for basalt that are required by observational data. Further, it is compatible with seismic images that show cold lithospheric plates sinking deep into the mantle. Despite its attractiveness, there is little

A.

B.

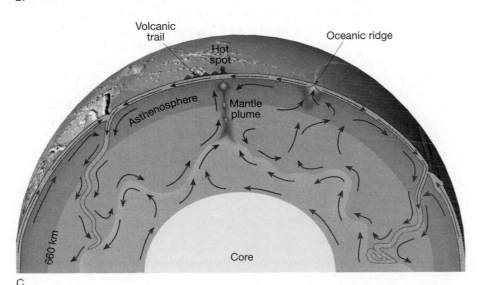

C.

Figure 7.27 Proposed models for mantle convection. **A.** The model shown in this illustration consists of two convection layers—a thin, convective layer above 660 kilometers and a thick one below. **B.** In this whole-mantle convection model, cold oceanic lithosphere descends into the lowermost mantle while hot mantle plumes transport heat toward the surface. **C.** This deep-layer model suggests that the mantle operates similar to a lava lamp on a low setting. Earth's heat causes these layers of convection to slowly swell and shrink in complex patterns without substantial mixing. Some material from the lower layer flows upward as mantle plumes.

seismic evidence to suggest that a deep mantle layer of this nature exists, except for the very thin layer located at the mantle-core boundary.

Although there is still much to be learned about the mechanisms that cause plates to move, some facts are clear. The unequal distribution of heat in Earth generates some type of thermal convection that ultimately drives plate-mantle motion. Furthermore, the descending lithospheric plates are active components of downwelling, and they serve to transport cold material into the mantle. Exactly how this convective flow operates is yet to be determined.

STUDENTS SOMETIMES ASK...
Will plate tectonics eventually "turn off" and cease to operate on Earth?

Because plate tectonic processes are powered by heat from within Earth (which is of a finite amount), the forces will slow sometime in the distant future to the point that the plates will cease to move. The work of external forces (wind, water, and ice), however, will continue to erode Earth's features, most of which will be eventually eroded flat. What a different world it will be—an Earth with no earthquakes, no volcanoes, and no mountains. Flatness will prevail!

Chapter Summary

• In the early 1900s *Alfred Wegener* set forth his *continental drift* hypothesis. One of its major tenets was that a supercontinent called *Pangaea* began breaking apart into smaller continents about 200 million years ago. The smaller continental fragments then drifted to their present positions. To support the claim that the now separate continents were once joined, Wegener and others used the *fit of South America and Africa, the distribution of ancient climates, fossil evidence,* and *rock structures.*

• One of the main objections to the continental drift hypothesis was its inability to provide an acceptable mechanism for the movement of continents.

• The theory of *plate tectonics,* a far more encompassing theory than continental drift, holds that Earth's rigid outer shell, called the *lithosphere,* consists of about seven large and numerous smaller segments called *plates* that are in motion relative to each other. Most of Earth's *seismic activity, volcanism,* and *mountain building* occur along the dynamic margins of these plates.

• A major departure of the plate tectonics theory from the continental drift hypothesis is that large plates contain both continental and ocean crust and the entire plate moves. By contrast, in continental drift, Wegener proposed that the sturdier continents drifted by breaking through the oceanic crust, much like ice breakers cut through ice.

• *Divergent plate boundaries* occur where plates move apart, resulting in upwelling of material from the mantle to create new seafloor. Most divergent boundaries occur along the axis of the oceanic ridge system and are associated with seafloor spreading, which occurs at rates of 2 to 15 centimeters per year. New divergent boundaries may form within a continent (for example, the East African rift valleys) where they may fragment a landmass and develop a new ocean basin.

• *Convergent plate boundaries* occur where plates move together, resulting in the subduction of oceanic lithosphere into the mantle along a deep oceanic trench. Convergence between an oceanic and continental block results in subduction of the oceanic slab and the formation of a *continental volcanic arc* such as the Andes of South America. Oceanic–oceanic convergence results in an arc-shaped chain of volcanic islands called a *volcanic island arc.* When two plates carrying continental crust converge, both plates are too buoyant to be subducted. The result is a "collision" resulting in the formation of a mountain belt such as the Himalayas.

• *Transform fault boundaries* occur where plates grind past each other without the production or destruction of lithosphere. Most transform faults join two segments of an oceanic ridge. Others connect spreading centers to subduction zones and thus facilitate the transport of oceanic crust created at a ridge crest to its site of destruction, at a deep-ocean trench. Still others, like the San Andreas Fault, cut through continental crust.

• The theory of plate tectonics is supported by (1) *paleomagnetism,* the direction and intensity of Earth's magnetism in the geologic past; (2) the global distribution of *earthquakes* and their close association with plate boundaries; (3) the ages of *sediments* from the floors of the deep-ocean basins; and (4) the existence of island groups that formed over *hot spots* and that provide a frame of reference for tracing the direction of plate motion.

• Three basic models for mantle convection are currently being evaluated. Mechanisms that contribute to this convective flow are slab-pull, ridge-push, and mantle plumes. *Slab-pull* occurs where cold, dense oceanic lithosphere is subducted and pulls the trailing lithosphere along. *Ridge-push* results when gravity sets the elevated slabs astride oceanic ridges in motion. Hot, buoyant *mantle plumes* are considered the upward flowing arms of mantle convection. One model suggests that mantle convection occurs in two layers separated at a depth of 660 kilometers. Another model proposes whole-mantle convection that stirs the entire 2900-kilometer-thick rocky mantle. Yet another model suggests that the bottom third of the mantle gradually bulges upward in some areas and sinks in others without appreciable mixing.

Key Terms

asthenosphere (p. 199)
continental drift (p. 192)
continental volcanic arc (p. 207)

convergent plate boundary (p. 199)
divergent plate boundary (p. 199)
hot spot (p. 216)

lithosphere (p. 199)
mantle plume (p. 219)
normal polarity (p. 213)

Review Questions

1. Who is credited with developing the continental drift hypothesis?

2. What was probably the first evidence that led some to suspect the continents were once connected?

3. What was Pangaea?

4. List the evidence that Wegener and his supporters gathered to substantiate the continental drift hypothesis.

5. Explain why the discovery of the fossil remains of *Mesosaurus* in both South America and Africa, but nowhere else, supports the continental drift hypothesis.

6. Early in this century, what was the prevailing view of how land animals migrated across vast expanses of ocean?

7. How did Wegener account for the existence of glaciers in the southern landmasses, while at the same time areas in North America, Europe, and Siberia supported lush tropical swamps?

8. On what basis were plate boundaries first established?

9. What are the three major types of plate boundaries? Describe the relative plate motion at each of these boundaries.

10. What is seafloor spreading? Where is active seafloor spreading occurring today?

11. What is a subduction zone? With what type of plate boundary is it associated?

12. Where is lithosphere being consumed? Why must the production and destruction of lithosphere be going on at approximately the same rate?

13. Briefly describe how the Himalaya Mountains formed.

14. Differentiate between transform faults and the other two types of plate boundaries.

15. Some predict that California will sink into the ocean. Is this idea consistent with the theory of plate tectonics?

16. Define the term *paleomagnetism*.

17. How does the continental drift hypothesis account for the apparent wandering of Earth's magnetic poles?

18. Describe the distribution of earthquake epicenters and foci depths as they relate to oceanic trench systems.

19. What is the age of the oldest sediments recovered by deep-ocean drilling? How do the ages of these sediments compare to the ages of the oldest continental rocks?

20. How do hot spots and the plate tectonics theory account for the fact that the Hawaiian Islands vary in age?

21. With what type of plate boundary are the following places or features associated (be as specific as possible): Himalayas, Aleutian Islands, Red Sea, Andes Mountains, San Andreas Fault, Iceland, Japan, Mount St. Helens?

22. Briefly describe the three models proposed for mantle-plate convection. What is lacking in each of these models?

Examining the Earth System

1. As an integral subsystem of the Earth system, plate tectonics has played a major role in determining the events that have taken place on Earth since its formation about 4.5 billion years ago. Briefly comment on the general effect that the changing positions of the continents and the redistribution of land and water over Earth's surface have had on the atmosphere, hydrosphere, and biosphere through time. You may want to review plate tectonics by investigating the topic on the Internet using a search engine such as HotBot (**http://www.hotbot.com/**) or Yahoo (**http://www.yahoo.com/**) and visiting the United States Geological Survey's (USGS) This Dynamic Earth: The Story of Plate Tectonics Website at **http://pubs.usgs.gov/publications/text/dynamic.html.**

2. Assume that plate tectonics did not cause the breakup of the supercontinent Pangaea. Using Figure 7.2, describe how the climate (atmosphere), vegetation and animal life (biosphere), and geological features (solid Earth) of those locations currently occupied by the cities of Seattle, WA; Chicago, IL; New York, NY; and your college campus location would be different from the conditions that exist today.

Web Resources

The *Earth Science* Website uses the resources and flexibility of the Internet to aid in your study of the topics in this chapter. Written and developed by Earth science instructors, this site will help improve your understanding of Earth science. Visit **http://www.prenhall.com/tarbuck** and click on the cover of *Earth Science 10e* to find:

- On-line review quizzes.
- Web-based critical thinking and writing exercises.
- Links to chapter-specific Web resources.
- Internet-wide key term searches.

http://www.prenhall.com/tarbuck

A recent eruption of Italy's Mount Etna. (Photo by Art Wolfe)

CHAPTER 8

VOLCANOES AND OTHER IGNEOUS ACTIVITY

The significance of igneous activity may not be obvious at first glance. However, because volcanoes extrude molten rock that formed at great depth, they provide the only windows we have for direct observation of processes that occur many kilometers below Earth's surface. Furthermore, the atmosphere and oceans are thought to have evolved from gases emitted during volcanic eruptions. Either of these facts is reason enough for igneous activity to warrant our attention. All volcanic eruptions are spectacular, but why are some destructive and others quiescent? Is the entire island of Hawaii a volcano as high as Mount Everest resting on the ocean floor? Do the large volcanoes of Washington and Oregon present a threat to human lives? This chapter considers these and other questions as we explore the formation and movement of magma.

On Sunday, May 18, 1980, one of the largest volcanic eruption to occur in North America in historic times transformed a picturesque volcano into a decapitated remnant (Figure 8.1). On this date in southwestern Washington State, Mount St. Helens erupted with tremendous force. The blast blew out the entire north flank of the volcano, leaving a gaping hole. In one brief moment, a prominent volcano whose summit had been more than 2900 meters (9500 feet) above sea level was lowered by more than 400 meters (1350 feet).

The event devastated a wide swath of timber-rich land on the north side of the mountain (Figure 8.2). Trees within a 400-square-kilometer area lay intertwined and flattened, stripped of their branches and appearing from the air like toothpicks strewn about. The accompanying mudflows carried ash, trees, and water-saturated rock debris 29 kilometers down the Toutle River. The eruption claimed 59 lives, some dying from the intense heat and the suffocating cloud of ash and gases, others from being hurled by the blast, and still others from entrapment in the mudflows.

The eruption ejected nearly a cubic kilometer of ash and rock debris. Following the devastating explosion, Mount St. Helens continued to emit great quantities of hot gases and ash, some of which were propelled more than 18,000 meters (over 11 miles) into the stratosphere. During the next few days, this very fine-grained mate-

Figure 8.1 Before and after photographs show the transformation of Mount St. Helens caused by the May 18, 1980, eruption. The dark area in the "after" photo is debris-filled Spirit Lake, partially visible in the "before" photo. (Photos courtesy of U.S. Geological Survey)

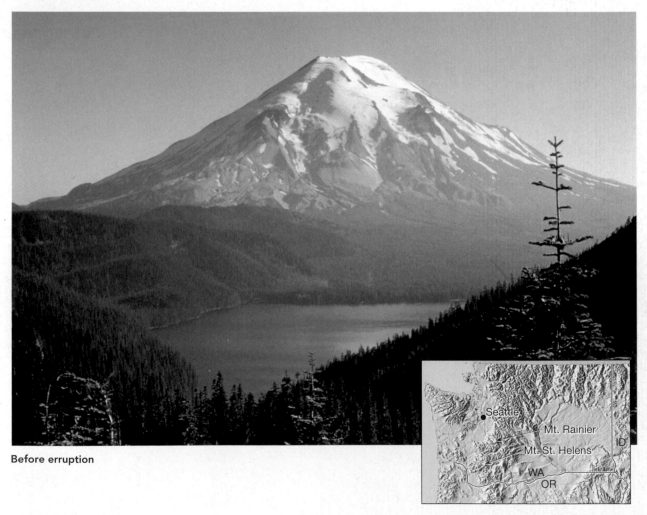

Before erruption

rial was carried around Earth by strong upper-air winds. Measurable deposits were reported in Oklahoma and Minnesota, with crop damage into central Montana. Meanwhile, ash fallout in the immediate vicinity exceeded 2 meters in depth. The air over Yakima, Washington (130 kilometers to the east), was so filled with ash that residents experienced midnight-like darkness at noon.

Mount St. Helens is one of 15 large volcanoes and innumerable smaller ones that comprise the Cascade Range, which extends from British Columbia to northern California. Eight of the largest volcanoes have been active in the past few hundred years. Of the remaining seven active volcanoes, the most likely to erupt again are Mount Baker and Mount Rainier in Washington, Mount Shasta and Lassen Peak in California, and Mount Hood in Oregon.

Not all volcanic eruptions are as violent as the 1980 Mount St. Helens event. Some volcanoes, such as Hawaii's Kilauea volcano, generate relatively quiet outpourings of fluid lavas. These "gentle" eruptions are not without some fiery displays; occasionally fountains of incandescent lava spray hundreds of meters into the air. Such events, however, typically pose minimal threat to human life and property.

Testimony to the quiet nature of Kilauea's eruptions is the fact that the Hawaiian Volcanoes Observatory has operated on its summit since 1912. This, despite the fact that Kilhauea has had more than 50 eruptive phases since record keeping began in 1823. Further, the longest and largest of Kilauea's eruptions began in 1983 and remains active, although it has received only modest media attention.

Why do volcanoes like Mount St. Helens erupt explosively, whereas others like Kilauea are relatively quiet? Why do volcanoes occur in chains like the Aleutian Islands or the Cascade Range? Why do some volcanoes form on the ocean floor, while others occur on the continents? This chapter will deal with these and other questions as we explore the nature and movement of magma and lava (Figure 8.3).

STUDENTS SOMETIMES ASK...

*After all the destruction during
the eruption of Mount St. Helens,
what does the area look like today?*

The area continues to make a slow recovery. Surprisingly, many organisms survived the blast, including animals that live underground and plants (particularly those protected by snow or near streams, where erosion quickly removed the ash). Others have adaptations that allow them to quickly repopulate devastated areas. Fully 20 years after the blast, plants have revegetated the area, first-growth forests are beginning

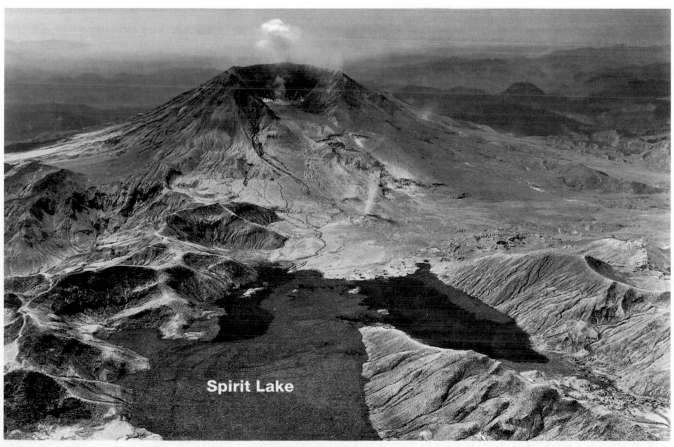

Spirit Lake

After erruption

Figure 8.2 Douglas fir trees were snapped off or uprooted by the lateral blast of Mount St. Helens on May 18, 1980. (Photo by Krafft-Explorer/Photo Researchers, Inc.)

to be established, and many animals have returned. Once the old-growth forest is complete (in a few hundred years), it might be hard to find much evidence of the destruction, other than a thick weathered ash layer in the soil.

The volcano itself is rebuilding, too. A large lava dome is forming inside the summit crater, suggesting that the mountain will build up again. Many volcanoes similar to Mount St. Helens exhibit this behavior: rapid destruction followed by slow rebuilding. If you really want to see what it looks like, go to the Mount St. Helens National Volcanic Home Page at http://www.fs.fed.us/gpnf/mshnvm/, where they have a "volcanocam" with real-time images of the mountain.

The Nature of Volcanic Eruptions

Forces Within
↳ Igneous Activity

Volcanic activity is commonly perceived as a process that produces a picturesque, cone-shaped structure that periodically erupts in a violent manner. Although some eruptions are very explosive, many others are not. What determines whether a volcano extrudes magma violently or gently? The primary factors include the

Figure 8.3 A river of basaltic lava from the January 17, 2002 erruption of Mount Nyiragongo destroyed many homes in Goma, Congo. (Photo by Marco Longari/CORBIS)

magma's *composition,* its *temperature,* and the amount of *dissolved gases* it contains. To varying degrees these factors affect the magma's **viscosity** (*viscos* = sticky). The more viscous the material, the greater its resistance to flow. For example, syrup is more viscous than water. Magma associated with an explosive eruption may be thousands of times more viscous than magma that is extruded in a quiescent manner.

Factors Affecting Viscosity

The effect of temperature on viscosity is easily seen. Just as heating syrup makes it more fluid (less viscous), the mobility of lava is strongly influenced by temperature. As a lava flow cools and begins to congeal, its mobility decreases and eventually the flow halts.

A more significant influence on volcanic behavior is the chemical composition of magmas. This was discussed in Chapter 2 along with the classification of igneous rocks. Recall that a major difference among various igneous rocks is their silica SiO$_2$) content. The same is true of the magmas from which rocks form. Magmas that produce basaltic rocks contain about 50 percent silica, whereas magmas that produce granitic rocks contain over 70 percent silica (Table 8.1).

A magma's viscosity is directly related to its silica content. In general, the more silica in magma, the greater its viscosity. The flow of magma is impeded because silicate structures link together into long chains even before crystallization begins. Consequently, because of their high silica content, rhyolitic (felsic) lavas are very viscous and tend to form comparatively short, thick flows. By contrast, basaltic (mafic) lavas, which contain less silica, tend to be more fluid and have been known to travel distances of 150 kilometers (90 miles) or more before congealing (Figure 8.3).

In Hawaiian eruptions, the magmas are hot and basaltic, so they are extruded with ease. By contrast, highly viscous magmas are more difficult to force through a vent. On occasion, the vent may become plugged with viscous magma, which can result in a buildup of gases that produce an explosive eruption.

However, a viscous magma is not explosive by itself. It is the gas content that puts the bang into a violent eruption.

Importance of Dissolved Gases in Magma

During explosive eruptions, it is the **volatiles** (the gaseous components of magma) that provide the force that extrudes molten rock from the vent. These gases are mostly water vapor and carbon dioxide. As magma moves into a near-surface environment, such as within a volcano, the confining pressure in the uppermost portion of the magma body is greatly reduced. The reduction in confining pressure allows dissolved gases to be released suddenly, just as opening a soda bottle allows dissolved carbon dioxide gas bubbles to escape.

At high temperatures and low near-surface pressures, these gases will expand to occupy hundreds of times their original volume. Very fluid basaltic magmas allow the expanding gases to bubble upward and escape from the vent with relative ease. As they escape, the gases will often propel incandescent lava hundreds of meters into the air, producing lava fountains. Although spectacular, such fountains are mostly harmless and not generally associated with major explosive events that cause great loss of life and property. Rather, eruptions of fluid basaltic lavas, such as those that occur in Hawaii, are relatively quiescent.

At the other extreme, highly viscous magmas impede the upward migration of expanding gases. The gases collect in bubbles and pockets that increase in size until they explosively eject the semimolten rock from the volcano (Figure 8.4). The result is a Mount St. Helens or Mt. Pinatubo.

To summarize, the viscosity of magma, plus the quantity of dissolved gases and the ease with which they can escape, determines the nature of a volcanic eruption. We can now understand the "gentle" volcanic eruptions of hot, fluid lavas in Hawaii and the explosive, violent, eruptions of viscous lavas from volcanoes such as Mount St. Helens.

Table 8.1 Magmas have different compositions, which cause their properties to vary

Composition	Silica Content	Viscosity	Gas Content	Tendency to Form Pyroclastics	Volcanic Landform
Basaltic (Mafic)	Least (~50%)	Least	Least (1–2%)	Least	Shield Volcanoes Basalt Plateaus Cinder Cones
Andesitic (Intermediate)	Intermediate (~60%)	Intermediate	Intermediate (3–4%)	Intermediate	Composite Cones
Rhyolitic (Felsic)	Most (~70%)	Greatest	Most (4–6%)	Greatest	Pyroclastic Flows Volcanic Domes

Figure 8.4 Mount St. Helens emitting volcanic ash on July 22, 1980, two months after the huge May erruption. (Photo by David Weintraub/Photo Researchers, Inc.)

What Is Extruded During Eruptions?

Internal Processes
↳ Igneous Activity

Lava may appear to be the primary material extruded from a volcano, but this is not always the case. Just as often, explosive eruptions eject huge quantities of broken rock, lava bombs, fine ash, and dust. Moreover, all volcanic eruptions emit large amounts of gas. In this section we will examine each of these materials associated with a volcanic eruption.

Lava Flows

Because of their low silica content, hot basaltic lavas are usually very fluid. They flow in thin, broad sheets or streamlike ribbons. On the island of Hawaii, such lavas have been clocked at speeds of 30 kilometers (20 miles) per hour down steep slopes. These velocities are rare, however, and flow rates of 10 to 300 meters per hour are more common. In contrast, the movement of silica-rich (rhyolitic) lava is on occasion too slow to be perceptible.

When fluid basaltic lavas of the Hawaiian type congeal, they commonly form a relatively smooth skin that wrinkles as the still-molten subsurface lava continues to advance (Figure 8.5A). These are known as **pahoehoe flows** (pronounced *pah-hoy-hoy*) and resemble the twisted braids in ropes.

Another common type of basaltic lava has a surface of rough, jagged blocks with dangerously sharp edges

and spiny projections (Figure 8.5B). The name **aa** (pronounced *ah-ah*) is given to these flows. In general, Hawaiian aa flows are relatively cool and thick and, depending on the slope, move more slowly than pahoehoe flows. Further, gases escaping from the surface produce numerous voids and sharp spines in the congealing lava. As the molten interior advances, the outer crust is broken further, giving the flow the appearance of an advancing mass of lava rubble.

Gases

Magmas contain varied amounts of dissolved gases (*volatiles*) held in the molten rock by confining pressure, just as carbon dioxide is held in soft drinks. As with soft drinks, as soon as the pressure is reduced, the gases begin to escape. Obtaining gas samples from an erupting volcano is often difficult and dangerous, so geologists can often only estimate the amount of gas originally contained within the magma.

The gaseous portion of most magmas is believed to make up from 1 to 6 percent of the total weight, with most of this being water vapor. Although the percentage may be small, the actual quantity of emitted gas can exceed thousands of tons each day.

The composition of volcanic gases is important to scientists because they have contributed significantly to gases that make up the atmosphere. Analysis of samples taken during Hawaiian eruptions indicates that the gases are about 70 percent water vapor, 15 percent carbon dioxide, 5 percent nitrogen, 5 percent sulfur, and lesser amounts of chlorine, hydrogen, and argon. Sulfur compounds are easily recognized by their pungent odor

A.   **B.**

Figure 8.5 **A.** Typical pahoehoe (ropy) lava flow, Kilauea Hawaii. (Photo by Doug Perrine/DRK) **B.** Typical slow-moving aa flow. (Photo by J. D. Griggs/U.S. Geological Survey)

and because they readily form sulfuric acid—a natural source of air pollution.

Pyroclastic Materials

When basaltic lava is extruded, dissolved gases escape quite freely and continually. These gases propel incandescent blobs of lava to great heights. Some of this ejected material may land near the vent and build a cone-shaped structure, whereas smaller particles will be carried great distances by the wind. By contrast, viscous (rhyolitic) magmas are highly charged with gases, and upon release they expand greatly as they blow pulverized rock, lava, and glass fragments from the vent. The particles produced in both of these situations are referred to as **pyroclastic materials** (*pyro* = fire, *clast* = fragment). These ejected fragments range in size from very fine dust and sand-sized volcanic ash (less than 2 millimeters) to pieces that weigh several tons.

Ash and *dust* particles are produced from gas-laden viscous magma during an explosive eruption (see Figure 8.4). As magma moves up in the vent, the gases rapidly expand, generating a froth of melt that might resemble froth that flows from a just opened bottle of champagne. As the hot gases expand explosively, the froth is blown into very fine glassy fragments. When the hot ash falls, the glassy shards often fuse to form a rock called *welded tuff*. Sheets of this material, as well as ash deposits that later consolidate, cover vast portions of the western United States.

Also common are pyroclasts that range in size from small beads to walnuts termed *lapilli* ("little stones"). These ejecta are commonly called *cinders* (2–64 millimeters). Particles larger than 64 millimeters (2.5 inches)

in diameter are called *blocks* when they are made of hardened lava and *bombs* when they are ejected as incandescent lava. Because bombs are semimolten upon ejection, they often take on a streamlined shape as they hurtle through the air (Figure 8.6). Because of their size, bombs and blocks usually fall on the slopes of a cone; however, they are occasionally propelled far from the volcano by the force of escaping gases. For instance, bombs 6 meters (20 feet) long and weighing about 200 tons were blown 600 meters (2000 feet) from the vent during an eruption of the Japanese volcano Asama.

Volcanic Structures and Eruptive Styles

 Forces Within
↳ Igneous Activity

The popular image of a volcano is that of a solitary, graceful, snowcapped cone, such as Mount Hood in Oregon or Japan's Fujiyama. These picturesque, conical mountains are produced by volcanic activity that occurred intermittently over thousands, or even hundreds, of thousands, of years. However, many volcanoes do not fit this image. Some volcanoes are only 30 meters (100 feet) high and formed during a single eruptive phase that may have lasted only a few days.

Volcanic landforms come in a wide variety of shapes and sizes, and each structure has a unique eruptive history. Nevertheless, volcanologists have been able to classify volcanic landforms and determine their eruptive patterns. In this section we will consider the general

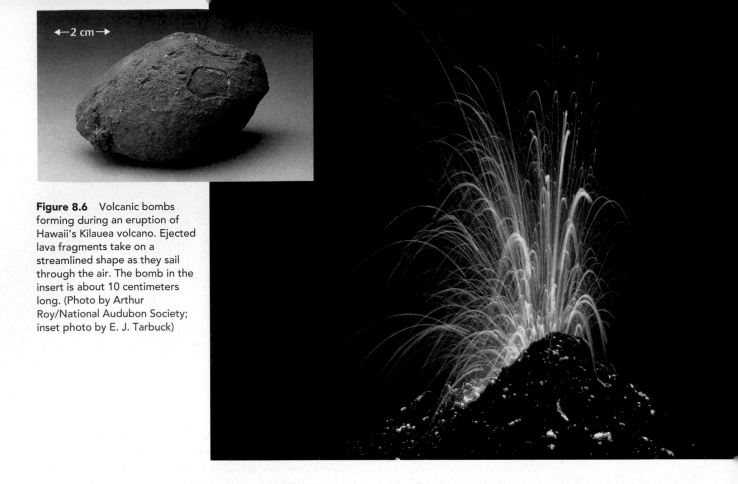

Figure 8.6 Volcanic bombs forming during an eruption of Hawaii's Kilauea volcano. Ejected lava fragments take on a streamlined shape as they sail through the air. The bomb in the insert is about 10 centimeters long. (Photo by Arthur Roy/National Audubon Society; inset photo by E. J. Tarbuck)

anatomy of a volcano and look at three major volcanic types: *shield volcanoes, cinder cones,* and *composite cones.* This discussion will be followed by an overview of other significant volcanic landforms.

Anatomy of a Volcano

Volcanic activity frequently begins when a **fissure** (crack) develops in the crust as magma moves forcefully toward the surface. As the gas-rich magma moves up this linear fissure, its path is usually localized into a circular **conduit,** or **pipe,** that terminates at a surface opening called a **vent** (Figure 8.7). Successive eruptions of lava, pyroclastic material, or frequently a combination of both often separated by long periods of inactivity eventually build the structure we call a **volcano.**

Located at the summit of many volcanoes is a steep-walled depression called a **crater** (*crater* = a bowl). Craters are constructional features that were built upward as ejected fragments collected around the vent to form a doughnut-like structure. Some volcanoes have multiple summit craters, whereas others have very large, more or less circular depressions called **calderas.** Calderas are large collapse structures that may or may not form in association with a volcano. (We will consider the formation of various types of calderas later.)

During early stages of growth most volcanic discharges come from a central summit vent. As a volcano matures, material also tends to be emitted from fissures that develop along the flanks, or base, of the volcano. Continued activity from a flank eruption may produce a small

Figure 8.7 Anatomy of a "typical" composite cone (see also Figures 8.8 and 8.11 for a comparison with a shield and cinder cone, respectively).

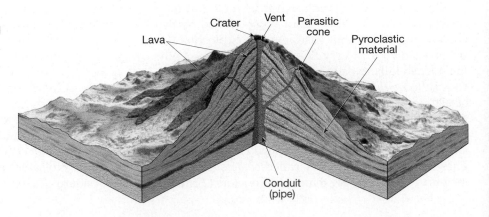

parasitic cone (*parasitus* = one who eats at the table of another). Mount Etna in Italy, for example, has more than 200 secondary vents, some of which have built cones. Many of these vents, however, emit only gases and are appropriately called **fumaroles** (*fumus* = smoke).

The form of a particular volcano is largely determined by the composition of the contributing magma. As you will see, fluid Hawaiian-type lavas tend to produce broad structures with gentle slopes, whereas more viscous silica-rich lavas (and some gas-rich basaltic lavas) tend to generate cones with moderate to steep slopes.

Shield Volcanoes

Shield volcanoes are produced by the accumulation of fluid basaltic lavas and exhibit the shape of a broad slightly domed structure that resembles a warrior's shield (Figure 8.8). Most (but not all) shield volcanoes have grown up from the deep-ocean floor to form islands or seamounts. For example, the islands of the Hawaiian chain, Iceland, and the Galapagos are either a single shield volcano or the coalescence of several shields. Extensive study of the Hawaiian Islands confirms that each shield was built from a myriad of basaltic lava flows averaging a few meters thick. Also, these islands consist of only about 1 percent pyroclastic ejecta.

Mauna Loa: Earth's Largest Volcano.   Mauna Loa is one of five overlapping shield volcanoes that together comprise the Big Island of Hawaii (Figure 8.8). From its base on the floor of the Pacific Ocean to its summit, Mauna Loa is over 9 kilometers (6 miles) high, exceeding that of Mount Everest. This massive pile of basaltic rock has a volume of 40,000 cubic kilometers that was extruded over a period of nearly a million years. For comparison, the volume of material composing Mauna Loa is roughly 200 times greater than the amount composing a large composite cone such as Mount Rainier (Figure 8.9). Most shields, however, are more modest in size. For example, the classic Icelandic shield, Skjalbreidur, rises to a height of only about 600 meters (2000 feet) and is 10 kilometers (6 miles) across its base.

Young shields, particularly those located in Iceland, emit very fluid lava from a central summit vent and have sides with gentle slopes that vary from 1 to 5 degrees. Mature shields, as exemplified by Mauna Loa, have steeper slopes in the middle sections (about 10 degrees), while their summits and flanks are comparatively flat. During the mature stage, lavas are discharged from the summit vents, as well as from rift zones that develop along the slopes. Most lava is discharged as the fluid pahoehoe type, but as these flows cool downslope, many change into a clinkery aa flow. Once an eruption is well established, a large fraction of

Figure 8.8   Shield volcanoes are built primarily of fluid basaltic lava flows and contain only a small percentage of pyroclastic materials. These broad, slightly domed structures, exemplified by the Hawaiian Islands, are the largest volcanoes on Earth. (Photo by Greg Vaughn)

A. Mauna Loa, Hawaii, a large shield volcano

Sea level

0   10   20 km

4 km

B. Mt. Rainier, Washington, a large composite cone   C. Sunset Crater, Arizona, a large cinder cone

Figure 8.9   Profiles of volcanic landforms. **A.** Profile of Mauna Loa, Hawaii, the largest shield volcano in the Hawaiian chain. Note size comparison with Mount Rainier, Washington, a large composite cone. **B.** Profile of Mount Rainier Washington. Note how it dwarfs a typical cinder cone. **C.** Profile of Sunset, Arizona, a typical steep-sided cinder-cone.

the lava (perhaps 80 percent) flows through a well-developed system of lava tubes. This greatly increases the distance lava can travel before it solidifies. Thus, lava emitted near the summit often reaches the sea, thereby adding to the width of the cone at the expense of its height.

Another feature common to a mature, active shield volcano is a large steep-walled caldera that occupies its summit. Calderas form when the roof of the volcano collapses as magma from the central magma reservoir migrates to the flanks, often feeding fissure eruptions. Mauna Loa's summit caldera measures 2.6 by 4.5 kilometers (1.6 by 2.8 miles) and has a depth that averages about 150 meters (500 feet).

Kilauea, Hawaii: Eruption of a Shield Volcano. Kilauea, the most active and intensely studied shield volcano in the world, is located on the island of Hawaii in the shadow of Mauna Loa. More than 50 eruptions have been witnessed here since record keeping began in 1823. Several months before each eruptive phase, Kilauea inflates as magma gradually migrates upward and accumulates in a central reservoir located a few kilometers below the summit. For up to 24 hours in advance of an eruption, swarms of small earthquakes warn of the impending activity.

Most of the activity on Kilauea during the past 50 years occurred along the flanks of the volcano in a region called the East Rift Zone. The longest and largest rift eruption ever recorded on Kilauea began in 1983 and continues to this day, with no signs of abating. The first discharge began along a 6-kilometer (4-mile) fissure where a 100-meter (300-foot) high "curtain of fire" formed as red-hot lava was ejected skyward (Figure 8.10). When the activity became localized, a cinder and spatter cone given the Hawaiian name *Puu Oo* was built. Over the next three years the general eruptive pattern consisted of short periods (hours to days) when fountains of gas-rich lava sprayed skyward. Each event was followed by nearly a month of inactivity.

By the summer of 1986 a new vent opened up 3 kilometers downrift. Here smooth-surfaced pahoehoe lava formed a lava lake. Occasionally the lake over-flowed, but more often lava escaped through tunnels to feed pahoehoe flows that moved down the southeastern flank of the volcano toward the sea. These flows destroyed nearly a hundred rural homes, covered a major roadway, and eventually reached the sea. Lava has been intermittently pouring into the ocean ever since, adding new land to the island of Hawaii.

Cinder Cones

As the name suggests, **cinder cones** (also called **scoria cones**) are built from ejected lava fragments that take on the appearance of cinders or clinkers as they begin to harden while in flight. These fragments range in size from fine ash to bombs but consist mostly of pea- to walnut-size lapilli. Usually a product of relatively gas-rich basaltic magma, cinder cones consist of rounded to irregular fragments that are markedly vesicular (containing voids) and have a black to reddish-brown color. Occasionally an eruption of silica-rich magma will generate a light-colored cinder cone composed of ash and pumice fragments. Although cinder cones are composed mostly of loose pyroclastic material, they sometimes extrude lava. On such occasions the discharges come from vents located at or near the base rather than from the summit crater.

Cinder cones have a very simple, distinctive shape determined by the slope that loose pyroclastic material maintains as it comes to rest (Figure 8.11). Because cinders have a high angle of repose (the steepest angle at which material remains stable), young cinder cones are steep-sided, having *stopes* between 30 and 40 degrees. In addition, cinder cones have large, deep craters in relation to the overall size of the structure. Although relatively symmetrical, many cinder cones are elongated, and higher on the side that was downwind during the eruptions.

Cinder cones are usually the product of a single eruptive episode that sometimes lasts only a few weeks and rarely exceeds a few years. Once this event ceases, the magma in the pipe connecting the vent to the magma chamber solidifies, and the volcano never erupts again. As a consequence of this short life span,

Figure 8.10   Lava extruded along the East Rift Zone, Kilauea, Hawaii. (Photo by Greg Vaughn)

Figure 8.11   SP Crater, a cinder cone north of Flagstaff, Arizona. (Photo by Michael Collier)

cinder cones are small, usually between 30 meters (100 feet) and 300 meters (1000 feet) and rarely exceed 700 meters (2100 feet) in height (see Figure 8.9).

Cinder cones are found by the thousands all around the globe. Some are located in volcanic fields like the one located near Flagstaff, Arizona, which consists of about 600 cones. Others are parasitic cones of larger volcanoes. Mount Etna, for example, has dozens of cinder cones dotting its flanks.

Parícutin: Life of a Garden-Variety Cinder Cone. One of the very few volcanoes studied by geologists from beginning to end is the cinder cone called Parícutin, located about 320 kilometers (200 miles) west of Mexico City. In 1943 its eruptive phase began in a cornfield owned by Dionisio Pulido, who witnessed the event as he prepared the field for planting.

For two weeks prior to the first eruption, numerous Earth tremors caused apprehension in the nearby village of Parícutin. Then on February 20 sulfurous gases began billowing from a small depression that had been in the cornfield for as long as people could remember. During the night, hot, glowing rock fragments were ejected from the vent, producing a spectacular fireworks display. Explosive discharges continued, throwing hot fragments and ash occasionally as high as 6000 meters (20,000 feet) above the crater rim. Larger fragments fell near the crater, some remaining incandescent as they rolled down the slope. These built an aesthetically pleasing cone, while finer ash fell over a much larger area, burning and eventually covering the village of Parícutin. In the first day the cone grew to 40 meters (130 feet), and by the fifth day it was over 100 meters (330 feet) high. Within the first year more than 90 percent of the total ejecta had been discharged.

The first lava flow came from a fissure that opened just north of the cone, but after a few months flows began to emerge from the base of the cone itself. In June 1944 a clinkery aa flow 10 meters (30 feet) thick moved over much of the village of San Juan Parangaricutiro, leaving only the church steeple exposed (Figure 8.12). After nine years of intermittent pyroclastic explosions and nearly continuous discharge of lava from vents at its base, the activity ceased almost as quickly as it had begun. Today, Parícutin is just another one of the scores of cinder cones dotting the landscape in this region of Mexico. Like the others, it will not erupt again.

Composite Cones

Earth's most picturesque yet potentially dangerous volcanoes are **composite cones** or **stratovolcanoes** (Figure 8.13). Most are located in a relatively narrow zone that rims the Pacific Ocean, appropriately called the *Ring of Fire* (see Figure 8.28). This active zone includes a chain of continental volcanoes that are distributed along the west coast of South and North America, including the large cones of the Andes and the Cascade Range of the western United States and Canada. The latter group includes Mount St. Helens, Mount Rainier, and Mount Garibaldi. The most active regions in the Ring of Fire are located along curved belts of volcanic islands situated adjacent to the deep ocean trenches of the northern and western Pacific. This nearly continuous chain of volcanoes stretches from the Aleutian Islands to Japan and the Philippines and ends on the North Island of New Zealand.

The classic composite cone is a large, nearly symmetrical structure composed of both lava and pyroclastic deposits. Just as shield volcanoes owe their shape to fluid basaltic lavas, composite cones reflect the nature of the erupted material. For the most part, composite cones are the product of gas-rich magma having an andesitic composition. Relative to shields, the silica-rich magmas typical of composite cones generate thick viscous lavas that travel short distances. In addition,

Figure 8.12 The village of San Juan Parangaricutiro engulfed by lava from Parícutin, shown in the background. Only the church towers remain. (Photo by Tad Nichols)

Figure 8.13 Mount Shasta, California, one of the largest composite cones in the Cascade Range. Shastina is the smaller parasitic cone on the left. (Photo by David Muench)

composite cones may generate explosive eruptions that eject huge quantities of pyroclastic material.

The growth of a "typical" composite cone begins with both pyroclastic material and lava being emitted from a central vent. As the structure matures, lavas tend to flow from fissures that develop on the lower flanks of the cone. This activity may alternate with explosive eruptions that eject pyroclastic material from the summit crater. Sometimes both activities occur simultaneously.

A conical shape, with a steep summit area and more gradually sloping flanks, is typical of many large composite cones. This classic profile, which adorns calendars and postcards, is partially a consequence of the way viscous lavas and pyroclastic ejecta contribute to the growth of the cone. Coarse fragments ejected from the summit crater tend to accumulate near their source. Because of their high angle of repose, coarse materials contribute to the steep slopes of the summit area. Finer ejecta, on the other hand, are deposited as a thin layer over a large area. This acts to flatten the flank of the cone. In addition, during the early stages of growth, lavas tend to be more abundant and flow greater distances from the vent than do later lavas. This contributes to the cone's broad base. As the volcano matures, the short flows that come from the central vent serve to armor and strengthen the summit area. Consequently, steep summit slopes exceeding 40 degrees are sometimes possible. Two of the most perfect cones—Mount Mayon in the Philippines and Fujiyama in Japan—exhibit the classic form we expect of a composite cone, with its steep summit and gently sloping flanks.

Despite their symmetrical form, most composite cones have a complex history. Huge mounds of volcanic debris surrounding many cones provide evidence that, in the distant past, a large section of the volcano slid downslope as a massive landslide. Others develop horseshoe-shaped depressions at their summits as a result of explosive eruptions, or as occurred during the 1980 eruption of Mount St. Helens, a combination of a landslide and the eruption of 0.6 cubic kilometer of magma left a gaping void on the north side of the cone. Often, so much rebuilding has occurred since these eruptions that no trace of the amphitheater-shaped scar remains.

Living in the Shadow of a Composite Cone

More than 50 volcanoes have erupted in the United States in the past 200 years (see Figure 8.30). Fortunately, the most explosive of these eruptions, except for Mount St. Helens in 1980, occurred in sparsely inhabited regions of Alaska. On a global scale numerous destructive eruptions have occurred during the past few thousand years, a few of which may have influenced the course of human civilization.

Nuée Ardente: A Deadly Pyroclastic Flow

One of the most devastating phenomena associated with composite cones are **pyroclastic flows,** which consist of hot gases infused with incandescent ash and larger rock fragments. The most destructive of these fiery flows, called **nuée ardentes** (also referred to as *glowing avalanches*), are capable of racing down steep volcanic slopes at speeds that can approach 200 kilometers (125 miles) per hour (Figure 8.14).

The ground-hugging portion of a glowing avalanche is rich in particulate matter, which is suspended by jets of buoyant gases passing upward through the flow.

Figure 8.14 Nuée ardente races down the slope of Mount St. Helens on August 7, 1980, at speeds in excess of 100 kilometers (60 miles) per hour. (Photo by Peter W. Lipman, U.S. Geological Survey)

Some of these gases have escaped from newly erupted volcanic fragments. In addition, air that is overtaken and trapped by an advancing flow may be heated sufficiently to provide buoyancy to the particulate matter of the nuée ardente. Thus, these flows, which can include large rock fragments in addition to ash, travel downslope in a nearly frictionless environment. This helps to explain why some nuée ardente deposits are found more than 100 kilometers (60 miles) from their source.

The pull of gravity is the force that causes these heavier-than-air flows to sweep downslope much like a snow avalanche. Some pyroclastic flows result when a powerful eruption blasts pyroclastic material laterally out the side of a volcano. Probably more often, nuée ardentes form from the collapse of tall eruption columns that form over a volcano during an explosive event. Once gravity overcomes the initial upward thrust provided by the escaping gases, the ejecta begin to fall. Massive amounts of incandescent blocks, ash, and pumice fragments that fall onto the summit area begin to cascade downslope under the influence of gravity. The largest fragments have been observed bouncing down the flanks of a cone, while the finer materials travel rapidly as an expanding tongue-shaped cloud.

The Destruction of St. Pierre. In 1902 an infamous nuée ardente from Mount Pelée, a small volcano on the Caribbean island of Martinique, destroyed the port town of St. Pierre. The destruction happened in moments and was so devastating that almost all of St. Pierre's 28,000 inhabitants were killed. Only one person on the outskirts of town—a prisoner protected in a dungeon—and a few people on ships in the harbor were spared (Figure 8.15).

Shortly after this calamitous eruption, scientists arrived on the scene. Although St. Pierre was mantled by only a thin layer of volcanic debris, they discovered that masonry walls nearly a meter thick were knocked over like dominoes; large trees were uprooted and cannons were torn from their mounts. A further reminder of the destructive force of this nuée ardente is preserved in the ruins of the mental hospital. One of the immense steel chairs that had been used to confine alcoholic patients can be seen today, contorted, as though it were made of plastic.

Lahars: Mudflows on Active and Inactive Cones

In addition to violent eruptions, large composite cones may generate a type of mudflow referred to by its Indonesian name **lahar.** The destructive mudflows occur when volcanic debris becomes saturated with water and rapidly moves down steep volcanic slopes, generally following gullies and stream valleys. Some lahars are triggered when large volumes of ice and snow melt during an eruption. Others are generated when heavy rainfall saturates weathered volcanic deposits. Thus, lahars can occur even when a volcano is *not* erupting.

When Mount St. Helens erupted in 1980, several lahars formed. These flows and accompanying flood waters raced down the valleys of the north and south forks of the Toutle River at speeds exceeding 30 kilometers per hour. Water levels in the river rose to 4 meters (13 feet) above flood stage, destroying or severely damaging nearly all the homes and bridges along the impacted area (Figure 8.16). Fortunately, the area was not densely populated.

In 1985 deadly lahars were produced during a small eruption of Nevado del Ruiz, a 5300 meter (17,400 foot) volcano in the Andes Mountains of Colombia. Hot pyroclastic material melted ice and snow that capped the mountain (Nevado means *snow* in Spanish) and sent torrents of ash and debris down three major river valleys that flank the volcano. Reaching speeds of 100 kilometers (60 miles) per hour, these mudflows tragically took 25,000 lives.

Mount Rainier, Washington, is considered by many to be America's most dangerous volcano because, like Nevado del Ruiz, it has a thick year-round mantle of snow and ice. Adding to the risk is the fact that 100,000 people live in the valleys around Rainier, and many homes are built on lahars that flowed down the volcano hundreds or thousands of years ago. A future eruption, or perhaps just a period of heavy rainfall, may produce lahars that will likely take similar paths.

Figure 8.15 St. Pierre as it appeared shortly after the eruption of Mount Pelée, 1902. (Reproduced from the collection of the Library of Congress)

STUDENTS SOMETIMES ASK...

*If volcanoes are so dangerous,
why do people live on or near them?*

Realize that many who live near volcanoes did not choose the location; they were simply born there. Their ancestors may have lived in the region for generations. Historically, many have been drawn to volcanic regions because of their fertile soils. Not all volcanoes have explosive eruptions, but all active volcanoes are dangerous. Certainly, choosing to live close to an active composite cone like Mount St. Helens or Italy's Mount Vesuvius has a high inherent risk. However, the time interval between successive eruptions might be several decades or more—plenty of time for generations of people to forget the last eruption and consider the volcano to be dormant (*dormin* = to sleep) and therefore safe. Many people that choose to live near an active volcano have the belief that the *relative* risk is no higher than in other hazard-prone places. In essence, they are gambling that they will be able to live out their lives before the next major eruption.

Figure 8.16 A house damaged by a lahar along the Toutle River, west-northwest of Mount St. Helens. The end section of the house was torn free and lodged against trees. (Photo by D. R. Crandell, U.S. Geological Survey)

BOX 8.1

PEOPLE AND THE ENVIRONMENT:
Eruption of Vesuvius A.D. 79

In addition to producing some of the most violent volcanic activity, composite cones can erupt unexpectedly. One of the best documented of these events was the A.D. 79 eruption of the Italian volcano we now call Vesuvius. Prior to this eruption, Vesuvius had been dormant for centuries and had vineyards adorning its sunny slopes. On August 24, however, the tranquility ended, and in less than 24 hours the city of Pompeii (near Naples) and more than 2000 of its 20,000 residents perished. Most were entombed beneath a layer of pumice nearly 3 meters (10 feet) thick. They remained this way for nearly 17 centuries, until the city was partially excavated, giving archaeologists a superbly detailed picture of ancient Roman life.

By reconciling historical records with detailed scientific studies of the region, volcanologists have pieced together the chronology of the destruction of Pompeii. The eruption most likely began as steam discharges on the morning of August 24. By early afternoon fine ash and pumice fragments formed a tall eruptive cloud emanating from Vesuvius. Shortly thereafter, debris from this cloud began to shower Pompeii, located 9 kilometers (6 miles) downwind of the volcano. Undoubtedly, many people fled during this early phase of the eruption. For the next several hours pumice fragments as large as 5 centimeters (2 inches) fell on Pompeii. One historical record of this eruption states that people located more distant than Pompeii tied pillows to their heads in order to fend off the flying fragments.

The pumice fall continued for several hours, accumulating at the rate of 12 to 15 centimeters (5 to 6 inches) per hour. Most of the roofs in Pompeii eventually gave way. Despite the accumulation of more than 2 meters of pumice, many of the people that had not evacuated Pompeii were probably still alive the morning of August 25. Then suddenly and unexpectedly a surge of searing hot dust and gas swept rapidly down the flanks of Vesuvius. This blast killed an estimated 2000 people who had somehow managed to survive the pumice fall. Some may have been killed by flying debris, but most died of suffocation as a result of inhaling ash-laden gases. Their remains were quickly buried by the falling ash, which rain cemented into a hard mass before their bodies had time to decay. The subsequent decomposition of the bodies produced cavities in the hardened ash that replicated the form of the entombed bodies, preserving in some cases even facial expressions. Nineteenth-century excavators found these cavities and created casts of the corpses by pouring plaster of Paris into the voids (Figure 8.A). Some of the plaster casts show victims trying to cover their mouth in an effort to draw what would be their last breath.

Volcanologists now realize that several destructive flows of hot, asphyxiating ash-laden gas swept over the countryside surrounding Vesuvius. Skeletons excavated from the nearby town of Herculaneum indicate that most of its inhabitants were probably killed by these flows. Further, many of those who evacuated Pompeii probably met a similar fate. An estimated 16,000 people may have perished in this tragic and unexpected event.

Figure 8.A Plaster casts of several victims of the A.D. 79 eruption of Mount Vesuvius that destroyed the Italian city of Pompeii. (Photo by Leonard von Matt/ Photo Researchers, Inc.)

Other Volcanic Landforms

The most obvious volcanic structure is a cone. But other distinctive and important landforms are also associated with volcanic activity.

Calderas

Calderas (*caldaria* = a cooking pot) are large collapse depressions having a more or less circular form. Their diameters exceed one kilometer, and many are tens of kilometers across. (Those less than a kilometer across are called *collapse pits*.) Most calderas are formed by one of the following processes: 1) the collapse of the summit of a large composite volcano following an explosive eruption of silica-rich pumice and ash fragments; 2) the collapse of the top of a shield volcano caused by subterranean drainage from a central magma chamber; and 3) the collapse of a large area caused by the discharge of colossal volumes of silica-rich pumice and ash along ring fractures.

Crater Lake–Type Calderas. Crater Lake, Oregon, is located in a caldera that has a maximum diameter of 10 kilometers (6 miles) and is 1175 meters (over 3800 feet) deep. This caldera formed about 7000 years ago when a composite cone, later named Mount Mazama, violently extruded 50 to 70 cubic kilometers of pyroclastic material (Figure 8.17). With the loss of support, 1500 meters (nearly a mile) of the summit of this once prominent cone collapsed. After the collapse, rainwater filled the caldera (Figure 8.18). Later volcanic activity built a small cinder cone in the lake. Today this cone, called Wizard Island, provides a mute reminder of past activity.

Hawaiian-Type Calderas. Although most calderas are produced by *collapse following an explosive eruption,*

Eruption of
Mount Mazama

Partialy emptied
magma chamber

Collapse of
Mount Mazama

Formation of Crater Lake and Wizard Island

Figure 8.17 Sequence of events that formed Crater Lake, Oregon. About 7000 years ago, the summit of former Mount Mazama collapsed following a violent eruption that partly emptied the magma chamber. Subsequent eruptions produced the cinder cone called Wizard Island. Rainfall and groundwater contributed to form the lake. (After H. Williams, *The Ancient Volcanoes of Oregon*, p. 47. Courtesy of the University of Oregon)

some are not. For example, Hawaii's active shield volcanoes, Mauna Loa and Kilauea, both have large calderas at their summits. Kilauea's measures 3.3 by 4.4 kilometers (about 2 by 3 miles) and is 150 meters (500 feet) deep. Each caldera formed by gradual subsidence of the summit as magma slowly drained laterally from the central magma chamber to a rift zone, often producing flank eruptions.

Yellowstone-Type Calderas. Although the 1980 eruption of Mount St. Helens was spectacular, it pales by comparison to what happened 630,000 years ago in the region now occupied by Yellowstone National Park. Here approximately 1000 cubic kilometers of pyroclastic material erupted, eventually producing a caldera 70 kilometers (43 miles) across. This event produced showers of ash as far away as the Gulf of Mexico. Vestiges of this activity are the many hot springs and geysers in the region.

The formation of a large Yellowstone-type caldera begins when a silica-rich (rhyolitic) magma body is emplaced near the surface, upwarping the overlying rocks. Next, ring fractures develop in the roof providing a pathway to the surface for the highly gas-rich magma. This initiates an explosive eruption of colossal proportions ejecting huge volumes (usually exceeding 100 cubic kilometers) of pyroclastic materials, mainly in the form of ash and pumice fragments. Typically, these materials form a pyroclastic flow that spreads across the landscape at speeds that may exceed 100 kilometers (60 miles) per hour, destroying most living things in its path. After coming to rest, the hot fragments of ash and pumice fuse together, forming a welded tuff that closely resembles a solidified lava flow. Finally, with the loss of support, the roof of the magma chamber collapses, generating a large caldera.

Calderas of the Yellowstone-type are the largest volcanic structures on Earth. Some geologists have compared their destructive force with that of the impact of a small asteroid. Fortunately, no eruption of this type has occurred in historic times. Other examples of large calderas located in the United States are California's Long Valley Caldera and Valles Caldera located west of Los Alamos, New Mexico.

Fissure Eruptions and Lava Plateaus

We think of volcanic eruptions as building a cone or shield from a central vent. But by far the greatest volume of volcanic material is extruded from fractures in the crust called **fissures** (*fissura* = to split). Rather than building a cone, these long, narrow cracks may emit a low-viscosity basaltic lava, blanketing a wide area.

The extensive Columbia Plateau in the northwestern United States was formed this way (Figure 8.19). Here, numerous **fissure eruptions** extruded very fluid basaltic lava (Figure 8.20). Successive flows, some 50 meters (160 feet) thick, buried the existing landscape as they built a lava plateau nearly a mile thick. The fluid nature of the lava is evident, because some remained molten long enough to flow 150 kilometers (90 miles) from its source. The term **flood basalts** appropriately describes these flows. Massive accumulations of basaltic lava, similar to those of the Columbia Plateau, occur worldwide. One of the largest is the Deccan Traps, a thick sequence of flat-lying basalt flows covering nearly 500,000 square kilometers (195,000 square miles) of

Figure 8.18 Crater Lake in Oregon occupies a caldera about 10 kilometers (6 miles) in diameter. Wizard Island is a cinder cone located in the caldera. (Photo by Greg Vaughn/Tom Stack and Associates)

west central India. When the Deccan Traps formed about 66 million years ago, nearly 2 million cubic kilometers of lava were extruded in less than 1 million years. Another huge deposit of flood basalts, called the Ontong Java Plateau, is found on the floor of the Pacific Ocean.

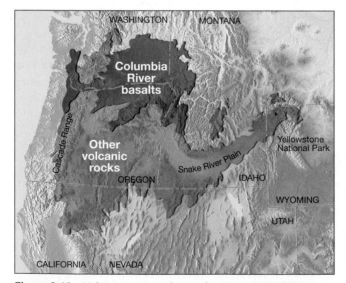

Figure 8.19 Volcanic areas in the northwestern United States. The Columbia River basalts cover an area of nearly 200,000 square kilometers (80,000 square miles). Activity here began about 17 million years ago as lava poured out of large fissures, eventually producing a basalt plateau with an average thickness of more than 1 kilometer. (After U.S. Geological Survey)

Volcanic Pipes and Necks

Most volcanoes are fed magma through short conduits, called *pipes*, that connect a magma chamber to the surface. In rare circumstances, pipes may extend tubelike to depths exceeding 200 kilometers (125 miles). When this occurs, the ultramafic magmas that migrate up these structures produce rocks that are thought to be samples of the mantle that have undergone very little alteration during their ascent. Geologists consider these unusually deep conduits to be "windows" into Earth, for they allow us to view rock normally found only at great depth.

The best-known volcanic pipes are the diamond-bearing structures of South Africa. Here, the rocks filling the pipes originated at depths of at least 150 kilometers (90 miles), where pressure is high enough to generate diamonds and other high-pressure minerals. The task of transporting essentially unaltered magma (along with diamond inclusions) through 150 kilometers of solid rock is exceptional. This fact accounts for the scarcity of natural diamonds.

Volcanoes on land are continually being lowered by weathering and erosion. Cinder cones are easily eroded, because they are composed of unconsolidated materials. However, all volcanoes will eventually succumb to relentless erosion over geologic time. As erosion progresses, the rock occupying the volcanic pipe is often more resistant and may remain standing above the surrounding terrain long after most of the cone has vanished. Shiprock, New Mexico, is such a feature and

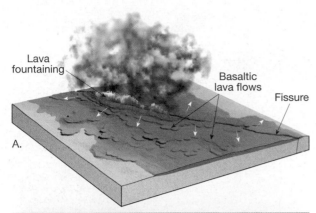

Lava
fountaining

Basaltic
lava flows

Fissure

A.

B.

Figure 8.20   Basaltic fissure eruption. **A.** Lava fountaining from a fissure forming fluid lava flows called flood basalts. **B.** Photo of basalt flows (dark) near Idaho Falls. (Photo by John S. Shelton)

is called a **volcanic neck** (Figure 8.21). This structure, higher than many skyscrapers, is but one of many such landforms that protrude conspicuously from the red desert landscapes of the American Southwest.

Intrusive Igneous Activity

 GEODe   Forces Within
↳ Igneous Activity

Although volcanic eruptions can be among the most violent and spectacular events in nature and therefore worthy of detailed study, most magma is emplaced at depth. Thus, an understanding of intrusive igneous activity is as important to geologists as the study of volcanic events.

The structures that result from the emplacement of igneous material at depth are called **plutons,** named for Pluto, the god of the lower world in classical mythology. Because all plutons form out of view beneath Earth's surface, they can be studied only after uplifting and erosion have exposed them. The challenge lies in reconstructing the events that generated these structures millions or even hundreds of millions of years ago.

Plutons are known to occur in a great variety of sizes and shapes. Some of the most common types are illustrated in Figure 8.22. Notice that some of these structures have a tabular (tabletop) shape, whereas others are quite massive. Also, observe that some of these bodies cut across existing structures, such as layers of sedimentary rock; others form when magma is injected between sedimentary layers. Because of these differences, intrusive igneous bodies are generally classified according to their shape as either *tabular* (*tabula* = table) or *massive* and by their orientation with respect to the host rock. Plutons are said to be *discordant* (*discordare* = to disagree) if they cut across existing structures and *concordant* (*concordare* = to agree) if they form parallel to

Figure 8.21   Ship Rock, New Mexico, is a volcanic neck. This structure, which stands over 420 meters (1380 feet) high, consists of igneous rock that crystallized in the vent of a volcano that has long since been eroded away. The tabular structure in the background is a dike that served to feed lava flows along the flanks of the once active volcano. (Photo by Tom Bean)

Figure 8.22 Illustrations showing basic igneous plutons. **A.** This block diagram shows the relationship between volcanism and intrusive igneous activity. **B.** This view illustrates the basic intrusive igneous structures, some of which have been exposed by erosion long after their formation. **C.** After millions of years of uplifting and erosion a batholith is exposed at the surface.

A. Implacement of magma

B. Crystallization of igneous plutons and erosion

C. Extensive uplift and erosion exposes batholith

features such as sedimentary strata. As you can see in Figure 8.22A, plutons are closely associated with volcanic activity.

Dikes

Dikes are tabular discordant bodies that are produced when magma is injected into fractures. The force exerted by the emplaced magma can be great enough to sep-arate the walls of the fracture further. Once crystallized, these sheetlike structures have thicknesses ranging from less than a centimeter to more than a kilometer. The largest have lengths of hundreds of kilometers. Most dikes, however, are a few meters thick and extend laterally for no more than a few kilometers.

Dikes are often found in groups that once served as vertically oriented pathways followed by molten rock that fed ancient lava flows. The parent pluton is generally not observable. Some dikes are found radiating, like spokes on a wheel, from an eroded volcanic neck. In these situations the active ascent of magma is thought to have generated fissures in the volcanic cone out of which lava flowed.

Sills and Laccoliths

Sills and laccoliths are concordant plutons that form when magma is intruded in a near-surface environment. They differ in shape and usually differ in composition.

Sills. **Sills** are tabular plutons formed when magma is injected along sedimentary bedding surfaces (Figure 8.23). Horizontal sills are the most common, although all orientations, even vertical, are known to exist. Because of their relatively uniform thickness and large areal extent, sills are likely the product of very fluid magmas. Magmas having a low silica content are more fluid, so most sills are composed of the rock basalt.

The emplacement of a sill requires that the overlying sedimentary rock be lifted to a height equal to the thickness of the sill. Although this is a formidable task, in shallow environments it often requires less energy than forcing the magma up the remaining distance to the surface. Consequently, sills form only at shallow depths, where the pressure exerted by the weight of overlying rock layers is low. Although sills are intruded between layers, they can be locally discordant. Large sills frequently cut across sedimentary layers and resume their concordant nature at a higher level.

One of the largest and most studied of all sills in the United States is the Palisades Sill. Exposed for 80 kilometers along the west bank of the Hudson River in southeastern New York and northeastern New Jersey, this sill is about 300 meters thick. Because of its resistant nature, the Palisades Sill forms an imposing cliff

that can be seen easily from the opposite side of the Hudson.

In many respects, sills closely resemble buried lava flows. Both are tabular and often exhibit columnar jointing (Figure 8.24). **Columnar joints** form as igneous rocks cool and develop shrinkage fractures that produce elongated, pillarlike columns. Further, because sills generally form in near-surface environments and may be only a few meters thick, the emplaced magma often cools quickly enough to generate a fine-grained texture.

Laccoliths. **Laccoliths** are similar to sills because they form when magma is intruded between sedimentary layers in a near-surface environment. However, the magma that generates laccoliths is more viscous. This less fluid magma collects as a lens-shaped mass that arches the overlying strata upward (see Figure 8.22). Consequently, a laccolith can occasionally be detected because of the dome-shaped bulge it creates at the surface.

Most large laccoliths are probably not much wider than a few kilometers. The Henry Mountains in southeastern Utah are largely composed of several laccoliths believed to have been fed by a much larger magma body emplaced nearby.

Batholiths

By far the largest intrusive igneous bodies are **batholiths** (*bathos* = depth, *lithos* = stone). Most often, batholiths occur in groups that form linear structures several hundreds of kilometers long and up to 100 kilometers wide. The Idaho batholith, for example, encompasses an area of more than 40,000 square kilometers and consists of many plutons. Indirect evidence gathered from gravitational studies indicates that batholiths are also very thick, possibly extending dozens of kilometers into the crust. Based on the amount exposed by erosion, some batholiths are at least several kilometers thick.

By definition, a plutonic body must have a surface exposure greater than 100 square kilometers (40 square miles) to be considered a batholith. Smaller plutons of this type are termed *stocks*. Many stocks appear to be portions of batholiths that are not yet fully exposed.

Batholiths may compose the core of mountain systems. Here uplifting and erosion have removed the surrounding rock, thereby exposing the resistant igneous body. Some of the highest peaks in the Sierra Nevada, such as Mount Whitney, are carved from such a granitic mass (Figure 8.25).

Large expanses of granitic rock also occur in the stable interiors of the continents, such as the Canadian Shield of North America. These relatively flat exposures are the remains of ancient mountains that have long

Figure 8.23 Salt River Canyon, Arizona. The dark, essentially horizontal band is a sill of basaltic composition that intruded into horizontal layers of sedimentary rock. (Photo by E. J. Tarbuck)

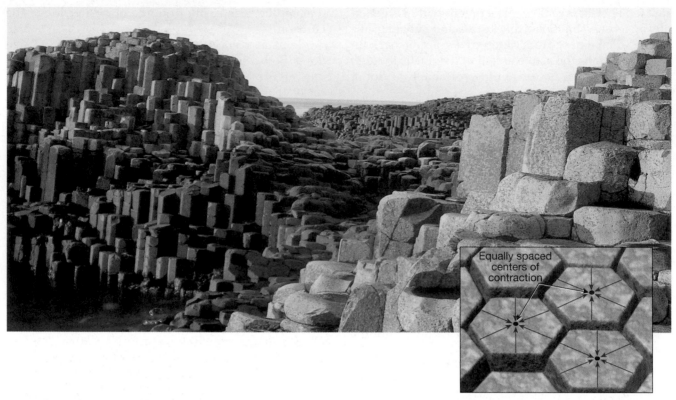

Figure 8.24   Columnar jointing in basalt, Giants Causeway National Park, Northern Ireland. These five-to-seven-sided columns are produced by contraction and fracturing that results as a lava flow or sill gradually cools. (Photo by Tom Till)

Figure 8.25   Mount Whitney California. This mountain makes up just a tiny portion of the Sierra Nevada batholith, a huge structure that extends for approximately 400 kilometers. (Photo by Art Wolfe)

since been leveled by erosion. Thus, the rocks that make up the batholiths of youthful mountain ranges, such as the Sierra Nevada, were generated near the top of a magma chamber, whereas in shield areas, the roots of former mountains and, thus, the lower portions of batholiths, are exposed. In Chapter 9 we will further consider the role of igneous activity as it relates to mountain building.

Origin of Magma

The origin of magma has been controversial in geology, almost from the beginning of the science. How do magmas of different compositions form? Why do volcanoes in the deep-ocean basins primarily extrude basaltic lava, whereas those adjacent to oceanic trenches extrude mainly andesitic lava? These are some of the questions we will address in the following sections.

Generating Magma from Solid Rock

Based on available scientific evidence, *Earth's crust and mantle are composed primarily of solid, not molten, rock*. Although the outer core is a fluid, its iron-rich material is very dense and remains deep within Earth. So, what is the source of magma that produces igneous activity?

Geologists conclude that magma originates when essentially solid rock, located in the crust and upper mantle, melts. The most obvious way to generate magma from solid rock is to raise the temperature above the level at which the rock begins to melt.

Role of Heat. What source of heat is sufficient to melt rock? Workers in underground mines know that temperatures get higher as they go deeper. Although the rate of temperature change varies from place to place, it *averages* between 20°C and 30°C per kilometer in the *upper* crust. This change in temperature with depth is known as the **geothermal gradient.** Estimates indicate that the temperature at a depth of 100 kilometers ranges between 1200°C and 1400°C. At these high temperatures, rocks in the lower crust and upper mantle are near but somewhat cooler than their melting point temperatures. Thus, they are very hot but still essentially solid.

There are several ways that enough additional heat can be generated within the crust or upper mantle to produce some magma. First, at subduction zones friction generates heat as huge slabs of crust slide past one another. Second, crustal rocks are heated as they descend into the mantle during subduction. Third, hot mantle rocks can rise and intrude crustal rocks. Although all of these processes generate some magma, the quantities produced are relatively small and the distribution is very limited. As you will see, the vast bulk of

magma forms without the aid of an additional heat source.

Role of Pressure. If temperature were the only factor that determined whether or not rock melts, our planet would be a molten ball covered with a thin, solid outer shell. This, of course, is not the case. The reason is that pressure also increases with depth.

Melting, which is accompanied by an increase in volume, *occurs at higher temperatures at depth* because of greater confining pressure. Consequently, an increase in confining pressure causes an increase in the rock's melting temperature. Conversely, reducing confining pressure lowers a rock's melting temperature. When confining pressure drops enough, **decompression melting** is triggered. This may occur when rock *ascends* as a result of convective upwelling, thereby moving into zones of lower pressure. (Recall that even though the mantle is a *solid, it does flow* at very slow rates over time scales of millions of years.) This process is responsible for generating magma along ocean ridges where plates are rifting apart (Figure 8.26).

Role of Volatiles. Another important factor affecting the melting temperature of rock is its water content. Water and other volatiles act as salt does to melt ice. That is, volatiles cause rock to melt at lower temperatures. Further, the effect of volatiles is magnified by increased pressure. Consequently, "wet" rock buried at depth has a much lower melting temperature than does "dry" rock of the same composition and under the same confining pressure. Therefore, in addition to a rock's composition, its temperature, depth (confining pressure), and water content determine whether it exists as a solid or liquid.

Volatiles play an important role in generating magma in regions where cool slabs of oceanic lithosphere descend into the mantle (Figure 8.27). As an oceanic plate sinks, both heat and pressure drive water from the subducting crustal rocks. These volatiles, which are very mobile, migrate into the wedge of hot mantle that lies above. This process is believed to lower the melting temperature of mantle rock sufficiently to generate some melt. Laboratory studies have shown that the melting point of basalt can be lowered by as much as 100°C by the addition of only 0.1 percent water.

When enough mantle-derived basaltic magma forms, it will buoyantly rise toward the surface. In a continental setting, basaltic magma may "pond" beneath crustal rocks, which have a lower density and are already near their melting temperature. This may result in some melting of the crust and the formation of a secondary, silica-rich magma.

In summary, magma can be generated under three sets of conditions: (1) *heat* may be added; for example,

Figure 8.26 As hot mantle rock ascends, it continually moves into zones of lower pressure. This drop in confining pressure can trigger melting, even without additional heat.

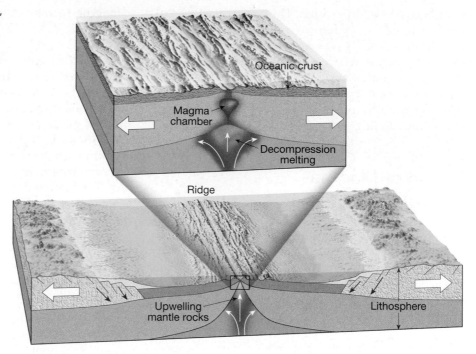

a magma body from a deeper source intrudes and melts crustal rock; (2) *a decrease in pressure* (without the addition of heat) can result in *decompression melting*; and (3) the *introduction of volatiles* (principally water) can lower the melting temperature of mantle rock sufficiently to generate magma.

Partial Melting and Magma Compositions

An important difference exists between the melting of a substance that consists of a single compound, such as ice, and melting igneous rocks, which are mixtures of several different minerals. Ice melts at a specific temperature, whereas igneous rocks melt over a tempera-

ture range of about 200°C. As rock is heated, minerals with the lowest melting points tend to start melting first. Should melting continue, minerals with higher melting points begin to melt, and the composition of the magma steadily approaches the overall composition of the rock from which it was derived. Most often, melting is not complete. This process, known as **partial melting,** produces most, if not all, magma.

An important consequence of partial melting is the *production of a magma with a higher silica content than the original rock.* Recall from the discussion of Bowen's reaction series that basaltic (mafic) rocks contain mostly high-melting-temperature minerals that are comparatively low in silica, whereas granitic (felsic) rocks are composed primarily of low-melting-temperature sili-

Figure 8.27 As an oceanic plate descends into the mantle, water and other volatiles are driven from the subducting crustal rocks. These volatiles lower the melting temperature of mantle rock sufficiently to generate magma.

Figure 8.28 Locations of some of Earth's major volcanoes. Note the concentration of volcanoes encircling the Pacific basin, known as the "Ring of Fire."

cates that are enriched in silica (see Chapter 2). Because silica-rich minerals melt first, magmas generated by partial melting are nearer to the granitic end of the compositional spectrum than are the rocks from which they formed (see Figure 2.10).

Plate Tectonics and Igneous Activity

Geologists have known for decades that the global distribution of volcanism is not random. Of the more than 800 active volcanoes that have been identified, most are located along the margins of the ocean basins—most notably within the circum-Pacific belt known as the *Ring of Fire* (Figure 8.28). This group of volcanoes consists mainly of composite cones that emit volatile-rich magma having an intermediate (andesitic) composition that occasionally produce awe-inspiring eruptions.

The volcanoes comprising a second group emit very fluid basaltic lavas and are confined to the deep-ocean basins, including well-known examples on Hawaii and Iceland. In addition, this group contains many active submarine volcanoes that dot the ocean floor; particularly notable are the innumerable small seamounts that lie along the axis of the oceanic ridge.

A third group includes those volcanic structures that are irregularly distributed in the interiors of the continents. Volcanism on continents is the most diverse, ranging from eruptions of very fluid basaltic lavas, like those that generated the Columbia River basalts to explosive eruptions of silica-rich rhyolitic magma as occurred in Yellowstone.

Until the late 1960s, geologists had no explanation for the apparently haphazard distribution of continental volcanoes, nor were they able to account for the almost continuous chain of volcanoes that circles the margin of the Pacific basin. With the development of the theory of plate tectonics, the picture was greatly clarified. Recall that most magma originates in the upper mantle and that the mantle is essentially solid, *not molten* rock. The basic connection between plate tectonics and volcanism is that *plate motions provide the mechanisms by which mantle rocks melt to generate magma.*

We will examine three zones of igneous activity and their relationship to plate boundaries (Figure 8.29). These active areas are located 1) along convergent plate boundaries where plates move toward each other and one sinks beneath the other; 2) along divergent plate boundaries, where plates move away from each other and new seafloor is created; and 3) areas within the plates proper that are not associated with any plate boundary.

Igneous Activity at Convergent Plate Boundaries

Recall that at convergent plate boundaries slabs of oceanic crust are bent as they descend into the mantle, generating an oceanic trench. As a slab sinks deeper into the mantle, the increase in temperature and pressure drives volatiles (mostly H_2O) from the oceanic crust. These mobile fluids migrate upward into the wedge-shaped piece of mantle located between the subducting slab and overriding plate (see Figure 8.27). Once

Convergent plate volcanism

Volcanic island arc

Trench

Oceanic crust

Marginal sea

Continental crust

Subducting oceanic lithosphere

Mantle rock melts

Water driven from plate

Asthenosphere

Oceanic crust

Hot spot

Hawaii

Decompression melting

Rising mantle plume

Intraplate volcanism

Continental volcanic arc

Trench

Oceanic crust

Continental crust

Subducting oceanic lithosphere

Mantle rock melts

Water driven from plate

Convergent plate volcanism

Figure 8.29 Three zones of volcanism. Two of these zones are plate boundaries, and the third is the interior area of the plates.

Figure 8.29 Continued

the sinking slab reaches a depth of about 100 to 150 kilometers, these water-rich fluids reduce the melting point of hot mantle rock sufficiently to trigger some melting. The partial melting of mantle rock generates magma with a basaltic composition. After a sufficient quantity of magma has accumulated, it slowly migrates upward.

Volcanism at a convergent plate margin results in the development of a linear or slightly curved chain of volcanoes called a *volcanic arc*. These volcanic chains develop roughly parallel to the associated trench—at distances of 200 to 300 kilometers (100 to 200 miles). Volcanic arcs can be constructed on oceanic, or continental, lithosphere. Those that develop within the ocean and grow large enough for their tops to rise above the surface are labeled *island archipelagos* in most atlases. Geologists prefer the more descriptive term **volcanic island arcs,** or simply **island arcs** (Figure 8.29 upper left). Several young volcanic island arcs of this type border the western Pacific basin, including the Aleutians, the Tongas, and the Marianas.

Volcanism associated with convergent plate boundaries may also develop where slabs at oceanic lithosphere are subducted under continental lithosphere to produce a **continental volcanic arc** (Figure 8.29 lower left). The mechanisms that generate these magmas are essentially the same as those operating at island arcs. The major difference is that continental crust is much thicker and is composed of rocks having a higher silica content than oceanic crust. Hence through the assimilation of silica-rich crustal rocks a magma body may change composition as it rises through continental crust. Stated another way, magmas generated in the mantle may change from a comparatively dry, fluid basaltic magma to a viscous andesitic or rhyolitic magma having a high concentration of volatiles as it moves up through the continental crust. The volcanic chain of the Andes Mountains along the western edge of South America is perhaps the best example of a mature continental volcanic arc.

Since the Pacific basin is essentially bordered by convergent plate boundaries (and associated subduction zones), it is easy to see why the irregular belt of explosive volcanoes we call the Ring of Fire formed in this region. The volcanoes of the Cascade Range in the northwestern United States, including Mount Hood, Mount Rainier, and Mount Shasta, are included in this group (Figure 8.30).

Igneous Activity at Divergent Plate Boundaries

The greatest volume of magma (perhaps 60 percent of Earth's total yearly output) is produced along the oceanic ridge system in association with seafloor spreading (Figure 8.29 upper right). Here, below the ridge axis where the lithospheric plates are being continually

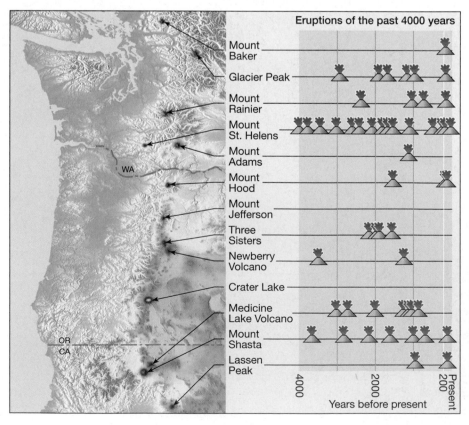

Figure 8.30 Of the 13 potentially active volcanoes in the Cascade Range, 11 have erupted in the past 4000 years and 7 in just the past 200 years. More than 100 eruptions, most of which were explosive, have occurred in the past 4000 years. Mount St. Helens is the most active volcano in the Cascades. Its eruptions have ranged from relatively quiet outflows of lava to explosive events much larger than that of May 18, 1980. Each eruption symbol in the diagram represents from one to several dozen eruptions closely spaced in time. (After U.S. Geological Survey)

BOX 8.2

EARTH AS A SYSTEM:
Volcanic Activity and Climate

One example of the interplay between different parts of the Earth system is the relationship between volcanic activity and changes in climate. We know that changes in the composition of the atmosphere can have a significant impact on climate. Moreover, we know that volcanic eruptions can emit large quantities of gases and particles into the atmosphere, thus altering its composition. So do volcanic eruptions actually influence Earth's climate?

The idea that explosive eruptions might alter Earth's climate goes back many years. In fact, Benjamin Franklin speculated that the eruption of a large volcano in Iceland might have been responsible for the unusually cold winter of 1783–84. The basic premise is that suspended volcanic material reflects a portion of the incoming solar radiation, which in turn drops temperatures in the lowest layer of the atmosphere (this layer, called the *troposphere*, extends from Earth's surface to a height of about 10 kilometers).

Perhaps the most notable cool period linked to a volcanic event is the "year without a summer" that followed the 1815 eruption of Indonesia's Mount Tambora. The cold temperatures of spring and summer were unprecedented and were followed by an early fall in many parts of the Northern Hemisphere, including eastern North America and Western Europe. The abnormal temperatures and shortened growing season led to crop failures, which in turn led to famine in some regions.

When Mount St. Helens erupted in 1980, there was almost immediate speculation: Can an eruption such as this change our climate? There is no doubt that the large quantity of volcanic ash emitted by the explosive eruption had

local and regional effects for a short period (Figure 8.B). Still, studies indicated that any longer-term lowering of hemispheric temperatures was negligible. The cooling was so slight, probably less than 0.1°C (0.2°F), that it could not be distinguished from other natural temperature fluctuations.

Two years of monitoring and studies following the 1982 eruption of Mexico's El Chichón volcano indicated that its cooling effect on global mean temperature was greater than that of Mount St. Helens—on the order of 0.3 to 0.5°C (0.5 to 0.9°F). The eruption of

Figure 8.B   When Mount St. Helens erupted on May 18, 1980, huge quantities of volcanic ash were blown into the atmosphere. This satellite image was taken less than eight hours after the eruption. The ash cloud had already spread as far as western Montana. Volcanic ash has little long-term impact on global climate because it settles quickly from the air. A more significant factor affecting climate is the quantity of sulfur dioxide gas emitted during an eruption. (Photo courtesy of the National Environmental Satellite Service)

El Chichón, was less explosive than the Mount St. Helens blast, so why did it have a greater impact on global temperatures? The reason is that the material emitted by Mount St. Helens was largely fine ash that settled out in a relatively short time. El Chichón, on the other hand, emitted far greater quantities of sulfur dioxide gas (an estimated 40 times more) than Mount St. Helens. This gas combines with water vapor to produce a dense cloud of tiny sulfuric-acid particles in the stratosphere (the atmospheric layer between about 10 and 50 kilometers above the surface). The particles, called *aerosols*, may take up to several years to settle out completely. They lower the troposphere's mean temperature because they reflect solar radiation back to space.

The Philippines volcano, Mount Pinatubo, erupted explosively in June 1991, injecting 25 million to 30 million tons of sulfur dioxide into the stratosphere. The event provided scientists with an opportunity to study the climatic impact of a major explosive volcanic eruption using NASA's spaceborne Earth Radiation Budget Experiment. During the next year the haze of tiny aerosols lowered global temperatures by 0.5°C (0.9°F).

We now understand that volcanic clouds that remain in the stratosphere for a year or more are composed largely of sulfuric-acid droplets and not of volcanic ash, as was once thought. Thus, the volume of ash emitted during an explosive event is *not* an accurate criterion for predicting the global atmospheric effects of an eruption. Furthermore, the preceding examples illustrate that the impact on climate of a single volcanic eruption, no matter how explosive, is relatively small and short-lived.

---

pulled apart, the solid yet mobile mantle responds to the decrease in overburden and rises upward to fill in the rift. Recall that as rock rises, it experiences a decrease in confining pressure and undergoes melting without the addition of heat. This process, called *decompression melting*, generates large quantities of magma.

Partial melting of mantle rock at spreading centers produces basaltic magma having a composition that is surprisingly similar to that generated along convergent plate boundaries. Because this newly formed basaltic magma is less dense than the mantle rock from which it was derived, it buoyantly rises.

Collecting in reservoirs located just beneath the ridge crest, about 10 percent of this magma eventually migrates upward along fissures to erupt as flows on the ocean floor. This activity continuously adds new basaltic rock to the plate margins, temporarily welding them together, only to break again as spreading continues. Along some ridges, outpouring of bulbous pillow lavas build numerous small seamounts. At other locations erupted lavas produce fluid flows that create more subdued topography.

Although most spreading centers are located along the axis of an oceanic ridge, some are not. In particular, the East African Rift is a site where continental crust is being ripped apart (Figure 8.29, lower right). Vast outpourings of fluid basaltic lavas are common in this region. The East African Rift zone also contains numerous small volcanoes and even a few large composite cones, as exemplified by Mount Kilimanjaro.

Intraplate Igneous Activity

We know why igneous activity is initiated along plate boundaries, but why do eruptions occur in the interiors of plates? Hawaii's Kilauea is considered the world's most active volcano, yet it is situated thousands of kilometers from the nearest plate boundary in the middle of the vast Pacific plate. Other sites of **intraplate volcanism** (meaning "within the plate") include the Canary Islands, Yellowstone, and several volcanic centers that you may be surprised to learn are located in the Sahara Desert of northern Africa.

We now recognize that most intraplate volcanism occurs where a mass of hotter than normal mantle material called a **mantle plume** ascends toward the surface (Figure 8.29 left). Although the depth at which (at least some) mantle plumes originate is still hotly debated, many appear to form deep within Earth at the core–mantle boundary. These plumes of solid yet mobile mantle rock rise toward the surface in a manner similar to the blobs that form within a lava lamp. (These are the gaudy lamps that contain two immiscible liquids in a glass container. As the base of the lamp is heated, the denser liquid at the bottom becomes buoyant and forms blobs that rise to the top.) Like the blobs in a lava lamp, a mantle plume has a bulbous head that draws out a narrow stalk beneath it as it rises. Once the plume head nears the top of the mantle, decompression melting generates basaltic magma that may eventually trigger volcanism at the surface. The result is a localized volcanic region a few hundred kilometers across called a **hot spot** (Figure 8.29 right). More than 100 hot spots have been identified, and most have persisted for millions of years. The land surface around hot spots is often elevated, showing that it is buoyed up by a plume of warm low-density material. Furthermore, by measuring the heat flow in

these regions, geologists have determined that the mantle beneath hot spots must be 100–150°C hotter than normal.

The volcanic activity on the island of Hawaii, with its outpourings of basaltic lava, is certainly the result of hot-spot volcanism (Figure 8.29 left). Where a mantle plume has persisted for long periods of time, a chain of volcanic structures may form as the overlying plate moves over it. In the Hawaiian Islands, hot-spot activity is currently centered on Kilauea. However, over the past 80 million years the same mantle plume generated a chain of volcanic islands (and seamounts) that extend thousands of kilometers from the Big Island in a northwestward direction across the Pacific.

Mantle plumes are also thought to be responsible for the vast outpourings of basaltic lava that create large basalt plateaus such as the Columbia Plateau in the northwestern United States, India's Deccan Plateau, and the Ontong Java Plateau in the western Pacific (Figure 8.29 right).

Although the plate tectonics theory has answered many questions regarding the distribution of igneous activity, many new questions have arisen: Why does seafloor spreading occur in some areas but not others? How do mantle plumes and associated hot spots originate? These and other questions are the subject of continuing geologic research.

?

STUDENTS SOMETIMES ASK...

Some of the larger volcanic eruptions, like the eruption of Krakatoa, must have been impressive. What was it like?

On August 27, 1883, in what is now Indonesia, the volcanic island of Krakatoa exploded and was nearly obliterated. The sound of the explosion was heard an incredible 4800 kilometers (3000 miles) away at Rodriguez Island in the western Indian Ocean. Dust from the explosion was propelled into the atmosphere and circled Earth on high-altitude winds. This dust produced unusual and beautiful sunsets for nearly a year.

Not many were killed directly by the explosion, because the island was uninhabited. However, the displacement of water from the energy released during the explosion was enormous. The resulting *seismic sea wave* or *tsunami* exceeded 35 meters (116 feet) in height. It devastated the coastal region of the Sunda Strait between the nearby islands of Sumatra and Java, drowning over 1000 villages and taking more than 36,000 lives. The energy carried by this wave reached every ocean basin and was detected by tide recording stations as far away as London and San Francisco.

Summary

- The primary factors that determine the nature of volcanic eruptions include the magma's *temperature*, its *composition*, and the *amount of dissolved gases it* contains. As lava cools, it begins to congeal, and as *viscosity* increases, its mobility decreases. *The viscosity of magma is directly related to its silica content. Rhyolitic* lava, with its high silica content, is very viscous and forms short, thick flows. *Basaltic* lava, with a lower silica content, is more fluid and may travel a long distance before congealing. Dissolved gases provide the force that propels molten rock from the vent of a volcano.

- The materials associated with a volcanic eruption include *lava flows* (*pahoehoe* and *aa* flows for basaltic lavas), *gases* (primarily in the form of *water vapor*), and *pyroclastic material* (pulverized rock and lava fragments blown from the volcano's vent, which include *ash, pumice, lapilli, cinders, blocks,* and *bombs*).

- Successive eruptions of lava from a central vent result in a mountainous accumulation of material known as a *volcano*. Located at the summit of many volcanoes is a steep-walled depression called a *crater. Shield cones* are broad, slightly domed volcanoes built primarily of fluid, basaltic lava. *Cinder cones* have steep slopes composed of pyroclastic material. *Composite cones*, or *stratovolcanoes*, are large, nearly symmetrical structures built of interbedded lavas and pyroclastic deposits. Composite cones produce some of the most violent volcanic activity. Often associated with a violent eruption is a *nuée ardente*, a fiery cloud of hot gases infused with incandescent ash that races down steep volcanic slopes. Large composite cones may also generate a type of mudflow known as a *lahar*.

- Most volcanoes are fed by *conduits* or *pipes*. As erosion progresses, the rock occupying the pipe is often more resistant and may remain standing above the surrounding terrain as a *volcanic neck*. The summits of some volcanoes have large, nearly circular depressions called *calderas* that result from collapse following an explosive eruption. Calderas also form on shield volcanos by subterranean drainage from a central magma chamber, and the largest calderas form by the discharge of colossal volumes of silica-rich pumice along ring fractures. Although volcanic eruptions from a central vent are the most familiar, by far the largest amounts of volcanic material are extruded from cracks in the crust called *fissures*. The term *flood basalts* describes the fluid, waterlike, basaltic lava flows that cover an extensive region in the northwestern United States known as the Columbia Plateau. When silica-rich magma is extruded, *pyroclastic flows* consisting largely of ash and pumice fragments usually result.

- Igneous intrusive bodies are classified according to their *shape* and by their *orientation with respect to the host rock*, generally sedimentary rock. The two general shapes are *tabular* (tablelike) and *massive*. Intrusive igneous bodies that cut across existing sedimentary beds are said to be *discordant*, whereas those that form parallel to existing sedimentary beds are *concordant*.

- *Dikes* are tabular, discordant igneous bodies produced when magma is injected into fractures that cut across rock layers. Tabular, concordant bodies called *sills* form when magma is injected along the bedding surfaces of sedimentary rocks. *Laccoliths* are similar to sills but form from less-fluid magma that collects as a lens-shaped mass that arches the overlying strata upward. *Batholiths*, the largest intrusive igneous bodies with surface exposures of more than 100 square kilometers (40 square miles), frequently make up the cores of mountains.

- Magma originates from essentially solid rock of the crust and mantle. In addition to a rock's composition, its temperature, depth (confining pressure), and water content determine whether it exists as a solid or liquid. Thus, magma can be generated by *raising a rock's temperature*, as occurs when a hot mantle plume "ponds" beneath crustal rocks. A *decrease in pressure* can cause *decompression melting*. Further, the *introduction of volatiles* (water) can lower a rock's melting point sufficiently to generate magma. Because melting is generally not complete, a process called *partial melting* produces a melt made of the lowest-melting-temperature minerals, which are higher in silica than the original rock. Thus, magmas generated by partial melting are nearer to the granitic (felsic) end of the compositional spectrum than are the rocks from which they formed.

- *Most active volcanoes are associated with plate boundaries.* Active areas of volcanism are found along oceanic ridges where seafloor spreading is occurring (*divergent plate boundaries*), in the vicinity of ocean trenches where one plate is being subducted beneath another (*convergent plate boundaries*), and in the interiors of plates themselves (*intraplate volcanism*). Rising plumes of hot mantle rock are the source of most intraplate volcanism.

Key Terms

aa flow (p. 230)
batholith (p. 245)
caldera (p. 232)
cinder cone (p. 234)
columnar joint (p. 245)
composite cone (p. 236)

conduit (p. 232)
continental volcanic arc (p. 252)
crater (p. 232)
decompression melting (p. 247)
dike (p. 244)
fissure (p. 232)

fissure eruption (p. 241)
flood basalt (p. 241)
furmarole (p. 233)
geothermal gradient (p. 247)
hot spot (p. 254)
intraplate volcanism (p. 254)

Review Questions

1. What is the difference between magma and lava?

2. What three factors determine the nature of a volcanic eruption? What role does each play?

3. Why is a volcano fed by highly viscous magma likely to be a greater threat than a volcano supplied with very fluid magma?

4. Describe pahoehoe and aa lava.

5. List the main gases released during a volcanic eruption.

6. Analysis of samples taken during Hawaiian eruptions on the island of Hawaii indicate that _____ was the most abundant gas released.

7. How do volcanic bombs differ from blocks of pyroclastic debris?

8. Compare a volcanic crater to a caldera.

9. Compare and contrast the three main types of volcanoes (size, shape, eruptive style, and so forth).

10. Name one example of each of the three types of volcanoes.

11. Compare the formation of Hawaii with that of Parícutin.

12. Describe the formation of Crater Lake. Compare it to the caldera formed during the eruption of Kilauea.

13. What are the largest volcanic structures on Earth?

14. What is Ship Rock, New Mexico, and how did it form?

15. How do the eruptions that created the Columbia Plateau differ from eruptions that create volcanic peaks?

16. Describe each of the four intrusive igneous features discussed in the text (dike, sill, laccolith, and batholith).

17. Why might a laccolith be detected at Earth's surface before being exposed by erosion?

18. What is the largest of all intrusive igneous bodies? Is it tabular or massive? Concordant or discordant?

19. Explain how most magma is thought to originate.

20. Spreading-center volcanism is associated with which rock type? What causes rocks to melt in regions of spreading-center volcanism?

21. What is the Ring of Fire?

22. Are volcanic eruptions in the Ring of Fire generally quiet or violent? Name a volcano that would support your answer.

23. The Hawaiian Islands and Yellowstone Park are thought to be associated with which of the three zones of volcanism?

24. Volcanic islands in the deep ocean are composed primarily of what igneous rock type?

25. Briefly describe the mechanism by which explosive volcanic eruptions are thought to influence Earth's climate (See Box 8.2).

Examining the Earth System

1. Describe some of the consequences that a great and prolonged increase in explosive volcanic activity might have on each of Earth's four spheres. An excellent Web site to begin volcano research is the University of California at Santa Barbara Volcano Information Center at **http://magic.geol.ucsb.edu/~fisher/**

2. Despite the potential for devastating destruction, humans live, work, and play on or near many active volcanoes. What are some of the benefits that a volcano or volcanic region might offer? (List some volcanoes and the assets they provide.)

Web Resources

 The *Earth Science* Website uses the resources and flexibility of the Internet to aid in your study of the topics in this chapter. Written and developed by Earth science instructors, this site will help improve your understanding of Earth science. Visit **http://www.prenhall.com/tarbuck** and click on the cover of *Earth Science* 10e to find:

- **On-line review quizzes.**
- **Web-based critical thinking and writing exercises.**
- **Links to chapter-specific Web resources.**
- **Internet-wide key term searches.**

http://www.prenhall.com/tarbuck

Mount Moran (on right) in Wyoming's Grand Tetons National Park. (Photo by Art Wolfe)

CHAPTER 9

MOUNTAIN BUILDING

Mountains provide some of the most spectacular scenery on our planet (Figure 9.1). This splendor has been captured by poets, painters, and songwriters alike. Geologists believe that at some time all continental regions were mountainous masses and have concluded that the continents grow by the addition of mountains to their flanks. Consequently, as geologists unravel the secrets of mountain formation, they also gain a deeper understanding of the evolution of Earth's continents. If continents do indeed grow by adding mountains to their flanks, how do geologists explain the existence of mountains (the Urals, for example) that are located in the interior of a landmass? To answer this and related questions, this chapter attempts to piece together the sequence of events believed to generate these lofty structures. We begin our look at mountain building by examining the process of rock deformation and the structures that result.

Rock Deformation

Every body of rock, no matter how strong, has a point at which it will fracture or flow. **Deformation** (*de* = out, *forma* = form) is a general term that refers to all changes in the original shape and/or size of a rock body. Most crustal deformation occurs along plate mar-

gins. Recall from Chapter 7 that the lithosphere consists of large segments (plates) that move relative to one another. Plate motions and the interactions along plate boundaries generate forces that cause rock to deform.

When rocks are subjected to forces (stresses) greater than their own strength, they begin to deform, usually by folding, flowing, or fracturing (Figure 9.2). It is easy to visualize how rocks break, because we normally think of them as being brittle. But how can rock masses be bent into intricate folds without being broken during the process? To answer this question, geologists performed laboratory experiments in which rocks were subjected to forces under conditions that simulated those existing at various depths within the crust.

Although each rock type deforms somewhat differently, the general characteristics of rock deformation were determined from these experiments. Geologists discovered that when stress is gradually applied, rocks first respond by deforming elastically. Changes that result from *elastic deformation* are recoverable; that is, like a rubber band, the rock will return to nearly its original size and shape when the force is removed. (As you saw in Chapter 6, the energy for most earthquakes comes from stored elastic energy that is released as rock snaps back to its original shape.)

Figure 9.1 This peak is part of the Karakoram Range in Pakistan. (Photo by Art Wolfe)

Figure 9.2 Folded sedimentary layers exposed on the face of Mount Kidd, Alberta, Canada. (Photo by Peter French/DRK Photo)

Once the elastic limit (strength) of a rock is surpassed, it either flows (*ductile deformation*) or fractures (*brittle deformation*). The factors that influence the strength of a rock and thus how it will deform include temperature, confining pressure, rock type, and time.

Temperature and Confining Pressure

Rocks near the surface, where temperatures and confining pressures are low, tend to behave like a brittle solid and fracture once their strength is exceeded. This type of deformation is called **brittle** (*bryttian* = to shatter) **failure** or **brittle deformation.** From our everyday experience, we know that glass objects, wooden pencils, china plates, and even our bones exhibit brittle failure once their strength is surpassed. By contrast at depth, where temperatures and confining pressures are high, rocks exhibit *ductile* behavior. **Ductile deformation** is a type of solid-state flow that produces a change in the size and shape of an object without fracturing. Ordinary objects that display ductile behavior include modeling clay, bee's wax, caramel candy, and most metals. For example, a copper penny placed on a railroad track will be flattened and deformed (without breaking) by the force applied by a passing train.

Ductile deformation of a rock—strongly aided by high temperature and high confining pressure—is somewhat similar to the deformation of a penny flattened by a train. Rocks that display evidence of ductile flow usually were deformed at great depth and may exhibit contorted folds that give the impression that the strength of the rock was akin to soft putty (Figure 9.2).

Rock Type

In addition to the physical environment, the mineral composition and texture of a rock greatly influence how it will deform. For example, crystalline rocks, such as granite, basalt, and quartzite, that are composed of minerals that have strong internal molecular bonds tend to fail by brittle fracture. By contrast, sedimentary rocks that are weakly cemented, or metamorphic rocks that contain zones of weakness, such as foliation, are more susceptible to ductile flow. Rocks that are weak and thus most likely to behave in a ductile manner when subjected to differential forces include rock salt, gypsum, and shale; limestone, schist, and marble are of intermediate strength. In fact, rock salt is so weak that it deforms under small amounts of differential stress and rises like stone pillars through beds of sediment that lie in and around the Gulf of Mexico.

Time

One key factor that researchers are unable to duplicate in the laboratory is how rocks respond to small amounts of force applied over long spans of *geologic time*. However, insights into the effects of time on deformation are provided in everyday settings. For example, marble benches have been known to sag under their own weight over a span of a hundred years or so, and wooden bookshelves may bend after being loaded with books for a relatively short period. In nature small stresses applied over long time spans surely play an important role in the deformation of rock. Forces that are unable to deform rock when initially applied may cause rock to flow if the force is maintained over an extended period of time.

Folds

During mountain building, flat-lying sedimentary and volcanic rocks are often bent into a series of wavelike undulations called **folds.** Folds in sedimentary strata are much like those that would form if you were to hold the ends of a sheet of paper and then push them to-

261

Normal limb

Overturned limb

Anticline   Syncline   Anticline   Syncline

Symmetrical fold ⟶ ⟵ Asymmetrical fold ⟶ ⟵ Overturned fold

Figure 9.3   Block diagram of principal types of folded strata. The upfolded or arched structures are anticlines. The downfolds or troughs are synclines. Notice that the limb of an anticline is also the limb of the adjacent syncline.

gether. In nature, folds come in a wide variety of sizes and configurations. Some folds are broad flexures in which rock units hundreds of meters thick have been slightly warped. Others are very tight microscopic structures found in metamorphic rocks. Size differences notwithstanding, most folds are the result of *compressional forces* that result in the shortening and thickening of the crust.

Types of Folds

The two most common types of folds are anticlines and synclines (Figure 9.3). An **anticline** is most commonly formed by the upfolding, or arching, of rock layers.* Anticlines are sometimes spectacularly displayed where highways have been cut through deformed strata (Figure 9.4). Often found in association with anticlines are downfolds, or troughs, called **synclines.** Notice in Figure 9.3 that the limb of an anticline is also a limb of the adjacent syncline.

Depending on their orientation, these basic folds are described as *symmetrical* when the limbs are mirror images of each other and *asymmetrical* when they are not. An asymmetrical fold is said to be *overturned* if one limb is tilted beyond the vertical (Figure 9.3). An overturned fold can also "lie on its side" so that a plane extending through the axis of the fold would be horizontal. These *recumbent* folds are common in mountainous regions such as the Alps.

Folds do not continue forever; rather, their ends die out much like the wrinkles in cloth. Some folds *plunge* because the axis of the fold penetrates into the ground (Figure 9.5). As the figure shows, both anticlines and synclines can plunge. Figure 9.6 shows an example of a plunging anticline and the pattern produced when ero-

sion removes the upper layers of the structure and exposes its interior. Note that the outcrop pattern of an anticline points in the direction it is plunging, whereas the opposite is true for a syncline. A good example of the kind of topography that results when erosional forces attack folded sedimentary strata is found in the Valley and Ridge Province of the Appalachians.

Although we have separated our discussion of folds and faults, in the real world folds are generally intimately coupled with faults. Examples of this close association are broad, regional features called *monoclines.* Particularly prominent features of the Colorado Plateau, **monoclines** are large, steplike folds in otherwise horizontal sedimentary strata (Figure 9.7). These folds appear to be the result of the reactivating of steeply

Figure 9.4   Anticline in folded sedimentary rocks, Devon, England. (Photo by Tom Bean/DRK Photo)

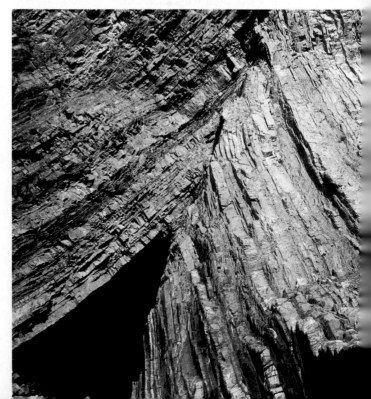

*By strict definition, an anticline is a structure in which the oldest strata are found in the center. This most typically occurs when strata are upfolded. Further, a syncline is strictly defined as a structure in which the youngest strata are found in the center. This occurs most commonly when strata are downfolded.

dipping fault zones located in basement rocks beneath the plateau. As large blocks of basement rock were displaced upward along ancient faults, the comparatively ductile sedimentary strata above responded by folding. On the Colorado Plateau, monoclines display a narrow zone of steeply inclined beds that flatten out to form the uppermost layers of large elevated areas, including the Zuni Uplift, Echo Cliffs Uplift, and San Rafael Swell. Displacement along these reactivated faults often exceeds 1 kilometer (0.6 mile).

Domes and Basins

Broad upwarps in basement rock may deform the overlying cover of sedimentary strata and generate large folds. When this upwarping produces a circular or elongated structure, the feature is called a **dome.** Downwarped structures having a similar shape are termed **basins.**

The Black Hills of western South Dakota is a large domed structure thought to be generated by upwarping. Here erosion has stripped away the highest portions of the upwarped sedimentary beds, exposing older igneous and metamorphic rocks in the center (Figure 9.8). Remnants of these once continuous sedimen-

tary layers are visible, flanking the crystalline core of these mountains.

Several large basins exist in the United States (Figure 9.9). The basins of Michigan and Illinois have very gently sloping beds similar to saucers. These basins are thought to be the result of large accumulations of sediment, whose weight caused the crust to subside.

Because large basins usually contain sedimentary beds sloping at very low angles, they are usually identified by the age of the rocks composing them. The youngest rocks are found near the center, and the oldest rocks are at the flanks. This is just the opposite order of a domed structure, such as the Black Hills, where the oldest rocks form the core.

Faults and Joints

Faults are fractures in the crust along which appreciable displacement has taken place. Occasionally, small faults can be recognized in road cuts where sedimentary beds have been offset a few meters, as shown in Figure 9.10. Faults of this scale usually occur as single discrete breaks. By contrast, large faults, like the San Andreas Fault in California, have displacements of hun-

Figure 9.5 Plunging folds. **A.** Idealized view of plunging folds in which a horizontal surface has been added. **B.** View of plunging folds as they might appear after extensive erosion. Notice that in a plunging anticline the outcrop pattern "points" in the direction of the plunge, while the opposite is true of plunging synclines.

Figure 9.6 Sheep Mountain, a doubly plunging anticline. Note that erosion has cut the flanking sedimentary beds into low ridges that make a "V" pointing in the direction of plunge. (Photo by John S. Shelton)

? STUDENTS SOMETIMES ASK...

How do geologists determine which side of a fault has moved?

Surprisingly, for many faults it cannot be definitively established. For example, in the picture of a fault shown in Figure 9.10, did the left side move down, or did the right side move up? Since the surface (at the top of the photo) has been eroded flat, either side could have moved; or both could have moved, with one side moving more than the other (for instance, they both may have moved up, it's just that the right side moved up more than the left side). That's why geologists talk about *relative* motion across faults. In this case, the left side moved down *relative to* the right side, and the right side moved up *relative to* the left side (note arrows on photo).

dreds of kilometers and consist of many interconnecting fault surfaces. These *fault zones* can be several kilometers wide and are often easier to identify from high-altitude photographs than at ground level.

Dip-Slip Faults

Faults in which the movement is primarily parallel to the inclination, or dip, of the fault surface are called **dip-slip faults.** Vertical displacements along dip-slip faults may produce long, low cliffs called **fault scarps** (*scarpe* = a slope). Fault scarps, such as the one shown in Figure 9.11, are produced by displacements that generate earthquakes.

It has become common practice to call the rock surface that is immediately above the fault the *hanging wall* and to call the rock surface below, the *footwall.* This

A.

B.

Figure 9.7 Monocline. **A.** Monocline located near Mexican Hat, Utah. (Photo by Tom Bean/DRK Photo) **B.** Monocline consisting of bent sedimentary beds that were deformed by faulting in the bedrock below.

nomenclature arose from prospectors and miners who excavated shafts and tunnels along fault zones because these are frequently sites of ore deposits. In these tunnels, the miners would walk on the rocks below the mineralized fault zone (the footwall) and hang their lanterns on the rocks above (the hanging wall).

Two major types of dip-slip faults are *normal faults* and *reverse faults.*

Normal Faults.   Dip-slip faults are classified as **normal faults** when the hanging wall block moves down relative to the footwall block (Figure 9.12A). Most normal faults have steep dips of about 60°, which tend to flatten out with depth. However, some dip-slip faults have much lower dips, with some approaching horizontal. Because of the downward motion of the hanging wall block, normal faults accommodate lengthening, or extension, of the crust.

Most normal faults are small, having displacements of only a meter or so, like the one shown in the road cut in Figure 9.10. Others extend for tens of kilometers where they may sinuously trace the boundary of a

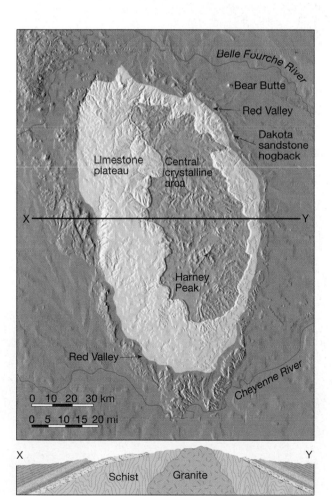

Figure 9.8   The Black Hills of South Dakota, a large domal structure with resistant igneous and metamorphic rocks exposed in the core.

mountain front. In the western United States, large-scale normal faults like these are associated with structures called **fault-block mountains** (see Figure 9.17).

Examples of fault-block mountains include the Teton Range of Wyoming and the Sierra Nevada of California. Both are faulted along their eastern flanks, which were uplifted as the blocks tilted downward to the west. These precipitous mountain fronts were produced over a period of 5 million to 10 million years by many irregularly spaced episodes of faulting. Each event was responsible for just a few meters of displacement.

Normal faulting is also prevalent at spreading centers where plate divergence occurs. Here, a central block called a **graben** (*graben* = ditch) is bounded by normal faults and drops as the plates separate (Figure 9.13). These grabens produce an elongated valley bounded by relatively uplifted structures called **horsts.** The Rift Valley of East Africa is made up of several large grabens, above which tilted horsts produce a linear mountainous topography. This valley, nearly 6000 kilometers (3600 miles) long, contains the excavation sites of some of the earliest human fossils.

An excellent example of horst and graben topography is found in the Basin and Range Province, a region that encompasses Nevada and portions of surrounding states (Figure 9.14). Here the crust has been elongated and broken to create more than 200 relatively small mountain ranges. Averaging about 80 kilometers in length, the ranges rise 900 to 1500 meters above the adjacent down-faulted basins. Notice in Figure 9.14 that slopes of the normal faults in the Basin and Range Province decrease with depth and join together to form a nearly horizontal fault called a *detachment fault.* These detachment faults extend for several kilometers below the surface. Here they form a major boundary between the rocks below, which exhibit ductile deformation, and the rocks above, which demonstrate brittle deformation via faulting.

Fault motion provides geologists with a method of determining the nature of the forces at work within

?

STUDENTS SOMETIMES ASK...
*Has anyone ever seen
a fault scarp forming?*

Amazingly, yes. There have been several instances where people have been at the fortuitously appropriate place and time to observe the creation of a fault scarp—and have lived to tell about it. In Idaho a large earthquake in 1983 created a 3-meter (10-foot) fault scarp that was witnessed by several people, many of whom were knocked off their feet. More often, though, fault scarps are noticed *after* they form. For example, a 1999 earthquake in Taiwan created a fault scarp that formed a new waterfall and destroyed a nearby bridge.

Figure 9.9 The bedrock geology of the Michigan Basin. Notice that the youngest rocks are centrally located, while the oldest beds flank this structure.

KEY

Upper Pennsylvanian

Lower Pennsylvanian

Upper Mississippian

Lower Mississippian

Devonian

Silurian

Ordovician

Earth. Normal faults indicate the existence of tensional forces that pull the crust apart. This "pulling apart" can be accomplished either by uplifting that causes the surfaces to stretch and break or by opposing horizontal forces.

Reverse and Thrust Faults. **Reverse faults** and **thrust faults** are dip-slip faults in which the hanging wall block moves up relative to the footwall block (Figure 9.12B and C). Reverse faults have dips greater than 45°, and thrust faults have dips less than 45°. Because the hanging wall block moves up and over the footwall block, reverse and thrust faults accommodate shortening of the crust.

Most high-angle reverse faults are small and accommodate local displacements in regions dominated by other types of faulting. Thrust faults, on the other hand, exist at all scales. In mountainous regions such as the Alps, Northern Rockies, Himalayas, and Appalachians, thrust faults have displaced strata as far as 50 kilometers over adjacent rock units. The result of this large-scale movement is that older strata end up overlying younger rocks.

Whereas normal faults occur in tensional environments, reverse and thrust faults result from strong compressional stresses. In these settings, crustal blocks are displaced *toward* one another, with the hanging wall being displaced upward relative to the footwall. Thrust faulting is most pronounced in subduction zones and other convergent boundaries where plates are collid-

ing. Compressional forces generally produce folds as well as faults and result in a thickening and shortening of the material involved.

Strike-slip Faults

Faults in which the dominant displacement is horizontal and parallel to the trend, or strike, of the fault surface are called **strike-slip faults** (Figure 9.12D). Because of their large size and linear nature, many strike-slip faults produce a trace that is visible over a great distance (Fig-

Figure 9.10 Faulting caused the vertical displacement of these beds located near Kanab, Utah. Arrows show relative motion of rock units. (Photo by Tom Bean/DRK Photo)

Figure 9.11   A fault scarp located near Joshua Tree National Monument, California. (Photo by A. P. Trujillo/APT Photos)

large earthquakes. One of the most noteworthy of these was the great San Francisco earthquake of 1906. During this strong earthquake, structures such as fences that were built across the San Andreas Fault were displaced as much as 4.7 meters (15 feet). Because the movement along the San Andreas causes the crustal block on the opposite side of the fault to move to the right as you face the fault, it is called a *right-lateral* strike-slip fault. The Great Glen fault in Scotland is a well-known example of a *left-lateral* strike-slip fault, which exhibits the opposite sense of displacement.

Many major strike-slip faults cut through the lithosphere and accommodate motion between two large crustal plates. This special kind of strike-slip fault is called a **transform** (*trans* = across, *forma* = form) **fault.** Numerous transform faults cut the oceanic lithosphere and link offset segments of oceanic ridges. Others accommodate displacement between continental plates that move horizontally with respect to each other. One of the best-known transform faults is California's San Andreas Fault (see Box 9.1). This plate-bounding fault can be traced for about 950 kilometers (600 miles) from the Gulf of California to a point along the Pacific Coast north of San Francisco, where it heads out to sea. Ever since its formation, about 29 million years ago, displacement along the San Andreas Fault has exceeded 560 kilometers (340 miles). This movement has accommodated the northward displacement of southwestern California and the Baja Peninsula of Mexico in relation to the remainder of North America.

ure 9.15). Rather than a single fracture along which movement takes place, large strike-slip faults consist of a zone of roughly parallel fractures. The zone may be up to several kilometers wide. The most recent movement, however, is often along a strand only a few meters wide, which may offset features such as stream channels (see Figure 9.A in Box 9.1). Furthermore, crushed and broken rocks produced during faulting are more easily eroded, often producing linear valleys or troughs that mark the locations of strike-slip faults.

The earliest scientific records of strike-slip faulting were made following surface ruptures that produced

A. Normal fault (tensional)

B. Reverse fault (compressional)

C. Thrust fault (compressional)

D. Strike-slip fault (shear)

Figure 9.12   Block diagrams of four types of faults. **A.** Normal fault. **B.** Reverse fault. **C.** Thrust fault. **D.** Strike-slip fault.

Figure 9.13 Diagrammatic sketch of downfaulted blocks (grabens and half-grabens) and upfaulted blocks (horsts).

Joints

Among the most common rock structures are fractures called **joints**. Unlike faults, joints are fractures along which *no appreciable-displacement* has occurred. Although some joints have a random orientation, most occur in roughly parallel groups (Figure 9.16).

We have already considered two types of joints. Earlier we learned that *columnar joints* form when igneous rocks cool and develop shrinkage fractures that produce elongated, pillarlike columns (see Figure 8.24). Also recall that sheeting produces a pattern of gently curved joints that develop more or less parallel to the

surface of large exposed igneous bodies such as batholiths. Here the jointing results from the gradual expansion that occurs when erosion removes the overlying load (see Figure 3.4).

In contrast to the situations just described, most joints are produced when rocks in the outermost crust are deformed. Here forces associated with crustal movements cause the rock to fail by brittle fracture. For example, when folding occurs, rocks situated at the axes of the folds are elongated and pulled apart to produce tensional joints. Extensive joint patterns can also develop in response to relatively subtle and often barely

Figure 9.14 Normal faulting in the Basin and Range Province. Here, tensional stresses have elongated and fractured the crust into numerous blocks. Movement along these fractures has tilted the blocks producing parallel mountain ranges called fault-block mountains. (Photo by Michael Collier)

Figure 9.15 Aerial view of strike-slip (right-lateral) fault in Southern Nevada. (Photo by Martin G. Miller)

perceptible regional upwarping and downwarping of the crust. In many cases, the cause for jointing at a particular locale is not readily apparent.

Many rocks are broken by two or even three sets of intersecting joints that slice the rock into numerous regularly shaped blocks. These joint sets often exert a strong influence on other geologic processes. For example, chemical weathering tends to be concentrated along joints, and in many areas groundwater movement and the resulting dissolution in soluble rocks is controlled by the joint pattern (Figure 9.16). Moreover, a system of joints can influence the direction that stream courses follow. The rectangular drainage pattern described in Chapter 4 is such a case.

Mountain Belts

Like other people, geologists have been inspired more by Earth's mountains than by any other landform (Figure 9.17). Through extensive scientific exploration over the last 150 years, much has been learned about the internal processes that generate these often spectacular terrains. The name for the processes that collectively produce a mountain belt is **orogenesis,** (*oros* = mountain, *genesis* = to come into being). The rocks comprising mountains provide striking visual evidence of the enormous compressional forces that have deformed large sections of Earth's crust and subsequently elevated them to their present positions. Although folding is often the most conspicuous sign of these forces, thrust faulting, metamorphism, and igneous activity are always present in varying degrees.

Mountain building has occurred during the recent geologic past in several locations around the world. These young mountainous belts include the American Cordillera, which runs along the western margin of the Americas from Cape Horn to Alaska and includes the Andes and Rocky mountains; the Alpine-Himalaya chain, which extends from the Medi-

Figure 9.16 Chemical weathering is enhanced along joints in sandstone, near Moab, Utah. (Photo by Michael Collier)

terranean through Iran to northern India and into Indochina; and the mountainous terrains of the western Pacific, which include volcanic island arcs such as Japan, the Philippines, and Sumatra. Most of these young mountain belts have come into existence within the last 100 million years. Some, including the Himalayas, began their growth as recently as 45 million years ago.

STUDENTS SOMETIMES ASK...

You mentioned that most mountains are the result of crustal deformation. Are there areas that exhibit mountainous topography but have been produced without crustal deformation?

Yes. Plateaus—areas of high-standing rocks that are essentially horizontal—are one example of a feature that can be deeply dissected by erosional forces into rugged, mountain-like landscapes. Although these highlands resemble mountains topographically, they lack the structures associated with orogenesis. The opposite situation also exists. For instance, the Piedmont section of the eastern Appalachians exhibits topography that is nearly as subdued as that seen in the Great Plains. Yet, because this region is composed of deformed metamorphic rocks, it is clearly part of the Appalachian Mountains.

BOX 9.1

PEOPLE AND THE ENVIRONMENT:
The San Andreas Fault System

The San Andreas, the best-known and largest fault system in North America, first attracted wide attention after the great 1906 San Francisco earthquake and fire. Following this devastating event, geologic studies demonstrated that a displacement of as much as 5 meters along the fault had been responsible for the earthquake. It is now known that this dramatic event is just one of many thousands of earthquakes that have resulted from repeated movements along the San Andreas throughout its 29-million-year history.

Where is the San Andreas fault system located? As shown in Figure 9.A, it trends in a northwesterly direction for nearly 1300 kilometers (780 miles) through much of western California. At its southern end, the San Andreas connects with a spreading center located in the Gulf of California. In the north, the fault enters the Pacific Ocean at Point Arena, where it is thought to continue its northwesterly trend, eventually joining the Mendocino fracture zone. In the central section, the San Andreas is relatively simple and straight. However, at its two extremities, several branches spread from the main trace, so that in some areas the fault

Figure 9.A Map showing the extent of the San Andreas fault system. Inset shows only a few of the many splinter faults that are part of this great fault system.

zone exceeds 100 kilometers (60 miles) in width.

Over much of its extent, a linear trough reveals the presence of the San Andreas Fault. When the system is viewed from the air, linear scars, offset stream channels, and elongated ponds mark the trace in a striking manner. On the ground, however, surface expressions of the faults are much more difficult to detect. Some of the most distinctive landforms include long, straight escarpments, narrow ridges, and sag ponds formed by settling of blocks within the fault zone. Further, many stream channels characteristically bend sharply to the right where they cross the fault (Figure 9.B).

With the development of the theory of plate tectonics, geologists began to realize the significance of this great fault system. The San Andreas Fault is a transform boundary separating two crustal plates that move very slowly. The Pacific plate, located to the west, moves northwestward relative to the North American plate, causing earthquakes along the fault (Table 9.A).

The San Andreas is undoubtedly the most studied of any fault system in the world. Although many questions remain unanswered, geologists have learned that each fault segment exhibits somewhat different behavior. Some portions of the San Andreas exhibit a slow creep with little noticeable seismic

Figure 9.B Aerial view showing offset stream channel across the San Andreas Fault on the Carrizo Plain west of Taft, California. (Photo by Michael Collier/DRK Photo)

activity. Other segments regularly slip, producing small earthquakes, whereas still other segments seem to store elastic energy for hundreds of years and rupture in great earthquakes. This knowledge is useful when assigning earthquake-hazard potential to a given segment of the fault zone.

Because of the great length and complexity of the San Andreas Fault, it is more appropriately referred to as a "fault system." This major fault system consists primarily of the San Andreas Fault and several major branches, including the Hayward and Calaveras faults of central California and the San Jacinto and Elsinore faults of southern California (Figure 9.A). These major segments, plus a vast number of smaller faults that include the Imperial Fault, San Fernando Fault, and the Santa Monica Fault, collectively accommodate the relative motion between the North American and Pacific plates.

Ever since the great San Francisco earthquake of 1906, when as much as 5 meters of displacement occurred, geologists have attempted to establish the cumulative displacement along this fault over its 29-million-year history. By matching rock units across the fault, geologists have determined that the total accumulated displacement from earthquakes and creep exceeds 560 kilometers (340 miles).

Table 9.A Major earthquakes on the San Andreas fault system.

Date	Location	Magnitude	Remarks
1812	Wrightwood, CA	7	Church at San Juan Capistrano collapsed, killing 40 worshipers.
1812	Santa Barbara channel	7	Churches and other buildings wrecked in and around Santa Barbara.
1838	San Francisco peninsula	7	At one time thought to have been comparable to the great earthquake of 1906.
1857	Fort Tejon, CA	8.25	One of the greatest U.S. earthquakes. Occurred near Los Angeles, then a city of 4000.
1868	Hayward, CA	7	Rupture of the Hayward fault caused extensive damage in San Francisco Bay area.
1906	San Francisco, CA	8.25	The great San Francisco earthquake. As much as 80 percent of the damage caused by fire.
1940	Imperial Valley	7.1	Displacement on the newly discovered Imperial fault.
1952	Kern County	7.7	Rupture of the White Wolf fault. Largest earthquake in California since 1906. Sixty million dollars in damages and 12 people killed.
1971	San Fernando Valley	6.5	One-half billion dollars in damages and 58 lives claimed.
1989	Santa Cruz Mountains	7.1	Loma Prieta earthquake. Six billion dollars in damages, 62 lives lost, and 3757 people injured.
1994	Northridge (Los Angeles area)	6.9	Over 15 billion dollars in damages, 51 lives lost, and over 5000 injured.

In addition to these relatively young mountain belts, several chains of older mountains exist on Earth as well. Although these older structures are deeply eroded and topographically less prominent, they clearly possess the same structural features found in younger mountains. Typical of this older group are the Appalachians in the eastern United States and the Urals in Russia.

Over the years, several hypotheses have been put forward regarding the formation of Earth's major mountain belts. One early proposal suggested that mountains are simply wrinkles in Earth's crust, produced as the planet cooled from its original semimolten state. As Earth lost heat, it contracted and shrank. In response to this process, the crust was deformed similar to when the peel of an orange wrinkles as the fruit dries out. However, neither this nor any other early hypothesis was able to withstand careful scrutiny and had to be discarded.

Mountain Building at Convergent Boundaries

With the development of the theory of plate tectonics, a widely accepted model for orogenesis became available. According to this model, most mountain building occurs at convergent plate boundaries. Here colliding plates provide the compressional forces that fold, fault, and metamorphose the thick accumulations of sediments that have been deposited along the flanks of landmasses. In addition, partial melting of mantle rock provides a source of magma that intrudes and further deforms these deposits.

To unravel the events that produce mountains, researchers examine ancient mountain structures as well as sites where orogenesis is currently active. Of particular interest are active subduction zones, where lithospheric plates are converging. At most modern-day subduction zones, volcanic arcs are forming. This situation is typified by Alaska's Aleutian Islands and by the Andean arc of western South America.

Aleutian-Type Mountain Building

Aleutian-type convergent zones occur where two oceanic plates converge and one is subducted beneath the other (Figure 9.18A). This activity results in partial melting of the mantle wedge located above the subducting plate and eventually leads to the growth of a volcanic island arc on the ocean floor. Because they are associated with subducting oceanic lithosphere, island arcs are typically found on the margins of a shrinking ocean basin, such as the Pacific—where the majority of volcanic island arcs are found. Examples of active island arcs include the Mariana, New Hebrides, Tonga, and Aleutian arcs.

Island arcs result from the steady subduction of oceanic lithosphere and tend to be relatively long-lived features. Here, somewhat sporadic volcanic activity and the emplacement of magma at depth, as well as the accumulation of sediment that is scraped off the subducting plate, increase the volume of the crust (Figure 9.18B). Some volcanic island arcs, such as Japan, appear to have been built up by two or three different episodes of subduction or possibly by the joining of two island arcs.

As illustrated by Japan, the continued development of a volcanic island arc can result in the formation of mountainous topography consisting of belts of igneous and metamorphic rocks. This activity, however, is viewed as just one phase in the development of a major mountain belt. As you will see later, these volcanic arcs eventually become positioned along the margin of a large continental block, where sooner or later they will be involved in a major mountain-building episode.

Andean-Type Mountain Building

Mountain building along continental margins involves the convergence of an oceanic plate and a plate whose leading edge contains continental crust. Exemplified by the Andes Mountains, an *Andean-type convergent zone* results in the formation of a volcanic arc and related tectonic features inland of the continental margin. Both Andean- and Aleutian-type plate margins have many broadly similar features, reflecting their comparable tectonic settings. There are a number of differences, however, related to the type of overriding plate.

Figure 9.17   The Grand Tetons of Wyoming are an example of fault-block mountains. (Photo by David Muench)

UNDERSTANDING EARTH:
Mountain Building away from Plate Margins

In the American West, extending from the Front Range of the southern Rocky Mountains across the Colorado Plateau and through the Basin and Range Province, the topography consists of lofty peaks and elevated plateaus (Figure 9.C). According to the plate tectonics model, mountain belts are produced along continental margins in association with convergent plate boundaries. But this region of mountainous topography extends inland almost 1600 kilometers (1000 miles), far from the nearest plate boundary.

The portion of the Rocky Mountains that extends from southern Montana to New Mexico was produced by a period of deformation known as the *Laramide Orogeny*. This event, which created some of the most picturesque scenery in the United States, peaked about 60 million years ago (Figure 9.C). The mountain ranges generated during the Laramide Orogeny include Colorado's Front Range, the Sangre de Cristo of New Mexico and Colorado, and Wyoming's Bighorns (see Figure 9.20).

The event that may have triggered this period of uplift started with the nearly horizontal subduction of the Farallon plate eastward beneath North America as far inland as the Black Hills of South Dakota (Figure 9.D). As the subducted slab scraped beneath the continent, compressional forces initiated a period of tectonic activity. About 65 million years ago the Farallon plate began to sink into the mantle. As this

Figure 9.D One possible mechanism for the uplift that led to the formation of the southern Rockies, Colorado Plateau, and Basin and Range Province. **A.** Nearly horizontal subduction of an oceanic plate initiated a period of tectonic activity. **B.** Sinking of this oceanic slab allowed for upwelling of hot mantle material that buoyantly raised the crust.

relatively cool plate gradually separated from the lithosphere above, it was replaced by hot rock that upwelled from the mantle (Figure 9.D). Thus, according to this scenario, the hot mantle provided the buoyancy to raise the southern Rockies, as well as the Colorado Plateau and the Basin and Range Province.

In the southern Rockies this event uplifted large blocks of ancient basement rocks along high-angle faults to produce mountainous topography separated by large basins that became filled with sediment as the mountains subsequently eroded. The upwelling that is associated with the Basin and Range Province started about 50 million years ago and remains active today. Here the buoyancy of the warm material caused upwarping and rifting that elongated the overlying crust by 200 to 300 kilometers. The lower crust is ductile and easily stretched. The upper crust, on the other hand, is brittle and deforms by faulting (see Figure 9.14). The extension and faulting broke the uplifted crust, causing individual blocks to shift. The high portions of these tilted blocks make up the mountain ranges, whereas their low areas form the basins, now partially filled with sediment.

Figure 9.C Fall colors in the Mount Sneffels Wilderness of the Colorado Rockies. (Photo by Gavrel Jecan/Art Wolfe, Inc.)

Figure 9.18 The development of a volcanic island arc by the convergence of two oceanic plates. Continuous subduction along these Aleutian-type convergent zones results in the development of thick units of continental-type crust.

Passive Margins.   The first stage in the development of an idealized Andean-type mountain belt occurs prior to the formation of the subduction zone. During this period the continental margin is a **passive continental margin;** that is, it is not a plate boundary but a part of the same plate as the adjoining oceanic crust. The East Coast of North America provides a present-day example of a passive continental margin. Here, as at other passive continental margins surrounding the Atlantic, deposition of sediment on the continental shelf is producing a thick wedge of shallow-water sandstones, limestones, and shales (Figure 9.19A). Beyond the continental shelf, turbidity currents are depositing sediments on the continental slope and rise (see Chapter 12).

Active Continental Margins.   At some point the continental margin becomes active. A subduction zone forms and the deformation process begins (Figure 9.19B). A good place to examine an **active continental**

margin is the west coast of South America. Here the Nazca plate is being subducted beneath the South American plate along the Peru-Chile trench. This subduction zone probably formed prior to the breakup of the supercontinent Pangaea.

In an idealized Andean-type subduction, convergence of the continental block and the subducting oceanic plate leads to deformation and metamorphism of the continental margin. Once the oceanic plate descends to about 100 kilometers, partial melting of mantle rock above the subducting slab generates magma that migrates upward (Figure 9.19B). During the development of this continental volcanic arc, sediment derived from the land and scraped from the subducting plate is plastered against the landward side of the trench. This chaotic accumulation of sedimentary and metamorphic rocks with occasional scraps of ocean crust is called an **accretionary** (*ad* = toward, *crecere* = to grow) **wedge** (Figure 9.19B). Prolonged subduction can build an accretionary wedge that is large enough to stand above sea level (Figure 9.19C).

Andean-type mountain belts are composed of two roughly parallel zones. The volcanic arc develops on the continental block. It consists of volcanoes and large intrusive bodies intermixed with high-temperature metamorphic rocks. The seaward segment is the accretionary wedge. It consists of folded, faulted sedimentary and metamorphic rocks (Figure 9.19C).

Sierra Nevada and Coast Ranges.   One of the best examples of an inactive Andean-type orogenic belt is found in the western United States. It includes the Sierra Nevada and the Coast Ranges in California (Figure 9.20). These parallel mountain belts were produced by the subduction of a portion of the Pacific Basin under the western edge of the North American plate. The Sierra Nevada batholith is a remnant of a portion of the continental volcanic arc that was produced by several surges of magma over tens of millions of years. Subsequent uplifting and erosion have removed most of the evidence of past volcanic activity and exposed a core of crystalline, igneous, and associated metamorphic rocks.

In the trench region, sediments scraped from the subducting plate, plus those provided by the eroding continental volcanic arc, were intensely folded and faulted into an accretionary wedge. This chaotic mixture of rocks presently constitutes the Franciscan Formation of California's Coast Ranges. Uplifting of the Coast Ranges took place only recently, as evidenced by the young unconsolidated sediments that still mantle portions of these highlands.

Continental Collisions

So far, we have discussed the formation of mountain belts where the leading edge of just one of the two lithospheric plates was capped by continental crust.

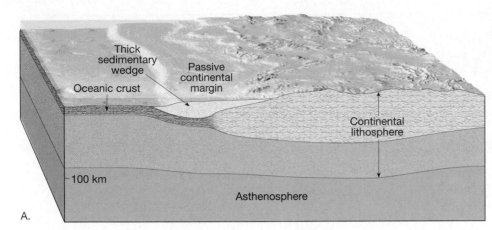

Figure 9.19 Mountain building along an Andean-type subduction zone. **A.** Passive continental margin with extensive wedge of sediments. **B.** Plate convergence generates a subduction zone, and partial melting produces a developing continental volcanic arc. **C.** Continued convergence and igneous activity further deform and thicken the crust, elevating the mountain belt, while the accretionary wedge grows.

However, it is possible that both plates may be carrying continental crust. Because continental crust is apparently too buoyant to undergo any appreciable subduction, a collision between the continental fragments results (see Figure 7.14).

An example of such a collision began about 45 million years ago when India collided with the Eurasian plate. Prior to this event, India, which was once part of Antarctica, had split from that continent and over the course of millions of years slowly moved a few thousand kilometers due north (see Figure 7.14A). The result of the collision was the formation of the spectacular Himalaya Mountains and the Tibetan Plateau (see Figure 7.14B).

Although most of the oceanic crust that separated these landmasses prior to the collision was subducted, some was caught up in the squeeze, along with sediment that lay offshore. Today these sedimentary rocks and slivers of oceanic crust are elevated high above sea level. Geologists believe that following such a collision,

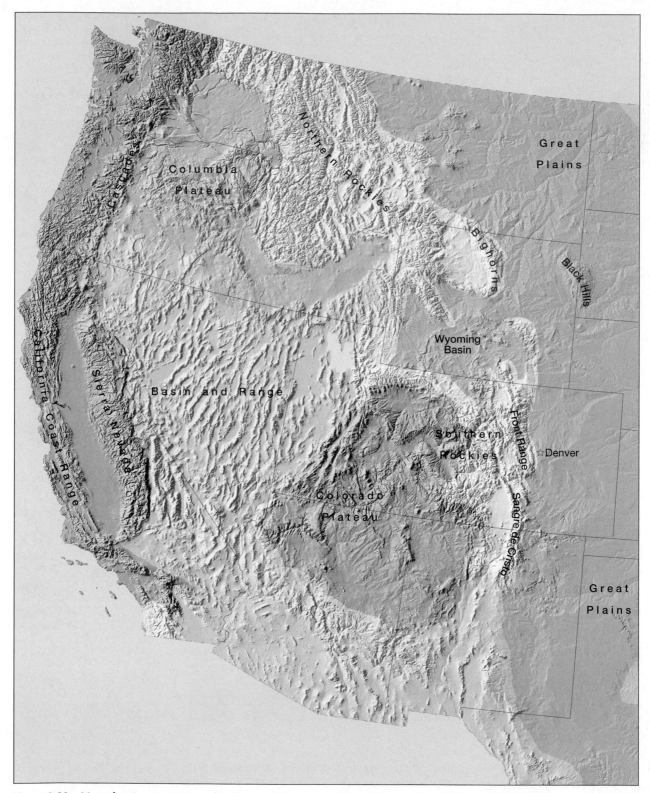

Figure 9.20   Map of major mountainous landforms of the western United States. (After Thelin and Pike, USGS)

the subducted oceanic plate decouples from the rigid continental plate and continues its descent into the mantle.

The spreading center that propelled India northward is still active; hence, India continues to be thrust into Asia at an estimated rate of a few centimeters each year. However, numerous earthquakes recorded off the southern coast of India indicate that a new subduction zone may be in the making. If formed, it would provide a subduction site for the floor of the Indian Ocean, which is continually being produced at a spreading center located to the southwest. Should this occur, India's northward journey, relative to Asia, would come to an end and the growth of the Himalayas would cease.

A similar but much older collision is believed to have taken place when the European continent collided with the Asian continent to produce the Ural Mountains, which extend in a north–south direction through Russia. Prior to the discovery of plate tectonics, geologists had difficulty explaining the existence of mountain ranges such as the Urals, which are located deep within continental interiors. How could thousands of meters of marine sediment be deposited and then become highly deformed while situated in the middle of a large stable landmass?

Continental Accretion and Mountain Building

Plate tectonics theory originally suggested two mechanisms for orogenesis: (1) Continental collisions were proposed to explain the formation of such mountainous terrains as the Alps, Himalayas, Appalachians, and Urals; and (2) subduction of oceanic lithosphere was thought to be the underlying tectonic process for many circum-Pacific mountain chains, as typified by the Andes, Aleutians, and Japan. Subsequent investigations, however, indicated a third mechanism of orogenesis: Smaller crustal fragments collide and merge with continental margins, and through this process of collision and accretion, many of the mountainous regions rimming the Pacific have been generated.

Geologists refer to these accreted crustal blocks as terranes. Simply, the term **terrane** designates any crustal fragment that has a geologic history distinct from that of the adjoining terranes. Terranes come in varied shapes and sizes; some are no larger than volcanic islands. Others, such as the one making up the entire Indian subcontinent, are very large.

What is the nature of these crustal fragments, and where do they come from? Research suggests that prior to their accretion to a continental block, some of the fragments may have been microcontinents similar to the present-day island of Madagascar, located in the Indian Ocean east of Africa. Many others were island arcs like Japan, the Philippines, and the Aleutian Islands. Still others may have been submerged crustal fragments, such as those occuring on the floor of the western Pacific (Figure 9.21). Over a hundred of these *"oceanic plateaus"* are presently known to exist. Their origins vary; some are thought to be submerged continental fragments, whereas others are extinct volcanic islands, such as those of the Hawaiian Island–Emperor Seamount chain. Others are submerged volcanic

Figure 9.21   Distribution of present-day oceanic plateaus and other submerged crustal fragments. (Data from Ben-Avraham and others)

plateaus created by massive outpourings of lava associated with hot-spot activity.

Accretion of Foreign Terranes.   The widely accepted view is that as oceanic plates move, they carry the embedded oceanic plateaus, volcanic island arcs, and microcontinents to an Andean-type subduction zone. Eventually a collision between the crustal fragment and the continent occurs. Relatively small crustal pieces are peeled from the subducting plate and thrust in relatively thin sheets upon the adjacent continental block. This newly added material increases the width and thickness of the continental crust. The material may later be overridden and/or displaced farther inland by the addition of other fragments.

Figure 9.22   This sequence illustrates the collision of an inactive volcanic island arc with an Andean-type plate margin.

The accretion of larger crustal fragments, such as a mature island arc, may result in a mountain range—although much smaller than those resulting from a continent–continent collision (Figure 9.22). Because of its buoyancy, an island arc will not subduct beneath the continental plate and will therefore plow into the continent, deforming both blocks. When no further convergence is possible, a new trench may develop on the seaward side of the accreted volcanic arc.

Accretion and Mountain Building.   The idea that mountain building occurs in association with the accretion of crustal fragments to a continental mass arose principally from studies conducted in the northern portion of the North American Cordillera (Figure 9.23). Here it was determined that some mountainous areas, principally those in the orogenic belts of Alaska and British Columbia, contain fossil and magnetic evidence indicating that these strata once lay nearer the equator.

It is now assumed that many other terranes found in the North American Cordillera were once scattered throughout the eastern Pacific, much as we find island arcs and oceanic plateaus distributed in the western Pacific today (Figure 9.21). Since the breakup of Pangaea 200 million years ago, North America has been migrating westward, overriding the Pacific Basin. Apparently, this activity resulted in the piecemeal addition of fragments to the entire Pacific margin of the continent, from the Baja Peninsula to northern Alaska (Figure 9.23). In a like manner, many modern microcontinents will eventually be accreted to active continental margins, producing new orogenic belts.

STUDENTS SOMETIMES ASK...

What's the difference between a terrane and a terrain?

The term *terrane* is used to designate a distinct and recognizable series of rock formations that has been transported by plate tectonic processes. Since geologists who mapped these rocks were unsure where they came from, these rocks were sometimes called "exotic," "suspect," "accreted," or "foreign" terranes. Don't confuse this with the term *terrain,* which describes the shape of the surface topography or "lay of the land."

Buoyancy and the Principle of Isostasy

In addition to the horizontal movements of lithospheric plates, gradual up-and-down motions of the crust are observed at many locations around the globe. Although much of this vertical movement occurs along plate mar-

Figure 9.23 Map showing terranes thought to have been added to western North America during the past 200 million years. (Redrawn after D. R. Hutchinson and others)

Figure 9.24 Former wave-cut platforms now exist as a series of elevated terraces on the west side of San Clemente Island off the southern California coast. Once at sea level, the highest terraces are now about 400 meters (1320 feet) above it. (Photo by John S. Shelton)

gravitational balance is called **isostasy** (*iso* = equal, *stasis* = standing). One way to grasp the concept of isostasy is to envision a series of wooden blocks of different heights floating in water, as shown in Figure 9.25. Note that the thicker wooden blocks float higher than the thinner blocks. In a similar manner, many mountain belts stand high above the surface because they have buoyant (less dense) crustal "roots" that extend deep into the (more dense) mantle that supports them from below.

Now visualize what would happen if another small block of wood were placed atop one of the blocks in Figure 9.25. The combined block would sink until a new isostatic balance was reached. However, the top of the combined block would actually be higher than before, and the bottom would be lower. This process of establishing a new level of gravitational equilibrium is called **isostatic adjustment.**

Applying the concept of isostasy, we should expect that when weight is added to the crust, it will respond by subsiding, and when weight is removed, the crust will rebound. (Visualize what happens to a ship as cargo is being loaded and unloaded.) Evidence for crustal subsidence followed by crustal rebound is pro-

Figure 9.25 This drawing illustrates how wooden blocks of different thicknesses float in water. In a similar manner, thick sections of crustal material float higher than thinner crustal slabs.

gins and is associated with episodes of mountain building, some of it is not. Rather, it occurs in the interiors of continents far from plate boundaries.

Evidence for crustal uplift occurs along the West Coast of the United States. When the elevation of a coastal area remains unchanged for an extended period, a wave-cut platform develops (see Figure 14.14). In parts of California, ancient wave-cut platforms can now be found as terraces hundreds of meters above sea level (Figure 9.24). Such evidence of crustal uplift is easy to find; unfortunately, the reason for uplift is not always as easy to establish.

Early workers discovered that Earth's less-dense crust floats on top of the denser and deformable rocks of the mantle. The concept of a floating crust in

279

vided by Ice Age glaciers. When continental ice sheets occupied portions of North America during the Pleistocene epoch, the added weight of 3-kilometer-thick masses of ice caused downwarping of Earth's crust by hundreds of meters. In the 8000 years since the last ice sheet melted, uplifting of as much as 330 meters (1000 feet) has occurred in Canada's Hudson Bay region, where the thickest ice had accumulated.

As this example illustrates, crustal buoyancy can account for considerable vertical movement. Recall that mountain building causes the crust to shorten and thicken. Thus, because of isostasy, deformed strata experience regional uplift both during mountain building and for an extended period afterward. Usually, little additional deformation is associated with the latter activity. As the crust rises, the processes of erosion accelerate, carving the deformed strata into a mountainous landscape.

One of the consequences of the isostatic adjustment is that as erosion lowers the summits of mountains, the crust will rise in response to the reduced load (Figure 9.26). However, each episode of isostatic uplift is somewhat less than the elevation loss due to erosion. The processes of uplifting and erosion will continue until the mountain block reaches "normal" crustal thickness. When this occurs, the mountains will be eroded to near sea level, and the once deeply buried interior of the mountain will be exposed at the surface. In addition, as mountains are worn down, the eroded sediment is deposited on adjacent landscapes, causing these areas to subside (Figure 9.26).

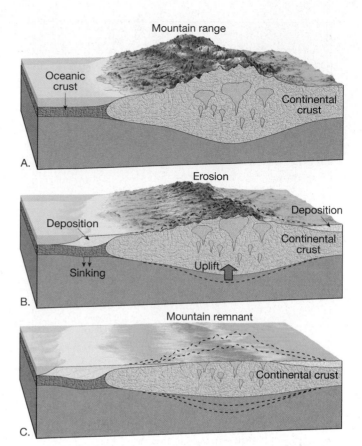

Figure 9.26 This sequence illustrates how the combined effect of erosion and isostatic adjustment results in a thinning of the crust in mountainous regions. **A.** When mountains are young, the continental crust is thickest. **B.** As erosion lowers the mountains, the crust rises in response to the reduced load. **C.** Erosion and uplift continue until the mountains reach "normal" crustal thickness.

Chapter Summary

- *Deformation* refers to changes in the shape and/or volume of a rock body. Rocks deform differently depending on the environment (temperature and confining pressure), the composition of the rock, and the length of time stress is maintained. Rocks first respond by deforming *elastically* and will return to their original shape when the stress is removed. Once their elastic limit (strength) is surpassed, rocks either deform by ductile flow or they fracture. *Ductile deformation* is a solid-state flow that results in a change in size and shape of rocks without fracturing. Ductile deformation occurs in a high-temperature/high-pressure environment. In a near-surface environment, most rocks deform by *brittle failure*.

- Among the most basic geologic structures associated with rock deformation are *folds* (flat-lying sedimentary and volcanic rocks bent into a series of wavelike undulations). The two most common types of folds are *anticlines*, formed by the upfolding, or arching, of rock layers, and *synclines*, which are downfolds. Most folds are the result of horizontal *compres-*

sional stresses. Domes (upwarped structures) and *basins* (downwarped structures) are circular or somewhat elongated folds formed by vertical displacements of strata.

- Faults are fractures in the crust along which appreciable displacement has occurred. Faults in which the movement is primarily vertical are called *dip-slip faults*. Dip-slip faults include both *normal* and *reverse faults*. Low-angle reverse faults are called *thrust faults*. Normal faults indicate *tensional stresses* that pull the crust apart. Along spreading centers, divergence can cause a central block called a *graben*, bounded by normal faults, to drop as the plates separate.

- Reverse and thrust faulting indicate that *compressional forces* are at work. Large *thrust faults* are found along subduction zones and other convergent boundaries where plates are colliding.

- *Strike-slip faults* exhibit mainly horizontal displacement parallel to the fault surface. Large strike-slip faults, called *transform faults*, accommodate displacement between plate boundaries. Most transform faults cut the oceanic lithosphere and link spreading centers. The San Andreas Fault cuts the continental lithosphere and accommodates the northward displacement of southwestern California.

• *Joints* are fractures along which no appreciable displacement has occurred. Joints generally occur in groups with roughly parallel orientations and are the result of brittle failure of rock units located in the outermost crust.

• The name for the processes that collectively produce a mountain system is *orogenesis*. Most mountains consist of roughly parallel ridges of folded and faulted sedimentary and volcanic rocks, portions of which have been strongly metamorphosed and intruded by younger igneous bodies.

• Major mountain systems form along *convergent plate boundaries*. *Andean-type mountain building* along continental margins involves the convergence of an oceanic plate and a plate whose leading edge contains continental crust. At some point in the formation of Andean-type mountains a *subduction zone* forms along with a *continental volcanic arc*.

• *Continental collisions*, in which both plates are carrying continental crust, have resulted in the formation of the Himalaya Mountains and the Tibetan Plateau. Continental collisions also formed many other mountain belts, including the Alps, Urals, and Appalachians.

• Recent investigations indicate that *accretion*, a third mechanism of orogenesis, takes place where *small crustal fragments collide and accrete to continental margins* along plate boundaries. The accreted crustal blocks are referred to as *terranes*. The mountainous topography of Alaska and the westernmost regions of Canada, the United States, and Mexico formed as the result of the accretion to terranes to North America.

• Earth's less dense crust floats on top of the denser and deformable rocks of the mantle, much like wooden blocks floating in water. The concept of a floating crust in gravitational balance is called *isostasy*. Most mountainous topography is located where the crust has been shortened and thickened. Therefore, mountains have deep, buoyant crustal roots that isostatically support them. As erosion lowers the peaks, *isostatic adjustment* gradually raises the mountains in response. The processes of erosion and uplifting will continue until the mountain block reaches "normal" crustal thickness.

Key Terms

accretionary wedge (p. 274)
active continental margin (p. 274)
anticline (p. 262)
basin (p. 263)
brittle failure (deformation) (p. 261)
deformation (p. 260)
dip-slip fault (p. 264)
dome (p. 263)
ductile deformation (p. 261)
fault (p. 263)

fault-block mountains (p. 264)
fault scarp (p. 264)
fold (p. 261)
graben (p. 264)
horst (p. 264)
isostasy (p. 279)
isostatic adjustment (p. 279)
joint (p. 266)
monocline (p. 262)
normal fault (p. 264)

orogenesis (p. 267)
passive continental margin (p. 274)
reverse fault (p. 265)
strike-slip fault (p. 266)
syncline (p. 262)
terrane (p. 277)
thrust fault (p. 265)
transform fault (p. 266)

Review Questions

1. What is rock deformation?

2. How is brittle deformation different from ductile deformation?

3. List three factors that determine how rocks will behave when exposed to stresses that exceed their strength. Briefly explain the role of each.

4. Distinguish between anticlines and synclines, domes and basins, anticlines and domes.

5. How is a monocline different from an anticline?

6. The Black Hills of South Dakota are a good example of what type of structural feature?

7. Contrast the movements that occur along normal and reverse faults. What type of force is indicated by each fault?

8. Is the fault shown in Figure 9.10 a normal or reverse fault?

9. Describe a horst and a graben. Explain how a graben valley forms, and name one.

10. What type of faults are associated with fault-block mountains?

11. How are reverse faults different from thrust faults? In what way are they the same?

12. The San Andreas Fault is an excellent example of a _____ fault.

13. How are joints different from faults?

14. In the plate tectonics model, which type of plate boundary is most directly associated with mountain building?

15. Briefly describe the development of a volcanic island arc.

16. The formation of mountainous topography at a volcanic island arc, such as Japan, is considered just one phase in the development of a major mountain belt. Explain.

17. What is an accretionary wedge? Briefly describe its formation.

18. What is a passive margin? Give an example. Give an example of an active continental margin.

19. In what way are the Sierra Nevada and the western Andes similar?

20. Compare mountain building along an Andean-type convergent boundary and a continent-continent convergent boundary.

21. How does the plate tectonics theory help explain the existence of fossil marine life in rocks atop the Ural Mountains?

22. Define the term *terrane*. How is it different from the term *terrain*?

23. List three structures that fall under the heading "oceanic plateaus." In addition to oceanic plateaus, what other features are thought to be carried by the oceanic lithosphere and eventually accreted to a continent?

24. List some evidence that supports the concept of crustal uplift.

25. What happens to a floating object when weight is added? Subtracted? How does this principal apply to changes in the elevation of mountains? What term is applied to the adjustment that causes crustal uplift of this type?

26. How does the formation and melting of Pleistocene ice sheets support the idea that the lithosphere tries to remain in isostatic balance?

Examining the Earth System

1. A good example of the interactions between Earth's spheres is the influence of mountains on climate. Examine the temperature graphs and annual precipitation amounts for the cities of Seattle and Spokane, Washington (Figure 9.27). Notice on the inset map that mountains (the Cascades) separate the two cities and that the general wind direction in the region is from west to east (the prevailing westerlies of the midlatitudes). Which city receives the greatest annual precipitation? Contrast the summer and winter temperatures that occur at each city. What influence have the mountains apparently had on Spokane's rainfall and temperature?

2. The Cascades have had a profound effect on the amount and type of plant and animal life (biosphere) that inhabit the region around Spokane and Seattle. Provide several specific examples to verify this statement.

Figure 9.27 Comparison of the average monthly temperatures and precipitation for Seattle and Spokane. Note that Seattle receives an average of 38.4 inches of precipitation per year while Spokane receives only 17 inches of precipitation on average.

Web Resources

The *Earth Science* Website uses the resources and flexibility of the Internet to aid in your study of the topics in this chapter. Written and developed by Earth science instructors, this site will help improve your understanding of Earth science. Visit **http://www.prenhall.com/tarbuck** and click on the cover of *Earth Science 10e* to find:

- **On-line review quizzes.**
- **Web-based critical thinking and writing exercises.**
- **Links to chapter-specific Web resources.**
- **Internet-wide key term searches.**

http://www.prenhall.com/tarbuck

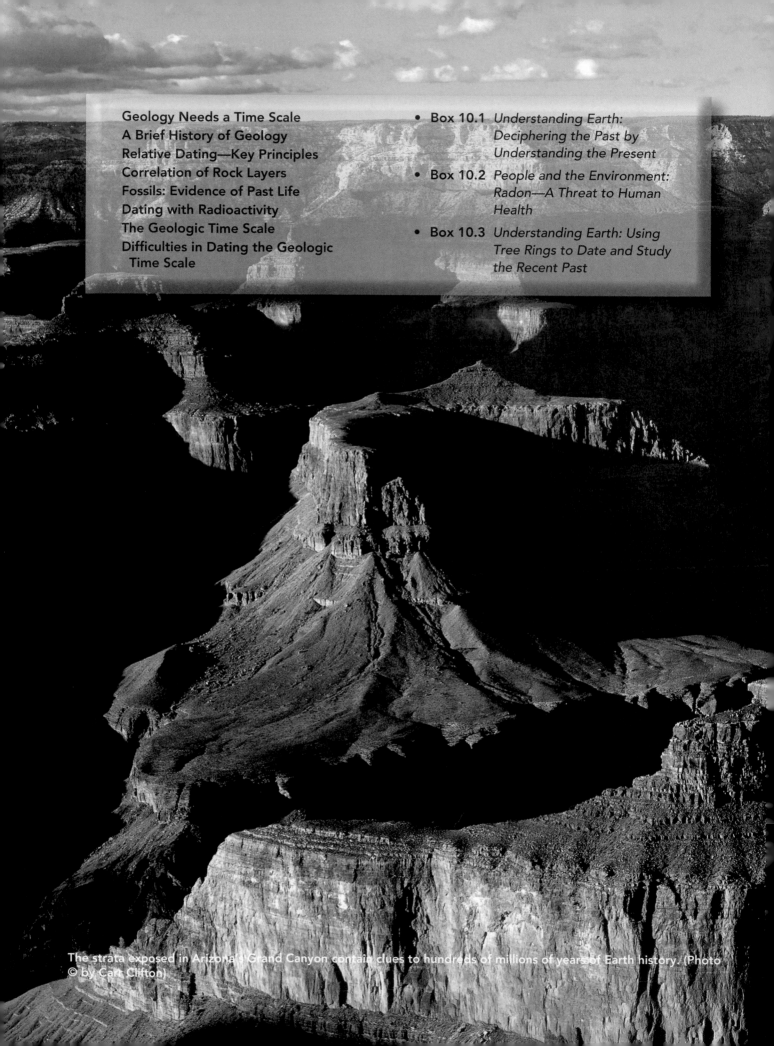

The strata exposed in Arizona's Grand Canyon contain clues to hundreds of millions of years of Earth history. (Photo © by Carr Clifton)

CHAPTER 10

GEOLOGIC TIME

In the eighteenth century, James Hutton recognized the immensity of Earth history and the importance of time as a component in all geological processes. In the nineteenth century, others effectively demonstrated that Earth had experienced many episodes of mountain building and erosion, which must have required great spans of geologic time. Although these pioneering scientists understood that Earth was very old, they had no way of knowing its true age. Was it tens of millions, hundreds of millions, or even billions of years old? Rather, a geologic time scale was developed that showed the sequence of events based on relative dating principles. What are these principles? What part do fossils play? With the discovery of radioactivity and the development of radiometric dating techniques, geologists now can assign fairly accurate dates to many of the events in Earth history. What is radioactivity? Why is it a good "clock" for dating the geologic past? In this chapter we shall answer these questions.

Geology Needs a Time Scale

In 1869 John Wesley Powell, who was later to head the U.S. Geological Survey, led a pioneering expedition down the Colorado River and through the Grand Canyon (Figure 10.1). Writing about the rock layers that were exposed by the downcutting of the river, Powell said that "the canyons of this region would be a Book of Revelations in the rock-leaved Bible of geology." He was undoubtedly impressed with the millions of years of Earth history exposed along the walls of the Grand Canyon (see chapter-opening photo).

Powell realized that the evidence for an ancient Earth is concealed in its rocks. Like the pages in a long and complicated history book, rocks record the geological events and changing life forms of the past. The book, however, is not complete. Many pages, especially in the early chapters, are missing. Others are tattered, torn, or smudged. Yet enough of the book remains to allow much of the story to be deciphered.

Interpreting Earth history is a prime goal of the science of geology. Like a modern-day sleuth, the geologist must interpret clues found preserved in the rocks. By studying rocks, especially sedimentary rocks, and the features they contain, geologists can unravel the complexities of the past.

Geological events by themselves, however, have little meaning until they are put into a time perspective. Studying history, whether it be the Civil War or the Age of Dinosaurs, requires a calendar. Among geology's major contributions to human knowledge is the *geologic time scale* and the discovery that Earth history is exceedingly long.

The geologists who developed the geologic time scale revolutionized the way people think about time and how they perceive our planet. They learned that Earth is much older than anyone had previously imagined and that its surface and interior have been changed

A.

B.

Figure 10.1   A. Start of the expedition from Green River station. A drawing from Powell's 1875 book. **B.** Major John Wesley Powell, pioneering geologist and the second director of the U.S. Geological Survey. (Courtesy of the U.S. Geological Survey, Denver)

over and over again by the same geological processes that operate today.

A Brief History of Geology

In the mid-1600s, James Ussher, Anglican Archbishop of Armagh, Primate of all Ireland, published a work that had immediate and profound influence on people's view of Earth's age. A respected scholar of the Bible, Ussher constructed a chronology of human and Earth history in which he determined that Earth was only a few thousand years old, having been created in 4004 B.C. Ussher's treatise earned widespread acceptance among Europe's scientific and religious leaders, and his chronology was soon printed in the margins of the Bible itself.

During the seventeenth and eighteenth centuries the doctrine of **catastrophism** strongly influenced people's thinking about Earth. Briefly stated, catastrophists believed that Earth's landscapes had been developed primarily by great catastrophes. Features such as mountains and canyons, which today we know take great periods of time to form, were explained as having been produced by sudden and often worldwide disasters triggered by unknowable causes that no longer operate. This philosophy was an attempt to fit the rate of

Earth processes to the prevailing ideas on the age of Earth.

Birth of Modern Geology

Modern geology began in the late 1700s when James Hutton, a Scottish physician and gentleman farmer, published his *Theory of the Earth.* In this work, Hutton put forth a fundamental principle that is a pillar of geology today: **uniformitarianism.** It simply states that *the physical, chemical, and biological laws that operate today have also operated in the geologic past.* This means that the forces and processes that we observe presently shaping our planet have been at work for a very long time. Thus, to understand ancient rocks, we must first understand present-day processes and their results. This idea is commonly expressed by saying "the present is the key to the past" (see Box 10.1).

Prior to Hutton's *Theory of the Earth,* no one had effectively demonstrated that geological processes occur over extremely long periods of time. However, Hutton persuasively argued that weak, slow-acting processes could, over long spans of time, produce effects just as great as those resulting from sudden catastrophic events. Unlike his predecessors, Hutton carefully cited verifiable observations to support his ideas.

BOX 10.1

UNDERSTANDING EARTH:
Deciphering the Past by
Understanding the Present

Louis Agassiz, a Swiss scientist born in 1807, was instrumental in formulating modern ideas about the Ice Age (Figure 10.A). The development of this knowledge provides an excellent example of the application of the principle of uniformitarianism.

In 1821 Agassiz heard another scientist present a paper in which he indicated that glacial features occurred in places that were a significant distance from existing glaciers in the Alps. This, of course, implied that the glaciers had once occupied areas considerably beyond their present limits. Agassiz was skeptical about this hypothesis and set out to invalidate it. Ironically, his fieldwork in the Alps convinced him of the merits of his colleague's hypothesis. Agassiz found the same unique deposits and features that can be seen forming in association with active glaciers in places far beyond the limits of the ice. Subsequent work led Agassiz to hypothesize that a great ice age had occurred in response to a period of worldwide climate change and had affected large parts of the globe. Agassiz's ideas eventually developed into our present-day glacial theory.

The proof of the glacial theory proposed by Agassiz and others constitutes a classic example of applying the principle of uniformitarianism. Realizing that certain landforms and other features are produced by no other known process but glacial activity, they were able to reconstruct the extent of now vanished ice sheets. Clearly, understanding the present was the key to deciphering the past.

Figure 10.A   Louis Agassiz (1807–1873) played a major role in the development of glacial theory. (Courtesy of Harvard University Archives)

For example, when he argued that mountains are sculpted and ultimately destroyed by weathering and the work of running water, and that their wastes are carried to the oceans by processes that can be observed, Hutton said, "We have a chain of facts which clearly demonstrates that the materials of the wasted mountains have traveled through the rivers"; and further, "There is not one step in all this progress that is not to be actually perceived." He then went on to summarize this thought by asking a question and immediately providing the answer: "What more can we require? Nothing but time."

Geology Today

Today the basic tenets of uniformitarianism are just as viable as in Hutton's day. Indeed, we realize more strongly than ever that the present gives us insight into the past and that the physical, chemical, and biological laws that govern geological processes remain unchanging through time. However, we also understand that the doctrine should not be taken too literally. To say that geological processes in the past were the same as those occurring today is not to suggest that they always had the same relative importance or that they operated at precisely the same rate. Moreover, some important geologic processes are not currently observable, but evidence that they occur is well established. For example, we know that Earth has experienced impacts from large meteorites even though we have no human witnesses. Such events altered Earth's crust, modified its climate, and strongly influenced life on the planet.

The acceptance of uniformitarianism meant the acceptance of a very long history for Earth. Although Earth's processes vary in intensity, they still take a very long time to create or destroy major landscape features.

For example, geologists have established that mountains once existed in portions of present-day Minnesota, Wisconsin, and Michigan. Today the region consists of low hills and plains. Erosion gradually destroyed these peaks. Estimates indicate that the North American continent is being lowered at a rate of about 3 centimeters per 1000 years. At this rate, it would take 100 million years for water, wind, and ice to lower mountains that were 300 meters (10,000 feet) high.

But even this time span is relatively short on the time scale of Earth history, for the rock record contains evidence that shows Earth has experienced many cycles of mountain building and erosion. Concerning the ever changing nature of Earth through great expanses of geologic time, Hutton made a statement that was to become his most famous. In concluding his classic 1788 paper published in the *Transactions of the Royal Society of Edinburgh,* he stated, "The results, therefore, of our present enquiry is, that we find no vestige of a beginning—no prospect of an end."

It is important to remember that although many features of our physical landscape may seem to be unchanging over our lifetimes, they are nevertheless changing, but on time scales of hundreds, thousands, or even many millions of years.

Relative Dating—Key Principles

 Geologic Time
↳ Relative Dating

During the late 1800s and early 1900s, various attempts were made to determine the age of Earth. Although some of the methods appeared promising at the time, none proved reliable. What these scientists were seeking was a **numerical date.** Such dates specify the actual number of years that have passed since an event occurred—for example, the extinction of the dinosaurs about 65 million years ago. Today our understanding of radioactivity allows us to accurately determine numerical dates for rocks that represent important events in Earth's distant past. We will study radioactivity later in this chapter. Prior to the discovery of radioactivity, geologists had no accurate and dependable method of numerical dating and had to rely solely on relative dating.

Relative dating means placing rocks in their proper *sequence of formation*—which ones formed first, second, third, and so on. Relative dating cannot tell us how long ago something took place, only that it followed one event and preceded another. The relative dating techniques that were developed are valuable and still widely used. Numerical dating methods did not replace these techniques; they simply supplemented them. To establish a relative time scale, a few basic principles or rules had to be discovered and applied. Although they may seem obvious to us today, they were major breakthroughs in thinking at the time, and their discovery and acceptance was an important scientific achievement.

STUDENTS SOMETIMES ASK...
You mentioned early attempts at determining Earth's age that proved unreliable. How did nineteenth-century scientists go about making such calculations?

One method that was attempted several times involved the rate at which sediment is deposited. Some reasoned that if they could determine the rate that sediment accumulates and could further ascertain the total thickness of sedimentary rock that had been deposited during Earth history, they could estimate the length of geologic time. All that was necessary was to divide the rate of sediment accumulation into the total thickness of sedimentary rock.

Estimates of Earth's age varied each time this method was attempted. The age of Earth as calculated by this method ranged from 3 million to 1.5 billion years! Obviously this method was riddled with difficulties. Can you suggest what some might have been?

Law of Superposition

Nicolaus Steno, a Danish anatomist, geologist, and priest (1636–1686), is credited with being the first to recognize a sequence of historical events in an outcrop of sedimentary rock layers. Working in the mountains of western Italy, Steno applied a very simple rule that has come to be the most basic principle of relative dating—the **law of superposition.** The law simply states that in an undeformed sequence of sedimentary rocks, each bed is older than the one above it and younger than the one below. Although it may seem obvious that a rock layer could not be deposited unless it had something older beneath it for support, it was not until 1669 that Steno clearly stated the principle.

This rule also applies to other surface-deposited materials, such as lava flows and beds of ash from volcanic eruptions. Applying the law of superposition to the beds exposed in the upper portion of the Grand Canyon (Figure 10.2), you can easily place the layers in their proper order. Among those that are shown, the sedimentary rocks in the Supai Group must be the oldest, followed in order by the Hermit Shale, Coconino Sandstone, Toroweap Formation, and Kaibab Limestone.

Principle of Original Horizontality

Steno is also credited with recognizing the importance of another basic principle, called the **principle of original horizontality.** Simply stated, it means that layers of sediment are generally deposited in a horizontal position. Thus, if we observe rock layers that are flat, it means they have not been disturbed and thus still have their *original* horizontality. The layers in the Grand Canyon illustrate this in the chapter-opening photo and

in Figure 10.2. But if they are folded or inclined at a steep angle, they must have been moved into that position by crustal disturbances sometime *after* their deposition (Figure 10.3).

Principle of Cross-Cutting Relationships

When a fault cuts through other rocks, or when magma intrudes and crystallizes, we can assume that the fault or intrusion is younger than the rocks affected. For example, in Figure 10.4, the faults and dikes clearly must have occurred *after* the sedimentary layers were deposited.

This is the **principle of cross-cutting relationships.** By applying the cross-cutting principle, you can see that fault *A* occurred *after* the sandstone layer was deposited, because it "broke" the layer. However, fault *A* occurred *before* the conglomerate was laid down, because that layer is unbroken.

We can also state that dike *B* and its associated sill are older than dike *A*, because dike *A* cuts the sill. In the same manner, we know that the batholith was emplaced after movement occurred along fault *B*, but before dike *B* was formed. This is true because the batholith cuts across fault *B*, and dike *B* cuts across the batholith.

Inclusions

Sometimes inclusions can aid the relative dating process. **Inclusions** are pieces of one rock unit that are contained within another. The basic principle is logical and straightforward. The rock mass adjacent to the one containing the inclusions must have been there first in order to provide the rock fragments. Therefore, the rock

A.

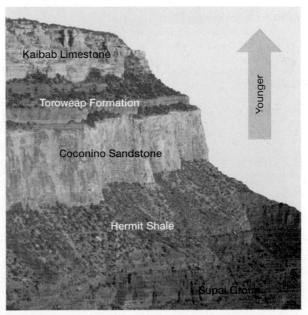

B.

Figure 10.2   Applying the law of superposition to these layers exposed in the upper portion of the Grand Canyon, the Supai Group is oldest and the Kaibab Limestone is youngest. (Photo by E. J. Tarbuck)

Figure 10.3 Most layers of sediment are deposited in a nearly horizontal position. Thus, when we see rock layers that are folded or tilted, we can assume that they must have been moved into that position by crustal disturbances *after* their deposition. These folded layers are exposed in the Namib Desert (southwestern Africa). (Photo by Michael Fogden/DRK Photo)

mass containing inclusions is the younger of the two. Figure 10.5 provides an example. Here the inclusions of intrusive igneous rock in the adjacent sedimentary layer indicate that the sedimentary layer was deposited on top of a weathered igneous mass rather than being intruded from below by magma that later crystallized.

Unconformities

When we observe layers of rock that have been deposited essentially without interruption, we call them **conformable.** Particular sites exhibit conformable beds representing certain spans of geologic time. However, no place on Earth has a complete set of conformable strata.

Throughout Earth history, the deposition of sediment has been interrupted again and again. All such breaks in the rock record are termed unconformities. An

unconformity represents a long period during which deposition ceased, erosion removed previously formed rocks, and then deposition resumed. In each case uplift and erosion are followed by subsidence and renewed sedimentation. Unconformities are important features because they represent significant geologic events in Earth history. Moreover, their recognition helps us identify what intervals of time are not represented by strata and thus are missing from the geologic record.

The rocks exposed in the Grand Canyon of the Colorado River represent a tremendous span of geologic history. It is a wonderful place to take a trip through time. The canyon's colorful strata record a long history of sedimentation in a variety of environments—advancing seas, rivers and deltas, tidal flats, and sand dunes. But the record is not continuous. Unconformities represent

Figure 10.4 Cross-cutting relationships are an important principle used in relative dating. An intrusive rock body is younger than the rocks it intrudes. A fault is younger than the rock layers it cuts.

A. Intrusive
igneous rock

B. Exposure and
weathering of intrusive igneous rock

C. Deposition of
sedimentary layers

Figure 10.5   These three diagrams illustrate one way that inclusions can form, and a type of unconformity termed a *nonconformity*. In the third diagram we know the igneous rock must be older because pieces of it are included in the overlying sedimentary bed. When older intrusive igneous rocks are overlain by younger sedimentary layers, a nonconformity is said to exist. The photo shows an inclusion of dark igneous rock in a lighter-colored and younger host rock. (Photo by Tom Bean)

vast amounts of time that have not been recorded in the canyon's layers. Figure 10.6 is a geologic cross section of the Grand Canyon. Refer to it as you read about the three basic types of unconformities: angular unconformities, disconformities, and nonconformities.

Angular Unconformity.   Perhaps the most easily recognized unconformity is an **angular unconformity.** It consists of tilted or folded sedimentary rocks that are overlain by younger, more flat-lying strata.

An angular unconformity indicates that during the pause in deposition, a period of deformation (folding or tilting) and erosion occurred (Figure 10.7).

When James Hutton studied an angular unconformity in Scotland more than 200 years ago, it was clear to him that it represented a major episode of geologic activity (Figure 10.7E). He also appreciated the immense time span implied by such relationships. When a companion later wrote of their visit to the site, he stated that "the mind seemed to grow giddy by looking so far into the abyss of time."

Disconformity.   When contrasted with angular unconformities, **disconformities** are more common, but usually far less conspicuous because the strata on either side are essentially parallel. For example, look at the disconformities in the cross section of the Grand Canyon in Figure 10.6. Many disconformities are difficult to identify because the rocks above and below are similar and there is little evidence of erosion. Such a break often resembles an ordinary bedding plane. Other disconformities are easier to identify because the ancient erosion surface is cut deeply into the older rocks below.

Nonconformity.   The third basic type of unconformity is a **nonconformity.** Here the break separates older metamorphic or intrusive igneous rocks from younger sedimentary strata (Figures 10.5 and 10.6). Just as angular unconformities and disconformities imply crustal movements, so too do nonconformities. Intrusive igneous masses and metamorphic rocks originate far below the surface. Thus, for a nonconformity to develop, there must be a period of uplift and the erosion of overlying rocks. Once exposed at the surface, the igneous or metamorphic rocks are subjected to weathering and erosion prior to subsidence and the renewal of sedimentation.

Using Relative Dating Principles

If you apply the principles of relative dating to the hypothetical geologic cross section in Figure 10.8 (p. 294), you can place in proper sequence the rocks and the events they represent. The statements within the figure summarize the logic used to interpret the cross section.

In this example, we establish a relative time scale for the rocks and events in the area of the cross section. Remember that this method gives us no indication as to how many years of Earth history are represented, for we have no numerical dates. Nor do we know how this area compares to any other.

Correlation of Rock Layers

To develop a geologic time scale that is applicable to the entire Earth, rocks of similar age in different regions must be matched up. Such a task is referred to as **correlation.**

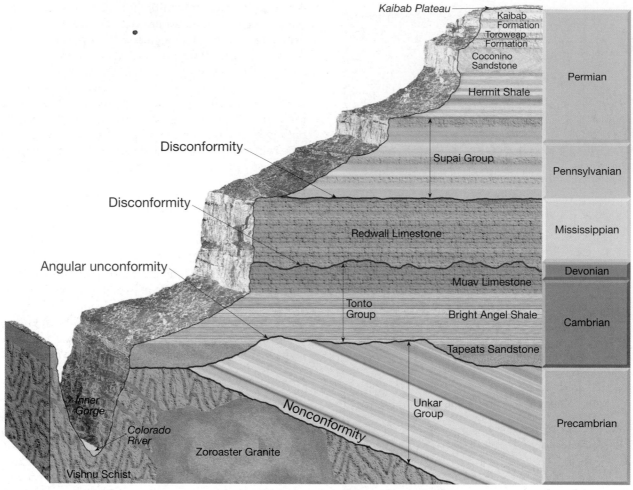

Figure 10.6 This cross section through the Grand Canyon illustrates the three basic types of unconformities. An angular unconformity can be seen between the tilted Precambrian Unkar Group and the Cambrian Tapeats Sandstone. Two disconformities are marked, above and below the Redwall Limestone. A nonconformity occurs between the igneous and metamorphic rocks of the Inner Gorge and the sedimentary strata of the Unkar Group.

Within a limited area, correlating the rocks of one locality with those of another may be done simply by walking along the outcropping edges. However, this might not be possible when the rocks are mostly concealed by soil and vegetation. Correlation over short distances is often achieved by noting the position of a distinctive rock layer in a sequence of strata. Or, a layer may be identified in another location if it is composed of very distinctive or uncommon minerals.

By correlating the rocks from one place to another, a more comprehensive view of the geologic history of a region is possible. Figure 10.9, for example, shows the correlation of strata at three sites on the Colorado Plateau in southern Utah and northern Arizona. No single locale exhibits the entire sequence, but correlation reveals a more complete picture of the sedimentary rock record.

Many geologic studies involve relatively small areas. Such studies are important in their own right, but their full value is realized only when the rocks are correlated with those of other regions. Although the methods just described are sufficient to trace a rock formation over rel-

atively short distances, they are not adequate for matching rocks that are separated by great distances. When correlation between widely separated areas or between continents is the objective, geologists must rely on fossils.

Fossils: Evidence of Past Life

Fossils, the remains or traces of prehistoric life, are important inclusions in sediment and sedimentary rocks. They are important tools for interpreting the geologic past. Knowing the nature of the life forms that existed at a particular time helps researchers understand past environmental conditions. Further, fossils are important time indicators and play a key role in correlating rocks of similar ages that are from different places.

Types of Fossils

Fossils are of many types. The remains of relatively recent organisms may not have been altered at all. Such objects as teeth, bones, and shells are common exam-

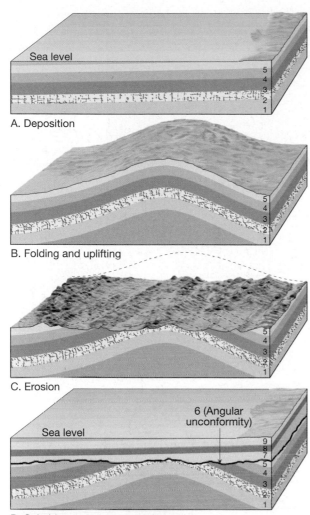

A. Deposition

B. Folding and uplifting

C. Erosion

D. Subsidence and renewed deposition

E.

Figure 10.7 Formation of an angular unconformity. An angular unconformity represents an extended period during which deformation and erosion occurred. Part E shows an angular unconformity at Siccar Point, Scotland, that was first described by James Hutton more than 200 years ago. (Photo by A. P. Trujillo/APT Photos)

ples. Far less common are entire animals, flesh included, that have been preserved because of rather unusual circumstances. Remains of prehistoric elephants called mammoths that were frozen in the Arctic tundra of Siberia and Alaska are examples, as are the mummified remains of sloths preserved in a dry cave in Nevada.

Given enough time, the remains of an organism are likely to be modified. Often fossils become *petrified* (literally, "turned into stone"), meaning that the small internal cavities and pores of the original structure are filled with precipitated mineral matter (Figure 10.10A, p. 296). In other instances *replacement* may occur. Here the cell walls and other solid material are removed and replaced with mineral matter. Sometimes the microscopic details of the replaced structure are faithfully retained.

Molds and casts constitute another common class of fossils. When a shell or other structure is buried in sediment and then dissolved by underground water, a *mold* is created. The mold faithfully reflects only the shape and surface marking of the organism; it does not reveal any information concerning its internal structure. If these hollow spaces are subsequently filled with mineral matter, *casts* are created (Figure 10.10B).

A type of fossilization called *carbonization* is particularly effective in preserving leaves and delicate animal forms. It occurs when fine sediment encases the remains of an organism. As time passes, pressure squeezes out the liquid and gaseous components and leaves behind a thin residue of carbon (Figure 10.10C). Black shales deposited as organic-rich mud in oxygen-poor environments often contain abundant carbonized remains. If the film of carbon is lost from a fossil preserved in fine-grained sediment, a replica of the surface, called an *impression*, may still show considerable detail (Figure 10.10D).

Delicate organisms, such as insects, are difficult to preserve, and consequently they are relatively rare in the fossil record. Not only must they be protected from decay but they must not be subjected to any pressure that would crush them. One way in which some insects have been preserved is in *amber*, the hardened resin of ancient trees. The fly in Figure 10.10E was preserved after being trapped in a drop of sticky resin. Resin sealed off the insect from the atmosphere and protected the remains from damage by water and air. As the resin hardened, a protective pressure-resistant case was formed.

In addition to the fossils already mentioned, there are numerous other types, many of them only traces of prehistoric life. Examples of such indirect evidence include:

1. Tracks—animal footprints made in soft sediment that was later lithified (Figure 10.10F).
2. Burrows—tubes in sediment, wood, or rock made by an animal. These holes may later become filled with mineral matter and preserved. Some of the oldest-known fossils are believed to be worm burrows.

Figure 10.8 Geologic cross section of a hypothetical region.

3. Coprolites—fossil dung and stomach contents that can provide useful information pertaining to food habits of organisms.

4. Gastroliths—highly polished stomach stones that were used in the grinding of food by some extinct reptiles.

Conditions Favoring Preservation

Only a tiny fraction of the organisms that have lived during the geologic past have been preserved as fossils. Normally the remains of an animal or plant are destroyed. Under what circumstances are they preserved? Two special conditions appear to be necessary: rapid burial and the possession of hard parts.

When an organism perishes, its soft parts usually are quickly eaten by scavengers or decomposed by bacteria. Occasionally, however, the remains are buried by sediment. When this occurs, the remains are protected from the environment, where destructive processes operate. Rapid burial therefore is an important condition favoring preservation.

In addition, animals and plants have a much better chance of being preserved as part of the fossil record if they have hard parts. Although traces and imprints of soft-bodied animals such as jellyfish, worms, and insects exist, they are not common. Flesh usually decays so rapidly that preservation is exceedingly unlikely. Hard parts such as shells, bones, and teeth predominate in the record of past life.

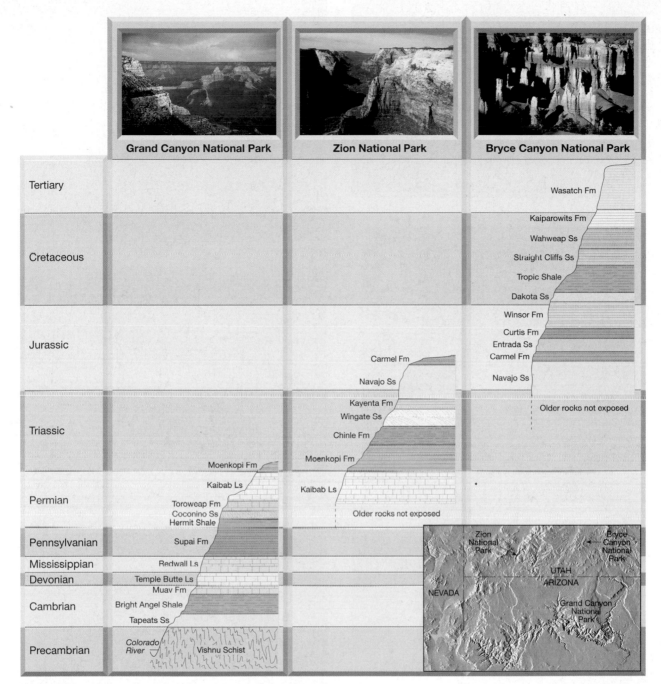

Figure 10.9 Correlation of strata at three locations on the Colorado Plateau reveals a more complete view of the extent of sedimentary rocks in the region. (After U.S. Geological Survey)

Because preservation is contingent on special conditions, the record of life in the geologic past is biased. The fossil record of those organisms with hard parts that lived in areas of sedimentation is quite abundant. However, we get only an occasional glimpse of the vast array of other life forms that did not meet the special conditions favoring preservation.

Fossils and Correlation

The existence of fossils had been known for centuries, yet it was not until the late 1700s and early 1800s that their significance as geologic tools was made evident.

During this period an English engineer and canal builder, William Smith, discovered that each rock formation in the canals he worked on contained fossils unlike those in the beds either above or below. Further, he noted that sedimentary strata in widely separated areas could be identified and correlated by their distinctive fossil content.

Based on Smith's classic observations and the findings of many geologists who followed, one of the most important and basic principles in historical geology was formulated: *Fossil organisms succeed one another in a definite and determinable order, and therefore any time period*

Figure 10.10 There are many types of fossilization. Six examples are shown here. **A.** Petrified wood in Petrified Forest National Park, Arizona. **B.** Natural casts of shelled invertebrates. **C.** A fossil bee preserved as a thin carbon film. **D.** Impressions are common fossils and often show considerable detail. **E.** Insect in amber. **F.** Dinosaur footprint in fine-grained limestone near Tuba City, Arizona. (Photo A by David Muench; Photos B, D, and F by E. J. Tarbuck; Photo C courtesy of the National Park Service; Photo E by Breck P. Kent)

can be recognized by its fossil content. This has come to be known as the **principle of fossil succession.** In other words, when fossils are arranged according to their age by applying the law of superposition to the rocks in which they are found, they do not present a random or haphazard picture. To the contrary, fossils show changes that document the evolution of life through time.

For example, an Age of Trilobites is recognized quite early in the fossil record. Then, in succession, paleontologists recognize an Age of Fishes, an Age of Coal Swamps, an Age of Reptiles, and an Age of Mammals. These "ages" pertain to groups that were especially plentiful and characteristic during particular time periods. Within each of the ages, there are many subdivisions based, for example, on certain species of trilobites and certain types of fish, reptiles, and so on. This same succession of dominant organisms, never out of order, is found on every continent.

Once fossils were recognized as time indicators, they became the most useful means of correlating rocks of similar age in different regions. Geologists pay particular attention to certain fossils called **index fossils.** These fossils are widespread geographically and are limited to a short span of geologic time, so their presence provides an important method of

matching rocks of the same age. Rock formations, however, do not always contain a specific index fossil. In such situations, groups of fossils are used to establish the age of the bed. Figure 10.11 illustrates how an assemblage of fossils can be used to date rocks more precisely than could be accomplished by the use of only one of the fossils.

In addition to being important and often essential tools for correlation, fossils are important environmental indicators. Although much can be deduced about past environments by studying the nature and characteristics of sedimentary rocks, a close examination of any fossils present can usually provide a great deal more information.

For example, when the remains of certain clam shells are found in limestone, the geologist can assume that the region was once covered by a shallow sea, because that is where clams live today. Also, by using what we know of living organisms, we can conclude that fossil animals with thick shells capable of withstanding pounding and surging waves must have inhabited shorelines. Conversely, animals with thin, delicate shells probably indicate deep, calm offshore waters. Hence, by looking closely at the types of fossils, the approximate position of an ancient shoreline may be identified.

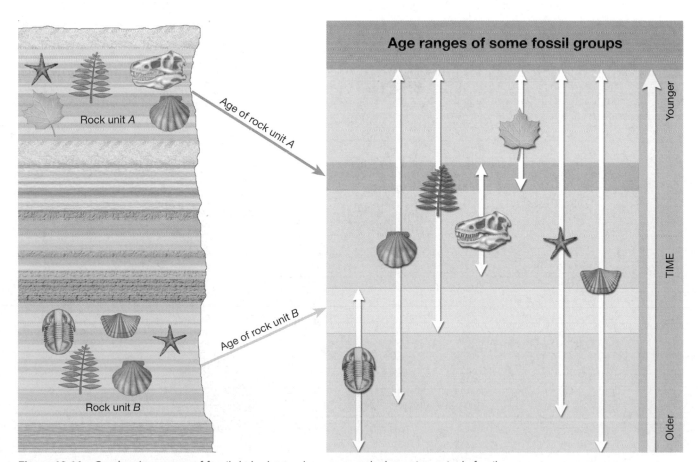

Figure 10.11   Overlapping ranges of fossils help date rocks more exactly than using a single fossil.

Further, fossils can indicate the former temperature of the water. Certain present-day corals require warm and shallow tropical seas like those around Florida and the Bahamas. When similar corals are found in ancient limestones, they indicate that a Florida-like marine environment must have existed when the corals were alive. These examples illustrate how fossils can help unravel the complex story of Earth history.

Dating with Radioactivity

 Geologic Time
↳ Radiometric Dating

In addition to establishing relative dates by using the principles described in the preceding sections, it is also possible to obtain reliable numerical dates for events in the geologic past. For example, we know that Earth is about 4.5 billion years old and that the dinosaurs became extinct about 65 million years ago. Dates that are expressed in millions and billions of years truly stretch our imagination because our personal calendars involve time measured in hours,

weeks, and years. Nevertheless, the vast expanse of geologic time is a reality, and it is radiometric dating that allows us to measure it. In this section you will learn about radioactivity and its application in radiometric dating.

Reviewing Basic Atomic Structure

Recall from Chapter 1 that each atom has a *nucleus* containing protons and neutrons and that the nucleus is orbited by electrons. *Electrons* have a negative electrical charge, and *protons* have a positive charge. A *neutron* is actually a proton and an electron combined, so it has no charge (it is neutral).

The *atomic number* (each element's identifying number) is the number of protons in the nucleus. Every element has a different number of protons and thus a different atomic number (hydrogen = 1, carbon = 6, oxygen = 8, uranium = 92, etc.). Atoms of the same element always have the same number of protons, so the atomic number stays constant.

Practically all of an atom's mass (99.9%) is in the nucleus, indicating that electrons have virtually no mass at all. So, by adding the protons and neutrons in an atom's nucleus, we derive the atom's *mass number*. The

Figure 10.12 Common types of radioactive decay. Notice that in each case the number of protons (atomic number) in the nucleus changes, thus producing a different element.

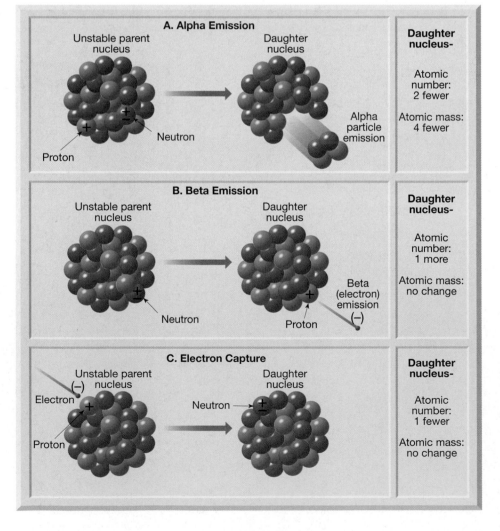

number of neutrons can vary, and these variants, or *isotopes*, have different mass numbers.

To summarize with an example, uranium's nucleus always has 92 protons, so its atomic number always is 92. But its neutron population varies, so uranium has three isotopes: uranium-234 (mass of protons + neutrons = 234), uranium-235, and uranium-238. All three isotopes are mixed in nature. They look the same and behave the same in chemical reactions.

Radioactivity

The forces that bind protons and neutrons together in the nucleus usually are strong. However, in some isotopes, the nuclei are unstable because the forces binding protons and neutrons together are not strong enough. As a result, the nuclei spontaneously break apart (decay), a process called **radioactivity.**

What happens when unstable nuclei break apart? Three common types of radioactive decay are illustrated in Figure 10.12. and are summarized as follows:

1. Alpha particles (α particles) may be emitted from the nucleus. An alpha particle consists of 2 protons and 2 neutrons. Consequently, the emission of an alpha particle means that the mass number of the isotope is reduced by 4 and the atomic number is decreased by 2.

2. When a beta particle (β particle), or electron, is given off from a nucleus, the mass number remains unchanged, because electrons have practically no mass. However, because the electron has come from a neutron (remember, a neutron is a combination of a proton and an electron), the nucleus contains one more proton than before. Therefore, the atomic number increases by 1.

3. Sometimes an electron is captured by the nucleus. The electron combines with a proton and forms an additional neutron. As in the last example, the mass number remains unchanged. However, as the nucleus now contains one less proton, the atomic number decreases by 1.

An unstable (radioactive) isotope of an element is called the *parent.* The isotopes resulting from the decay of the parent are the *daughter products.* Figure 10.13 provides an example of radioactive decay. Here it can be seen that when the radioactive parent, uranium-238 (atomic number 92, mass number 238), decays, it follows a number of steps, emitting 8 alpha particles and 6 beta particles before finally becoming the stable daughter product lead-206 (atomic number 82, mass number 206). One of the unstable daughter products produced during this decay series is radon. Box 10.2 examines the hazards associated with this radioactive gas.

Certainly among the most important results of the discovery of radioactivity is that it provided a reliable means of calculating the ages of rocks and minerals that contain particular radioactive isotopes. The procedure is called **radiometric dating.** Why is radiometric dating reliable? Because the rates of decay for many isotopes have been precisely measured and do not vary under the physical conditions that exist in Earth's outer layers. Therefore, each radioactive isotope used for dating has been decaying at a fixed rate since the formation of the rocks in which it occurs, and the products of decay have been accumulating at a corresponding rate. For example, when uranium is incorporated into a mineral that crystallizes from magma, there is no lead (the stable daughter product) from previous decay. The radiometric "clock" starts at this point. As the uranium in this newly formed mineral disintegrates, atoms of the daughter product are trapped, and measurable amounts of lead eventually accumulate.

Half-Life

The time required for one half of the nuclei in a sample to decay is called the **half-life** of the isotope. Half-life is a common way of expressing the rate of radioactive disintegration. Figure 10.14 illustrates what occurs when a radioactive parent decays directly into its stable daughter product. When the quantities of parent and daughter are equal (ratio 1:1), we know that one half-life has transpired. When one-quarter of the original parent

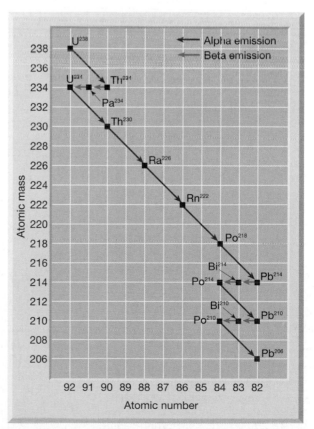

Figure 10.13   The most common isotope of uranium (U-238) is an example of a radioactive decay series. Before the stable end product (Pb-206) is reached, many different isotopes are produced as intermediate steps.

atoms remain and three-quarters have decayed to the daughter product, the parent/daughter ratio is 1:3 and we know that two half-lives have passed. After three half-lives, the ratio of parent atoms to daughter atoms is 1:7 (one parent for every seven daughter atoms).

If the half-life of a radioactive isotope is known and the parent/daughter ratio can be measured, the age of the sample can be calculated. For example, assume that the half-life of a hypothetical unstable isotope is 1 million years and the parent/daughter ratio in a sample is 1:15. Such a ratio indicates that four half-lives have passed and that the sample must be 4 million years old.

Radiometric Dating

Notice that the *percentage* of radioactive atoms that decay during one half-life is always the same: 50 percent. However, the *actual number* of atoms that decay with the passing of each half-life continually decreases. Thus, as the percentage of radioactive parent atoms declines, the proportion of stable daughter atoms rises, with the increase in daughter atoms just matching the drop in parent atoms. This fact is the key to radiometric dating.

Of the many radioactive isotopes that exist in nature, five have proved particularly useful in providing radiometric ages for ancient rocks (Table 10.1). Rubidium-87, thorium-232, and the two isotopes of uranium

---

BOX 10.2

PEOPLE AND THE ENVIRONMENT:
Radon—A Threat to Human Health

Richard L. Hoffman*

Radioactivity is defined as the spontaneous emission of atomic particles and/or electromagnetic waves from unstable atomic nuclei. For example, in a sample of uranium-238, unstable nuclei decay and produce a variety of radioactive progeny or "daughter" products as well as energetic forms of radiation (Table 10.A). One of its radioactive decay products is radon—a colorless, odorless, invisible gas.

Radon gained public attention in 1984 when a worker in a Pennsylvania nuclear power plant set off radiation alarms not when he left work, but when he first arrived. His clothing and hair were contaminated with radon decay products. Investigation revealed that his basement at home had a radon level 2800 times the average level in indoor air. The home was located along a geological formation known as the Reading Prong—a mass of uranium-bearing black rock that runs from near Reading, Pennsylvania, to near Trenton, New Jersey.

Originating in the radio decay of traces of uranium and thorium found in almost all soils, radon isotopes (Rn-222 and Rn-220) are continually renewed in an ongoing, natural process. Geologists estimate that the top six feet of soil from an average acre of land contains about 50 pounds of uranium (about 2 to 3 parts per million); some types of rocks contain more. Radon is

*Dr. Hoffman is Professor of Chemistry, Emeritus, Illinois Central College.

Table 10.A Decay Products of Uranium-238

Some Decay Products of Uranium-238	Decay Particle Produced	Half-Life
Uranium-238	alpha	4.5 billion years
Radium-226	alpha	1600 years
Radon-222	**alpha**	**3.82 days**
Polonium-218	alpha	3.1 minutes
Lead-214	beta	26.8 minutes
Bismuth-214	beta	19.7 minutes
Polonium-214	alpha	1.6×10^{-4} second
Lead-210	beta	20.4 years
Bismuth-210	beta	5.0 days
Polonium-210	alpha	138 days
Lead-206	none	stable

continually generated by the gradual decay of this uranium. Because uranium has a half-life of about 4.5 billion years, radon will be with us forever.

Radon itself decays, having a half-life of only about four days. Its decay products (except lead-206) are all radioactive solids that adhere to dust particles, many of which we inhale. During prolonged exposure to a radon-contaminated environment, some decay will occur while the gas is in the lungs, thereby placing the radioactive radon progeny in direct contact with delicate lung tissue. Steadily accumulating evidence indicates radon to be a significant cause of lung cancer second only to smoking.

A house with a radon level of 4.0 picocuries per liter of air has about eight to nine atoms of radon decaying every minute in every liter of air. The EPA suggests indoor radon levels be kept below this level. EPA risk estimates are conservative; they are based on an assumption that one would spend 75 percent of a 70-year time span (about 52 years) in the contaminated space, which most people would not.

Once radon is produced in the soil, it diffuses throughout the tiny spaces between soil particles. Some radon ultimately reaches the soil surface, where it dissipates into the air. Radon enters buildings and homes through holes and cracks in basement floors and walls. Radon's density is greater than air, so it tends to remain in basements during its short decay cycle.

The source of radon is as enduring as its generation mechanism within Earth; radon will never go away. However, cost-effective mitigation strategies are available to reduce radon to acceptable levels, generally without great expense.

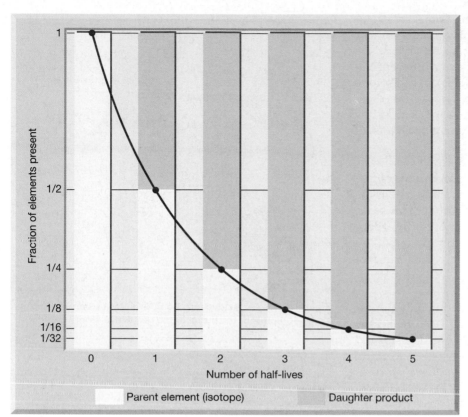

Figure 10.14   The radioactive decay curve shows change that is exponential. Half of the radioactive parent remains after one half-life. After a second half-life one-quarter of the parent remains, and so forth.

are used only for dating rocks that are millions of years old, but potassium-40 is more versatile. Although the half-life of potassium-40 is 1.3 billion years, analytical techniques make possible the detection of tiny amounts of its stable daughter product, argon-40, in some rocks that are younger than 100,000 years.

It is important to realize that an accurate radiometric date can be obtained only if the mineral remained a closed system during the entire period since its formation. A correct date is not possible unless there was neither the addition nor loss of parent or daughter isotopes. This is not always the case. In fact, an important limitation of the potassium-argon method arises from the fact that argon is a gas, and it may leak from minerals, throwing off measurements. Cross-checking of samples, using two different radiometric methods, is done where possible to ensure accurate age determinations.

?

S T U D E N T S   S O M E T I M E S   A S K . . .

With radioactive decay, is there ever
a time that all of the parent material
is converted to the daughter product?

Theoretically, no. During each half-life, half of the parent material is converted to daughter product. Then half again is converted after another half-life, and so on. (Figure 10.14 shows how this logarithmic relationship works—notice that the red line becomes nearly parallel to the horizontal axis after several half-lives.) By converting only half of the remaining parent material to daughter product, there is never a time when all the parent material would be converted. Think about it this

way. If you kept cutting a cake in half and eating only half, would you ever eat all of it? (The answer is no, assuming you had a sharp enough knife to slice the cake at an atomic scale!) However, after many half-lives, the parent material can exist in such small amounts that it is essentially undetectable.

Dating with Carbon-14

To date very recent events, carbon-14 is used. Carbon-14 is the radioactive isotope of carbon. The process is often called **radiocarbon dating.** Because the half-life of carbon-14 is only 5730 years, it can be used for dating events from the historic past as well as those from very recent geologic history. In some cases carbon-14 can be used to date events as far back as 75,000 years.

Carbon-14 is continuously produced in the upper atmosphere as a consequence of cosmic-ray bombardment. Cosmic rays, which are high-energy particles, shatter the

Table 10.1 Radioactive isotopes frequently used in radiometric dating.

Radioactive Parent	Stable Daughter Product	Currently Accepted Half-Life Values
Uranium-238	Lead-206	4.5 billion years
Uranium-235	Lead-207	713 million years
Thorium-232	Lead-208	14.1 billion years
Rubidium-87	Strontium-87	47.0 billion years
Potassium-40	Argon-40	1.3 billion years

nuclei of gas atoms, releasing neutrons. Some of the neutrons are absorbed by nitrogen atoms (atomic number 7), causing their nuclei to emit a proton. As a result, the atomic number decreases by 1 (to 6), and a different element, carbon-14, is created (Figure 10.15A). This isotope of carbon quickly becomes incorporated into carbon dioxide, which circulates in the atmosphere and is absorbed by living matter. As a result, all organisms contain a small amount of carbon-14, including yourself.

While an organism is alive, the decaying radiocarbon is continually replaced, and the proportions of carbon-14 and carbon-12 remain constant. Carbon-12 is the stable and most common isotope of carbon. However, when any plant or animal dies, the amount of carbon-14 gradually decreases as it decays to nitrogen-14 by beta emission (Figure 10.15B). By comparing the proportions of carbon-14 and carbon-12 in a sample, radiocarbon dates can be determined.

Although carbon-14 is useful in dating only the last small fraction of geologic time, it has become a very valuable tool for anthropologists, archaeologists, and historians, as well as for geologists who study very recent Earth history. (Box 10.3 explores another method of studying and dating recent events.) In fact, the de-

Figure 10.15 **A.** Production and **B.** decay of carbon-14. These sketches represent the nuclei of the respective atoms.

BOX 10.3

UNDERSTANDING EARTH:
Using Tree Rings to Date and Study the Recent Past

If you look at the top of a tree stump or at the end of a log, you will see that it is composed of a series of concentric rings. Each of these *tree rings* becomes larger in diameter outward from the center (Figure 10.B). Every year in temperate regions trees add a layer of new wood under the bark. Characteristics of each tree ring, such as size and density, reflect the environmental conditions (especially climate) that prevailed during the year when the ring formed. Favorable growth conditions produce a wide ring; unfavorable ones produce a narrow ring. Trees growing at the same time in the same region show similar tree-ring patterns.

Because a single growth ring is usually added each year, the age of the tree when it was cut can be determined by counting the rings. If the year of cutting is known, the age of the tree and the year in which each ring formed can be determined by counting back from the outside ring.* This procedure can be used to determine the dates of recent geologic events. For example, the minimum number of years since a new land surface was created by a landslide or a flood. The dating and study of annual rings in trees is called *dendrochronology*.

To make the most effective use of tree rings, extended patterns known as ring chronologies are established. They are produced by comparing the patterns of rings among trees in an area. If the same pattern can be identified in two samples, one of which has been dated, the second sample can be dated from the first by matching the ring pattern common to both. This technique, called *cross dating*, is illustrated in Figure 10.C. Tree-ring chronologies extending back for thousands of years

Figure 10.B Each year a growing tree produces a layer of new cells beneath the bark. If the tree is felled and the trunk examined (or if a core is taken, to avoid cutting the tree), each year's growth can be seen as a ring. Because the amount of growth (thickness of a ring) depends upon precipitation and temperature, tree rings are useful records of past climates. (Photo by Stephen J. Krasemann/DRK Photo)

*Scientists are not limited to working with trees that have been cut down. Small, nondestructive core samples can be taken from living trees.

velopment of radiocarbon dating was considered so important that the chemist who discovered this application, Willard F. Libby, received a Nobel prize.

Importance of Radiometric Dating

Bear in mind that although the basic principle of radiometric dating is simple, the actual procedure is quite complex. The analysis that determines the quantities of parent and daughter must be painstakingly precise. In addition, some radioactive materials do not decay directly into the stable daughter product. As you saw in Figure 10.13, uranium-238 produces 13 intermediate unstable daughter products before the fourteenth and final daughter product, the stable isotope lead-206, is produced.

Radiometric dating methods have produced literally thousands of dates for events in Earth history. Rocks from several localities have been dated at more than 3 billion years, and geologists realize that still older rocks exist. For example, a granite from South Africa has been dated at 3.2 billion years, and it contains inclusions of quartzite. (Remember that inclusions are older than the rock containing them.) Quartzite itself is a metamorphic rock that originally was the sedimentary rock sandstone. Sandstone, in turn, is the product of the lithification of sediments produced by the weathering of existing rocks. Thus, we have a positive indication that much older rocks existed.

Radiometric dating has vindicated the ideas of James Hutton, Charles Darwin, and others who inferred that geologic time must be immense. Indeed, modern dating methods have proved that there has been enough time for the processes we observe to have accomplished tremendous tasks.

The Geologic Time Scale

GEODe   Geologic Time
↳ Geologic Time Scale

Geologists have divided the whole of geologic history into units of varying magnitude. Together they comprise the **geologic time scale** of Earth history (Figure 10.16). The major units of the time scale were delineated during the nineteenth century, principally by scientists working in Western Europe and Great Britain. Because radiometric dating was unavailable at that time, the entire time scale was created using methods of rela-

Figure 10.C   Cross dating is a basic principle in dendrochronology. Here it was used to date an archaeological site by correlating tree-ring patterns for wood from trees of three different ages. First, a tree-ring chronology for the area is established using cores extracted from living trees. This chronology is extended further back in time by matching overlapping patterns from older, dead trees. Finally, cores taken from beams inside the ruin are dated using the chronology established from the other two sites.

have been established for some regions. To date a timber sample of unknown age, its ring pattern is matched against the reference chronology.

Tree-ring chronologies are unique archives of environmental history and have important applications in such disciplines as climate, geology, ecology, and archaeology. For example, tree rings are used to reconstruct climate variations within a region for spans of thousands of years prior to human historical records. Knowledge of such long-term variations is of great value in making judgments regarding the recent record of climate change.

In summary, dendrochronology provides useful numerical dates for events in the historic and recent prehistoric past. Moreover, because tree rings are a storehouse of data, they are a valuable tool in the reconstruction of past environments.

tive dating. It was only in the twentieth century that radiometric dating permitted numerical dates to be added.

Structure of the Time Scale

The geologic time scale subdivides the 4.5-billion-year history of Earth into many different units and provides a meaningful time frame within which the events of the geologic past are arranged. As shown in Figure 10.16, **eons** represent the greatest expanses of time. The eon that began about 540 million years ago is the **Phanerozoic,** a term derived from Greek words meaning *visible life.* It is an appropriate description because the rocks and deposits of the Phanerozoic eon contain abundant fossils that document major evolutionary trends.

Another glance at the time scale reveals that the Phanerozoic eon is divided into **eras.** The three eras within the Phanerozoic are the **Paleozoic** (*paleo* = ancient, *zoe* = life), the **Mesozoic** (*meso* = middle,

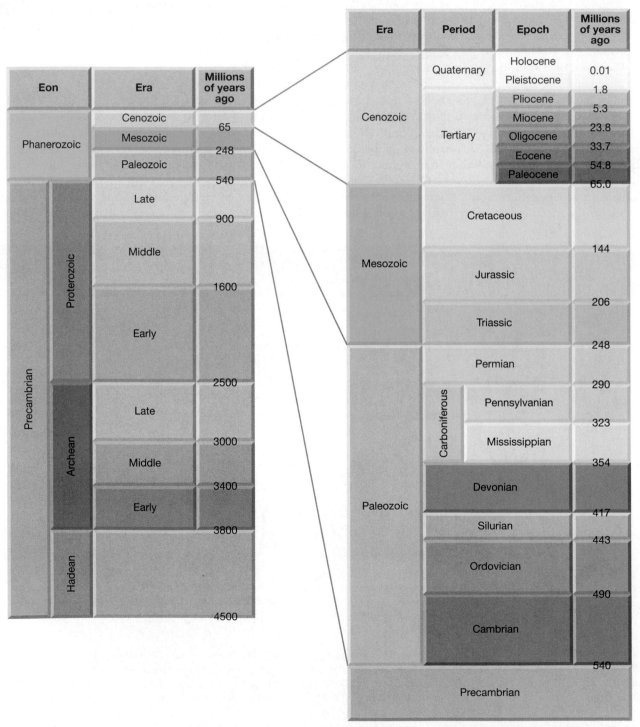

Figure 10.16 The geologic time scale. The numerical dates were added long after the time scale had been established using relative dating techniques. (Data from Geological Society of America)

zoe = life), and the **Cenozoic** (*ceno* = recent, *zoe* = life). As the names imply, the eras are bounded by profound worldwide changes in life forms. Each era is subdivided into **periods.** The Paleozoic has seven, the Mesozoic three, and the Cenozoic two. Each of these 12 periods is characterized by a somewhat less profound change in life forms as compared with the eras.

Finally, periods are divided into still smaller units called **epochs.** As you can see in Figure 10.16, seven epochs have been named for the periods of the Cenozoic. The epochs of other periods, however, are not usually referred to by specific names. Instead, the terms *early, middle,* and *late* are generally applied to the epochs of these earlier periods.

Precambrian Time

Notice that the detail of the geologic time scale does not begin until about 540 million years ago, the date for the beginning of the Cambrian period. The more than 4 billion years prior to the Cambrian is divided into three eons, the *Hadean,* the *Archean,* and the *Proterozoic.* It is also common for this vast expanse of time to simply be referred to as the **Precambrian.** Although it represents about 88 percent of Earth history, the Precambrian is not divided into nearly as many smaller time units as is the Phanerozoic eon.

The quantity of information geologists have deciphered about Earth's past is somewhat analogous to the detail of human history. The further back we go, the less we know. Certainly more data and information exist about the past 10 years than for the first decade of the twentieth century; the events of the nineteenth century have been documented much better than the events of the first century A.D., and so on. Thus it is with Earth history. The more recent past has the freshest, least disturbed, and most ob-

servable record. The further back in time the geologist goes, the more fragmented the record and clues become.

Difficulties in Dating the Geologic Time Scale

Although reasonably accurate numerical dates have been worked out for the periods of the geologic time scale (see Figure 10.16), the task is not without difficulty. The primary problem in assigning numerical dates to units of time is the fact that not all rocks can be dated by radiometric methods. Recall that for a radiometric date to be useful, all minerals in the rock must have formed at about the same time. For this reason, radioactive isotopes can be used to determine when minerals in an igneous rock crystallized and when pressure and heat created new minerals in a metamorphic rock.

However, samples of sedimentary rock can only rarely be dated directly by radiometric means. A sedimentary rock may include particles that contain radioactive isotopes, but the rock's age cannot be accurately determined because the grains making up the rock are not the same age as the rock in which they occur. Rather, the sediments have been weathered from rocks of diverse ages.

Radiometric dates obtained from metamorphic rocks may also be difficult to interpret, because the age of a particular mineral in a metamorphic rock does not necessarily represent the time when the rock initially formed. Instead, the date may indicate any one of a number of subsequent metamorphic phases.

If samples of sedimentary rocks rarely yield reliable radiometric ages, how can numerical dates be assigned to sedimentary layers? Usually the geologist must relate them to datable igneous masses, as in Figure 10.17. In

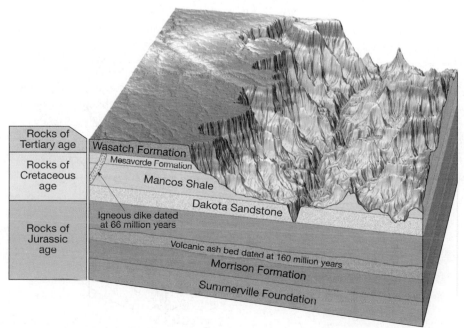

Figure 10.17 Numerical dates for sedimentary layers are usually determined by examining their relationship to igneous rocks. (After U.S. Geological Survey)

this example, radiometric dating has determined the ages of the volcanic ash bed within the Morrison Formation and the dike cutting the Mancos Shale and Mesaverde Formation. The sedimentary beds below the ash are obviously older than the ash, and all the layers above the ash are younger (principle of superposition). The dike is younger than the Mancos Shale and the Mesaverde Formation but older than the Wasatch Formation because the dike does not intrude the Tertiary rocks (cross-cutting relationships).

From this kind of evidence, geologists estimate that a part of the Morrison Formation was deposited about 160 million years ago, as indicated by the ash bed. Further, they conclude that the Tertiary period began after the intrusion of the dike, 66 million years ago. This is one example of literally thousands that illustrates how datable materials are used to *bracket* the various episodes in Earth history within specific time periods. It shows the necessity of combining laboratory dating methods with field observations of rocks.

Chapter Summary

• During the seventeenth and eighteenth centuries, *catastrophism* influenced the formulation of explanations about Earth. Catastrophism states that Earth's landscapes have been developed primarily by great catastrophes. By contrast, *uniformitarianism,* one of the fundamental principles of modern geology advanced by *James Hutton* in the late 1700s, states that the physical, chemical, and biological laws that operate today have also operated in the geologic past. The idea is often summarized as "the present is the key to the past." Hutton argued that processes that appear to be slow-acting could, over long spans of time, produce effects that were just as great as those resulting from sudden catastrophic events.

• The two types of dates used by geologists to interpret Earth history are (1) *relative dates*, which put events in their *proper sequence of formation*, and (2) *numerical dates*, which pinpoint the *time in years* when an event took place.

• Relative dates can be established using the *law of superposition, principle of original horizontality, principle of cross-cutting relationships, inclusions*, and *unconformities.*

• *Correlation,* the matching up of two or more geologic phenomena in different areas, is used to develop a geologic time scale that applies to the entire Earth.

• *Fossils* are the remains or traces of prehistoric life. The special conditions that favor preservation are *rapid burial* and the possession of *hard parts* such as shells, bones, or teeth.

• Fossils are used to *correlate* sedimentary rocks from different regions by using the rocks' distinctive fossil content and applying the *principle of fossil succession*. It states that fossil organisms succeed one another in a definite and determinable order, and therefore any time period can be recognized by its fossil content.

• Each atom has a nucleus containing *protons* (positively charged particles) and *neutrons* (neutral particles). Orbiting the nucleus are negatively charged *electrons*. The *atomic number* of an atom is the number of protons in the nucleus. The *mass number* is the number of protons plus the number of neutrons in an atom's nucleus. *Isotopes* are variants of the same atom, but with a different number of neutrons and hence a different mass number.

• *Radioactivity* is the spontaneous breaking apart (decay) of certain unstable atomic nuclei. Three common types of radioactive decay are (1) emission of alpha particles from the nucleus, (2) emission of beta particles (electrons) from the nucleus, and (3) capture of electrons by the nucleus.

• An unstable *radioactive isotope*, called the *parent*, will decay and form stable *daughter products*. The length of time for half of the nuclei of a radioactive isotope to decay is called the *half-life* of the isotope. If the half-life of the isotope is known and the parent/daugher ratio can be measured, the age of a sample can be calculated.

• The *geologic time scale* divides Earth's history into units of varying magnitude. It is commonly presented in chart form, with the oldest time and event at the bottom and the youngest at the top. The principal subdivisions of the geologic time scale, called *eons*, include the *Hadean, Archean, Proterozoic* (together, these three eons are commonly referred to as the *Precambrian*), and, beginning about 540 million years ago, the *Phanerozoic*. The Phanerozoic (meaning "visible life") eon is divided into the following *eras*: *Paleozoic* ("ancient life"), *Mesozoic* ("middle life"), and *Cenozoic* ("recent life").

• A significant problem in assigning numerical dates to units of time is that *not all rocks can be dated radiometrically*. A sedimentary rock may contain particles of many ages that have been weathered from different rocks that formed at various times. One way geologists assign numerical dates to sedimentary rocks is to relate them to datable igneous masses, such as dikes and volcanic ash beds.

Key Terms

angular unconformity (p. 291)
catastrophism (p. 287)
Cenozoic era (p. 305)
conformable (p. 290)
correlation (p. 291)
cross-cutting relationships, principle of (p. 289)
disconformity (p. 291)

eon (p. 304)
epoch (p. 305)
era (p. 304)
fossil (p. 292)
fossil succession, principle of (p. 297)
geologic time scale (p. 303)
half-life (p. 299)
inclusions (p. 289)

index fossil (p. 297)
Mesozoic era (p. 304)
nonconformity (p. 291)
numerical date (p. 288)
original horizontality, principle of (p. 289)
Paleozoic era (p. 304)
period (p. 305)

Phanerozoic eon (p. 304)
Precambrian (p. 305)
radioactivity (p. 299)

radiocarbon dating (p. 301)
radiometric dating (p. 299)
relative dating (p. 288)

superposition, law of (p. 289)
unconformity (p. 290)
uniformitarianism (p. 287)

Review Questions

1. Contrast catastrophism and uniformitarianism. How did the proponents of each perceive the age of Earth?

2. Distinguish between numerical and relative dating.

3. What is the law of superposition? How are cross-cutting relationships used in relative dating?

4. When you observe an outcrop of steeply inclined sedimentary layers, what principle allows you to assume that the beds became tilted *after* they were deposited?

5. Refer to Figure 10.4 to answer the following questions:
   (a) Is fault A older or younger than the sandstone layer?
   (b) Is dike A older or younger than the sandstone layer?
   (c) Was the conglomerate deposited before or after fault A?
   (d) Was the conglomerate deposited before or after fault B?
   (e) Which fault is older, A or B?
   (f) Is dike A older or younger than the batholith?

6. A mass of granite is in contact with a layer of sandstone. Using a principle described in this chapter, explain how you might determine whether the sandstone was deposited on top of the granite or the granite was intruded from below after the sandstone was deposited.

7. Distinguish among angular unconformity, disconformity, and nonconformity.

8. What is meant by the term *correlation?*

9. List and briefly describe at least five different types of fossils.

10. List two conditions that improve an organism's chances of being preserved as a fossil.

11. Why are fossils such useful tools in correlation?

12. In addition to being important aids in dating and correlating rocks, how else are fossils helpful in geologic investigations?

13. If a radioactive isotope of thorium (atomic number 90, mass number 232) emits 6 alpha particles and 4 beta particles during the course of radioactive decay, what are the atomic number and mass number of the stable daughter product?

14. Why is radiometric dating the most reliable method of dating the geologic past?

15. Assume that a hypothetical radioactive isotope has a half-life of 10,000 years. If the ratio of radioactive parent to stable daughter product is 1:3, how old is the rock containing the radioactive material? What if the ratio were 1:15?

16. To make calculations easier, let us round the age of Earth to 5 billion years.
    (a) What fraction of geologic time is represented by recorded history (assume 5000 years for the length of recorded history)?
    (b) The first abundant fossil evidence does not appear until the beginning of the Cambrian period (approximately 550 million years ago). What percentage of geologic time is represented by abundant fossil evidence?

17. What subdivisions make up the geologic time scale? What is the primary basis for differentiating the eras?

18. Briefly describe the difficulties in assigning numerical dates to layers of sedimentary rock.

Examining the Earth System

1. Figure 10.10A is a large petrified log in Arizona's Petrified Forest National Park. Describe the transition of this tree from being part of the biosphere to being a component of the solid Earth. How might the hydrosphere and/or atmosphere have played a role in the transition?

2. The famous angular unconformity at Scotland's Siccar Point was originally studied by James Hutton in the late

1700s (see Figure 10.7E, p. 293). Can you describe in a general way what occurred to produce this feature? Could all of the spheres of the Earth system have been involved? The Earth system is powered by energy from two sources. How are both sources represented here?

Web Resources

The *Earth Science* Website uses the resources and flexibility of the Internet to aid in your study of the topics in this chapter. Written and developed by Earth science instructors, this site will help improve your understanding of Earth science. Visit **http://www.prenhall.com/tarbuck** and click on the cover of *Earth Science* 10e to find:

- **Online review quizzes.**
- **Web-based critical thinking and writing exercises.**
- **Links to chapter-specific Web resources.**
- **Internet-wide key term searches.**

http://www.prenhall.com/tarbuck

Weathering and erosion have exposed these petrified logs in the Triassic Chinle Formation in Arizona's Petrified Forest National Park. (Photo by David Muench)

CHAPTER 11

EARTH'S HISTORY: A BRIEF SUMMARY

Earth has a long and complex history. The splitting and colliding of continents has resulted in the formation of new ocean basins and the creation of Earth's great mountain ranges. Furthermore, the nature of life on our planet has experienced dramatic changes through time.

Many of the changes on planet Earth occur at a "snail's pace," generally too slow for people to perceive. Thus, human awareness of evolutionary change is fairly recent. Evolution is not confined to life forms, for all of Earth's "spheres" have evolved together: the atmosphere, hydrosphere, solid Earth, and biosphere. Examples are evolutionary changes in the air we breathe, evolution of the world oceans, the rise of mountains, ponderous movements of crustal plates, the comings and goings of vast ice sheets, and the evolution of a vast array of life forms. As each facet of Earth has evolved, it has powerfully influenced the others.

As we saw in Chapter 10, geologists have many tools at their disposal for interpreting the clues about Earth's past. Using these tools, and clues that are contained in the rock record, geologists have been able to unravel many of the complex events of the geological past. The goal of this chapter is to provide a brief overview of the history of our planet and its life forms (Figure 11.1). We will describe how our physical world assumed its present form and how Earth's inhabitants changed through time.*

Precambrian Time: Vast and Enigmatic

The Precambrian encompasses immense geological time, from Earth's distant beginnings 4.5 billion years ago until the start of the Cambrian period, over 4 billion years later. Thus, the Precambrian spans about 88 percent of Earth's history. To get a visual sense of this proportion, look at the right side of Figure 11.2, which shows the relative time span of eras. Our knowledge of this ancient time is sketchy, for much of the early rock record has been obscured by the very Earth processes you have been studying, especially plate tectonics, erosion, and deposition.

Untangling the long, complex Precambrian rock record is a formidable task, and it is far from complete. Most Precambrian rocks are devoid of fossils, which hinders correlation of rocks. Rocks of this great age are metamorphosed and deformed, extensively eroded, and obscured by overlying strata. Consequently, this least-understood span of Earth's history has not been successfully divided into briefer time units, as have later intervals. Indeed, Precambrian history is written in scattered, speculative episodes, like a long book with many missing chapters.

Precambrian Rocks

Looking at Earth from the Space Shuttle, astronauts see plenty of ocean (71 percent) and much less land area (29 percent). Over large expanses of the continents, the orbiting space scientists gaze upon many Paleozoic, Mesozoic, and Cenozoic rock surfaces, but fewer Precambrian surfaces. This demonstrates the law of superposition: Precambrian rocks in these regions are buried from view beneath varying thicknesses of more recent rocks. Here, Precambrian rocks peek through the surface where younger strata are extensively eroded, as in the Grand Canyon and in some mountain ranges. However, on each continent, large core areas of Precambrian rocks dominate the surface, mostly as deformed metamorphic rocks. These areas are called **shields** because they roughly resemble a warrior's shield in shape.

Figure 11.3 shows these shield areas of Precambrian rocks worldwide. In North America (including Greenland), the Canadian Shield encompasses 7.2 million square kilometers (2.8 million square miles), the equivalent of about 10 states of Texas put together.

Much of what we know about Precambrian rocks comes from mining the ores contained in such rocks. The mining of iron, nickel, gold, silver, copper, chromium, uranium, and diamonds has provided Precambrian rock samples for study, and surveys to locate valuable ore deposits have revealed much about the rocks.

Noteworthy are extensive iron-ore deposits. Rocks from the middle Precambrian (1.2 billion to 2.5 billion years ago) contain most of Earth's iron ore, mainly as the mineral hematite (Fe_2O_3). These iron-rich sedimentary rocks probably represent the time when oxygen became sufficiently abundant to react with iron dissolved in shallow lakes and seas. Later, after much of the iron was oxidized and deposited on lake and sea bottoms, formation of these iron-rich deposits declined and oxygen levels in the ocean and atmosphere began to increase. Because most of Earth's free oxygen results from plant photosynthesis, the formation of extensive Precambrian iron-ore deposits is linked to life in the sea.

Notably absent in the Precambrian are fossil fuels (coal, oil, natural gas). The reason is clear—a virtual absence of land plants to form coal swamps and of certain animals to form petroleum. Fossil fuels are from a later time.

Earth's Atmosphere Evolves

Earth's atmosphere is unlike that of any other body in the solar system. No other planet has the same life-sustaining mixture of gases as Earth.

Today, the air you breathe is a stable mixture of 78 percent nitrogen, 21 percent oxygen, about 1 percent

---

*You may want to refer to the section on the "Early Evolution of Earth" in the Introduction (p. 2). It sets the stage by providing a brief but useful summary of the formation of Earth and the other members of our solar system.

Figure 11.1 Paleontologists study ancient life. This researcher is examining the partially cleaned skull of the *Tyrannosaurus rex* known as "Sue" at Chicago's Field Museum of Natural History. The aim of historical geology is to understand the development of Earth and its life through time. Fossils are essential tools in that quest. (Photo by Ira Block)

argon (an inert gas), and trace gases like carbon dioxide and water vapor. But our planet's original atmosphere, several billion years ago, was far different.

Earth's very earliest atmosphere probably was swept into space by the *solar wind*, a vast stream of particles emitted by the Sun. As Earth slowly cooled, a more enduring atmosphere formed. The molten surface solidified into a crust, and gases that had been dissolved in the molten rock were gradually released, a process called **outgassing.** Outgassing continues today from hundreds of active volcanoes worldwide. Thus, geologists hypothesize that Earth's original atmosphere was made up of gases similar to those released in volcanic emissions today: water vapor, carbon dioxide, nitrogen, and several trace gases.

As the planet continued to cool, the water vapor condensed to form clouds, and great rains commenced. At first, the water evaporated in the hot air before reaching the ground, or quickly boiled away upon contacting the surface, just like water sprayed on a hot grill. This accelerated the cooling of Earth's crust. When the surface had cooled below water's boiling point (100°C or 212°F), torrential rains slowly filled low areas, forming the oceans. This reduced not only the water vapor in the air but also the amount of carbon dioxide, for it became dissolved in the water. What remained was a nitrogen-rich atmosphere.

If Earth's primitive atmosphere resulted from volcanic outgassing, we have a problem, because volcanoes do not emit free oxygen. Where did the very significant percentage of oxygen in our present atmosphere (nearly 21 percent) come from?

The major source of oxygen is green plants. Put another way, *life itself* has strongly influenced the composition of our present atmosphere. Plants did not just adapt to their environment; they actually influenced it, dramatically altering the composition of the entire planet's atmosphere by using carbon dioxide and releasing oxygen. This is a good example of how Earth operates as a giant system in which living things interact with their environment.

How did plants come to alter the atmosphere? The key is the way in which plants create their own food. They employ *photosynthesis*, in which they use light energy to synthesize food sugars from carbon dioxide and water. The process releases a waste gas: oxygen. Those of us in the animal kingdom rely on oxygen to metabolize our food, and we in turn exhale carbon dioxide as a waste gas. The plants use this carbon dioxide for more photosynthesis, and so on, in a continuing system.

The first life forms on Earth, probably bacteria, did not need oxygen. Their life processes were geared to the earlier, oxygenless atmosphere. Even today, many *anaerobic* bacteria thrive in environments that lack free oxygen. Later, primitive plants evolved that used photosynthesis and released oxygen. Slowly, the oxygen content of Earth's atmosphere increased. The Precambrian rock record suggests that much of the first free oxygen did not remain free because it combined with (oxidized) other substances dissolved in water, especially iron. Iron has tremendous affinity for oxygen, and the two elements combine to form iron oxides (rust) at any opportunity.

Then, once the available iron satisfied its need for oxygen, substantial quantities of oxygen accumulated in the atmosphere. By the beginning of the Paleozoic era, about 4 billion years into Earth's existence (after seven-eighths of Earth's history had transpired), the fossil record reveals abundant ocean-dwelling organisms that require oxygen to live. Hence, the composition of Earth's atmosphere has evolved together with its life forms, from an oxygenless envelope to today's oxygen-rich environment.

Precambrian Fossils

A century ago, the earliest fossils known dated from the Cambrian period, about 540 million years ago. None were known from the Precambrian. This created a major problem for science at that time: How could complex organisms like trilobites abruptly appear in the geologic record? The answer, or course, was that there were fossils in Precambrian rocks, but they were rare, small, obscure, and simply had not been discovered yet. Today,

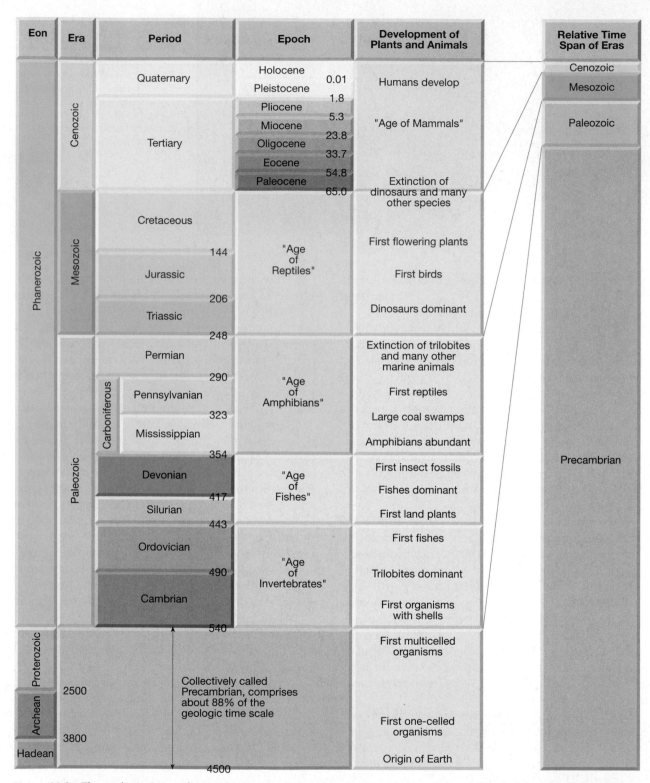

Figure 11.2 The geologic time scale. Numbers on the time scale represent time in millions of years before the present. These dates were added long after the time scale had been established using relative dating techniques. The Precambrian accounts for more than 88 percent of geologic time. (Data from Geological Society of America)

Figure 11.3   Today's continents look very different from the Precambrian. Remnants of Precambrian rocks are the continental shields, composed largely of metamorphosed igneous and sedimentary rocks.

our knowledge of Precambrian life, although far from complete, is quite extensive.

Precambrian fossils are disappointing if you are expecting to see fascinating plants and large animals, for these had not yet evolved. Instead, the most common **Precambrian fossils are stromatolites.** These are distinctively layered mounds or columns of calcium carbonate (Figure 11.4). Stromatolites are not the remains of actual organisms but are material deposited by algae. They are indirect evidence of algae because they closely resemble similar deposits made by modern algae.

Stromatolites did not become common until the middle Precambrian, around 2 billion years ago. Stro-

matolites are large, but most actual organisms preserved in Precambrian rocks are microscopic. Well-preserved remains of many tiny organisms have been discovered, extending the record of life back beyond 3.5 billion years.

Many of these most ancient fossils are preserved in *chert*, a hard, dense chemical sedimentary rock. Chert must be very thinly sliced and studied under powerful microscopes to observe bacteria and algae fossils within it.

Microfossils have been found at several locations worldwide. Two notable areas are in southern Africa, where the rocks date to more than 3.1 billion years, and

A.   **B.**

Figure 11.4   Stromatolites are among the most common Precambrian fossils. **A.** These Precambrian fossil stromatolites from Argentina are more than 2 billion years old. (Photo by Sinclair Stammers/Photo Researchers, Inc.) **B.** Modern stromatolites growing in saline area, Western Australia. (Photo by Bill Bachman/Photo Researchers, Inc.)

in the Gunflint Chert (named for its use in flintlock rifles) of Lake Superior, which dates to 1.7 billion years. In both places, bacteria and blue-green algae have been discovered. The fossils are of the most primitive organisms, *prokaryotes.* Their cells lack organized nuclei, and they reproduce asexually.

More advanced organisms, *eukaryotes,* have cells that contain nuclei. Eukaryotes are among billion-year-old fossils discovered at Bitter Springs in Australia, such as green algae. Unlike prokaryotes, eukaryotes reproduce sexually, which means that genetic material is exchanged between organisms. This reproductive mode permits greatly increased genetic variation. Thus, development of eukaryotes may have dramatically increased the rate of evolutionary change.

Plant fossils date from the middle Precambrian, but animal fossils came a bit later, in the late Precambrian. Many of these fossils are *trace fossils,* meaning that they are not of the animals themselves but of their activities, such as trails and worm holes. Areas in Australia and Newfoundland have yielded hundreds of fossil impressions of soft-bodied creatures. Most, if not all, of the Precambrian fauna lacked shells, which would develop as protective armor during the Paleozoic.

As the Precambrian came to a close, the fossil record disclosed diverse and complete multicelled organisms. This set the stage for more complex plants and animals to evolve at the dawn of the Paleozoic era.

Paleozoic Era: Life Explodes

Following the long Precambrian, the most recent 540 million years of Earth history are divided into three eras: Paleozoic, Mesozoic, and Cenozoic. The Paleozoic era encompasses about 292 million years and is by far the longest of the three. Seven periods make up the Paleozoic era (see Figure 11.2).

Before the Paleozoic, life forms possessed no hard parts—shells, scales, bones, or teeth. The beginning of the Paleozoic is marked by the appearance of the first life forms with hard parts (Figure 11.5). Hard parts greatly enhanced their chance of being preserved as part of the fossil record. Therefore, our knowledge of life's diversification improves greatly from the Paleozoic onward. This diversity is demonstrated in Figure 11.6.

Abundant Paleozoic fossils have allowed geologists to construct a far more detailed time scale for the last one-eighth of geologic time than for the preceding seven-eighths, the Precambrian. Moreover, because every organism is associated with a particular environment, the greatly improved fossil record provided invaluable information for deciphering ancient environments. To facilitate our brief tour of the Paleozoic, we divide it into Early Paleozoic (Cambrian, Ordovician, Silurian periods) and Late Paleozoic (Devonian, Mississippian, Pennsylvanian, Permian periods).

STUDENTS SOMETIMES ASK...

I know that era names refer to "ancient," "middle," and "recent" life. What is the origin of period names?

There is no overall scheme for naming the periods; rather, these names have diverse origins. Several names refer to places that have prominent strata of that age. For example, the Cambrian period is taken from the Roman name for Wales

A.

B.

Figure 11.5 Fossils of common Paleozoic life forms. **A.** Natural cast of a trilobite. Trilobites dominated the Paleozoic ocean, scavenging food from the bottom. **B.** Extinct coiled cephalopods. Like their modern descendants, these were highly developed marine organisms. (Photos by Edward J. Tarbuck)

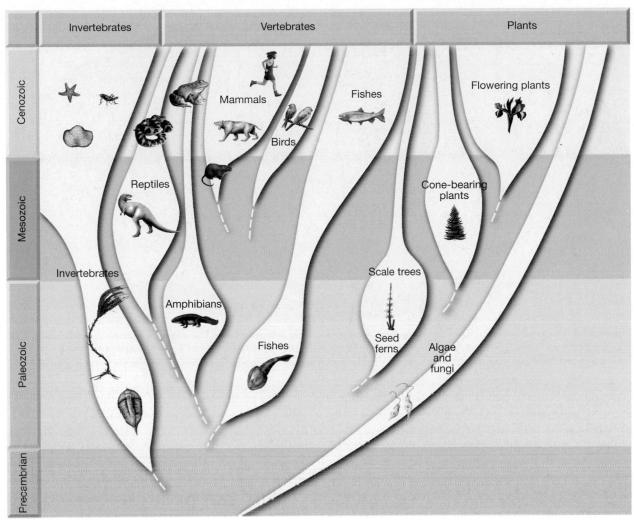

Figure 11.6 This chart indicates the times of appearance and relative abundance of major groups of organisms. The wider the band, the more dominant the group.

(Cambria). The Permian is named for the province of Perm in Russia, while the Jurassic period gets its name from the Jura Mountains located between France and Switzerland.

Early Paleozoic History

The early Paleozoic consists of a 123-million-year span that embraces the Cambrian, Ordovician, and Silurian periods. Anyone approaching Earth from space at this time would have seen the familiar blue planet with plentiful white clouds, but the arrangement of continents would have looked very different from today (Figure 11.7). At this time, the vast southern continent of Gondwanaland encompassed five continents (South America, Africa, Australia, Antarctica, India, and perhaps China). Evidence of an extensive continental glaciation places western Africa near the South Pole!

Landmasses that were not part of Gondwanaland existed as five separate units and some scattered fragments. Although the exact position of these northern continents is uncertain, ancestral North America and Europe are thought to have been near the equator and separated by a narrow sea, as shown on the map in Figure 11.7.

As the Paleozoic opened, North America was a land with no living things, plant or animal. There were no Appalachian or Rocky Mountains; the continent was largely a barren lowland. Several times during the Cambrian and Ordovician periods, shallow seas moved inland and then receded from the interior of the continent. Deposits of clean sandstones, used today to make glass, mark the edge of these shallow seas in the mid-continent.

Early in the Paleozoic, a mountain-building event affected eastern North America from the present-day central Appalachians to Newfoundland. The mountains produced during this event, known as the Taconic Orogeny, have since eroded away, leaving behind deformed strata and a large volume of detrital sedimentary rocks that were derived from the weathering of these mountains.

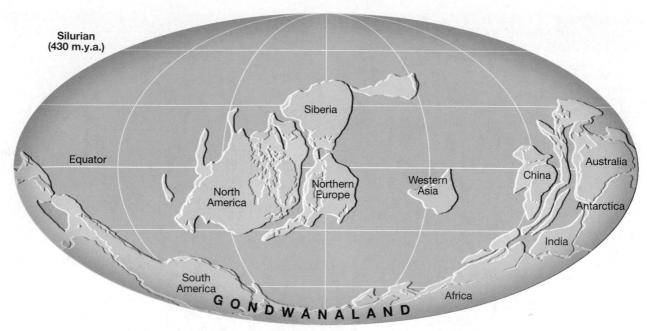

Figure 11.7   Reconstruction of Earth as it may have appeared in early Paleozoic time. The southern continents were joined into a single landmass called Gondwanaland. Four of the five landmasses that would later join to form the northern continent of Laurasia lay scattered roughly along the equator. (After C. Scotese, R. K. Bambach, C. Barton, R. VanderVoo, and A. Ziegler)

During the Silurian period, much of North America was once again inundated by shallow seas. This time large barrier reefs restricted circulation between shallow marine basins and the open ocean. Water in these basins evaporated, causing deposition of large quantities of rock salt and gypsum. Today these thick *evaporite beds* are important resources for the chemical, rubber, plasterboard, and photographic industries in Ohio, Michigan, and western New York State.

Early Paleozoic Life

Life in early Paleozoic time was restricted to the seas. Vertebrates had not yet evolved, so life consisted of several invertebrate groups (see Box 11.1). The Cambrian period was the golden age of *trilobites*. More than 600 genera of these mud-burrowing scavengers flourished worldwide. By Ordovician times, *brachiopods* outnumbered the trilobites. Brachiopods are among the most widespread Paleozoic fossils and, except for one modern group, are now extinct. Although the adults lived attached to the seafloor, the young larvae were free-swimming. This mobility accounts for the group's wide geographic distribution.

The Ordovician also marked the appearance of abundant *cephalopods*—mobile, highly developed mollusks that became the major predators of their time. The descedents of cephalopods include the modern squid, octopus, and nautilus. Cephalopods were the first truly large organisms on Earth (Figure 11.8). Whereas the largest trilobites seldom exceeded 30 centimeters (12 inches) in length and the biggest brachiopods were no more than about 20 centimeters (8 inches) across, one species of cephalopod reached a length of nearly 10 meters (30 feet).

The beginning of the Cambrian period marks an important event in animal evolution. For the first time, organisms appeared that secreted material that formed *hard parts*, such as shells. Why several diverse life forms began to develop hard parts about the same time remains unanswered. One proposal suggests that because an external skeleton provides protection from predators, hard parts evolved for survival. Yet, the fossil record does not seem to support this hypothesis. Organisms with hard parts were plentiful in the Cambrian period, whereas predators such as cephalopods were not abundant until the Ordovician period, some 70 million years later.

Whatever the answer, hard parts clearly served many useful purposes and aided adaptations to new ways of life. Sponges, for example, developed a network of fine interwoven silica spicules that allowed them to grow larger and more erect, capable of extending above the surface in search for food. Mollusks (clams and snails) secreted external shells of calcium carbonate that protected them and allowed body organs to function in a more controlled environment. The successful trilobites developed an exoskeleton of a protein called *chitin*, which permitted them to burrow through soft sediment in search of food (see Figure 11.5A).

Late Paleozoic History

The late Paleozoic consists of four periods—the Devonian, Mississippian, Pennsylvanian, and Permian—that span about 160 million years. Tectonic forces reorganized Earth's landmasses during this time, culminating with the formation of the supercontinent *Pangaea* (Figure 11.9).

BOX 11.1

UNDERSTANDING EARTH:
The Burgess Shale

The possession of hard parts greatly enhances the likelihood of organisms being preserved in the fossil record. Nevertheless, there have been rare occasions in geologic history when large numbers of soft-bodied organisms have been preserved. The Burgess Shale is one well-known example. Located in the Canadian Rockies near the town of Field in southeastern British Columbia, the site was discovered in 1909 by Charles D. Walcott of the Smithsonian Institution.

The Burgess Shale is a site of exceptional fossil preservation and records a diversity of animals found nowhere else (Figure 11.A). The animals of the Burgess Shale lived shortly after the *Cambrian explosion,* a time when there had been a huge expansion of marine biodiversity. Its beautifully preserved fossils represent our most complete and authoritative snapshot of Cambrian life, far better than deposits containing only fossils of organisms with hard parts. To date, more than 100,000 unique fossils have been found.

The animals preserved in the Burgess Shale inhabited a warm shallow sea adjacent to a large reef that was part of the continental margin of North America. During the Cambrian, the North American continent was in the tropics astride the equator. Life was restricted to the ocean, and the land was barren and uninhabited.

What were the circumstances that led to the preservation of the many life forms found in the Burgess Shale? The animals lived in and on underwater mud banks that formed as sediment accumulated on the outer margins of a reef adjacent to a steep escarpment (cliff). Periodically the accumulation of muds became unstable and the slumping and sliding sediments moved down the escarpment as turbidity currents. These flows transported the animals in a turbulent cloud of sediment to the base of the reef where they were buried. Here, in an environment lacking oxygen, the buried carcasses were protected from scavengers and decomposing bacteria. This process occurred again and again, building a thick sequence of fossil-rich sedimentary layers. Beginning about 175 million years ago, mountain-building forces elevated these strata from the seafloor and moved them many kilometers eastward along huge faults to their present location in the Canadian Rockies.

The Burgess Shale is one of the most important fossil discoveries of the twentieth century. Its layers preserve for us an intriguing glimpse of early animal life that is more than a half billion years old.

Figure 11.A   Two examples of Burgess Shale Fossils. *Thaumaptilon walcotti* (at left) was a relatively large (up to 20 centimeters, or 8 inches, long) leaf-like animal. (Photo by National Museum of Natural History) *Aysheaia pedunculata* (on right) was an ancient relative of modern velvet worms and may have clung to soft sponges with tiny hooks on its feet. (Photo by Royal Tyrrell Museum)

Forming Pangaea, the Supercontinent.   As ancestral North America collided with Africa, the narrow sea that separated these landmasses began to close slowly (compare Figure 11.9B and 11.9C). Strong compressional forces from this collision deformed the rocks to produce the original northern Appalachian Mountains of eastern North America.

During the fusion of North America and Africa, the other northern continents began to converge (Figure 11.9). By the Permian period, this newly formed landmass had collided with western Asia and the Siberian landmass along the line of the Ural Mountains. Through this union, the northern continent of *Laurasia* was born, encompassing present-day North America, Europe, western Asia, Siberia, and perhaps China.

As Laurasia was forming, Gondwanaland migrated northward. By the Pennsylvanian period, Gondwanaland collided with Laurasia, forming a mountainous belt through central Europe. Simultaneously, a collision between the African fragment of Gondwanaland and the southeastern edge of North America created the southern Appalachian Mountains.

By the close of the Paleozoic, all the continents had fused into the supercontinent of Pangaea (Figure 11.9). With only a single vast continent, the world's climate became very seasonal, having extremes far greater than those we experience today. As illustrated in Box 11.2, these altered climatic conditions may have contributed to one of the most dramatic biological declines in all of Earth history.

Figure 11.8 During the Ordovician period (490–443 million years ago), the shallow waters of an inland sea over central North America contained an abundance of marine invertebrates. Shown in this reconstruction are straight-shelled cephalopods, trilobites, brachiopods, snails, and corals. (© The Field Museum, Neg. #GEO80820c, Chicago)

Late Paleozoic Life

During most of the late Paleozoic, organisms diversified dramatically. Some 400 million years ago, plants that had adapted to survive at the water's edge began to move inland, becoming *land plants.* These earliest land plants were leafless vertical spikes about the size of your index finger. However, by the end of the Devonian, 40 million years later, the fossil record indicates the existence of forests with trees tens of meters high.

In the oceans, armor-plated fishes that had evolved during the Ordovician continued to adapt. Their armor plates thinned to lightweight scales that increased their speed and mobility (Figure 11.10). Other fishes evolved during the Devonian, including primitive sharks that had a skeleton made of cartilage and bony fishes, the groups to which virtually all modern fishes belong. Because of this, the Devonian period is often called the "age of fishes."

By late Devonian time, two groups of bony fishes—the lung fish and the lobe-finned fish—became adapted to land environments. Not unlike their modern relatives, these fishes had primitive lungs that supplemented their breathing through gills. Lobe-finned fish likely occupied tidal flats or small ponds. In times of drought, they may have used their bony fins to "walk" from dried-up pools in search of other ponds. Through time, the lobe-finned fish began to rely more on their lungs and less on their gills. By late Devonian time, they had evolved into true air-breathing amphibians with fishlike heads and tails. It should be noted that insects had already invaded the land.

Modern amphibians, like frogs, toads, and salamanders, are small and occupy limited biological niches. But conditions during the remainder of the Paleozoic were ideal for these newcomers to the land. Plants and insects, which were their main diet, already were very abundant and large. Having only minimal competition from other land dwellers, the amphibians rapidly diversified. Some groups took on roles and forms that were more similar to modern reptiles, such as crocodiles, than to modern amphibians.

By the Pennsylvanian period, large tropical swamps extended across North America, Europe, and Siberia (Figure 11.11). Trees approached 30 meters (100 feet), with trunks over a meter across. The coal deposits that fueled the Industrial Revolution, and which provide a substantial portion of our electric power today, originated in these vast swamps. Further, it was in the lush coal swamp environment of the late Paleozoic that the amphibians evolved quickly into a variety of species.

Mesozoic Era: Age of the Dinosaurs

Spanning about 183 million years, the Mesozoic era is divided into three periods: the Triassic, Jurassic, and Cretaceous. The Mesozoic era witnessed the beginning

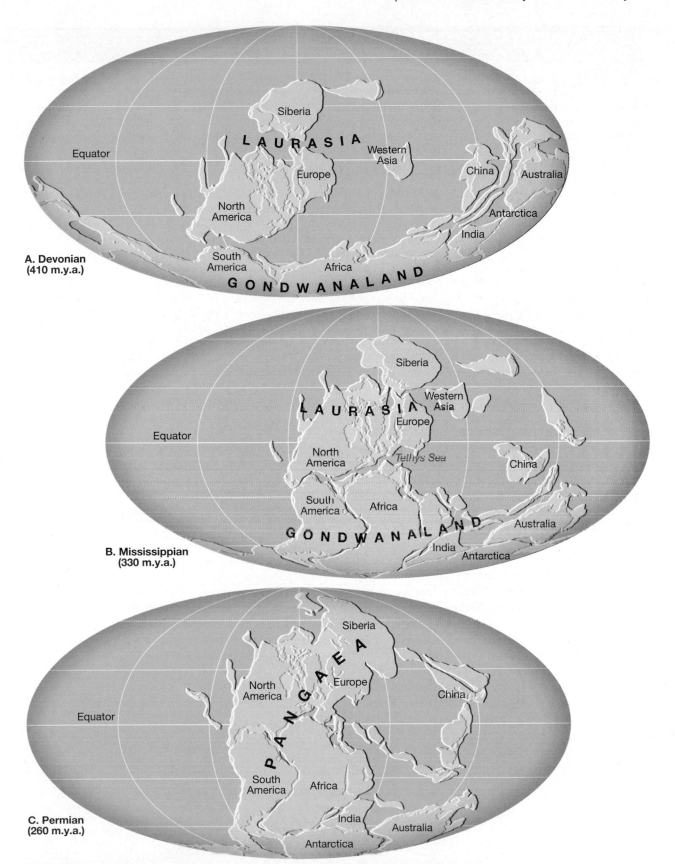

**A. Devonian
(410 m.y.a.)**

**B. Mississippian
(330 m.y.a.)**

**C. Permian
(260 m.y.a.)**

Figure 11.9   During the late Paleozoic, plate movements were joining together the major landmasses to produce the supercontinent of Pangaea. (After C. Scotese, R. K. Bambach, C. Barton, R. VanderVoo, and A. Ziegler)

BOX 11.2

EARTH AS A SYSTEM:
The Great Paleozoic Extinction

The Paleozoic ended with the Permian period, a time when Earth's major landmasses joined to form the supercontinent Pangaea. This redistribution of land and water and changes in the elevations of landmasses brought pronounced changes in world climates. Broad areas of the northern continents became elevated above sea level, and the climate grew drier. These changes apparently triggered extinctions of many species on land and sea.

By the close of the Permian, 75 percent of the amphibian families had disappeared, and plants had declined in number and variety. Although many amphibian groups became extinct, their descendants, the reptiles, would become the most successful and advanced animals on Earth. Marine life was not spared. At least 80 percent, and perhaps as much as 95 percent, of marine life disappeared. Many marine invertebrates that had been dominant during the Paleozoic, including all the remaining trilobites as well as some types of corals and brachiopods, failed to adapt to the widespread environmental changes.

The late Paleozoic extinction was the greatest of at least five mass extinctions to occur over the past 500 million years. Each extinction wreaked havoc with the existing biosphere, wiping out large numbers of species. In each case, however, the survivors formed new biological communities that were more diverse than their predecessors. Thus, mass extinctions actually invigorated life on Earth, as the few hardy survivors eventually filled more niches than the ones left by the victims.

The cause of the great Paleozoic extinction is uncertain. The climate changes from the formation of Pangaea and the associated drop in sea level undoubtedly stressed many species. In addition, at least 2 million cubic kilometers of lava flowed across Siberia to produce what is called the Siberian Traps. Perhaps debris from these eruptions blocked incoming sunlight, or perhaps enough sulfuric acid was emitted to make the seas virtually uninhabitable. Some recent research suggests that an impact with an extraterrestrial body may have contributed. Whatever caused the late Paleozoic extinction, it is clear that without it a very different population of organisms would today inhabit this planet.

---

of the breakup of the supercontinent Pangaea. Also in this era, organisms that had survived the great Permian extinction described in Box 11.2 began to diversify in spectacular ways. On land, dinosaurs became dominant and remained unchallenged for over 100 million years. Because of this, the Mesozoic era is often called the "age of dinosaurs."

Early geologists recognized a profound difference between the fossils in Permian strata and those in younger Triassic rocks, as if someone had drawn a bold line to separate the two time periods. Clearly one half of the fossil groups that occurred in late Paleozoic rocks were missing in Mesozoic rocks. On this basis, it was decided to separate the Paleozoic and Mesozoic at the Permian–Triassic boundary.

Mesozoic History

The Mesozoic era began with much of the world's land above sea level. In fact, in North America no period exhibits a more meager marine sedimentary record than does the Triassic period. Of the exposed Triassic strata, most are red sandstones and mudstones that lack fossils and contain features indicating a terrestrial environment.

As the second period opened, the Jurassic, the sea invaded western North America. Adjacent to this shallow sea, extensive continental sediments were deposited on what is now the Colorado Plateau. The most prominent is the Navajo Sandstone, a windblown, white quartz sandstone that in places approaches a thickness of 300 meters (1000 feet). These massive dunes indicate

Figure 11.10 These placoderms or "plate-skinned" fish were abundant during the Devonian (417–354 million years ago). (Drawing after A. S. Romer)

Figure 11.11 Restoration of a Pennsylvania coal swamp (323–290 million years ago). Shown are scale trees (left), seed ferns (lower left), and scouring rushes (right). Also note the large dragonfly. (© The Field Museum, Neg. #GEO85637, Chicago. Photographer: John Weinstein.)

that a major desert occupied much of the American Southwest during early Jurassic times.

A well-known Jurassic deposit is the Morrison Formation, within which is preserved the world's richest storehouse of dinosaur fossils. Included are fossilized bones of huge dinosaurs such as *Apatosaurus* (formerly *Brontosaurus*), *Brachiosaurus*, and *Stegosaurus*.

As the Jurassic period gave way to the Cretaceous, shallow seas once again invaded much of western North America, the Atlantic, and Gulf coastal regions. This created great swamps like those of the Paleozoic era, forming Cretaceous coal deposits that are very important economically in the western United States and Canada. For example, on the Crow Indian reservation in Montana, there exists nearly 20 billion tons of high-quality coal of Cretaceous age.

A major event of the Mesozoic era was the breakup of Pangaea (Figure 11.12). A rift developed between what is now the eastern United States and western Africa, marking the birth of the Atlantic Ocean. It also represents the beginning of the breakup of Pangaea, a process that continued for 200 million years, through the Mesozoic and into the Cenozoic.

As Pangaea fragmented, the westward-moving North American plate began to override the Pacific plate. This tectonic event began a continuous wave of deformation that moved inland along the entire western margin of the continent. By Jurassic times, subduction of the Pacific plate had begun to produce the chaotic mixture of rocks that exists today in the Coast Ranges of California. Further inland, igneous activity was widespread, and for nearly 60 million years huge masses of magma rose to within a few miles of the surface, where

they cooled and solidified. The remnants of this intrusive activity include the granitic rocks of the Sierra Nevada, the Idaho batholith, and the Coast Range batholith of British Columbia.

Tectonic activity that began in the Jurassic continued throughout the Cretaceous, ultimately forming the vast mountains of western North America (Figure 11.13). Compressional forces moved huge rock units in a shinglelike fashion toward the east. Throughout much of the western margin of North America, older rocks were thrust eastward over younger strata, for a distance exceeding 150 kilometers (90 miles).

Toward the end of the Mesozoic, the middle and southern ranges of the Rocky Mountains formed. This mountain-building event, called the Laramide Orogeny, resulted when tectonic forces uplifted large blocks of Precambrian rocks in what is today Colorado and Wyoming.

Mesozoic Life

As the Mesozoic era dawned, its life forms were survivors of the great Paleozoic extinction. These survivors diversified in many new ways to fill the biological voids created at the close of the Paleozoic. On land, conditions favored those that could adapt to drier climates. Among plants, the gymnosperms were one such group. Unlike the first plants to invade the land, the seed-bearing gymnosperms did not depend on free-standing water for fertilization. Consequently, these plants were not restricted to a life near the water's edge.

The gymnosperms quickly became the dominant trees of the Mesozoic. They included the cycads, the

A. 200 Million Years Ago (Early Jurassic)

B. 150 Million Years Ago (Late Jurassic)

C. 100 Million Years Ago (Cretaceous)

D. 50 Million Years Ago (Tertiary)

Figure 11.12 The breakup of the supercontinent of Pangaea began and continued throughout the Mesozoic era. (After D. Walsh and C. Scotese)

Figure 11.13 Uplifted and deformed sedimentary strata in the Canadian Rockies near Lake Louise, Alberta, Canada. (Photo by Carr Clifton/Minden Pictures)

conifers, and the ginkgoes. The cycads resembled a large pineapple plant. The ginkgoes had fan-shaped leaves, much like their modern relatives. Largest were the conifers, whose modern descendants include the pines, firs, and junipers. The best-known fossil occurrence of these ancient trees is in northern Arizona's Petrified Forest National Park. Here, huge petrified logs lie exposed at the surface, having been weathered from rocks of the Triassic Chinle Formation (see chapter-opening photo).

The Shelled Egg. Among the animals, reptiles readily adapted to the drier Mesozoic environment. They were the first true terrestrial animals. Unlike amphibians, reptiles have shell-covered eggs that can be laid on land. The elimination of a water-dwelling stage (like the tadpole stage in frogs) was an important evolutionary step. Of interest is the fact that the watery fluid within the reptilian egg closely resembles seawater in chemical composition. Because the reptile embryo develops in this watery environment, the shelled egg has been characterized as a "private aquarium" in which the embryos of these land vertebrates spend their water-dwelling stage of life.

Dinosaurs Dominate. With the perfection of the shelled egg, reptiles quickly became the dominant land animals. They continued this dominance for more than 160 million years (Figure 11.14). Most awesome of the Mesozoic reptiles were the dinosaurs. Some of the huge dinosaurs were carnivorous (*Tyrannosaurus*), whereas others were herbivorous (like the ponderous *Apatosaurus*, formerly *Brontosaurus*). The extremely long neck of *Apatosaurus* may have been an adaptation for feeding on tall conifer trees. However, not all dinosaurs were large. In fact,

certain small forms closely resembled modern fleet-footed lizards. Further, evidence indicates that some dinosaurs, unlike their present-day reptile relatives, were warm blooded.

The reptiles made one of the most spectacular adaptive radiations in all of Earth history (Figure 11.15). One group, the pterosaurs, took to the air. These "dragons of

Figure 11.14 Fossil skull of *Sarcosuchus imperator*, or "flesh crocodile emperor." Remains of this large crocodile have been uncovered in the desert of Niger. These animals were river dwellers, indicating that the climate of this region during the Cretaceous period was very different from today. For more about this amazing animal, see the *Students Sometimes Ask* on page 324. (Photo by Project Exploration)

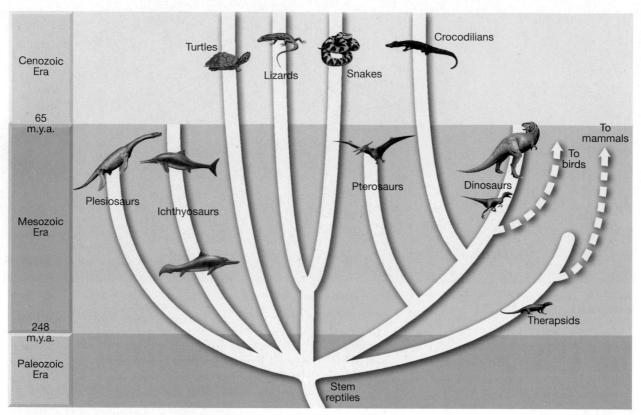

Figure 11.15   Origin and development of the major reptile groups and their descendants. Dinosaurs, pterosaurs (flying reptiles), and large marine reptiles all became extinct by the end of the Mesozoic.

the sky" possessed huge membranous wings that allowed them rudimentary flight (Figure 11.16). Another group of reptiles, exemplified by the fossil *Archaeopteryx,* led to more successful flyers: the birds. Whereas some reptiles took to the skies, others returned to the sea, including the fish-eating plesiosaurs and ichthyosaurs. These reptiles became proficient swimmers but retained their reptilian teeth and breathed by means of lungs.

At the close of the Mesozoic, many reptile groups became extinct. Only a few types survived to recent times, including the turtles, snakes, crocodiles, and lizards. The huge land-dwelling dinosaurs, the marine plesiosaurs, and the flying pterosaurs all are known only through the fossil record. What caused this great extinction? (See Box 11.3.)

STUDENTS SOMETIMES ASK...
Many dinosaurs were very large.
Were they the only large reptiles?

No, indeed. One well-publicized example is a crocodile known as *Sarcosuchus imperator* (see Figure 11.14). This huge river dweller lived in Africa about 110 million years ago (Cretaceous period). At maturity (50–60 years) the animal weighed 8 metric tons (more than 17,600 pounds!) and was about 12 meters (40 feet) long—as long as *Tyrannosaurus rex* and much heavier. Its jaws were roughly as long as an adult human. This animal has appropriately been dubbed "supercroc." Paleontologists indicate that the teeth and jaw suggest a diet of large vertebrates, including fish and dinosaurs.

Figure 11.16   Fossils of the great flying reptile Pteranodon have been recovered from Cretaceaous chalk deposits located in Kansas. Pteranodon had a wingspan of 7 meters (22 feet), but flying reptiles with twice this wingspan have been discovered in strata of west Texas.

BOX 11.3

EARTH AS A SYSTEM:
Demise of the Dinosaurs

Review for a moment the geologic time scale, Figure 11.2. You can see that its divisions represent times of significant geological and/or biological change. Of special interest is the boundary between the Mesozoic era ("middle life") and Cenozoic era ("recent life"), about 65 million years ago. Around this time, more than half of all plant and animal species died out in a *mass extinction*. This boundary marks the end of the era in which dinosaurs and other reptiles dominated the landscape, and the beginning of the era when mammals become very important (Figure 11.B). Because the last period of the Mesozoic is the Cretaceous (abbreviated K to avoid confusion with other "C" periods), and the first period of the Cenozoic is the Tertiary (abbreviated T), the time of this mass extinction is called the *Cretaceous-Tertiary* or *KT boundary*.

The extinction of the dinosaurs is generally attributed to this group's inability to adapt to some radical change in environmental conditions. What event could have triggered the rapid extinction of the dinosaurs—one of the most successful groups of land animals ever to have lived?

The most strongly supported hypothesis proposes that about 65 million years ago a large meteorite about 10 kilometers in diameter collided with Earth (Figure 11.C). The speed at impact was about 70,000 kilometers per hour! The force of the impact vaporized the meteorite and trillions of tons of Earth's crust. Huge quantities of dust and other metamorphosed debris was blasted high into the atmosphere. For months the encircling dust cloud would have greatly restricted the sunlight reaching Earth's surface. With insufficient sunlight for photosynthesis, delicate food chains would have collapsed. Large plant-eating dinosaurs would be affected more adversely than smaller life forms because of the tremendous volume of vegetation they consumed. Next would come the demise of large carnivores. When the sunlight returned, more than half of the species on Earth, including numerous marine organisms, had become extinct.

This period of mass extinction appears to have affected all land animals larger than dogs. Supporters of this catastrophic event scenario suggest that small ratlike mammals were able to survive a breakdown in their food chains lasting perhaps several months. Large dinosaurs, they contend, could not survive this event. The loss of these large reptiles opened habitats for the small mammals that remained. These new habitats, along with evolutionary

Figure 11.C   Employing gravitational measurements, scientists have located a giant impact crater that formed about 65 million years ago and has since been filled with sediment. About 180 kilometers in diameter, Chicxulub crater is regarded by some researchers as the impact site that resulted in the demise of the dinosaurs.

forces, led to the development of the large mammals that occupy our modern world.

What evidence points to such a catastrophic collision 65 million years ago? First, a thin layer of sediment nearly 1 centimeter thick has been discovered at the KT boundary, worldwide. This sediment contains a high level of the element *iridium*, rare in Earth's crust but found in high proportions in stony meteorites. Could this layer be the scattered remains of the meteorite that was responsible for the environmental changes that led to the demise of many reptile groups?

Despite its growing support, some scientists disagree with the impact hypothesis. They suggest instead that huge volcanic eruptions led to the breakdown in the food chain. To support their hypothesis, they cite enormous outpourings of lavas in the Deccan Plateau of northern India about 65 million years ago.

Whatever caused the KT extinction, we now have a greater appreciation of the role of catastrophic events in shaping the history of our planet and the life that occupies it. Could a catastrophic event having similar results occur today? This possibility may explain why an event that occurred 65 million years ago has captured the interest of so many.

Figure 11.B   Dinosaurs dominated the Mesozoic landscape until their extinction at the close of the Cretaceous period. This skeleton of *Tyrannosaurus* stands on display in Chicago's Field Museum of Natural History. (Photo by Don and Pat Valenti/DRK Photo)

Cenozoic Era: Age of Mammals

The Cenozoic era, or "era of recent life," encompasses the past 65 million years of Earth history. It is the "post-dinosaur" era, the time of mammals, including humans. It is during this span that the physical landscapes and life forms of our modern world came into being. The Cenozoic era represents a much smaller fraction of geologic time than either the Paleozoic or the Mesozoic. Although shorter, it nevertheless possesses a rich history, because the completeness of the geologic record improves as time approaches the present. The rock formations of this time span are more widespread and less disturbed than those of any preceding time.

The Cenozoic era is divided into two periods of very unequal duration, the Tertiary period and the Quaternary period. The Tertiary period includes five epochs and embraces about 63 million years, practically all of the Cenozoic era. The Quaternary period consists of two epochs that represent only the last 2 million years of geologic time.

Cenozoic North America

Most of North America was above sea level throughout the Cenozoic era. However, the eastern and western margins of the continent experienced markedly contrasting events, because of their different relationships with plate boundaries. The Atlantic and Gulf coastal regions, far removed from an active plate boundary, were tectonically stable. In contrast, western North America was the leading edge of the North American plate. As a result, plate interactions during the Cenozoic gave rise to many events of mountain building, volcanism, and earthquakes in the West.

Eastern North America.   The stable continental margin of eastern North America was the site of abundant marine sedimentation. The most extensive deposition surrounded the Gulf of Mexico, from the Yucatan Peninsula in present-day Mexico to Florida. Here, the great buildup of sediment caused the crust to downwarp and produced numerous faults. In many instances, the faults created traps in which oil and natural gas accumulated. Today, these and other petroleum traps are the most economically important resource in Cenozoic strata of the Gulf Coast, as evidenced by the Gulf's well-known off-shore drilling platforms.

By early Cenozoic time, most of the original Appalachians had been eroded to a low plain. Then, by the mid-Cenozoic, isostatic adjustments raised the region once again, changing its orientation to base level and rejuvenating its rivers.* Streams eroded with renewed vigor, gradually sculpting the surface into its present-day topography (Figure 11.17). Sediments from all of

this erosion were deposited along the eastern margin of the continent, where they attained a thickness of many kilometers. Today, portions of the strata deposited during the Cenozoic are exposed as the gently sloping Atlantic and Gulf coastal plains.

Western North America.   In the West, the formation of the Rocky Mountains was coming to an end. As erosion lowered the mountains, the basins between uplifted ranges filled with sediments. Eastward, a great wedge of sediment from the eroding Rockies was building the Great Plains.

Beginning in the Miocene epoch, a broad region from northern Nevada into Mexico experienced crustal movements that formed more than 150 fault-block mountain ranges. They rise abruptly above the adjacent basins, creating the Basin and Range Province (see Chapter 9).

As the Basin and Range Province was forming, the entire western interior of the continent was gradually uplifted. This uplift reelevated the Rockies and caused many of the West's major rivers to vigorously downcut. As the rivers became entrenched, many spectacular gorges were formed, including the Grand Canyon of the Colorado River, the Grand Canyon of the Snake River, and the Black Canyon of the Gunnison. The present topography of the Rocky Mountains is in large measure the result of this late Tertiary uplift and the subsequent excavation of the early Tertiary basin deposits by reinvigorated streams.

Volcanic activity was common in the West during much of the Cenozoic. Beginning in the Miocene epoch, great volumes of fluid basaltic lava flowed from fissures in portions of present-day Washington, Oregon, and Idaho. These eruptions built the extensive (1.3 million square kilometers) Columbia Plateau (see Figure 8.19). Immediately west of the Columbia Plateau, volcanic activity was quite different. Here, thicker magmas with a higher silica content erupted explosively, creating the Cascade Range, a chain of stratovolcanoes from northern California to the Canadian border, some of which remain active—like Washington's Mount St. Helens (see Figures 8.1 and 8.30).

A final episode of folding occurred in the West in late Tertiary time, creating the Coast Ranges, which stretch along the Pacific Coast. Meanwhile, the Sierra Nevada became fault-block mountains as they were uplifted along their eastern flank, creating the imposing mountain front we know today.

As the Tertiary period drew to a close, the effects of mountain building, volcanic activity, isostatic adjustments, and extensive erosion and sedimentation had created a physical landscape very similar to the configuration of today. All that remained of Cenozoic time was the final 2-million-year episode called the Quaternary period. During this most recent (and current) phase of Earth history, in which humans evolved, the

---

*To review the concept of base level, see Chapter 4.

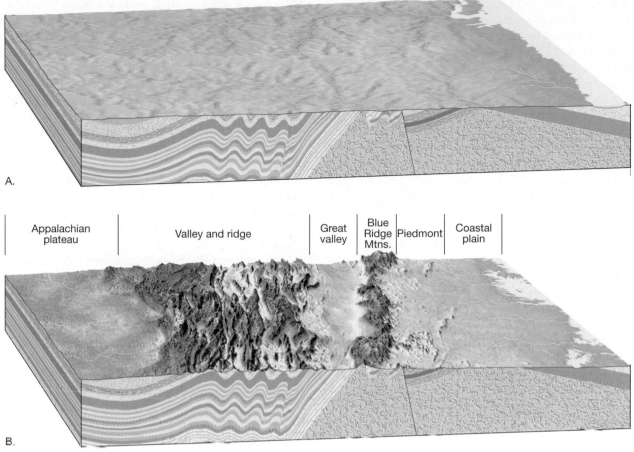

A.

| Appalachian plateau | Valley and ridge | Great valley | Blue Ridge Mtns. | Piedmont | Coastal plain |

B.

Figure 11.17 The formation of the modern Appalachian Mountains. **A.** The original Appalachians eroded to a low plain. **B.** The recent upwarping and erosion of the Appalachians began nearly 30 million years ago to produce the present topography.

action of glacial ice and other erosional agents added the finishing touches.

Cenozoic Life

Mammals replaced reptiles as the dominant land animals in the Cenozoic. Angiosperms (flowering plants with covered seeds) replaced gymnosperms as the dominant land plants. Marine invertebrates took on a modern look. Microscopic animals called *foraminifera* became especially important. Today foraminifera are among the most intensely studied of all fossils because their widespread occurrence makes them invaluable in correlating Tertiary sediments. Tertiary strata are very important to the modern world, for they yield more oil than do rocks of any other age.

The Cenozoic is often called the "age of mammals," because these animals came to dominate land life. It could also be called the "age of flowering plants," for the angiosperms enjoy a similar status in the plant world. As a result of advances in seed fertilization and dispersal, angiosperms experienced rapid development and expansion as the Mesozoic drew to a close. Thus, as the Cenozoic era began, angiosperms were already the dominant land plants.

Development of the flowering plants, in turn, strongly influenced the evolution of both birds and mammals. Birds that feed on seeds and fruits, for example, evolved rapidly during the Cenozoic in close association with the flowering plants. During the middle Tertiary, grasses developed rapidly and spread over the plains. This fostered the emergence of herbivorous (plant-eating) mammals that were mainly grazers. In turn, the development and spread of grazing animals established the setting for the evolution of the carnivorous mammals that preyed upon them.

Mammals Replace Reptiles. Back in the Mesozoic, an important evolutionary event was the appearance of primitive mammals in the late Triassic, about the same time the dinosaurs emerged. Yet throughout the period of dinosaur dominance, mammals remained in the background as small and inconspicuous animals. By the close of the Mesozoic era, dinosaurs and other reptiles no longer dominated the land. It was only after these large reptiles became extinct that mammals came into their own as the dominant land animals. The transition is a major example in the fossil record of the replacement of one large group by another.

Mammals are distinct from reptiles in important respects. Mammalian young are born live, and mammals maintain a steady body temperature—that is, they are "warm-blooded." This latter adaptation allowed mammals to lead more active and diversified lives than reptiles because they could survive in cold regions and search for food during any season or time of day. Other mammalian adaptations included the development of insulating body hair and more efficient heart and lungs.

It is worth noting that perfect boundaries cannot be drawn between mammalian and reptilian traits. One minor group of animals, the monotremes, still lay eggs. The two species in this group, the duck-billed platypus and the spiny anteater, are found only in Australia. Moreover, although modern reptiles are "cold-blooded," some paleontologists believe that dinosaurs may have been "warm-blooded."

With the demise of most Mesozoic reptiles, Cenozoic mammals diversified rapidly. The many forms that exist today evolved from small primitive mammals that were characterized by short legs, flat five-toed feet, and small brains. Their development and specialization took four principal directions: (1) increase in size, (2) increase in brain capacity, (3) specialization of teeth to better accommodate a particular diet, and (4) specialization of limbs to better equip the animal for life in a particular environment.

Marsupials, Placentals, and Diversity. Following the reptilian extinctions at the close of the Mesozoic, two groups of mammals, the marsupials and the placentals, evolved and expanded to dominate the Cenozoic. The groups differ principally in their modes of reproduction. Young marsupials are born live but at a very early stage of development. After birth, the tiny and immature young crawl into the mother's external stomach pouch to complete their development. Examples are kangaroos, opossums, and koala bears.

Placental mammals, in contrast, develop within the mother's body for a much longer period, so that birth occurs after the young are relatively mature and independent. Most mammals are placental, including humans.

Today marsupials are found primarily in Australia, where they went through a separate evolutionary expansion during the Cenozoic, largely isolated from placental mammals.

In South America, both primitive marsupials and placentals coexisted before that landmass became completely isolated during the breakup of Pangaea. Evolution and specialization of both groups continued undisturbed for approximately 40 million years until the close of the Pliocene epoch, when the Central American land bridge emerged, connecting the two American

A.

B.

Figure 11.18 La Brea tar pits in Los Angeles, California. **A.** Fossils being excavated from the thick tar in 1914. **B.** Skeleton of an extinct saber-toothed cat. These bones came from the La Brea tar pits. (Photo courtesy of The George C. Page Museum)

continents. Then an invasion of advanced carnivores from North America brought the extinction of many hoofed mammals that had persisted in South America for millions of years. The marsupials, except for opossums, also could not compete and became extinct. Both Australia and South America provide excellent examples of how isolation caused by the separation of continents increased the diversity of animals in the world.

Large Mammals and Extinction.   As we have seen, mammals diversified quite rapidly during the Cenozoic era. One tendency was for some groups to became very large. For example, by the Oligocene epoch a hornless rhinoceros that stood nearly 5 meters (16 feet) high had evolved. It is the largest land mammal known to have existed. As time approached the present, many other types evolved to a large size as well—more, in fact, than now exist. Many of these large forms were common as recently as 11,000 years ago. However, a wave of late Pleistocene extinctions rapidly eliminated these animals from the landscape.

In North America, the mastodon and mammoth, both huge relatives of the elephant, became extinct. In addition, saber-toothed cats, giant beavers, large ground sloths, horses, camels, giant bison, and others died out

(Figure 11.18). In Europe, late Pleistocene extinctions included woolly rhinos, large cave bears, and the Irish elk. The reason for this recent wave of large animal extinctions puzzles scientists. These animals had survived several major glacial advances and interglacial periods, so it is difficult to ascribe these extinctions to climatic change. Some scientists believe that early humans hastened the decline of these mammals by selectively hunting large forms. Although this hypothesis is preferred by many, it is not yet accepted by all.

STUDENTS SOMETIMES ASK...
What are the La Brea tar pits?

The La Brea tar pits, located in downtown Los Angeles, are famous because they contain a rich and very well preserved assemblage of Pleistocene vertebrate fossils (see Figure 11.18). These organisms roamed southern California from 40,000 to 8000 years ago. The collection includes 59 species of mammals and more than 130 species of birds. Hundreds of invertebrate and plant fossils are also preserved. Tar pits form when crude oil seeps to the surface and the light portion evaporates, leaving behind sticky pools of heavy tar.

Chapter Summary

• The *Precambrian* spans about 88% of Earth history, beginning with the formation of Earth about 4.5 billion years ago and ending 540 million years ago with the diversification of life that marks the start of the Paleozoic era. It is the least understood span of Earth's history because most Precambrian rocks are buried from view. However, on each continent there is a "core area" of Precambrian rocks called the *shield*. The iron-ore deposits of Precambrian age represent the time when oxygen became abundant in the atmosphere and combined with iron to form iron oxide.

• Earth's primitive atmosphere consisted of such gases as water vapor, carbon dioxide, nitrogen, and several trace gases that were released in volcanic emissions, a process called *outgassing*. The first life forms on Earth, probably *anaerobic bacteria*, did not require oxygen. As life evolved, plants, through the process of *photosynthesis,* used carbon dioxide and water and released oxygen into the atmosphere. Once the available iron on Earth was oxidized (combined with oxygen), substantial quantities of oxygen accumulated in the atmosphere. About 4 billion years into Earth's existence, the fossil record reveals abundant ocean-dwelling organisms that require oxygen to live.

• The most common middle Precambrian fossils are *stromatolites*. Microfossils of bacteria and blue-green algae, both primitive *prokaryotes* whose cells lack organized nuclei, have been found in chert, a hard, dense, chemical sedimentary rock in southern Africa (3.1 billion years of age) and near Lake Superior (1.7 billion years of age). *Eukaryotes,* with cells containing organized nuclei, are among billion-year-old fossils discovered in Australia. Plant fossils date from the middle

Precambrian, but animal fossils came a bit later, in the late Precambrian. Many of these fossils are *trace fossils,* and not of the animals themselves.

• The *Paleozoic era* extends from 540 million years ago to about 248 million years ago. The beginning of the Paleozoic is marked by the *appearance of the first life forms with hard parts,* such as shells. Therefore, abundant Paleozoic fossils occur, and a far more detailed record of Paleozoic events can be constructed. During the early Paleozoic (the Cambrian, Ordovician, and Silurian periods) the vast southern continent of *Gondwanaland* existed. Seas inundated and receded from North America several times, leaving thick evaporite beds of rock salt and gypsum. Life in the early Paleozoic was restricted to the seas and consisted of several invertebrate groups. During the late Paleozoic (the Devonian, Mississippian, Pennsylvanian, and Permian periods), ancestral North America collided with Africa to produce the original northern Appalachian Mountains, and the northern continent of *Laurasia* formed. By the close of the Paleozoic, all the continents had fused into the supercontinent of *Pangaea.* During most of the Paleozoic, organisms diversified dramatically. Insects and plants moved onto the land, and amphibians evolved and diversified quickly. By the Pennsylvanian period, large tropical swamps, which became the major coal deposits of today, extended across North America, Europe, and Siberia. At the close of the Paleozoic, altered climatic conditions caused one of the most dramatic biological declines in all of Earth history.

• The *Mesozoic era,* often called the *"age of dinosaurs,"* began about 248 million years ago and ended approximately 65 million years ago. Early in the Mesozoic much of the land was above sea level. However, by the middle Mesozoic, seas invaded western North America. As Pangaea began to break up, the westward-moving North American plate began to override the Pacific plate, causing crustal deformation along the entire western margin of the continent. Organisms that had survived extinction at the end of the Paleozoic began to diversify in spectacular ways. *Gymnosperms* (cycads, conifers, and ginkgoes) quickly became the dominant trees of the Mesozoic because they could adapt to the drier climates. Reptiles quickly became the dominant land animals, with one group eventually becoming the birds. The most awesome of the Mesozoic reptiles were the *dinosaurs.* At the close of the Mesozoic, many reptile groups, including the dinosaurs, became extinct.

• The *Cenozoic era,* or *"era of recent life,"* began approximately 65 million years ago and continues today. It is *the time of mammals,* including humans. The widespread, less disturbed rock formations of the Cenozoic provide a rich geologic record. Most of North America was above sea level throughout the Cenozoic. Because of their different relations with tectonic plate boundaries, the eastern and western margins of the North American continent experienced contrasting events. The stable eastern margin was the site of abundant sedimentation as isostatic adjustment raised the eroded Appalachians, causing the streams to downcut with renewed vigor and deposit their sediment along the continental margin. In the west, building of the Rocky Mountains was coming to an end, the Basin and Range Province was forming, and volcanic activity was extensive. The Cenozoic is often called *"the age of mammals"* because these animals replaced the reptiles as the dominant land life. Two groups of mammals, the *marsupials* and the *placentals,* evolved and expanded to dominate the era. One tendency was for some mammal groups to become very large. However, a wave of late *Pleistocene* extinctions rapidly eliminated these animals from the landscape. Some scientists believe that humans hastened the decline of these animals by selectively hunting the larger species. The Cenozoic could also be called the *"age of flowering plants."* As a source of food, flowering plants strongly influenced the evolution of both birds and herbivorous (plant-eating) mammals throughout the Cenozoic era.

Key Terms

outgassing (p. 311)   shields (p. 310)   stromatolites (p. 313)

Review Questions

1. Explain why Precambrian history is more difficult to decipher than more recent geological history.

2. What is the major source of free oxygen in Earth's atmosphere?

3. Match the following words and phrases to the most appropriate time span. Select among the following: Precambrian, early Paleozoic, late Paleozoic, Mesozoic, Cenozoic.
   (a) Pangaea came into existence.
   (b) Bacteria and blue-green algae preserved in chert.
   (c) The era that encompasses the least amount of time.
   (d) Shields.
   (e) Age of dinosaurs.
   (f) Formation of the original northern Appalachian mountains.
   (g) Mastodons and mammoths.
   (h) Extensive deposits of rock salt.
   (i) Triassic, Jurassic, and Cretaceous.
   (j) Coal swamps extended across North America, Europe, and Siberia.
   (k) Gulf Coast oil deposits formed.
   (l) Formation of most of the world's major iron-ore deposits.

   (m) Massive sand dunes covered large portions of the Colorado Plateau region.
   (n) The "age of fishes" occurred during this span.
   (o) Cambrian, Ordovician, and Silurian.
   (p) Pangaea began to break apart.
   (q) "Age of mammals."
   (r) Animals with hard parts first appeared in abundance.
   (s) Gymnosperms were the dominant trees.
   (t) Columbia Plateau formed.
   (u) Stromatolites are among its more common fossils.
   (v) "Golden age of trilobites" occurred during this span.
   (w) Fault-block mountains form in the Basin and Range Province.

4. Briefly discuss two proposals that attempt to explain why several groups developed hard parts at the beginning of the Cambrian period. Do these proposals appear to provide a satisfactory explanation? Why or why not?

5. List some differences between amphibians and reptiles. List differences between reptiles and mammals.

6. Describe a hypothesis that attempts to explain the extinction of the dinosaurs.

7. Contrast the eastern and western margins of North America during the Cenozoic era in terms of their relationships to plate boundaries.

Examining the Earth System

1. The Earth system has been responsible for both the conditions that favored the evolution of life on this planet and for the mass extinctions that have occurred throughout geological time. Describe the role of the biosphere, hydrosphere, and solid Earth in forming the current level of atmospheric oxygen. How did Earth's near-space environment interact with the atmosphere and biosphere to contribute to the great mass extinction that marked the end of the dinosaurs?

2. Most of the vast North American coal resources located from Pennsylvania to Illinois began forming during the Pennsylvanian and Mississippian periods of Earth history. (This time period is also referred to as the Carboniferous period.) Using Figure 11.11, a restoration of a Pennsylvanian period coal swamp, describe the climatic and biological conditions associated with this unique environment. Next, examine Figure 11.9B, C. The maps show the formation of the supercontinent of Pangaea and illustrate the geographic position of North America during the period of coal formation. Where, relative to the equator, was North America located during the time of coal formation? What role did plate tectonics play in determining the conditions that eventually produced North America's eastern coal reserves? Why is it unlikely that the coal-forming environment will repeat itself in North America in the near future? (You may find it helpful to visit the Illinois State Museum Mazon Creek Fossils exhibit at **http://www.museum.state.il.us/exhibits/mazon_creek**, and/or the University of California Time Machine Exhibit at **http://www.ucmp.berkeley.edu/help/timeform.html**.

Web Resources

The *Earth Science* Website uses the resources and flexibility of the Internet to aid in your study of the topics in this chapter. Written and developed by Earth science instructors, this site will help improve your understanding of Earth science. Visit **http://www.prenhall.com/tarbuck** and click on the cover of *Earth Science* 10e to find:

- **Online review quizzes.**
- **Web-based critical thinking and writing exercises.**
- **Links to chapter-specific Web resources.**
- **Internet-wide key term searches.**

http://www.prenhall.com/tarbuck

A view of the North Atlantic near Cape Canaveral, Florida. (Photo by Mark Muench)

CHAPTER 12

THE OCEAN FLOOR*

*This chapter was revised by Professor Alan P. Trujillo, Palomar College.

How deep is the ocean? How much of Earth is covered by the global sea? What does the seafloor look like? Humans have long been interested in finding answers to these questions, but it was not until rather recently that these seemingly simple questions could be answered. Suppose, for example, that all of the water were drained from the ocean. What would we see? Plains? Mountains? Canyons? Plateaus? Indeed, the ocean conceals all of these features, and more. And what about the carpet of sediment that covers much of the seafloor? Where did it come from, and what can be learned by examining it? This chapter provides answers to these questions.

Calling Earth the "water planet" is certainly appropriate, because nearly 71 percent of its surface is covered by the global ocean. Although the ocean comprises a much greater percentage of Earth's surface than the continents, it has only been in the relatively recent past that the ocean became an important focus of study. Recently there has been a virtual explosion of data about the oceans, and with it oceanography has grown dramatically. **Oceanography** is an interdisciplinary science that draws on the methods and knowledge of geology, chemistry, physics, and biology to study all aspects of the world ocean.

The Vast World Ocean

A glance at a globe or a view of Earth from space reveals a planet dominated by an interconnected single world ocean (Figure 12.1). Indeed, it is for this reason that Earth is often referred to as the *blue planet.*

Geography of the Oceans

The area of Earth is about 510 million square kilometers (197 million square miles). Of this total, approximately 360 million square kilometers (140 million square miles), or 71 percent, is represented by oceans and marginal seas (meaning seas around the ocean's margin, like the Mediterranean Sea and Caribbean Sea). Continents and islands comprise the remaining 29 percent, or 150 million square kilometers (58 million square miles).

Even though oceans dominate Earth's surface, is the distribution of land and water similar in the Northern and Southern hemispheres? By studying a world map or globe (Figure 12.1), it is readily apparent that the continents and oceans are not evenly divided between the two hemispheres. In the Northern Hemisphere, for instance, nearly 61 percent of the surface is water, whereas about 39 percent is land. In the Southern Hemisphere, on the other hand, almost 81 percent of the surface is water, and only 19 percent is land. It is no wonder then that the Northern Hemisphere is called the *land hemisphere* and the Southern Hemisphere the *water hemisphere.*

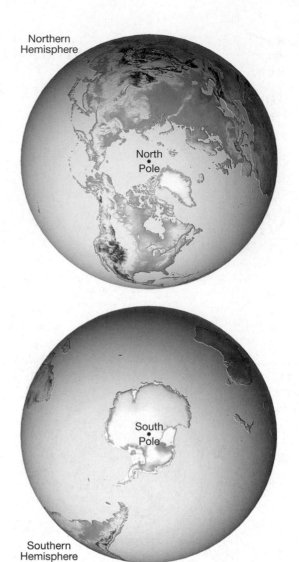

Figure 12.1 These views of Earth show the uneven distribution of land and water between the Northern and Southern hemispheres. Almost 81 percent of the Southern Hemisphere is covered by the oceans—20 percent more than the Northern Hemisphere.

Figure 12.2A shows the distribution of land and water in the Northern and Southern hemispheres. Between latitudes 45 degrees north and 70 degrees north, there is actually more land than water, whereas between 40 degrees south and 65 degrees south there is almost no land to interrupt the oceanic and atmospheric circulation.

The world ocean can be divided into four main ocean basins (Figure 12.2B):

1. The *Pacific Ocean,* which is the largest ocean (and the largest single geographic feature on the planet), covering over half of the ocean surface area on Earth. In fact, the Pacific Ocean is so large that all of the continents could fit into the space occupied by it—with room left over! It is also the world's deepest ocean, with an average depth of 3940 meters (12,927 feet).

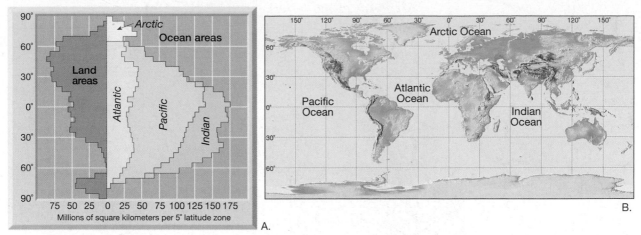

Figure 12.2 Distribution of land and water. **A.** The graph shows the amount of land and water in each 5-degree latitude belt. **B.** The world map provides a more familiar view.

2. The *Atlantic Ocean*, which is about half the size of the Pacific Ocean but is not quite as deep. It is a relatively narrow ocean as compared to the Pacific and is bounded by almost parallel continental margins.

3. The *Indian Ocean*, which is slightly smaller than the Atlantic Ocean but has about the same average depth. Unlike the Pacific and Atlantic oceans, it is largely a Southern Hemisphere water body.

4. The *Arctic Ocean*, which is about 7 percent the size of the Pacific Ocean and is only a little more than one-quarter as deep as the rest of the oceans.

Comparing the Oceans to the Continents

A major difference between continents and the ocean basins is their relative levels. The average elevation of the continents above sea level is about 840 meters (2756 feet), whereas the average depth of the oceans is nearly 4 and a half times this amount—3729 meters (12,234 feet). The volume of ocean water is so large that if Earth's solid mass were perfectly smooth (level) and spherical, the oceans would cover Earth's entire surface to a uniform depth of more than 2000 meters (1.2 miles)!

Mapping the Ocean Floor

 Oceans
 ↳ Floor of the Oceans

If all water were drained from the ocean basins, a great variety of features would be seen, including linear chains of volcanoes, tall mountain ranges, deep canyons, rift valleys, and large submarine plateaus. In fact, the scenery would be nearly as diverse as that on the continents.

An understanding of seafloor features came with the development of techniques to measure the depth of the oceans. **Bathymetry** (*bathos* = depth, *metry* = measurement) is the measurement of ocean depths and

the charting of the shape or topography of the ocean floor.

Bathymetric Techniques

The first understanding of the ocean floor's varied topography did not unfold until the historic three-and-a-half-year voyage of the HMS *Challenger* (Figure 12.3). From December 1872 to May 1876, the *Challenger* expedition made the first—and perhaps still most comprehensive—study of the global ocean ever attempted by one agency. The 127,500-kilometer (79,200-mile) trip took the ship and its crew of scientists to every ocean except the Arctic. Throughout the voyage, they sampled various ocean properties, including water depth, which was accomplished by laboriously lowering a long, weighted line overboard. Not many years later the knowledge gained by the *Challenger* of the ocean's great depth and varied topography was further expanded with the laying of transatlantic communication cables, especially in the North Atlantic Ocean. However, as long as a weighted line was the only way to measure ocean depths, knowledge of seafloor features remained limited.

In the 1920s a technological breakthrough occurred with the invention of electronic depth-sounding equipment (**echo sounder**). (The echo sounder is also referred to as *sonar*, an acronym for *so*und *na*vigation *a*nd *r*anging.) The echo sounder works by transmitting sound waves toward the ocean bottom (Figure 12.4A). A sensitive receiver intercepts the echo reflected from the bottom, and a clock precisely measures the time interval to fractions of a second. By knowing the speed of sound waves in water—about 1500 meters (4900 feet) per second—and the time required for the energy pulse to reach the ocean floor and return, depth can be calculated. The depths determined from continuous monitoring of these echoes are plotted so that a profile of the ocean floor is obtained. By laboriously combining profiles from several adjacent traverses, a chart of the seafloor can be produced.

Figure 12.3 The H.M.S. *Challenger.* (From C.W. Thomson and Sir John Murray, *Report on the Scientific Results of the Voyage of the H.M.S. Challenger,* Vol. 1. Great Britain: Challenger Office, 1895, Plate 1)

In the last few decades, researchers have designed even more sophisticated echo sounders to map the ocean floor. In contrast to simple echo sounders, *multibeam sonar* employs an array of sound sources and listening devices. Thus, rather than obtaining the depth of a single point every few seconds, this technique obtains a profile of a narrow strip of seafloor (Figure 12.4B). These profiles are recorded every few seconds as the research vessel advances, building a continuous swath of relatively detailed coverage. When a ship uses multibeam sonar to make a map of a section of seafloor, it travels through the area in a regularly spaced back-and-forth pattern—called, appropriately enough, "mowing the lawn."

Despite their greater efficiency and enhanced detail, research vessels equipped with multibeam sonar travel at a mere 10 to 20 kilometers (6 to 12 miles) per hour. It would take at least 100 vessels outfitted with this equipment hundreds of years to map the entire seafloor. This explains why only about 5 percent of the seafloor has been mapped in detail—and why large areas of the seafloor have not yet been mapped with sonar at all.

Figure 12.4 Echo sounders. **A.** An echo sounder determines the water depth by measuring the time interval required for an acoustic wave to travel from a ship to the seafloor and back. The speed of sound in water is 1500 m/sec. Therefore, depth = 1/2 (1500 m/sec x echo travel time). **B.** Modern multibeam sonar obtains a profile of a narrow swath of seafloor every few seconds.

Viewing the Ocean Floor from Space

Another technological breakthrough that has led to an enhanced understanding of the seafloor involves measuring the shape of the ocean surface from space. After compensating for waves, tides, currents, and atmospheric effects, it was discovered that the ocean surface is not perfectly flat. This is because gravity attracts water toward regions where massive seafloor features occur. Therefore, mountains and ridges produce elevated areas on the ocean surface, and, conversely, canyons and trenches cause slight depressions. Satellites equipped with *radar altimeters* are able to measure these subtle differences by bouncing microwaves off the sea surface (Figure 12.5). These devices can measure variations as small as 3 to 6 centimeters. Such data have added greatly to the knowledge of ocean-floor topography. Cross-checked with traditional sonar depth measurements, the data are used to produce detailed ocean-floor maps, such as the one shown in Figure 12.6.

Provinces of the Ocean Floor

Oceanographers studying the topography of the ocean floor have delineated three major units: *continental margins,* the *ocean basin floor,* and the *oceanic (mid-ocean) ridge.* The map in Figure 12.7 outlines these provinces

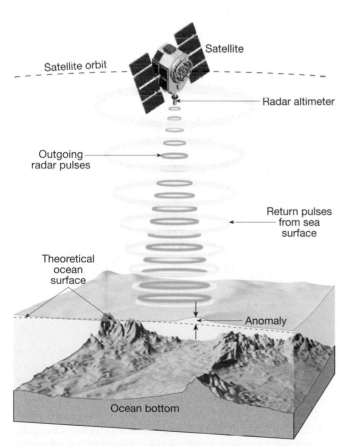

Figure 12.5   A satellite altimeter measures the variation in sea surface elevation, which is caused by gravitational attraction and mimics the shape of the seafloor. The sea surface anomaly is the difference between the measured and theoretical ocean surface.

for the North Atlantic Ocean, and the profile at the bottom of the illustration shows the varied topography. Such profiles usually have their vertical dimension exaggerated many times—40 times in this case—to make topographic features more conspicuous. Vertical exaggeration, however, makes slopes shown in seafloor profiles appear to be *much* steeper than they actually are.

Continental Margins

 Oceans
  ↳ Floor of the Oceans

Two main types of **continental margins** have been identified—*passive* and *active.* Passive margins are found along most of the coastal areas that surround the Atlantic Ocean, including the east coasts of North and South America, as well as the coastal areas of Western Europe and Africa. Passive margins are *not* associated with plate boundaries and therefore experience very little volcanism and few earthquakes. Here, weathered materials eroded from the adjacent landmass accumulate to form a thick, broad wedge of relatively undisturbed sediments.

By contrast, active continental margins occur where oceanic lithosphere is being subducted beneath the edge of a continent. The result is a relatively narrow margin, consisting of highly deformed sediments that were scraped from the descending lithospheric slab. Active continental margins are common around the Pacific Rim, where they parallel deep oceanic trenches.

Passive Continental Margins

The features comprising a **passive continental margin** include the continental shelf, the continental slope, and the continental rise (Figure 12.8).

Continental Shelf.   The **continental shelf** is a gently sloping submerged surface extending from the shoreline toward the ocean basin floor. Because it is underlain by continental crust, it is clearly a flooded extension of the continents.

The continental shelf varies greatly in width. Almost nonexistent along some continents, the shelf may extend seaward as far as 1500 kilometers (930 miles) along others. On average, the continental shelf is about 80 kilometers (50 miles) wide and 130 meters (425 feet) deep at its seaward edge. The average inclination of the continental shelf is only about one-tenth of 1 degree, a drop of only about 2 meters per kilometer (10 feet per mile). The slope is so slight that it would appear to an observer to be a horizontal surface.

Although continental shelves represent only 7.5 percent of the total ocean area, they have economic and political significance because they contain important mineral deposits, including large reservoirs of oil and natural gas, as well as huge sand and gravel deposits. The waters of

Figure 12.6   The topography of Earth's solid surface is shown on these two pages.

the continental shelf also contain many important fishing grounds, which are significant sources of food.

Even though the continental shelf is relatively featureless, some areas are mantled by extensive glacial deposits and are thus quite rugged. The most profound features are long valleys running from the coastline into deeper waters. Many of these *shelf valleys* are the seaward extensions of river valleys on the adjacent landmass. Such valleys appear to have been excavated during the Pleistocene epoch (Ice Age).

During this time great quantities of water were stored in vast ice sheets on the continents. This caused sea level to drop by 100 meters (330 feet) or more, exposing large areas of the continental shelves (see Figure 5.3, p. 133). Because of this drop in sea level, rivers extended their courses, and land-dwelling plants and animals inhabited the newly exposed portions of the continents. Dredging off the coast of North America has retrieved the ancient remains of numerous land dwellers, including mammoths, mas-

Puerto-Rico
Trench

Mid-Atlantic Ridge

Peru-Chile
trench

Mid-Indian Ridge

South Sandwich
Trench

Southwest Indian Ridge

Southeast Indian Ridge

Figure 12.6   Continued

todons, and horses, adding to the evidence that portions of the continental shelves were once above sea level.

Most passive continental shelves, such as those along the East Coast of the United States, consist of shallow-water deposits that can reach several kilometers in thickness. Such deposits have led researchers to conclude that these thick accumulations of sediment are produced along a gradually subsiding continental margin.

Continental Slope.   Marking the seaward edge of the continental shelf is the **continental slope,** a relatively steep structure (as compared with the shelf) that marks the boundary between continental crust and oceanic crust (Figure 12.8). Although the inclination of the continental slope varies greatly from place to place, it averages about 5 degrees and in places may exceed 25 degrees. Further, the continental slope is a relatively narrow feature, averaging only about 20 kilometers (12 miles) in width.

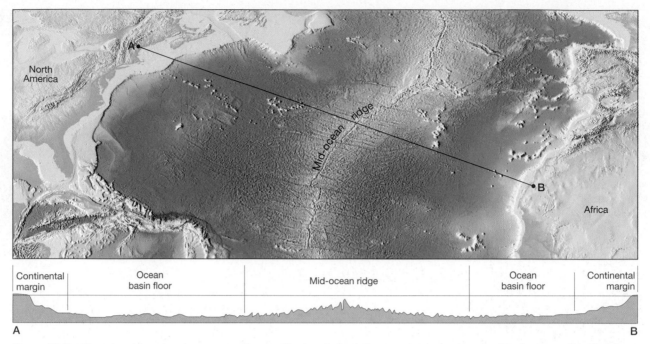

Figure 12.7 Map view (*above*) and corresponding profile view (*below*) showing the major topographic divisions of the North Atlantic Ocean. On the profile, the vertical scale has been expanded (exaggerated) by 40 times to make topographic features more conspicuous.

Submarine Canyons and Turbidity Currents. Deep, steep-sided valleys known as **submarine canyons** are cut into the continental slope and may extend across the entire continental rise to the ocean basin floor (Figure 12.9). Although some of these canyons appear to be the seaward extensions of river valleys such as the Hudson Valley and Amazon River, others are not directly associated with existing river systems. Furthermore, because these canyons extend to depths far below the lowest level of the sea during the Ice Age, their formation cannot be attributed to stream erosion. These features must be created by some process that operates far below the ocean surface.

Most available information suggests that submarine canyons have been eroded, at least in part, by turbidity currents (Figure 12.9). **Turbidity currents** are periodic downslope movements of dense, sediment-laden water. They are created when sand and mud on the continental shelf and slope are dislodged, perhaps by an earthquake, and are thrown into suspension. Because such mud-choked water is denser than normal seawater, it flows downslope, eroding and accumulating more sediment. The erosional work accomplished by these muddy torrents is thought to be the major force in the excavation of most submarine canyons (Box 12.1).

Figure 12.8 Features of a passive continental margin, including the continental shelf, continental slope, and continental rise. Note that the slopes shown for the continental shelf and continental slope are greatly exaggerated. The continental shelf has an average slope of one-tenth of 1 degree, whereas the continental slope has an average slope of about 5 degrees.

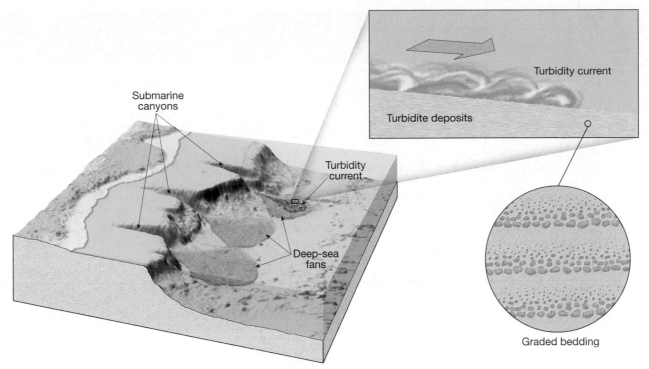

Figure 12.9   Turbidity currents are high-density flows of rock material mixed with water that move downslope, eroding the continental margin and enlarging submarine canyons. Turbidity currents eventually lose momentum and deposit their loads of sediment as deep sea fans. Beds deposited by these currents are called *turbidites*. Each event produces a single bed characterized by a decrease in sediment size from bottom to top, a feature known as a *graded bed*.

Narrow continental margins, such as the one located along the California coast, are dissected by numerous submarine canyons. Here erosion has extended many of these canyons landward into shallow water. Sediments carried to the coasts by rivers are transported along the shore by wave activity until they reach a submarine canyon. This steady supply of sediment collects until it becomes unstable and moves as a massive underwater landslide (turbidity current) to the deep-ocean floor.

Turbidity currents eventually lose momentum and come to rest along the floor of the ocean basin (Figure 12.9). As these currents slow, the suspended sediments begin to settle out. First, the coarser, heavier sand is dropped, followed by successively finer deposits of silt and then clay. These deposits, called **turbidites,** are characterized by a decrease in sediment grain size from bottom to top, a phenomenon known as **graded bedding** (Figure 12.9, *inset*).

Although there is still more to be learned about the complex workings of turbidity currents, it has been well established that they are an important mechanism of sediment transport in the ocean. By the action of turbidity currents, submarine canyons are eroded and sediments are deposited on the deep-ocean floor.

Continental Rise.   In regions where trenches do not exist, the steep continental slope merges into a more gradual incline known as the **continental rise.** Here the slope drops to about one-third degree, or about 6 meters per kilometer (32 feet per mile). Whereas the width of the continental slope averages about 20 kilometers (12 miles), the continental rise may extend for hundreds of kilometers into the deep-ocean basin.

The continental rise consists of a thick accumulation of sediment that moved downslope from the continental shelf to the deep-ocean floor. The sediments are delivered to the base of the continental slope by turbidity currents that periodically flow down submarine canyons. When these muddy currents emerge from the mouth of a canyon onto the relatively flat ocean floor, they deposit sediment that forms a **deep-sea fan** (Figure 12.9). Deep-sea fans have the same basic shape as the *alluvial fans* that form at the foot of steep mountain slopes on land. As fans from adjacent submarine canyons grow, they merge laterally with one another to produce a continuous covering of sediment at the base of the continental slope that is called the continental rise.

Active Continental Margins

Along some coasts the continental slope descends abruptly into a deep-ocean trench located between the continent and ocean basin. Thus, the landward wall of the trench and the continental slope are essentially the same feature. In such locations, the continental shelf is very narrow, if it exists at all.

Active continental margins are located primarily around the Pacific Ocean in areas where the leading edge of a continent is overrunning oceanic lithosphere (Figure 12.10). Here sediments from the ocean floor and pieces of

BOX 12.1

UNDERSTANDING EARTH:
A Grand Break—Evidence
for Turbidity Currents

How do earthquakes and telephone cables help explain how turbidity currents move across the ocean floor and carve submarine canyons? In 1929, the Grand Banks earthquake in the North Atlantic Ocean severed some of the trans-Atlantic telephone and telegraph cables that lay across the seafloor south of Newfoundland near the earthquake epicenter (Figure 12.A). At first it was assumed that seafloor movement caused all of these breaks. However, analysis of the data revealed that the cables closest to the epicenter broke simultaneously during the earthquake, but cables that crossed the slope and deeper ocean floor at greater distances from the epicenter

were broken progressively later in time. It seemed unusual that certain cables were affected by the failure of the slope due to ground shaking, but others were broken several hours later.

Following a reanalysis of the data several years later, a hypothesis was formulated to explain the distribution of cable breaks. The hypothesis suggested that a turbidity current moving down the continental slope and rise could account for the pattern of cable breaks. Based on the sequence of breaks, the turbidity current must have reached speeds approaching 80 kilometers (50 miles) per hour on the steep portions of the continental slope, and about 24 kilometers (15 miles) per hour

on the more gently sloping continental rise. Thus, turbidity currents reach high speeds and are strong enough to break underwater cables, suggesting that they must be powerful enough to erode submarine canyons.

Further evidence of turbidity currents comes from several studies that have documented turbidity currents using sonar. For instance, a study of Rupert Inlet in British Columbia, Canada, monitored turbidity currents moving through an underwater channel. Such studies indicate that submarine canyons are gradually eroded by turbidity currents over long periods of time, just as canyons on land are carved by running water.

Figure 12.A Profile of the seafloor showing the events of the November 18, 1929, earthquake off the coast of Newfoundland. The arrows point to cable breaks; the numbers show times of breaks in hours and minutes after the earthquake. Vertical scale is greatly exaggerated. (After B. C. Heezen and M. Ewing, "Turbidity Currents and Submarine Slump and the 1929 Grand Banks Earthquake," *American Journal of Science* 250:867)

oceanic crust are scraped from the descending oceanic plate and plastered against the edge of the overriding continent. This chaotic accumulation of deformed sediment and scraps of oceanic crust is called an *accretionary* (*ad* = toward, *crescere* = to grow) *wedge*. Prolonged plate subduction, along with the accretion of sediments on the landward side of the trench, can produce a large accumulation of sediments along a continental margin. A large accretionary wedge, for example, is found along the northern coast of Japan's Honshu Island.*

Some subduction zones have little or no accumulation of sediments, indicating that ocean sediments are

being carried into the mantle with the subducting plate. Here the continental margin is very narrow, as the trench may lie a mere 50 kilometers (31 miles) offshore.

The Ocean Basin Floor

 Oceans
  ↳ Floor of the Oceans

Between the continental margin and the oceanic ridge lies the **ocean basin floor** (see Figure 12.7). The size of this region—almost 30 percent of Earth's surface—is roughly comparable to the percentage of land above sea level. This region includes remarkably flat areas known

---

*For more on accretionary wedges and subduction, see Chapter 9.

Continental volcanic arc

Trench

Accretionary
wedge

Oceanic crust

Subducting oceanic lithosphere

Asthenosphere

Melting

Continental
lithosphere

100 km

200 km

Figure 12.10 Active continental margin.
Here sediments from the ocean floor are
scraped from the descending plate and
added to the continental crust as an
accretionary wedge.

as *abyssal plains,* tall volcanic peaks called *seamounts* and
guyots, and *deep-ocean trenches,* which are extremely
deep linear depressions in the ocean floor.

Deep-Ocean Trenches

Deep-ocean trenches are long, relatively narrow creas-
es in the seafloor that form the deepest parts of the
ocean. Most trenches are located along the margins of
the Pacific Ocean (see Figure 12.6) where many exceed
10,000 meters (33,000 feet) in depth. A portion of one
trench—the Challenger Deep in the Mariana Trench—
has been measured at a record 11,022 meters (36,163
feet) below sea level, making it the deepest known part
of the world ocean.

STUDENTS SOMETIMES ASK...

*Have humans ever explored
the deepest ocean trenches?
Could anything live there?*

Humans have indeed visited the deepest part of the oceans—
where there is crushing high pressure, complete darkness,
and near-freezing water temperatures—over 40 years ago! In
January 1960, U.S. Navy Lt. Don Walsh and explorer Jacques
Piccard descended to the bottom of the Challenger Deep region of the Mariana Trench in the deep diving bathyscaphe
Trieste. At 9906 meters (32,500 feet), the men heard a loud
cracking sound that shook the cabin. They were unable to see
that a 7.6-centimeter (3-inch) Plexiglas viewing port had
cracked (miraculously, it held for the rest of the dive). More
than five hours after leaving the surface, they reached the bottom at 10,912 meters (35,800 feet)—a record depth of human
descent that has not been broken since. They did see some life
forms that are adapted to life in the deep: a small flatfish, a
shrimp, and some jellyfish.

Although deep-ocean trenches represent only a
very small portion of the area of the ocean floor, they
are nevertheless significant geologic features. Trench-

es are sites of plate convergence where a moving
crustal plate subducts and plunges back into the mantle*. In addition to earthquakes being created as one
plate descends beneath another, volcanic activity is
also associated with these regions. The release of
volatiles—especially water—from a descending plate
triggers melting in the wedge of asthenosphere above
it. This buoyant material slowly migrates upward and
gives rise to volcanic activity at the surface. Thus,
trenches are often paralleled by an arc-shaped row of
active volcanoes called a **volcanic island arc.** Furthermore, **continental volcanic arcs,** such as those making up portions of the Andes, are located parallel to
trenches that lie adjacent to continental margins (see
Figure 12.10). The large number of trenches and associated volcanic activity along the margins of the Pacific Ocean is why the region is known as the *Ring
of Fire.*

Abyssal Plains

Abyssal (*a* = without, *byssus* = bottom) **plains** are
deep, incredibly flat features; in fact, these regions are
likely the most level places on Earth. The abyssal plain
found off the coast of Argentina, for example, has less
than 3 meters (10 feet) of relief over a distance exceeding
1300 kilometers (800 miles). The monotonous topography of abyssal plains is occasionally interrupted by the
protruding summit of a partially buried volcanic peak.

Using *seismic profilers* (instruments that generate
signals designed to penetrate far below the ocean floor),
researchers have determined that abyssal plains owe
their relatively featureless topography to thick accumulations of sediment that have buried an otherwise
rugged ocean floor (Figure 12.11). The nature of the sediment indicates that these plains consist primarily of
fine sediments transported far out to sea by turbidity
currents or deposited through suspension settling.

*For more on convergent plate boundaries, see Chapter 7.

Abyssal plains are found in all oceans of the world. However, the Atlantic Ocean has the most extensive abyssal plains because it has few trenches to act as traps for sediment carried down the continental slope.

Seamounts and Guyots

Dotting the ocean floor are isolated volcanic peaks called **seamounts,** which may rise hundreds of meters above the surrounding topography. Although these steep-sided conical peaks are found on the floors of all the oceans, the greatest number have been identified in the Pacific.

Some, like the Emperor Seamount chain in the Pacific that stretches from the Hawaiian Islands to the Aleutian trench, form in association with volcanic hot spots (see Figure 12.6). Others are born near oceanic ridges, divergent plate boundaries where plates move apart.* If the volcano grows large enough before being carried from the crest of the mid-ocean ridge, the structure emerges as an island. Examples of volcanic islands in the Atlantic include the Azores, Ascension, Tristan da Cunha, and St. Helena.

During the time they exist as islands, some of these volcanoes are eroded to near sea level by running water and wave action. Over a span of millions of years, the islands gradually sink and disappear below the water surface as the moving plate slowly carries them away from the elevated oceanic ridge or hot spot where they originated (see Box 12.2). These submerged, flat-topped seamounts are called **guyots** or **tablemounts.** †

---

*For more on hot spots and divergent plate boundaries, see Chapter 7.

†The term *guyot* is named after Princeton University's first geology professor. It is pronounced "GEE-oh" with a hard *g* as in "give."

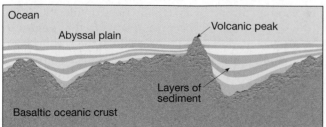

Figure 12.11 Seismic cross section and matching sketch across a portion of the Madeira abyssal plain in the eastern Atlantic Ocean, showing the irregular oceanic crust buried by sediments. (Image courtesy of Charles Hollister, Woods Hole Oceanographic Institution)

The Oceanic Ridge

Oceans
↳ Floor of the Oceans

The **oceanic ridge,** also called the **mid-ocean ridge,** is a topographically elevated feature that is found near the center of most ocean basins (see Figures 12.6 and 12.7). It is characterized by extensive faulting and numerous volcanic structures that have developed on the newly formed crust.

The interconnected oceanic ridge system is the longest topographic feature on Earth's surface, exceeding 70,000 kilometers (43,000 miles) in length. Representing 23 percent of Earth's surface, the mid-ocean ridge winds through all major oceans in a manner similar to the seam on a baseball. The term *ridge* may be misleading, as these features are not narrow but have widths of from 1000 to 4000 kilometers (620 to 2500 miles) and, in places, may occupy as much as one half the total area of the ocean floor. A close examination of Figure 12.6 shows that the mid-ocean ridge is broken into segments that are offset by large transform faults. Further, along the axis of some segments are deep down-faulted structures called **rift valleys.** The name *rift valley* has been applied to these features because they are so strikingly similar to continental rift valleys, such as the East African Rift Valley.

Features associated with rift valleys include active volcanoes, recent underwater lava flows, and black smokers, which are a type of *hydrothermal* (*hydro* = water, *thermos* = hot) *vent* that spews dark mineral-rich water. Unusual life forms reliant on the hot water are found near many black smokers (see Box 13.2, p. 367).

On average, the oceanic ridge rises 2.5 kilometers (1.5 miles) above the adjacent ocean basin floor. It is, however, much different from mountains found on the continents. Rather than consisting of thick sequences of folded and faulted sedimentary rocks, the oceanic ridge consists of layer upon layer of basaltic rocks that have been faulted and uplifted. The abundant volcanic activity along the crest of the ridge is associated with seafloor spreading, which occurs at divergent plate boundaries where two lithospheric plates are moving apart.

The far ends of the flanks of most ridges are topographically relatively featureless and rise very gradually (with a slope less than 1 degree) toward the ridge crest. Approaching the ridge crest, however, the topography begins to exhibit greater local relief as volcanic structures and faulted valleys become more prominent. The most rugged topography is found on those ridges that exhibit large rift valleys.

Partly because of its accessibility to both American and European scientists, the Mid-Atlantic Ridge (see Figure 12.6) has been studied more thoroughly than ridge systems in other ocean basins. The Mid-Atlantic

Ridge is a broad, submerged structure standing 2500 to 3000 meters (8200 to 9800 feet) above the adjacent ocean basin floor. In a few places, such as Iceland, the ridge has actually grown above sea level. Throughout most of its length, however, this divergent plate boundary lies 2500 meters (8200 feet) below sea level. Another prominent feature of the Mid-Atlantic Ridge is its deep linear rift valley extending along the ridge axis. In places this rift valley is deeper than the Grand Canyon of the Colorado River and two or three times as wide.

Seafloor Sediments

Except for steep areas of the continental slope and areas near the crest of the mid-ocean ridge, most of the ocean floor is covered with sediment. Part of this material has been deposited by turbidity currents, and the rest has slowly settled onto the seafloor from above. The thickness of this carpet of debris varies greatly. In some trenches, which act as traps for sediment originating on the continental margin, accumulations may approach 10 kilometers (6 miles) in thickness. In general, however, sediment accumulations are considerably less. In the Pacific Ocean, for example, sediment thickness is about 600 meters (2000 feet) or less, whereas on the floor of the Atlantic, the thickness varies from about 500 to 1000 meters (1500 to 3000 feet).

Although deposits of sand-size particles are found on the deep-ocean floor, *mud* (composed of fine clay-sized particles) is the most common sediment covering this region. Muds also predominate on the continental shelf and slope, but the sediments in these areas are coarser overall because of greater quantities of sand.

Types of Seafloor Sediments

Seafloor sediments can be classified according to their origin into three broad categories: (1) **terrigenous** (*terra* = land, *generare* = to produce); (2) **biogenous** (*bio* = life, *generare* = to produce); and (3) **hydrogenous** (*hydro* = water, *generare* = to produce). Although each category is discussed separately, seafloor sediments usually come from a variety of sources and are thus mixtures of the various sediment types.

Terrigenous Sediment.   Terrigenous sediment consists primarily of mineral grains that were weathered from continental rocks and transported to the ocean. Larger particles (gravel and sand) usually settle rapidly near shore whereas finer particles (microscopic clay-size particles) can take years to settle to the ocean floor and may be carried thousands of kilometers by ocean currents. As a consequence, virtually every part of the ocean receives some terrigenous sediment. The rate at which this sediment accumulates on the deep-ocean floor, though, is very slow. To form a 1-centimeter (0.4-inch) *abyssal clay* layer, for example, requires as much as 50,000 years. Conversely, on the continental

margins near the mouths of large rivers, terrigenous sediment accumulates rapidly and forms thick deposits. In the Gulf of Mexico, for instance, the sediment has reached a depth of many kilometers.

Because fine particles usually remain suspended in the water for a very long time, ample opportunity exists for chemical reactions to occur. Because of this, the colors of the deep-sea sediments are often red or brown. This results when iron in the particle or in the water reacts with dissolved oxygen in the water and produces a coating of iron oxide (rust).

Biogenous Sediment.   Biogenous sediment consists of shells and skeletons of marine animals and algae (Figure 12.12). This debris is produced mostly by microscopic organisms living in sunlit waters near the ocean surface. Once these organisms die, their hard *tests* (*testa* = shell) constantly "rain" down and accumulate on the seafloor.

The most common biogenous sediment is *calcareous* ($CaCO_3$) *ooze*, which, as its name implies, has the consistency of thick mud. This sediment is produced from the tests of organisms such as *coccolithophores* (single-celled algae) and *foraminifers* (small animals) that inhabit warm surface waters. When calcareous tests slowly sink into deeper parts of the ocean, they begin to dissolve. This results because the deeper, cold seawater is rich in carbon dioxide and is thus more acidic than warm water. In seawater deeper than about 4500 meters (15,000 feet), calcareous tests will completely dissolve before they reach bottom. Consequently, calcareous ooze does not accumulate in the deep-ocean basins.

Other biogenous sediments include *siliceous* (SiO_2) *ooze* and phosphate-rich material. The former is composed primarily of tests of *diatoms* (single-celled algae) and *radiolarians* (single-celled animals) that prefer cooler surface waters, whereas the latter is derived from the bones, teeth, and scales of fish and other marine organisms.

? STUDENTS SOMETIMES ASK...

Are diatoms an ingredient in diatomaceous earth, which is used in swimming pool filters?

Not only are diatoms used as swimming pool filters, they are also used in a variety of everyday products, including toothpaste (yes, you're brushing your teeth with the remains of dead microscopic organisms!). Diatoms secrete walls of silica in a great variety of forms that accumulate as sediments in enormous quantities. Because it is light weight, chemically stable, has high surface area, and is highly absorbent, diatomaceous earth has many practical uses. The main uses of diatoms include: filters (for refining sugar, straining yeast from beer, and filtering swimming pool water); mild abrasives (in household cleaning and polishing compounds and facial scrubs); and absorbents (for chemical spills).

BOX 12.2

UNDERSTANDING EARTH:
Explaining Coral Atolls—Darwin's Hypothesis

Coral *atolls* are ring-shaped structures that often extend several thousand meters below sea level (Figure 12.B). What causes atolls to form, and how do they attain such a great thickness?

Corals are colonial animals about the size of an ant that feed with stinging tentacles and are related to jellyfish. Most corals protect themselves by growing a hard external skeleton made of calcium carbonate. Where corals reproduce and grow over many centuries, their skeletons fuse into large structures called *coral reefs*. Other corals—as well as sponges and algae—begin to attach to the reef, enlarging it further. Eventually fishes, sea slugs, octopus, and other organisms are attracted to these diverse and productive habitats.

Corals require specific environmental conditions to grow. For example, reef-building corals grow best in waters with an average annual temperature of about 24°C (75°F). They cannot survive prolonged exposure to temperature below 18°C (64°F) or above 30°C (86°F). In addition, reef-builders require an attachment site (usually other corals) and clear sunlit water. Consequently, the limiting depth of most active reef growth is only about 45 meters (150 feet).

The restricted environmental conditions required for coral growth create an interesting paradox: How can corals—which require warm, shallow, sunlit water no deeper than a few dozen meters to live—create thick structures such as coral atolls that extend into deep water?

Figure 12.B   An aerial view of Tetiaroa Atoll in the Pacific. The light blue waters of the relatively shallow lagoon contrast with the dark blue color of the deep ocean surrounding the atoll. (Photo by Douglas Peebles Photography)

Hydrogenous Sediment.   Hydrogenous sediment consists of minerals that crystallize directly from seawater through various chemical reactions. Hydrogenous sediments represent a relatively small portion of the overall sediment in the ocean. They do, however, have many different compositions and are distributed in diverse environments of deposition.

Some of the most common types of hydrogenous sediment include:

- *Manganese nodules*, which are rounded, hard lumps of manganese, iron, and other metals that precipitate in concentric layers around a central object (such as a volcanic pebble or a grain of sand). The nodules can be up to 20 centimeters (8 inches) in diameter and are often littered across large areas of the deep seafloor.

- *Calcium carbonates*, which form by precipitation directly from seawater in warm climates. If this material is buried and hardened, it forms limestone. Most limestone, however, is composed of biogenous sediment.

- *Metal sulfides*, which are usually precipitated as coatings on rocks near black smokers associated with the crest of the mid-ocean ridge. These deposits contain iron, nickel, copper, zinc, silver, and other metals in varying proportions.

- *Evaporites*, which form where evaporation rates are high and there is restricted open-ocean circulation. As water evaporates from such areas, the remaining seawater becomes saturated with dissolved minerals, which then begin to precipitate. Heavier than seawater, they sink to the bottom or form a characteristic white crust of evaporite minerals around the edges of these areas. Collectively termed "salts," some evaporite minerals taste salty, such as *halite* (common table salt, NaCl), and some do not, such as the calcium sulfate minerals *anhydrite* ($CaSO_4$) and *gypsum* ($CaSO_4 \cdot 2H_2O$).

STUDENTS SOMETIMES ASK...
Has a large body of seawater ever evaporated?

Yes—in fact, several have. The evidence for this includes thick and extensive deposits of evaporite minerals. A large-scale example of evaporite deposition occurred in the Mediterranean Sea about 6 million years ago. Evidence suggests that the Mediterranean Sea was cut off from the Atlantic Ocean at the narrow Strait of Gibraltar—in effect, creating a dam. Because of the warm and dry climate in this region, the Mediterranean Sea nearly evaporated in a few thousand years, leaving a thick deposit of evaporites on the hot, dry basin floor. A half million years later, erosion, further tectonic activity, or a rise

The naturalist Charles Darwin was one of the first to formulate a hypothesis on the origin of atolls. From 1831 to 1836 he sailed aboard the British ship HMS *Beagle* during its famous circumnavigation of the globe. In various places that Darwin visited, he noticed a progression of stages in coral reef development from (1) a *fringing reef* along the margins of a volcano to (2) a *barrier reef* with a volcano in the middle to (3) an *atoll*, which consists of a continuous or broken ring of coral reef surrounded by a central lagoon (Figure 12.C). The essence of Darwin's hypothesis was that as a volcanic island slowly sinks, the corals continue to build the reef complex upward.

Darwin's hypothesis explained how coral reefs, which are restricted to shallow water, can build structures that now exist in much deeper water. During Darwin's time, however, there was no plausible mechanism to account for how an island might sink through time.

Today, plate tectonic movement of the seafloor helps explain how a volcanic island can become extinct and experience a change in elevation over long periods of time. Further, drilling through atolls has revealed that volcanic rock does indeed underlie the oldest (and deepest) coral reef structures, confirming Darwin's hypothesis. Thus, atolls owe their existence to the gradual sinking of volcanic islands containing coral reefs that build upward through time.

Figure 12.C   Formation of a coral atoll due to the gradual sinking of oceanic crust and upward growth of the coral reef. **A.** A fringing coral reef forms around a volcanic island. **B.** As the volcanic island sinks, the fringing reef gradually becomes a barrier reef. **C.** Eventually, the volcano is completely submerged and an atoll remains.

in worldwide sea level caused the dam to breach at Gibraltar. A huge waterfall spilled into the Mediterranean and filled the sea with seawater once again.

Distribution of Seafloor Sediments

Figure 12.13 shows the distribution of seafloor sediments. Coarse-grained terrigenous deposits dominate continental margin areas, whereas fine-grained terrigenous material (abyssal clay) is common in deeper areas of the ocean basins. However, deep-ocean deposits are dominated by calcareous oozes, which are found on the shallower portions of deep-ocean areas along the mid-ocean ridge. Siliceous oozes are found beneath areas of unusually high biologic productivity such as the Antarctic and the equatorial Pacific Ocean. Hydrogenous sediment comprises only a small proportion of deposits in the ocean.

Various types of sediment accumulate on nearly all areas of the ocean floor in the same way dust accumulates in all parts of your home (which is why seafloor sediment is often referred to as "marine dust"). Even the deep-ocean floor far from land receives small amounts of windblown material, microscopic biogenous particles, and even space dust.

There are a few places in the ocean, however, where very little sediment accumulates. One such place is along the continental slope, where there is active erosion by turbidity and other deep-ocean currents. Another place where very little sediment can be found is along the oceanic ridge. Here, the seafloor along the crest of the ridge is so young (because of seafloor spreading), and the rates of sediment accumulation far from land are so slow that there has not been enough time for sediments to accumulate.

Seafloor Sediments and Climate Change

Reliable climate records go back only a couple of hundred years, at best. How do scientists learn about climates and climate changes prior to that time? The answer is that they must reconstruct past climates from *indirect* evidence; that is, they must examine and analyze clues from phenomena that responded to and reflect changing atmospheric conditions. An important technique for analyzing Earth's climate history is the study of seafloor sediments.

Most seafloor sediments contain the remains of microscopic organisms that once lived near the sea surface (the ocean–atmosphere interface). When such

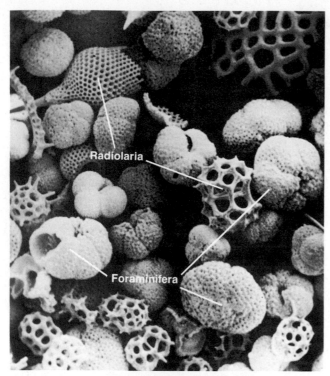

Figure 12.12 Microscopic hard parts of radiolarians and foraminifers are examples of biogenous sediments. This photomicrograph has been enlarged hundreds of times. (Photo courtesy of Deep Sea Drilling Project, Scripps Institution of Oceanography, University of California, San Diego)

. . . [W]e would expect that in any area of the ocean/atmosphere interface the average annual temperature of the surface water of the ocean would approximate that of the contiguous atmosphere. The temperature equilibrium established between surface seawater and the air above it should mean that . . . changes in climate should be reflected in changes in organisms living near the surface of the deep sea. . . . When we recall that the seafloor sediments in vast areas of the ocean consist mainly of shells of [oceanic] foraminifers, and that these animals are sensitive to variations in water temperature, the connection between such sediments and climatic change becomes obvious.*

In seeking to understand climate and other environmental changes, scientists are tapping the huge reservoir of data in seafloor sediments. Sediment cores gathered by drilling ships and other research vessels have provided invaluable data that have significantly expanded scientific knowledge and understanding of past climates (Box 12.3). Analysis of seafloor sediments, for instance, has revealed periods of Ice Ages and global warming, ocean circulation changes, the timing of major extinction events, and the movement of Earth's plates.

Resources from the Seafloor

The seafloor is rich in mineral and organic resources. Most, however, are not easily accessible, and recovery involves technological challenges and high cost. Nevertheless, certain resources have high value and thus make appealing exploration targets.

near-surface organisms die, their tests slowly settle onto the ocean floor, where they can become buried and preserved over time. Thus, the deep-ocean floor has become a repository for sediment representing millions of years of Earth history.

Seafloor sediments are useful recorders of worldwide climate change because the numbers and types of organisms living near the sea surface change with the climate:

---

*Richard F. Flint, *Glacial and Quaternary Geology* (New York: Wiley, 1971), p. 718.

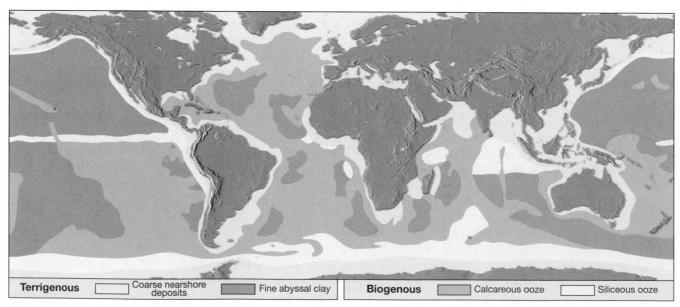

| **Terrigenous** | ☐ Coarse nearshore deposits | ▨ Fine abyssal clay | **Biogenous** | ▨ Calcareous ooze | ☐ Siliceous ooze |

Figure 12.13   Distribution of marine sediment.

Energy Resources

Among the nonliving resources extracted from the oceans, more than 95 percent of the economic value comes from energy products. The main energy products are oil and natural gas, which are currently being extracted, and gas hydrates, which are not yet utilized but have vast potential.

Oil and Natural Gas.   The ancient remains of microscopic organisms, buried within marine sediments before they could decompose, are the source of today's deposits of oil and natural gas. The percentage of world oil produced from offshore regions has increased from trace amounts in the 1930s to over 30 percent today. Most of this increase results from continuing

BOX 12.3
UNDERSTANDING EARTH:
Collecting the Geologic History of the Deep-Ocean Floor

A fundamental aspect of scientific inquiry is the gathering of basic facts through observation and measurement. Formulating and testing hypotheses requires reliable data. The acquisition of such information is no easy task when it comes to sampling the vast storehouse of data contained in seafloor sediments and oceanic crust. Acquiring samples is technically challenging and very expensive.

The *Ocean Drilling Program (ODP)* uses the ship *JOIDES Resolution* to drill into the seafloor and collect long cylinders (cores) of sediment and rock. The "*JOIDES*" in the ship's name stands for *Joint Oceanographic Institutions for Deep Earth Sampling* and reflects the international commitment from the program's 22 participating countries. The "*Resolution*" honors the ship HMS *Resolution*, commanded more than 200 years ago by the prolific English explorer Captain James Cook.

The *JOIDES Resolution* has a tall metal derrick that is used to conduct *rotary drilling,* while the ship's thrusters hold it in a fixed position at sea (Figure 12.D). Individual sections of drill pipe are fitted together to make a single string of pipe up to 8200 meters (27,000 feet) long. The drill bit, located at the end of the pipe string, rotates as it is pressed against the ocean bottom and

can drill up to 2100 meters (6900 feet) below the seafloor. Like twirling a soda straw into a layer cake, the drilling operation crushes the rock around the outside and retains a cylinder of rock (a core sample) on the inside of the hollow pipe, which can then be raised on board the ship and analyzed in state-of-the-art laboratory facilities.

Since 1985, the ship has drilled more than 2000 holes worldwide. The result has been the recovery of more than 168,000 meters (more than 104 miles!) of core samples that represent millions of years of Earth history and are used by scientists to study many aspects of Earth science, including changes in global climate.

The ODP is scheduled to be replaced in 2003 by the Integrated Ocean Drilling Program (IODP), which will be led by the United States and Japan. Two new drill ships will be constructed: one for drilling shallow high-resolution cores, and one with advanced technology that can drill up to 4000 meters (13,000 feet) below the seafloor. The primary objective of the new program is to collect cores that will allow scientists to better understand Earth history and Earth system processes, including the properties of the deep crust, climate change patterns, earthquake mechanisms, and the microbiology of the deep-ocean floor.

Figure 12.D   *JOIDES Resolution* drills into the ocean floor and collects cores of sediment and rock for analysis. The ship's dynamic positioning system consists of powerful thrusters (small propellers) that allow it to remain stationary above a drill site. Previous drill sites can be reused years later and are located by bouncing sound waves between the ship's hydrophones and sonar beacons. A remote television camera aids in positioning the drill pipe into the reeentry cone.

technological advancements employed by offshore drilling platforms (Figure 12.14).

Major offshore reserves exist in the Persian Gulf, in the Gulf of Mexico, off the coast of southern California, in the North Sea, and in the East Indies. Additional reserves are probably located off the north coast of Alaska and in the Canadian Arctic, Asian seas, Africa, and Brazil. Because the likelihood of finding major new reserves on land is small, future offshore exploration will continue to be important, especially in deeper waters of the continental margins. A major environmental concern about offshore petroleum exploration is the possibility of oil spills caused by inadvertent leaks or blowouts during the drilling process.

Gas Hydrates. **Gas hydrates** are unusually compact chemical structures made of water and natural gas. The most common type of natural gas is methane, which produces *methane hydrate.* Gas hydrates occur beneath permafrost areas on land and under the ocean floor at depths below 525 meters (1720 feet).

Most oceanic gas hydrates are created when bacteria break down organic matter trapped in seafloor sediments, producing methane gas with minor amounts of ethane and propane. These gases combine with water in deep-ocean sediments (where pressures are high and temperatures are low) in such a way that the gas is trapped inside a latticelike cage of water molecules.

Vessels that have drilled into gas hydrates have retrieved cores of mud mixed with chunks or layers of gas hydrates (Figure 12.15A) that fizzle and evaporate quickly when they are exposed to the relatively warm, low-pressure conditions at the ocean surface. Gas hydrates resemble chunks of ice but ignite when lit by a flame because methane and other flammable gases are released as gas hydrates vaporize (Figure 12.15B).

Some estimates indicate that as much as 20 quadrillion cubic meters (700 quadrillion cubic feet) of methane are locked up in sediments containing gas hydrates. This is equivalent to about *twice* as much carbon as Earth's coal, oil, and conventional gas reserves combined, so gas hydrates have great potential. One major drawback in exploiting reserves of gas hydrate is that they rapidly decompose at surface temperatures and pressures. In the future, however, these vast seafloor reserves of energy may help power modern society.

Other Resources

Other major resources from the seafloor include sand and gravel, evaporative salts, and manganese nodules.

Sand and Gravel. The offshore sand-and-gravel industry is second in economic value only to the petroleum industry. Sand and gravel, which includes rock fragments that are washed out to sea and shells of marine organisms, is mined by offshore barges using suction dredges. Sand and gravel is primarily used as an aggregate in concrete, as a fill material in grading projects, and on recreational beaches.

In some cases, materials of high economic value are associated with offshore sand and gravel deposits. Gem-quality diamonds, for example, are recovered from gravels on the continental shelf offshore of South Africa and Australia. Sediments rich in tin have been mined from some offshore areas of Southeast Asia. Platinum and gold have been found in deposits in gold-mining areas throughout the world, and some Florida beach sands are rich in titanium.

Evaporative Salts. When seawater evaporates, the salts increase in concentration until they can no longer remain dissolved, so they precipitate out of solution

Figure 12.14 Offshore drilling rigs are used to tap the oil and natural gas reserves of the continental shelf. These platforms are near Santa Barbara, California. (Photo by Gregory Ochocki/Photo Researchers, Inc.)

A.

B.

Figure 12.15 Gas hydrates. **A.** A sample retrieved from the ocean floor shows layers of white icelike gas hydrate mixed with mud. **B.** Gas hydrates evaporate when exposed to surface conditions and release natural gas, which can be ignited. (Photos courtesy of GEOMAR Research Center, Kiel, Germany)

and form salt deposits, which can then be harvested. The most economically important salt is *halite* (common table salt). Halite is widely used for seasoning, curing, and preserving foods. It is also used in water conditioners, in agriculture, in the clothing industry for dying fabric and to de-ice roads.

Ever since ancient times, the ocean has been an important source of salt for human consumption, and the sea remains a significant supplier. Today, about 30 percent of the world's salt is produced by evaporating seawater (Figure 12.16).

Manganese Nodules. Manganese nodules contain significant concentrations of manganese, iron, and smaller concentrations of copper, nickel, and cobalt, all of which have a variety of economic uses. Cobalt, for example, is deemed "strategic" (essential to United States national security) because it is required

to produce dense, strong alloys with other metals and is used in high-speed cutting tools, powerful permanent magnets, and jet engine parts. Technologically, mining the deep-ocean floor for manganese nodules is possible but economically not profitable (Figure 12.17).

Nodules are widely distributed, but not all regions have the same potential for mining. Good locations have abundant nodules that contain the economically opti-

Figure 12.17 Manganese nodules photographed at a depth of 2909 fathoms (5323 meters) beneath the *Robert Conrad*, south of Tahiti. (Photo courtesy of Lawrence Sullivan, Lamont-Doherty Earth Observatory/Columbia University))

Figure 12.16 Harvesting salt produced by the evaporation of seawater. (Photo by William E. Townsend, Jr./Photo Researchers, Inc.)

mum mix of copper, nickel, and cobalt. Sites meeting these criteria, however, are relatively limited. Additionally, there are political problems of establishing mining rights far from land and environmental concerns about disturbing large portions of the deep-ocean floor.

Chapter Summary

• *Oceanography is an interdisciplinary science* that draws on the methods and knowledge of geology, chemistry, physics, and biology to study all aspects of the world ocean.

• *Earth is a planet dominated by oceans.* Seventy-one percent of Earth's area consists of oceans and marginal seas. In the Southern Hemisphere, often called the *water hemisphere*, about 81 percent of the surface is water. The world ocean can be divided into four main ocean basins: the *Pacific Ocean* (largest and deepest ocean), the *Atlantic Ocean* (about half the size of the Pacific), the *Indian Ocean* (slightly smaller than the Atlantic and mostly in the Southern Hemisphere), and the *Arctic Ocean* (smallest and shallowest ocean). The average depth of the oceans is 3729 meters (12,234 feet).

• *Ocean bathymetry is determined using echo sounders and multibeam sonars,* which bounce sonic signals off the ocean floor. Ship-based receivers record the reflected echoes and accurately measure the time interval of the signals. With this information, ocean depths are calculated and plotted to produce maps of ocean-floor topography. Recently, *satellite measurements* of the shape of the ocean surface have added data for mapping ocean-floor features.

• The zones that collectively make up a *passive continental margin* include the *continental shelf* (a gently sloping, submerged surface extending from the shoreline toward the deep-ocean basin); the *continental slope* (the true edge of the continent, which has a steep slope that leads from the continental shelf into deep water); and in regions where trenches do not exist, the steep continental slope merges into a more gradual incline known as the *continental rise* (which consists of sediments that have moved downslope from the continental shelf to the deep-ocean floor).

• *Submarine canyons* are deep, steep-sided valleys that originate on the continental slope and may extend to the deep-ocean basin. Many submarine canyons have been excavated by *turbidity currents,* which are downslope movements of dense, sediment-laden water.

• *Active continental margins* are located primarily around the Pacific Rim in areas where the leading edge of a continent is overrunning oceanic lithosphere. Here sediment scraped from the descending oceanic plate is plastered against the continent to form a collection of sediments called an *accretionary wedge.* An active continental margin generally has a narrow continental shelf, which grades into a steep continental slope and deep-ocean trench.

• The *ocean basin floor* lies between the continental margin and the oceanic ridge system. The features of the ocean basin floor include *deep-ocean trenches* (the deepest parts of the ocean, where moving crustal plates descend into the mantle), *abyssal plains* (the most level places on Earth, consisting of thick accumulations of sediments that were deposited atop the low, rough portions of the ocean floor), and *seamounts* and *guyots* (isolated volcanic peaks on the ocean floor that originate near the mid-ocean ridge or in association with volcanic hot spots).

• *Atolls* form from corals that grow on the flanks of sinking volcanic islands, where the corals continue to build the reef complex upward as the island sinks.

• The *oceanic (mid-ocean) ridge* winds through the middle of most ocean basins. Seafloor spreading occurs along this broad feature, which is characterized by an elevated position, extensive faulting, and volcanic structures that have developed on newly formed oceanic crust. Most of the geologic activity associated with ridges occurs along a narrow region on the ridge crest, called the *rift valley,* where magma moves upward to create new slivers of oceanic crust.

• *There are three broad categories of seafloor sediments. Terrigenous sediment* consists primarily of mineral grains that were weathered from continental rocks and transported to the ocean; *biogenous sediment* consists of shells and skeletons of marine animals and plants; and *hydrogenous sediment* includes minerals that crystallize directly from seawater through various chemical reactions. The global distribution of marine sediments is affected by proximity to source areas and water temperatures that favor the growth of certain marine organisms.

• *Seafloor sediments are helpful when studying worldwide climate changes* because they often contain the remains of organisms that once lived near the sea surface. The numbers and types of these organisms change as the climate changes, and their remains in seafloor sediments record these changes.

• *Energy resources* from the seafloor include *oil and natural gas* and large untapped deposits of *gas hydrates.* Other seafloor resources include *sand and gravel, evaporative salts,* and metals within *manganese nodules.*

Key Terms

abyssal plain (p. 343)
active continental margin (p. 341)
bathymetry (p. 335)
biogenous sediment (p. 345)
continental margin (p. 337)
continental rise (p. 341)

continental shelf (p. 337)
continental slope (p. 339)
continental volcanic arc (p. 343)
deep-ocean trench (p. 343)
deep-sea fan (p. 341)
echo sounder (p. 335)

gas hydrate (p. 350)
graded bedding (p. 341)
guyot (p. 344)
hydrogenous sediment (p. 345)
mid-ocean ridge (p. 344)
ocean basin floor (p. 342)

Review Questions

1. How does the area of Earth's surface covered by the oceans compare with that of the continents? Describe the distribution of land and water on Earth.

2. Name the four main oceans basins. Of the four:
   (a) Which one is the largest in area? Which one is smallest?
   (b) Which one is the deepest? Which one is shallowest?
   (c) Which one is almost entirely within the Southern Hemisphere?
   (d) Which one is exclusively in the Northern Hemisphere?

3. How does the average depth of the ocean compare to the average elevation of the continents?

4. Assuming that the average speed of sound waves in water is 1500 meters per second, determine the water depth in meters if a signal sent out by an echo sounder requires 6 seconds to strike bottom and return to the recorder (see Figure 12.4).

5. Describe how satellites orbiting Earth can determine features on the seafloor without being able to directly observe them beneath several kilometers of seawater.

6. List the three major subdivisions of a passive continental margin. Which subdivision is considered a flooded extension of the continent? Which has the steepest slope?

7. Describe the differences between active and passive continental margins. Be sure to include how various features relate to plate tectonics and give a geographic example of each type of margin.

8. Defend or rebut the statement "Most submarine canyons found on the continental slope and rise were formed during the Ice Age when rivers extended their valleys seaward."

9. What are turbidity currents? How do they differ from turbidites? What is meant by the term *graded bedding?*

10. What are differences between a submarine canyon and a deep-ocean trench?

11. Describe the process by which abyssal plains are created. Why are abyssal plains more extensive on the floor of the Atlantic than on the floor of the Pacific?

12. Discuss how seamounts and guyots are created and why they have a different shape.

13. Describe the environmental conditions required for the development of coral reefs.

14. What paradox was there concerning atolls? Describe Darwin's proposal on the origin of atolls. How was his proposal confirmed or disproved?

15. Describe features associated with the oceanic ridge.

16. How are oceanic ridges and deep-ocean trenches related to plate tectonics?

17. Distinguish among the three basic types of seafloor sediment, and give an example of each.

18. Describe how seafloor sediments are useful in studying past climates. What kind of information can sediments reveal?

19. Discuss the present importance, future prospects, and potential problems for production of offshore petroleum; gas hydrates; sand and gravel; evaporative salts; and manganese nodules.

Examining the Earth System

1. Describe some of the material and energy exchanges that take place at the interface between the (a) ocean surface and atmosphere, (b) ocean water and ocean floor, and (c) ocean biosphere and ocean water.

2. Sediment on the seafloor often leaves clues about various conditions that existed during deposition. What do the following layers in a seafloor core tell about the environment in which each layer was deposited?

- Layer #5 (top): Red layer composed of fine clays
- Layer #4: Siliceous ooze
- Layer #3: Calcareous ooze
- Layer #2: Fragments of coral reef
- Layer #1 (bottom): Rocks of basalt composition with some metal sulfide coatings

  Explain how one area of the seafloor could experience such varied conditions of deposition.

Web Resources

The *Earth Science* Website uses the resources and flexibility of the Internet to aid in your study of the topics in this chapter. Written and developed by Earth science instructors, this site will help improve your understanding of Earth science. Visit **http://www.prenhall.com/tarbuck** and click on the cover of *Earth Science 10e* to find:

- **On-line review quizzes.**
- **Web-based critical thinking and writing exercises.**
- **Links to chapter-specific Web resources.**
- **Internet-wide key term searches.**

http://www.prenhall.com/tarbuck

An underwater seascape in the south Pacific near Fiji. (Photo by Gary Bell/Seapics.com)

CHAPTER 13

OCEAN WATER
AND OCEAN LIFE*

*This chapter was prepared by Professor Alan P.
Trujillo, Palomar College.

What is the difference between pure water and seawater? One of the most obvious differences is that seawater contains dissolved substances that give it a distinctly salty taste. These dissolved substances are not simply sodium chloride (table salt)—they include various other salts, metals, and even dissolved gases. In fact, every known naturally occurring element is found dissolved in at least trace amounts in seawater. Unfortunately, the salt content of seawater makes it unsuitable for drinking or for irrigating most crops and causes it to be highly corrosive to many materials. Yet, many parts of the ocean are teeming with life that is superbly adapted to the marine environment.

There is an amazing variety of life in the ocean, from microscopic bacteria and algae to the largest organism alive today (the blue whale). Water is the major component of nearly every life form on Earth, and our own body fluid chemistry is remarkably similar to the chemistry of seawater.

Figure 13.1 Relative proportions of water and dissolved components in seawater. Components shown by chemical symbol are chlorine (Cl^-), sodium (Na^+), sulfate (SO_4^{2-}), magnesium (Mg^{2+}), calcium (Ca^{2+}), potassium (K^+), strontium (Sr^{2+}), bromine (Br^-), and carbon (C).

Composition of Seawater

Seawater consists of about 3.5 percent (by weight) dissolved mineral substances that are collectively termed "salts." Although the percentage of dissolved components may seem small, the actual quantity is huge because the ocean is so vast.

Salinity

Salinity (*salinus* = salt) is the total amount of solid material dissolved in water. More specifically, it is the ratio of the mass of dissolved substances to the mass of the water sample. Many common quantities are expressed in percent (%), which is really *parts per hundred.* Because the proportion of dissolved substances in seawater is such a small number, oceanographers typically express salinity in *parts per thousand (‰).* Thus, the average salinity of seawater is 3.5% or 35‰.

Figure 13.1 shows the principal elements that contribute to the ocean's salinity. If one wanted to make artificial seawater, it could be approximated by following the recipe shown in Table 13.1. From this table it is evident that most of the salt in seawater is sodium chloride—common table salt. Sodium chloride together with the next four most abundant salts comprise over 99 percent of all dissolved substances in the sea. Although only eight elements make up these five most abundant salts, seawater contains all of Earth's other naturally occurring elements. Despite their presence in minute quantities, many of these elements are very important

in maintaining the necessary chemical environment for life in the sea.

Sources of Sea Salts

What are the primary sources for the vast quantities of dissolved substances in the ocean? Chemical weathering of rocks on the continents is one source. These dissolved materials are delivered to the oceans by streams at an estimated rate of more than 2.3 billion metric tons (2.5 billion short tons) annually.* The second major source of elements found in seawater is from Earth's interior. Through volcanic eruptions, large quantities of water vapor and other gases have been emitted during much of geologic time. This process, called *outgassing,* is the principal source of water in the oceans. Certain elements—notably chlorine, bromine, sulfur, and boron—were outgassed along with water and exist in the ocean in much greater abundance than could be explained by weathering of rocks alone.

Although rivers and volcanic activity continually contribute dissolved substances to the oceans, the salinity of seawater is not increasing. In fact, evidence suggests that the composition of seawater has been relatively stable for millions of years. Why doesn't the sea get saltier? The answer is because material is being removed just as rapidly as it is added. For example, some dissolved components are withdrawn from seawater by organisms as they build hard parts. Other components are removed when they chemically precipitate out of the water as sediment. Still others are exchanged at the oceanic ridge by *hydrothermal (hydro =*

---

*A metric ton equals 1000 kilograms or 2205 pounds. Thus, it is larger than the "ton" that most Americans are familiar with (called a *short ton*). There are 2000 pounds in a short ton.

Table 13.1 Recipe for artificial seawater.	
To make seawater, combine:	**Amount (grams)**
Sodium chloride (NaCl)	23.48
Magnesium chloride ($MgCl_2$)	4.98
Sodium sulfate (Na_2SO_4)	3.92
Calcium chloride ($CaCl_2$)	1.10
Potassium chloride (KCl)	0.66
Sodium bicarbonate ($NaHCO_3$)	0.192
Potassium bromide (KBr)	0.096
Hydrogen borate (H_3BO_3)	0.026
Strontium chloride ($SrCl_2$)	0.024
Sodium fluoride (NaF)	0.003
Then add:	
Pure water (H_2O) to form 1000 grams of solution.	

water, *thermos* = hot) *activity*. The net effect is that the overall makeup of seawater has remained relatively constant through time.

Processes Affecting Seawater Salinity

Because the ocean is well mixed, the relative abundances of the major components in seawater are essentially constant, no matter where the ocean is sampled.

Variations in salinity, therefore, are primarily a consequence of changes in the water content of the solution.

Various surface processes alter the amount of water in seawater, thereby affecting salinity (Figure 13.2). Processes that add large amounts of fresh water to seawater—and thereby decrease salinity—include precipitation, runoff from land, icebergs melting, and sea ice melting. Processes that remove large amounts of fresh water from seawater—and thereby increase seawater salinity—include evaporation and the formation of sea ice. High salinities, for example, are found where evaporation rates are high, as is the case in the dry subtropical regions (roughly between 25 and 35 degrees north or south latitude). Conversely, where large amounts of precipitation dilute ocean waters, as in the midlatitudes (between 35 and 60 degrees north or south latitude) and near the equator, lower salinities prevail (Figure 13.3).

Surface salinity in polar regions varies seasonally due to the formation and melting of sea ice. When seawater freezes in winter, sea salts do not become part of the ice. Therefore, the salinity of the remaining seawater increases. In summer when sea ice melts, the addition of the relatively fresh water dilutes the solution and salinity decreases.

Surface salinity variation in the open ocean normally ranges from 33‰ to 38‰. Some marginal seas, however, demonstrate extraordinary extremes. For ex-

Icebergs

Sea ice

Runoff

Evaporation

Figure 13.2   Processes affecting seawater salinity. Processes that *decrease* seawater salinity include precipitation, runoff, icebergs melting, and sea ice melting. Processes that *increase* seawater salinity include formation of sea ice and evaporation.

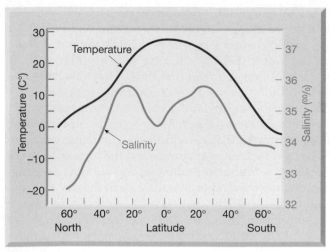

Figure 13.3 This graph shows variations in the ocean's surface temperature (top curve) and surface salinity (lower curve) with latitude. As you might expect, average temperatures are highest near the equator and get colder moving poleward. An important factor influencing differences in surface salinity is variations in rainfall and evaporation rates. For example, in the dry subtropics in the vicinity of the Tropics of Cancer and Capricorn, high evaporation rates remove more water than is replaced by the meager rainfall, resulting in high surface salinities. In the wet equatorial region, abundant rainfall reduces surface salinities.

STUDENTS SOMETIMES ASK...
*What would happen to a person
if he or she drank seawater?*

It depends on the quantity. The salinity of seawater is about four times greater than that of your body fluids. In your body, seawater causes your internal membranes to lose water through *osmosis*, which transports water molecules from higher concentrations (the normal body chemistry of your internal fluids) to areas of lower concentrations (your digestive tract containing seawater). Thus, your natural body fluids would move into your digestive tract and eventually be expelled, causing dehydration if seawater is consumed in large amounts. If you've inadvertently swallowed a little bit of seawater, don't worry too much.

ample, in the restricted waters of the Middle East's Persian Gulf and Red Sea—where evaporation far exceeds precipitation—salinity may exceed 42‰. Conversely, very low salinities occur where large quantities of fresh water are supplied by rivers and precipitation. Such is the case for northern Europe's Baltic Sea, where salinity is often below 10‰.

Ocean Temperature Variation

The ocean's surface water temperature varies with the amount of solar radiation received, which is primarily a function of latitude (Figure 13.3). The intensity of solar radiation in high latitudes is significantly less than that

STUDENTS SOMETIMES ASK...
*Where is the saltiest
water in the world?*

Some of the most highly saline water in the world is found in arid regions that have inland lakes, which are often called "seas" because they are so salty. The Great Salt Lake in Utah, for example, has a salinity of 280‰, and the Dead Sea on the border of Israel and Jordan has a salinity of 330‰. The water in the Dead Sea, therefore, contains 33 percent dissolved solids and is almost *10 times saltier than seawater*. As a result, these waters have such high density and are so buoyant that while lying down in the water, you can easily float—with arms and legs sticking up above water level!

received in tropical latitudes.* Therefore, much lower sea surface temperatures are found in high-latitude regions, and much higher sea surface temperatures are found in low-latitude regions.

Temperature Variation with Depth

If a thermometer were lowered from the surface of the ocean into deeper water, what temperature pattern would be found? Surface waters are warmed by the Sun, so they generally have higher temperatures than deeper waters. However, the observed temperature pattern depends on the latitude.

Figure 13.4 shows two graphs of temperature versus depth: one for high-latitude regions and one for low-latitude regions. The low-latitude curve begins at the surface with high temperature, but the temperature

---

*For more on this subject, see the section on "Earth-Sun Relationships" in Chapter 15.

Figure 13.4 Variations in ocean water temperature with depth for low- and high-latitude regions. The layer of rapidly changing temperature, called the *thermocline*, is present only in the low latitudes.

PEOPLE AND THE ENVIRONMENT:
Desalination of Seawater—
Fresh Water from the Sea

Earth's growing population uses fresh water in greater volumes each year. As fresh water becomes increasingly scarce, several countries have begun using the ocean as a source of water. The removal of salts and other chemicals to extract low-salinity ("fresh") water from seawater is termed *desalination.*

Worldwide, there are more than 12,500 desalination plants (Figure 13.A), with the majority located in arid regions of the Middle East, Caribbean, and Mediterranean. The United States produces only about 10 percent of the world's desalted water, with the majority produced in Florida and California. Because desalinated water is expensive to produce, most desalination plants are small-scale operations. In fact, desalination plants provide only about 1 percent of the world's drinking water. More than half of the world's desalination plants use *distillation* to purify water, while most of the remaining plants use *membrane processes.*

In distillation, saltwater is evaporated, and the resulting water vapor is collected and condensed to produce fresh water. This simple procedure is very efficient at purifying seawater. For instance, distillation of 35‰ seawater produces fresh water with a salinity of only 0.03‰, which is about 10 times fresher than bottled water.

Membrane processes such as *electrolysis* or *reverse osmosis* use specialized semipermeable membranes to separate dissolved components from water molecules, thereby purifying water. Worldwide, at least 30 countries are operating reverse osmosis units. Santa Barbara, California, for example, operates a reverse osmosis plant that produces up to 34 million liters (9 million gallons) of fresh water daily, which supplies up to 60 percent of its municipal water needs. This method is also used in many household water purification units and aquariums.

Other methods of desalination include freeze separation, crystallization of dissolved components directly from seawater, solvent demineralization using chemical catalysts, and even making use of salt-eating bacteria!

Although fresh water produced by various desalination methods is becoming more important as a source of water for human and even industrial use, it is unlikely to be an important supply for agricultural purposes because of the enormous quantities of water necessary to support agriculture. Consequently, making the deserts "bloom" by irrigating them with fresh water produced by desalination is only a dream and, for economic reasons, is likely to remain so for the foreseeable future.

Figure 13.A Desalination plants such as this one in Saudi Arabia make fresh water from seawater. (Photo by Anthony Howarth)

decreases rapidly with depth because of the inability of the Sun's rays to penetrate very far into the ocean. At a depth of about 1000 meters (3300 feet), the temperature remains just a few degrees above freezing and is relatively constant from this level down to the ocean floor. The layer of ocean water between about 300 meters (980 feet) and 1000 meters (3300 feet), where there is a rapid change of temperature with depth, is called the **thermocline** (*thermo* = heat, *cline* = slope). The thermocline is a very important structure in the ocean because it creates a vertical barrier to many types of marine life.

The high-latitude curve in Figure 13.4 displays a pattern quite different from the low-latitude curve. Surface water temperatures in high latitudes are much cooler than in low latitudes, so the curve begins at the surface with low temperature. Deeper in the ocean, the temperature of the water is similar to that at the surface (just a few degrees above freezing), so the curve remains vertical and there is no rapid change of temperature with depth. A thermocline is not present in high latitudes; instead, the water column is *isothermal* (*iso* = same, *thermo* = heat).

Some high-latitude waters can experience minor warming during the summer months. Thus, certain high-latitude regions experience an extremely weak seasonal thermocline. Midlatitude waters, on the other hand, experience a more dramatic seasonal thermocline and exhibit characteristics intermediate between high- and low-latitude regions.

Ocean Temperature Change over Time

Seawater has many unique thermal properties that make it resistant to changes in temperature. Marine species are well adapted to life in the ocean, and many cannot withstand rapid changes in temperature.* In fact, researchers have concluded that the ocean's stability as a habitat has been instrumental in the development of life on Earth.

Scientists who study the effects of global warming on the ocean indicate that atmospheric warming would eventually be transferred to the ocean, potentially impacting marine organisms. How can we determine if the average temperature of the global ocean is actually changing? Accurately measuring such changes with traditional thermometers would be extremely difficult. One group of researchers, led by Walter Munk of Scripps Institution of Oceanography, has initiated an experiment to determine the amount of ocean warming by using sound as a thermometer.

The scientists have used a worldwide sound channel—called the *SOFAR channel* (an acronym for *SOund Fixing And Ranging*)—to transmit low frequency sound across an ocean basin to a distant receiver. The experiment—which is called *ATOC* (*Acoustic Thermometry of Ocean Climate*)—is designed to accurately measure the travel time of similar sound signals through the SOFAR channel now and in the future. The speed of sound in seawater increases as temperature increases, so sound should take less time to travel the same distance in the future if, in fact, the oceans are warming.

The group successfully tested the experiment by transmitting signals across the Pacific Ocean in 1991 and 1995. The transmissions were halted, however, because of concern about the sound's effect on whales, some of which may also use the SOFAR channel for communication. Nonetheless, the transmissions that have been conducted have established an important baseline for comparison with future measurements.

Ocean Density Variation

Density is defined as mass per unit volume but can be thought of as a measure of *how heavy something is for its size*. For instance, an object that has low density is lightweight for its size, such as a dry sponge, foam packing, or a surfboard. Conversely, an object that has high density is heavy for its size, such as cement, most metals, or a large container full of water.

Density is an important property of ocean water because it determines the water's vertical position in the ocean. Furthermore, density differences cause large areas of ocean water to sink or float. For example, when high-density seawater is added to low-density

fresh water, the denser seawater sinks below the fresh water.

Factors Affecting Seawater Density

Seawater density is influenced by two main factors: *salinity* and *temperature.* An increase in salinity adds dissolved substances and results in an increase in seawater density. An increase in temperature, on the other hand, causes thermal expansion and results in a decrease in seawater density. Such a relationship where one variable decreases as a result of another variable's increase is known as an *inverse relationship,* where one variable is *inversely proportional* with the other.

Temperature has the greatest influence on surface seawater density because variations in surface seawater temperature are greater than salinity variations. In fact, only in the extreme polar areas of the ocean, where temperatures are low and remain relatively constant, does salinity significantly affect density. Cold water that also has high salinity is some of the highest-density water in the world.

Density Variation with Depth

By extensively sampling ocean waters, oceanographers have learned that temperature and salinity—and the water's resulting density—vary with depth. Figure 13.5 shows two graphs of density versus depth: one for high-latitude regions and one for low-latitude regions. Not surprisingly, the curves in Figure 13.5 are a mirror image of the temperature curves in Figure 13.4. This similarity demonstrates that temperature is the most important factor affecting seawater density and that temperature is inversely proportional with density.

The low-latitude curve in Figure 13.5 begins at the surface with low density (related to high surface water

Figure 13.5 Variations in ocean water density with depth for low- and high-latitude regions. The layer of rapidly changing density, called the *pycnocline,* is present in the low latitudes but absent in the high latitudes.

---

*This is discussed more fully in the section on "Controls of Temperature" in Chapter 15.

temperatures). However, density increases rapidly with depth because the water temperature is getting colder. At a depth of about 1000 meters (3300 feet), seawater density reaches a maximum value related to the water's low temperature. From this depth to the ocean floor, density remains constant and high. The layer of ocean water between about 300 meters (980 feet) and 1000 meters (3300 feet), where there is a rapid change of density with depth, is called the **pycnocline** (*pycno* = density, *cline* = slope). A pycnocline has a high gravitational stability and presents a significant barrier to mixing between low-density water above and high-density water below.

The high-latitude curve in Figure 13.5 is also related to the temperature curve for high latitudes shown in Figure 13.4. Figure 13.5 shows that in high latitudes, there is high-density (cold) water at the surface and high-density (cold) water below. Thus, the high-latitude density curve remains vertical, and there is no rapid change of density with depth. A pycnocline is not present in high latitudes; instead, the water column is *isopycnal* (*iso* = same, *pycno* = density).

Ocean Layering

The ocean, like Earth's interior, is layered according to density. Low-density water exists near the surface, and higher-density water occurs below. Except for some shallow inland seas with a high rate of evaporation, the highest-density water is found at the greatest ocean depths. Oceanographers generally recognize a three-layered structure in most parts of the open ocean: a shallow surface mixed zone, a transition zone, and a deep zone (Figure 13.6).

Because solar energy is received at the ocean surface, it is here that water temperatures are warmest. The mixing of these waters by waves as well as the turbulence from currents and tides creates a rapid vertical heat transfer. Hence, this *surface mixed zone* has nearly uniform temperatures. The thickness and temperature of this layer vary, depending on latitude and season. The zone usually extends to about 300 meters (980 feet)

but may attain a thickness of 450 meters (1500 feet). The surface mixed zone accounts for only about 2 percent of ocean water.

Below the Sun-warmed zone of mixing, the temperature falls abruptly with depth (see Figure 13.4). Here, a distinct layer called the *transition zone* exists between the warm surface layer above and the deep zone of cold water below. The transition zone includes a prominent thermocline and associated pycnocline and accounts for about 18 percent of ocean water.

Below the transition zone is the *deep zone,* where sunlight never reaches and water temperatures are just a few degrees above freezing. As a result, water density remains constant and high. Remarkably, the deep zone includes about 80 percent of ocean water, indicating the immense depth of the ocean (the average depth of the ocean is 3729 meters, or 12,234 feet).

In high latitudes, the three-layer structure of ocean layering does not exist (Figure 13.6). This is because the water column is isothermal (and isopycnal), which indicates that there is no rapid change in temperature (or density) with depth. Consequently, good vertical mixing between surface and deep waters can occur in high-latitude regions. Here, cold high-density water forms at the surface, sinks, and initiates deep-ocean currents, which are discussed in Chapter 14.

The Diversity of Ocean Life

A wide variety of organisms inhabit the marine environment. These organisms range in size from microscopic bacteria and algae to blue whales, which are as long as three buses lined up end to end. Marine biologists have identified over 250,000 marine species, a number that is constantly increasing as new organisms are discovered.

Most marine organisms live within the sunlit surface waters of the ocean. Strong sunlight supports **photosynthesis** (*photo* = light, *syn* – with, *thesis* = an arranging) by marine algae, which either directly or indirectly provide food for the vast majority of marine

Figure 13.6   Oceanographers recognize three main "layers" of the ocean based on characteristic water density, which varies with temperature and salinity. The warm surface mixed layer accounts for only 2 percent of ocean water; the transition zone includes the thermocline and pycnocline and accounts for 18 percent of ocean water; the deep zone contains cold, high-density water that accounts for 80 percent of ocean water.

organisms. All marine algae live near the surface because they need sunlight; most marine animals also live near the surface because this is where food can be obtained. In shallow water areas close to land, sunlight reaches all the way to the bottom, resulting in an abundance of marine life on the ocean floor.

There are advantages and disadvantages to living in the marine environment. One advantage is that there is an abundance of water available, which is necessary for supporting all types of life. One disadvantage is that maneuvering in water, which has high density and impedes movement, can be difficult. The individual success of species depends on their ability to avoid predators, find food, and cope with the physical challenges of their environment.

Classification of Marine Organisms

Marine organisms can be classified according to where they live (their habitat) and how they move (their mobility). Organisms that inhabit the water column can be classified as either *plankton* (floaters) or *nekton* (swimmers). All other organisms are *benthos* (bottom dwellers).

Plankton (Floaters) **Plankton** (*planktos* = wandering) include all organisms—algae, animals, and bacteria—that drift with ocean currents. Just because plankton drift does not mean they are unable to swim. Many plankton can swim but either move very weakly or move only vertically within the water column.

Among plankton, the algae (photosynthetic cells, most of which are microscopic) are called **phytoplank-**ton (*phyto* = plant, *planktos* = wandering) and the animals are called **zooplankton** (*zoo* = animal, *planktos* = wandering). Representative members of each group are shown in Figure 13.7.

Plankton are extremely abundant and very important within the marine environment. In fact, most of Earth's **biomass**—the mass of all living organisms—consists of plankton adrift in the oceans. Even though 98 percent of marine species are bottom dwelling, the vast majority of the ocean's biomass is planktonic.

Nekton (Swimmers) **Nekton** (*nektos* = swimming) include all animals capable of moving independently of the ocean currents, by swimming or other means of propulsion. They are capable not only of determining their position within the ocean but also, in many cases, of long migrations. Nekton include most adult fish and squid, marine mammals, and marine reptiles (Figure 13.8).

Although nekton move freely, they are unable to move throughout the breadth of the ocean. Gradual changes in temperature, salinity, density, and availability of nutrients effectively limit their lateral range. The deaths of large numbers of fish, for example, can be caused by temporary shifts of water masses in the ocean. High water pressure at depth normally limits the vertical range of nekton.

Fish may appear to exist everywhere in the oceans, but they are more abundant near continents and islands and in colder waters. Some fish, such as salmon, ascend fresh water rivers to spawn. Many eels do just the reverse, growing to maturity in fresh water and then descending the streams to breed in the great depths of the ocean.

A.
B.

Figure 13.7 **A.** A variety of live *phytoplankton* from the Atlantic Ocean. They include various diatoms and dinoflagellates. (Copyright N. T. Nicoll) **B.** This image of *zooplankton* includes the most common forms known as copepods and their larvae. The picture also includes larval forms of other common marine organisms. (Copyright N. T. Nicoll)

Figure 13.8 Necton includes all animals capable of moving independently of ocean currents. **A.** Gray reef shark, Bikini Atoll. (Photo by Doug Perrine/DRK Photo); **B.** California market squid. (Photo by Tom McHugh/Photo Researchers, Inc.); **C.** School of grunts, Florida Keys. (Photo by Larry Lipsky/DRK Photo); **D.** Yellow head moray eel. (Photo by David B. Fleetham/Innerspace Visions)

Benthos (Bottom Dwellers) The term **benthos** (*benthos* = bottom) describes organisms living on or in the ocean bottom. *Epifauna* (*epi* = upon, *fauna* = animal) live on the surface of the seafloor, either attached to rocks or moving along the bottom. *Infauna* (*in* = inside, *fauna* = animal) live buried in the sand, shells, or mud. Some benthos, called *nektobenthos,* live on the bottom but also swim or crawl through the water above the ocean floor. Examples of benthos are shown in Figure 13.9.

The shallow coastal ocean floor contains a wide variety of physical conditions and nutrient levels, both of which have allowed a great number of species to evolve. Moving across the bottom from the shore into deeper water, the number of benthos species may remain relatively constant, but the biomass of benthos organisms decreases. In addition, shallow coastal areas are the only locations where large marine algae (often called "seaweeds") are found attached to the bottom. This is the case because these are the only areas of the seafloor that receive sufficient sunlight.

Throughout most of the deeper parts of the seafloor, animals live in perpetual darkness, where photosynthesis cannot occur. They must feed on each other, or on whatever nutrients fall from the productive surface waters.

The deep-sea bottom is an environment of coldness, stillness, and darkness. Under these conditions, life progresses slowly, and organisms that live in the deep sea usually are widely distributed because physical conditions vary little on the deep-ocean floor, even over great distances.

Marine Life Zones

The distribution of marine organisms is affected by the chemistry, physics, and geology of the oceans. Marine organisms are influenced by a variety of physical oceanographic factors. Some of these factors—such as availability of sunlight, distance from shore, and water depth—are used to divide the ocean into distinct marine life zones (Table 13.2 and Figure 13.10).

Availability of Sunlight The upper part of the ocean into which sunlight penetrates is called the **photic** (*photos* = light) **zone.** The clarity of seawater is affected by many factors, such as the amount of plankton, suspended sediment, and decaying organic particles in the water. In addition, the amount of sunlight varies with atmospheric conditions, time of day, season of the year, and latitude.

Figure 13.9 Benthos describes organisms living on or in the ocean bottom. **A.** Sea star. (Photo by David Hall/Photo Researchers, Inc.); **B.** Yellow tube sponge. (Photo by Andrew Martinez/Photo Researchers, Inc.); **C.** Green sea urchin. (Photo by Andrew Martinez/Photo Researchers, Inc.); **D.** Coral crab. (Photo by Fred Bavendam/Peter Arnold, Inc.)

The **euphotic** (*eu* = good, *photos* = light) **zone** is the portion of the photic zone near the surface where light is strong enough for photosynthesis to occur. In the open ocean, this zone can reach a depth of 100 meters (330 feet), but the zone will be much shallower close to shore where water clarity is typically reduced. In the euphotic zone, phytoplankton use sunlight to produce food molecules and become the basis of most oceanic food webs.

Different wavelengths of sunlight are absorbed as they pass through seawater. The longer wavelength red and orange colors are absorbed first, the greens and yellows next, and the shorter wavelength blues and violets penetrate the farthest. In fact, faint traces of blue and violet light can still be measured in extremely clear water at depths of 1000 meters (3300 feet).

Although photosynthesis cannot occur much below 100 meters (330 feet), there is enough light in the lower photic zone for marine animals to avoid predators, find food, recognize their species, and locate mates. Even deeper is the **aphotic** (*a* = without, *photos* = light) **zone,** where there is no sunlight.

Distance from Shore Marine life zones can also be subdivided based on distance from shore. The area where the land and ocean meet and overlap is called

the **intertidal zone.** This narrow strip of land between high and low tides is alternately covered and uncovered by seawater with each tidal change. Even though it appears to be a harsh place to live with crashing waves, periodic drying out, and rapid changes in temperature, salinity, and oxygen concentrations, many species live here that are superbly adapted to the dramatic environmental changes.

Seaward from the low-tide line is the **neritic** (*neritos* = of the coast) **zone.** This covers the gently

Basis	Marine Life Zone	Subdivision	Characteristics
Table 13.2 Marine life zones.			
Available sunlight	Photic		Sunlit surface waters.
		Euphotic	Has enough sunlight to support photosynthesis.
	Aphotic		No sunlight; many organisms have bioluminescent capabilities.
Distance from shore	Intertidal		Narrow strip of land between high and low tides; dynamic area.
	Neritic		Above continental shelf; high biomass and diversity of species.
	Oceanic		Open ocean beyond the shelf break; low nutrient concentrations.
Depth	Pelagic		All water above the ocean floor; organisms swim or float.
	Benthic		Bottom of ocean; organisms attach to, burrow into, or crawl on seafloor.
		Abyssal	Deep-sea bottom; dark, cold, high pressure; sparse life.

Water Depth   A third method of classifying marine habitats is based on water depth. Open ocean of *any* depth is called the **pelagic** (*pelagios* = of the sea) **zone.** Animals in this zone swim or float freely. The photic part of the pelagic zone is home to phytoplankton, zooplankton, and nekton, such as tuna, sea turtles, and dolphins. The aphotic part has strange species like viperfish and giant squid that are adapted to life in deep water.

Benthos organisms such as giant kelp, sponges, crabs, sea anemones, sea stars, and marine worms that attach to, crawl upon, or burrow into the seafloor occupy parts of the **benthic** (*benthos* = bottom) **zone.** The benthic zone includes any sea-bottom surface regardless of its distance from shore and is mostly inhabited by benthos organisms.

The **abyssal** (*a* = without, *byssus* = bottom) **zone** is a subdivision of the benthic zone and includes the deep-ocean floor, such as *abyssal plains*. This zone is characterized by extremely high water pressure, consistently low temperature, no sunlight, and sparse life. Two food sources typically exist at abyssal depths: (1) tiny decaying particles steadily "raining" down from above, which provide food for filter-feeders, brittle stars, and burrowing worms; and (2) large fragments or entire dead bodies falling at scattered sites, which supply meals for actively searching fish, such as the grenadier, tripodfish, and hagfish, which locate food by chemical sensing. However, a third source of food has recently been discovered that is associated with hot springs on the seafloor, called *hydrothermal vents* (Box 13.2).

Oceanic Productivity

Why are some regions of the ocean teeming with life, while other areas seem barren? The answer is related to the amount of primary productivity in various parts of the ocean. **Primary productivity** is the amount of carbon fixed by organisms through the synthesis of organic matter using energy derived from solar radiation (*photosynthesis*) or chemical reactions (*chemosynthesis*). Although chemosynthesis supports hydrothermal vent biocommunities along the oceanic ridge (Box 13.2), it is much less significant than photosynthesis in worldwide oceanic productivity.

Two factors influence a region's photosynthetic productivity: *availability of nutrients* (such as nitrates, phosphorous, iron, and silica) and the *amount of solar radiation* (sunlight). Thus, the most abundant marine life exists where there are ample nutrients and good sunlight. Oceanic productivity, however, varies dramatically because of the uneven distribution of nutrients throughout the photosynthetic zone and seasonal changes in the availability of solar energy.

A permanent *thermocline* (and resulting *pycnocline*) develops nearly everywhere in the oceans (see Fig-

sloping continental shelf out to the *shelf break* (see Figure 12.8, p. 340). This zone can be very narrow or may extend hundreds of kilometers from shore. The neritic zone is often shallow enough for sunlight to reach all the way to the ocean floor, putting it entirely within the photic zone.

Although the neritic zone covers only about 5 percent of the world ocean, it is rich in both biomass and number of species. Many organisms find the conditions here ideal because photosynthesis occurs readily, nutrients wash in from the land, and the bottom provides shelter and habitat. This zone is so rich, in fact, that it supports 90 percent of the world's commercial fisheries.

Beyond the continental shelf is the **oceanic zone.** The open ocean reaches great depths, and as a result, surface waters typically have lower nutrient concentrations because nutrients tend to sink out of the photic zone to the deep-ocean floor. This low nutrient concentration usually results in much smaller populations than the more productive neritic zone.

Figure 13.10 The ocean is divided into marine life zones, based on availability of light, distance from shore, and water depth.

ure 13.6). This layer forms a barrier to vertical mixing and prevents the resupply of nutrients to sunlit surface waters. In the midlatitudes, a thermocline develops only during the summer season, and in polar regions a thermocline does not usually develop at all. The degree to which waters develop a thermocline profoundly affects the amount of productivity observed at different latitudes.

Productivity in Polar Oceans

Polar regions such as the Arctic Ocean's Barents Sea, which is off the northern coast of Europe, experience continuous darkness for about three months of winter and continuous illumination for about three months during summer. Productivity of phytoplankton—mostly single-celled algae called *diatoms*—peaks there during May (Figure 13.11), when the Sun rises high enough in the sky so that there is deep penetration of sunlight into the water. As soon as the diatoms develop, zooplankton—mostly small crustaceans called *copepods* (Figure 13.11) and larger *krill*—begin feeding on them. The zooplankton biomass peaks in June and continues at a relatively high level until winter darkness begins in October.

Recall that density and temperature change very little with depth in polar regions (see Figures 13.4 and 13.5), so these waters are *isothermal* and there is no barrier to mixing between surface waters and deeper, nutrient-rich waters. In the summer, however, melting ice creates a thin, low-salinity layer that does not readily mix with the deeper waters. This stratification is crucial to summer production, because it helps prevent phytoplankton from being carried into deeper, darker waters. Instead, they are concentrated in the sunlit surface waters where they reproduce continuously.

Because of the constant supply of nutrients rising from deeper waters below, high-latitude surface waters typically have high nutrient concentrations. The availability of solar energy, however, is what limits photosynthetic productivity in these areas.

Productivity in Tropical Oceans

You may be surprised to learn that productivity is low in tropical regions of the open ocean. Because the Sun is more directly overhead, light penetrates much deeper into tropical oceans than in temperate and polar waters, and solar energy is available year-round. However, productivity is low in tropical regions of the open ocean because a permanent thermocline produces a stratification of water masses that prevents mixing between surface waters and nutrient-rich deeper waters (Figure 13.12). In essence, the thermocline is a barrier that

BOX 13.2

EARTH AS A SYSTEM:
Deep-Sea Hydrothermal Vent Biocommunities—Earth's First Life?

Deep-sea hydrothermal vents form along many active rift zones of the oceanic ridge. Here seawater percolates down into the hot and newly formed oceanic crust. During its journey, the water can become saturated with minerals before it spews back into the ocean as a *black smoker* (Figure 13.B). Black smokers usually emit from tall chimneys composed of metal sulfide minerals that have precipitated as the hot ventwater is exposed to cold seawater.

Water temperatures at some vents reach as high as 350°C (660°F), which is far too hot for anything to live. At some vents, however, water temperatures of 100°C (212°F) or lower nourish dense, exotic *hydrothermal vent biocommunities* of organisms found nowhere else in the world. In fact, hundreds of new species—and even new genera and families—have been discovered surrounding these deep-sea habitats since their discovery by scientists along the Galápagos Rift in 1977. Additional hydrothermal vent biocommunities exist in discrete patches along the oceanic ridge and have been visited by scientists in deep-sea submersibles

along the East Pacific Rise, Mid-Atlantic Ridge, Mid-Indian Ocean Ridge, and the Juan de Fuca Ridge.

How do these organisms survive in this dark, hot, sulfur-rich environment where photosynthesis cannot occur? Study of hydrothermal vent organisms reveals that microscopic bacteria-like organisms called *archaea* (archaeo = ancient) living in and near the vents perform *chemosynthesis* (chemo = chemistry, *syn* = with, *thesis* = an arranging) and constitute the base of the food web. Hydrothermal vents provide heat energy for archaea to oxidize hydrogen sulfide (H_2S), which is formed by the reaction of hot water with dissolved sulfate (SO_4^{-2}). Through chemosynthesis, archaea produce sugars and other foods that enable them and many other organisms to live in this very unusual and extreme environment.

Some archaea live symbiotically inside giant gutless tube-dwelling worms (Figure 13.C). These archaea provide food for the tube worms to grow as rapidly as 1 meter (3.3 feet) each year and up to 3 meters (10 feet) in length. Other archaea are consumed by specialized yellow mussels, giant white

Figure 13.C Tube worms up to 3 meters (10 feet) in length are among the organisms found in the extreme environment of hydrothermal vents along the crest of the oceanic ridge where sunlight is nonexistent. These organisms obtain their food from internal microscopic bacteria-like archaea, which acquire their nourishment and energy through the processes of chemosynthesis. (Photo by Al Giddings Images, Inc.)

clams, and pink sea urchins. These in turn are eaten by unique crabs and fishes. Thus, archaea are the foundation of a living ecosystem that does not require sunlight.

It is very likely that environments similar to those of hydrothermal vents were present during the early history of the planet. Some scientists have suggested that the uniformity of conditions and abundant energy of the vents would have provided an ideal habitat for the origin of life. In fact, hydrothermal vents may represent one of the oldest life-sustaining environments, because hydrothermal activity occurs wherever there are both volcanoes and water. An additional line of evidence in support of hydrothermal vents harboring some of Earth's first life exists in the fact that archaea contain ancient genetic makeup.

Figure 13.B View from the submersible *Alvin* of a black smoker spewing hot, mineral-rich water along the East Pacific Rise. When heated solutions meet cold seawater, metal sulfides precipitate and form mounds of minerals around these hydrothermal vents. (Photo by Dudley Foster, Woods Hole Oceanographic Institution)

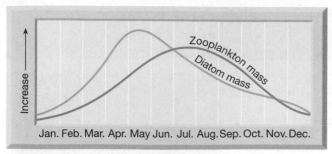

Figure 13.11 One example of productivity in polar oceans is illustrated by the Barents Sea. A springtime increase in diatom mass is followed closely by an increase in zooplankton abundance.

Figure 13.12 Productivity in tropical oceans. Although tropical regions receive adequate sunlight year-round, a permanent thermocline prevents the mixing of surface and deep water. As phytoplankton consume nutrients in the surface layer, productivity is limited because the thermocline prevents replenishment of nutrients from deeper water. Thus, productivity remains at a steady, low level.

eliminates the supply of nutrients from deeper waters below. So, productivity in tropical regions is limited by the lack of nutrients (unlike polar regions, where productivity is limited by the lack of sunlight). In fact, these areas have so few organisms that they are considered biological deserts.

Productivity in Temperate Oceans

Productivity is limited by available sunlight in polar regions and by nutrient supply in the tropics. In temperate (midlatitude) regions, a combination of these two limiting factors controls productivity as shown in Figure 13.13 (which shows the pattern for the Northern Hemisphere; in the Southern Hemisphere, the seasons are reversed).

Winter   Productivity in temperate oceans is very low during winter, even though nutrient concentration is highest at this time. The reason is that solar energy is limited because days are short and the Sun angle is low. As a result, the depth at which photosynthesis can occur is so shallow that phytoplankton do not grow much.

Spring   The Sun rises higher in the sky during spring, creating a greater depth at which photosynthesis can occur. A *spring bloom* of phytoplankton occurs because solar energy and nutrients are available, and a seasonal thermocline develops (due to increased solar heating) that traps algae in the euphotic zone (Figure 13.13). This creates a tremendous demand for nutrients in the euphotic zone, so the supply is quickly depleted, causing productivity to decrease sharply. Even though the days are lengthening and sunlight is increasing, productivity during the spring bloom is limited by the lack of nutrients.

Summer   The Sun rises even higher in the summer, so surface waters in temperate parts of the ocean continue to warm. A strong seasonal thermocline is created that in turn prevents vertical mixing, so nutrients depleted from surface waters cannot be replaced by those from deeper waters. Throughout summer,

the phytoplankton population remains relatively low (Figure 13.13).

Fall   Solar radiation diminishes in the fall as the Sun moves lower in the sky, so surface temperatures drop and the summer thermocline breaks down. Nutrients return to the surface layer as increased wind strength mixes surface waters with deeper waters. These conditions create a *fall bloom* of phytoplankton, which is much less dramatic than the spring bloom (Figure 13.13). The fall bloom is very short-lived because sunlight (not nutrient supply, as in the spring bloom) becomes the limiting factor as winter approaches to repeat the seasonal cycle.

Figure 13.14 compares the seasonal variation in phytoplankton biomass of tropical, north polar, and north temperate regions, where the total area under each curve represents photosynthetic productivity. The figure shows the dramatic peak in productivity in polar oceans during the summer; the steady, low rate of productivity year-round in the tropical oceans; and the seasonal productivity that occurs in temperate oceans. It also shows that the highest overall productivity occurs in temperate regions.

Oceanic Feeding Relationships

Marine algae, plants, bacteria, and bacteria-like archaea are the main oceanic producers. As these producers make food (organic matter) available to the consuming animals of the ocean, it passes from one feeding population to the next. Only a small percentage of the energy taken in at any level is passed on to the next because energy is consumed and lost at each level. As a result, the producers' biomass in the ocean is many times greater than the mass of the top consumers, such as sharks or whales.

PEOPLE AND THE ENVIRONMENT:
The Iron Hypothesis—Fertilizing the Ocean to Reduce Global Warming

The documented increase in Earth's average surface temperature over the last 130 years—likely as a result of human activities that have increased the amount of carbon dioxide in the atmosphere—has been a topic of much debate and discussion.* Nevertheless, scientists are seeking ways to prevent further alteration of Earth's climate by reducing the amount of atmospheric carbon dioxide. Oceanographer John Martin was one of the first to propose increasing oceanic productivity as a way to remove carbon dioxide from the atmosphere, thereby reducing global warming.

How does an increase in ocean productivity remove carbon dioxide from the atmosphere? Through photosynthesis, microscopic marine algae con-

*This topic is explored more fully in "Human Impact on Global Climate" in Chapter 19.

vert carbon dioxide dissolved in the ocean to carbohydrates and oxygen gas. As the number of photosynthetic marine organisms increases, additional carbon dioxide is removed from the ocean, which in turn causes the ocean to absorb more carbon dioxide from the atmosphere.

Areas of the ocean that have relatively low productivity, such as the tropical oceans, are a good place to stimulate productivity and thus increase the amount of carbon dioxide removed from the atmosphere. In 1987, Martin determined that the absence of iron limited productivity in tropical oceans, so he proposed fertilizing the ocean with iron—the only element that seemed to be lacking—in order to increase its productivity (the "iron hypothesis"). In 1993, Martin's associates added finely ground iron to a test area of the ocean near the Galápagos Islands in the Pacific Ocean. Their results and

the results of other "Iron Ex" open-ocean experiments in 1995 and 1999 showed that adding iron to the ocean increased productivity up to 30 times. Although these results are promising, there were problems grinding the iron fine enough, dispersing the iron, and keeping it in suspension for long periods of time.

An even larger problem may be the long-term global environmental effects of adding additional iron and large amounts of carbon dioxide to the ocean. Many scientists fear that increased amounts of carbon dioxide in the ocean would upset the ocean's natural chemical balance and alter the entire global marine ecosystem. Proponents of iron fertilization, however, claim that stimulating productivity mimics what is done naturally in the ocean and that iron fertilization may be one of the most promising solutions for reducing atmospheric carbon dioxide.

STUDENTS SOMETIMES ASK...
Are all sharks dangerous?

Of the more than 350 shark species, about 80 percent are unable to hurt people or rarely encounter people. The largest shark—also the largest fish in the world—is the whale shark (*Rhincodon typus*), which reaches lengths of up to 15 meters (50 feet) but eats only plankton and is therefore not considered dangerous.

Trophic Levels

Chemical energy stored in the mass of the ocean's algae (the "grass of the sea") is transferred to the animal community mostly through feeding. Zooplankton are *herbivores* (*herba* = grass, *vora* = eat), so they eat diatoms and other microscopic marine algae. Larger herbivores feed on the larger algae and marine plants that grow attached to the ocean bottom near shore.

The herbivores (grazers) are then eaten by larger animals, the *carnivores* (*carni* = meat, *vora* = eat). They in turn are eaten by another population of larger carnivores, and so on. Each of these feeding stages is called a **trophic** (*tropho* = nourishment) **level.**

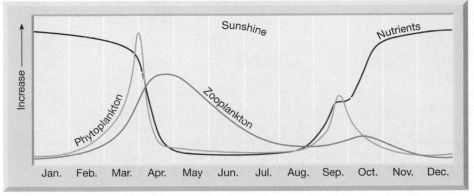

Figure 13.13 Productivity in temperate oceans (Northern Hemisphere). The graph shows the relationship among phytoplankton, zooplankton, amount of sunshine, and nutrient levels for surface waters.

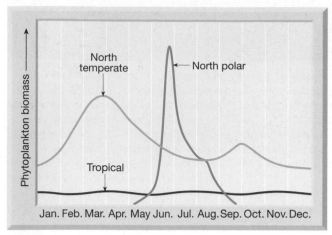

Figure 13.14 Comparison of productivity in tropical, temperate, and polar oceans (Northern Hemisphere), showing seasonal variation in phytoplankton biomass. The total area under each curve represents annual photosynthetic productivity.

Generally, individual members of a feeding population are larger—but not too much larger—than the organisms they eat. There are conspicuous exceptions, however, such as the blue whale. Up to 30 meters (100 feet) long, it is possibly the largest animal that has ever existed on Earth, yet it feeds mostly on krill, which have a maximum length of only 6 centimeters (2.4 inches).

Transfer Efficiency

The transfer of energy between trophic levels is very inefficient. The efficiencies of different algal species vary, but the average is only about *2 percent*, which means that 2 percent of the light energy absorbed by algae is ultimately synthesized into food and made available to herbivores.

Figure 13.15 shows the passage of energy between trophic levels through an entire ecosystem, from the solar energy assimilated by phytoplankton through all trophic levels to the ultimate carnivore—humans. Because energy is lost at each trophic level, it takes thousands of smaller marine organisms to produce a single fish that is so easily consumed during a meal!

Food Chains and Food Webs

A **food chain** is a sequence of organisms through which energy is transferred, starting with an organism that is the primary producer, then an herbivore, then one or more carnivores, finally culminating with the "top carnivore," which is not usually preyed upon by any other organism.

Because energy transfer between trophic levels is inefficient, it is advantageous for fishers to choose a population that feeds as close to the primary producing population as possible. This increases the biomass available for food and the number of individuals avail-

Figure 13.15 Ecosystem energy flow and efficiency. For every 500,000 units of radiant energy input available to the producers (phytoplankton), only one unit of mass is added to the fifth trophic level (humans). Average phytoplankton transfer efficiency is 2 percent (98 percent loss) and all other trophic levels average 10 percent efficiency (90 percent loss). The ultimate effect of energy transfer between trophic levels is that the number of individuals and the total biomass decrease at successive trophic levels because the amount of available energy decreases.

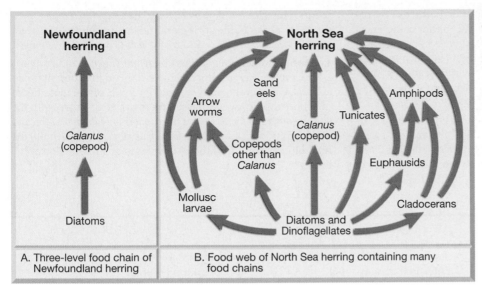

Figure 13.16   Comparison between a food chain and a food web. **A.** A food chain is the passage of energy along a single path, such as from diatoms to copepods to Newfoundland herring. Feeding relationships are rarely this simple. **B.** A food web, showing multiple paths for food sources of the North Sea herring.

able to be taken by the fishery. Newfoundland herring, for example, are an important fishery that usually represents the third trophic level in a food chain. They feed primarily on small copepods that feed, in turn, upon diatoms (Figure 13.16A).

Feeding relationships are rarely as simple as that of the Newfoundland herring. More often, top carnivores in a food chain feed on a number of different animals, each of which has its own simple or complex feeding relationships. This constitutes a **food web,** as shown in Figure 13.16B for North Sea herring.

Animals that feed through a food web rather than a food chain are more likely to survive because they have alternative foods to eat should one of their food sources diminish in quantity or even disappear. Newfoundland herring, on the other hand, eat only copepods, so the disappearance of copepods would catastrophically affect their population.

Chapter Summary

- *Salinity* is the amount of dissolved substances in water, usually expressed in parts per thousand (‰). Seawater salinity in the open ocean *averages* 35‰. The principal elements that contribute to the ocean's salinity are *chlorine* (55 percent) and *sodium* (31 percent), which combine to produce table salt. The primary *sources of the elements in sea salt* in the ocean are *chemical weathering* of rocks on the continents and *volcanic outgassing*.

- *Variations in seawater salinity* are primarily caused by changing the *water content*. Natural processes that add large amounts of fresh water to seawater and *decrease salinity* include *precipitation, runoff from land, icebergs melting,* and *sea ice melting*. Processes that remove large amounts of fresh water from seawater and *increase salinity* include the *formation of sea ice* and *evaporation*. Seawater salinity in the open ocean *ranges from 33‰ to 38‰*, with some marginal seas experiencing considerably more variation.

- The ocean's *surface temperature* is related to the amount of solar energy received and *varies as a function of latitude*. Low-latitude regions have distinctly colder water at depth, creating a *thermocline*, which is a layer of rapidly changing temperature. No thermocline exists in high-latitude regions because the water column is *isothermal*.

- Water's unique thermal properties have caused the *ocean's temperature to remain stable* for long periods of time, facilitating the development of life on Earth. Experiments have been conducted that *send sound through the ocean* to determine if the ocean's temperature is increasing as a result of global warming.

- Seawater *density is mostly affected by water temperature* but also by salinity. Low-latitude regions have distinctly denser (colder) water at depth, creating a *pycnocline,* which is layer of rapidly changing density. No pycnocline exists in high-latitude regions because the water column is *isopycnal*.

- Most open-ocean regions exhibit a *three-layered structure based on water density*. The shallow *surface mixed zone* has warm and nearly uniform temperatures. The *transition zone* includes a prominent thermocline and associated pycnocline. The *deep zone* is continually dark and cold and accounts for 80 percent of the water in the ocean. In high latitudes, the three-layered structure does not exist.

- *Marine life is superbly adapted to the oceans*. Marine organisms can be *classified* into one of three groups based on *habitat* and *mobility*. *Plankton* are free-floating forms with little power of locomotion, *nekton* are swimmers, and *benthos* are bottom dwellers. *Most of the ocean's biomass is planktonic*.

- Three criteria are frequently used to establish *marine life zones*. Based on *availability of sunlight*, the ocean can be divided into the *photic zone* (which includes the *euphotic zone*) and the *aphotic zone*. Based on *distance from shore*, the ocean can be divided into the *intertidal zone,* the *neritic zone,* and the *oceanic zone*. Based on *water depth*, the ocean can be divided into the *pelagic zone* and the *benthic zone* (which includes the *abyssal zone*).

- *Primary productivity* is the amount of carbon fixed by organisms through the synthesis of organic matter using energy derived from solar radiation (*photosynthesis*) or chemical reactions (*chemosynthesis*). Chemosynthesis is much less significant than photosynthesis in worldwide oceanic productivity. Photosynthetic productivity in the ocean varies due to the *availability of nutrients* and *amount of solar radiation*.

- *Oceanic photosynthetic productivity* varies at different latitudes because of *seasonal changes* and the *development of a thermocline*. In *polar oceans*, the availability of *solar radiation limits productivity* even though nutrient levels are high. In *tropical oceans*, a strong *thermocline* exists year-round, so the *lack of nu-* trients generally limits productivity. In *temperate oceans*, productivity peaks in the spring and fall and is *limited by the lack of solar radiation in winter* and by the *lack of nutrients in summer*.

- The Sun's energy is utilized by *phytoplankton* and converted to *chemical energy*, which is passed through different *trophic levels*. On average, only about *10 percent* of the mass taken in at one trophic level is passed on to the next. As a result, the *size of individuals increases* but the *number of individuals decreases* with each trophic level of the *food chain* or *food web*. Overall, the total biomass of populations decreases at successive trophic levels.

Key Terms

abyssal zone (p. 365)
aphotic zone (p. 364)
benthic zone (p. 365)
benthos (p. 363)
biomass (p. 362)
density (p. 360)
euphotic zone (p. 364)
food chain (p. 370)

food web (p. 371)
intertidal zone (p. 364)
nekton (p. 364)
neritic zone (p. 362)
oceanic zone (p. 365)
pelagic zone (p. 365)
photic zone (p. 363)
photosynthesis (p. 361)

phytoplankton (p. 362)
plankton (p. 362)
primary productivity (p. 365)
pycnocline (p. 361)
salinity (p. 356)
thermocline (p. 359)
trophic level (p. 369)
zooplankton (p. 362)

Review Questions

1. Define salinity. What is the average salinity of the ocean? Why do oceanographers typically express salinity in parts per thousand?

2. What are the six most abundant components (elements) dissolved in seawater? What is produced when the two most abundant elements are combined?

3. What are the two primary sources for the materials that comprise the dissolved components in seawater?

4. Describe the processes that affect seawater salinity. For each process, indicate whether water is added or removed and if it decreases or increases salinity. What physical conditions create high salinity water in the Red Sea and low salinity water in the Baltic Sea?

5. Why is desalination not likely to be a significant source of water for agriculture in the foreseeable future?

6. What one factor is primarily responsible for influencing seawater temperature?

7. Describe temperature variation with depth in both high and low latitudes. Why do high-latitude waters generally lack a thermocline?

8. What is the goal of the ATOC experiment? How do scientists propose achieving their goal? What problems could the experiment create?

9. Which two factors influence seawater density? Of the two, which one has the greatest influence on surface seawater density?

10. Describe density variation with depth in both high and low latitudes. Why do high-latitude waters generally lack a pycnocline?

11. Describe the ocean's layered structure. Why does the three-layer structure not exist in high latitudes?

12. Describe the lifestyles of plankton, nekton, and benthos and give examples of each.

13. Of plankton, nekton, and benthos, which group comprises the largest biomass? Explain why.

14. List three physical factors that are used to divide the ocean into marine life zones. Describe how each factor influences the abundance and distribution of marine life.

15. Why are there greater numbers and types of organisms in the neritic zone than in the oceanic zone?

16. What is the base of the food web for hydrothermal vent biocommunities? What is unusual about this food source?

17. List the two methods in which primary productivity is accomplished in the ocean. Of these two methods, which one is the most significant? What two factors influence it?

18. Compare the biological productivity of polar, temperate, and tropical regions of the oceans. Consider seasonal changes, the development of a thermocline, the availability of nutrients, and variations in solar radiation.

19. What is the average efficiency of energy transfer between trophic levels? Use this efficiency to determine how much phytoplankton mass is required to add *1 gram* of new mass to a killer whale, which is a third-level carnivore. Include a diagram that shows the different trophic levels and the relative size and abundance of organisms at different levels. How would your answer change if the efficiency were half the average rate, or twice the average rate?

20. Describe the advantage that a top carnivore gains by eating from a food web as compared to a single food chain.

Examining the Earth System

1. Discuss the ocean's importance in the hydrologic cycle (see Figures 4.1 and 4.2). If Earth did not have oceans, how might this affect (a) the global biosphere, and (b) the rock cycle?

2. Describe the iron hypothesis and discuss the relative merits and dangers of undertaking a project that could cause dramatic changes in the global environment. For details about the iron hypothesis and its inventor, oceanographer John Martin, visit NASA's Earth Observatory site at:

   http://earthobservatory.nasa.gov/Library/Giants/Martin/

Web Resources

 The *Earth Science* Website uses the resources and flexibility of the Internet to aid in your study of the topics in this chapter. Written and developed by Earth science instructors, this site will help improve your understanding of Earth science. Visit **http://www.prenhall.com/tarbuck** and click on the cover of *Earth Science* 10e to find:

- **Online review quizzes.**
- **Web-based critical thinking and writing exercises.**
- **Links to chapter-specific Web resources.**
- **Internet-wide key term searches.**

http://www.prenhall.com/tarbuck

Waves breaking along California's rocky Big Sur coast. (Photo by Carr Clifton)

CHAPTER 14

THE DYNAMIC OCEAN*

*This chapter was revised by Professor Alan P. Trujillo, Palomar College.

The restless waters of the ocean are constantly in motion, powered by many different forces. Winds, for example, generate surface currents, which influence coastal climate and provide nutrients that affect the abundance of algae and other marine life in surface waters. Winds also produce waves that carry energy from storms to distant shores, where their impact erodes the land. In some areas, density differences create deep-ocean circulation, which is important for ocean mixing and recycling nutrients. In addition, the Moon and the Sun produce tides, which periodically raise and lower average sea level. This chapter examines these movements of ocean waters and their effect upon coastal regions (Figure 14.1).

Surface Circulation

Ocean currents are masses of ocean water that flow from one place to another. The amount of water can be large or small, currents can be at the surface or deep below, and the phenomena that create them can be simple or complex. In all cases, however, the currents that are generated involve water masses in motion.

Surface currents develop from friction between the ocean and the wind that blows across its surface. Some of these currents are short-lived and affect only small areas. Such water movements are responses to local or seasonal influences. Other surface currents are relatively permanent phenomena that extend over large portions of the oceans. These major horizontal movements of surface waters are closely related to the general circulation pattern of the atmosphere. This is clearly illustrated by comparing the pattern of global winds shown in Figure 17.13 (p. 481) and the position of Earth's principal surface ocean currents shown in Figure 14.2.

Ocean Circulation Patterns

Huge circular-moving current systems dominate the surfaces of the oceans. These large whirls of water within an ocean basin are called **gyres** (*gyros* = a circle). Figure 14.2 shows the world's five main gyres: the *North Pacific Gyre*, the *South Pacific Gyre*, the *North Atlantic Gyre*, the *South Atlantic Gyre*, and the *Indian Ocean Gyre* (which exists mostly within the Southern Hemisphere). The center of each gyre coincides with the subtropics at about 30 degrees north or south latitude, so they are often called *subtropical gyres*. Comparing Figure 17.13 (p. 481) to Figure 14.2 shows that there is a striking correspondence between the direction of surface-current flow and the major wind belts of the world.

Figure 14.1 Wind is not only responsible for creating waves, but it also provides the force that drives the ocean's surface circulation. (Photo by Marc Muench)

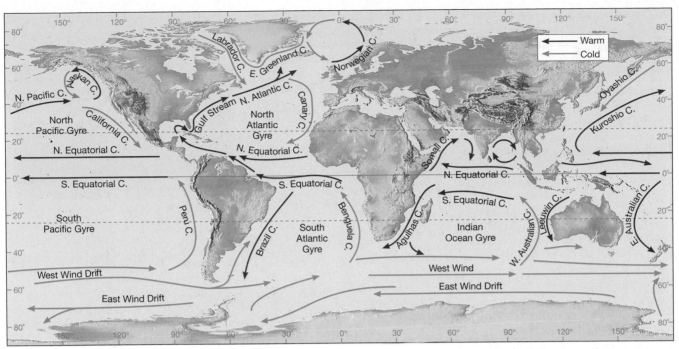

Figure 14.2   Average ocean surface currents in February–March. The ocean's circulation is organized into five major current gyres (large circular-moving loops of water), which exist in the North Pacific, South Pacific, North Atlantic, South Atlantic, and Indian Oceans.

As shown in Figure 14.2, subtropical gyres rotate clockwise in the Northern Hemisphere and counterclockwise in the Southern Hemisphere. Why do the gyres flow in different directions in the two hemispheres? Although wind is the force that generates surface currents, other factors also influence the movement of ocean waters. The most significant of these is the **Coriolis effect.** Because of Earth's rotation, currents are deflected to the *right* in the Northern Hemisphere and to the *left* in the Southern Hemisphere. (The Coriolis effect is more fully explained in Chapter 17.) As a consequence, gyres flow in opposite directions in the two different hemispheres.

Four main currents generally exist within each gyre (Figure 14.2). The North Pacific Gyre, for example, consists of the North Equatorial Current, the Kuroshio* Current, the North Pacific Current, and the California Current. The tracking of floating objects that are released into the ocean intentionally or accidentally reveal that it takes about six years for the objects to go all the way around the loop (see Box 14.1).

In the North Atlantic, the North Equatorial Current is deflected northward through the Caribbean, where it becomes the Gulf Stream (Figure 14.2). As the Gulf Stream moves along the East Coast of the United States, it is strengthened by the prevailing westerly winds and is deflected to the east (to the right) between the Car-

olinas and New England. As it continues northeastward, it gradually widens and slows until it becomes a vast, slowly moving current known as the North Atlantic Current, which, because of its sluggish nature, is also known as the North Atlantic Drift.

As the North Atlantic Current approaches Western Europe, it splits, part of it moving northward past Great Britain, Norway, and Iceland, carrying heat to these otherwise chilly areas. The other part is deflected southward as the cool Canary Current. As the Canary Current moves southward, it eventually merges into the North Equatorial Current, completing the gyre. Because the North Atlantic Ocean basin is about half the size of the North Pacific, it takes floating objects about three years to go completely around this gyre.

The circular motion of gyres leaves a large central area that has no well-defined currents. In the North Atlantic, this zone of calmer waters is known as the Sargasso Sea, named for the large quantities of *Sargassum*, a type of floating seaweed encountered there.

The ocean basins in the Southern Hemisphere exhibit a similar pattern of flow as the Northern Hemisphere basins, with surface currents that are influenced by wind belts, the position of continents, and the Coriolis effect. In the South Atlantic and South Pacific, for example, surface ocean circulation is very much the same as in their Northern Hemisphere counterparts except that the direction of flow is counterclockwise (Figure 14.2).

The Indian Ocean exists mostly in the Southern Hemisphere, so it follows a surface circulation pattern similar to other Southern Hemisphere ocean basins

---

*Kuroshio is pronounced "kuhr-ROH-shee-oh" and is sometimes known as the Japan Current. The term Kuroshio is Japanese for "black tide" in reference to its clear, lifeless waters.

BOX 14.1

UNDERSTANDING EARTH:
Running Shoes as Drift Meters—Just Do It

Any floating object can serve as a makeshift drift meter, as long as it is known where the object entered the ocean and where it was retrieved. The path of the object can then be inferred, providing information about the movement of surface currents. If the times of release and retrieval are known, the speed of currents can also be determined. Oceanographers have long used *drift bottles* (a floating "message in a bottle" or a radio-transmitting device set adrift in the ocean) to track the movement of currents and, more recently, to refine computer models of ocean circulation.

Many objects have inadvertently become drift meters when ships have lost some (or all) of their cargo at sea. In this way, Nike athletic shoes and colorful floating bathtub toys (Figure 14A, *inset*) have helped oceanographers advance the understanding of surface circulation in the North Pacific Ocean.

In May 1990 the container vessel *Hansa Carrier* was en route from Korea to Seattle, Washington, when it encountered a severe North Pacific storm. During the storm the ship lost 21 deck containers overboard, including five that held Nike athletic shoes. The shoes that were released from their containers floated and were carried east by the North Pacific Current. Within six months thousands of the shoes began to wash up along the beaches of Alaska, Canada, Washington, and Oregon (Figure 14.A), over 2400 kilometers (1500 miles) from the site of the spill. A few shoes were found on beaches in northern California, and over two years later shoes from the spill were even recovered from the north end of the Big Island of Hawaii!

With help from the beachcombing public and remotely based lighthouse operators, information on the location and number of shoes collected was compiled during the months following the spill. Serial numbers inside the shoes were traced to individual containers, indicating that only four of the

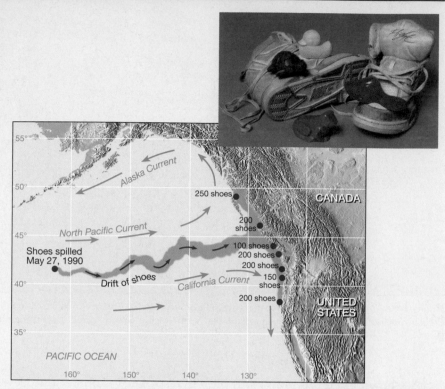

Figure 14.A Large ships crossing the ocean have lost entire containers overboard. If the containers release floating items, inadvertent float meters are launched that help oceanographers track ocean surface currents. Map shows path of drifting shoes and recovery locations from a spill in 1990; inset shows recovered shoes from the 1990 spill and plastic bathtub toys from a spill in 1992. (Courtesy of *Eos Transactions AGU* 75:37, p. 425, 1994)

five containers had released their shoes; evidently, one entire container sank without opening. Thus, a maximum of 30,910 pairs of shoes (61,820 individual shoes) were released. Before the shoe spill, the largest number of drift bottles purposefully released at one time by oceanographers was about 30,000. Although only 2.6 percent of the shoes were recovered; this compares favorably with the 2.4 percent recovery rate of drift bottles released by oceanographers conducting research.

In January 1992 another cargo ship lost 12 containers overboard during a storm to the north of where the shoes had previously spilled. One of these containers held 29,000 packages of small, floatable, colorful plastic bathtub toys in the shapes of blue turtles, yellow ducks, red beavers, and green frogs

(Figure 14.A, *inset*). Even though the toys were housed in plastic packaging glued to a cardboard backing, studies showed that after 24 hours in seawater, the glue deteriorated, thereby releasing over 100,000 individual floating toys.

The floating bathtub toys began to come ashore in southeast Alaska 10 months later, verifying computer models of North Pacific circulation. The models indicate that many of the bathtub toys will continue to be carried by the Alaska Current, eventually dispersing throughout the North Pacific Ocean.

Since 1992 oceanographers have continued to study ocean currents by tracking other floating items spilled from cargo ships, including 34,000 hockey gloves, 5 million plastic Lego pieces, and an unidentified number of small plastic doll parts.

(Figure 14.2). The small portion of the Indian Ocean in the Northern Hemisphere, however, is influenced by the seasonal wind shifts known as the summer and winter *monsoons* (*mausim* = season). When the winds change direction, the surface currents also reverse direction.

The West Wind Drift is the only current that completely encircles Earth (Figure 14.2). It flows around the ice-covered continent of Antarctica, where no large landmasses are in the way, so its cold surface waters circulate in a continuous loop. It moves in response to the Southern Hemisphere prevailing westerly winds, and portions of it split off into the adjoining southern ocean basins.

The Gulf Stream

Anyone who navigates the oceans needs to be aware of currents. By understanding the direction and strength of ocean currents, sailors soon realize that their voyage time can be reduced if they travel with a current, or increased if they travel against a current. Such is the case for the strong Gulf Stream current, which is the best known and most studied of all ocean currents.

The Gulf Stream flows northward along the East Coast of the United States. It was given the name Gulf Stream because it carries warm water from the Gulf of Mexico and because it was narrow and well defined—similar to a stream, but in the ocean. As deputy postmaster general for the new colonies, Benjamin Franklin was the first to produce a map of the Gulf Stream (Figure 14.3). His interest in the North Atlantic began when it was brought to his attention that ships carrying the mail took two weeks longer going from England to America than in the other direction. P. L. Richardson[*] describes how Franklin discovered that an ocean current was causing the delay:

> While he was in London as Deputy Postmaster General for the American colonies, Franklin was consulted on the question of why the mail packets took a fortnight longer to sail to America than the merchant ships. In October 1768 Franklin discussed this problem with his cousin Timothy Folger, a Nantucket ship captain then visiting London. Folger told him the packet captains were ignorant of the Gulf Stream and frequently sailed in this current.... Folger sketched the Gulf Stream on a chart and added written notes on how to avoid the Gulf Stream and Franklin had the chart printed in 1769 or 1770.

Even though the Franklin-Folger chart of the Gulf Stream was widely distributed, it was initially ignored by the captains of the mail ships because they thought that simple American fishers could not possess superior knowledge of the sea.

Nearly 100 years later Matthew Fontaine Maury, the founder of physical oceanography, described the Gulf Stream in his 1855 book, *The Physical Geography of the Sea*:

> There is a river in the ocean. In the severest droughts it never fails, and in the mightiest floods . . . it never overflows. Its banks and its bottoms are of cold water, while its current is of warm. The Gulf of Mexico is its fountain, and its mouth is in the Arctic Sea. It is the Gulf Stream. There is in the world no other such majestic flow of waters.

Today the warm waters of the Gulf Stream can be mapped by Earth-orbiting satellites (Figure 14.3). Such satellite maps reveal that the Gulf Stream has many complexities, including prominent bends called meanders, which produce large circular eddies that spin off from the main current and can last for up to two years before dissipating. Although the Franklin-Folger chart was created long before the advent of modern technology, which allows navigators to map ocean currents from space, it still remains a good summary of the average path of the Gulf Stream.

Ocean Currents and Climate

Ocean currents have an important effect on climates. When currents from low-latitude regions move into higher latitudes, they transfer heat from warmer to cooler areas on Earth. In fact, the North Atlantic Current—an extension of the warm Gulf Stream—keeps Great Britain and much of northwestern Europe warmer during the winter than one would expect for their latitudes, which are similar to the latitudes of Alaska and Labrador. The prevailing westerly winds carry the moderating effects far inland. For example, Berlin, Germany (52 degrees north latitude), has an average January temperature similar to that experienced at New York City, which lies 12 degrees latitude farther south.

In contrast to warm ocean currents whose effects are felt mostly in the middle latitudes in winter, the influence of cold currents is most pronounced in the tropics or during summer months in the middle latitudes. Cold currents originate in cold high-latitude regions. As these currents travel equatorward, they tend to moderate the warm temperatures of adjacent land areas. Such is the case for the Benguela Current along western Africa, the Peru Current along the west coast of South America, and the California Current (see Figure 14.2).

In addition to influencing temperatures of adjacent land areas, cold currents have other climatic influences. For example, where tropical deserts exist along the west coasts of continents, cold ocean currents have a dramatic impact. The principal west coast deserts are the Atacama in Peru and Chile, and the Namib in southwestern Africa. The aridity along these coasts is intensified because the lower atmosphere is chilled by cold offshore waters. When this occurs, the air becomes very stable and resists the upward movement necessary to create precipitation-producing clouds. In addition, the presence of cold currents causes temperatures to approach and often reach the dew point, the temperature

[*]P. L. Richardson, 1980, "Benjamin Franklin and Timothy Folger's First Printed Chart of the Gulf Stream," *Science* 207 (4431): 643.

Figure 14.3 To aid ships crossing the Atlantic, Benjamin Franklin, with the assistance of his cousin, Timothy Folger, produced the first detailed chart of the Gulf Stream about 1769 or 1770. Today satellite images provide views of the Gulf Stream's complexities such as this false-color image of sea surface temperature. The warm waters of the Gulf Stream are shown in red and orange; colder waters are shown in green, blue, and purple. As the Gulf Stream meanders northward, some of its meanders pinch off to form large circular eddies. (Courtesy of O. Brown, R. Evans, and M. Carle/University of Miami Rosenstiel School of Marine and Atmospheric Science)

at which water vapor condenses. As a result, these areas are characterized by high relative humidities and much fog. Thus, not all tropical deserts are hot with low humidities and clear skies. Rather, the presence of cold currents transforms some tropical deserts into relatively cool, damp places that are often shrouded in fog.

Ocean currents also play a major role in maintaining Earth's heat balance. They accomplish this task by transferring heat from the tropics, where there is an excess of heat, to the polar regions, where a heat deficit exists. Ocean water movement accounts for about a quarter of this heat transport, and winds transport the remaining three-quarters.

Upwelling

In addition to producing surface currents, winds can also cause *vertical* water movements. **Upwelling,** the rising of cold water from deeper layers to replace warmer surface water, is a common wind-induced vertical movement. One type of upwelling, called *coastal upwelling,* is most characteristic along the west coasts

of continents, most notably along California, western South America, and West Africa.

Coastal upwelling occurs in these areas when winds blow toward the equator and parallel to the coast (Figure 14.4). Coastal winds combined with the Coriolis effect cause surface water to move away from shore. As the surface layer moves away from the coast, it is replaced by water that "upwells" from below the surface. This slow upward movement of water from depths of 50 to 300 meters (165 to 1000 feet) brings water that is cooler than the original surface water and results in lower surface water temperatures near the shore.

For swimmers who are accustomed to the warm waters along the mid-Atlantic shore of the United States, a swim in the Pacific off the coast of central California can be a chilling surprise. In August, when temperatures in the Atlantic are 21°C (70°F) or higher, central California's surf is only about 15°C (60°F).

Upwelling brings greater concentrations of dissolved nutrients, such as nitrates and phosphates, to the ocean surface. These nutrient-enriched waters from below promote the growth of microscopic plankton,

Figure 14.4   Coastal upwelling occurs along the west coasts of continents where winds blow toward the equator and parallel to the coast. Owing to the Coriolis effect (deflection to the left in the Southern Hemisphere), surface water moves away from the shore, which brings cold, nutrient-rich water to the surface. This image from the SeaStar satellite shows chlorophyll concentration along the southwest coast of Africa (February 21, 2001). An instrument aboard the satellite detects changes in seawater color caused by changing concentrations of chlorophyll. High chlorophyll concentrations indicate high amounts of photosynthesis, which is linked to the upwelling nutrients. Red indicates high concentrations and blue low concentrations. (Provided by the SeaWiFS Project, NASA/Goddard Space Flight Center and ORBIMAGE)

which in turn support extensive populations of fish and other marine organisms. Figure 14.4 is a satellite image that shows high productivity due to coastal upwelling off the southwest coast of Africa.

Deep-Ocean Circulation

In contrast to the largely horizontal movements of surface currents, deep-ocean circulation has a significant vertical component and accounts for the thorough mixing of deep-water masses. This component of ocean circulation is a response to density differences among water masses that cause denser water to sink and slowly spread out beneath the surface. Because the density variations that cause deep-ocean circulation are caused by differences in temperature and salinity, deep-ocean circulation is also referred to as **thermohaline** (*thermo* = heat, *haline* = salt) **circulation.**

Recall from Chapter 13 that an increase in seawater density can be caused by a decrease in temperature or an increase in salinity. Density changes due to salinity variations are important in very high latitudes, where water temperature remains low and relatively constant.

Most water involved in deep-ocean currents (thermohaline circulation) begins in high latitudes at the surface. In these regions, surface water becomes cold and its salinity increases as sea ice forms (Figure 14.5). When this surface water becomes dense enough, it sinks, initiating deep-ocean currents. Once this water sinks, it is removed from the physical processes that increased its density in the first place, and so its temperature and salinity remain largely unchanged for the duration it spends in the deep ocean.

Near Antarctica, surface conditions create the highest density water in the world. This cold saline brine slowly sinks to the seafloor, where it moves throughout the ocean basins in sluggish currents. After sinking from the surface of the ocean, deep waters will not reappear at the surface for an average of 500 to 2000 years.

A simplified model of ocean circulation is similar to a conveyor belt that travels from the Atlantic Ocean through the Indian and Pacific Oceans and back again (Figure 14.6). In this model, warm water in the ocean's upper layers flows poleward, converts to dense water, and returns equatorward as cold deep water that eventually upwells to complete the circuit. As this "conveyor belt" moves around the globe, it influences global climate by converting warm water to cold and liberating heat to the atmosphere.

Waves

GEODe   Oceans
↳ Coastal Processes

Ocean waves are energy traveling along the interface between ocean and atmosphere, often transferring energy from a storm far out at sea over distances of several thousand kilometers. That's why even on calm days the ocean still has waves that travel across its surface. When observing waves, always remember that you are watching *energy* travel through a medium (water). If you make waves by tossing a pebble into a pond, or by splashing in a pool, or by blowing across the surface of

Figure 14.5 Sea ice in the Arctic Ocean. When seawater freezes, sea salts do not become part of the ice. Consequently, the salt content of the remaining seawater becomes more concentrated, which makes it denser and prone to sink. (Photo by Wayne Lynch/DRK Photo)

a cup of coffee, you are imparting *energy* to the water, and the waves you see are just the visible evidence of the energy passing through.

The power of waves is most noticeable along the shore—the dynamic interface between land and sea where waves are constantly rolling in and breaking. Sometimes the waves are low and gentle; at other times they pound the shore with an awesome fury. The conflict that results is never-ending and often dramatic.

Wave Characteristics

Most ocean waves derive their energy and motion from the wind. When a breeze is less than 3 kilometers (2 miles) per hour, only small wavelets appear. At greater wind speeds, more stable waves gradually form and advance with the wind.

Characteristics of ocean waves are illustrated in Figure 14.7, which shows a simple, nonbreaking wave-

form. The tops of the waves are the *crests,* which are separated by *troughs.* Halfway between the crests and troughs is the *still water level,* which is the level that the water would occupy if there were no waves. The vertical distance between trough and crest is called the **wave height,** and the horizontal distance between successive crests (or troughs) is the **wavelength.** The time it takes one full wave—one wavelength—to pass a fixed position is the **wave period.**

The height, length, and period that are eventually achieved by a wave depend on three factors: (1) wind speed; (2) length of time the wind has blown; and (3) **fetch,** the distance that the wind has traveled across open water. As the quantity of energy transferred from the wind to the water increases, both the height and steepness of the waves also increase. Eventually, a critical point is reached where waves grow so tall that they topple over, forming ocean breakers called *whitecaps.*

Figure 14.6 Idealized "conveyor belt" model of ocean circulation, which is initiated in the North Atlantic Ocean when warm water transfers its heat to the atmosphere, cools, and sinks below the surface. This water moves southward as a subsurface flow and joins water that encircles Antarctica. From here, this deep water spreads into the Indian and Pacific Oceans, where it slowly rises and completes the conveyer as it travels along the surface into the North Atlantic Ocean.

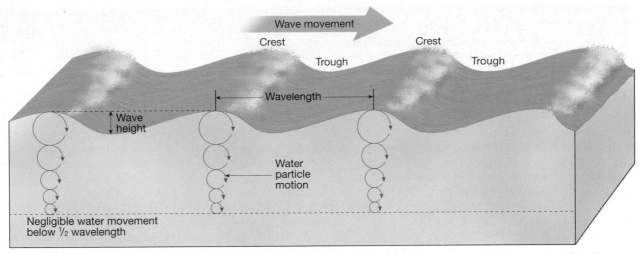

Figure 14.7   Diagrammatic view of an idealized non-breaking ocean wave showing the basic parts of a wave as well as the movement of water particles at depth. Negligible water movement occurs below a depth equal to one half the wavelength (*lower dashed line*).

When the wind stops or changes direction, or the waves leave the storm area where they were created, they continue on without relation to local winds. The waves also undergo a gradual change to *swells,* which are lower in height and longer in length and may carry a storm's energy to distant shores. Because many independent wave systems exist at the same time, the sea surface acquires a complex and irregular pattern, sometimes producing very large waves (Box 14.2). The sea waves that are seen from shore are usually a mixture of swells from faraway storms and waves created by local winds.

Circular Orbital Motion

Waves can travel great distances across ocean basins. In one study, waves generated near Antarctica were tracked as they traveled through the Pacific Ocean basin. After more than 10,000 kilometers (over 6000 miles), the waves finally expended their energy a week

later along the shoreline of the Aleutian Islands of Alaska. The water itself doesn't travel the entire distance, but the waveform does. As the wave travels, the water passes the energy along by moving in a circle. This movement is called *circular orbital motion.*

Observation of an object floating in waves reveals that it moves not only up and down but also slightly forward and backward with each successive wave. Figure 14.8 shows that a floating object moves up and backward as the crest approaches, up and forward as the crest passes, down and forward after the crest, down and backward as the trough approaches, and rises and moves backward again as the next crest advances. When the movement of the toy boat shown in Figure 14.8 is traced as a wave passes, it can be seen that the boat moves in a circle and it returns to essentially the same place. Circular orbital motion allows a waveform (the wave's shape) to move forward *through the water* while the individual water particles that transmit the wave move around in a circle. Wind moving across a field of wheat causes a similar phenomenon: The wheat itself doesn't travel across the field, but the waves do.

The energy contributed by the wind to the water is transmitted not only along the surface of the sea but also downward. However, beneath the surface the circular motion rapidly diminishes until, at a depth equal to one half the wavelength measured from still water level, the movement of water particles becomes negligible. This depth is known as the *wave base.* The dramatic decrease of wave energy with depth is shown by the rapidly diminishing diameters of water-particle orbits in Figure 14.7.

Waves in the Surf Zone

As long as a wave is in deep water, it is unaffected by water depth (Figure 14.9, *left*). However, when a wave approaches the shore, the water becomes shallower and

PEOPLE AND THE ENVIRONMENT:
The Move of the Century—Relocating
the Cape Hatteras Lighthouse

In spite of efforts to protect structures that are too close to the shore, they can still be in danger of being destroyed by receding shorelines and the destructive power of waves. Such was the case for one of the nation's most prominent landmarks, the candy-striped lighthouse at Cape Hatteras, North Carolina, which is 21 stories tall—the nation's tallest lighthouse.

The lighthouse was built in 1870 on the Cape Hatteras barrier island 457 meters (1500 feet) from the shoreline to guide mariners through the dangerous offshore shoals known as the "Graveyard of the Atlantic." As the barrier island began migrating toward land, its beach narrowed. When the waves began to lap just 37 meters (120 feet) from its brick and granite base, there was concern that even a moderate-strength hurricane could trigger beach erosion sufficient to topple the lighthouse.

In 1970 the U.S. Navy built three groins in front of the lighthouse in an effort to protect the lighthouse from further erosion. The groins initially slowed erosion but disrupted sand flow in the surf zone, which caused the flattening of nearby dunes and the formation of a bay south of the lighthouse. Attempts to increase the width of the beach in front of the lighthouse included beach nourishment and artificial offshore beds of seaweed, both of which failed to widen the beach substantially. In the 1980s the Army Corps of Engineers proposed building a massive stone seawall around the lighthouse but decided the eroding coast would eventually move out from under the structure, leaving it stranded at sea on its own island. In 1988 the National Academy of Sciences determined that the shoreline

in front of the lighthouse would retreat so far as to destroy the lighthouse and recommended relocation of the tower as had been done with smaller lighthouses. In 1999 the National Park Service, which owns the lighthouse, finally authorized moving the structure to a safer location.

Moving the lighthouse, which weighs 4395 metric tons (4830 short tons), was accomplished by severing it from its foundation and carefully hoisting it onto a platform of steel beams fitted with roller dollies. Once on the platform, it was slowly rolled along a specially designed steel track using a series of hydraulic jacks. A strip of vegetation was cleared to make a runway along which the lighthouse traveled 1.5

meters (5 feet) at a time, with the track picked up from behind and reconstructed in front of the tower as it moved. In less than a month, the lighthouse was gingerly transported 884 meters (2900 feet) from its original location, making it one of the largest structures ever successfully moved.

After its $12 million move, the lighthouse now resides in a scrub oak and pine woodland (Figure 14.B). Although it now stands farther inland, the light's slightly higher elevation makes it visible just as far out to sea, where it continues to warn mariners of the hazardous shoals. At the current rate of shoreline retreat, the lighthouse should be safe from the threat of waves for at least another century.

Figure 14.B   When North Carolina's Cape Hatteras Lighthouse was threatened by shoreline erosion in 1999, it was relocated 488 meters (1600 feet) from the shore. (Photo by Drew Wilson © 1999, *Virginian–Pilot*)

influences wave behavior. The wave begins to "feel bottom" at a water depth equal to its wave base. Such depths interfere with water movement at the base of the wave and slow its advance (Figure 14.9, *center*).

As a wave advances toward the shore, the slightly faster waves farther out to sea catch up, decreasing the wavelength. As the speed and length of the wave diminish, the wave steadily grows higher. Finally, a critical point is reached when the wave is too steep to

support itself and the wave front collapses, or *breaks* (Figure 14.9, *right*), causing water to advance up the shore.

The turbulent water created by breaking waves is called **surf.** On the landward margin of the surf zone, the turbulent sheet of water from collapsing breakers, called *swash*, moves up the slope of the beach. When the energy of the swash has been expended, the water flows back down the beach toward the surf zone as *backwash*.

Figure 14.8   The movements of the toy boat show that the wave form advances, but the water does not advance appreciably from its original position. In this sequence, the wave moves from left to right as the boat (and the water in which it is floating) rotates in an imaginary circle.

Beaches and Shoreline Processes

 GEODe
Oceans
↳ Coastal Processes

Beaches are composed of whatever material is locally available. When this material—sediment—comes from the erosion of beach cliffs or nearby coastal mountains, beaches are composed of mineral particles from these

STUDENTS SOMETIMES ASK...
What are tidal waves?

Tidal waves, more accurately known as *tsunami* (*tsu* = harbor, *nami* = wave), have *nothing* to do with the tides. They are long-wavelength, fast-moving, often large, and sometimes destructive waves that originate from sudden changes in the topography of the seafloor. Tsunami are initiated by underwater faulting, landslides, or volcanic eruptions. Since the mechanisms that trigger these dangerous waves are frequently seismic events, tsunami are appropriately termed *seismic sea waves*. For more information about tsunami and their destructive effects, see "Tsunami" in Chapter 6.

---

rocks and may be relatively coarse in texture. When the sediment comes primarily from rivers that drain lowland areas, beaches are finer in texture.

Some beaches have a significant biological component. For example, in low-latitude areas such as southern Florida, where there are no mountains or other sources of rock-forming minerals nearby, most beaches are composed of shell fragments and the remains of organisms that live in coastal waters. Some beaches on volcanic islands in the open ocean are composed of black or green fragments of the basaltic lava that comprise the islands, or of coarse debris from coral reefs that develop around islands in low latitudes.

Regardless of the composition, though, the material that comprises the beach does not stay in one place. Instead, the waves that crash along the shoreline are constantly moving it. Thus, beaches can be thought of as material in transit along the shoreline.

Wave Erosion

During calm weather, wave action is minimal. During storms, however, waves are capable of causing much erosion. The impact of large, high-energy waves against the shore can be awesome in its violence. Each breaking wave may hurl thousands of tons of water against the land, sometimes causing the ground literally to tremble. The pressures exerted by Atlantic waves in wintertime, for example, average nearly 10,000 kilograms per square meter (more than 2000 pounds per square foot).

It is no wonder that cracks and crevices are quickly opened in cliffs, coastal structures, and anything else that is subjected to these enormous shocks (Figure 14.10). Water is forced into every opening, causing air in the cracks to become highly compressed by the thrust of crashing waves. When the wave subsides, the air expands rapidly, dislodging rock fragments and enlarging and extending preexisting fractures.

In addition to the erosion caused by wave impact and pressure, **abrasion,** the sawing and grinding action of the water armed with rock fragments, is also impor-

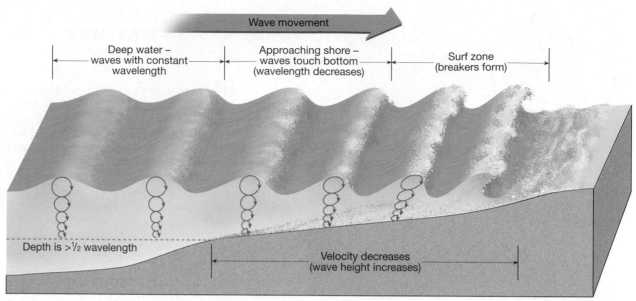

Figure 14.9   Changes that occur when a wave moves onto shore. The waves touch bottom as they encounter water depths less than half a wavelength. The wave speed decreases, and the waves stack up against the shore, causing the wavelength to decrease. This results in an increase in wave height to the point where the waves pitch forward and break in the surf zone.

tant. In fact, abrasion is probably more intense in the surf zone than in any other environment. Smooth, rounded stones and pebbles along the shore are evidence of the relentless grinding action of rock against rock in the surf zone (Figure 14.11A). Such fragments are also used as "tools" by the waves as they cut horizontally into the land (Figure 14.11B). Moreover, waves are very effective at breaking down rock material and supplying sand to beaches.

Wave Refraction

The bending of waves, called **wave refraction** (*refringere* = to break up), plays an important part in shoreline processes. It affects the distribution of ener-

gy along the shore and thus strongly influences where and to what degree erosion, sediment transport, and deposition will take place.

Waves seldom approach the shore straight on. Rather, most waves move toward the shore at a slight angle. However, when they reach the shallow water of a smoothly sloping bottom, the wave crests are refracted (bent) and tend to line up nearly parallel to the shore. Such bending occurs because the part of the wave nearest the shore touches bottom and slows first, whereas the part of the wave that is still in deep water continues forward at its full speed. This causes wave crests to become nearly parallel to the shore regardless of their original orientation.

Figure 14.10   The force of breaking waves can be powerful. Here waves batter a seawall at Sea Bright, New Jersey. A seawall 5 to 6 meters (16 to 20 feet) high and 8 kilometers (5 miles) long was built to protect the town and the railroad that brought tourists to the beach. As you can see, after the wall was built, the beach narrowed dramatically. (Photo by Rafael Macia/Photo Researchers, Inc.)

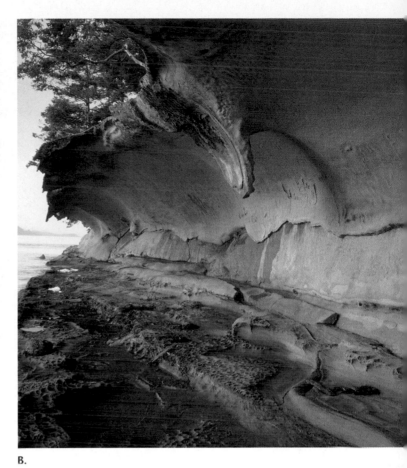

A.   B.

Figure 14.11   **A.** Abrasion can be intense in the surf zone. Smooth rounded stones along the shore are an obvious reminder of this fact. Garrapata State Park, California (Photo by Car Clifton) **B.** Sandstone cliff undercut by wave erosion at Gabriola Island, British Columbia, Canada. (Photo by Fletcher and Baylis/Photo Researchers, Inc.)

Because of refraction, wave energy is concentrated against the sides and ends of headlands that project into the water, whereas wave attack is weakened in bays. This differential wave attack along irregular coastlines is illustrated in Figure 14.12. Because the waves reach the shallow water in front of the headland sooner than they do in adjacent bays, they are bent more nearly parallel to the protruding land and strike it from all three sides. By contrast, refraction in the bays causes waves to diverge and expend less energy. In these zones of weakened wave activity, sediments can accumulate and form sandy beaches. Over a long period, erosion of the headlands and deposition in the bays will straighten an irregular shoreline.

Longshore Transport

Most streams supply large amounts of sediment to the shore, where it is reworked and transported by waves. With each breaking wave, sand is moved a little farther up or down the coast, making beaches analogous to "rivers of sand."

Although waves are refracted, most still reach the shore at a slight angle. Consequently, the uprush of water from each breaking wave (the swash) is at an oblique angle to the shoreline. However, the backwash is straight down the slope of the beach. The effect of this pattern of water movement is to transport sediment in a zigzag pattern along the beach face (Figure 14.13). This movement is called **beach drift,** and it can transport sand and pebbles hundreds or even thousands of meters daily. However, a more typical rate is 5 to 10 meters per day.

Oblique waves also produce currents within the surf zone that flow parallel to the shore and move substantially more sediment than beach drift (Figure 14.13). Because the water here is turbulent, these **longshore currents** easily move the fine suspended sand and roll larger sand and gravel along the bottom. When the sediment transported by longshore currents is added to the quantity moved by beach drift, the total amount can be very large. At Sandy Hook, New Jersey, for example, the quantity of sand transported along the shore over a 48-year period averaged almost 680,000 metric tons (750,000 short tons) annually. For a 10-year period at Oxnard, California, more than 1.4 million metric tons (1.5 million short tons) of sediment moved along the shore each year.

Figure 14.12 Wave refraction along an irregular coastline. As waves first touch bottom in the shallows off the headlands, they are slowed, causing the waves to refract and align nearly parallel to the shoreline. This causes wave energy to be concentrated at headlands (resulting in erosion) and dispersed in bays (resulting in deposition).

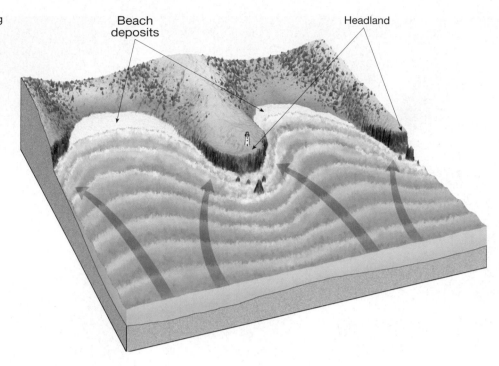

Beach deposits

Headland

Both rivers and coastal zones move water and sediment from one area (*upstream*) to another (*downstream*). As a result, the beach has often been characterized as a "river of sand." Beach drift and longshore currents, however, move in a zigzag pattern, whereas rivers flow mostly in a turbulent, swirling fashion. Additionally, the direction of flow of longshore currents along a shoreline can change, whereas rivers flow in the same direction (downhill). Longshore currents change direction because the direction that waves approach the beach changes seasonally. Nevertheless, longshore currents generally flow southward along both the Atlantic and Pacific shores of the United States.

STUDENTS SOMETIMES ASK...

Are rip currents the same thing as longshore currents?

No. Longshore currents occur in the surf zone and move roughly parallel to the shore. By contrast, rip currents occur perpendicular to the coast and move in the opposite direction of breaking waves. Most of the backwash from spent waves finds its way back to the open ocean as an unconfined flow across the ocean bottom called *sheet flow*. However, a portion of the returning water moves seaward in more concentrated surface *rip currents*. Rip currents do not travel far beyond the surf zone before breaking up and can be recognized by the

Figure 14.13 Beach drift and longshore currents are created by obliquely breaking waves. Beach drift occurs as incoming waves carry sand obliquely up the beach, while the water from spent waves carries it directly down the slope of the beach. Similar movements occur offshore in the surf zone to create the longshore current. These processes transport large quantities of material along the beach and in the surf zone.

Net movement of sand grains

Path of sand particles

Longshore current

way they interfere with incoming waves or by the sediment that is often held in suspension within the rip current. They can also be a hazard to swimmers, who, if caught in them, can be carried out away from the shore.

Shoreline Features

Oceans
↳ Coastal Processes

A fascinating assortment of shoreline features can be observed along the world's coastal regions. These shoreline features vary depending on the type of rocks exposed along the shore, the intensity of waves, the nature of coastal currents, and whether the coast is stable, sinking, or rising. Features that owe their origin primarily to the work of erosion are called *erosional features*, while deposits of sediment produce *depositional features*.

Erosional Features

Many coastal landforms owe their origin to erosional processes. Such erosional features are common along the rugged and irregular New England coast and along the steep shorelines of the West Coast of the United States.

Wave-Cut Cliffs, Wave-Cut Platforms, and Marine Terraces. **Wave-cut cliffs,** as the name implies, originate by the cutting action of the surf against the base of coastal land. As erosion progresses, rocks overhanging the notch at the base of the cliff crumble

into the surf and the cliff retreats. A relatively flat, benchlike surface, called a **wave-cut platform,** is left behind by the receding cliff (Figure 14.14, *left*). The platform broadens as wave attack continues. Some debris produced by the breaking waves remains along the water's edge as sediment on the beach, while the remainder is transported farther seaward. If a wave-cut platform is uplifted above sea level by tectonic forces, it becomes a **marine terrace** (Figure 14.14, *right*). Marine terraces are easily recognized by their gentle seaward-sloping shape and are often desirable sites for coastal roads, buildings, or agriculture.

Sea Arches and Sea Stacks. Headlands that extend into the sea are vigorously attacked by waves because of refraction. The surf erodes the rock selectively, wearing away the softer or more highly fractured rock at the fastest rate. At first, sea caves may form. When two caves on opposite sides of a headland unite, a **sea arch** results (Figure 14.15). Eventually, the arch falls in, leaving an isolated remnant, or **sea stack,** on the wave-cut platform (Figure 14.15). In time, it too will be consumed by the action of the waves.

Depositional Features

Sediment eroded from the beach is transported along the shore and deposited in areas where wave energy is low. Such processes produce a variety of depositional features.

Spits, Bars, and Tombolos. Where beach drift and longshore currents are active, several features related to the movement of sediment along the shore may

Figure 14.14 Wave-cut platform and marine terrace. A wave-cut platform is exposed at low tide along the California coast at Bolinas Point near San Francisco. An elevated wave-cut platform, called a marine terrace, has been uplifted to the right. (Photo by John S. Shelton)

Figure 14.15 Sea arch and sea stack at the tip of Mexico's Baja Peninsula. (Photo by Mark A. Johnson/The Stock Market)

develop. A **spit** (*spit* = spine) is an elongated ridge of sand that projects from the land into the mouth of an adjacent bay. Often the end in the water hooks landward in response to the dominant direction of the longshore current (Figure 14.16). The term **baymouth bar** is applied to a sandbar that completely crosses a bay, sealing it off from the open ocean (Figure 14.16). Such a feature tends to form across bays where currents are weak, allowing a spit to extend to the other side. A **tombolo** (*tombolo* = mound), a ridge of sand that connects an island to the mainland or to another island, forms in much the same manner as a spit.

Barrier Islands.   The Atlantic and Gulf Coastal Plains are relatively flat and slope gently seaward. The shore zone is characterized by **barrier islands.** These low ridges of sand parallel the coast at distances from 3 to 30 kilometers (1.9 to 19 miles) offshore. From Cape Cod, Massachusetts, to Padre Island, Texas, nearly 300 barrier islands rim the coast (Figure 14.17).

Most barrier islands are from 1 to 5 kilometers (0.6 to 3 miles) wide and between 15 and 30 kilometers (9 to 18 miles) long. The tallest features are sand dunes, which usually reach heights of 5 to 10 meters (16 to 33 feet). The lagoons that separate these narrow islands from the shore are zones of relatively quiet water that allow small craft traveling between New York and northern Florida to avoid the rough waters of the North Atlantic.

Barrier islands probably formed in several ways. Some originated as spits that were subsequently severed from the mainland by wave erosion or by the

Figure 14.16   High-altitude image of a well-developed spit and baymouth bar along the coast of Martha's Vineyard, Massachusetts. Also notice the tidal delta in the lagoon adjacent to the inlet through the baymouth bar. (Photo courtesy of USDA-ASCS)

Figure 14.17 Nearly 300 barrier islands rim the Gulf and Atlantic Coasts. The islands along the south Texas coast and along the coast of North Carolina are excellent examples.

general rise in sea level following the last episode of glaciation. Others were created when turbulent waters in the line of breakers heaped up sand that had been scoured from the bottom. Finally, some barrier islands may be former sand-dune ridges that originated along the shore during the last glacial period, when sea level was lower. As the ice sheets melted, sea level rose and flooded the area behind the beach-dune complex.

The Evolving Shore

A shoreline continually undergoes modification regardless of its initial configuration. At first most coastlines are irregular, although the degree of and reason for the irregularity may differ considerably from place to place. Along a coastline that is characterized by varied geology, the pounding surf may initially increase its irregularity because the waves will erode the weaker rocks more easily than the stronger ones. However, if a shoreline remains stable, marine erosion and deposition will eventually produce a straighter, more regular coast.

Figure 14.18 illustrates the evolution of an initially irregular coast that remains relatively stable and shows many of the coastal features discussed in the previous section. As headlands are eroded and erosional features such as wave-cut cliffs and wave-cut platforms are created, sediment is produced that is carried along the shore by beach drift and longshore currents. Some material is deposited in the bays, while other debris is formed into depositional features such as spits and baymouth bars. At the same time, rivers fill the bays with sediment. Ultimately, a smooth coast results.

Stabilizing the Shore

  Oceans
↳ Coastal Processes

Today the coastal zone teems with human activity. Unfortunately, people often treat the shoreline as if it were a stable platform on which structures can be built safely. This approach jeopardizes both people and the shoreline because many coastal landforms are relatively fragile, short-lived features that are easily damaged by development. As anyone who has endured a strong coastal storm knows, the shoreline is not always a safe place to live (Figure 14.19).

Compared with other natural hazards, such as earthquakes, volcanic eruptions, and landslides, shoreline erosion appears to be a more continuous and predictable process that causes relatively modest damage to limited areas. In reality, the shoreline is one of Earth's

Figure 14.18 These diagrams illustrate the changes that can take place through time along an initially irregular coastline that remains relatively stable. The coastline shown in part **A** gradually evolves to **B**, then **C**, and then **D**. The diagrams also serve to illustrate many of the features described in the section on shoreline features. (Photos by E. J. Tarbuck except photo of wave-cut cliff by A. P. Trujillo/APT Photos)

Spit

Sea stack

Wave-cut cliff

Beach deposits

Baymouth bar

Figure 14.18   Continued

most dynamic places that changes rapidly in response to natural forces. Storms, for example, are capable of eroding beaches and cliffs at rates that far exceed the long-term average. Such bursts of accelerated erosion not only have a significant impact on the natural evolution of a coast but can also have a profound impact on people who reside in the coastal zone. Erosion along the coast causes significant property damage. Huge sums are spent annually not only to repair damage but also to prevent or control erosion. Already a problem at many sites, shoreline erosion is certain to become increasingly serious as extensive coastal development continues.

Although the same processes cause change along every coast, not all coasts respond in the same way. Interactions among different processes and the relative importance of each process depend on local factors. The factors include: (1) the proximity of a coast to sediment-laden rivers, (2) the degree of tectonic activity, (3) the topography and composition of the land, (4) prevailing winds and weather patterns, and (5) the configuration of the coastline and nearshore areas.

During the past 100 years, growing affluence and increasing demands for recreation have brought unprecedented development to many coastal areas. As both the number and the value of buildings have increased, so too have efforts to protect property from storm waves by stabilizing the shore.

Hard Stabilization

Structures built to protect a coast from erosion or to prevent the movement of sand along a beach are known as **hard stabilization.** Hard stabilization can take many forms and often results in predictable yet unwanted outcomes. Hard stabilization includes groins, breakwaters, and seawalls.

Groins.   To maintain or widen beaches that are losing sand, groins are sometimes constructed. A **groin** (*groin* = ground) is a barrier built at a right angle to the beach to trap sand that is moving parallel to the shore. Groins are usually constructed of large rocks but may also be composed of wood. These structures often do their job so effectively that the longshore current beyond the groin becomes sand-starved. As a result, the current erodes sand from the beach on the downstream side of the groin.

To offset this effect, property owners downstream from the structure may erect a groin on their property. In this manner, the number of groins multiplies, resulting in a *groin field* (Figure 14.20). An example of such proliferation is the shoreline of New Jersey, where hundreds of these structures have been built. Because it has been shown that groins often do not provide a satisfactory solution, they are no longer the preferred method of keeping beach erosion in check.

STUDENTS SOMETIMES ASK...
*Is it true that sea level is
rising due to global warming?*

Yes. One probable impact of a human-induced global warming is a rise in sea level. Some of the rise comes from water added by shrinking glaciers. A more significant factor is that a warmer atmosphere causes an increase in ocean volume due to thermal expansion. Higher air temperatures warm the upper layers of the ocean, which causes the water to expand and sea level to rise. For more on this, see the discussion of "Human Impact on Global Climate" in Chapter 19.

Figure 14.19   Erosion caused by strong storms forced abondonment of this highly developed shoreline area in Long Island, New York. Coastal areas are dynamic places that can change rapidly in response to natural forces. (Photo by Mark Wexler/Woodfin Camp & Associates)

Breakwaters and Seawalls. Hard stabilization can also be built parallel to the shoreline. One such structure is a **breakwater,** the purpose of which is to protect boats from the force of large breaking waves by creating a quiet water zone near the shore. However, when this is done, the reduced wave activity along the shore behind the structure may allow sand to accumulate. If this happens, the boat anchorage will eventually fill with sand while the downstream beach erodes and retreats. At Santa Monica, California, where the building of a breakwater created such a problem, the city uses a dredge to remove sand from the protected quiet water zone and deposit it farther downstream where longshore currents continue to move the sand down the coast (Figure 14.21).

Another type of hard stabilization built parallel to the shore is a **seawall,** which is designed to armor the coast and defend property from the force of breaking waves (see Figure 14.10). Waves expend much of their energy as they move across an open beach. Seawalls cut this process short by reflecting the force of unspent waves seaward. As a consequence, the beach to the seaward side of the seawall experiences significant erosion and may, in some instances, be eliminated entirely. Once the width of the beach is reduced, the seawall is subjected to even greater pounding by the waves. Eventually this battering takes its toll on the seawall such that the seawall will fail and a larger, more expensive structure must be built to take its place.

The wisdom of building temporary protective structures along shorelines is increasingly questioned.

The opinion of many coastal scientists and engineers is expressed in the following excerpt from a position paper that was developed as a result of a conference on America's Eroding Shoreline:

> It is now clear that halting the receding shoreline with protective structures benefits only a few and seriously degrades or destroys the natural beach and the value it holds for the majority. Protective structures divert the ocean's energy temporarily from private properties, but usually refocus the energy on the adjacent natural beaches. Many interrupt the natural sand flow in coastal currents, robbing many beaches of vital sand replacement.*

Alternatives to Hard Stabilization

Armoring the coast with hard stabilization has several potential drawbacks, including the cost of the structure and the loss of sand on the beach. Alternatives to hard stabilization include beach nourishment and relocation.

Beach Nourishment. **Beach nourishment** represents an approach to stabilizing shoreline sands without hard stabilization. As the term implies, this practice simply involves the addition of large quantities of sand to the beach system (Figure 14.22). By building the beaches seaward, beach quality and storm protection are both improved. Beach nourishment, however, is not a permanent solution to the problem of shrinking beaches because the same processes that removed the sand in the first place will eventually remove the replacement sand as well. In addition, it can be very expensive because huge volumes of sand must be

---

*"Strategy for Beach Preservation Proposed," *Geotimes* 30 (No. 12, December 1985): 15.

Figure 14.20 A series of groins along the shoreline near Chichester, Sussex, England. Because groins trap sand on the upcurrent side, the movement of sand along this coast caused by the longshore current must be from lower right toward upper left. (Photo by Sandy Stockwell/London Aerial Photo Library/CORBIS)

Figure 14.21 Aerial view of a breakwater at Santa Monica, California. The breakwater appears as a faint line in the water behind which many boats are anchored. The construction of the breakwater disrupted longshore transport and caused the seaward growth of the beach. (Photo by John S. Shelton)

Breakwater

A.

B.

Figure 14.22   Miami Beach. **A.** Before beach nourishment and **B.** After beach nourishment. (Courtesy of the U.S. Army Corps of Engineers, Vicksburg District)

transported to the beach from offshore areas, nearby rivers, or other source areas for sand.

In some instances, beach nourishment can lead to unwanted environmental effects. For example, beach replenishment at Waikiki Beach, Hawaii, involved replacing the natural coarse calcareous beach sand with softer, muddier sand. Destruction of the softer sand by breaking waves increased the water's turbidity and killed offshore coral reefs. Similar damage to local coral communities occurred after beach replenishment at Miami Beach.

Beach nourishment appears to be an economically viable long-range solution to the beach-preservation problem only in areas where there exists dense development, large supplies of sand, relatively low wave energy, and reconcilable environmental issues. Unfortunately, few areas possess all these attributes.

Relocation.   Instead of building structures such as groins and seawalls to hold the shoreline in place or adding sand to replenish eroding beaches, another option is also available. Many coastal scientists and planners are calling for a policy shift from defending and rebuilding beaches and coastal property in high hazard areas to *relocating* storm-damaged or at-risk buildings (Box 14.2) and letting nature reclaim the beach. This approach is similar to that adopted by the federal government for river floodplains following the devastating 1993 Mississippi River floods in which vulnerable structures are abandoned and relocated on higher, safer ground.

Such proposals, of course, are controversial. People with significant shoreline investments shudder at

the thought of not rebuilding and defending coastal developments from the erosional wrath of the sea. Others, however, argue that with sea level rising, the impact of coastal storms will only get worse in the decades to come. This group advocates that oft-damaged structures be relocated or abandoned to improve personal safety and to reduce costs. Such ideas will no doubt be the focus of much study and debate as states and communities evaluate and revise coastal land-use policies.

Erosion Problems Along U.S. Coasts

The shoreline along the Pacific Coast of the United States is strikingly different from that characterizing the Atlantic and Gulf Coast regions. Some of the differences are related to plate tectonics. The West Coast represents the leading edge of the North American plate, and because of this, it experiences active uplift and deformation. By contrast, the East Coast is a tectonically quiet region that is far from any active plate margin. Because of this basic geological difference, the nature of shoreline erosion problems along America's opposite coasts is different.

Atlantic and Gulf Coasts.   Much of the coastal development along the Atlantic and Gulf coasts has occurred on barrier islands. Typically, barrier islands consist of a wide beach that is backed by dunes and separated from the mainland by marshy lagoons. The broad expanses of sand and exposure to the ocean have made barrier islands exceedingly attractive sites for development. Unfortunately, development has taken place more rapidly than has our understanding of barrier island dynamics.

Because barrier islands face the open ocean, they receive the full force of major storms that strike the coast. When a storm occurs, the barriers absorb the energy of the waves primarily through the movement of sand. This process and the dilemma that results have been described as follows:

> Waves may move sand from the beach to offshore areas or, conversely, into the dunes; they may erode the dunes, depositing sand onto the beach or carrying it out to sea; or they may carry sand from the beach and the dunes into the marshes behind the barrier, a process known as overwash. The common factor is movement. Just as a flexible reed may survive a wind that destroys an oak tree, so the barriers survive hurricanes and nor'easters not through unyielding strength but by giving before the storm.
>
> This picture changes when a barrier is developed for homes or a resort. Storm waves that previously rushed harmlessly through gaps between the dunes now encounter buildings and roadways. Moreover, since the dynamic nature of the barriers is readily perceived only during storms, homeowners tend to attribute damage to a particular storm, rather than to the basic mobility of coastal barriers. With their homes or investments at stake, local residents are more likely to seek to hold the sand in place and the waves at bay than to admit that development was improperly placed to begin with.*

Pacific Coast. In contrast to the broad, gently sloping coastal plains of the Atlantic and Gulf Coasts, much of the Pacific Coast is characterized by relatively narrow beaches that are backed by steep cliffs and mountain ranges. Recall that America's western margin is a more rugged and tectonically active region than the eastern margin. Because uplift continues, the rise in sea level in the West is not so readily apparent. Nevertheless, like the shoreline erosion problems facing the East's barrier islands, West Coast difficulties also stem largely from the alteration of natural systems by people.

A major problem facing the Pacific shoreline—particularly along southern California—is a significant narrowing of many beaches. The bulk of the sand on many of these beaches is supplied by rivers that transport it from the mountainous regions to the coast. Over the years this natural flow of material to the coast has been interrupted by dams built for irrigation and flood control. The reservoirs effectively trap the sand that would otherwise nourish the beach environment. When the beaches were wider, they served to protect the cliffs behind them from the force of storm waves. Now, however, the waves move across the narrowed beaches without losing much energy and cause more rapid erosion of the sea cliffs.

Although the retreat of the cliffs provides material to replace some of the sand impounded behind dams, it also endangers homes and roads built on the bluffs.

In addition, development atop the cliffs aggravates the problem. Urbanization increases runoff, which, if not carefully controlled, can result in serious bluff erosion. Watering lawns and gardens adds significant quantities of water to the slope. This water percolates downward toward the base of the cliff, where it may emerge in small seeps. This action reduces the slope's stability and facilitates mass-wasting events.

Shoreline erosion along the Pacific Coast varies considerably from one year to the next, largely because of the sporadic occurrence of storms. As a consequence, when the infrequent but serious episodes of erosion occur, the damage is often blamed on the unusual storms and not on coastal development or the sediment-trapping dams that may be great distances away.

Coastal Classification

The great variety of shorelines demonstrates their complexity. Indeed, to understand any particular coastal area, many factors must be considered, including rock types, size and direction of waves, frequency of storms, tidal range, and offshore topography. In addition, practically all coastal areas were affected by the worldwide rise in sea level that accompanied the melting of Ice Age glaciers at the close of the Pleistocene epoch. Finally, tectonic events that uplift or downdrop the land or change the volume of ocean basins must be taken into account. The large number of factors that influence coastal areas make shoreline classification difficult.

Nonetheless, many geologists classify coasts based on changes that have occurred with respect to sea level. This commonly used classification system divides coasts into two very general categories: emergent and submergent. **Emergent coasts** develop either because an area experiences uplift or as a result of a drop in sea level. Conversely, **submergent coasts** are created when sea level rises or the land adjacent to the sea subsides.

Emergent Coasts

In some areas, the coast is clearly emergent because rising land or falling water levels expose wave-cut cliffs and marine terraces above sea level. Excellent examples include portions of coastal California where uplift has occurred in the recent geological past (see Figures 14.14 and 9.24). In the case of the Palos Verdes Hills, south of Los Angeles, California, seven different terrace levels exist, indicating at least seven episodes of uplift. The ever persistent sea is now cutting a new platform at the base of the cliff. If uplift follows, it too will become an elevated marine terrace.

Other examples of emergent shores include regions that were once buried beneath great ice sheets. When glaciers were present, their weight depressed the crust, and when the ice melted, the crust began gradually to spring back. Consequently, prehistoric shoreline features today are found high above sea level. The Hudson Bay region of

*Frank Lowenstein, "Beaches or Bedrooms—The Choice as Sea Level Rises," *Oceanus* 28 (No. 3, Fall 1985): 22.

Canada is one such area, portions of which are still rising at a rate of more than 1 centimeter (0.4 inch) annually.

Submergent Coasts

In contrast to the preceding examples, other coastal areas show definite signs of submergence. Shorelines that have been submerged in the relatively recent past are often highly irregular because the sea typically floods the lower reaches of river valleys flowing into the ocean. The ridges separating the valleys, however, remain above sea level and project into the sea as headlands. These drowned river mouths, which are called **estuaries** (*aestus* = tide), characterize many coasts today. Along the Atlantic coastline, Chesapeake and Delaware bays are examples of estuaries created by submergence (Figure 14.23). The picturesque coast of Maine, particularly in the vicinity of Acadia National Park, is another excellent example of an area that was flooded by the postglacial rise in sea level and transformed into a highly irregular submerged coastline.

Keep in mind, however, that most coasts have a complicated geologic history. With respect to sea level, many coasts have at various times emerged and then submerged again. Each time they retain some of the features created during the previous situation.

Tides

Tides are daily changes in the elevation of the ocean surface. Their rhythmic rise and fall along coastlines have been known since antiquity. Other than waves, they are the easiest ocean movements to observe (Figure 14.24).

Although known for centuries, tides were not explained satisfactorily until Sir Isaac Newton applied the law of gravitation to them. Newton showed that there is a mutual attractive force between two bodies, as between Earth and the Moon. Because both the atmosphere and the ocean are fluids and are free to move, both are deformed by this force. Hence, ocean tides result from the gravitational attraction exerted upon Earth by the Moon and, to a lesser extent, by the Sun.

Causes of Tides

To illustrate how tides are produced, consider the Earth as a rotating sphere covered to a uniform depth with water (Figure 14.25). Further, ignore the effect of the Sun for now. It is easy to see how the Moon's gravitational force can cause the water to bulge on the side of Earth nearest the Moon. In addition, however, an equally large tidal bulge is produced on the side of Earth directly *opposite* the Moon.

Both tidal bulges are caused, as Newton discovered, by the pull of gravity. Gravity is inversely proportional to the square of the distance between two objects, meaning simply that it quickly weakens with distance. In this case, the two objects are the Moon and Earth. Because the force of gravity decreases with distance, the

Figure 14.23   Major estuaries along the East Coast of the United States. The lower portions of many river valleys were flooded by the rise in sea level that followed the end of the last Ice Age, creating large estuaries such as Chesapeake and Delaware bays.

STUDENTS SOMETIMES ASK...
Where are the world's largest tides?

The world's largest *tidal range* (the difference between successive high and low tides) is found in the northern end of Nova Scotia's 258-kilometer- (160-mile-) long Bay of Fundy. During maximum spring tide conditions, the tidal range at the mouth of the bay (where it opens to the ocean) is only about 2 meters (6.6 feet). However, the tidal range progressively increases from the mouth of the bay northward because the natural geometry of the bay concentrates tidal energy. In the northern end of Minas Basin, the maximum spring tidal range is about 17 meters (56 feet). This extreme tidal range leaves boats high and dry during low tide (see Figure 14.24).

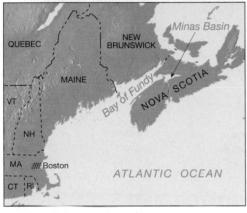

Figure 14.24   High tide and low tide on Nova Scotia's Minas Basin in the Bay of Fundy. The areas exposed during low tide and flooded during high tide are called *tidal flats*. Tidal flats here are extensive. (Courtesy of Nova Scotia Department of Tourism and Culture)

Moon's gravitational pull on Earth is slightly greater on the near side of Earth than on the far side. The result of this differential pulling is to stretch (elongate) the "solid" Earth very slightly. In contrast, the world ocean, which is mobile, is deformed quite dramatically by this effect to produce the two opposing tidal bulges.

Because the position of the Moon changes only moderately in a single day, the tidal bulges remain in place while Earth rotates "through" them. For this reason, if you stand on the seashore for 24 hours, Earth will rotate you through alternating areas of higher and lower water. As you are carried into each tidal bulge, the tide rises, and as you are carried into the intervening troughs between the tidal bulges, the tide falls. Therefore, most places on Earth experience two high tides and two low tides each day.

In addition, the tidal bulges migrate as the Moon revolves around Earth. As a result, the tides—like the time of moonrise—shift about 50 minutes later each day. In essence, the tidal bulges exist in fixed positions relative to the Moon, which is slowly moving eastward.

In many locations, there may be an inequality between the high tides during a given day. Depending on the Moon's position, the tidal bulges may be inclined to the equator as in Figure 14.25. This figure illustrates that one high tide experienced by an observer in the Northern Hemisphere is considerably higher than the high tide half a day later. In contrast, a Southern Hemisphere observer would experience the opposite effect.

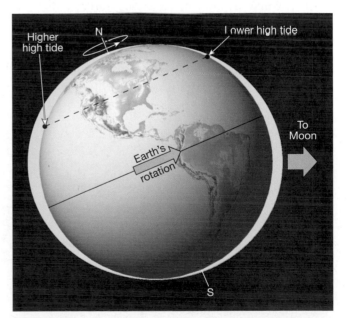

Figure 14.25   Idealized tidal bulges on Earth caused by the Moon. If Earth were covered to a uniform depth with water, there would be two tidal bulges: one on the side of Earth facing the Moon (*right*) and the other on the opposite side of Earth (*left*). Depending on the Moon's position, tidal bulges may be inclined relative to Earth's equator. In this situation, Earth's rotation causes an observer to experience two unequal high tides during a day.

Monthly Tidal Cycle

The primary body that influences the tides is the Moon, which makes one complete revolution around Earth every 29 and a half days. The Sun, however, also influences the tides. It is far larger than the Moon, but because it is much farther away, its effect is considerably less. In fact, the Sun's tide-generating effect is only about 46 percent that of the Moon's.

Near the times of new and full moons, the Sun and Moon are aligned and their forces are added together (Figure 14.26A). Accordingly, the combined gravity of these two tide-producing bodies causes larger tidal bulges (higher high tides) and larger tidal troughs (lower low tides), producing a large tidal range. These are called the **spring** (*springen* = to rise up) **tides,** which have no connection with the spring season but occur twice a month during the time when the Earth–Moon–Sun system is aligned. Conversely, at about the time of the first and third quarters of the Moon, the gravitational forces of the Moon and Sun act on Earth at right angles, and each partially offsets the influence of the other (Figure 14.26B). As a result, the daily tidal range is less. These are called **neap** (*nep* = scarcely or barely touching) **tides,** which also occur twice each month. Each month, then, there are two spring tides and two neap tides, each about one week apart.

Tidal Patterns

So far, the basic causes and types of tides have been explained. Keep in mind, however, that these theoretical considerations cannot be used to predict either the height or the time of actual tides at a particular place. This is because many factors—including the shape of the coastline, the configuration of ocean basins, and water depth—greatly influence the tides. Consequently, tides at various locations respond differently to the

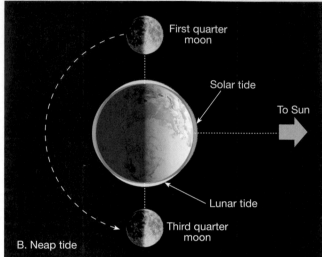

Figure 14.26 Earth–Moon–Sun positions and the tides. **A.** When the Moon is in the full or new position, the tidal bulges created by the Sun and Moon are aligned, there is a large tidal range on Earth, and spring tides are experienced. **B.** When the Moon is in the first- or third-quarter position, the tidal bulges produced by the Moon are at right angles to the bulges created by the Sun. Tidal ranges are smaller, and neap tides are experienced.

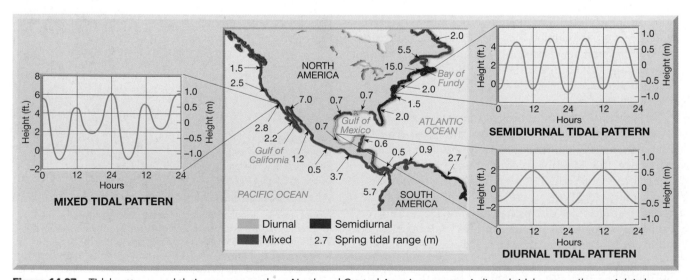

Figure 14.27 Tidal patterns and their occurrence along North and Central American coasts. A diurnal tidal pattern (*lower right*) shows one high and low tide each tidal day. A semidiurnal pattern (*upper right*) shows two highs and lows of approximately equal heights during each tidal day. A mixed tidal pattern (*left*) shows two highs and lows of unequal heights during each tidal day.

Tidal   Tidal   Barrier
flats   delta   island

Lagoon

Figure 14.28   As a rapidly moving tidal current moves through a barrier island's inlet into the quiet waters of the lagoon, the current slows and deposits sediment, creating a tidal delta. Because this tidal delta has developed on the landward side of the inlet, it is called a *flood delta*.

tide-producing forces. This being the case, the nature of the tide at any coastal location can be determined most accurately by actual observation. The predictions in tidal tables and tidal data on nautical charts are based on such observations.

Three main tidal patterns exist worldwide. A **diurnal** (*diurnal* = daily) **tidal pattern** is characterized by a single high tide and a single low tide each tidal day (Figure 14.27). Tides of this type occur along the northern shore of the Gulf of Mexico, among other locations. A **semidiurnal** (*semi* = twice, *diurnal* = daily) **tidal pattern** exhibits two high tides and two low tides each tidal day, with the two highs about the same height and the two lows about the same height (Figure 14.27). This type of tidal pattern is common along the Atlantic Coast of the United States. A **mixed tidal pattern** is similar to a semidiurnal pattern except that it is characterized by a large inequality in high water heights, low water heights, or both (Figure 14.27). In this case, there are usually two high and two low tides each day, with high tides of dif-

ferent heights and low tides of different heights. Such tides are prevalent along the Pacific Coast of the United States and in many other parts of the world.

Tidal Currents

Tidal current is the term used to describe the *horizontal* flow of water accompanying the rise and fall of the tides. These water movements induced by tidal forces can be important in some coastal areas. Tidal currents that advance into the coastal zone as the tide rises are called *flood currents*. As the tide falls, seaward-moving water generates *ebb currents*. Periods of little or no current, called *slack water*, separate flood and ebb. The areas affected by these alternating tidal currents are called **tidal flats** (see Figure 14.24). Depending on the nature of the coastal zone, tidal flats vary from narrow strips seaward of the beach to zones that may extend for several kilometers.

Although tidal currents are not important in the open sea, they can be rapid in bays, river estuaries, straits, and other narrow places. Off the coast of Brittany in France, for example, tidal currents that accompany a high tide of 12 meters (40 feet) may attain a speed of 20 kilometers (12 miles) per hour. Tidal currents are not generally considered to be major agents of erosion and sediment transport, but notable exceptions occur where tides move through narrow inlets. Here they scour the narrow entrances to many harbors that would otherwise be blocked.

Sometimes deposits called **tidal deltas** are created by tidal currents (Figure 14.28). They may develop either as *flood deltas* landward of an inlet or as *ebb deltas* on the seaward side of an inlet. Because wave activity and longshore currents are reduced on the sheltered landward side, flood deltas are more common and actually more prominent (see Figure 14.16). They form after the tidal current moves rapidly through an inlet. As the current emerges into more open waters from the narrow passage, it slows and deposits its load of sediment.

Chapter Summary

• *The ocean's surface currents follow the general pattern of the world's major wind belts.* Surface currents are parts of huge, slowly moving loops of water called *gyres* that are centered in the subtropics of each ocean basin. The *positions of the continents* and the *Coriolis effect* also influence the movement of ocean water within gyres. Because of the Coriolis effect, subtropical gyres move *clockwise in the Northern Hemisphere* and *counterclockwise in the Southern Hemisphere.* Generally, *four main currents* comprise each subtropical gyre.

• *Ocean currents are important in navigation and for the effect they have on climates.* Poleward-moving *warm ocean currents moderate winter temperatures* in the middle latitudes. Cold currents exert their greatest influence during summer in middle latitudes and year-round in the tropics. In addition to cooler

temperatures, *cold currents are associated with greater fog frequency and drought.*

• *Upwelling,* the rising of colder water from deeper layers, is a wind-induced movement that brings cold, nutrient-rich water to the surface. *Coastal upwelling* is most characteristic along the west coasts of continents.

• In contrast to surface currents, *deep-ocean circulation* is governed by *gravity* and driven by *density differences.* The two factors that are most significant in creating a dense mass of water are *temperature* and *salinity,* so the movement of deep-ocean water is often termed *thermohaline circulation.* Most water involved in thermohaline circulation begins in high latitudes at the surface when the salinity of the cold water increases due to sea ice formation. This dense water sinks initiating deep-ocean currents.

- *Waves are moving energy* and *most ocean waves are initiated by the wind.* The three factors that influence the *height, wavelength,* and *period* of a wave are (1) *wind speed,* (2) *length of time the wind has blown,* and (3) *fetch,* the distance that the wind has traveled across open water. Once waves leave a storm area, they are termed *swells,* which are symmetrical, longer-wavelength waves.

- As waves travel, *water particles transmit energy by circular orbital motion,* which extends to a depth equal to one half the wavelength. When a wave travels into shallow water, it experiences physical changes that can cause the wave to collapse, or *break,* and form *surf.*

- *Beaches are composed of any locally derived material* that is in transit along the shore. Wave erosion is caused by *wave impact pressure* and *abrasion* (the sawing and grinding action of water armed with rock fragments). The bending of waves is called *wave refraction.* Owing to refraction, wave impact is concentrated against the sides and ends of headlands and dispersed in bays.

- Most waves reach the shore at an angle. The uprush (swash) and backwash of water from each breaking wave moves the sediment in a zigzag pattern along the beach. This movement is called *beach drift.* Oblique waves also produce *longshore currents* within the surf zone that flow parallel to the shore and transport more sediment than beach drift.

- *Erosional features* include *wave-cut cliffs* (which originate from the cutting action of the surf against the base of coastal land), *wave-cut platforms* (relatively flat, benchlike surfaces left behind by receding cliffs), and *marine terraces* (uplifted wave-cut platforms). Erosional features also include *sea arches* (formed when a headland is eroded and two sea caves from opposite sides unite), and *sea stacks* (formed when the roof of a sea arch collapses).

- *Depositional features* include *spits* (elongated ridges of sand that project from the land into the mouth of an adjacent bay), *baymouth bars* (sandbars that completely cross a bay), and *tombolos* (ridges of sand that connect an island to the mainland or to another island). Along the Atlantic and Gulf Coastal Plains, the coastal region is characterized by offshore *barrier islands,* which are low ridges of sand that parallel the coast.

- Local *factors that influence shoreline erosion* are (1) the proximity of a coast to sediment-laden rivers, (2) the degree of tectonic activity, (3) the topography and composition of the land, (4) prevailing winds and weather patterns, and (5) the configuration of the coastline and nearshore areas.

- *Hard stabilization* involves building hard, massive structures in an attempt to protect a coast from erosion or prevent the movement of sand along the beach. Hard stabilization includes *groins* (short walls constructed at a right angle to the shore to trap moving sand), *breakwaters* (structures built parallel to the shore to protect it from the force of large breaking waves), and *seawalls* (armoring the coast to prevent waves from reaching the area behind the wall). *Alternatives to hard stabilization* include *beach nourishment,* which involves the addition of sand to replenish eroding beaches, and *relocation* of damaged or threatened buildings.

- Because of basic geological differences, the *nature of shoreline erosion problems along America's Pacific and Atlantic/Gulf Coasts is very different.* Much of the development along the Atlantic and Gulf Coasts has occurred on barrier islands, which receive the full force of major storms. Much of the Pacific Coast is characterized by narrow beaches backed by steep cliffs and mountain ranges. A major problem facing the Pacific shoreline is the narrowing of beaches caused by irrigation and flood control dams that interrupt the natural flow of sand to the coast.

- One commonly used classification of coasts is based upon changes that have occurred with respect to sea level. *Emergent coasts* often exhibit wave-cut cliffs and marine terraces and develop either because an area experiences uplift or as a result of a drop in sea level. Conversely, *submergent coasts* commonly display drowned river mouths called *estuaries* and are created when sea level rises or the land adjacent to the sea subsides.

- *Tides,* the daily rise and fall in the elevation of the ocean surface at a specific location, are caused by the *gravitational attraction* of the Moon and, to a lesser extent, the Sun. The Moon and the Sun each produce a pair of *tidal bulges* on Earth. These tidal bulges remain in fixed positions relative to the generating bodies as Earth rotates through them, resulting in alternating high and low tides. *Spring tides* occur near the times of new and full moons when the Sun and Moon are aligned and their bulges are added together to produce especially high and low tides (a *large daily tidal range*). Conversely, *neap tides* occur at about the times of the first and third quarters of the Moon when the bulges of the Moon and Sun are at right angles, producing a *smaller daily tidal range.*

- *Three main tidal patterns exist worldwide.* A *diurnal tidal pattern* exhibits one high and low tide daily; a *semidiurnal tidal pattern* exhibits two high and low tides daily of about the same height; and a *mixed tidal pattern* usually has two high and low tides daily of different heights.

- *Tidal currents* are horizontal movements of water that accompany the rise and fall of the tides. *Tidal flats* are the areas that are affected by the advancing and retreating tidal currents. When tidal currents slow after emerging from narrow inlets, they deposit sediment that may eventually create *tidal deltas.*

Key Terms

abrasion (p. 385)
barrier island (p. 390)
baymouth bar (p. 390)
beach drift (p. 387)
beach nourishment (p. 395)
breakwater (p. 395)
Coriolis effect (p. 377)
diurnal tidal pattern (p. 401)
emergent coast (p. 397)

estuary (p. 398)
fetch (p. 382)
groin (p. 394)
gyre (p. 376)
hard stabilization (p. 394)
longshore current (p. 387)
marine terrace (p. 389)
mixed tidal pattern (p. 401)
neap tide (p. 400)

sea arch (p. 389)
sea stack (p. 389)
seawall (p. 395)
semidiurnal tidal pattern (p. 401)
spit (p. 390)
spring tide (p. 400)
submergent coast (p. 397)
surf (p. 384)
thermohaline circulation (p. 381)

Review Questions

1. What is the primary driving force of surface ocean currents? How do the distribution of continents on Earth and the Coriolis effect influence these currents?

2. What is a gyre? How many currents exist within each gyre? Name the five subtropical gyres, and identify the primary surface currents that comprise each gyre.

3. Why did Benjamin Franklin want to know about the surface current pattern in the North Atlantic Ocean?

4. How do ocean currents influence climate? Give at least one example.

5. Describe the process of coastal upwelling. Why is an abundance of marine life associated with these areas?

6. What is the driving force of deep-ocean circulation? Why is the movement of deep currents often termed "thermohaline circulation"?

7. List three factors that determine the height, length, and period of a wave.

8. Describe the motion of a floating object as a wave passes (see Figure 14.8).

9. Describe the physical changes that occur to a wave's speed, wavelength, and height as a wave moves into shallow water and breaks on the shore.

10. How do waves cause erosion?

11. What is wave refraction? What is the effect of this process along irregular coastlines?

12. What is beach drift, and how is it related to a longshore current? Why are beaches often called "rivers of sand"?

13. Describe the formation of the following shoreline features: wave-cut cliff, wave-cut platform, marine terrace, sea stack, spit, baymouth bar, tombolo.

14. List three ways that barrier islands may form.

15. List the types of hard stabilization and describe what each is intended to do. What effect does each one have on the distribution of sand on the beach?

16. List two alternatives to hard stabilization, indicating potential problems with each one.

17. Relate the damming of rivers to the shrinking of beaches at some locations along the West Coast of the United States.

18. What observable features would lead you to classify a coastal area as emergent?

19. Are estuaries associated with submergent or emergent coasts? Explain.

20. Discuss the origin of ocean tides. Explain why the Sun's influence on Earth's tides is only about half that of the Moon's, even though the Sun is so much more massive than the Moon.

21. Explain how an observer can experience two unequal high tides during a day (see Figure 14.25).

22. How do diurnal, semidiurnal, and mixed tidal patterns differ?

23. Distinguish between a flood current and an ebb current. Of flood current, ebb current, high slack water, and low slack water, when is the best time to navigate a boat in a shallow, rocky harbor?

Examining the Earth System

1. Describe the exchanges of energy between the ocean and the atmosphere that are responsible for (1) surface ocean currents and (2) the ocean's deep circulation.

2. If the warm North Atlantic Drift were to cease, how might the climate of Western Europe change? Speculate about how such a climate change might influence the biosphere.

3. It is believed that global temperatures will be increasing in the decades to come. How can a warmer atmosphere lead to a rise in sea level? How could geologic events associated with the solid Earth contribute to a rise in sea level? Describe some possible changes in coastal areas if sea level were to rise. You may find some interesting and useful information at this Environmental Protection Agency Website: **http://www.epa.gov/globalwarming/impacts/coastal/index.html**.

Web Resources

The *Earth Science* Website uses the resources and flexibility of the Internet to aid in your study of the topics in this chapter. Written and developed by Earth science instructors, this site will help improve your understanding of Earth science. Visit **http://www.prenhall.com/tarbuck** and click on the cover of *Earth Science 10e* to find:

• **Online review quizzes.**
• **Web-based critical thinking and writing exercises.**
• **Links to chapter-specific Web resources.**
• **Internet-wide key term searches.**

http://www.prenhall.com/tarbuck

Climber in Antarctica's Ellsworth Mountains. Even summertime here is frigid. Latitude and altitude are two important controls of temperature. (Photo by Gordon Wiltsie/Peter Arnold, Inc.)

CHAPTER 15

THE ATMOSPHERE: COMPOSITION, STRUCTURE, AND TEMPERATURE

Earth's atmosphere is unique. No other planet in our solar system has an atmosphere with the exact mixture of gases or the heat and moisture conditions necessary to sustain life as we know it. The gases that make up Earth's atmosphere and the controls to which they are subject are vital to our existence. In this chapter we begin our examination of the ocean of air in which we all must live. We will try to answer a number of basic questions. What is the composition of the atmosphere? At what point do we leave the atmosphere and enter outer space? What causes the seasons? How is air heated? What factors control temperature variations over the globe?

Weather influences our everyday activities, our jobs, and our health and comfort (Figure 15.1). Many of us pay little attention to the weather unless we are inconvenienced by it or when it adds to our enjoyment outdoors. Nevertheless, there are few other aspects of our physical environment that affect our lives more than the phenomena we collectively call the weather.

The United States has the greatest variety of weather of any country in the world. Severe weather events such as tornadoes, flash floods, and intense thunderstorms, as well as hurricanes and blizzards, are collectively more frequent and more damaging in the United States than in any other nation. Beyond its direct impact on the lives of individuals, the weather has a strong effect on the world economy by influencing agriculture, energy use, water resources, transportation, and industry.

Weather clearly influences our lives a great deal. Yet it is also important to realize that people influence the atmosphere and its behavior as well. There are, and will continue to be, significant political and scientific decisions that must be made involving these impacts. Answers to questions regarding air pollution and its control and the effects of various emissions on global climate and the atmosphere's ozone layer are important examples (see Box 15.1). So there is a need for increased awareness and understanding of our atmosphere and its behavior.

Weather and Climate

Acted on by the combined effects of Earth's motions and energy from the Sun, our planet's formless and invisible envelope of air reacts by producing an infinite variety of weather, which in turn creates the basic pattern of global climates. Although not identical, weather and climate have much in common.

Weather is constantly changing, sometimes from hour to hour and at other times from day to day. It is a term that refers to the state of the atmosphere at a given

Figure 15.1 Digging out after a major lake-effect snowstorm buried Buffalo, New York, in late December 2001. The city was under a state of emergency after receiving more than 6 feet of snow. (Photo by Mike Groll/Getty Images)

time and place. Whereas changes in the weather are continuous and sometimes seemingly erratic, it is nevertheless possible to arrive at a generalization of these variations. Such a description of aggregate weather conditions is termed **climate.** It is based on observations that have been accumulated over many years. Climate is often defined simply as "average weather," but this is an inadequate definition. In order to more accurately portray the character of an area, variations and extremes must also be included, as well as the probabilities that such departures will take place. For example, it is not only necessary for farmers to know the average rainfall during the growing season, but it is also important to know the frequency of extremely wet and extremely dry years. Thus, climate is the sum of all statistical weather information that helps describe a place or region.

The nature of weather and climate is expressed in terms of the same basic **elements,** those quantities or properties that are measured regularly. The most important are (1) air temperature, (2) humidity, (3) type and amount of cloudiness, (4) type and amount of precipitation, (5) air pressure, and (6) the speed and direction of the wind. These elements are the major variables from which weather patterns and climate types are deciphered. Although you will study these elements separately at first, keep in mind that they are very much interrelated. A change in any one of the elements will often bring about changes in the others.

Composition of the Atmosphere

Sometimes the term *air* is used as if it were a specific gas, which, of course, it is not. Rather, **air** is a *mixture* of many discrete gases, each with its own physical properties, in which varying quantities of tiny solid and liquid particles are suspended.

Major Components

The composition of air is not constant; it varies from time to time and from place to place. If the water vapor, dust, and other variable components were removed from the atmosphere, we would find that its makeup is very stable worldwide up to an altitude of about 80 kilometers (50 miles).

As you can see in Figure 15.2, two gases—nitrogen and oxygen—make up 99 percent of the volume of clean, dry air. Although these gases are the most plentiful components of air and are of great significance to life on Earth, they are of minor importance in affecting weather phenomena. The remaining 1 percent of dry air is mostly the inert gas argon (0.93 percent) plus tiny quantities of a number of other gases. Carbon dioxide, although present in only minute amounts (0.036 percent), is nevertheless an important constituent of air.

Carbon dioxide is of great interest to meteorologists because it is an efficient absorber of energy emitted by Earth and thus influences the heating of the atmosphere.

Variable Components

Air includes many gases and particles that vary significantly from time to time and place to place. Important examples include water vapor, dust particles, and ozone. Although usually present in small percentages, they can have significant effects on weather and climate.

Water Vapor.   The amount of water vapor in the air varies considerably, from practically none at all up to about 4 percent by volume. Why is such a small fraction of the atmosphere so significant? Certainly the fact that water vapor is the source of all clouds and precipitation would be enough to explain its importance. However, water vapor has other roles. Like carbon dioxide, it has the ability to absorb heat given off by Earth as well as some solar energy. It is therefore important when we examine the heating of the atmosphere.

When water changes from one state to another (see Figure 16.2, p. 435), it absorbs or releases heat. This energy is termed *latent heat,* which means "hidden heat." As we shall see in later chapters, water vapor in the atmosphere transports this latent heat from one region to another, and it is the energy source that helps drive many storms.

Aerosols.   The movements of the atmosphere are sufficient to keep a large quantity of solid and liquid particles suspended within it. Although visible dust sometimes clouds the sky, these relatively large particles are too heavy to stay in the air for very long. Still, many particles are microscopic and remain suspended for considerable periods of time. They may originate from many sources, both natural and human made, and include sea salts from breaking waves, fine soil blown into the air, smoke and soot from fires, pollen and microorganisms lifted by the wind, ash and dust from volcanic eruptions, and more. Collectively, these tiny solid and liquid particles are called **aerosols.**

From a meteorological standpoint, these tiny, often invisible particles can be significant. First, many act as surfaces on which water vapor can condense, an important function in the formation of clouds and fog. Second, aerosols can absorb, reflect, and scatter incoming solar radiation. Thus, when an air-pollution episode is occurring or when ash fills the sky following a volcanic eruption, the amount of sunlight reaching Earth's surface can be measurably reduced. Finally, aerosols contribute to an optical phenomenon we have all observed—the varied hues of red and orange at sunrise and sunset (Figure 15.3).

Ozone.   Another important component of the atmosphere is **ozone.** It is a form of oxygen that combines

BOX 15.1

PEOPLE AND THE ENVIRONMENT:
Altering the Atmosphere's Composition— Sources and Types of Air Pollution

Air pollutants are airborne particles and gases that occur in concentrations that endanger the health and well-being of organisms or disrupt the orderly functioning of the environment. One category of pollutants, the *primary pollutants*, are emitted directly from identifiable sources. They pollute the air immediately upon being emitted. Figure 15.A shows the major primary pollutants and the sources that produce them. When the sources are examined, the significance of the transportation category is obvious. It accounts for nearly half of our air pollution (by weight). In addition to highway vehicles, this category includes trains, ships, and airplanes. Still, the tens of millions of cars and trucks on U.S. roads are, without a doubt, the greatest contributors (Figure 15.B).

Sometimes the direct impact of primary pollutants on human health and the environment is less severe than the effects of the secondary pollutants they

form. *Secondary pollutants* are not emitted directly into the air, but form in the atmosphere when reactions take place among primary pollutants. The chemicals that make up smog are important examples, as is the sulfuric acid that falls as acid precipitation. After the primary pollutant, sulfur dioxide, is emitted into the atmosphere, it combines with oxygen to produce sulfur trioxide, which then combines with water to create this irritating and corrosive acid.

Many reactions that produce secondary pollutants are triggered by strong sunlight and so are called *photochemical reactions*. One common example occurs when nitrogen oxides absorb solar radiation, initiating a chain of complex reactions. When certain volatile organic compounds are present, the result is the formation of a number of undesirable secondary products that are very reactive, irritating, and toxic. Collectively, this noxious mixture of gases and particles is called *photochemical smog*.

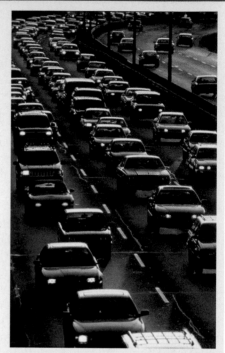

Figure 15.B This crowded freeway reminds us that the transportation category accounts for nearly half of the air pollution (by weight) in the United States. (Photo by Derek Trask/The Stock Market)

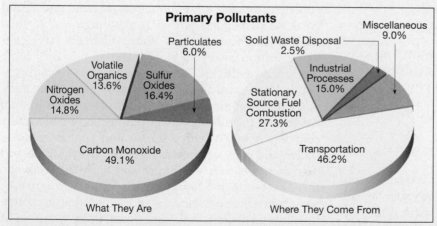

Figure 15.A Major primary pollutants and their sources. Percentages are calculated on the basis of weight. (Data from the U.S. Environmental Protection Agency)

Air-quality standards have not yet been met in a large number of places. But significant improvements in air quality have occurred ever since the U.S. Environmental Protection Agency initiated air-pollution control programs. Between 1970 and 2000 national total emissions of primary pollutants declined by about one-third. This occurred during a span when the U.S. population increased by about one-third and vehicle miles traveled rose nearly 130 percent. Despite continued improvements in air quality, more than 100 million people lived in counties with unhealthy air in 2000.

three oxygen atoms into each molecule (O_3). Ozone is not the same as oxygen we breathe, which has two atoms per molecule (O_2). There is very little of this gas in the atmosphere, and its distribution is not uniform. In the lowest portion of the atmosphere, ozone represents less than one part in 100 million. It is concentrated well above the surface in a layer called the *stratosphere*, between 10 and 50 kilometers (6 and 31 miles).

In this altitude range, oxygen molecules (O_2) are split into single atoms of oxygen (O) when they absorb ultraviolet radiation emitted by the Sun. Ozone is then created when a single atom of oxygen (O) and a molecule of oxygen (O_2) collide. This must happen in the presence of a third, neutral molecule that acts as a *catalyst* by allowing the reaction to take place without itself being consumed in the process. Ozone is concentrated in the 10- to 50-kilometer height range because a cru-

Figure 15.2 Proportional volume of gases composing dry air. Nitrogen and oxygen clearly dominate.

cial balance exists there: The ultraviolet radiation from the Sun is sufficient to produce single atoms of oxygen, and there are enough gas molecules to bring about the required collisions.

The presence of the ozone layer in our atmosphere is crucial to those of us who dwell on Earth. The reason is that ozone absorbs the potentially harmful ultraviolet (UV) radiation from the Sun. If ozone did not filter a great deal of the ultraviolet radiation, and if the Sun's UV rays reached the surface of Earth undiminished, our planet would be uninhabitable for most life as we know it. Thus, anything that reduces the amount of

Figure 15.3 Dust in the air can cause sunsets to be especially colorful. (Photo by Steve Elmore/The Stock Market)

ozone in the atmosphere could affect the well-being of life on Earth. Just such a problem exists and is described in Box 15.2.

Height and Structure of the Atmosphere

 Atmosphere
  ↳ Heating the Atmosphere

To say that the atmosphere begins at Earth's surface and extends upward is obvious. However, where does the atmosphere end and outer space begin? There is no sharp boundary; the atmosphere rapidly thins as you travel away from Earth, until there are too few gas molecules to detect.

Pressure Changes

To understand the vertical extent of the atmosphere, let us examine the changes in the atmospheric pressure with height. Atmospheric pressure is simply the weight of the air above. At sea level, the average pressure is slightly more than 1000 millibars. This corresponds to a weight of slightly more than 1 kilogram per square centimeter (14.7 pounds per square inch). Obviously the pressure at higher altitudes is less (Figure 15.4).

One half of the atmosphere lies below an altitude of 5.6 kilometers (3.5 miles). At about 16 kilometers (10 miles), 90 percent of the atmosphere has been traversed, and above 100 kilometers (62 miles), only 0.00003 percent of all the gases making up the atmosphere remains. Even so, traces of our atmosphere extend far beyond this altitude, gradually merging with the emptiness of space.

Temperature Changes

By the early twentieth century, much had been learned about the lower atmosphere. The upper atmosphere was partly known from indirect methods. Data from balloons and kites had revealed that the air temperature dropped with increasing height above Earth's surface. This phenomenon is felt by anyone who has climbed a high mountain and is obvious in pictures of snow-capped mountaintops rising above snow-free lowlands (Figure 15.5). We divide the atmosphere vertically into four layers on the basis of temperature (Figure 15.6).

Troposphere.   The bottom layer in which we live, where temperature decreases with an increase in altitude, is the **troposphere**. The term literally means the region where air "turns over," a reference to the appreciable vertical mixing of air in this lowermost zone. The troposphere is the chief focus of meteorologists, because it is in this layer that essentially all important weather phenomena occur.

BOX 15.2

PEOPLE AND THE ENVIRONMENT:
Ozone Depletion—A Global Issue

The loss of ozone high in the atmosphere as a consequence of human activities is a serious global-scale environmental problem (Figure 15.C). For nearly a billion years Earth's ozone layer has protected life on the planet. However, over the past half century, people have unintentionally placed the ozone layer in jeopardy by polluting the atmosphere. The offending chemicals are known as chlorofluorocarbons (CFCs for short). They are versatile compounds that are chemically stable, odorless, nontoxic, noncorrosive, and inexpensive to produce. Over the decades many uses were developed for CFCs, including as coolants for air-conditioning and refrigeration equipment, cleaning solvents for electronic components, propellants for aerosol sprays, and the production of certain plastic foams.

No one worried about how CFCs might affect the atmosphere until three scientists, Paul Crutzen, F. Sherwood Rowland, and Mario Molina, studied the relationship. In 1974 they alerted the world when they reported that CFCs were probably reducing the average concentration of ozone in the stratosphere. In 1995 these scientists were awarded the Nobel Prize in chemistry for their pioneering work.

They discovered that because CFCs are practically inert (that is, not chemically active) in the lower atmosphere, a portion of these gases gradually makes its way to the ozone layer, where sunlight separates the chemicals into their constituent atoms. The chlorine atoms released this way, through a complicated series of reactions, have the net effect of removing some of the ozone.

Because ozone filters out most of the ultraviolet (UV) radiation from the Sun, a decrease in its concentration permits more of these harmful wavelengths to reach Earth's surface. The most serious threat to human health is an increased risk of skin cancer. An increase in damaging UV radiation also can impair the human immune system as well as promote cataracts, a clouding of the eye lens that reduces vision and may cause blindness if not treated.

In response to this problem, an international agreement known as the *Montreal Protocol* was developed under the sponsorship of the United Nations to eliminate the production and use of CFCs.

Although relatively strong action was taken, CFC levels in the atmosphere will not drop rapidly. Once in the atmosphere, CFC molecules can take many years to reach the ozone layer and once there, they can remain active for decades. This does not promise a near-term reprieve for the ozone layer. Even after CFC levels begin to decline, the ozone layer probably will not cleanse itself of most of the pollutants until sometime in the second half of the twenty-first century.

100   150   200   250   300   350   400   450
Dobson Units

Figure 15.C   This satellite image shows ozone distribution in the Southern Hemisphere in October 1998. The "hole" forms over Antarctica during the Southern Hemisphere spring. It is that region where ozone concentrations are less than 220 Dobson units. The 1998 Antarctic ozone hole was one of the largest ever observed, extending over about 26 million square kilometers (10 million square miles). (Data from NOAA)

The temperature decrease in the troposphere is called the **environmental lapse rate.** Its average value is 6.5°C per kilometer (3.5°F per 1000 feet), a figure known as the *normal lapse rate.* It should be emphasized, however, that the environmental lapse rate is not a constant, but rather can be highly variable, and must be regularly measured. To determine the actual environmental lapse rate as well as gather information about vertical changes in pressure, wind, and humidity, radiosondes are used. The *radiosonde* is an instrument package that is attached to a balloon and transmits data by radio as it ascends through the atmosphere (Figure 15.7).

The thickness of the troposphere is not the same everywhere; it varies with latitude and the season. On

Figure 15.4   Atmospheric pressure variation with altitude. The rate of pressure decrease with an increase in altitude is not constant. Rather, pressure decreases rapidly near Earth's surface and more gradually at greater heights.

Figure 15.5   Temperatures drop with an increase in altitude in the troposphere. Therefore, it is possible to have snow on a mountaintop and warmer, snow-free lowlands below. Mount Kerkeslin, Jasper National Park, Alberta, Canada. (Photo by Carr Clifton/Minden Pictures)

the average, the temperature drop continues to a height of about 12 kilometers (7.4 miles). The outer boundary of the troposphere is the *tropopause.*

Stratosphere.   Beyond the tropopause is the **stratosphere.** In the stratosphere, the temperature remains constant to a height of about 20 kilometers (12 miles) and then begins a gradual increase that continues until the *stratopause,* at a height of nearly 50 kilometers (30 miles) above Earth's surface. Below the tropopause, atmospheric properties like temperature and humidity are readily transferred by large-scale turbulence and mixing. Above the tropopause, in the stratosphere, they are not. Temperatures increase in the stratosphere because it is in this layer that the atmosphere's ozone is concentrated. Recall that ozone absorbs ultraviolet radiation from the Sun. As a consequence, the stratosphere is heated.

Mesosphere.   In the third layer, the **mesosphere,** temperatures again decrease with height until, at the *mesopause,* more than 80 kilometers (50 miles) above the surface, the temperature approaches −90°C.

Thermosphere.   The fourth layer extends outward from the mesopause and has no well-defined upper limit. It is the **thermosphere,** a layer that contains only a tiny fraction of the atmosphere's mass. In the extremely rarefied air of this outermost layer, temperatures again increase, owing to the absorption of very short-wave, high-energy solar radiation by atoms of oxygen and nitrogen.

Temperatures rise to extremely high values of more than 1000°C in the thermosphere. But such temperatures are not comparable to those experienced near Earth's surface. Temperature is defined in terms of the average speed at which molecules move. Because the gases of the thermosphere are moving at very high speeds, the temperature is very high. But the gases are so sparse that, collectively, they possess only an insignificant quantity of heat. For this reason, the temperature of a satellite orbiting Earth in the thermosphere is determined chiefly by the amount of solar radiation it absorbs and not by the high temperature of the almost nonexistent surrounding air. If an astronaut inside were to expose his or her hand, it would not feel hot.

Earth–Sun Relationships

Always remember that nearly all of the energy that drives Earth's variable weather and climate comes from the Sun. Earth intercepts only a minute percentage of the energy given off by the Sun—less than one two-billionth. This may seem to be an insignificant amount until we realize that it is several hundred thousand times the electrical-generating capacity of the United States.

Figure 15.6 Thermal structure of the atmosphere.

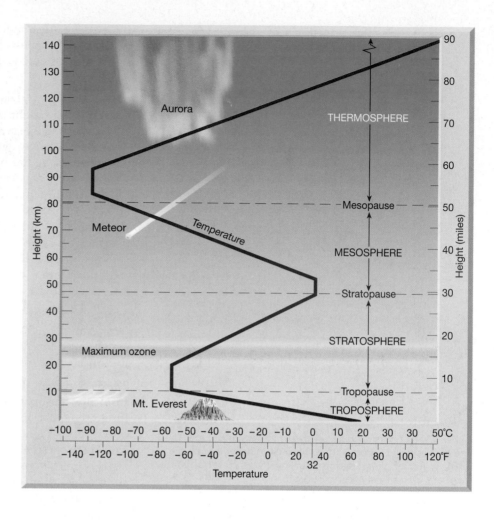

Solar energy is not distributed evenly over Earth's land–sea surface. The amount of energy received varies with latitude, time of day, and season of the year. Contrasting images of polar bears on ice rafts and palm trees along a remote tropical beach serve to illustrate the extremes. It is the unequal heating of Earth that creates winds and drives the ocean's currents. These movements, in turn, transport heat from the tropics toward the poles in an unending attempt to balance energy inequalities. The consequences of these processes are the phenomena we call weather. If the Sun were "turned off," global winds and ocean currents would quickly cease. Yet as long as the Sun shines, the winds *will* blow and weather *will* persist. So to understand how the atmosphere's dynamic weather machine works, we must first know why different latitudes receive varying quantities of solar energy and why the amount of solar energy changes to produce the seasons. As you will see, the variations in solar heating are caused by the motions of Earth relative to the Sun and by variations in Earth's land–sea surface.

Earth's Motions

Earth has two principal motions—rotation and revolution. **Rotation** is the spinning of Earth about its axis. The axis is an imaginary line running through the poles. Our planet rotates once every 24 hours, producing the daily cycle of daylight and darkness. At any moment, half of Earth is experiencing daylight, and the other half darkness. The line separating the dark half of Earth from the lighted half is called the **circle of illumination.**

Revolution refers to the movement of Earth in its orbit around the Sun. Hundreds of years ago, most people believed that Earth was stationary in space and that the Sun and stars revolved around our planet. Today, we know that Earth is traveling at nearly 113,000 kilometers (70,000 miles) per hour in an eliptical orbit about the Sun.

Seasons

We know that it is colder in winter than in summer. But why? Length of daylight certainly accounts for some of the difference. Long summer days expose us to more solar radiation, whereas short winter days expose us to less.

Furthermore, a gradual change in the angle of the noon Sun above the horizon is quite noticeable (Figure 15.8). At midsummer, the noon Sun is seen high above the horizon. But as summer gives way to autumn, the noon Sun appears lower in the sky and sunset occurs

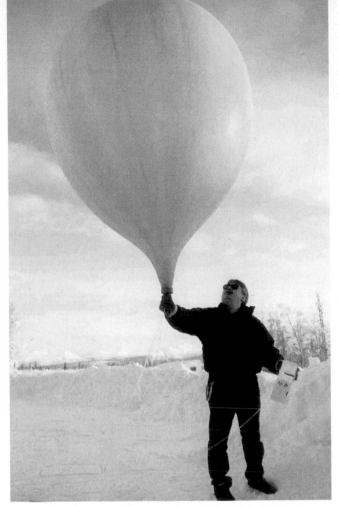

Figure 15.7 A lightweight package of instruments, *the radiosonde*, is carried aloft by a small weather balloon. Radiosondes supply data on vertical changes in temperature, pressure, and humidity. (Photo by AP/Wide World Photos)

earlier each evening. What we observe here is the annual shifting of the solar angle or *altitude* of the Sun.

The seasonal variation in the altitude of the Sun affects the amount of energy received at Earth's surface in two ways. First, when the Sun is high in the sky, the solar rays are most concentrated (you can see this in Figure 15.9A). The lower the angle, the more spread out

Figure 15.8 Daily paths of the Sun for June and December for an observer in the middle latitudes in the Northern Hemisphere. Notice that the angle of the Sun above the horizon is much greater in the summer than in the winter.

and less intense is the solar radiation reaching the surface (Figure 15.9B, C). To illustrate this principle, hold a flashlight at a right angle to a surface and then change the angle.

Second, and of lesser importance, the angle of the Sun determines the amount of atmosphere the rays must penetrate (Figure 15.10). When the Sun is directly overhead, the rays pass through a thickness of only 1 atmosphere, whereas rays entering at a 30-degree angle travel through twice this amount, and 5-degree rays travel through a thickness roughly equal to 11 atmospheres. The longer the path, the greater the chances for absorption, reflection, and scattering by the atmosphere, all of which reduce the intensity at the surface. The same effects account for the fact that we cannot look directly at the midday Sun, but we can enjoy gazing at a sunset.

It is also important to remember that Earth has a spherical shape. Hence, on any given day only places located at a particular latitude receive vertical (90-degree) rays from the Sun. As we move either north or south of this location, the Sun's rays strike at an ever decreasing angle. Thus, the nearer a place is to the latitude receiving vertical rays of the Sun, the higher will be its noon Sun and the more intense will be the radiation it receives.

Earth's Orientation

What causes the fluctuations in the Sun angle and length of daylight that occur during the course of a year? They occur because *Earth's orientation to the Sun continually changes as it travels along its orbit*. Earth's axis is not perpendicular to the plane of its orbit around the Sun. Instead it is tilted 23 1/2 degrees from the perpendicular, as shown in Figure 15.11. This is called the **inclination of the axis.** As you will see, if the axis were not inclined, we would have no seasonal changes. In addition, because the axis remains pointed in the same direction (toward the North Star) as Earth journeys around the Sun, the orientation of Earth's axis to the Sun's rays is constantly changing (Figure 15.11).

For example, on one day in June each year the axis is such that the Northern Hemisphere is "leaning" 23 1/2 degrees *toward* the Sun. Six months later, in December, when Earth has moved to the opposite side of its orbit, the Northern Hemisphere "leans" 23 1/2 degrees *away* from the Sun. On days between these extremes, Earth's axis is leaning at amounts less than 23 1/2 degrees to the rays of the Sun. This change in orientation causes the spot where the Sun's rays are vertical to make an annual migration from 23 1/2 degrees north of the equator to 23 1/2 degrees south of the equator.

In turn, this migration causes the angle of the noon Sun to vary by up to 47 degrees (23 1/2 + 23 1/2) for many locations during the year. For example, a midlat-

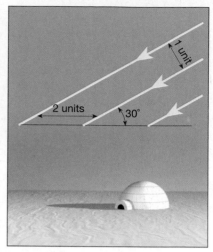

Figure 15.9 Changes in the Sun's angle causes variations in the amount of solar energy reaching Earth's surface. The higher the angle, the more intense the solar radiation.

itude city like New York (about 40 degrees north latitude) has a maximum noon Sun angle of 73 1/2 degrees when the Sun's vertical rays reach their farthest northward location in June and a minimum noon Sun angle of 26 1/2 degrees six months later.

Solstices and Equinoxes

Historically, four days a year have been given special significance based on the annual migration of the direct rays of the Sun and its importance to the yearly weather cycle. On June 21 or 22, Earth is in a position where the axis in the Northern Hemisphere is tilted 23 1/2 degrees toward the Sun (Figure 15.12A). At this time the vertical rays of the Sun strike 23 1/2 degrees north latitude (23 1/2 degrees north of the equator), a latitude known as the **Tropic of Cancer.** For people in the North-

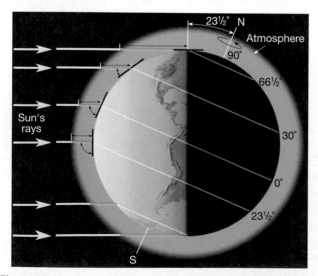

Figure 15.10 Notice that rays striking Earth at a low angle (toward the poles) must travel through more of the atmosphere than rays striking at a high angle (around the equator) and thus are subject to greater depletion by reflection and absorption.

ern Hemisphere, June 21 or 22 is known as the **summer solstice,** the first "official" day of summer.

Six months later, on about December 21 or 22, Earth is in an opposite position, with the Sun's vertical rays striking at 23 1/2 degrees south latitude (Figure 15.12B). This parallel is known as the **Tropic of Capricorn.** For those in the Northern Hemisphere, December 21 and 22 is the **winter solstice,** the first day of winter. However, at the same time in the Southern Hemisphere, people are experiencing just the opposite: the summer solstice.

The equinoxes occur midway between the solstices. September 22 or 23 is the date of the **autumnal equinox** in the Northern Hemisphere, and March 21 or 22 is the date of the **spring equinox.** On these dates, the vertical rays of the Sun strike the equator (0 degrees latitude) because Earth is in such a position in its orbit that the axis is tilted neither toward nor away from the Sun (Figure 15.12C).

The length of daylight versus darkness is also determined by Earth's position in orbit. The length of daylight on June 21, the summer solstice in the Northern Hemisphere, is greater than the length of night. This fact can be established from Figure 15.12A by comparing the fraction of a given latitude that is on the "day" side of the circle of illumination with the fraction on the "night" side. The opposite is true for the winter solstice, when the nights are longer than the days. Again for comparison let us consider New York City, which has 15 hours of daylight on June 21 and only nine hours on December 21 (you can see this in Figure 15.12 and Table 15.1). Also note from Table 15.1 that the farther you are north of the equator on June 21, the longer the period of daylight. When you reach the Arctic Circle (66 1/2° N), the length of daylight is 24 hours. This is the land of the "midnight Sun," which does not set for about six months at the North Pole (Figure 15.13).

Figure 15.11   Earth–Sun relationships.

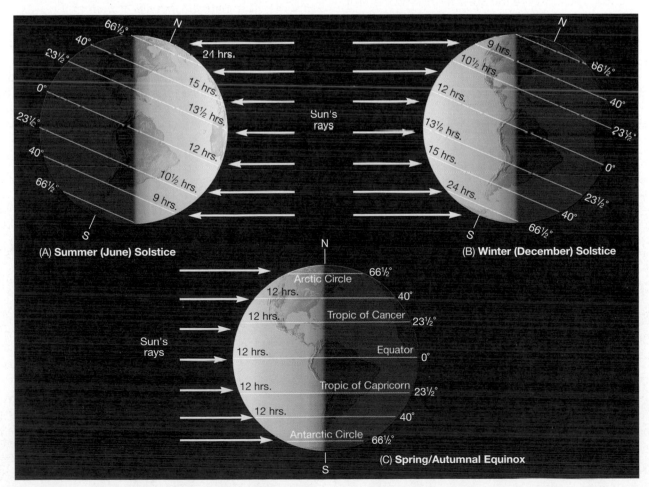

(A) **Summer (June) Solstice**

(B) **Winter (December) Solstice**

(C) **Spring/Autumnal Equinox**

Figure 15.12   Characteristics of the solstices and equinoxes.

Latitude (degrees)	Summer Solstice	Winter Solstice	Equinoxes
0	12 h	12 h	12 h
10	12 h 35 min	11 h 25 min	12
20	13 12	10 48	12
30	13 56	10 04	12
40	14 52	9 08	12
50	16 18	7 42	12
60	18 27	5 33	12
70	24 h (for 2 mo)	0 00	12
80	24 h (for 4 mo)	0 00	12
90	24 h (for 6 mo)	0 00	12

Table 15.1 Length of daylight.

During an equinox (meaning "equal night"), the length of daylight is 12 hours everywhere on Earth, because the circle of illumination passes directly through the poles, dividing the latitudes in half.

As a review of the characteristics of the summer solstice for the Northern Hemisphere, examine Figure 15.12A and Table 15.1 and consider the following facts:

1. The solstice occurs on June 21 or 22.
2. The vertical rays of the Sun are striking the Tropic of Cancer (23 1/2 degrees north latitude).
3. Locations in the Northern Hemisphere are experiencing their greatest length of daylight (opposite for the Southern Hemisphere).
4. Locations north of the Tropic of Cancer are experiencing their highest noon Sun angles (opposite for places south of the Tropic of Capricorn).

5. The farther you are north of the equator, the longer the period of daylight, until the Arctic Circle is reached, where daylight lasts for 24 hours (opposite for the Southern Hemisphere).

The facts about the winter solstice are just the opposite. It should now be apparent why a midlatitude location is warmest in the summer. It is then that the days are longest and the Sun's altitude is highest.

In summary, seasonal variations in the amount of solar energy reaching places on Earth's surface are caused by the migrating vertical rays of the Sun and the resulting variations in Sun angle and length of daylight.

All places at the same latitude have identical Sun angles and lengths of daylight. If the Earth–Sun relationships just described were the only controls of temperature, we would expect these places to have identical temperatures as well. Obviously this is not the case. Although the altitude of the Sun is an extremely important control of temperature, it is not the only control, as you will see.

STUDENTS SOMETIMES ASK...

The "official" first day of winter isn't until December 21 or 22. Realistically, doesn't winter start sooner?

The most publicized dates for the seasons are based on the astronomical definition, which uses the dates of the solstices and equinoxes as the "first day" of each of the seasons. Because the weather phenomena we normally associate with each season do not coincide well with this definition, mete-

Figure 15.13 Multiple exposures of the midnight Sun in late June or July in high northern latitudes—Alaska, Scandinavia, northern Canada, etc. (Photo by Brian Stablyk/Tony Stone Images)

orologists prefer to divide the year into four three-month periods based primarily on temperature. Thus, winter is defined as December, January, and February, the three coldest months of the year in the Northern Hemisphere. Summer is defined as the three warmest months, June, July, and August. Spring and autumn are the transitional periods between these two seasons. This definition is more useful for meteorological purposes.

Heat and Temperature

The concepts of *heat* and *temperature* are often confused. The phrase "in the heat of the day" is one common expression in which the word "heat" is misused to describe the concept of temperature. *Heat* is a form of energy. Recall that all matter is composed of atoms or molecules that are constantly vibrating and therefore possess kinetic energy, which is the energy of motion. **Heat** is defined as the *total* kinetic energy of all the atoms and molecules that make up a substance. By contrast, *temperature* refers to intensity—that is, the degree of "hotness." **Temperature** is a measure of the *average* kinetic energy of the individual atoms or molecules in a substance. When heat is added to a substance, its atoms move faster and faster and its temperature rises. When heat is removed, its atoms move slower and its temperature drops.

Although the ideas of temperature and heat are distinct, they are nevertheless related. Certainly, the addition or subtraction of heat causes temperatures to increase or decrease. Moreover, differences in temperature determine the direction of heat flow. When two bodies having different temperatures are in contact, *heat always moves from the higher-temperature body to the lower-temperature body.* For example, when you touch a hot stove, heat enters your hand because the stove is much warmer than your hand. By contrast, when you hold an ice cube, heat is transferred from your hand to the ice cube. Thus, if two objects of different temperature are in contact, the warmer object will become cooler and the cooler object will become warmer until they both reach the same temperature.

Mechanisms of Heat Transfer

 GEODe
EARTH SCIENCE

Atmosphere
↳ Heating the Atmosphere

Three mechanisms of heat transfer are recognized: conduction, convection, and radiation. Although we present them separately, all three processes go on simultaneously in the atmosphere. In addition, these mechanisms operate to transfer heat between Earth's surface (both land and water) and the atmosphere.

Conduction

Conduction is familiar to all of us. Anyone who has touched a metal spoon that was left in a hot pan has discovered that heat was conducted through the spoon. **Conduction** is *the transfer of heat through matter by molecular activity.* The energy of molecules is transferred through collisions from one molecule to another, with the heat flowing from the higher temperature to the lower temperature.

The ability of substances to conduct heat varies considerably. Metals are good conductors, as those of us who have touched hot metal have quickly learned (Figure 15.14). Air, conversely, is a very poor conductor of heat. Consequently, conduction is important only between Earth's surface and the air directly in contact with the surface. As a means of heat transfer for the atmosphere as a whole, conduction is the least significant.

Convection

Much of the heat transport that occurs in the atmosphere is carried on by convection. **Convection** is *the transfer of heat by mass movement or circulation within a substance.* It takes place in fluids (for example, liquids like the ocean and gases like air) where the atoms and molecules are free to move about.

The pan of water in Figure 15.14 illustrates the nature of simple convective circulation. Radiation from the fire warms the bottom of the pan, which conducts heat to the water near the bottom of the container. As the water is heated, it expands and becomes less dense than the water above. Because of this new buoyancy, the warmer water rises. At the same time, cooler, denser water near the top of the pan sinks to the bottom, where it becomes heated. As long as the water is heated unequally—that is, from the bottom up—the water will continue to "turn over," producing a *convective circulation.* In a similar manner, most of the heat acquired in the lowest layer of the atmosphere by way of radiation and conduction is transferred by convective flow.

Figure 15.14   The three mechanisms of heat transfer: conduction, convection, and radiation.

On a global scale, convection in the atmosphere creates a huge, worldwide air circulation. This is responsible for the redistribution of heat between hot equatorial regions and the frigid poles. This important process will be discussed in detail in Chapter 17.

Radiation

The third mechanism of heat transfer is radiation. As shown in Figure 15.14, radiation travels out in all directions from its source. Unlike conduction and convection, which need a medium to travel through, radiant energy readily travels through the vacuum of space. Thus, radiation is the heat-transfer mechanism by which solar energy reaches our planet.

Because the Sun is the ultimate source of energy that creates our weather, we will consider the nature of solar radiation in more detail. From our everyday experience we know that the Sun emits light and heat as well as the ultraviolet rays that cause suntan. Although these forms of energy comprise a major portion of the total energy that radiates from the Sun, they are only part of a large array of energy called **radiation** or **electromagnetic radiation.** This array or spectrum of electromagnetic energy is shown in Figure 15.15. All radiation, whether X rays, radio waves, or heat waves, travel through the vacuum of space at 300,000 kilometers (186,000 miles) per second and only slightly slower through our atmosphere.

It helps to understand radiant energy by picturing ripples made in a pond by tossing in a pebble or by observing ocean waves. Like these waves, *electromagnetic waves,* as waves of radiant energy are called, come in various sizes. For our purpose, the most important difference among electromagnetic waves is their *wavelength,* or the distance from one crest to the next. Radio waves have the longest wavelengths, ranging to tens of

kilometers. Gamma waves are the shortest, being less than a billionth of a centimeter long.

Visible light, as the name implies, is the only portion of the spectrum we can see. We often refer to visible light as "white" light because it appears "white" in color. However, it is easy to show that white light is really a mixture of colors, each corresponding to a specific wavelength (Figure 15.16). By using a prism, white light can be divided into the colors of the rainbow, from violet with the shortest wavelength—0.4 micrometer (1 micrometer is 0.0001 centimeter)—to red with the longest wavelength—0.7 micrometer.

Located adjacent to red, and having a longer wavelength, is **infrared** radiation, which we cannot see but which we can detect as heat. The closest invisible waves to violet are called **ultraviolet** rays. They are responsible for the sunburn that can occur after an intense exposure to the Sun. Although we divide radiant energy into groups based on our ability to perceive them, all forms of radiation are basically the same. When any form of radiant energy is absorbed by an object, the result is an increase in molecular motion, which causes a corresponding increase in temperature.

To understand better how the atmosphere is heated, it is useful to have a general understanding of the basic laws governing radiation.

1. *All objects, at whatever temperature, emit radiant energy.* Thus, not only hot objects like the Sun but also Earth, including its polar ice caps, continually emit energy.

2. *Hotter objects radiate more total energy per unit area than do colder objects.*

3. *The hotter the radiating body, the shorter the wavelength of maximum radiation.* The Sun, with a surface tem-

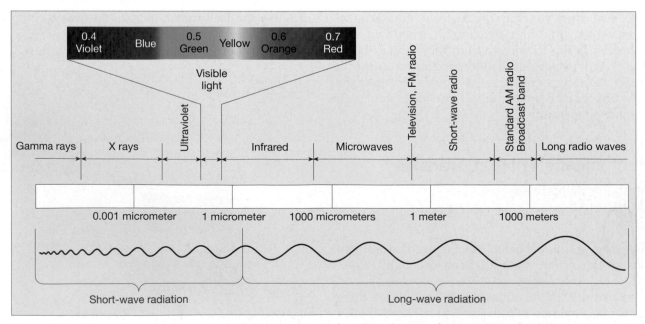

Figure 15.15 The electromagnetic spectrum, illustrating the wavelengths and names of various types of radiation.

Figure 15.16 Visible light consists of an array of colors we commonly call the "colors of the rainbow." Rainbows are relatively common optical phenomena produced by the bending and reflection of light by drops of water. Denali National Park, Alaska. (Photo by Carr Clifton)

perature of nearly 6000°C, radiates maximum energy at 0.5 micrometer, which is in the visible range. The maximum radiation for Earth occurs at a wavelength of 10 micrometers, well within the infrared (heat) range. Because the maximum Earth radiation is roughly 20 times longer than the maximum solar radiation, it is often called long-wave radiation, and solar radiation is called short-wave radiation.

4. *Objects that are good absorbers of radiation are good emitters as well.* Earth's surface and the Sun are nearly perfect radiators because they absorb and radiate with nearly 100 percent efficiency for their respective temperatures. On the other hand, *gases are selective absorbers and radiators.* Thus, the atmosphere, which is nearly transparent to (does not absorb) certain wavelengths of radiation, is nearly opaque (a good absorber) to others. Our experience tells us that the atmosphere is transparent to visible light; hence, it readily reaches Earth's surface. This is not the case for the longer wavelength radiation emitted by Earth.

The Fate of Incoming Solar Radiation

GEODe
Atmosphere
↳ Heating the Atmosphere

When radiation strikes an object, there are usually three different results. First, some of the energy is *absorbed* by the object. Recall that when radiant energy is absorbed, it is converted to heat, which causes an increase in temperature. Second, substances such as water and air are transparent to certain wavelengths of radiation. Such materials simply *transmit* this energy. Radiation that is transmitted does not contribute energy to the object. Third, some radiation may "bounce off" the object without being absorbed or transmitted. *Reflection and scattering* are responsible for redirecting incoming solar radiation. In summary, *radiation may be absorbed, transmitted, or redirected (reflected or scattered).*

Figure 15.17 shows the fate of incoming solar radiation averaged for the entire globe. Notice that the atmosphere is quite transparent to incoming solar radiation. On average, about 50 percent of the solar energy reaching the top of the atmosphere is absorbed at Earth's surface. Another 30 percent is reflected back to space by the atmosphere, clouds, and reflective surfaces such as snow and water. The remaining 20 percent is absorbed by clouds and the atmosphere's gases. What determines whether solar radiation will be transmitted to the surface, scattered, or reflected outward? It depends greatly on the wavelength of the energy being transmitted, as well as on the nature of the intervening material.

Reflection and Scattering

Reflection is the process whereby light bounces back from an object at the same angle at which it encounters a surface and with the same intensity (Figure 15.18A). By contrast, **scattering** produces a larger number of weaker rays that travel in different directions. Although scattering disperses light both forward and backward (*backscattering*), more energy is dispersed in the forward direction (Figure 15.18B).

Figure 15.17 Average distribution of incoming solar radiation by percentage. More solar energy is absorbed by Earth's surface than by the atmosphere. Consequently, the air is not heated directly by the Sun, but is heated indirectly from Earth's surface.

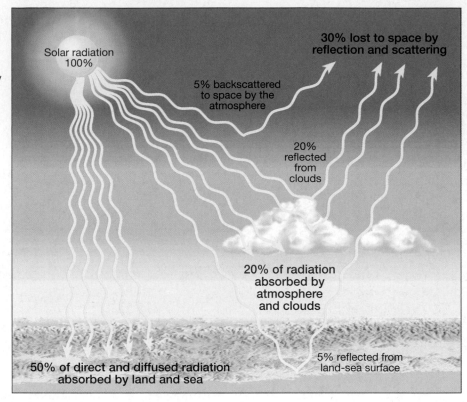

Solar radiation 100%

30% lost to space by reflection and scattering

5% backscattered to space by the atmosphere

20% reflected from clouds

20% of radiation absorbed by atmosphere and clouds

5% reflected from land-sea surface

50% of direct and diffused radiation absorbed by land and sea

Reflection. About 30 percent of the solar energy reaching the outer atmosphere is reflected back to space. Included in this figure is the amount sent skyward by backscattering. This energy is lost to Earth and does not play a role in heating the atmosphere.

The fraction of the total radiation that is reflected by a surface is called its **albedo.** Thus, the albedo for Earth as a whole (the *planetary albedo*) is 30 percent. However, the albedo from place to place as well as from time to time in the same locale varies considerably, depending on the amount of cloud cover and particulate matter in the air, as well as on the angle of the Sun's rays and the nature of the surface. A lower Sun angle means that more atmosphere must be penetrated, thus making the "obstacle course" longer and therefore the loss of solar radiation greater (see Figure 15.10). Table 15.2 gives the albedo for various surfaces. Note that the angle at which the Sun's rays strike a water surface greatly affects its albedo.

Scattering. Although incoming solar radiation travels in a straight line, small dust particles and gas molecules in the atmosphere scatter some of this energy in all directions. The result, called *diffused light*, explains how light reaches into the area beneath a shade tree, and how a room is lit in the absence of direct sunlight. Further, scattering accounts for the brightness and even the blue color of the daytime sky. In contrast, bodies like the Moon and Mercury, which are without atmospheres, have dark skies and "pitch-black" shadows, even during daylight hours. Overall, about half of the solar radiation that is absorbed at Earth's surface arrives as diffused (scattered) light.

Absorption

As stated earlier, gases are selective absorbers, meaning that they absorb strongly in some wavelengths, moderately in others, and only slightly in still others. When a gas molecule absorbs light waves, this energy is transformed into internal molecular motion, which is detectable as a rise in temperature.

Nitrogen, the most abundant constituent in the atmosphere, is a poor absorber of all types of incoming radiation. Oxygen and ozone are efficient absorbers of ultraviolet radiation. Oxygen removes most of the shorter ultraviolet radiation high in the atmosphere,

A. Reflection   B. Scattering

Figure 15.18 Reflection and scattering. **A.** Reflected light bounces back from a surface at the same angle at which it strikes that surface and with the same intensity. **B.** When a beam of light is scattered, it results in a larger number of weaker rays, traveling in all different directions. Usually more energy is scattered in the forward direction than is backscattered.

Table 15.2 Albedo of various surfaces.	
Surface	**Percent Reflected**
Clouds, stratus	
<150 meters thick	25–63
150–300 meters thick	45–75
300–600 meters thick	59–84
Average of all types and thicknesses	50–55
Concrete	17–27
Crops, green	5–25
Forest, green	5–10
Meadows, green	5–25
Ploughed field, moist	14–17
Road, blacktop	5–10
Sand, white	30–60
Snow, fresh-fallen	80–90
Snow, old	45–70
Soil, dark	5–15
Soil, light (or desert)	25–30
Water	8*

*Typical albedo value for a water surface. The albedo of a water surface varies greatly depending upon the Sun angle. If the Sun angle is greater than 30 degrees, the albedo is less than 5 percent. When the Sun is near the horizon (Sun angle less than 3 degrees), the albedo is more than 60 percent.

and ozone absorbs most of the remaining ultraviolet rays in the stratosphere. The absorption of UV radiation in the stratosphere accounts for the high temperatures experienced there. The only other significant absorber of incoming solar radiation is water vapor, which, along with oxygen and ozone, accounts for most of the solar radiation absorbed within the atmosphere.

For the atmosphere as a whole, none of the gases are effective absorbers of visible radiation. This explains why most visible radiation reaches Earth's surface and why we say that the atmosphere is *transparent* to incoming solar radiation. Thus, the atmosphere does not acquire the bulk of its energy directly from the Sun. Rather, it is heated chiefly by energy that is first absorbed by Earth's surface and then reradiated to the sky.

Heating the Atmosphere: The Greenhouse Effect

 GEODe

Atmosphere
  ↳ Heating the Atmosphere

Approximately 50 percent of the solar energy that strikes the top of the atmosphere reaches Earth's surface and is absorbed. Most of this energy is then reradiated skyward. Because Earth has a much lower surface temperature than the Sun, the radiation that it emits has longer wavelengths than solar radiation.

The atmosphere as a whole is an efficient absorber of the longer wavelengths emitted by Earth (*terrestrial*

radiation). Water vapor and carbon dioxide are the principal absorbing gases. Water vapor absorbs roughly five times more terrestrial radiation than do all other gases combined and accounts for the warm temperatures found in the lower troposphere, where it is most highly concentrated. Because the atmosphere is quite transparent to shorter-wavelength solar radiation and more readily absorbs longer-wavelength terrestrial radiation, the atmosphere is heated from the ground up rather than vice versa. This explains the general drop in temperature with increasing altitude experienced in the troposphere. The farther from the "radiator," the colder it becomes.

When the gases in the atmosphere absorb terrestrial radiation, they warm, but they eventually radiate this energy away. Some travels skyward, where it may be reabsorbed by other gas molecules, a possibility less likely with increasing height because the concentration of water vapor decreases with altitude. The remainder travels Earthward and is again absorbed by Earth. For this reason, Earth's surface is continually being supplied with heat from the atmosphere as well as from the Sun. Without these absorptive gases in our atmosphere, Earth would not be a suitable habitat for humans and numerous other life forms.

This very important phenomenon has been termed the **greenhouse effect** because it was once thought that greenhouses were heated in a similar manner (Figure 15.19). The gases of our atmosphere, especially water vapor and carbon dioxide, act very much like the glass in the greenhouse. They allow shorter-wavelength solar radiation to enter, where it is absorbed by the objects inside. These objects in turn reradiate the heat, but at longer wavelengths, to which glass is nearly opaque. The heat therefore is trapped in the greenhouse. However, a more important factor in keeping a greenhouse warm is the fact that the greenhouse itself prevents mixing of air inside with cooler air outside. Nevertheless, the term *greenhouse effect* is still used.

? STUDENTS SOMETIMES ASK...

Isn't the greenhouse effect responsible for global warming?

The popular press frequently points to the greenhouse effect as the "villain" of the global warming problem. It is important to note that the greenhouse effect and global warming *are not* the same thing. Without the greenhouse effect, Earth would be uninhabitable. We do have mounting evidence that human activity (particularly the release of carbon dioxide into the atmosphere) is responsible for a rise in global temperatures (see Chapter 19). Thus, human activities seem to be enhancing an otherwise natural process (the greenhouse effect) to increase Earth's temperature. Nevertheless, to equate the greenhouse effect, which makes life possible, with undesirable changes to our atmosphere caused by human activity is incorrect.

Figure 15.19 The heating of the atmosphere. Most of the short-wavelength radiation from the Sun passes through the atmosphere and is absorbed by Earth's land-sea surface. This energy is then emitted from the surface as longer-wavelength radiation, much of which is absorbed by certain gases in the atmosphere. Some of the energy absorbed by the atmosphere will be reradiated Earthward. This so-called greenhouse effect is responsible for keeping Earth's surface much warmer than it would be otherwise.

1. Much of the incoming, short wavelength, solar radiation penetrates the atmosphere and heats Earth's surface.

2. Objects on Earth's surface emit long wavelength radiation skyward.

3. Greenhouse gases absorb outgoing, long wavelength, radiation and reradiate some of this energy Earthward, thus trapping heat in the lower atmosphere.

Temperature Measurement and Data

Temperature is one of the basic elements of weather and climate. When someone asks what it is like outside, air temperature is often the first element we mention. At a weather station, the temperature is read on a regular basis from instruments mounted in an instrument shelter (Figure 15.20). The shelter protects the instruments from direct sunlight and allows a free flow of air. In addition to a standard mercury thermometer, the shelter is likely to contain a thermograph to continuously record temperature and a set of maximum–minimum thermometers. As their name implies, these thermometers record the highest and lowest temperatures during a measurement period, usually 24 hours.

The daily maximum and minimum temperatures are the bases for many of the temperature data compiled by meteorologists:

1. By adding the maximum and minimum temperatures and then dividing by two, the *daily mean temperature* is calculated.

2. The *daily range* of temperature is computed by finding the difference between the maximum and minimum temperatures for a given day.

3. The *monthly mean* is calculated by adding together the daily means for each day of the month and dividing by the number of days in the month.

4. The *annual mean* is an average of the 12 monthly means.

5. The *annual temperature range* is computed by finding the difference between the highest and lowest monthly means.

Mean temperatures are particularly useful for making comparisons, whether on a daily, monthly, or annual basis (see Box 15.3). It is common to hear a weather reporter state, "Last month was the hottest July on record," or "Today Chicago was 10 degrees warmer than Miami." Temperature ranges are also useful statistics, because they give an indication of extremes.

Many different applications of temperature data have been devised. One familiar use of temperature data relates to human perception of temperature. Television weather reports use indices that attempt to portray levels of human comfort and discomfort. Such indices are based on the fact that our sensation of temperature is often different from the actual temperature of the air as recorded by a thermometer.

The human body is a heat engine that continually releases energy. Anything that influences the rate of heat loss from the body also influences our sensation of temperature, thereby affecting our feeling of comfort. Several factors control the thermal comfort of the human body, and certainly air temperature is a major factor. Other environmental conditions are also significant, such as relative humidity, wind speed, and sunshine. For example, many of us know that a cold, windy winter day can feel much colder than the air temperature would seem to indicate.

Why Temperatures Vary: The Controls of Temperature

A *temperature control* is any factor that causes temperature to vary from place to place and from time to time. Earlier in this chapter we examined the most important cause for temperature variations—differences in the receipt of solar radiation. Because variations in Sun angle and length of daylight depend on latitude, they are responsible for warm temperatures in the tropics and colder temperatures at more poleward locations. Of

A.

B.

Figure 15.20   **A.** Standard instrument shelter. A shelter protects instruments from direct sunlight and allows for the free flow of air. (Courtesy of Qualimetrics, Inc.) **B.** This modern shelter contains an electrical thermometer called a *thermistor*. (Photo by Bobbé Christopherson)

course, seasonal temperature changes at a given latitude occur as the Sun's vertical rays migrate toward and away from a place during the year.

However, latitude is not the only control of temperature. If it were, we would expect all places along the same parallel of latitude to have identical temperatures. This is clearly not the case. For example, Eureka, California, and New York City are both coastal cities at about the same latitude and both have an annual mean temperature of 11°C (52°F). However, New York City is 9°C (16°F) warmer than Eureka in July and 10°C cooler in January. In another example, two cities in Ecuador—Quito and Guayaquil—are relatively close to each other, yet the annual mean temperatures of these two cities differ by 12°C (21°F). To explain these situations and countless others, we must realize that factors other than latitude also exert a strong influence on temperature. In the next sections, we will examine these other controls, which include differential heating of land and water, altitude, geographic position, cloud cover and albedo, and ocean currents.*

---

*For a discussion of the effects of ocean currents on temperature, see Chapter 14.

STUDENTS SOMETIMES ASK...
What are the highest and lowest temperatures ever recorded at Earth's surface?

The world's record-high temperature is nearly 59°C(136°F)! It was recorded on September 13, 1922, at Azizia, Libya, in North Africa's Sahara Desert. The lowest recorded temperature is −89°C(−129°F). It should come as no surprise that this incredibly frigid temperature was recorded in Antarctica, at the Russian Vostok Station, on August 24, 1960.

Land and Water

The heating of Earth's surface controls the heating of the air above it. Therefore, to understand variations in air temperature, we must examine the nature of the surface. Different land surfaces absorb varying amounts of incoming solar energy, which in turn cause variations in the temperature of the air above. The largest contrast, however, is not between different land surfaces, but between land and water. *Land heats more rapidly and to higher temperatures than water and cools more rapidly and to lower temperatures than water.* Temperature variations,

BOX 15.3

UNDERSTANDING EARTH:
Showing Temperature Distribution Using Isotherms

To examine the distribution of air temperatures over large areas, isotherms are commonly used. An *isotherm* is a line that connects points on a map that have the same temperature (*iso* = equal, *therm* = temperature). Therefore, all points through which an isotherm passes have identical temperatures for the time period indicated. Generally, isotherms representing 5° or 10° temperature differences are used, but any interval may be chosen. Figure 15.D illustrates how isotherms are drawn on a map. Notice that most

isotherms do not pass directly through the observing stations, because the station readings may not coincide with the values chosen for the isotherms. Only an occasional station temperature will be exactly the same as the value of the isotherm, so it is usually necessary to draw the lines by estimating the proper position between stations.

Isothermal maps are valuable tools because they clearly make temperature distribution visible at a glance. Areas of low and high temperatures are easy to pick out. In addition, the amount of temperature change per unit of distance,

called the *temperature gradient,* is easy to visualize. Closely spaced isotherms indicate a rapid rate of temperature change, whereas more widely spaced lines indicate a more gradual rate of change. You can see this in Figure 15.D. The isotherms are closer in Colorado and Utah (steeper temperature gradient), whereas the isotherms are spread farther in Texas (gentler temperature gradient). Without isotherms, a map would be covered with numbers representing temperatures at dozens or hundreds of places, which would make patterns difficult to see.

Figure 15.D Temperature distribution using isotherms. Isotherms are lines that connect points of equal temperature. Showing temperature distribution in this way makes patterns easier to see. On television, and in many newspapers, temperature maps are in color. Rather than labeling isotherms, the area *between* isotherms is labeled. For example, the zone between the 60° and 70° isotherms is labeled "60s."

therefore, are considerably greater over land than over water.

Among the reasons for the differential heating of land and water are the following:

1. The *specific heat* (amount of energy needed to raise 1 gram of a substance 1°C) is far greater for water than for land. Thus, water requires a great deal more heat to raise its temperature the same amount as an equal quantity of land.

2. Land surfaces are opaque, so heat is absorbed only at the surface. Water, being more transparent, allows some solar radiation to penetrate to a depth of many meters.

3. The water that is heated often mixes with water below, thus distributing the heat through an even larger mass.

4. Evaporation (a cooling process) from water bodies is greater than that from land surfaces.

All these factors collectively cause water to warm more slowly, store greater quantities of heat, and cool more slowly than land.

Monthly temperature data for two cities will demonstrate the moderating influence of a large water body and the extremes associated with land (Figure 15.21). Vancouver, British Columbia, is located along the windward Pacific coast, whereas Winnipeg, Manitoba, is in a continental position far from the influence of water. Both cities are at about the same latitude and thus experience similar Sun angles and lengths of daylight. Winnipeg, however, has a mean January temperature that is 20°C lower than Vancouver's. Conversely, Winnipeg's July mean is 2.6°C higher than Vancouver's. Although their latitudes are nearly the same, Winnipeg, which has no water influence, experiences much greater temperature extremes than does Vancouver, which does. The key to Vancouver's moderate year-round climate is the Pacific Ocean.

On a different scale, the moderating influence of water may also be demonstrated when temperature variations in the Northern and Southern hemispheres are compared. In the Northern Hemisphere, 61 percent is covered by water, and land accounts for the remaining 39 percent. However, in the Southern Hemisphere, 81 percent is covered by water and 19 percent by land. The Southern Hemisphere is correctly called the *water hemisphere* (see Figure 12.1, p. 334). Table 15.3 portrays the considerably smaller annual temperature variations in the water-dominated Southern Hemisphere as compared with the Northern Hemisphere.

Altitude

The two cities in Ecuador mentioned earlier—Quito and Guayaquil—demonstrate the influence of altitude on mean temperature (Figure 15.22). Both cities are near the equator and relatively close to one another, but the annual mean temperature at Guayaquil is 25°C (77°F) compared to Quito's mean of 13°C (55°F). The difference may be understood when the cities' elevations are noted. Guayaquil is only 12 meters (40 feet) above sea level, whereas Quito is high in the Andes Mountains at 2800 meters (9200 feet).

Recall that temperatures drop an average of 6.5°C per kilometer in the troposphere; thus, cooler temperatures are to be expected at greater heights (see Figure 15.5). Yet, the magnitude of the difference is not explained completely by the normal lapse rate. If the normal lapse rate is used, we would expect Quito to be about 18°C cooler than Guayaquil, but the difference is only 12°C. The fact that high-altitude places such as

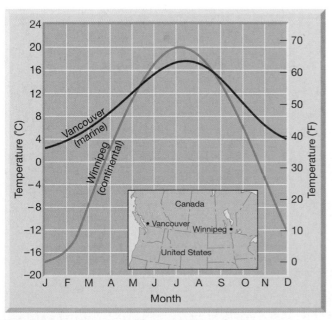

Figure 15.21 Mean monthly temperatures for Vancouver, British Columbia, and Winnipeg, Manitoba. Vancouver has a much smaller annual temperature range owing to the strong marine influence of the Pacific Ocean. Winnipeg illustrates the greater extremes associated with an interior location.

Latitude	Northern Hemisphere	Southern Hemisphere
0	0	0
15	3	4
30	13	7
45	23	6
60	30	11
75	32	26
90	40	31

Table 15.3 Variation in annual mean temperature range (°C) with latitude.

Figure 15.22 Because Quito is high in the Andes Mountains, it experiences much cooler temperatures than Guayaquil, which is near sea level. Because both cities are near the equator, they have a negligible annual temperature range.

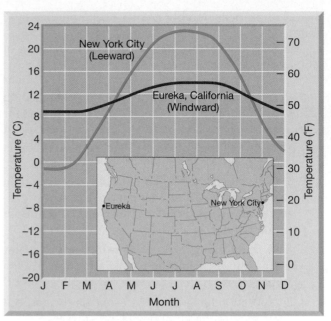

Figure 15.23 Monthly mean temperatures for Eureka, California, and New York City. Both cities are coastal and located at about the same latitude. Because Eureka is strongly influenced by prevailing winds from the ocean and New York City is not, the annual temperatures range at Eureka is much smaller.

Quito are warmer than the value calculated using the normal lapse rate results from the absorption and reradiation of solar energy by the ground surface.

Geographic Position

The geographic setting can greatly influence the temperatures experienced at a specific location. A coastal location where prevailing winds blow from the ocean onto the shore (a *windward* coast) experiences considerably different temperatures than does a coastal location where the prevailing winds blow from the land toward the ocean (a *leeward* coast). In the first situation, the windward coast will experience the full moderating influence of the ocean—cool summers and mild winters, compared to an inland station at the same latitude.

A leeward coast, on the other hand, will have a more continental temperature pattern because the winds do not carry the ocean's influence onshore. Eureka, California, and New York City, the two cities mentioned earlier, illustrate this aspect of geographic position. The annual temperature range at New York City is 19°C (34°F) greater than Eureka's (Figure 15.23).

Seattle and Spokane, both in the state of Washington, illustrate a second aspect of geographic position—mountains that act as barriers. Although Spokane is only about 360 kilometers (220 miles) east of Seattle, the towering Cascade Range separates the cities. Consequently, Seattle's temperatures show a marked marine influence, but Spokane's are more typically

continental (Figure 15.24). Spokane is 7°C (13°F) cooler than Seattle in January and 4°C (7°F) warmer than Seattle in July. The annual range at Spokane is 11°C (20°F) greater than at Seattle. The Cascade Range effectively cuts Spokane off from the moderating influence of the Pacific Ocean.

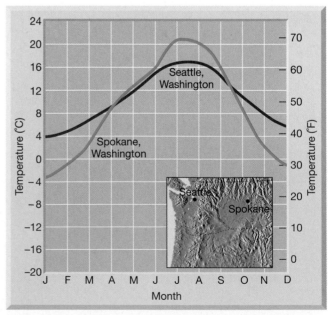

Figure 15.24 Monthly mean temperatures for Seattle and Spokane, Washington. Because the Cascade Mountains cut off Spokane from the moderating influence of the Pacific Ocean, its annual temperature range is greater than Seattle's.

Cloud Cover and Albedo

The extent of cloud cover is a factor that influences temperatures in the lower atmosphere. Cloud cover is important because many clouds have a high albedo and therefore reflect a significant portion of the sunlight that strikes them back to space. By reducing the amount of incoming solar radiation, daytime temperatures will be lower than if the clouds were absent and the sky were clear (Figure 15.25A).

At night, clouds have the opposite effect as during daylight. They act as a blanket by absorbing outgoing terrestrial radiation and reradiating a portion of it back to the surface (Figure 15.25B). Consequently, some of the heat that otherwise would have been lost remains near the ground. Thus, nighttime air temperatures do not drop as low as they would on a clear night. The effect of cloud cover is to reduce the daily temperature range by lowering the daytime maximum and raising the nighttime minimum.

A.

B.

Figure 15.25 Clouds reduce the daily temperature range. **A.** During daylight hours, clouds reflect solar radiation back to space. Therefore, the maximum temperature is lower than if the sky were clear. **B.** At night, the minimum temperature will not fall as low because clouds retard the loss of heat.

Clouds are not the only phenomenon that increase albedo and thereby reduce air temperatures. We also recognize that snow- and ice-covered surfaces have high albedos. This is one reason why mountain glaciers do not melt away in the summer and why snow may still be present on a mild spring day. In addition, during the winter, when snow covers the ground, daytime maximums on a sunny day are less than they otherwise would be because energy that the land would have absorbed and used to heat the air is reflected and lost.

World Distribution of Temperature

Take a moment to study the two world isothermal maps (Figures 15.26 and 15.27). From hot colors near the equator to cool colors toward the poles, these maps portray sea-level temperatures in the seasonally extreme months of January and July. On these maps you can study global temperature patterns and the effects of the controlling factors of temperature, especially latitude, the distribution of land and water, and ocean currents. Like most isothermal maps of large regions, all temperatures on these world maps have been reduced to sea level to eliminate the complications caused by differences in altitude.

On both maps, the isotherms generally trend east and west and show a decrease in temperatures poleward from the tropics. They illustrate one of the most fundamental aspects of world temperature distribution: that the effectiveness of incoming solar radiation in heating Earth's surface and the atmosphere above it is largely a function of latitude.

Moreover, there is a latitudinal shifting of temperatures caused by the seasonal migration of the Sun's vertical rays. To see this, compare the color bands by latitude on the two maps. For example, on the January map, the hot spots of 30°C are *south* of the equator, but in July they have shifted *north* of the equator.

If latitude were the only control of temperature distribution, our analysis could end here, but this is not the case. The added effect of the differential heating of land and water is clearly reflected on the January and July temperature maps. The warmest and coldest temperatures are found over land—note the coldest area, a purple oval in Siberia, and the hottest areas, the deep orange ovals—all over land. Consequently, because temperatures do not fluctuate as much over water as over land, the north-south migration of isotherms is greater over the continents than over the oceans.

In addition, it is clear that the isotherms in the Southern Hemisphere, where there is little land and where the oceans predominate, are much more regular than in the Northern Hemisphere, where they bend sharply northward in July and southward in January over the continents.

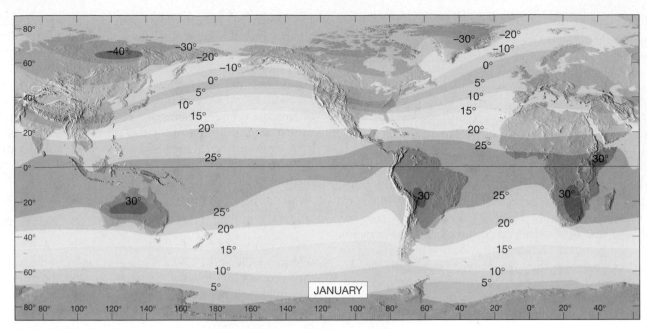

Figure 15.26   World mean sea-level temperatures in January in degrees Celsius.

Isotherms also reveal the presence of ocean currents. Warm currents cause isotherms to be deflected toward the poles, whereas cold currents cause an equatorward bending. The horizontal transport of water poleward warms the overlying air and results in air temperatures that are higher than would otherwise be expected for the latitude. Conversely, currents moving toward the equator produce cooler than expected air temperatures.

Figures 15.26 and 15.27 show the seasonal extremes of temperature, so comparing them enables us to see the annual range of temperature from place to place. Comparing the two maps shows that a station near the equator has a very small annual range because it experiences little variation in the length of daylight and it always has a relatively high Sun angle. A station in the middle latitudes, however, experiences wide variations in Sun angle and length of daylight and hence large variations in temperature. Therefore, we can state that the annual temperature range increases with an increase in latitude.

Moreover, land and water also affect seasonal temperature variations, especially outside the tropics. A

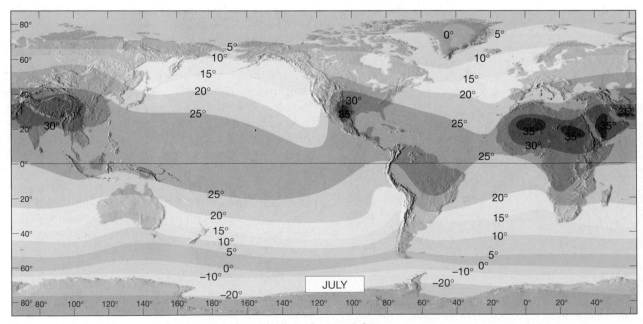

Figure 15.27   World mean sea-level temperatures in July in degrees Celsius.

continental location must endure hotter summers and colder winters than a coastal location. Consequently, outside the tropics the annual temperature range will increase with an increase in continentality.

A classic example of the effect of latitude and continentality on annual temperature range is Yakutsk,

Russia. This city in the heart of Siberia is just a few degrees south of the Arctic Circle and far from the influence of water. As a result, Yakutsk has an average annual temperature range of 62.2°C (112° F), among the greatest ranges in the world.

Chapter Summary

- *Weather* is the state of the atmosphere at a particular place for a short period of time. *Climate*, on the other hand, is a generalization of the weather conditions of a place over a long period of time.

- The most important *elements*, those quantities or properties that are measured regularly, of weather and climate are (1) *air temperature*, (2) *humidity*, (3) type and amount of *cloudiness*, (4) type and amount of *precipitation*, (5) *air pressure*, and (6) the speed and direction of the *wind*.

- If water vapor, dust, and other variable components of the atmosphere were removed, clean, dry air would be composed almost entirely of *nitrogen* (N), about 78% of the atmosphere by volume, and *oxygen* (O_2), about 21%. *Carbon dioxide* (CO_2), although present only in minute amounts (0.036%), is important because it has the ability to absorb heat radiated by Earth and thus helps keep the atmosphere warm. Among the variable components of air, *water vapor* is very important because it is the source of all clouds and precipitation and, like carbon dioxide, it is also a heat absorber.

- *Ozone* (O_3), the triatomic form of oxygen, is concentrated in the 10- to 50-kilometer altitude range of the atmosphere, and is important to life because of its ability to absorb potentially harmful ultraviolet radiation from the Sun.

- Because the atmosphere gradually thins with increasing altitude, it has no sharp upper boundary but simply blends into outer space. Based on temperature, the atmosphere is divided vertically into four layers. The *troposphere* is the lowermost layer. In the troposphere, temperature usually decreases with increasing altitude. This *environmental lapse rate* is variable, but averages about 6.5°C per kilometer (3.5°F per 1000 feet). Essentially all important weather phenomena occur in the troposphere. Above the troposphere is the *stratosphere*, which exhibits warming because of absorption of ultraviolet radiation by ozone. In the *mesosphere*, temperatures again decrease. Upward from the mesosphere is the *thermosphere*, a layer with only a tiny fraction of the atmosphere's mass and no well-defined upper limit.

- The two principal motions of Earth are (1) *rotation*, the spinning of Earth about its axis, which produces the daily cycle of daylight and darkness; and (2) *revolution*, the movement of Earth in its orbit around the Sun.

- *Several factors act together to cause the seasons.* Earth's axis is inclined 23 1/2° from the perpendicular to the plane of its orbit around the Sun and remains pointed in the same direc-

tion (toward the North Star) as Earth's journeys around the Sun. As a consequence, Earth's orientation to the Sun continually changes. The yearly fluctuations in the angle of the Sun and length of daylight brought about by Earth's changing orientation to the Sun cause the seasons.

- The three mechanisms of heat transfer are (1) *conduction*, the transfer of heat through matter by molecular activity; (2) *convection*, the transfer of heat by the movement of a mass or substance from one place to another; and (3) *radiation*, the transfer of heat by electromagnetic waves.

- *Electromagnetic radiation* is energy emitted in the form of rays, called *electromagnetic waves*. All radiation is capable of transmitting energy through the vacuum of space. An important difference among electromagnetic waves is their *wavelengths*, which range from very long *radio waves* to very short *gamma rays*. *Visible light* is the only portion of the electromagnetic spectrum we can see. Some of the basic laws that govern radiation as it heats the atmosphere are (1) all objects emit radiant energy; (2) hotter objects radiate more total energy than do colder objects; (3) the hotter the radiating body, the shorter the wavelengths of maximum radiation; and (4) objects that are good absorbers of radiation are good emitters as well.

- The general drop in temperature with increasing altitude in the troposphere supports the fact that the *atmosphere is heated from the ground up*. Approximately 50 percent of the solar energy that strikes the top of the atmosphere is ultimately absorbed at Earth's surface. Earth emits the absorbed radiation in the form of long-wave radiation. The atmospheric absorption of this long-wave *terrestrial radiation*, primarily by water vapor and carbon dioxide, is responsible for heating the atmosphere.

- The factors that cause temperature to vary from place to place, also called the *controls of temperature*, are (1) *differences in the receipt of solar radiation*—the greatest single cause; (2) the unequal heating and cooling of *land and water*, in which land heats more rapidly and to higher temperatures than water and cools more rapidly and to lower temperatures than water; (3) *altitude*; (4) *geographic position*; (5) *cloud cover and albedo*; and (6) *ocean currents*.

- Temperature distribution is shown on a map by using *isotherms*, which are lines that connect equal temperatures. Differences between January and July temperatures around the world can be explained in terms of the basic controls of temperature.

Key Terms

aerosols (p. 407)
air (p. 407)
albedo (p. 420)
autumnal equinox (p. 414)
circle of illumination (p. 412)
climate (p. 407)
conduction (p. 417)
convection (p. 417)
electromagnetic radiation (p. 418)
element (of weather and climate)
    (p. 407)
environmental lapse rate (p. 410)

greenhouse effect (p. 421)
heat (p. 417)
inclination of the axis (p. 413)
infrared (p. 418)
mesosphere (p. 411)
ozone (p. 407)
radiation (p. 418)
reflection (p. 419)
revolution (p. 412)
rotation (p. 412)
scattering (p. 419)
spring equinox (p. 414)

stratosphere (p. 411)
summer solstice (p. 414)
temperature (p. 417)
thermosphere (p. 411)
Tropic of Cancer (p. 414)
Tropic of Capricorn (p. 414)
troposphere (p. 409)
ultraviolet (p. 418)
visible light (p. 418)
weather (p. 406)
winter solstice (p. 414)

Review Questions

1. Distinguish between weather and climate.

2. List the basic elements of weather and climate.

3. What are the two major components of clean, dry air? Are they important meteorologically?

4. Why are water vapor and aerosols important constituents of our atmosphere?

5. What source is responsible for the most pollution? (See Figure 15.A.)

6. What is the difference between primary and secondary pollutants? (See Box 15.1)

7. (a) Why is ozone important to life on Earth? (b) What is the most serious threat to human health of a decrease in the stratosphere's ozone? (See Box 15.2)

8. The atmosphere is divided vertically into four layers on the basis of temperature. List the names of these layers in order (from lowest to highest) and describe how temperature changes in each layer.

9. If the temperature at sea level were 23°C, what would the air temperature be at a height of 2 kilometers *under average conditions*?

10. Why do temperatures increase in the stratosphere?

11. Briefly explain the primary cause of the seasons.

12. After examining Table 15.1, write a general statement that relates the season, the latitude, and the length of daylight.

13. Describe the relationship between the temperature of a radiating body and the wavelengths it emits.

14. Distinguish between heat and temperature.

15. Describe the three basic mechanisms of heat transfer.

16. Figure 15.17 illustrates what happens to incoming solar radiation. The percentages shown, however, are only global averages. In particular, the amount of solar radiation reflected (albedo) may vary considerably. What factors might cause variations in albedo?

17. Describe the process by which Earth's atmosphere is heated.

18. How are the following temperature data computed: daily mean, daily range, monthly mean, annual mean, annual range?

19. Quito, Ecuador, is located on the equator and is not a coastal city. It has an average annual temperature of only 13°C (55°F). What is the likely cause for this low average temperature?

20. In what ways can geographic position be considered a control of temperature?

21. How does cloud cover influence the maximum temperature on an overcast day? How is the nighttime minimum influenced by clouds?

22. Yakutsk is located in Siberia at about 60 degrees north latitude. This Russian city has one of the highest average annual temperature ranges in the world: 62.2°C (112°F). Explain the reasons for the very high annual temperature range.

Examining the Earth System

1. Give an example or two of how the Earth system might be affected if Earth's axis were perpendicular to the plane of its orbit instead of being inclined 23 1/2°.

2. Speculate on the changes in global temperatures that might occur if Earth had substantially more land area and less ocean area than at present. How might such changes influence the biosphere?

Web Resources

 The *Earth Science* Website uses the resources and flexibility of the Internet to aid in your study of the topics in this chapter. Written and developed by Earth science instructors, this site will help improve your understanding of Earth science. Visit **http://www.prenhall.com/tarbuck** and click on the cover of *Earth Science* 10e to find:

- Online review quizzes.
- Web-based critical thinking and writing exercises.
- Links to chapter-specific Web resources.
- Internet-wide key term searches.

http://www.prenhall.com/tarbuck

Towering cumulonimbus cloud develops over San Carlos, Mexico. (Photo by Gary Gray/DRK Photo)

CHAPTER 16

MOISTURE, CLOUDS, AND PRECIPITATION

As you observe day-to-day weather changes, you witness the powerful role of water in the air. This raises questions: What is humidity, and how is it measured? Why do clouds form on some occasions but not on others? What processes produce clouds and precipitation? Why are some clouds thin, white, and harmless, while others are towering, gray, and ominous? What is the difference between sleet and hail? What is fog? Are all fogs alike? This chapter investigates the remarkable phenomena associated with moisture in the air.

Scientists agree: When it comes to understanding atmospheric processes, *water vapor is the most important gas in the atmosphere.* Water vapor constitutes only a small fraction of the gases in the atmosphere, varying from nearly 0 to about 4 percent by volume. But the importance of water in the air vastly exceeds what these small percentages would indicate.

Recall from Chapter 15 that water vapor is an important heat-absorbing gas. This property is critical to the heating of the atmosphere. Water vapor is also the source of all condensation and precipitation. Clouds and fog, as well as rain, snow, sleet, and hail, are among the most conspicuous and observable of weather phenomena (Figure 16.1).

We all expect rain to fall when certain cloud types are present, yet most of us have no idea of the complex processes that must take place in a cloud to produce rain. The formation of a single average raindrop requires water from nearly a million microscopic cloud droplets. What mechanisms foster the creation of raindrops? The wide variations in the amount of precipitation from place to place, as well as local differences from time to time, have a significant impact on the nature of the physical landscape and on our lifestyles.

Water's Changes of State

Atmosphere
↳ Moisture and Cloud Formation

Water vapor is an odorless, colorless gas that mixes freely with the other gases of the atmosphere. Unlike oxygen and nitrogen—the two most abundant components of the atmosphere—water can change from one state of matter to another (solid, liquid, or gas) at the temperatures and pressures experienced on Earth. (By contrast, nitrogen will not condense to a liquid unless its temperature is lowered to −196°C [−371°F]). Because of this unique property water freely leaves the oceans as a gas and returns again as a liquid, producing the vital hydrologic cycle.

Latent Heat

The process of changing state requires that heat be absorbed or released, as shown in Figure 16.2. The heat energy involved is often measured in calories. One **calorie** is the amount of heat required to raise the temperature of 1 gram of water 1°C (1.8°F). Thus, when 10 calories of heat are added to 1 gram of water, a 10°C (18°F) temperature rise occurs.

Under certain conditions, heat may be added to a substance without an accompanying rise in tempera-

Figure 16.1 People caught in a downpour. (Photo by Mary Fulton/Liaison Agency, Inc.)

Figure 16.2   Changes of state.

ture. This situation occurs, for example, when ice changes to water. How can this be?

When heat is supplied to a glass of ice water the temperature of the ice water remains a constant 0°C (32°F) until all the ice has melted. If adding heat does not raise the temperature, then where does this heat go?

In this case, the added heat went to disrupt the internal crystalline structure of the ice cubes. Stated another way, the water molecules in the ice crystals were dislodged to form the noncrystalline substance liquid water. We call this process *melting.*

Because the heat used to melt ice does not produce a temperature change, it is referred to as **latent heat.** (Latent means *hidden,* like the latent fingerprints hidden at a crime scene.) This energy becomes stored in the liquid water and is not released as heat until the liquid returns to the solid state.

As you will see later, latent heat plays a crucial role in many atmospheric processes. For example, the release of latent heat aids the formation of the towering clouds often seen on warm summer days, and it is the major source of energy for violent thunderstorms, tornadoes, and hurricanes.

Liquid/Gas Change of State.   The process of converting a liquid to a gas is termed **evaporation.** It takes approximately 600 calories of energy to convert 1 gram of water to water vapor (Figure 16.2). The energy absorbed by the water molecules during evaporation is used solely to give them the motion needed to escape the surface of the liquid and become a gas. This energy is referred to as *latent heat of vaporization.* During the process of evaporation, it is the higher-temperature (faster-moving) molecules that escape the surface. As a result, the average molecular motion

(temperature) of the remaining water is reduced, causing a cooling effect—hence, the common expression that, "evaporation is a cooling process." You have undoubtedly experienced this cooling effect when stepping dripping wet from a swimming pool or bathtub.

Condensation is the process whereby water vapor changes to the liquid state. In the atmosphere, condensation generates phenomena such as clouds and fog. For condensation to occur, the water molecules must release their stored heat energy (*latent heat of condensation*) equal to what was absorbed during evaporation. This released energy plays an important role in producing violent weather and can transfer great quantities of heat from tropical oceans to more poleward locations. When condensation occurs in the atmosphere, it results in the formation of fog and clouds.

Solid/Liquid Change of State.   **Melting** is the process by which a solid is changed to a liquid. It requires absorption of approximately 80 calories of heat per gram of water (*latent heat of melting*). **Freezing,** the reverse process, releases these 80 calories per gram as *latent heat of fusion.*

Solid/Gas Change of State.   The final processes illustrated in Figure 16.2 are sublimation and deposition. **Sublimation** is the conversion of a solid directly to a gas, without passing through the liquid state. You may have observed this change in watching the sublimation of "dry ice" (frozen carbon dioxide) into white, wispy vapor, sometimes used to generate "fog" or "smoke" in theatrical productions. Water ice also sublimates, which is why unused ice cubes gradually shrink in a freezer.

S T U D E N T S   S O M E T I M E S   A S K . . .
What is freezer burn?

"Freezer burn" is a common expression used to describe foods that have been left in a frost-free refrigerator for extended periods of time. Frost-free refrigerators circulate comparatively dry air through the freezer compartments. This causes ice on the freezer walls to sublimate (change from a solid to a gas) so it can be removed by the circulating air. Unfortunately, this process also removes moisture from those frozen foods that are not in airtight containers. Consequently, over a period of a few months these foods begin to dry out rather than actually being burned.

The term **deposition** is used to denote the reverse process, the conversion of a vapor directly to a solid. This change occurs, for example, when water vapor is deposited as ice on cold objects such as grass or windows (Figure 16.3). These deposits are termed *white frost* or *hoar frost* but are generally just called *frost*. A household example of deposition is the "frost" that accumulates in a freezer compartment. As shown in Figure 16.2, sublimation and deposition involve an amount of energy equal to the total of the other two processes.

Humidity: Water Vapor in the Atmosphere

GEODe Atmosphere
↳ Moisture and Cloud Formation

Water vapor constitutes only a small fraction of the atmosphere, varying from as little as one-tenth of 1 percent up to about 4 percent by volume. But the impor-tance of water in the air is far greater than these small percentages would indicate. Indeed, scientists agree that *water vapor* is the most important gas in the atmosphere when it comes to understanding atmospheric processes.

Humidity is the general term for the amount of water vapor in air. Meteorologists employ several methods to express the water-vapor content of the air; we will examine three: 1. mixing ratio, 2. relative humidity and 3. dew-point temperature.

Saturation

Before we consider these humidity measures further, you need to understand the concept of **saturation.** Imagine a closed jar half full of water and half full of dry air, both at the same temperature. As the water begins to evaporate from the water surface, a small increase in pressure can be detected in the air above. This increase is the result of the motion of the water-vapor molecules that were added to the air through evaporation. In the open atmosphere, this pressure is termed **vapor pressure** and is defined as that part of the total atmospheric pressure that can be attributed to the water-vapor content.

In the closed container, as more and more molecules escape from the water surface, the steadily increasing vapor pressure in the air above forces more and more of these molecules to return to the liquid. Eventually the number of vapor molecules returning to the surface will balance the number leaving. At that point, the air is said to be *saturated*, or filled to capacity. However, if we add heat to the container, increasing the temperature of the water and air, more water will evaporate before a balance is reached. Consequently, at higher temperatures, more moisture is required for saturation. Stated another way, the water-

Figure 16.3 Frost on a window pane is an example of deposition. (Photo by D. Cavagnaro/DRK Photo)

vapor capacity of air is *temperature* dependent, with warm air having a much greater capacity than cold air. The amount of water vapor required for saturation at various temperatures varies surprisingly, as shown in Table 16.1

Mixing Ratio

Not all air is saturated, of course. Thus, we need ways to express how humid a parcel of air is. One method specifies the amount of water vapor contained in a unit of air. The **mixing ratio** is the mass of water vapor in a unit of air compared to the remaining mass of dry air.

$$\text{mixing ratio} = \frac{\text{mass of water vapor (grams)}}{\text{mass of dry air (kilograms)}}$$

Table 16.1 shows the mixing ratios of saturated air at various temperatures. For example, at 25°C (77°F) a saturated parcel of air (one kilogram) would contain 20 grams of water vapor.

Because the mixing ratio is expressed in units of mass (usually in grams per kilogram), it is not affected by changes in pressure or temperature. However, the mixing ratio is time consuming to measure by direct sampling. Thus, other methods are employed to express the moisture content of the air. These include relative humidity and dew-point temperature.

Relative Humidity

The most familiar and, unfortunately, the most misunderstood term used to describe the moisture content of air is relative humidity. **Relative humidity** *is a ratio of the air's actual water-vapor content compared with the amount of water vapor required for saturation at that temperature (and pressure).* Thus, relative humidity indicates how near the air is to saturation, rather than the actual quantity of water vapor in the air.

Table 16.1 Saturation Mixing Ratio (at Sea-Level Pressure)	
Temperature **°C (°F)**	**Saturation Mixing** **Ratio (g/kg)**
−40 (−40)	0.1
−30 (−22)	0.3
−20 (−4)	0.75
−10 (14)	2
0 (32)	3.5
5 (41)	5
10 (50)	7
15 (59)	10
20 (68)	14
25 (77)	20
30 (86)	26.5
35 (95)	35
40 (104)	47

To illustrate, we see from Table 16.1 that at 25°C, air is saturated when it contains 20 grams of water vapor per kilogram of air. Thus, if the air contains 10 grams per kilogram on a 25°C day, the relative humidity is expressed as 10/20, or 50 percent. Further, if air with a temperature of 25°C had a water-vapor content of 20 grams per kilogram, the relative humidity would be expressed as 20/20 or 100 percent. On those occasions when the relative humidity reaches 100 percent, the air is said to be saturated.

Because relative humidity is based on the air's water-vapor content, as well as the amount of moisture required for saturation, it can be changed in either of two ways. First, relative humidity can be changed by the addition or removal of water vapor. Second, because the amount of moisture required for saturation is a function of air temperature, relative humidity varies with temperature. (Recall that the saturation vapor pressure is temperature dependent, such that at higher temperatures it takes more water vapor to saturate air than at lower temperatures.)

Adding or Subtracting Moisture. Notice in Figure 16.4 that when water vapor is added to a parcel of air, its relative humidity increases until saturation occurs (100 percent relative humidity). What if even more moisture is added to this parcel of saturated air? Does the relative humidity exceed 100 percent? Normally, this situation does not occur. Instead, the excess water vapor condenses to form liquid water.

In nature, moisture is added to the air mainly via evaporation from the oceans. However, plants, soil, and smaller bodies of water do make substantial contributions.

Changes with Temperature. The second condition that affects relative humidity is air temperature. Examine Figure 16.5 carefully. Note that in Figure 16.5A when air at 20°C contains 7 grams of water vapor per kilogram, it has a relative humidity of 50 percent. This can be verified by referring to Table 16.1. Here we can see that at 20°C, air is saturated when it contains 14 grams of water vapor per kilogram of air. Because the air in Figure 16.5A contains 7 grams of water vapor, its relative humidity is 7/14, or 50 percent.

How does cooling affect relative humidity? When the flask in Figure 16.5A is cooled from 20° to 10°C, as shown in Figure 16.5B, the relative humidity increases from 50 to 100 percent. We can conclude from this that when the water-vapor content remains constant, a decrease in temperature results in an increase in relative humidity.

But there is no reason to assume that cooling would cease the moment the air reached saturation. What happens when the air is cooled below the temperature at which saturation occurs? Figure 16.5C illustrates this situation. Notice from Table 16.1 that when the flask is

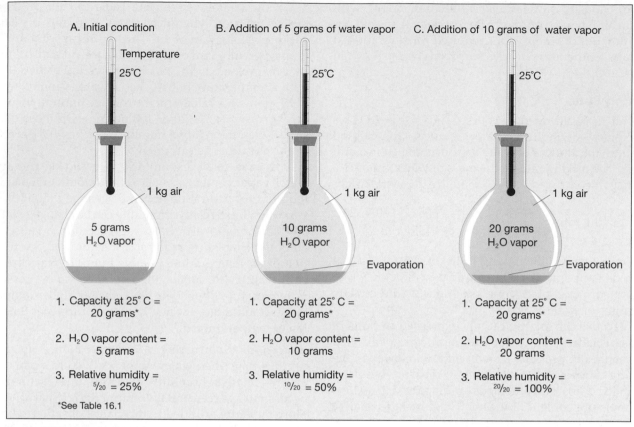

A. Initial condition   B. Addition of 5 grams of water vapor   C. Addition of 10 grams of water vapor

Temperature
25°C   25°C   25°C

1 kg air   1 kg air   1 kg air

5 grams   10 grams   20 grams
H₂O vapor   H₂O vapor   H₂O vapor

                             Evaporation   Evaporation

1. Capacity at 25° C =   1. Capacity at 25° C =   1. Capacity at 25° C =
   20 grams*   20 grams*   20 grams*

2. H₂O vapor content =   2. H₂O vapor content =   2. H₂O vapor content =
   5 grams   10 grams   20 grams

3. Relative humidity =   3. Relative humidity =   3. Relative humidity =
   5/20 = 25%   10/20 = 50%   20/20 = 100%

*See Table 16.1

Figure 16.4   Relative humidity. At a constant temperature, the relative humidity will increase as water vapor is added to the air. Here, the capacity remains constant at 20 grams per kilogram and the relative humidity rises from 25 to 100 percent as the water-vapor content increases.

cooled to 0°C, the air is saturated at 3.5 grams of water vapor per kilogram of air. Because this flask originally contained 7 grams of water vapor, 3.5 grams of water vapor will condense to form liquid droplets that collect on the walls of the container. In the meantime, the relative humidity of the air inside remains at 100 percent. This raises an important concept. When air aloft is cooled below its saturation level, some of the water vapor condenses to form clouds. As clouds are made of liquid droplets, this moisture is no longer part of the *water-vapor* content of the air.

We can summarize the effects of temperature on relative humidity as follows. When the water-vapor content of air remains at a constant level, a decrease in air temperature results in an increase in relative humidity and an increase in temperature causes a decrease in relative humidity. In Figure 16.6 the variations in temperature and relative humidity during a typical day demonstrate the relationship just described.

Dew-Point Temperature

Another important measure of humidity is the dew-point temperature. The **dew-point temperature** or simply the **dew point** is the temperature to which a parcel of air would need to be cooled to reach saturation. Note that in Figure 16.5, unsaturated air at 20°C must be cooled to 10°C before saturation occurs. Therefore, 10°C would be the dew-point temperature for this air. If the same air were cooled further, the air's saturation mixing ratio would be exceeded and the excess water vapor would condense, typically as dew, fog, or clouds. The term *dew point* stems from the fact that during night-time hours, objects near the ground often cool below the dew-point temperature and become coated with dew* (Figure 16.7).

Unlike relative humidity, which is a measure of how near the air is to being saturated, dew-point temperature is a measure of its *actual moisture* content. Because the dew-point temperature is directly related to the amount of water vapor in the air, and because it is easy to determine, it is one of the most widely used measures of humidity.

Recall that the saturation vapor pressure is temperature dependent and that for every 10°C (18°F) increase in temperature, the amount of water vapor needed for saturation doubles. Therefore, relatively cold

*Normally, we associate dew with grass. Because of transpiration by the blades of grass, the relative humidity on a calm night is much higher near the grass than a few inches above the surface. Consequently, dew forms on grass before it does on most other objects.

Figure 16.5   Relative humidity varies with temperature. When the water-vapor content (mixing ratio) remains constant, the relative humidity can be changed by increasing or decreasing the air temperature. In this example, when the temperature of the air in the flask was lowered from 20° to 10°C, the relative humidity increased from 50 to 100 percent. Further cooling (from 10° to 0°C) causes one-half of the water vapor to condense. In nature, cooling of air below its saturated mixing ratio generally causes condensation in the form of clouds, dew, or fog.

saturated air (0°C or 32°F) contains about half the water vapor of *saturated air* having a temperature of 10°C (50°F) and roughly one-fourth that of hot *saturated air* with a temperature of 20°C (68°F). Because the dew point is the temperature at which saturation occurs, we can conclude that high dew-point temperatures equate to moist air, and low dew-point temperatures indicate dry air. More precisely, based on what we have learned about vapor pressure and saturation, we can state that for every 10°C (18°F) increase in the dew-point temperature, the air contains about twice as much water vapor. Therefore, we know that air over Fort Myers,

Figure 16.6   Typical daily variations in temperature and relative humidity during a spring day at Washington, D.C. When temperature increases, relative humidity drops (see mid-afternoon) and vice versa.

Figure 16.7 Dew on a spider web. (Photo by Wolfgang Kaehler)

Florida, with a dew point of 25°C (77°F) contains about twice the water vapor of air situated over St. Louis, Missouri, with a dew point of 15°C (59°F) and four times that of Tucson, Arizona, with a dew point of 5°C (41°F).

Measuring Humidity

Relative humidity is commonly measured using a **hygrometer** (*hygro* = moisture, *metron* = measuring instrument). One type of hygrometer, called a **psychrometer,** consists of two identical thermometers mounted side by side (Figure 16.8). One thermometer, the *dry-bulb*, gives the present air temperature. The other, called the *wet-bulb* thermometer, has a thin muslin wick tied around the end (see end of thermometers in the photo).

To use the psychrometer, the cloth sleeve is saturated with water and a continuous current of air is passed over the wick. This is done either by swinging the instrument freely in the air or by fanning air past it. As a consequence, water evaporates from the wick, and the heat absorbed by the evaporating water makes the temperature of the wet bulb drop. The loss of heat that was required to evaporate water from the wet bulb lowers the thermometer reading.

The amount of cooling that takes place is directly proportional to the dryness of the air. The drier the air, the more moisture evaporates. The more heat the evaporating water absorbs, the greater the cooling. Therefore, the larger the difference that is observed between the thermometer readings, the lower the relative humidity; the smaller the difference, the higher the relative humidity. If the air is saturated, no evaporation will occur, and the two thermometers will have identical readings.

To determine the precise relative humidity from the thermometer readings, a standard table is used (Refer to Appendix D, Table D.1). With the same information but using a different table (Table D.2), the dew-point temperature can also be calculated.

Another instrument used for measuring relative humidity, the *hair hygrometer*, can be read directly, without

the use of tables. The hair hygrometer operates on the principle that hair or certain synthetic fibers change their length in proportion to changes in the relative humidity, lengthening as relative humidity increases, and shrinking as the relative humidity drops. The tension of a bundle of hairs is linked mechanically to an indicator that is calibrated between 0 and 100 percent. Thus,

Figure 16.8 Sling psychrometer. This instrument is used to determine both relative humidity and dew point. The dry-bulb thermometer gives the current air temperature. The web-bulb thermometer is covered with a cloth wick that is dipped in water. The thermometers are spun until the temperature of the wet-bulb thermometer stops declining. Then the thermometers are read and the data used in conjunction with the tables in Appendix D. (Photo by E. J. Tarbuck)

we need only glance at the dial to determine the relative humidity. Unfortunately, the hair hygrometer is less accurate than the psychrometer. Further, it requires frequent calibration and is slow in responding to changes in humidity, especially at low temperatures.

A different type of hygrometer is used in remote-sensing instrument packages such as radiosondes that transmit upper-air observations back to ground stations. The *electric hygrometer* contains an electrical conductor coated with a moisture-absorbing chemical. It works on the principle that the passage of current varies as the relative humidity varies.

The Basis of Cloud Formation: Adiabatic Cooling

GEODe Atmosphere
↳ Moisture and Cloud Formation

Up to this point, we have considered basic properties of water vapor and how its variability is measured. We are now ready to examine some of the important roles that water vapor plays in weather, especially the formation of clouds.

Fog and Dew versus Cloud Formation

Recall that condensation occurs when water vapor changes to a liquid. Condensation may form dew, fog, or clouds. Although these three forms are different, all require saturated air to develop. As indicated earlier, saturation occurs either when sufficient water vapor is added to the air or, more commonly, when the air is cooled to its dew point.

Near Earth's surface, heat is readily exchanged between the ground and the air above. During evening hours, the surface radiates heat away, causing the surface and adjacent air to cool rapidly. This radiation cooling accounts for the formation of dew and some types of fog. Thus, surface cooling that occurs after sunset accounts for some condensation. However, cloud formation often takes place during the warmest part of the day. Clearly some other mechanism must operate aloft that cools air sufficiently to generate clouds.

Adiabatic Temperature Changes

The process that is responsible for most cloud formation is easily visualized if you have ever pumped up a bicycle tire and noticed that the pump barrel became quite warm. The heat you felt was the consequence of the work you did on the air to compress it. When energy is used to compress air, the motion of the gas molecules increases and therefore the temperature of the air rises. Conversely, air that is allowed to escape from a bicycle tire *expands and cools*. This results because the expanding air pushes (does work on) the surrounding air and must cool by an amount equivalent to the energy expended.

You have probably felt the cooling effect of the propellant gas expanding as you applied hair spray or spray deodorant. As the compressed gas in the aerosol can is released, it quickly expands and cools. This drop in temperature occurs *even though heat is neither added nor subtracted*. Such variations are known as **adiabatic temperature changes** and result when air is compressed or allowed to expand. In summary, *when air is allowed to expand, it cools, and when it is compressed, it warms.*

Adiabatic Cooling and Condensation

To simplify the following discussion, it helps if we imagine a volume of air enclosed in a thin elastic cover. Meteorologists call this imaginary volume of air a **parcel.** Typically, we consider a parcel to be a few hundred cubic meters in volume, and we assume that it acts independently of the surrounding air. It is also assumed that no heat is transferred into, or out of, the parcel. Although highly idealized, over short time spans a parcel of air behaves in a manner much like an actual volume of air moving vertically in the atmosphere.

Dry Adiabatic Rate. As you travel from Earth's surface upward through the atmosphere, the atmospheric pressure rapidly diminishes, because there are fewer and fewer gas molecules. Thus, any time a parcel of air moves upward, it passes through regions of successively lower pressure. As a result, the ascending air expands. As it expands, it cools adiabatically. Unsaturated air cools at the constant rate of 10°C for every 1000 meters of ascent (5.5°F per 1000 feet).

Conversely, descending air comes under increasingly higher pressures, compresses, and is heated 10°C for every 1000 meters of descent. This rate of cooling or heating applies only to *unsaturated air* and is known as the **dry adiabatic rate.**

Wet Adiabatic Rate. If a parcel of air rises high enough, it will eventually cool to its dew point. Here the process of condensation begins. From this point on along its ascent, *latent heat of condensation* stored in the water vapor will be liberated. Although the air will continue to cool after condensation begins, the released latent heat works against the adiabatic process, thereby reducing the rate at which the air cools. This slower rate of cooling caused by the addition of latent heat is called the **wet adiabatic rate** of cooling. Because the amount of latent heat released depends on the quantity of moisture present in the air, the wet adiabatic rate varies from 5°C per 1000 meters for air with a high moisture content to 9°C per 1000 meters for dry air.

Figure 16.9 illustrates the role of adiabatic cooling in the formation of clouds. Note that from the surface up to the condensation level the air cools at the dry adiabatic rate. The wet adiabatic rate commences at the condensation level.

Processes that Lift Air

 GEODe Atmosphere
↳ Moisture and Cloud Formation

To review, when air rises, it expands and cools adiabatically. If air is lifted sufficiently, it will eventually cool to its dew-point temperature, saturation will occur, and clouds will develop. But why does air rise on some occasions and not on others?

It turns out that, in general, the tendency is for air to resist vertical movement. Therefore, air located near the surface tends to stay near the surface, and air aloft tends to remain aloft. Exceptions to this rule, as we shall see, include conditions in the atmosphere that give air sufficient buoyancy to rise without the aid of outside forces. In many situations, however, when you see clouds forming there is some mechanical phenomenon at work that forces the air to rise (at least initially).

We will be looking at four mechanisms that cause air to rise. These are:

1. *Orographic lifting*—air is forced to rise over a mountainous barrier.
2. *Frontal wedging*—warmer, less dense air, is forced over cooler, denser air.
3. *Convergence*—a pileup of horizontal air flow results in upward movement.
4. *Localized convective lifting*—unequal surface heating causes localized pockets of air to rise because of their buoyancy.

Orographic Lifting

Orographic lifting occurs when elevated terrains, such as mountains, act as barriers to the flow of air (Figure 16.10A). As air ascends a mountain slope, adiabatic cooling often generates clouds and copious precipitation. In fact, many of the rainiest places in the world are located on windward mountain slopes.

By the time air reaches the leeward side of a mountain, much of its moisture has been lost. If the air descends, it warms adiabatically, making condensation

Figure 16.9 Rising air cools at the dry adiabatic rate of 10° per 1000 meters, until the air reaches the dew point and condensation (cloud formation) begins. As air continues to rise, the latent heat released by condensation reduces the rate of cooling. The wet adiabatic rate is therefore always less than the dry adiabatic rate.

A. Orographic lifting

B. Frontal wedging

C. Convergence

D. Localized convective lifting

Figure 16.10   Four processes that lift air. **A.** *Orographic lifting*: Air is forced over a topographic barrier. **B.** *Frontal wedging*: Cold, dense air displaces warm, less dense air along their boundary. **C.** *Convergence*: When surface air converges, it increases in height to allow for the decreased area it occupies. **D.** *Localized convective lifting*: Unequal heating of Earth's surface causes pockets of air to be warmed more than the surrounding air.

and precipitation even less likely. As shown in Figure 16.10A, the result can be a **rain shadow desert.** The Great Basin Desert of the western United States lies only a few hundred kilometers from the Pacific Ocean, but it is effectively cut off from the ocean's moisture by the imposing Sierra Nevada (Figure 16.11). The Gobi Desert of Mongolia, the Takla Makan of China, and the Patagonia Desert of Argentina are other examples of deserts that exist because they are on the leeward sides of mountains (for a map showing deserts, see Figure 19.6).

Frontal Wedging

If orographic lifting were the only mechanism that forced air aloft, the relatively flat central portion of North America would be an expansive desert instead of the nation's breadbasket. Fortunately, this is not the case.

In central North America, masses of warm and cold air collide producing a **front.** Here the cooler, denser air acts as a barrier over which the warmer, less dense air rises. This process, called **frontal wedging,** is illustrated in Figure 16.10B.

It should be noted that weather-producing fronts are associated with storm systems called *middle-latitude cyclones.* Because these storms are responsible for producing a high percentage of the precipitation in the middle latitudes, we will examine them closely in Chapter 18.

Convergence

We saw that the collision of contrasting air masses forces air to rise. In a more general sense, whenever air in the lower atmosphere flows together, lifting results. This phenomenon is called **convergence.** When air flows in from more than one direction, it must go some-

Figure 16.11   Rain shadow desert. The arid conditions in California's Death Valley can be partially attributed to the adjacent mountains, which orographically remove the moisture from air originating over the Pacific. (Photo by James E. Patterson)

where. As it cannot go down, it goes up (Figure 16.10C). This, of course, leads to adiabatic cooling and possibly cloud formation.

The Florida peninsula provides an excellent example of the role that convergence can play in initiating cloud development and precipitation. On warm days, the airflow is from the ocean to the land along both coasts of Florida. This leads to a pileup of air along the coasts and general convergence over the peninsula. This pattern of air movement and the uplift that results is aided by intense solar heating of the land. The result is that the peninsula of Florida experiences the greatest frequency of midafternoon thunderstorms in the United States.

Localized Convective Lifting

On warm summer days, unequal heating of Earth's surface may cause pockets of air to be warmed more than the surrounding air. For instance, air above a paved parking lot will be warmed more than the air above an adjacent wooded park. Consequently, the parcel of air above the parking lot, which is warmer (less dense) than the surrounding air, will be buoyed upward (16.10D). These rising parcels of warmer air are called *thermals.* Birds such as hawks and eagles use these thermals to carry them to great heights where they can gaze down on unsuspecting prey. People have learned to employ these rising parcels using hang gliders as a way to "fly."

BOX 16.1

PEOPLE AND THE ENVIRONMENT:
Atmospheric Stability and Air Pollution

Air quality is not just a function of the quantity and types of pollutants emitted into the air, but it is also closely linked to the atmosphere's ability to disperse these noxious substances. Perhaps you have heard this following well-known phrase: "The solution to pollution is dilution." To a significant degree this is true. If the air into which the pollution is released is not dispersed, the air will become more toxic. Two of the most important atmospheric conditions affecting the dispersion of pollutants are the strength of the wind and the stability of the air. These factors are critical because they determine how rapidly pollutants are diluted by mixing with the surrounding air after leaving the source.

The way in which wind speed influences the concentration of pollutants is straightforward. When winds are weak or calm, the concentration of pollutants is higher than when winds are strong. High wind speeds mix polluted air into a greater volume of surrounding air and therefore cause the pollution to be more diluted. When winds are light, there is less turbulence and mixing, so the concentration of pollutants is higher.

Whereas wind speed governs the amount of air into which pollutants are initially mixed, atmospheric stability determines the extent to which vertical motions will mix the pollution with cleaner air above. The distance between Earth's surface and the height to which vertical air movements extend is termed the *mixing depth.* Generally, the greater the mixing depth, the better the air quality. When the mixing depth is several kilometers, pollutants are mixed through a large volume of cleaner air and dilute rapidly. When the mixing depth is shallow, pollutants are confined to a much smaller volume of air and concentrations can reach unhealthy levels. When air is stable, vertical motions are suppressed and mixing depths are small. Conversely, an unstable atmosphere promotes vertical air movements and greater mixing depths. Because heating of Earth's surface by the Sun enhances convectional movements, mixing depths are usually greater during the afternoon hours. For the same reason, mixing depths during the summer months are typically greater than during the winter months.

Temperature inversion represents a situation in which the atmosphere is very stable and the mixing depth is significantly restricted. Warm air overlying cooler air acts as a lid and prevents upward movement, leaving the pollutants trapped in a relatively narrow zone near the ground (Figure 16.A).

Inversions are generally classified into one of two categories—those that form near the ground and those that form aloft. A *surface inversion* develops close to the ground on clear and rela-

Figure 16.A Air pollution in downtown Los Angeles. Temperature inversions act as lids to trap pollutants below. (Photo by Ted Spiegel/Black Star)

The phenomenon that produces rising thermals is called **localized convective lifting.** When these warm parcels of air rise above the lifting condensation level, clouds form, which on occasion produce midafternoon rain showers. The accompanying rains, although occasionally heavy, are of short duration and widely scattered.

The Weathermaker: Atmospheric Stability

Atmosphere
↳ Moisture and Cloud Formation

When air rises, it cools and eventually produces clouds. Why do clouds vary so much in size, and why does the resulting precipitation vary so much? The answers are closely related to the *stability* of the air.

Recall that a parcel of air can be thought of as having a thin flexible cover that allows it to expand but prevents it from mixing with the surrounding air (picture a hot-air balloon). If this parcel were forced to rise, its temperature would decrease because of expansion. By comparing the parcel's temperature to that of the surrounding air, we can determine its stability. If the parcel were *cooler* than the surrounding environment, it would be more dense; and if allowed to do so, it would sink to its original position. Air of this type, called **stable air,** resists vertical movement.

If, however, our imaginary rising parcel were *warmer* and hence less dense than the surrounding air, it would continue to rise until it reached an altitude

Figure 16.B Generalized temperature profile for a surface inversion.

tively calm nights. It forms because the ground is a more effective radiator than the air above. This being the case, radiation from the ground to the clear night sky causes more rapid cooling at the surface than higher in the atmosphere. The result is that the air close to the ground is cooled more than the air above, yielding a temperature profile similar to the one shown in Figure 16.B. After sunrise the ground is heated and the inversion disappears.

Although surface inversions are usually shallow, they may be quite thick in regions where the land surface is uneven. Because cold air is denser (heavier) than warm air, the chilled air near the surface gradually drains from the uplands and slopes into adjacent lowlands and valleys. As might be expected, these thicker surface inversions

will not dissipate as quickly after sunrise.

Many extensive and long-lived air-pollution episodes are linked to temperature inversions that develop in association with the sinking air that characterizes slow-moving centers of high pressure (Figure 16.C). As the air sinks to lower altitudes, it is compressed and so its temperature rises. Because turbulence is almost always present near the ground, this lowermost portion of the atmosphere is generally prevented from participating in the general subsidence. Thus, an inversion develops aloft between the lower turbulent zone and the subsiding warmed layers above.

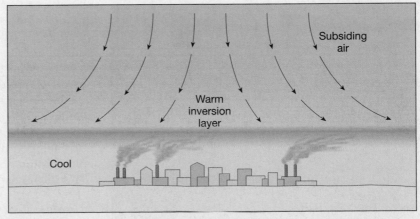

Figure 16.C Inversions aloft frequently develop in association with slow-moving centers of high pressure where the air aloft subsides and warms by compression. The turbulent surface zone does not subside as much. Thus, an inversion often forms between the lower turbulent zone and the subsiding layers above.

where its temperature equalled that of its surroundings. This is exactly how a hot-air balloon works, rising as long as it is warmer and less dense than the surrounding air (Figure 16.12). This type of air is classified as **unstable air.** In summary, stability is a property of air that describes its tendency to remain in its original position (stable), or to rise (unstable).

? STUDENTS SOMETIMES ASK...

What are the wettest places on Earth?

Many of the rainiest places in the world are located on windward mountain slopes. A station at Mount Waialeale, Hawaii, records the highest average annual rainfall, some 1234 centimeters (486 inches). The greatest recorded rainfall for a single 12-month period occurred at Cherrapunji, India, where an astounding 2647 centimeters (1042 inches, over 86 feet) fell. Much of this rainfall occurred in the month of July—a record 930 centimeters (366 inches, over 30 feet). This is 10 times more rain than Chicago receives in an average year.

Types of Stability

The stability of air is determined by measuring the temperature of the atmosphere at various heights. Recall from Chapter 15 that this measure is termed the *environmental lapse rate.* The environmental lapse rate is the temperature of the atmosphere as determined from observations made by radiosondes and aircraft. It is im-

Figure 16.12 As long as air is warmer than its surroundings, it will rise. Hot-air balloons rise up through the atmosphere for this reason. (Photo by Barbara Cushman Rowell)

portant not to confuse this with *adiabatic temperature changes,* which are changes in temperature caused by expansion or compression as a parcel of air rises or descends.

For illustration, we will examine a situation in which the environmental lapse rate is 5°C per 1000 meters (Figure 16.13). Under this condition, when air at the surface has a temperature of 25°C, the air at 1000 meters will be 5 degrees cooler, or 20°C, the air at 2000 meters will have a temperature of 15°C, and so forth. At first glance it appears that the air at the surface is less dense than the air at 1000 meters, because it is 5 degrees warmer. However, if the air near the surface is unsaturated and were to rise to 1000 meters, it would expand and cool at the dry adiabatic rate of 10°C per 1000 meters. Therefore, upon reaching 1000 meters, its temperature would have dropped 10°C. Being 5 degrees cooler than its environment, it would be denser and tend to sink to its original position. Hence, we say that the air near the surface is potentially cooler than the air aloft and therefore will not rise on its own. The air just described is *stable* and resists vertical movement.

With this as background, we will now look at three fundamental conditions of the atmosphere: absolute stability, absolute instability, and conditional instability.

Absolute Stability Stated quantitatively, **absolute stability** prevails when the environmental lapse rate is less than the wet adiabatic rate. Figure 16.14 depicts this situation using an environmental lapse rate of 5°C per 1000 meters and a wet adiabatic rate of 6°C per 1000 meters. Note that at 1000 meters the temperature of the surrounding air is 15°C, while the rising parcel of air has cooled to 10°C and is therefore the denser air. Even if this stable air were to be forced above the condensation level, it would remain cooler and denser than its environment, and thus it would tend to return to the surface.

The most stable conditions occur when the temperature in a layer of air actually increases with altitude. When such a reversal occurs, a *temperature inversion* is said to exist. Temperature inversions frequently occur on clear nights as a result of radiation cooling of Earth's surface. Under these conditions, an inversion is created because the ground and the air immediately above will cool more rapidly than the air aloft. When warm air overlies cooler air, it acts as a lid and prevents appreciable vertical mixing. Because of this, temperature inversions are responsible for trapping pollutants in a narrow zone near Earth's surface. This idea is explored more fully in Box 16.1.

Absolute Instability. At the other extreme, air is said to exhibit **absolute instability** when the environmental lapse rate is greater than the dry adiabatic rate. As shown in Figure 16.15, the ascending parcel of air is always warmer than its environment and will

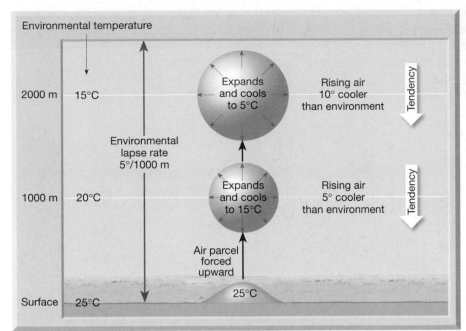

Figure 16.13   In a stable atmosphere, as an unsaturated parcel of air is lifted, it expands and cools at the dry adiabatic rate of 10°C per 1000 meters. Because the temperature of the rising parcel of air is lower than that of the surrounding environment, it will be heavier and, if allowed to do so, will sink to its original position.

continue to rise because of its own buoyancy. However, absolute instability is generally limited to near Earth's surface. On hot, sunny days the air above some surfaces, like paved shopping malls, is heated more than the air over adjacent surfaces. These invisible pockets of more intensely heated air, being less dense than the air aloft, will rise like a hot-air balloon. This phenomenon produces the small fluffy clouds we associate with fair weather. Occasionally, when the surface air is considerably warmer than the air aloft, clouds with considerable vertical development can form.

Figure 16.14   Absolute stability prevails when the environmental lapse rate is less than the wet adiabatic rate. The rising parcel of air is therefore always cooler and denser than the surrounding air. When stable air is forced to rise, it spreads out, producing flat, layered clouds.

BOX 16.2

UNDERSTANDING EARTH:
Science and Serendipity*

*S*erendipity is defined by Nobel Laureate Irving Langmuir as "the art of profiting from unexpected occurrences." In other words, if you are observing something and the entirely unexpected happens, and if you see in this accident a new and meaningful discovery, then you have experienced serendipity. Most nonscientists, some scientists, and, alas, many teachers are not aware that many of the great discoveries in science were serendipitous.

An excellent example of serendipity in science occurred when Tor Bergeron, the great Swedish meteorologist, discovered the importance of ice crystals in the initiation of precipitation in supercooled clouds. Bergeron's discovery occurred when he spent several weeks at a health resort at an altitude of 430 meters (1400 feet) on a hill near Oslo. During his stay, Bergeron noted that this hill was often "fogged-in" by a layer of supercooled clouds. As he walked along a narrow road in the fir forest along the hillside, he noticed that the "fog" did not enter the "road tunnel" at temperatures below −5°C, but did enter it when the temperature was warmer than 0°C. (Profiles of the hill, trees, and fog for the two temperature regimes is shown in Figure 16.D.)

Bergeron immediately concluded that at temperatures below about −5°C the branches of the firs acted as freez-

*Based on material prepared by Duncan C. Blanchard.

Figure 16.D Distribution of fog when the temperature is above freezing and when the temperature falls to −10°C.

ing nuclei upon which some of the supercooled droplets crystallized. Once the ice crystals developed, they grew rapidly at the expense of the remaining water droplets (see Figure 16.D). The result was the growth of ice crystals (rime) on the branches of the firs accompanied by a "clearing-off" between the trees and along the "road-tunnel."

From this experience, Bergeron realized that, if ice crystals somehow were to appear in the midst of a cloud of supercooled droplets, they would grow rapidly as water molecules diffused toward them from the evaporating cloud droplets. This rapid growth forms snow crystals that, depending on the air temperature beneath the cloud, fall to the ground as snow or rain. Bergeron had thus discovered one way that minuscule cloud droplets can grow large enough

to fall as precipitation (see the discussion entitled "Precipitation from Cold Clouds: The Bergeron Process," p. 457).

Serendipity influences the entire realm of science. Can we conclude that anyone who makes observations will necessarily make a major discovery? Not at all. A perceptive and inquiring mind is required, a mind that has been searching for order in a labyrinth of facts. As Langmuir said, the unexpected occurrence is not enough; you must know how to profit from it. Louis Pasteur observed that "In the field of observation, chance favors only the prepared mind." The discoverer of vitamin C, Nobel Laureat Albert Szent-Gyorgyi, remarked that discoveries are made by those who "see what everybody else has seen, and think what nobody else has thought." Serendipity is at the heart of science itself.

Conditional Instability. A more common type of atmospheric instability is called **conditional instability.** This situation prevails when moist air has an environmental lapse rate between the dry and wet adiabatic rates (between 5 and 10°C per 1000 meters). Simply, the atmosphere is said to be conditionally unstable when it is *stable* for an *unsaturated* parcel of air, but *unstable* for a *saturated* parcel of air. Notice in Figure 16.16 that the rising parcel of air is cooler than the surrounding air for the first 4000 meters. With the addition of latent heat above the lifting condensation level, the parcel becomes warmer than the surrounding air. From this point along its ascent the parcel will continue to rise because of its own buoyancy, without an outside force. Thus, conditional instability depends

on whether or not the rising air is saturated. The word *conditional* is used because the air must be forced upward, such as over mountainous terrain, before it becomes unstable and rises because of its own buoyancy.

Summary. In summary, the stability of air is determined by measuring the temperature of the atmosphere at various heights. In simple terms, a column of air is deemed unstable when the air near the bottom of this layer is significantly warmer (less dense) than the air aloft, indicating a steep environmental lapse rate. Under these conditions the air actually turns over, as the warm air below rises and displaces the colder air aloft. Conversely, the air is considered to be stable when the temperature drops gradually with

Figure 16.15   Absolute instability illustrated using an environmental lapse rate of 12°C per 1000 meters. The rising air is always warmer and therefore lighter than the surrounding air.

increasing altitude. The most stable conditions occur during a temperature inversion when the temperature actually increases with height. Under these conditions, there is very little vertical air movement.

Stability and Daily Weather

From the previous discussion, we can conclude that stable air resists vertical movement, whereas unstable air ascends freely because of its own buoyancy. But how do these facts manifest themselves in our daily weather?

Because stable air resists upward movement, we might conclude that clouds will not form when stable conditions prevail in the atmosphere. Although this seems reasonable, recall that processes exist that *force* air aloft. These include orographic lifting, frontal wedging, and convergence. On those occasions when stable air is forced aloft, the clouds that form are widespread and have little vertical thickness when compared to their horizontal dimension, and precipitation, if any, is light to moderate.

By contrast, clouds associated with the lifting of unstable air are towering and often generate thunderstorms and occasionally even a tornado. For this reason, we can conclude that on a dreary, overcast day with light drizzle, stable air has been forced aloft. On the other hand, during a day when cauliflower-shaped clouds appear to be growing as if bubbles of hot air are surging upward, we can be fairly certain that the ascending air is unstable (Figure 16.17).

In summary, the role of stability in determining our daily weather is very important. To a large degree, stability determines the type of clouds that develop and whether precipitation will come as a gentle shower or a heavy downpour.

Condensation and Cloud Formation

To review briefly, condensation occurs when water vapor in the air changes to a liquid. The result of this process may be dew, fog, or clouds. For any of these forms of condensation to occur, the air must be saturated. Saturation occurs most commonly when air is cooled to its dew point, or less often when water vapor is added to the air.

Generally, there must be a surface on which the water vapor can condense. When dew occurs, objects at or near the ground serve this purpose, like grass and car windows. But when condensation occurs in the air above the ground, tiny bits of particulate matter, known as **condensation nuclei,** serve as surfaces for water-vapor condensation. These nuclei are very important, for in their absence a relative humidity well in excess of 100 percent is needed to produce clouds.

Condensation nuclei such as microscopic dust, smoke, and salt particles (from the ocean) are profuse in the lower atmosphere. Because of this abundance of particles, relative humidity rarely exceeds 101 percent. Some particles, such as ocean salt, are particularly good nuclei because they absorb water. These particles are termed **hygroscopic** (*hygro* = moisture, *scopic* = to seek) **nuclei.** When condensation takes place, the initial growth rate of cloud droplets is rapid. It diminishes quickly because the excess water vapor is quickly ab-

Figure 16.16 Conditional instability illustrated using an environmental lapse rate of 8°C per 1000 meters, which lies between the dry adiabatic rate and the wet adiabatic rate. The rising parcel of air is cooler than the surrounding air below 4000 meters and warmer above 4000 meters.

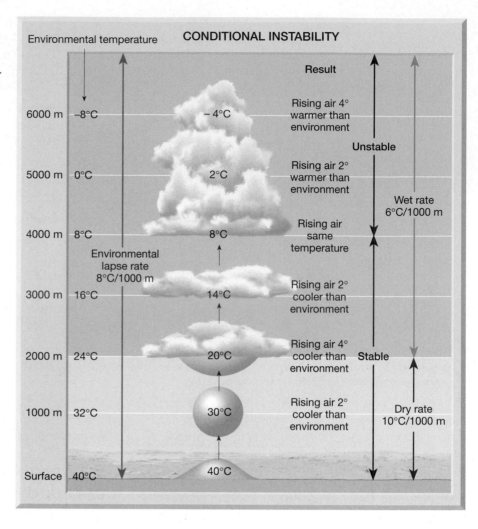

sorbed by the numerous competing particles. This results in the formation of a cloud consisting of millions upon millions of tiny water droplets, all so fine that they remain suspended in air. When cloud formation occurs at below-freezing temperatures, tiny ice crystals form. Thus, a cloud might consist of water droplets, ice crystals, or both.

The slow growth of cloud droplets by additional condensation and the immense size difference between cloud droplets and raindrops suggest that condensation alone is not responsible for the formation of drops large enough to fall as rain. We will first examine clouds and then return to the questions of how precipitation forms.

Types of Clouds

Clouds are among the most conspicuous and observable aspects of the atmosphere and its weather. **Clouds** are a form of condensation best described as *visible aggregates of minute droplets of water or tiny crystals of ice.* In addition to being prominent and sometimes spectacular features in the sky, clouds are of continual interest to meteorologists, because they provide a visible indication of what is going on in the atmosphere. Anyone who observes clouds with the hope of recognizing

different types often finds that there is a bewildering variety of these familiar white and gray masses streaming across the sky. Still, once one comes to know the basic classification scheme for clouds, most of the confusion vanishes.

Clouds are classified on the basis of their *form* and *height* (Figure 16.18). Three basic forms are recognized: cirrus, cumulus, and stratus.

- **Cirrus** (*cirrus* = a curl of hair) clouds are high, white, and thin. They can occur as patches or as delicate veil-like sheets or extended wispy fibers that often have a feathery appearance.
- **Cumulus** (*cummulus* = a pile) clouds consist of globular individual cloud masses. Normally they exhibit a flat base and have the appearance of rising domes or towers. Such clouds are frequently described as having a cauliflower structure.
- **Stratus** (*stratum* = a layer) clouds are best described as sheets or layers that cover much or all of the sky. While there may be minor breaks, there are no distinct individual cloud units.

All other clouds reflect one of these three basic forms or are combinations or modifications of them.

Figure 16.17 Cauliflower-shaped clouds provide evidence of unstable conditions in the atmosphere. (Photo by Dick Canby/DRK Photo)

Three levels of cloud heights are recognized: high, middle, and low (Figure 16.18). **High clouds** normally have bases above 6000 meters (20,000 feet), **middle clouds** generally occupy heights from 2000 to 6000 meters (6500 to 20,000 feet), and **low clouds** form below 2000 meters (6,500 feet). The altitudes listed for each height category are not hard and fast. There is some seasonal as well as latitudinal variation. For example, at high latitudes or during cold winter months in the midlatitudes, high clouds are often found at lower altitudes.

High Clouds. Three cloud types make up the family of high clouds (above 6000 meters): cirrus, cirrostratus, and cirrocumulus. *Cirrus* clouds are thin and delicate and sometimes appear as hooked filaments called "mares tails" (Figure 16.19A). As the names suggest, *cirrocumulus* clouds consist of fluffy masses (Figure 16.19B), whereas *cirrostratus* clouds are flat layers (Figure 16.19C). Because of the low temperatures and small quantities of water vapor present at high altitudes, all high clouds are thin and white and are made up of ice crystals. Further, these clouds are not considered precipitation makers. However, when cirrus clouds are followed by cirrocumulus clouds and increased sky coverage, they may warn of impending stormy weather.

Middle Clouds. Clouds that appear in the middle range (2000 to 6000 meters) have the prefix *alto* as part of their name. *Altocumulus* clouds are composed of globular masses that differ from cirrocumulus clouds in that they are larger and denser (Figure 16.19D). *Altostratus* clouds create a uniform white to grayish sheet covering the sky with the Sun or Moon visible as a bright spot (Figure 16.19E). Infrequent light snow or drizzle may accompany these clouds.

Low Clouds. There are three members in the family of low clouds: stratus, stratocumulus, and nimbostratus. *Stratus* are a uniform foglike layer of clouds that frequently covers much of the sky. On occasions these clouds may produce light precipitation. When stratus clouds develop a scalloped bottom that appears as long parallel rolls or broken globular patches, they are called *stratocumulus* clouds.

Nimbostratus clouds derive their name from the Latin *nimbus,* which means rainy cloud, and *stratus,* which means to cover with a layer (Figure 16.19F). As the name suggests, nimbostratus clouds are one of the chief precipitation producers. Nimbostratus clouds form in association with stable conditions. We might not expect clouds to grow or persist in stable air, yet cloud growth of this type is common when air is forced to rise, as occurs along a mountain range, a front, or near the center of a cyclone where converging winds cause air to ascend. Such forced ascent of stable air leads to the formation of a stratified cloud layer that is large horizontally compared to its depth.

Clouds of Vertical Development. Some clouds do not fit into any one of the three height categories mentioned. Such clouds have their bases in the low height range but often extend upward into the middle or high altitudes. Consequently, clouds in this category are called **clouds of vertical development.** They are all related to one another and are associated with unstable air. Although *cumulus* clouds are often connected with fair

451

weather (Figure 16.19G), they may grow dramatically under the proper circumstances. Once upward movement is triggered, acceleration is powerful, and clouds with great vertical extent form. The end result is often a towering cloud, called a *cumulonimbus*, that may produce rain showers or a thunderstorm (Figure 16.19H).

Definite weather patterns can often be associated with particular clouds or certain combinations of cloud types, so it is important to become familiar with cloud descriptions and characteristics. Table 16.2 lists the 10 basic cloud types that are recognized internationally and gives some characteristics of each.

Fog

Fog is generally considered to be an atmospheric hazard. When it is light, visibility is reduced to 2 or 3 kilometers (1 or 2 miles). However, when it is dense, visibility may be cut to a few dozen meters or less, making travel by any mode not only difficult but often dangerous (Figure 16.20). Officially, visibility must be reduced to 1 kilometer or less before fog is reported. While this figure is arbitrary, it does provide an objective criterion for comparing fog frequencies at different locations.

Fog is defined as *a cloud with its base at or very near the ground.* Physically, there is no basic difference between a fog and a cloud; their appearance and structure are the same. The essential difference is the method and place of formation. Whereas clouds result when air rises and cools adiabatically, most fogs are the consequence of radiation cooling or the movement of air over a cold surface. (The exception is upslope fog.) In other circumstances, fogs form when enough water vapor is added to the air to bring about saturation (evaporation fogs). We will examine both of these types.

Fogs Caused by Cooling

There are three common fogs that form when air at Earth's surface is chilled below its dew point. Often a fog is a combination of these types.

Advection Fog. When warm, moist air moves over a cool surface, the result might be a blanket of fog called **advection fog** (Figure 16.21). Examples of such fogs are very common. The foggiest location in the United States, and perhaps in the world, is Cape Disappointment, Washington. The name is indeed appropriate, because this station averages 2552 hours of fog each year (there are about 8760 hours in a year). The fog experienced at Cape Disappointment, and at other West Coast locations, is produced when warm, moist air from the Pacific Ocean moves over the cold California Current and is then carried onshore by the prevailing winds. Advection fogs are also relatively common in the winter season when warm air from the Gulf of Mexico moves across cold, often snow-covered surfaces of the Midwest and East.

Figure 16.18 Classification of clouds according to height and form.

Radiation Fog. **Radiation fog** forms on cool, clear, calm nights, when Earth's surface cools rapidly by radiation. As the night progresses, a thin layer of air in contact with the ground is cooled below its dew point. As the air cools and becomes denser, it drains into low areas, resulting in "pockets" of fog. The largest "pockets" are often river valleys, where thick accumulations may occur.

Upslope Fog As its name implies, **upslope fog** is created when relatively humid air moves up a gradually sloping plain or up the steep slopes of a mountain. As a result of this upward movement, air expands and cools adiabatically. If the dew point is reached, an extensive layer of fog may form. In the United States, the Great Plains offers an excellent example. When humid easterly or southeasterly winds blow westward from the Mississippi River upslope toward the Rocky Mountains, the air gradually rises, resulting in an adiabatic temperature decrease of about 13°C. When the difference between the air temperature and dew point of westward-moving air is less than 13°C, an extensive fog can result in the western plains.

Evaporation Fogs

When the saturation of air occurs primarily because of the addition of water vapor, the resulting fogs are called *evaporation fogs*. Two types of evaporation fogs are recognized: steam fog and frontal, or precipitation, fog.

Steam Fog.   When cool air moves over warm water, enough moisture may evaporate from the water surface to produce saturation. As the rising water vapor meets the cold air, it immediately recondenses and rises with the air that is being warmed from below. Because the water has a steaming appearance, the phenomenon is called **steam fog** (Figure 16.22). Steam fog is fairly common over lakes and rivers in the fall and early winter, when the water may still be relatively warm and the air is rather crisp. Steam fog is often shallow because as the steam rises, it reevaporates in the unsaturated air above.

Frontal Fog   When frontal wedging occurs, warm air is lifted over colder air. If the resulting clouds yield rain, and the cold air below is near the dew point, enough rain will evaporate to produce fog. A fog formed in this manner is called **frontal fog,** or **precip-**

itation fog. The result is a more or less continuous zone of condensed water droplets reaching from the ground up through the clouds.

In summary, both steam fog and frontal fog result from the addition of moisture to a layer of air. As you saw, the air is usually cool or cold and already near saturation. Because air's capacity for water vapor at low temperatures is small, only a relatively modest amount of evaporation is necessary to produce saturated conditions and fog.

How Precipitation Forms

Although all clouds contain water, why do some produce precipitation and others drift placidly overhead? This seemingly simple question perplexed meteorologists for many years. Before examining the processes that generate precipitation, we need to examine a couple of facts.

First, cloud droplets are very tiny, averaging under 20 micrometers (0.02 millimeter) in diameter (Figure 16.23). For comparison, a human hair is about 75 micrometers in diameter. The small size of cloud droplets results mainly because condensation nuclei

A. Cirrus

B. Cirrocumulus

C. Cirrostratus

D. Altocumulus

Figure 16.19 These photos depict common forms of several different cloud types. (Photos A, B, D, E, F, and G by E. J. Tarbuck. Photo C by A. and J. Verkaik/The Stock Market. Photo H by Doug Millar/Science Source/Photo Researchers, Inc.)

E. Altostratus

F. Nimbostratus

G. Cumulus

H. Cumulonimbus

Table 16.2 Cloud types and characteristics.

Cloud Family and Height	Cloud Type	Characteristics
High clouds— above 6000 meters (20,000 feet)	Cirrus	Thin, delicate, fibrous, ice-crystal clouds. Sometimes appear as hooked filaments called "mares' tails." (Figure 16.19A)
	Cirrocumulus	Thin, white, ice-crystal clouds in the form of ripples, waves, or globular masses all in a row. May produce a "mackerel sky." Least common of the high clouds. (Figure 16.19B)
	Cirrostratus	Thin sheet of white, ice-crystal clouds that may give the sky a milky look. Sometimes produce halos around the Sun or Moon. (Figure 16.19C).
Middle clouds— 2000–6000 meters (6500–20,000 feet)	Altocumulus	White to gray clouds often composed of separate globules; "sheep-back" clouds. (Figure 16.19D)
	Altostraatus	Stratified veil of clouds that are generally thin and may produce very light precipitation. When thin, the Sun or Moon may be visible as a bright spot, but no halos are produced. (Figure 16.19E)
Low clouds— below 2000 meters (6500 feet)	Stratocumulus	Soft, gray clouds in globular patches or rolls. Rolls may join together to make a continuous cloud.
	Stratus	Low uniform layer resembling fog but not resting on the ground. May produce drizzle.
	Nimbostratus	Amorphous layer of dark gray clouds. One of the chief precipitation-producing clouds. (Figure 16.19F)
Clouds of vertical development— 500–18,000 meters (1600–60,000 feet)	Cumulus	Dense, billowy clouds often characterized by flat bases. May occur as isolated clouds or closely packed. (Figure 16.19G)
	Cumulonimbus	Towering cloud sometimes spreading out on top to form an "anvil head." Associated with heavy rainfall, thunder, lightning, hail, and tornadoes. (Figure 16.19H)

are usually very abundant and the available water is distributed among numerous droplets rather than concentrated into fewer large droplets.

Second, because of their small size, the rate at which cloud droplets fall is incredibly slow. An average cloud droplet falling from a cloud base at 1000 meters would require several hours to reach the ground. However, it would never complete its journey. This cloud droplet would evaporate before it fell a few meters from the cloud base into the unsaturated air below.

How large must a droplet grow in order to fall as precipitation? A typical raindrop has a diameter of about 2000 micrometers (2 millimeters) or 100 times that of the average cloud droplet having a diameter of 20 micrometers (0.02 millimeter). However, the *volume* of a typical raindrop is a million times that of a cloud droplet. Thus, for precipitation to form, cloud droplets must grow in volume by roughly one million times. Two mechanisms that give rise to these "massive" drops are: the Bergeron process and the collision-coalescence process.

Figure 16.20 When dense, fog can reduce visibility to a few dozen meters or less. (Photo by Jeremy Walker/Tony Stone Images)

Figure 16.21 Advection fog rolling into San Francisco Bay. (Photo by Ed Pritchard/Tony Stone Worldwide)

Precipitation from Cold Clouds: The Bergeron Process

You have probably watched a TV documentary in which mountain climbers brave intense cold and a ferocious snowstorm to scale an ice-covered peak. Although it is hard to imagine, very similar conditions exist in the upper portions of towering cumulonimbus clouds, even on sweltering summer days. It turns out that the frigid conditions high in the troposphere provide an ideal environment to initiate precipitation. In fact, in the middle latitudes much of the rain that falls begins with the birth of snowflakes high in the cloud tops where temperatures are considerably below freezing. Obviously, in the winter, even low clouds are cold enough to trigger precipitation.

The process that generates much of the precipitation in the middle latitudes is named the **Bergeron process** for its discoverer, the highly respected Swedish meteorologist, Tor Bergeron (see Box 16.2). The process relies on two interesting phenomena: supercooling and supersaturation.

Figure 16.22 Steam fog rising from upper St. Regis Lake, Adirondack Mountains, New York. (Photo by Jim Brown/The Stock Market)

457

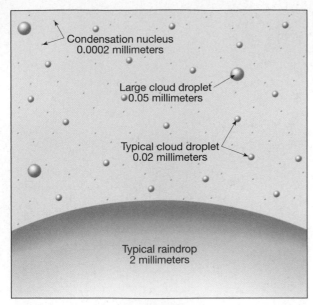

Figure 16.23 Comparative diameters of particles involved in condensation and precipitation processes.

Table 16.3 Relative humidity with respect to ice when relative humidity with respect to water is 100 percent.

Temperature (°C)	Relative Humidity with Respect to:	
	Water (%)	Ice (%)
0	100	100
−5	100	105
−10	100	110
−15	100	116
−20	100	121

the surrounding droplets. In turn, the water droplets evaporate to replenish the diminishing water vapor, thereby providing a continual source of vapor for the growth of the ice crystals (Figure 16.24).

Because the level of supersaturation with respect to ice can be quite great, the growth of ice crystals is generally rapid enough to generate crystals large enough to fall. During their descent, these ice crystals enlarge as they intercept cloud drops, which freeze upon them. Air movement will sometimes break up these delicate crystals and the fragments will serve as freezing nuclei. A chain reaction develops, producing many ice crystals, which by accretion form into large crystals called snowflakes.

In summary, the Bergeron process can produce precipitation throughout the year in the middle latitudes, provided at least the upper portions of clouds are cold enough to generate ice crystals. The type of precipitation (snow, sleet, rain or freezing rain) that reaches the ground depends on the temperature profile in the lower few kilometers of the atmosphere. When the surface temperature is above 4°C (39°F), snowflakes usually melt before they reach the ground and continue their descent as rain. Even on a hot summer day, a heavy downpour may have begun as a snowstorm high in the clouds overhead.

Precipitation from Warm Clouds: The Collision-Coalescence Process

A few decades ago, meteorologists believed that the Bergeron process was responsible for the formation of most precipitation. However, it was discovered that copious rainfall can be associated with clouds located well below the freezing level (*warm clouds*), particularly in the tropics. This led to the proposal of a second mechanism thought to produce precipitation—the **collision-coalescence process.**

Research has shown that clouds composed entirely of liquid droplets must contain some droplets larger than 20 micrometers (0.02 millimeters) if precipitation is to form (see Figure 16.23). These large droplets form when "giant" condensation nuclei are present, or when

Supercooling. Cloud droplets do not freeze at 0°C (32°F) as expected. In fact, pure water suspended in air does not freeze until it reaches a temperature of nearly −40°C (−40°F). Water in the liquid state below 0°C is referred to as **supercooled.** Supercooled water will readily freeze if it impacts an object. This explains why airplanes collect ice when they pass through a cloud composed of supercooled droplets.

In addition, supercooled droplets will freeze upon contact with solid particles that have a crystal form closely resembling that of ice. Such materials are termed **freezing nuclei.** The need for freezing nuclei to initiate the freezing process is similar to the requirement for condensation nuclei in the process of condensation.

In contrast to condensation nuclei, freezing nuclei are very sparse in the atmosphere and do not generally become active until the temperature reaches −10°C (14°F) or less. Only at temperatures well below freezing will ice crystals begin to form in clouds, and even at that, they will be few and far between. Once ice crystals form, they are in direct competition with the supercooled droplets for the available water vapor.

Supersaturation. When air is saturated (100 percent relative humidity) with respect to water, it is **supersaturated** (relative humidity is greater than 100 percent) with respect to ice. Table 16.3 shows that at −10°C (14°F), when the relative humidity is 100 percent with respect to water, the relative humidity with respect to ice is nearly 110 percent. Thus, ice crystals cannot coexist with water droplets, because the air always "appears" supersaturated to the ice crystals. Hence, the ice crystals begin to consume the "excess" water vapor, which lowers the relative humidity near

Figure 16.24   The Bergeron process. Ice crystals grow at the expense of cloud droplets until they are large enough to fall. The size of these particles has been greatly exaggerated.

Water molecule

Ice crystal

Cloud droplet

Snow crystal

hygroscopic particles such as sea salt exist. Hygroscopic particles begin to remove water vapor from the air at relative humidities under 100 percent and can grow quite large. Because the rate at which drops fall is size-dependent, these "giant" droplets fall most rapidly. As they plummet, they collide with smaller, slower droplets and coalesce. Growing larger in the process, they fall even more rapidly (or in an updraft, rise more slowly), increasing their chances of collision and rate of growth (Figure 16.25). After a great many such collisions they are large enough to fall to the surface without completely evaporating.

Because of the number of collisions required for growth to raindrop size, droplets in clouds that have great vertical thickness and abundant moisture have a better chance of reaching the required size. Updrafts also aid this process, for they allow the droplets to traverse the cloud repeatedly.

Raindrops can grow to a maximum size of 5 millimeters, at which point they fall at the rate of 33 kilometers (20 miles) per hour. At this size and speed, the water's surface tension, which holds the drop together, is overcome by the drag imposed by the air, which in turn pulls the drops apart. The resulting breakup of a large raindrop produces numerous smaller drops that begin anew the task of sweeping up cloud droplets. Drops that are less than 0.5 millimeter upon reaching

the ground are termed *drizzle* and require about 10 minutes to fall from a cloud 1000 meters (3300 feet) overhead.

The collision-coalescence process is not quite as simple as described. First, as the large droplets descend, they produce an air stream around them similar to that produced by a fast-moving automobile. If an automobile is driven at night and we use the bugs that are often out to be analogous to the cloud droplets, it is easy to visualize how most cloud droplets are swept aside. The larger the cloud droplet (or bug), the better its chance of colliding with the giant droplet (or car). Second, collision does not guarantee coalescence. Experiments indicate that atmospheric electricity may hold these droplets together once they collide. If a droplet with a negative charge should collide with a positively charged droplet, electrical attraction of opposite charges may bind them together.

In summary, two mechanisms are known to generate precipitation: the *Bergeron process* and the *collision-coalescence process.* The Bergeron process is dominant in the middle latitudes where cold clouds (or cold cloud tops) are the rule. In the tropics, abundant water vapor and comparatively few condensation nuclei are the norm. This leads to the formation of fewer, larger drops with fast fall velocities that grow by collision and coalescence.

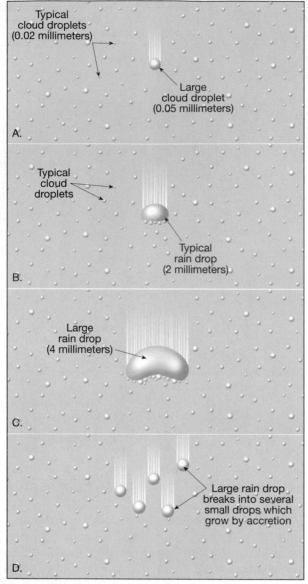

A.
Typical cloud droplets (0.02 millimeters)

Large cloud droplet (0.05 millimeters)

B.
Typical cloud droplets

Typical rain drop (2 millimeters)

C.
Large rain drop (4 millimeters)

D.
Large rain drop breaks into several small drops which grow by accretion

Figure 16.25 The collision–coalescence process. **A.** Most cloud droplets are so small that the motion of the air keeps them suspended. Because large cloud droplets fall more rapidly than smaller droplets, they are able to sweep up the smaller ones in their path and grow. **B.** As these drops increase in size, their fall velocity increases, resulting in increased air resistance, which causes the raindrop to flatten. **C.** As the raindrop approaches 4 millimeters in size it develops a depression in the bottom. **D.** Finally, when the diameter exceeds about 5 millimeters, the depression grows upward almost explosively, forming a donutlike ring of water that immediately breaks into smaller drops. (Note that the drops are not drawn to scale—a typical raindrop has a volume equal to roughly 1 million cloud droplets.)

Forms of Precipitation

Much of the world's precipitation begins as snow crystals or other solid forms, such as hail or graupel (Table 16.4). Entering the warmer air below the cloud, these ice particles often melt and reach the ground as raindrops. In some parts of the world, particularly the subtropics, precipitation often forms in clouds that are warmer than 0°C (32°F). These rains frequently occur over the ocean where cloud condensation nuclei are not plentiful and those that exist vary in size. Under such conditions, cloud droplets can grow rapidly by the collision-coalescence process to produce copious amounts of rain.

Because atmospheric conditions vary greatly from place to place as well as seasonally, several different forms of precipitation are possible. Rain and snow are the most common and familiar, but the other forms of precipitation listed in Table 16.4 are important as well. The occurrence of sleet, glaze, or hail is often associated with important weather events. Although limited in occurrence and sporadic in both time and space, these forms, especially glaze and hail, can cause considerable damage.

Rain

We are familiar with rain. In meteorology, the term **rain** is restricted to drops of water that fall from a cloud and have a diameter of at least 0.5 millimeter (0.002 inch). Most rain originates either in nimbostratus clouds or in towering cumulonimbus clouds that are capable of producing unusually heavy rainfalls known as *cloudbursts*. Raindrops rarely exceed about 5 millimeters (0.2 inch) in diameter. Larger drops do not survive, because surface tension, which holds the drops together, is exceeded by the frictional drag of the air. Consequently, large raindrops regularly break apart into smaller ones.

Fine, uniform drops of water having a diameter less than 0.5 millimeter (0.002 inch) are called *drizzle*. Drizzle can be so fine that the tiny drops appear to float and their impact is almost imperceptible. Drizzle and small raindrops generally are produced in stratus or nimbostratus clouds where precipitation can be continuous for several hours, or on rare occasions for days.

Snow

Snow is precipitation in the form of ice crystals (snowflakes) or, more often, aggregates of crystals. The size, shape, and concentration of snowflakes depend to a great extent on the temperature at which they form.

Recall that at very low temperatures, the moisture content of air is small. The result is the formation of very light, fluffy snow made up of individual six-sided ice crystals. This is the "powder" that downhill skiers talk so much about. By contrast, at temperatures warmer than about −5°C (23°F), the ice crystals join together into larger clumps consisting of tangled aggregates of crystals. Snowfalls composed of these composite snowflakes are generally heavy and have a high

Table 16.4 Forms of precipitation.

Type	Appropriate Size	State of Matter	Description
Mist	0.005 to 0.05 mm	Liquid	Droplets large enough to be felt on the face when air is moving 1 meter/second. Associated with stratus clouds.
Drizzle	Less than 0.5 mm	Liquid	Small uniform drops that fall from stratus clouds, generally for several hours.
Rain	0.5 to 5 mm	Liquid	Generally produced by nimbostratus or cumulonimbus clouds. When heavy, it can show high variability from one place to another.
Sleet	0.5 to 5 mm	Solid	Small, spherical to lumpy ice particles that form when raindrops freeze while falling through a layer of sub-freezing air. Because the ice particles are small, damage, if any, is generally minor. Sleet can make travel hazardous.
Glaze	Layers 1 mm to 2 cm thick	Solid	Produced when supercooled raindrops freeze on contact with solid objects. Glaze can form a thick coating of ice having sufficient weight to seriously damage trees and power lines
Rime	Variable	Solid	Deposits usually consisting of ice feathers that point into the wind. These delicate, frostlike accumulations form as supercooled cloud or fog droplets encounter objects and freeze on contact.
Snow	1 mm to 2 cm	Solid	The crystalline nature of snow allows it to assume many shapes including six-sided crystals, plates, and needles. Produced in supercooled clouds where water vapor is deposited as ice crystals that remain frozen during their descent.
Hail	5 mm to 10 cm	Solid	Precipitation in the form of hard, rounded pellets or irregular lumps of ice. Produced in large convective, cumulonimbus clouds, where frozen ice particles and supercooled water coexist.
Graupel	2 to 5 mm	Solid	Sometimes called soft hail, graupel forms when rime collects on snow crystals to produce irregular masses of "soft" ice. Because these particles are softer than hailstones, they normally flatten out upon impact.

moisture content, which makes them ideal for making snowballs.

STUDENTS SOMETIMES ASK...

*What is the record
snowfall in the U.S.?*

Although the lake effect snows of upstate New York are legendary, the greatest snowfalls usually occur in sparsely inhabited mountainous regions. The record for the largest snowfall from a single snowstorm in the United States goes to the Mt. Shasta Ski Bowl in California. This storm, which raged from February 13–19, 1959, dropped nearly 5 meters (16 feet) of snow!

Sleet and Glaze

Sleet is a wintertime phenomenon and refers to the fall of small particles of ice that are clear to translucent. For sleet to be produced, a layer of air with temperatures above freezing must overlie a subfreezing layer near the ground. When raindrops, which are often melted snow, leave the warmer air and encounter the colder air below, they freeze and reach the ground as small pellets of ice the size of the raindrops from which they formed.

On some occasions, when the vertical distribution of temperatures is similar to that associated with the formation of sleet, freezing rain or **glaze** results instead. In such situations, the subfreezing air near the ground is not thick enough to allow the raindrops to freeze. The raindrops, however, do become supercooled as they fall through the cold air and turn to ice upon colliding with solid objects. The result can be a thick coating of ice having sufficient weight to break tree limbs, down power lines, and make walking or driving extremely hazardous (Figure 16.26).

Hail

Hail is precipitation in the form of hard, rounded pellets or irregular lumps of ice. Moreover, large hailstones often consist of a series of nearly concentric shells of differing densities and degrees of opaqueness (Figure 16.27). Most hailstones have diameters between 1 centimeter (pea size) and 5 centimeters (golf ball–size), although some can be as big as an orange or larger.

Figure 16.26 Glaze forms when supercooled raindrops freeze on contact with objects. (Photo by Annie Griffiths Belt/DRK Photo)

Occasionally, hailstones weighing a pound or more have been reported. Many of these were probably composites of several stones frozen together.

The heaviest authenticated hailstone on record fell on Coffeyville, Kansas, September 3, 1970 (Figure 16.27). With a 14-centimeter (5.5-inch) diameter, this "giant"

Figure 16.27 A cross section of the Coffeyville hailstone. This largest recorded hailstone fell over Kansas in 1970 and weighed 766 grams (1.67 pounds). (University Corporation for Atmospheric Research/National Center for Atmospheric Research/National Science Foundation)

weighed 766 grams (1.67 pounds). It is estimated that this stone hit the ground at a speed in excess of 160 kilometers (100 miles) per hour.

The destructive effects of large hailstones are well known, especially to farmers whose crops have been devastated in a few minutes and to people whose windows and roofs have been damaged (Figure 16.28). In the United States, hail damage each year is in the hundreds of millions of dollars.

Hail is produced only in large cumulonimbus clouds where updrafts can sometimes reach speeds approaching 160 kilometers (100 miles) per hour, and where there is an abundant supply of supercooled water. Hailstones begin as small embryonic ice pellets that grow by collecting supercooled water droplets as they fall through the cloud. If they encounter a strong updraft, they may be carried upward again and begin the downward journey anew. Each trip through the supercooled portion of the cloud may be represented by an additional layer of ice. Hailstones can also form from a single descent through an updraft. Either way, the process continues until the hailstone encounters a downdraft or grows too heavy to remain suspended by the thunderstorm's updraft.

Rime

Rime is a deposit of ice crystals formed by the freezing of supercooled fog or cloud droplets on objects whose surface temperature is below freezing. When rime forms on trees, it adorns them with its characteristic ice feathers, which can be spectacular to behold (Figure 16.29). In these situations, objects such as pine needles act as freezing nuclei, causing the supercooled droplets to freeze on contact (see Box 16.2). On occasions when the wind is blowing, only the windward surfaces of objects will accumulate the layer of rime.

Measuring Precipitation

The most common form of precipitation—rain—is the easiest to measure. Any open container having a consistent cross section throughout can be a rain gauge. In general practice, however, more sophisticated devices are used so that small amounts of rainfall can be measured more accurately, and to reduce losses from evaporation. The *standard rain gauge* (Figure 16.30) has a diameter of about 20 centimeters (8 inches) at the top. Once the water is caught, a funnel conducts the rain into a cylindrical measuring tube that has a cross-sectional area only one-tenth as large as the receiver. Consequently, rainfall depth is magnified 10 times, which allows for accurate measurements to the nearest 0.025 centimeter (0.01 inch). The narrow opening also minimizes evaporation. When the amount of rain is less than 0.025 centimeter, it is reported as a *trace* of precipitation.

Figure 16.28 Hail damage to a used car lot in Fort Worth, Texas, May 6, 1995. This storm, which packed high winds and hail the size of baseballs, killed at least nine people and injured more than 100 as it swept through the northern part of the state. (Photo by Ron Heflin/AP Wide World Photos)

Measurement Errors

In addition to the standard rain gauge, several types of recording gauges are routinely used. These instruments record not only the amount of rain but also its time of occurrence and intensity (amount per unit of time).

No matter which rain gauge is used, proper exposure is critical. Errors arise when the gauge is shielded from obliquely falling rain by buildings, trees, or other high objects. Hence, the instrument should be at least as far away from such obstructions as the objects are high. Another cause of error is the wind. It has been shown that with increasing wind and turbulence, it becomes more difficult to collect a representative quantity of rain.

Measuring Snowfall

When snow records are kept, two measurements are normally taken: depth and water equivalent. Usually the depth of snow is measured with a calibrated stick. The actual measurement is simple, but choosing a *representative* spot often poses a dilemma. Even when winds are light or moderate, snow drifts freely. As a rule, it is best to take several measurements in an open place away from trees and obstructions and then aver-

Figure 16.29 Rime consists of delicate ice crystals that form when supercooled fog or cloud droplets freeze on contact with objects. (Photo by John Cancalosi/Stock Boston)

Figure 16.30 Precipitation measurement. The standard rain gauge allows for accurate rainfall measurement to the nearest 0.025 centimeter (0.01 inch). Because the cross-sectional area of the measuring tube is only one-tenth as large as the collector, rainfall is magnified 10 times.

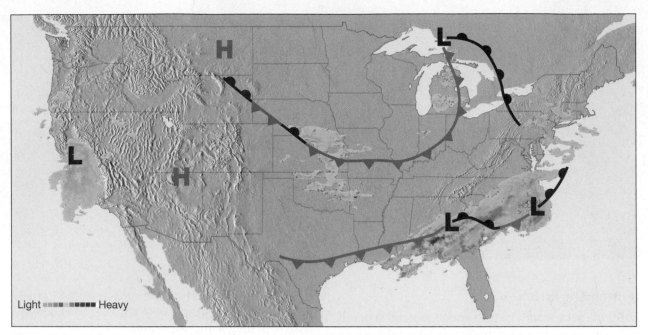

Figure 16.31   Color weather radar display, commonly seen on The Weather Channel and local TV weathercasts. Colors indicate different intensities of precipitation. Note the area of heavy precipitation in southeastern Alabama (red color). This map is a composite that uses data acquired from several radar installations.

age them. To obtain the water equivalent, samples can be melted and then weighed or measured as rain.

The quantity of water in a given volume of snow is not constant. A general ratio of 10 units of snow to 1 unit of water is often used when exact information is not available, but the actual water content of snow may deviate widely from this figure. It may take as much as 30 centimeters of light, fluffy dry snow or as little as 4 centimeters of wet snow to produce 1 centimeter of water.

Precipitation Measurement by Weather Radar

Today's TV weathercasts show helpful maps like the one in Figure 16.31 to depict precipitation patterns. The instrument that produces these images is the *weather radar.*

The development of radar has given meteorologists an important tool to probe storm systems that may be up to a few hundred kilometers away. All radar units have a transmitter that sends out short pulses of radio waves. The specific wavelengths that are used depend on the objects the user wants to detect. When radar is used to monitor precipitation, wavelengths between 3 and 10 centimeters are employed.

These wavelengths can penetrate small cloud droplets, but are reflected by larger raindrops, ice crystals, or hailstones. The reflected signal, called an *echo*, is received and displayed on a TV monitor. Because the echo is "brighter" when the precipitation is more intense, modern radar is able to depict not only the regional extent of the precipitation but also the rate of rainfall. Figure 16.31 is a typical radar display in which colors show precipitation intensity. Weather radar is also an important tool for determining the rate and direction of storm movement.

Chapter Summary

- *Water vapor,* an odorless, colorless gas, changes from one state of matter (solid, liquid, or gas) to another at the temperatures and pressures experienced near Earth's surface. The processes involved are *evaporation, condensation, melting, freezing, sublimation,* and *deposition.* During each change, *latent* (hidden) *heat* is either absorbed or released.

- *Humidity* is the general term to describe the amount of water vapor in the air. The methods used to express humidity quantitatively include (1) *mixing ratio,* the mass of water vapor in a unit of air compared to the remaining mass of dry air; (2) *vapor pressure,* that part of the total atmospheric pressure attributable to its water-vapor content; (3) *relative humidity,* the ratio of the air's actual water-vapor content compared with the amount of water vapor required for saturation at that temperature; and (4) *dew point,* that temperature to which a parcel of air would need to be cooled to reach saturation. When air is saturated, the pressure exerted by the water vapor, called the *saturation vapor pressure,* produces a balance between the number of water molecules leaving the surface of the water and the number returning. Because the saturation vapor pressure is temperature-dependent, at higher temperatures more water vapor is required for saturation to occur.

- *Relative humidity can be changed in two ways.* One is by *adding or subtracting water vapor.* The second is by *changing the air's temperature.* When air is cooled, its relative humidity increases.

- The cooling of air as it rises and expands owing to successively lower air pressure is the basic cloud-forming process. Temperature changes in air brought about by compressing or expanding the air are called *adiabatic temperature changes.* Unsaturated air warms by compression and cools by expansion at the rather constant rate of 10°C per 1000 meters of altitude change, a figure called the *dry adiabatic rate.* If air rises high enough, it will cool sufficiently to cause condensation and form a cloud. From this point on, air that continues to rise will cool at the *wet adiabatic rate,* which varies from 5°C to 9°C per 1000 meters of ascent. The difference in the wet and dry adiabatic rates is caused by the condensing water vapor releasing *latent heat,* thereby reducing the rate at which the air cools.

- Four mechanisms that can initiate the vertical movement of air are (1) *orographic lifting,* which occurs when elevated terrains, such as mountains, act as barriers to the flow of air; (2) *frontal wedging,* when cool air acts as a barrier over which warmer, less dense air rises; (3) *convergence,* which happens when air flows together and a general upward movement of air occurs; and (4) *localized convective lifting,* which occurs when unequal surface heating causes pockets of air to rise because of their buoyancy.

- The *stability of air* is determined by examining the temperature of the atmosphere at various altitudes. Air is said to be *unstable* when the *environmental lapse rate* (the rate of temperature decrease with increasing altitude in the troposphere) is greater than the *dry adiabatic rate.* Stated differently, a column of air is unstable when the air near the bottom is significantly warmer (less dense) than the air aloft.

- *For condensation to occur, air must be saturated.* Saturation takes place either when air is cooled to its dew point, which most commonly happens, or when water vapor is added to the air. There must also be a surface on which the water vapor can condense. In cloud and fog formation, tiny particles called *condensation nuclei* serve this purpose.

- *Clouds* are classified on the basis of their *appearance* and *height.* The three basic forms are *cirrus* (high, white, thin, wispy fibers), *cumulus* (globular, individual cloud masses), and *stratus* (sheets or layers that cover much or all of the sky). The four categories based on height are *high clouds* (bases normally above 6000 meters), *middle clouds* (from 2000 to 6000 meters), *low clouds* (below 2000 meters), and *clouds of vertical development.*

- *Fog* is defined as a cloud with its base at or very near the ground. Fogs form when air is cooled below its dew point or when enough water vapor is added to the air to bring about saturation. Various types of fog include *advection fog, radiation fog, upslope fog, steam fog,* and *frontal* (or *precipitation*) *fog.*

- For *precipitation* to form, millions of cloud droplets must somehow join together into large drops. Two mechanisms for the formation of precipitation have been proposed. (1) In clouds where the temperatures are below freezing, ice crystals form and fall as snowflakes. At lower altitudes the snowflakes melt and become raindrops before they reach the ground. (2) *Large* droplets form in warm clouds that contain large *hygroscopic* ("water-seeking") *nuclei,* such as salt particles. As these big droplets descend, they collide and join with smaller water droplets. After many collisions, the droplets are large enough to fall to the ground as rain.

- The forms of precipitation include *rain, snow, sleet, freezing rain (glaze), hail,* and *rime.*

Key Terms

absolute instability (p. 446)
absolute stability (p. 446)
adiabatic temperature change (p. 441)
advection fog (p. 452)
Bergeron process (p. 457)
calorie (p. 434)
cirrus (p. 450)
cloud (p. 450)
cloud of vertical development (p. 451)
collision-coalescence process (p. 458)
condensation (p. 435)
condensation nuclei (p. 449)
conditional instability (p. 448)
convergence (p. 443)
cumulus (p. 450)
deposition (p. 436)
dew-point temperature (p. 438)
dry adiabatic rate (p. 442)
evaporation (p. 435)
fog (p. 452)

freezing (p. 435)
freezing nuclei (p. 458)
frontal fog (p. 453)
front (p. 443)
frontal wedging (p. 443)
glaze (p. 461)
hail (p. 461)
high cloud (p. 451)
humidity (p. 436)
hygrometer (p. 440)
hygroscopic nuclei (p. 449)
latent heat (p. 435)
localized convective lifting (p. 445)
low cloud (p. 451)
melting (p. 435)
middle cloud (p. 451)
mixing ratio (p. 437)
orographic lifting (p. 442)
parcel (p. 441)
precipitation fog (p. 453)

psychrometer (p. 440)
radiation fog (p. 452)
rain (p. 460)
rain shadow desert (p. 443)
relative humidity (p. 437)
rime (p. 462)
saturation (p. 436)
sleet (p. 461)
snow (p. 460)
stable air (p. 445)
steam fog (p. 453)
stratus (p. 450)
sublimation (p. 435)
supercooled (p. 458)
supersaturation (p. 458)
unstable air (p. 446)
upslope fog (p. 452)
vapor pressure (p. 436)
wet adiabatic rate (p. 442)

Review Questions

1. Summarize the processes by which water changes from one state of matter to another. Indicate whether heat energy is absorbed or liberated.

2. After studying Table 16.1, write a generalization relating temperature and the amount of water vapor needed to saturate the air.

3. How do relative humidity and mixing ratio differ?

4. Referring to Figure 16.6, answer the following questions and then write a generalization relating changes in air temperature to changes in relative humidity.
   (a) During a typical day, when is the relative humidity highest? Lowest?
   (b) At what time of day would dew most likely form?

5. If the temperature remains unchanged and the mixing ratio decreases, how will relative humidity change?

6. On a cold winter day when the temperature is −10°C and the relative humidity is 50 percent, what is the mixing ratio (refer to Table 16.1)? What is the mixing ratio for a day when the temperature is 20°C and the relative humidity is 50 percent?

7. Explain the principle of the sling psychrometer and the hair hygrometer.

8. Using the standard tables in Appendix D, determine the relative humidity and dew-point temperature if the dry-bulb thermometer reads 16°C and the wet-bulb thermometer reads 12°C. How would the relative humidity and dew point change if the wet-bulb thermometer read 8°C?

9. On a warm summer day when the relative humidity is high, it may seem even warmer than the thermometer indicates. Why do we feel so uncomfortable on a muggy day?

10. Why does air cool when it rises through the atmosphere?

11. Explain the difference between environmental lapse rate and adiabatic cooling.

12. If unsaturated air at 23°C were to rise, what would its temperature be at 500 meters? If the dew-point temperature at the condensation level were 13°C, at what altitude would clouds begin to form?

13. Why does the adiabatic rate of cooling change when condensation begins? Why is the wet adiabatic rate not a constant figure?

14. The contents of an aerosol can are under very high pressure. When you push the nozzle on such a can, the spray feels cold. Explain.

15. How do orographic lifting and frontal wedging act to force air to rise?

16. Explain why the Great Basin area of the western United States is so dry. What term is applied to such a situation?

17. How does stable air differ from unstable air? Describe the general nature of the clouds and precipitation expected with each.

18. How do temperature inversions influence air pollution? (See Box 16.1.)

19. What is the function of condensation nuclei in cloud formation? The function of the dew point?

20. As you drink an ice-cold beverage on a warm day, the outside of the glass or bottle becomes wet. Explain.

21. What is the basis for the classification of clouds?

22. Why are high clouds always thin?

23. Which cloud types are associated with the following characteristics: thunder, halos, precipitation, hail, mackerel sky, lightning, mares' tails?

24. List five types of fog and discuss the details of their formation.

25. What is the difference between precipitation and condensation?

26. List the forms of precipitation and the circumstances of their formation.

27. Sometimes, when rainfall is light, the amount is reported as a trace. When this occurs, how much (or little) rain has fallen?

Examining the Earth System

1. The interrelationships among Earth's spheres have produced the Great Basin area of the western United States, which includes some of the driest areas in the world. Examine the map of the region in Figure 9.20. Although the area is only a few hundred miles from the Pacific Ocean, it is a desert. Why? Did any geologic factor(s) contribute to the formation of this desert environment? Do any major rivers have their source in the Great Basin? Explain. (For information about deserts in the United States, visit the United States Geological Survey (USGS) Deserts: Geology and Resources Website at **http://pubs.usgs.gov/gip/deserts/**)

2. Phoenix and Flagstaff, Arizona, are located nearby each other in the southwestern United States. Using the climate diagrams for the two cities found in Figure 19.13, describe the impact that elevation has on the precipitation and temperature of each city. Compare the natural vegetation of these nearby locations. (To examine the current weather conditions and forecasts for Phoenix and Flagstaff, Arizona, contact *USA Today* at **http://www.usatoday.com/weather/** and/or *The Weather Channel* at **http://www.weather.com/twc/homepage.twc**)

3. The amount of precipitation that falls at any particular place and time is controlled by the quantity of moisture in the air and many other factors. How might each of the following alter the precipitation at a particular locale? (1) An increase in the elevation of the land. (b) A decrease in the area covered by forests and other types of vegetation. (c) Lowering of average ocean-surface temperatures. (d) An increase in the percentage of time that the winds blow from an adjacent body of water. (e) A major episode of global volcanism lasting for decades.

Web Resources

The *Earth Science* Website uses the resources and flexibility of the Internet to aid in your study of the topics in this chapter. Written and developed by Earth science instructors, this site will help improve your understanding of Earth science. Visit **http://www.prenhall.com/tarbuck** and click on the cover of *Earth Science 10e* to find:

- **On-line review quizzes.**
- **Web-based critical thinking and writing exercises.**
- **Links to chapter-specific Web resources.**
- **Internet-wide key term searches.**

http://www.prenhall.com/tarbuck

Sailboats in a Norwegian fiord. (Photo by Imagebank/Getty Images)

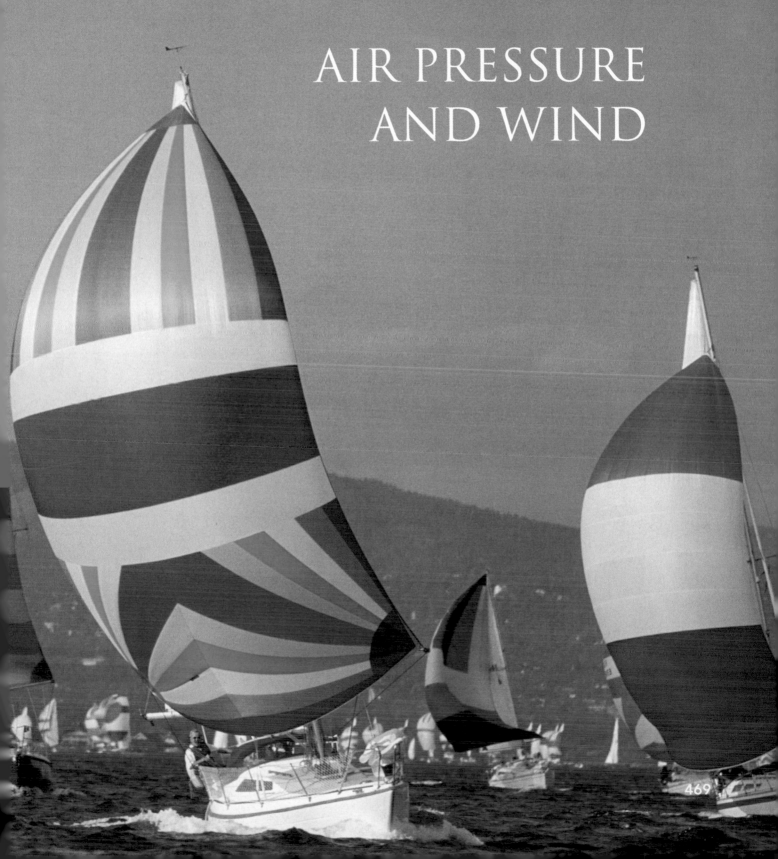

CHAPTER 17

AIR PRESSURE
AND WIND

469

Of the various elements of weather and climate, changes in air pressure are the least noticeable. In listening to a weather report, generally we are interested in moisture conditions (humidity and precipitation), temperature, and perhaps wind. It is the rare person, however, who wonders about air pressure. Although the hour-to-hour and day-to-day variations in air pressure are not perceptible to human beings, they are very important in producing changes in our weather. For example, it is variations in air pressure from place to place that generate winds that in turn can bring changes in temperature and humidity (Figure 17.1). Air pressure is one of the basic weather elements and is a significant factor in weather forecasting. As you will see, air pressure is closely tied to the other elements of weather in a cause-and-effect relationship.

Understanding Air Pressure

In Chapter 15 we noted that **air pressure** is simply the pressure exerted by the weight of air above. Average air pressure at sea level is about 1 kilogram per square centimeter, or 14.7 pounds per square inch. This is roughly the same pressure that is produced by a column of water 10 meters (33 feet) in height. With some simple arithmetic you can calculate that the air pressure exerted on the top of a small (50 centimeter by 100 centimeter) school desk exceeds 5000 kilograms (11,000 pounds), or about the weight of a 50-passenger school bus. Why doesn't the desk collapse under the weight of the ocean of air above? Simply, air pressure is exerted in all directions—down, up, and sideways. Thus, the air pressure pushing down on the desk exactly balances the air pressure pushing up on the desk.

You might be able to visualize this phenomenon better if you imagine a tall aquarium that has the same dimensions as the desktop. When this aquarium is filled to a height of 10 meters (33 feet), the water pressure at the bottom equals 1 atmosphere (14.7 pounds per square inch). Now, imagine what will happen if this aquarium is placed on top of our student desk so that all the force is directed downward. Compare this to what results when the desk is placed inside the aquarium and allowed to sink to the bottom. In the latter situation the desk survives because the water pressure is exerted in all directions, not just downward as

Figure 17.1   Palm trees buffeted by hurricane-force winds, Corpus Christi, Texas. (Photo by *National Geographic*/Getty Images)

in our earlier example. The desk, like your body, is "built" to withstand the pressure of 1 atmosphere. It is important to note that although we do not generally notice the pressure exerted by the ocean of air around us, except when ascending or descending in an elevator or airplane, it is nonetheless substantial. The pressurized suits used by astronauts on space walks are designed to duplicate the atmospheric pressure experienced at Earth's surface. Without these protective suits to keep body fluids from boiling away, astronauts would perish in minutes.

Measuring Air Pressure

 Atmosphere
↳ Air Pressure and Wind

When meteorologists measure atmospheric pressure, they employ a unit called the *millibar*. Standard sea-level pressure is 1013.2 millibars. Although millibars are used almost exclusively by meteorologists, you might be better acquainted with the expression "inches of mercury," which is used by the media to describe atmospheric pressure.

This expression dates from 1643, when Torricelli, a student of the famous Italian scientist Galileo, invented the **mercury barometer** (*bar* = pressure, *metron* = measuring intrument). Torricelli correctly described the atmosphere as a vast ocean of air that exerts pressure on us and all objects about us. To measure this force, he filled a glass tube, which was closed at one end, with mercury. He then inverted the tube into a dish of mercury (Figure 17.2A). Torricelli found that the mercury flowed out of the tube until the weight of the column was balanced by the pressure that the atmosphere exerted on the surface of the mercury in the dish. In other words, the weight of mercury in the column equaled the weight of the same diameter column of air that extended from the ground to the top of the atmosphere.

When air pressure increases, the mercury in the tube rises. Conversely, when air pressure decreases, so does the height of the mercury column. With some refinements the mercurial barometer invented by Torricelli is still the standard pressure-measuring instrument used today. Standard atmospheric pressure at sea level equals 29.92 inches of mercury.

The need for a smaller and more portable instrument for measuring air pressure led to the development of the **aneroid** (*an* = without, *ner* = fluid) **barometer.** Instead of having a mercury column held up by air pressure, the aneroid barometer uses a partially evacuated metal chamber. The chamber, being very sensitive to variations in air pressure, changes shape, compressing

Figure 17.2   A. Simple mercury barometer. The weight of the column of mercury is balanced by the pressure exerted on the dish of mercury by the air above. If the pressure decreases, the column of mercury falls; if the pressure increases, the column rises. B. An aneroid barograph makes a continuous record of pressure changes. One important advantage of the aneroid barometer is that it is easily adapted to a recording mechanism.

as the pressure increases and expanding as the pressure decreases. An advantage of the aneroid barometer is that it can be easily connected to a recording mechanism. The resulting instrument is a **barograph,** which provides a continuous record of pressure changes with the passage of time (Figure 17.2B).

STUDENTS SOMETIMES ASK...

*What is the lowest barometric
pressure ever recorded?*

All of the lowest recorded barometric pressures have been associated with strong hurricanes. The record for the United States is 888 millibars (26.20 inches) measured during Hurricane Gilbert in September 1988. The world's record, 870 millibars (25.70 inches), occurred during Typhoon Tip (a Pacific hurricane), in October 1979. Although tornadoes undoubtedly have produced even lower pressures, none have been accurately measured.

Factors Affecting Wind

Atmosphere
↳ Air Pressure and Wind

In Chapter 16 we examined the upward movement of air and its role in cloud formation. As important as vertical motion is, far more air moves horizontally, the phenomenon we call **wind.** What causes wind?

Simply stated, wind is the result of horizontal differences in air pressure. *Air flows from areas of higher pressure to areas of lower pressure.* You may have experienced this when opening a vacuum-packed can of coffee. The noise you hear is caused by air rushing from the higher pressure outside the can to the lower pressure inside. Wind is nature's attempt to balance such inequalities in air pressure. Because unequal heating of Earth's surface generates these pressure differences, *solar radiation is the ultimate energy source for most wind.*

If Earth did not rotate, and if there were no friction between moving air and Earth's surface, air would flow in a straight line from areas of higher pressure to areas of lower pressure. But because both factors exist, wind is controlled by a combination of forces, including: (1) the pressure-gradient force, (2) Coriolis effect, and (3) friction. We will now examine each of these factors.

Pressure-Gradient Force

Pressure differences create wind, and the greater these differences, the greater the wind speed. Over Earth's surface, variations in air pressure are determined from barometric readings taken at hundreds of weather stations. These pressure data are shown on a weather map using **isobars,** lines that connect places of equal air pressure (Figure 17.3). The spacing of isobars indicates the amount of pressure change occurring over a given

ff	Miles per hour
◎	Calm
	1–2
	3–8
	9–14
	15–20
	21–25
	26–31
	32–37
	38–43
	44–49
	50–54
	55–60
	61–66
	67–71
	72–77
	78–83
	84–89
	119–123

Figure 17.3 Isobars are lines connecting places of equal sea-level pressure. They are used to show the distribution of pressure on daily weather maps. Isobars are seldom straight, but usually form broad curves. Concentric rings of isobars indicate cells of high and low pressure. The "wind flags" indicate the expected airflow surrounding pressure cells and are plotted as "flying" with the wind (that is, the wind blows toward the station circle). Notice on this map that the isobars are more closely spaced and the wind speed is faster around the low-pressure center than around the high.

distance and is expressed as the **pressure gradient** (*gradus* = slope).

You might find it easier to visualize a pressure gradient if you think of it as being similar to the slope of a hill. A steep pressure gradient, like a steep hill, causes greater acceleration of an air parcel than does a weak pressure gradient (a gentle hill). Thus, the relationship between wind speed and the pressure gradient is straightforward: *Closely spaced isobars indicate a steep pressure gradient and high winds, whereas widely spaced isobars indicate a weak pressure gradient and light winds.* Figure 17.3 illustrates the relationship between the spacing of isobars and wind speed. Notice that wind speeds are greater in Ohio, Kentucky, Michigan, and Illinois, where isobars are more closely spaced, than in the western states where isobars are more widely spaced (see Box 17.1).

The pressure gradient is the driving force of wind, and it has both magnitude and direction. Its magnitude is determined from the spacing of isobars. The direction of force is always from areas of higher pressure to areas of lower pressure and at right angles to the isobars. Once the air starts to move, the Coriolis effect and friction come into play, but then only to modify the movement, not to produce it.

Coriolis Effect

The weather map in Figure 17.3 shows the typical air movements associated with high- and low-pressure systems. As expected, the air moves out of the regions of higher pressure and into the regions of lower pressure. However, the wind does not cross the isobars at right angles as the pressure-gradient force directs it. This deviation is the result of Earth's rotation and has been named the **Coriolis effect** after the French scientist who first thoroughly described it.

All free-moving objects or fluids, including the wind, are deflected to the *right* of their path of motion in the Northern Hemisphere and to the *left* in the Southern Hemisphere. The reason for this deflection can be illustrated by imagining the path of a rocket launched from the North Pole toward a target located on the equator (Figure 17.4). If the rocket took an hour to reach its target, Earth would have rotated 15 degrees to the east during its flight. To someone standing on Earth it would look as if the rocket veered off its path and hit Earth 15 degrees west of its target. The true path of the rocket is straight and would appear so to someone out in space looking down at Earth. It was Earth turning under the rocket that gave it its *apparent* deflection.

Note that the rocket was deflected to the right of its path of motion because of the counterclockwise rotation of the Northern Hemisphere. In the Southern Hemisphere, the effect is reversed. Clockwise rotation produces a similar deflection, but to the left of the path

of motion. The same deflection is experienced by wind regardless of the direction it is moving.

We attribute the apparent shift in wind direction to the Coriolis effect. This deflection (1) is always directed at right angles to the direction of airflow; (2) affects only wind direction, not wind speed; (3) is affected by wind speed (the stronger the wind, the greater the deflection); and (4) is strongest at the poles and weakens equatorward, becoming nonexistent at the equator.

It is of interest to point out that any "free-moving" object will experience a deflection caused by the Coriolis effect. This fact was dramatically discovered by the United States navy in World War II. During target practice long-range guns on battleships continually missed their targets by as much as several hundred yards until ballistic corrections were made for the changing position of a seemingly stationary target. Over a short distance, however, the Coriolis effect is relatively small.

Surface Winds versus Winds Aloft

The effect of friction on wind is important only within a few kilometers of Earth's surface. Friction acts to slow air movement and, as a consequence, alters wind direction. To illustrate friction's effect on wind direction, let us look at a situation in which it has no role. Above the friction layer, the pressure-gradient force and Coriolis effect work together to direct the flow of air. Under these conditions, the pressure-gradient force causes air to start moving across the isobars. As soon as the air starts to move, the Coriolis effect acts at right angles to this motion. The faster the wind speed, the greater the deflection.

Eventually, the Coriolis effect will balance the pressure-gradient force, and the wind will blow parallel to the isobars (Figure 17.5). Upper-air winds generally take this path and are called **geostrophic winds.** Because of the lack of friction with Earth's surface, geostrophic winds travel at higher speeds than do surface winds. This can be observed in Figure 17.6 by noting the wind flags, many of which indicate winds of 50 to 100 miles per hour.

The most prominent features of upper-level flow are the **jet streams.** First encountered by high-flying bombers during World War II, these fast-moving rivers of air travel between 120 and 240 kilometers (75 and 150 miles) per hour in a west-to-east direction. One such stream is situated over the polar front, which is the zone separating cool polar air from warm subtropical air.

Below 600 meters (2000 feet), friction complicates the airflow just described. Recall that the Coriolis effect is proportional to wind speed. Friction lowers the wind speed, so it reduces the Coriolis effect. Because the pressure-gradient force is not affected by wind speed, it wins the tug of war shown in Figure 17.7. The result

PEOPLE AND THE ENVIRONMENT:
Wind Energy

Approximately 0.25 percent (one-quarter of 1 percent) of the solar energy that reaches the lower atmosphere is transformed into wind. Although it is just a minuscule percentage, the absolute amount of energy is enormous. According to one estimate, if the winds of North and South Dakota could be harnessed, they would provide 80 percent of the electrical energy used in the United States.

Wind has been used for centuries as an almost free and nonpolluting source of energy. Sailing ships and wind-powered grist mills represent two of the early ways that this renewable resource was harnessed. Further, as rural America was settled, there was a strong reliance on wind power to pump water and later to generate small amounts of electricity.

Following the "energy crisis" that was brought about by an oil embargo in the 1970s, interest in wind power as well as other alternative forms of energy increased dramatically. In 1980 the federal government initiated a program to develop wind-power systems. A project sponsored by the U.S. Department of Energy involved setting up experimental wind farms in mountain passes known to have strong persistent winds (Figure 17.A). One of these facilities, located at Altamont Pass near San Francisco, now employs more than 7000 wind turbines.

In September 1998, an array of 143 20-story-high wind turbines went on line near Lake Benton in southwestern Minnesota. The 107-megawatt project was one of the largest wind farms in the world. Less than a year later, an additional 103 megawatts of wind-generated electrical power became available. When the wind is blowing, the Lake Benton farm can power nearly 80,000 homes. When the project is completed, the facility's output will be 425 megawatts, nearly four times the 1998 output.

As technology has improved, efficiency has increased and the costs of wind-generated electricity have become more competitive. Between 1983 and 2001 technological advances cut the cost of wind power by more than 85 percent. As a result, the growth of installed capacity in the United States (and elsewhere) has grown dramatically (Figure 17.B). Some estimate that in the next 50 to 60 years wind power could meet between 5 and 10 percent of the country's demand for electrical energy. One

Figure 17.B U.S.-installed wind-power capacity (in megawatts). Growth in recent years has been dramatic. (Data from U.S. Department of Energy and American Wind Energy Association)

area for the expansion of wind energy will likely be islands and other regions far from electrical grids that must import fuel for generating power.

Although the future for wind power appears promising, it is not without difficulties. In addition to technical advances that must continue to be made, noise pollution and the costs of large tracts of land in populated areas present significant obstacles to development. In the Netherlands, where windmills have a long history, proposals to locate wind turbines atop the country's dike system have been met with strong opposition. Part of the problem is that the Netherlands, like much of Europe, is densely populated and, hence, has few remote locations.

Another significant limitation of wind energy is that it is intermittent. If wind were to constitute a high proportion of the total energy supply, any shortfall could cause severe economic disruptions. Better means of storage would allow wind to contribute a significantly higher percentage of our energy needs. One proposal would use wind energy to produce hydrogen by electrolysis of water. This combustible gas would then be distributed and stored in a manner similar to natural gas.

Figure 17.A These wind turbines are operating near Palm Springs, California. (Photo by John Mead/Science Photo Library/Photo Researchers, Inc.)

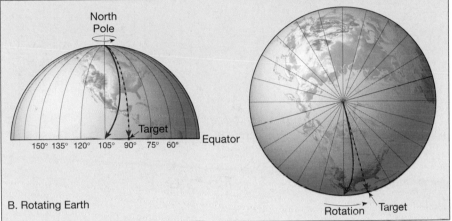

Figure 17.4 The Coriolis effect illustrated using a 1-hour flight of a rocket traveling from the North Pole to a location on the equator. **A.** On a nonrotating Earth, the rocket would travel straight to its target. **B.** However, Earth rotates 15° each hour. Thus, although the rocket travels in a straight line, when we plot the path of the rocket on Earth's surface, it follows a curved path that veers to the right of the target.

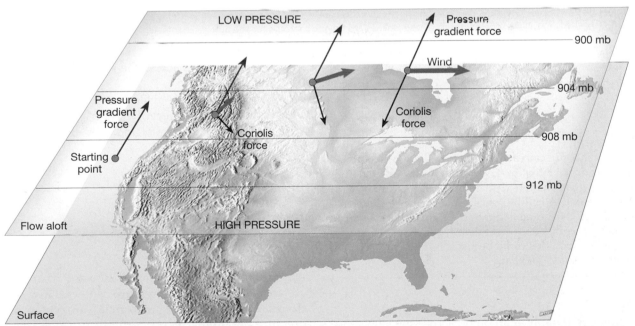

Figure 17.5 The geostrophic wind. The only force acting on a stationary parcel of air is the pressure-gradient force. Once the air begins to accelerate, the Coriolis effect deflects it to the right in the Northern Hemisphere. Greater wind speeds result in a stronger Coriolis effect (deflection) until the flow is parallel to the isobars. At this point the pressure-gradient force and Coriolis effect are in balance and the flow is called a *geostrophic wind*. It is important to note that in the "real" atmosphere, airflow is continually adjusting for variations in the pressure field. As a result, the adjustment to geostrophic equilibrium is much more irregular than shown.

A. Upper-level weather chart

ff	Miles per hour
◎	Calm
—	1–2
⌐	3–8
⌐	9–14
⌐	15–20
⌐	21–25
⌐	26–31
⌐	32–37
⌐	38–43
⌐	44–49
⌐	50–54
▲	55–60
▲	61–66
▲	67–71
▲	72–77
▲	78–83
▲	84–89
▲	119–123

B. Representation of upper-level chart

Figure 17.6 Upper-air winds. This map shows the direction and speed for the upper-air wind for a particular day. Note that the airflow is nearly parallel to the contours. These isolines are height contours for the 500-millibar level.

is a movement of air at an angle across the isobars toward the area of lower pressure.

The roughness of the terrain determines the angle of airflow across the isobars. Over the smooth ocean surface, friction is low and the angle is small. Over rugged terrain, where friction is higher, the angle that air makes as it flows across the isobars can be as great as 45 degrees.

In summary, upper airflow is nearly parallel to the isobars, whereas the effect of friction causes the surface winds to move more slowly and cross the isobars at an angle.

Highs and Lows

GEODe Atmosphere
↳ Air Pressure and Winds

Among the most common features on any weather map are areas designated as pressure centers. **Lows,** or **cyclones** (*kyklon* = moving in a circle), are centers of low pressure, and **highs,** or **anticyclones,** are high-pressure centers. As Figure 17.8 illustrates, the pressure decreases from the outer isobars toward the center in a low. In a high, just the opposite is the case—the values of the isobars increase from the outside toward the center. By knowing just a few basic facts about centers of high and

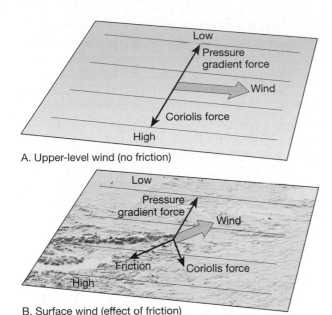

A. Upper-level wind (no friction)

B. Surface wind (effect of friction)

Figure 17.7   Comparison between upper-level winds and surface winds showing the effects of friction on airflow. Friction slows surface wind speed, which weakens the Coriolis effect, causing the winds to cross the isobars and move toward the lower pressure.

low pressure, you can greatly increase your understanding of current and forthcoming weather.

Cyclonic and Anticyclonic Winds

From the preceding section, you learned that the two most significant factors that affect wind are the pressure-gradient force and the Coriolis effect. Winds move from higher pressure to lower pressure and are deflected to the right or left by Earth's rotation. When these controls of airflow are applied to pressure centers in the Northern Hemisphere, the result is that winds blow inward and counterclockwise around a low. Around a high, they blow outward and clockwise (Figure 17.8).

In the Southern Hemisphere the Coriolis effect deflects the winds to the left, and therefore winds around a low blow clockwise, and winds around a high move counterclockwise. In either hemisphere, friction causes a net inflow (**convergence**) around a cyclone and a net outflow (**divergence**) around an anticyclone.

Weather Generalizations About Highs and Lows

Rising air is associated with cloud formation and precipitation, whereas subsidence produces clear skies. In this section you will learn how the movement of air can itself create pressure change and hence generate winds. In addition, you will examine the relationship between horizontal and vertical flow and its affect on the weather.

Let us first consider the situation around a surface low-pressure system where the air is spiraling inward. Here the net inward transport of air causes a shrinking of the area occupied by the air mass, a process that is termed *horizontal convergence*. Whenever air converges horizontally, it must pile up, that is, increase in height to allow for the decreased area it now occupies. This generates a taller and therefore heavier air column. Yet a surface low can exist only as long as the column of air above exerts less pressure than that occurring in surrounding regions. We seem to have encountered a paradox—a low-pressure center causes a net accumulation of air, which increases its pressure. Consequently, a surface cyclone should quickly eradicate itself in a manner

Figure 17.8   Cyclonic and anticyclonic winds in the Northern Hemisphere. Arrows show that winds blow into and counterclockwise around a low. By contrast, around a high, winds blow outward and clockwise.

not unlike what happens when a vacuum-packed can is opened.

You can see that for a surface low to exist for very long, compensation must occur aloft. For example, surface convergence could be maintained if divergence (spreading out) aloft occurred at a rate equal to the inflow below. Figure 17.9 shows the relationship between surface convergence (inflow) and divergence (outflow) aloft that is needed to maintain a low-pressure center.

Note that surface convergence about a cyclone causes a net upward movement. The rate of this vertical movement is slow, generally less than 1 kilometer per day. Nevertheless, because rising air often results in cloud formation and precipitation, a low-pressure center is generally related to unstable conditions and stormy weather (Figure 17.10A).

As often as not, it is divergence aloft that creates a surface low. Spreading out aloft initiates upflow in the atmosphere directly below, eventually working its way to the surface, where inflow is encouraged.

Like their cyclonic counterparts, anticyclones must be maintained from above. Outflow near the surface is accompanied by convergence aloft and general subsidence of the air column (Figure 17.9). Because descending air is compressed and warmed, cloud formation and precipitation are unlikely in an anticyclone. Thus, "fair" weather can usually be expected with the approach of a high (Figure 17.10B).

For reasons that should now be obvious, it has been common practice to print on household barometers the words "stormy" at the low-pressure end and "fair" on the high-pressure end. By noting whether the pressure is rising, falling, or steady, we have a good indication of what the forthcoming weather will be. Such a determination, called the **pressure,** or **barometric tendency,** is a very useful aid in short-range weather prediction.

In conclusion, you should now be better able to understand why television weather reporters emphasize the locations and projected paths of cyclones and anticyclones. The villain on these weather programs is always the low-pressure center, which produces "bad" weather in any season. Lows move in roughly a west-to-east direction across the United States and require a few days to more than a week for the journey. Their paths can be somewhat erratic; thus, accurate prediction of their movement is difficult, although essential, for short-range forecasting.

Meteorologists must also determine if the flow aloft will intensify an embryo storm or act to suppress its development. Because of the close tie between conditions at the surface and those aloft, a great deal of emphasis has been placed on the importance and understanding of the total atmospheric circulation, particularly in the midlatitudes. We will now examine the workings of Earth's general atmospheric circulation, and then again consider the structure of the cyclone in light of this knowledge.

General Circulation of the Atmosphere

As noted, the underlying cause of wind is unequal heating of Earth's surface. In tropical regions, more solar radiation is received than is radiated back to space. In polar regions the opposite is true: Less solar energy is

Figure 17.9 Airflow associated with surface cyclones and anticyclones. A low, or cyclone, has converging surface winds and rising air causing cloudy conditions. A high, or anticyclone, has diverging surface winds and descending air, which leads to clear skies and fair weather.

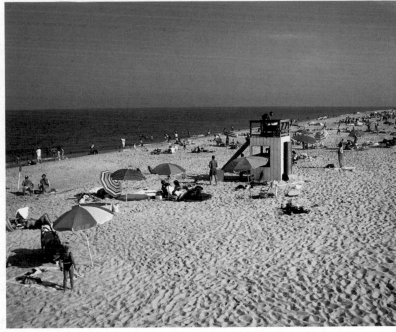

A.   B.

Figure 17.10   These two photographs illustrate the basic weather generalizations associated with pressure centers. **A.** A sea of umbrellas on a rainy day in Shanghai, China. Centers of low pressure are frequently associated with cloudy conditions and precipitation. (Photo by Getty Images) **B.** By contrast, clear skies and "fair" weather may be expected when an area is under the influence of high pressure. Sunbathers on a beach at Cape Henlopen, Delaware. (Photo by Mark Gibson/DRK Photo)

received than is lost. Attempting to balance these differences, the atmosphere acts as a giant heat-transfer system, moving warm air poleward and cool air equatorward. On a smaller scale, but for the same reason, ocean currents also contribute to this global heat transfer. The general circulation is very complex, and there is a great deal that has yet to be explained. We can, however, develop a general understanding by first considering the circulation that would occur on a nonrotating Earth having a uniform surface. We will then modify this system to fit observed patterns.

Circulation on a Nonrotating Earth

On a hypothetical nonrotating planet with a smooth surface of either all land or all water, two large thermally produced cells would form (Figure 17.11). The heated equatorial air would rise until it reached the tropopause, which, acting like a lid, would deflect the air poleward. Eventually, this upper-level airflow would reach the poles, sink, spread out in all directions at the surface, and move back toward the equator. Once there, it would be reheated and start its journey over again. This hypothetical circulation system has upper-level air flowing poleward and surface air flowing equatorward.

If we add the effect of rotation, this simple convection system will break down into smaller cells. Figure 17.12 illustrates the three pairs of cells proposed to

carry on the task of heat redistribution on a rotating planet. The polar and tropical cells retain the characteristics of the thermally generated convection described earlier. The nature of the midlatitude circulation is more complex and will be discussed in more detail in a later section.

Figure 17.11   Global circulation on a nonrotating Earth. A simple convection system is produced by unequal heating of the atmosphere.

Figure 17.12   Idealized global circulation proposed for the three-cell circulation model of a rotating Earth.

Idealized Global Circulation

Near the equator, the rising air is associated with the pressure zone known as the **equatorial low**—a region marked by abundant precipitation. As the upper-level flow from the equatorial low reaches 20 to 30 degrees latitude, north or south, it sinks back toward the surface. This subsidence and associated adiabatic heating produce hot, arid conditions. The center of this zone of subsiding dry air is the **subtropical high,** which encircles the globe near 30 degrees latitude, north and south (Figure 17.12). The great deserts of Australia, Arabia, and North Africa exist because of the stable dry conditions associated with the subtropical highs.

At the surface, airflow is outward from the center of the subtropical high. Some of the air travels equatorward and is deflected by the Coriolis effect, producing the reliable **trade winds.** The remainder travels poleward and is also deflected, generating the prevailing **westerlies** of the midlatitudes. As the westerlies move poleward, they encounter the cool **polar easterlies** in the region of the **subpolar low.** The interaction of these warm and cool winds produces the stormy belt known as the **polar front.** The source region for the variable polar easterlies is the **polar high.** Here, cold polar air is subsiding and spreading equatorward.

In summary, this simplified global circulation is dominated by four pressure zones. The subtropical

and polar highs are areas of dry subsiding air that flows outward at the surface, producing the prevailing winds. The low-pressure zones of the equatorial and subpolar regions are associated with inward and upward airflow accompanied by clouds and precipitation.

Influence of Continents

Up to this point, we have described the surface pressure and associated winds as continuous belts around Earth. However, the only truly continuous pressure belt is the subpolar low in the Southern Hemisphere. Here the ocean is uninterrupted by landmasses. At other latitudes, particularly in the Northern Hemisphere where landmasses break up the ocean surface, large seasonal temperature differences disrupt the pattern. Figure 17.13 shows the resulting pressure and wind patterns for January and July. The circulation over the oceans is dominated by semipermanent cells of high pressure in the subtropics and cells of low pressure over the subpolar regions. The subtropical highs are responsible for the trade winds and westerlies, as mentioned earlier.

The large landmasses, on the other hand, particularly Asia, become cold in the winter and develop a seasonal high-pressure system from which surface flow is directed off the land (Figure 17.13A). In the summer, the opposite occurs; the landmasses are heated and develop low-pressure cells, which permit air

Figure 17.13   Average surface pressure and associated global circulation for **A.** January and **B.** July.

to flow onto the land (Figure 17.13B). These seasonal changes in wind direction are known as the **monsoons.** During warm months, areas such as India experience a flow of warm, water-laden air from the Indian Ocean, which produces the rainy summer monsoon. The winter monsoon is dominated by dry continental air. A similar situation exists, but to a lesser extent, over North America.

In summary, the general circulation is produced by semipermanent cells of high and low pressure over the oceans and is complicated by seasonal pressure changes over land.

No. Regions that experience monsoons typically have both a wet and a dry season. Monsoon refers to a wind system that exhibits a pronounced season reversal in direction. In general, winter is associated with winds that blow predominantly off the continents and produce a dry winter monsoon. By contrast, in summer, warm moisture-laden air blows from the sea toward the land. Thus, the summer monsoon, which is usually associated with abundant precipitation, is the source of the misconception.

The Westerlies

The circulation in the midlatitudes, the zone of the westerlies, is complex and does not fit the convection system proposed for the tropics. Between about 30 and 60 degrees latitude, the general west-to-east flow is interrupted by migrating cyclones and anticyclones. In the Northern Hemisphere these cells move from west to east around the globe, creating an anticyclonic (clockwise) flow or a cyclonic (counterclockwise) flow in their area of influence. A close correlation exists between the paths taken by these surface pressure systems and the position of the upper-level airflow, indicating that the upper air strongly influences the movement of cyclonic and anticyclonic systems.

Among the most obvious features of the flow aloft are the seasonal changes. The steep temperature gradient across the middle latitudes in the winter months corresponds to a stronger flow aloft. In addition, the polar jet stream fluctuates seasonally such that its average position migrates southward with the approach of winter

and northward as summer nears. By midwinter, the jet core may penetrate as far south as central Florida.

Because the paths of low-pressure centers are guided by the flow aloft, we can expect the southern tier of states to experience more of their stormy weather in the winter season. During the hot summer months, the storm track is across the northern states, and some cyclones never leave Canada. The northerly storm track associated with summer applies also to Pacific storms, which move toward Alaska during the warm months, thus producing an extended dry season for much of our western coast. The number of cyclones generated is seasonal as well, with the largest number occurring in the cooler months when the temperature gradients are greatest. This fact is in agreement with the role of cyclonic storms in the distribution of heat across the midlatitudes.

Local Winds

Having examined Earth's large-scale circulation, let us turn briefly to winds that influence much smaller areas. Remember that all winds are produced for the same reason: pressure differences that arise because of temperature differences that are caused by unequal heating of Earth's surface. *Local winds* are simply small-scale winds produced by a locally generated pressure gradient. Those described here are caused either by topographic effects or variations in surface composition in the immediate area.

Land and Sea Breezes

In coastal areas during the warm summer months, the land surface is heated more intensely during the daylight hours than is the adjacent body of water (see the section on Land and Water in Chapter 15). As a result,

A. Sea breeze

B. Land breeze

Figure 17.14 Illustration of a sea breeze and a land breeze. **A.** During the daylight hours the air above the land heats and expands, creating an area of lower pressure. Cooler and denser air over the water moves onto the land, generating a sea breeze. **B.** At night the land cools more rapidly than the sea, generating an offshore flow called a *land breeze*.

A. Valley breeze

B. Mountain breeze

Figure 17.15 Valley and mountain breezes. **A.** Heating during the daylight hours warms the air along the mountain slopes. This warm air rises, generating a valley breeze. **B.** After sunset, cooling of the air near the mountain can result in cool air drainage into the valley, producing the mountain breeze.

the air above the land surface heats, expands, and rises, creating an area of lower pressure. A **sea breeze** then develops, because cooler air over the water (higher pressure) moves toward the warmer land (lower pressure) (Figure 17.14A). The sea breeze begins to develop shortly before noon and generally reaches its greatest intensity during the mid- to late afternoon. These relatively cool winds can be a significant moderating influence on afternoon temperatures in coastal areas.

At night, the reverse may take place. The land cools more rapidly than the sea, and the **land breeze** develops (Figure 17.14B). Small-scale sea breezes can also develop along the shores of large lakes. People who live in a city near the Great Lakes, such as Chicago, recognize this lake effect, especially in the summer. They are reminded daily by weather reports of the cool temperatures near the lake as compared to warmer outlying areas.

Mountain Valley and Breezes

A daily wind similar to land and sea breezes occurs in many mountainous regions. During daylight hours, the air along the slopes of the mountains is heated more intensely than the air at the same elevation over the valley floor. Because this warmer air is less dense, it glides up along the slope and generates a **valley breeze** (Figure 17.15A). The occurrence of these daytime upslope breezes can often be identified by the cumulus clouds that develop on adjacent mountain peaks.

After sunset, the pattern may reverse. Rapid radiation cooling along the mountain slopes produces a layer of cooler air next to the ground. Because cool air is denser than warm air, it drains downslope into the valley. Such a movement of air is called a **mountain breeze** (Figure 17.15B). The same type of cool air drainage can occur in places that have very modest slopes. The result is that the coldest pockets of air are usually found in the lowest spots. Like many other winds, mountain

and valley breezes have seasonal preferences. Although valley breezes are most common during the warm season when solar heating is most intense, mountain breezes tend to be more dominant in the cold season.

Chinook and Santa Ana Winds

Warm, dry winds are common on the eastern slopes of the Rockies, where they are called **chinooks.** Such winds are created when air descends the leeward (sheltered) side of a mountain and warms by compression. Because condensation may have occurred as the air ascended the windward side, releasing latent heat, the air descending the leeward slope will be warmer and drier than it was at a similar elevation on the windward side. Although the temperature of these winds is generally less than 10°C (50°F), which is not particularly warm, they occur mostly in the winter and spring when the affected areas may be experiencing below-freezing temperatures. Thus, by comparison, these dry, warm winds often bring a drastic change. When the ground has a snow cover, these winds are known to melt it in short order.

A chinooklike wind that occurs in southern California is the **Santa Ana.** These hot, desiccating winds greatly increase the threat of fire in this already dry area.

STUDENTS SOMETIMES ASK...
A friend who lives in Colorado talks about "snow eaters." What are they?

"Snow eaters" is a local term for chinooks, the warm, dry winds that descend the eastern slopes of the Rockies. These winds have been known to melt more than a foot of snow in a single day. A chinook that moved through Granville, North Dakota, on February 21, 1918, caused the temperature to rise from −33°F to 50°F, an increase of 83°F!

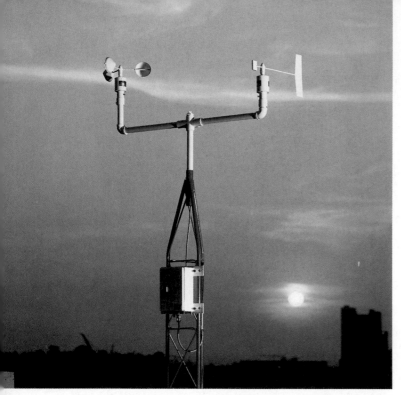

Figure 17.16 Wind vane and cup anemometer. The wind vane shows wind direction and the anemometer measures wind speed. (Photo by Belfort Instrument Company)

How Wind Is Measured

Two basic wind measurements, direction and speed, are particularly significant to the weather observer. *Winds are always labeled by the direction from which they blow.* A north wind blows *from* the north *toward* the south, an east wind *from* the east *toward* the west. The instrument most commonly used to determine wind direction is the **wind vane** (Figure 17.16, upper right). This instrument, a common sight on many buildings, always points *into* the wind. Often the wind direction is shown on a dial that is connected to the wind vane. The dial indicates wind direction, either by points of the compass (N, NE, E, SE, etc.) or by a scale of 0° to 360°. On the latter scale, 0 degrees or 360 degrees are both north, 90 degrees is east, 180 degrees is south, and 270 degrees is west.

When the wind consistently blows more often from one direction than from any other, it is called a **prevailing wind.** You may be familiar with the prevailing westerlies that dominate the circulation in the midlatitudes. In the United States, for example, these winds consistently move the "weather" from west to east across the continent. Embedded within this general eastward flow are cells of high and low pressure with the characteristic clockwise and counterclockwise flow. As a result, the winds associated with the westerlies, as measured at the surface, often vary considerably from day to day and from place to place. By contrast, the direction of airflow associated with the belt of trade winds is much more consistent, as can be seen in Figure 17.17.

Wind speed is commonly measured using a **cup anemometer** (*anemo* = wind, *metron* = measuring instrument) (Figure 17.16, upper left). The wind speed is read from a dial much like the speedometer of an automobile. Places where winds are steady and speeds are relatively high are potential sites for tapping wind energy (see Box 17.1).

By knowing the locations of cyclones and anticyclones in relation to where you are, you can predict the changes in wind direction that will be experienced as a pressure center moves past. Because changes in wind direction often bring changes in temperature and moisture conditions, the ability to predict the winds can be very useful. In the Midwest, for example, a north wind may bring cool, dry air from Canada, whereas a south wind may bring warm, humid air from the Gulf of Mexico. Sir Francis Bacon summed it up nicely when he wrote, "Every wind has its weather."

El Niño and La Niña

As can be seen in Figure 17.18A, the cold Peruvian current flows equatorward along the coast of Ecuador and Peru. This flow encourages upwelling of cold nutrient-filled waters that serve as the primary food source for millions of fish, particularly anchovies. Near the end of

Figure 17.17 Wind roses showing the percentage of time airflow is from various directions. **A.** Wind frequency for the winter in the eastern United States. **B.** Wind frequency for the winter in northern Australia. Note the reliability of the southeast trades in Australia as compared to the westerlies in the eastern United States. (Data from G. T. Trewartha)

A. Westerlies (winter)

B. Southeast Trades (winter)

A. Normal conditions

B. El Niño

Figure 17.18   The relationship between the Southern Oscillation and El Niño is illustrated on these simplified maps. **A.** Normally, the trade winds and strong equatorial currents flow toward the west. At the same time, the strong Peruvian current causes upwelling of cold water along the west coast of South America. **B.** When the Southern Oscillation occurs, the pressure over the eastern and western Pacific flip-flops. This causes the trade winds to diminish, leading to an eastward movement of warm water along the equator. As a result, the surface waters of the central and eastern Pacific warm, with far-reaching consequences to weather patterns.

each year, however, a warm current that flows southward along the coasts of Ecuador and Peru replaces the cold Peruvian current. During the nineteenth century the local residents named this warm countercurrent El Niño ("the child") after the Christ child because it usually appeared during the Christmas season. Normally, these warm countercurrents last for at most a few weeks when they again give way to the cold Peruvian flow. However, at irregular intervals of three to seven years, these countercurrents become unusually strong and replace normally cold offshore waters with warm equatorial waters (Figure 17.18B). Today, scientists use the

term **El Niño** for these episodes of ocean warming that affect the eastern tropical Pacific.

The onset of El Niño is marked by abnormal weather patterns that drastically affect the economies of Ecuador and Peru. As shown in Figure 17.18B, these unusually strong undercurrents amass large quantities of warm water that block the upwelling of colder, nutrient-filled water. As a result, the anchovies starve, devastating the fishing industry. At the same time, some inland areas that are normally arid receive an abnormal amount of rain. Here, pastures and cotton fields have yields far above the average. These climatic fluctuations have been known for years, but they were originally considered local phenomena. Today, we know that El Niño is part of the global circulation and affects the weather at great distances from Peru and Ecuador.

Two of the strongest El Niño events on record occurred between 1982–83 and 1997–98, and were responsible for weather extremes of a variety of types in many parts of the world. The 1997–98 El Niño brought ferocious storms that struck the California coast, causing unprecedented beach erosion, landslides, and floods. In the southern United States, heavy rains also brought floods to Texas and the Gulf states. The same energized jet stream that produced storms in the South, upon reaching the Atlantic, sheared off the northern portions of hurricanes, destroying the storms. It was one of the quietest Atlantic hurricane seasons in years.

Major El Niño events, such as the one in 1997 and 1998, are intimately related to the large-scale atmospheric circulation. Each time an El Niño occurs, the barometric pressure drops over large portions of the southeastern Pacific, whereas in the western Pacific, near Indonesia and northern Australia, the pressure rises (Figure 17.19). Then, as a major El Niño event comes to an end, the pressure difference between these two regions swings back in the opposite direction. This see-saw pattern of atmospheric pressure between the eastern and western Pacific is called the **Southern Oscillation.** It is an inseparable part of the El Niño warmings that

occur in the central and eastern Pacific every three to seven years. Therefore, this phenomenon is often termed El Niño/Southern Oscillation, or ENSO for short.

Winds in the lower atmosphere are the link between the pressure change associated with the Southern Oscillation and the extensive ocean warming associated with El Niño (see Box 17.2). During a typical year, the trade winds converge near the equator and flow westward toward Indonesia (Figure 17.19A). This steady westward flow creates a warm surface current that moves from east to west along the equator. The result is a "piling up" of a thick layer of warm surface water that produces higher sea levels (by 30 centimeters) in the western Pacific. Meanwhile, the eastern Pacific is characterized by a strong Peruvian current, upwelling of cold water, and lower sea levels.

Then when the Southern Oscillation occurs, the normal situation just described changes dramatically. Barometric pressure rises in the Indonesian region, causing the pressure gradient along the equator to weaken or even to reverse. As a consequence, the once-steady trade winds diminish and may even change direction. This reversal creates a major change in the equatorial current system, with warm water flowing eastward (Figure 17.19B). With time, water temperatures in the central and eastern Pacific increase and sea level in the region rises. This eastward shift of the warmest surface water marks the onset of El Niño and sets up changes

Figure 17.19 Simplified illustration of the see-saw pattern of atmospheric pressure between the eastern and western Pacific, called the *Southern Oscillation.* **A.** During average years, high pressure over the eastern Pacific causes surface winds and warm equatorial waters to flow westward. The result is a pileup of warm water in the western Pacific, which promotes the lowering of pressure. **B.** An El Niño event begins as surface pressure increases in the western Pacific and decreases in the eastern Pacific. This air pressure reversal weakens, or may even reverse the trade winds, and results in an eastward movement of the warm waters that had accumulated in the western Pacific.

A. Normal years

B. El Niño years

BOX 17.2

UNDERSTANDING EARTH:
Tracking El Niño from Space

The images in Figure 17.C show the progression of the 1997–98 El Niño as derived from the TOPEX/Poseidon satellite.* This satellite uses radar altimeters to bounce radar signals off the ocean surface to obtain precise measurements of the distance between the satellite and the sea surface. This information is combined with high-precision orbital data from the Global Positioning System (GPS) of satellites to produce maps of sea-surface height. Such maps show the topography of the sea surface. Elevated topography (hills) indicates warmer-than-average water, whereas areas of low topography (valleys) indicate cooler-than-normal water. With this detailed information, scientists can determine the speed and direction of surface ocean currents.

The colors represented in these images show sea-level height relative to average. Remember, hills are warm and valleys are cool. The white and red areas indicate places of higher-than-normal sea-surface heights (hills). In the white areas, the sea surface is between 14 and 32 centimeters above normal; in the red areas, sea level is elevated by about 10 centimeters. Green areas indicate average conditions, whereas purple shows zones that are at least 18 centimeters below average sea level.

The images depict the progression of the large warm-water mass from west

*This box is based on material provided by NASA's Goddard Space Flight Center.

April 25, 1997

July 25, 1997

November 10, 1997

March 14, 1998

Figure 17.C The image for April 25, 1997, shows the Pacific on the eve of the 1997–98 El Niño event. By examining the images for July 25 and November 10, you can see the dramatic buildup of warm water (white and red areas) in the eastern Pacific and the enlarging region of cool water (purple) in the western Pacific. By the time of the March 14, 1998 image, the area of warm water in the eastern Pacific was *much* smaller. (Courtesy of NASA Goddard Space Flight Center)

to east across the equatorial Pacific. At its peak in November 1997, the surface area covered by the warm water mass was about one and a half times the size of the contiguous 48 states. The added amount of warm water in the eastern Pacific with a temperature between 21° and 30°C was about 30 times the combined volume of the water in all of the U.S. Great Lakes.

---

in atmospheric circulation that affect areas far outside the tropical Pacific.

When an El Niño began in the summer of 1997, forecasters predicted that the pool of warm water over the Pacific would displace the paths of both the subtropical and midlatitude jet streams, which steer weather systems across North America (Figure 17.18). As predicted, the subtropical jet brought rain to the Gulf Coast, where Tampa, Florida, received more than three times its normal winter precipitation. Furthermore, the midlatitude jet pumped warm air far north into the continent. As a result, winter temperatures west of the Rockies were significantly above normal.

The effects of El Niño are somewhat variable depending in part on the temperatures and size of the warm pools. Nevertheless, some locales appear to be affected more consistently. In particular, during most El Niños, warmer-than-normal winters occur in the northern United States and Canada. In addition, normally arid portions of Peru and Ecuador, as well as the eastern United States, experience wet conditions. By contrast, drought conditions are generally observed in Indonesia, Australia, and the Philippines. One major benefit of the circulation associated with El Niño is a suppression of the number of Atlantic hurricanes.

The opposite of El Niño is an atmospheric phenomenon known as **La Niña.** Once thought to be the normal conditions that occur between two El Niño events, meteorologists now consider La Niña an important atmospheric phenomenon in its own right.

Figure 17.20   Average annual precipitation in millimeters. (Note, 400 mm is equal to 15.6 inches.)

Researchers have come to recognize that when surface temperatures in the eastern Pacific are *colder than average,* a La Niña event is triggered that has a distinctive set of weather patterns. A typical La Niña winter blows colder than normal air over the Pacific Northwest and the northern Great Plains while warming much of the rest of the United States. Further, greater precipitation is expected in the Northwest. During the La Niña winter of 1998–99, a world record snowfall for one season occurred in Washington State. Another La Niña impact is greater hurricane activity. A recent study concluded that the cost of hurricane damages in the United States is 20 times greater in La Niña years as compared to El Niño years.

In summary, the effects of El Niño and La Niña on world climate are widespread and variable. There is no place on Earth where the weather is indifferent to air and ocean conditions in the tropical Pacific. Events associated with El Niño and La Niña are now understood to have a significant influence on the state of weather and climate almost everywhere.

Global Distribution of Precipitation

A casual glance at Figure 17.20 shows a relatively complex pattern for the distribution of precipitation. Although the map appears to be complicated, the general features of the map can be explained by applying our knowledge of global winds and pressure systems.

In general, regions influenced by high pressure, with its associated subsidence and diverging winds, experience relatively dry conditions. On the other hand, regions under the influence of low pressure and its converging winds and ascending air receive ample precipitation. This pattern is illustrated by noting that the tropical region dominated by the equatorial low is the rainiest region on Earth. It includes the rain forests of the Amazon basin in South America and the Congo basin in Africa. Here the warm, humid trade winds converge to yield abundant rainfall throughout the year. By way of contrast, areas dominated by the subtropical high-pressure cells clearly receive much smaller amounts of precipitation. These are regions of extensive deserts. In the Northern Hemisphere the largest is the Sahara. Examples in the Southern Hemisphere include the Kalahari in southern Africa and the dry lands of Australia.

If Earth's pressure and wind belts were the only factors controlling precipitation distribution, the pattern shown in Figure 17.20 would be simpler. The inherent nature of the air is also an important factor in determining precipitation potential. Because cold air has a low capacity for moisture compared with warm air, we would expect a latitudinal variation in precipitation, with low latitudes receiving the greatest amounts of precipitation and high latitudes receiving the smallest amounts. Figure 17.20 indeed reveals heavy rainfall in equatorial regions and meager precipitation in high-latitude areas. Recall that the dry region in the warm subtropics is explained by the presence of the subtropical high.

In addition to latitudinal variations in precipitation, the distribution of land and water complicates the precipitation pattern. Large landmasses in the middle latitudes commonly experience decreased precipitation toward their interiors. For example, central North America and central Eurasia receive considerably less precipitation than do coastal regions at the same latitude. Furthermore, the effects of mountain barriers alter the idealized precipitation patterns we would expect solely from global wind and pressure systems. Windward mountain slopes receive abundant rainfall resulting from orographic lifting, whereas leeward slopes and adjacent lowlands are usually deficient in moisture.

Chapter Summary

- *Air has weight*: At sea level it exerts a pressure of 1 kilogram per square centimeter (14.7 pounds per square inch). *Air pressure* is the force exerted by the weight of air above. With increasing altitude there is less air above to exert a force, and thus air pressure decreases with altitude, rapidly at first, then much more slowly. The unit used by meteorologists to measure atmospheric pressure is the *millibar*. *Standard sea-level pressure* is expressed as 1013.2 millibars. *Isobars* are lines on a weather map that connect places of equal air pressure.

- A *mercury barometer* measures air pressure using a column of mercury in a glass tube that is sealed at one end and inverted in a dish of mercury. As air pressure increases, the mercury in the tube rises; conversely, when air pressure decreases, so does the height of the column of mercury. A mercury barometer measures atmospheric pressure in *inches of mercury*, the height of the column of mercury in the barometer. Standard atmospheric pressure at sea level equals 29.92 inches of mercury. *Aneroid* (without liquid) *barometers* consist of partially evacuated metal chambers that compress as air pressure increases and expand as pressure decreases.

- *Wind* is the horizontal flow of air from areas of higher pressure toward areas of lower pressure. Winds are controlled by the following combination of forces: (1) the *pressure-gradient force* (amount of pressure change over a given distance); (2) *Coriolis effect* (deflective effect of Earth's rotation to the right in the Northern Hemisphere and to the left in the Southern Hemisphere); and (3) *friction* with Earth's surface (slows the movement of air and alters wind direction).

- Upper-air winds, called *geostrophic winds*, blow parallel to the isobars and reflect a balance between the pressure-gradient force and the Coriolis effect. Upper-air winds are faster than surface winds because friction is greatly reduced aloft. Friction slows surface winds, which in turn reduces the Coriolis effect. The result is air movement at an angle across the isobars toward the area of lower pressure.

- The two types of pressure centers are (1) *cyclones,* or *lows* (centers of low pressure), and (2) *anticyclones,* or *highs* (high-pressure centers). In the Northern Hemisphere, winds around a low (cyclone) are counterclockwise and inward. Around a high (anticyclone), winds are clockwise and outward. In the Southern Hemisphere, the Coriolis effect causes winds to move clockwise around a low and counterclockwise around a high. Because air rises and cools adiabatically in low-pressure centers, cloudy conditions and precipitation are often associated with their passage. In high-pressure centers, descending air is compressed and warmed; therefore, cloud formation and precipitation are unlikely, and "fair" weather is usually expected.

- Earth's *global pressure zones* include the *equatorial low, subtropical high, subpolar low,* and *polar high.* The *global surface winds* associated with these pressure zones are the *trade winds, westerlies,* and *polar easterlies.*

- Particularly in the Northern Hemisphere, large seasonal temperature differences over continents disrupt the idealized, or zonal, global patterns of pressure and wind. In winter, large, cold landmasses develop a seasonal high-pressure system from which surface airflow is directed off the land. In summer, landmasses are heated and low pressure develops over them, which permits air to flow onto the land. The seasonal changes in wind direction are known as *monsoons.*

- In the middle latitudes, between 30 and 60 degrees latitude, the general west-to-east flow of the westerlies is interrupted by migrating cyclones and anticyclones. The paths taken by these pressure systems are closely related to upper-level airflow and the polar *jet stream.* The average position of the polar jet stream, and hence the paths followed by cyclones, migrates equatorward with the approach of winter and poleward as summer nears.

- *Local winds* are small-scale winds produced by a locally generated pressure gradient. Local winds include *sea* and *land breezes* (formed along a coast because of daily pressure differences caused by the differential heating of land and water); *valley* and *mountain breezes* (daily wind similar to sea and land breezes except in a mountainous area where the air along slopes heats differently from the air at the same elevation over the valley floor); and *chinook* and *Santa Ana winds* (warm, dry winds created when air descends the leeward side of a mountain and warms by compression).

- The two basic wind measurements are *direction* and *speed.* Winds are always labeled by the direction *from* which they blow. Wind direction is measured with a *wind vane,* and wind speed is measured using a *cup anemometer.*

- *El Niño* is the name given to the periodic warming of the ocean that occurs in the central and eastern Pacific. It is associated with periods when a weakened pressure gradient causes the trade winds to diminish. A major El Niño event triggers extreme weather in many parts of the world. When surface temperatures in the eastern Pacific are colder than average, a *La Niña* event is triggered. A typical La Niña winter blows colder-than-normal air over the Pacific Northwest and the northern Great Plains while warming much of the rest of the United States.

- The global distribution of precipitation is strongly influenced by the global pattern of air pressure and wind, latitude, and distribution of land and water.

Key Terms

air pressure (p. 470)
aneroid barometer (p. 471)
anticyclone (p. 476)
barograph (p. 471)
barometric tendency (p. 478)
chinook (p. 483)
convergence (p. 477)
Coriolis effect (p. 473)
cup anemometer (p. 484)
cyclone (p. 476)
divergence (p. 477)
El Niño (p. 485)
equatorial low (p. 480)

geostrophic wind (p. 473)
high (p. 476)
isobar (p. 472)
jet stream (p. 473)
land breeze (p. 483)
La Niña (p. 487)
low (p. 476)
mercury barometer (p. 471)
monsoon (p. 481)
mountain breeze (p. 483)
polar easterlies (p. 480)
polar front (p. 480)
polar high (p. 480)

pressure gradient (p. 473)
pressure tendency (p. 478)
prevailing wind (p. 484)
Santa Ana (p. 483)
sea breeze (p. 483)
Southern Oscillation (p. 486)
subpolar low (p. 480)
subtropical high (p. 480)
trade winds (p. 480)
valley breeze (p. 483)
westerlies (p. 480)
wind (p. 472)
wind vane (p. 484)

Review Questions

1. What is standard sea-level pressure in millibars? In inches of mercury? In pounds per square inch?

2. Mercury is 13 times heavier than water. If you built a barometer using water rather than mercury, how tall would it have to be to record standard sea-level pressure (in centimeters of water)? In feet?

3. Describe the principle of the aneroid barometer.

4. What force is responsible for generating wind?

5. Write a generalization relating the spacing of isobars to the speed of wind.

6. How does the Coriolis effect modify air movement?

7. Contrast surface winds and upper-air winds in terms of speed and direction.

8. Describe the weather that usually accompanies a drop in barometric pressure and a rise in barometric pressure.

9. Sketch a diagram (isobars and wind arrows) showing the winds associated with surface cyclones and anticyclones in both the Northern and Southern hemispheres.

10. If you live in the Northern Hemisphere and are directly west of the center of a cyclone, what most probably will be the wind direction? What will the wind direction be if you are west of an anticyclone?

11. The following questions relate to the global pattern of air pressure and winds.
    (a) The trade winds diverge from which pressure zone?
    (b) Which prevailing wind belts converge in the stormy region known as the polar front?
    (c) Which pressure belt is associated with the equator?

12. What influence does upper-level airflow seem to have on surface-pressure systems?

13. Describe the monsoon circulation of India.

14. What is a local wind? List three examples.

15. A northeast wind is blowing *from* the _____ (direction) *toward* the _____ (direction).

16. Describe the relationship between the Southern Oscillation and a major El Niño event.

17. Which global pressure system is responsible for deserts such as the Sahara and Kalahari in Africa? With which global pressure belt are the tropical rain forests of the Amazon and Congo basins associated?

18. Other than Earth's pressure and wind belts, list two other factors that exert a significant influence on the global distribution of precipitation.

Examining the Earth System

1. Examine the image of Africa in Figure I.3B (p. 4), and pick out the region dominated by the equatorial low and the areas influenced by the subtropical highs in each hemisphere. What clue(s) did you use? Speculate on the differences in the biosphere between the regions dominated by high pressure and the zone influenced by low pressure.

2. How are global winds related to surface ocean currents? (Try comparing Figure 17.13 with Figure 14.2). What is the *ultimate* source of energy that drives both of these circulations?

3. Winds and ocean currents change in the tropical Pacific during an El Niño event. How might this impact the biosphere and solid Earth in Peru and Ecuador? How about in Indonesia? (Two excellent Websites that deal with El Niño are the University of Southern California site at **http://www.usc.edu/org/seagrant/elnino/link.html** and NOAA's *El Niño Theme Page* at **http://www.pmel.noaa.gov/toga-tao/el-nino/nino-home.html**)

Web Resources

 The *Earth Science* Website uses the resources and flexibility of the Internet to aid in your study of the topics in this chapter. Written and developed by Earth science instructors, this site will help improve your understanding of Earth science. Visit **http://www.prenhall.com/tarbuck** and click on the cover of *Earth Science 10e* to find:

- **On-line review quizzes.**
- **Web-based critical thinking and writing exercises.**
- **Links to chapter-specific Web resources.**
- **Internet-wide key term searches.**

http://www.prenhall.com/tarbuck

Lighting from an isolated cumulonimbus cloud in the desert Southwest. (Photo by Kent Wood/Photo Researchers, Inc.)

CHAPTER 18

WEATHER PATTERNS AND SEVERE STORMS

Tornadoes and hurricanes rank among nature's most destructive forces. Each spring, newspapers report the death and destruction left in the wake of a band of tornadoes. During late summer and fall we hear occasional reports about hurricanes in the news. Storms with names like Andrew, Camille, Fran, and Mitch have made front-page headlines (Figure 18.1). Thunderstorms, although less intense and far more common than tornadoes and hurricanes, will also be part of our discussion on severe weather in this chapter. Before looking at violent weather, however, we will study those atmospheric phenomena that most often affect our day-to-day weather: air masses, fronts, and traveling middle-latitude cyclones. Here we will see the interplay of the elements of weather discussed in Chapters 15, 16, and 17.

Air Masses

Atmosphere
↳ Basic Weather Patterns

For many people who live in the middle latitudes, including much of the United States, summer heat waves and winter cold spells are familiar experiences. In the first instance, several days of high temperatures and oppressive humidities may finally end when a series of thunderstorms pass through the area, followed by a few days of relatively cool relief. By contrast, the clear skies that often accompany a span of frigid subzero days may be replaced by thick gray clouds and a period of snow as temperatures rise to levels that seem mild when compared to those that existed just a day earlier. In both examples, what was experienced was a period of generally constant weather conditions followed by a relatively short period of change, and then the reestablishment of a new set of weather conditions that remained for perhaps several days before changing again.

What Is an Air Mass?

The weather patterns just described result from movements of large bodies of air, called air masses. An **air mass,** as the term implies, is an immense body of air, usually 1600 kilometers (1000 miles) or more across and perhaps several kilometers thick, that is characterized by a similarity of temperature and moisture at any given altitude. When this air moves out of its region of origin, it will carry these temperatures and moisture conditions with it, eventually affecting a large portion of a continent.

An excellent example of the influence of an air mass is illustrated in Figure 18.2. Here a cold, dry mass from northern Canada moves southward. With a beginning temperature of −46°C (−51°F), the air mass warms 13 degrees, to −33°C (−27°F), by the time it reaches Winnipeg. It continues to warm as it moves southward through the Great Plains and into Mexico. Throughout its southward journey, the air mass becomes warmer. But it also brings some of the coldest weather of the winter to the places in its path. Thus, the air mass is

Figure 18.1 Three-dimensional computer image, based on satellite data, of Hurricane Mitch off the coast of Honduras, in October 1998. The storm took thousands of lives and caused an estimated $5 billion in damages. (Photo by NASA/Goddard Space Flight Center/Science Photo Library/Photo Researchers, Inc.)

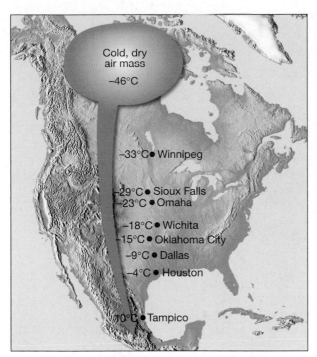

Figure 18.2 As this frigid Canadian air mass moved southward, it brought some of the coldest weather of the winter. (After Tom L. McKnight, *Physical Geography*, 5th ed. Upper Saddle River, NJ: Prentice Hall, 1996, p. 174)

modified, but it also modifies the weather in the areas over which it moves.

The horizontal uniformity of an air mass is not complete because it may extend through 20 degrees or more of latitude and cover hundreds of thousands to millions of square kilometers. Consequently, small differences in temperature and humidity are to be expected from one point to another at the same level. Still, the differences observed within an air mass are small in comparison to the rapid changes experienced across air-mass boundaries.

Because it may take several days for an air mass to traverse an area, the region under its influence will probably experience fairly constant weather, a situation called **air-mass weather.** Certainly some day-to-day variations may occur, but the events will be very unlike those in an adjacent air mass. For this reason the boundary between two adjoining air masses that have contrasting characteristics, called a *front*, marks a change in weather.

Source Regions

When a portion of the lower atmosphere moves slowly or stagnates over a relatively uniform surface, the air will assume the distinguishing features of that area, particularly with regard to temperature and moisture conditions.

The area where an air mass acquires its characteristic properties of temperature and moisture is called its **source region.** The source regions that produce

air masses influencing North America are shown in Figure 18.3

Air masses are classified according to their source region. **Polar (P) air masses** originate in high latitudes toward Earth's poles, whereas those that form in low latitudes are called **tropical (T) air masses.** The designation *polar* or *tropical* gives an indication of the temperature characteristics of the air masses. *Polar* indicates cold, and *tropical* indicates warm.

In addition, air masses are classified according to the nature of the surface in the source region. **Continental (c) air masses** form over land, and **maritime (m) air masses** originate over water. The designation *continental* or *maritime* thus suggests the moisture characteristics of the air mass. Continental air is likely to be dry, and maritime air, humid.

The four basic types of air masses according to this scheme of classification are continental polar (cP), continental tropical (cT), maritime polar (mP), and maritime tropical (mT).

Weather Associated with Air Masses

Continental polar and maritime tropical air masses influence the weather of North America most, especially east of the Rocky Mountains. Continental polar air masses originate in northern Canada, interior Alaska, and the Arctic areas that are uniformly cold and dry in winter and cool and dry in summer. In winter, an invasion of continental polar air brings the clear skies and cold temperatures we associate with a cold wave as it moves southward from Canada into the United States. In summer, this air mass may bring a few days of cooling relief.

Although cP air masses are not, as a rule, associated with heavy precipitation, those that cross the Great Lakes during late autumn and winter sometimes bring snow to the leeward shores (Figure 18.4). These localized storms often form when the surface weather map indicates no apparent cause for a snowstorm to occur. Known as **lake-effect snows,** they make Buffalo and Rochester, New York, among the snowiest cities in the United States (see Figure 15.1, p. 406). What causes lake-effect snow? During late autumn and early winter, the temperature contrast between the lakes and adjacent land areas can be large.* The temperature contrast can be especially great when a very cold cP air mass pushes southward across the lakes. When this occurs, the air acquires large quantities of heat and moisture from the relatively warm lake surface. By the time it reaches the opposite shore, the air mass is humid and unstable, and heavy snow showers are likely.

Maritime tropical air masses affecting North America most often originate over the warm waters of the

---

*Recall that land cools more rapidly and to lower temperatures than water. See the discussion of "Land and Water" in the section on "Why Temperatures Vary: The Controls of Temperature" in Chapter 15.

Figure 18.3 Air masses are classified on the basis of their source region. The designation continental (c) or maritime (m) gives an indication of moisture content, whereas polar (P) and tropical (T) indicate temperature conditions.

Gulf of Mexico, the Caribbean Sea, or the adjacent Atlantic Ocean. As you might expect, these air masses are warm, moisture-laden, and usually unstable. Maritime tropical air is the source of much, if not most, of the precipitation received in the eastern two-thirds of the United States. In summer, when an mT air mass invades the central and eastern United States, and occasionally southern Canada, it brings the high temperatures and oppressive humidity typically associated with its source region.

Of the two remaining air masses, maritime polar and continental tropical, the latter has the least influence on the weather of North America. Hot, dry continental tropical air, originating in the Southwest and Mexico during the summer, only occasionally affects the weather outside its source region.

During the winter, maritime polar air masses coming from the North Pacific often originate as continental polar air masses in Siberia. The cold, dry cP air is transformed into relatively mild, humid, unstable mP air during its long journey across the North Pacific (Figure 18.5). As this mP air arrives at the western shore of North America, it is often accompanied by low clouds and shower activity. When this air advances inland against the western mountains, orographic uplift pro-

duces heavy rain or snow on the windward slopes of the mountains. Maritime polar air also originates in the North Atlantic off the coast of eastern Canada and occasionally influences the weather of the northeastern United States. In winter, when New England is on the northern or northwestern side of a passing low-pressure center, the counterclockwise cyclonic winds draw in maritime polar air. The result is a storm characterized by snow and cold temperatures, known locally as a *nor'easter*.

Fronts

 Atmosphere
↳ Basic Weather Patterns

Fronts are boundaries that separate air masses of different densities. One air mass is usually warmer and contains more moisture than the other. Fronts can form between any two contrasting air masses. Considering the vast size of the air masses involved, fronts are relatively narrow, being 15- to 200-kilometer- (9- to 120-mile-) wide bands of discontinuity. On the scale of a

A.

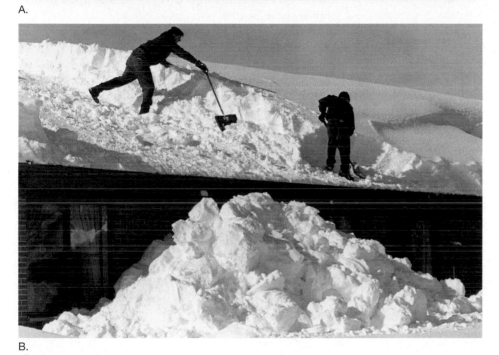

B.

Figure 18.4 **A.** The snowbelts of the Great Lakes region are the zones that most frequently experience lake-effect snowstorms. As continental polar air crosses the lakes in winter, it acquires moisture and is made unstable because of warming from below. Snow showers on the lee (downwind) side of the lakes are often the consequence of this air-mass modification. Compare the locations of Thunder Bay, Ontario, and Marquette, Michigan. In an average winter, Marquette, which is on the lee side of Lake Superior, receives two and a half times more snow than Thunder Bay. **B.** A remarkable six-day lake-effect snowstorm in November 1996 dropped 175 centimeters (nearly 69 inches) of snow on Chardon, Ohio, setting a new state record. (Photo by Tony Dejak/AP/Wide World Photos)

weather map, they are generally narrow enough to be represented by a broad line (as in Figure 18.10).

Above Earth's surface, the frontal surface slopes at a low angle so that warmer air overlies cooler air, as shown in Figure 18.6. In the ideal case, the air masses on both sides of the front move in the same direction and at the same speed. Under this condition, the front acts simply as a barrier that travels along with the air masses and that neither mass can penetrate.

Generally, however, the distribution of pressure across a front is such that one air mass moves faster than the other. Thus, one air mass actively advances into another and "clashes" with it. In fact, the boundaries were

tagged *fronts* during World War I by Norwegian meteorologists who visualized them as analogous to battle lines between two armies.

As one air mass moves into another, some mixing does occur along the frontal surface, but for the most part the air masses retain their distinct identities as one is displaced upward over the other. No matter which air mass is advancing, *it is always the warmer, less dense air that is forced aloft, whereas the cooler, denser air acts as the wedge upon which lifting takes place.* The term **overrunning** is generally applied to warm air gliding up along a cold air mass. We will now take a look at different types of fronts.

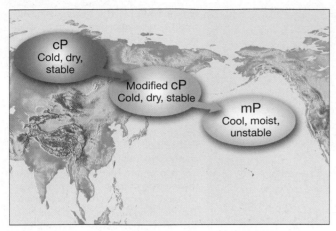

Figure 18.5 During winter, maritime polar (mP) air masses in the North Pacific usually begin as continental polar (cP) air masses in Siberia. The cP air is modified to mP as it slowly crosses the ocean.

Warm Fronts

When the surface (ground) position of a front moves so that warm air occupies territory formerly covered by cooler air, it is called a **warm front** (Figure 18.6). On a weather map, the surface position of a warm front is shown by a red line with semicircles extending into the cooler air.

East of the Rockies, warm tropical air often enters the United States from the Gulf of Mexico and overruns receding cool air. As the cold air retreats, friction with the ground greatly slows the advance of the surface position of the front compared to its position aloft. Stated another way, less dense, warm air has a hard time displacing denser cold air. For this reason, the

boundary separating these air masses acquires a very gradual slope. The average slope of a warm front is about 1:200, which means that if you are 200 kilometers (120 miles) ahead of the surface location of a warm front, you will find the frontal surface at a height of 1 kilometer (0.6 mile).

As warm air ascends the retreating wedge of cold air, it cools adiabatically to produce clouds, and frequently, precipitation. The sequence of clouds shown in Figure 18.6 typically precedes a warm front. The first sign of the approaching warm front is the appearance of cirrus clouds. These high clouds form where the overrunning warm air has ascended high up the wedge of cold air, 1000 kilometers (600 miles) or more ahead of the surface front.

As the front nears, cirrus clouds grade into cirrostratus, which blend into denser sheets of altostratus. About 300 kilometers (180 miles) ahead of the front, thicker stratus and nimbostratus clouds appear, and rain or snow begins.

Because of their slow rate of advance and very low slope, warm fronts usually produce light-to-moderate precipitation over a large area for an extended period. Warm fronts, however, are occasionally associated with cumulonimbus clouds and thunderstorms. This occurs when the overrunning air is unstable and the temperatures on opposite sides of the front contrast sharply. At the other extreme, a warm front associated with a dry air mass could pass unnoticed at the surface.

A gradual increase in temperature occurs with the passage of a warm front. As you would expect, the increase is most apparent when a large contrast exists between adjacent air masses. Moreover, a wind shift from the east to the southwest is generally noticeable. (The

Figure 18.6 Warm front produced as warm air glides up over a cold air mass. Precipitation is moderate and occurs within a few hundred kilometers of the surface front.

reason for this shift will become evident later.) The moisture content and stability of the encroaching warm air mass largely determine when clear skies will return. During the summer, cumulus, and occasionally cumulonimbus, clouds are embedded in the warm unstable air mass that follows the front. These clouds can produce precipitation, but it is usually restricted in extent and of short duration.

Cold Fronts

When cold air actively advances into a region occupied by warmer air, the boundary is called a **cold front** (Figure 18.7). As with warm fronts, friction tends to slow the surface position of a cold front more so than its position aloft. However, because of the relative positions of the adjacent air masses, the cold front steepens as it moves. On the average, cold fronts are about twice as steep as warm fronts, having a slope of perhaps 1:100. In addition, cold fronts advance more rapidly than do warm fronts. These two differences—rate of movement and steepness of slope—largely account for the more violent nature of cold-front weather compared to the weather generally accompanying a warm front.

The forceful lifting of air along a cold front is often so rapid that the latent heat released when water vapor condenses will increase the air's buoyancy appreciably. The heavy downpours and vigorous wind gusts associated with mature cumulonimbus clouds frequently result. Because a cold front produces roughly the same amount of lifting as a warm front, but over a shorter distance, the precipitation intensity is greater, but of shorter duration.

As a cold front approaches, generally from the west or northwest, towering clouds often can be seen in the distance. Near the front, a dark band of ominous clouds foretells the ensuing weather. A marked temperature drop and a wind shift usually accompany the passage of the front. (Again, the reason for this shift, from the south or southwest to the west or northwest, will be explained later.)

The sometimes violent weather and sharp temperature contrast along the cold front are symbolized on a weather map by a blue line with triangle-shaped points that extend into the warmer air mass (Figure 18.7).

The weather behind a cold front is dominated by a subsiding and relatively cold air mass. Thus, clearing usually begins soon after the front passes. Although the compression of air due to subsidence causes some adiabatic heating, the effect on surface temperatures is minor. In winter, the long, cloudless nights that often follow the passage of a cold front allow for abundant radiation cooling that reduces surface temperatures. When a cold front moves over a relatively warm area, surface heating of the air can produce shallow convection. This in turn may generate low cumulus or stratocumulus clouds behind the front.

Stationary Fronts and Occluded Fronts

Occasionally the flow on both sides of a front is neither toward the cold air mass nor toward the warm air mass, but almost parallel to the line of the front. Thus, the surface position of the front does not move. This condition is called a **stationary front**. On a weather map, stationary fronts are shown with blue triangular points on one side of the front and red semicircles on the other. At times, some overrunning occurs along a stationary

Figure 18.7 Fast-moving cold front and cumulonimbus clouds. Often thunderstorms occur if the warm air is unstable.

front. On these occasions, gentle to moderate precipitation is most likely.

The fourth type of front is the **occluded front.** Here an active cold front overtakes a warm front, as shown in Figure 18.8. As the advancing cold air wedges the warm front upward, a new front emerges between the advancing cold air and the air over which the warm front is gliding. The weather of an occluded front is generally complex. Most precipitation is associated with the warm air being forced aloft. When conditions are suitable, however, the newly formed front is capable of initiating precipitation of its own.

A word of caution is in order concerning the weather associated with various fronts. Although the preceding discussion will help you recognize the weather patterns associated with fronts, remember that these descriptions are generalizations. The weather generated along any individual front may or may not conform fully to this idealized picture. Fronts, like all aspects of nature, never lend themselves to classification as nicely as we would like.

The Middle-Latitude Cyclone

Atmosphere
↳ Basic Weather Patterns

So far we have examined the basic elements of weather as well as the dynamics of atmospheric motions. We are now ready to apply our knowledge of these diverse phenomena to an understanding of day-to-day weather patterns in the middle latitudes. For our purposes, *middle latitudes* refers to the region between southern Florida and Alaska. The primary weather producers here are **middle-latitude** or **midlatitude cyclones.** On weather maps, such as those used on the Weather Channel, they are shown by an **L,** meaning *low-pressure center.*

Middle-latitude cyclones are large centers of low pressure that generally travel from west to east. Lasting from a few days to more than a week, these weather systems have a counterclockwise circulation with an airflow inward toward their centers. Most middle-latitude cyclones also have a cold front extending from the central area of low pressure, and frequently a warm front as well. Convergence and forceful lifting initiate cloud development and often cause abundant precipitation.

Life Cycle

As early as the 1800s it was known that cyclones were the bearers of precipitation and severe weather. But it was not until the early part of the twentieth century that a model was developed that described how cyclones form and evolve. It was formulated by a group of Norwegian scientists and published in 1918. The model was created primarily from near-surface observations.

Years later, as data from the middle and upper troposphere and from satellite images became available, modifications were necessary. Yet this model is still an accepted working tool for interpreting the weather. If you keep this model in mind when you observe changes in the weather, the changes will no longer come as a surprise. You should begin to see some order in what had appeared to be disorder, and you might even occasionally "predict" the impending weather (see Box 18.1).

Formation: The Clash of Two Air Masses. Cyclones form along fronts and proceed through a generally predictable life cycle. Figure 18.9 shows six stages in the life of a typical midlatitude cyclone. In part A, the stage is set for cyclone formation. Here two air masses of different densities (temperatures) are moving

A.

B.

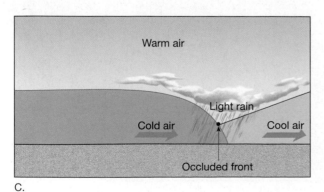

C.

Figure 18.8   Stages in the formation of an occluded front.

BOX 18.1

UNDERSTANDING EARTH:
A Brief Overview of the Weather Business

In the United States, the governmental agency responsible for gathering and disseminating weather-related information is the *National Weather Service (NWS)*. The mission of the NWS is as follows:

> The National Weather Service (NWS) provides weather, hydrologic and climate forecasts and warnings for the United States, its territories, adjacent waters and ocean areas, for the protection of life and property and the enhancement of the national economy. NWS data and products form a national information database and infrastructure that can be used by other governmental agencies, the private sector, the public and the global community.

Perhaps the most important services provided by the NWS are forecasts and warnings of hazardous weather including thunderstorms, flooding, hurricanes, tornadoes, winter weather, and extreme heat. According to the Federal Emergency Management Agency, 80 percent of all declared emergencies are weather related. In a similar vein, the Department of Transportation reports that more than 6000 fatalities per year can be attributed to the weather.

As the population grows, the economic impact from weather-related phenomena also escalates. Each year weather-related losses are in the billions of dollars. An estimated 90 percent of the U.S. public consults weather forecasts every day. As a result, the pressure on the NWS to provide more accurate and longer-range forecasts continues to grow.

To produce even a short-range forecast is an enormous task. It involves numerous steps, including collecting weather data, transmitting it, and compiling it on a global scale. These data must then be analyzed so that an accurate assessment of the current conditions can be made. From current weather patterns, various methods are used to determine the future state of the atmosphere, a task called *weather forecasting*.

The final phase in the weather business is the dissemination of a wide variety of forecasts. Each of the 119 Weather Forecast Offices operated by the NWS issues regional forecasts, aviation forecasts, and warnings covering their forecast area. The *local forecast* seen on The Weather Channel is an unedited version of a forecast issued by one of these offices. Further, all the data and products (maps, charts, and forecasts) produced by the NWS are available at no cost to the general public and to private forecasting services.

The demand for highly visual forecasts containing computer-generated graphics has grown along with the use of personal computers and the Internet. Because it is outside the mission of the National Weather Service, this publicly funded entity is not the source of the animated depictions of the weather that appear on most local newscasts. Instead, the private sector has taken over this task. In addition, private forecast services customize the NWS products to create a variety of specialized weather reports that are tailored for specific audiences. In a farming community, for example, the weather reports might include frost warnings, while winter forecasts in Denver, Colorado, include the snow conditions at ski resorts. Hence, the public receives most of its weather information through local TV stations and newspapers and/or through more widely distributed sources such as The Weather Channel, CNN, and USA Today (Figure 18.A).

It is important to note that, despite the valuable role that the private sector plays in disseminating weather-related information to the public, the NWS is the *official* voice in the United States for issuing warnings during life-threatening weather situations. Two major weather centers operated by the NWS serve critical functions in this regard. The Storm Prediction Center in Norman, Oklahoma, maintains a constant vigil for severe thunderstorms and tornadoes. Hurricane watches and warnings for the Atlantic, Caribbean, Gulf of Mexico, and eastern Pacific are issued by the National Hurricane Center/Tropical Prediction Center in Miami, Florida.

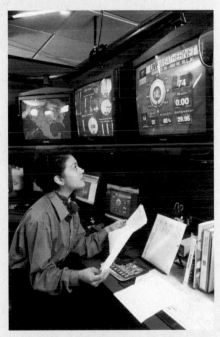

Figure 18.A Weather broadcaster previewing graphics. (Photo by Bob Daemmrich/Stock Boston)

roughly parallel to the front, but in opposite directions. (In the classic model, this would be continental polar air associated with the polar easterlies on the north side of the front and maritime tropical air driven by the westerlies on the south side of the front.)

Under suitable conditions, the frontal surface that separates these two contrasting air masses will take on a wave shape that is usually several hundred kilometers long. These waves are analogous to the waves produced on water by moving air, except that the scale is different. Some waves tend to dampen, or die out, whereas others grow in amplitude. Those storms that intensify develop waves that change in shape over time, much like a gentle ocean swell does as it moves into shallow water and becomes a tall, breaking wave.

Development of Cyclonic Flow. As the wave develops, warm air advances poleward invading the area

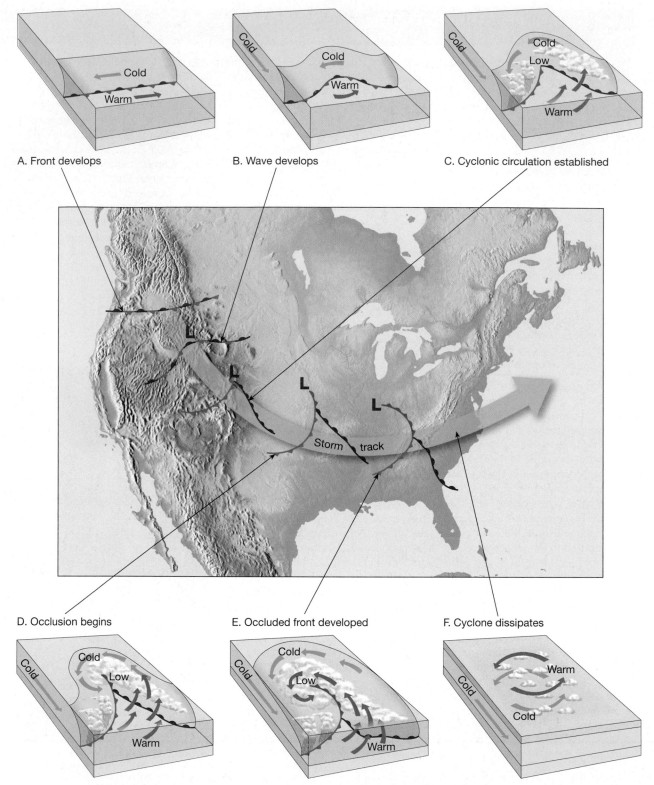

A. Front develops

B. Wave develops

C. Cyclonic circulation established

D. Occlusion begins

E. Occluded front developed

F. Cyclone dissipates

Figure 18.9 Stages in the life cycle of a middle-latitude cyclone, as proposed by J. Bjerknes.

formerly occupied by colder air, while cold air moves equatorward. This change in the direction of the surface flow is accompanied by a readjustment in the pressure pattern that results in nearly circular isobars, with the low pressure centered at the apex of the wave. The resulting flow is a counterclockwise cyclonic circulation that can be seen clearly on the weather map shown in Figure 18.10. Once the cyclonic circulation develops, we would expect general convergence to result in vertical lifting, especially where warm air is overrunning colder air. You can see in Figure 18.10 that the air in the warm sector (over the

Figure 18.10   Cloud patterns typically associated with a mature middle-latitude cyclone. The middle section is a map view. Note the cross-section lines (F–G, A–E). Above the map is a vertical cross section along line F–G. Below the map is a section along A–E. For cloud abbreviations, refer to Figures 18.6 and 18.7.

southern states) is flowing northeastward toward colder air that is moving toward the northwest. Because the warm air is moving in a direction perpendicular to this front, we can conclude that the warm air is invading a region formerly occupied by cold air. Therefore, this must be a warm front. Similar reasoning indicates that to the left (west) of the cyclonic disturbance, cold air from the northwest is displacing the air of the warm sector and generating a cold front.

Occlusion: The Beginning of the End.   Usually, the position of the cold front advances faster than the warm front and begins to close (lift) the warm front, as shown in Figure 18.9D, E. This process, known as **occlusion,** forms an *occluded front,* which grows in length as it displaces the warm sector aloft. As occlusion begins, the storm often intensifies. Pressure at the storm's center falls, and wind speeds increase. In the winter, heavy snowfalls and blizzardlike conditions are possible during this phase of the storm's evolution.

As more of the sloping discontinuity (front) is forced aloft, the pressure gradient weakens. In a day or two, the entire warm sector is displaced and cold air surrounds the cyclone at low levels (Figure 18.9F). Thus, the horizontal temperature (density) difference that existed between the two contracting air masses has been eliminated. At this point, the cyclone has exhausted its source of energy. Friction slows the surface flow, and the once highly organized counterclockwise flow ceases to exist.

Idealized Weather

The middle-latitude cyclone model provides a useful tool for examining the weather patterns of the middle latitudes. Figure 18.10 illustrates the distribution of clouds and thus the regions of possible precipitation associated with a mature system. Compare this drawing to the satellite image of a cyclone shown in Figure 18.11

Guided by the westerlies aloft, cyclones generally move eastward across the United States. Therefore, we can expect the first signs of a cyclone's arrival to appear in the western sky. In the region of the Mississippi Valley, however, cyclones often begin a more northeasterly path and occasionally move directly northward. Typically, a midlatitude cyclone requires two to four days to pass over a region. During that brief period, abrupt changes in atmospheric conditions may occur, particularly in the spring of the year when the greatest temperature contrasts occur across the midlatitudes.

Using Figure 18.10 as a guide, let us examine these weather producers and the changes we can expect as they pass in the spring of the year. To facilitate our discussion, Figure 18.10 includes two profiles along lines *A–E* and *F–G.*

- Imagine the change in weather as you move from right to left along profile *A–E* (bottom of figure). At point *A,* the sighting of high cirrus clouds is the first sign of the approaching cyclone. These high clouds can precede the surface front by 1000 kilometers (600 miles) or more, and they are normally accompanied by falling pressure. As the warm front advances, a lowering and thickening of the cloud deck is noticed.

- Within 12 to 24 hours after the first sighting of cirrus clouds, light precipitation begins (point *B*). As the front nears, the rate of precipitation increases, a rise in temperature is noticed, and winds begin to change from east or southeast to south or southwest.

- With the passage of the warm front, an area is under the influence of a maritime tropical air mass (point *C*). Usually the region affected by this sector of the cyclone experiences warm temperatures, south or southwest winds, and generally clear skies, although fair-weather cumulus or altocumulus clouds are not uncommon here.

- The relatively warm, humid weather of the warm sector passes quickly in the spring of the year and is replaced by gusty winds and precipitation generated along the cold front. The approach of a rapidly advancing cold front is marked by a wall of dark clouds (point *D*). Severe weather accompanied by heavy precipitation, hail, and an occasional tornado is a definite possibility at this time of year. The passage of the cold front is easily detected by a wind shift; the southwest winds are replaced by winds from the west to northwest and by a pronounced drop in temperature. Also, rising pressure hints of the subsiding cool, dry air behind the front.

Figure 18.11 Satellite view of a mature cyclone over the eastern United States. It is easy to see why we often refer to the cloud pattern of a cyclone as having a "comma" shape. (Courtesy of John Jensenius/National Weather Service)

- Once the front passes, skies clear as cooler air invades the region (point *E*). Often a day or two of almost cloudless deep-blue skies occurs unless another cyclone is edging into the region.

A very different set of weather conditions will prevail in those regions north of the storm's center along profile *F–G* of Figure 18.10. Often the storm reaches its greatest intensity in this zone, and the area along profile *F–G* receives the brunt of the storm's fury. Here temperatures remain cold during the passage of the system, and heavy snow, sleet, and/or freezing rain may develop during the winter months.

The Role of Airflow Aloft

When the earliest studies of cyclones were made, little was known about the nature of the airflow in the middle and upper troposphere. Since then, a close relationship has been established between surface disturbances and the flow aloft. Airflow aloft plays an important role in maintaining cyclonic and anticyclonic circulation. In fact, more often than not, these rotating surface wind systems are actually generated by upper-level flow.

Recall that the airflow around a surface low is inward, a fact that leads to mass convergence, or coming together (Figure 18.12). The resulting accumulation of air must be accompanied by a corresponding increase in surface pressure. Consequently, we might expect a low-pressure system to "fill" rapidly and be eliminated, just as the vacuum in a coffee can is quickly dissipated when we open it. However, this does not occur.

On the contrary, cyclones often exist for a week or longer. For this to happen, surface convergence must be offset by a mass outflow at some level aloft (Figure 18.12). As long as divergence (spreading out) aloft is equal to, or greater than, the surface inflow, the low pressure and its accompanying convergence can be sustained.

Because cyclones are bearers of stormy weather, they have received far more attention than anticyclones. Nevertheless, a close relation exists, which makes it difficult to separate any discussion of these two types of pressure systems. The surface air that feeds a cyclone, for example, generally originates as air flowing out of an anticyclone. Consequently, cyclones and anticyclones typically are found adjacent to one another. Like the cyclone, an anticyclone depends on the flow far above to maintain its circulation. In this instance, divergence at the surface is balanced by convergence aloft and general subsidence of the air column (Figure 18.12).

What's in a Name?

Up to now we have examined the middle-latitude cyclones which play such an important role in causing day-to-day weather changes. Yet the use of the term

Figure 18.12   Idealized diagram depicting the support that divergence and convergence aloft provide to cyclonic and anticyclonic circulation at the surface. Divergence aloft initiates upward air movement, reduced surface pressure, and cyclonic flow. In contrast, convergence along the jet stream results in general subsidence of the air column, increased surface pressure, and anticyclonic surface winds.

cyclone is often confusing. To many people, the term implies only an intense storm, such as a tornado or a hurricane. When a hurricane unleashes its fury on India or Bangladesh, for example, it is usually reported in the media as a cyclone (the term denoting a hurricane in that part of the world).

Similarly, tornadoes are referred to as cyclones in some places. This custom is particularly common in portions of the Great Plains of the United States. Recall that in the *Wizard of Oz,* Dorothy's house was carried from her Kansas farm to the land of Oz by a cyclone. Indeed, the nickname for the athletic teams at Iowa State University is the *Cyclones.* Although hurricanes and tornadoes are, in fact, cyclones, the vast majority of cyclones are *not* hurricanes or tornadoes. The term *cyclone* simply refers to the circulation around any low-pressure center, no matter how large or intense it is.

Tornadoes and hurricanes are both smaller and more violent than middle-latitude cyclones. Middle-latitude cyclones can have a diameter of 1600 kilometers (1000 miles) or more. By contrast, hurricanes average only 600 kilometers (375 miles) across, and tornadoes, with a diameter of just 0.25 kilometer (0.16 mile) are much too small to show up on a weather map.

The thunderstorm, a much more familiar weather event, hardly needs to be distinguished from tornadoes, hurricanes, and midlatitude cyclones. Unlike the flow of air about these latter storms, the circulation associated with thunderstorms is characterized by strong up-and-down movements. Winds in the vicinity of a thunderstorm do not follow the inward spiral of a cyclone, but they are typically variable and gusty.

Although thunderstorms form "on their own" away from cyclonic storms, they also form in conjunction with cyclones. For instance, thunderstorms are frequently spawned along the cold front of a midlatitude cyclone, where on rare occasions a tornado may descend from the thunderstorm's cumulonimbus tower. Hurricanes also generate widespread thunderstorm activity. Thus, thunderstorms are related in some manner to all three types of cyclones mentioned here.

Thunderstorms

This is the first of three severe weather types we will examine in this chapter. Sections on tornadoes and hurricanes follow this look at thunderstorms.

Severe weather has a fascination that everyday weather phenomena cannot provide. The lightning display and booming thunder generated by a severe thunderstorm can be a spectacular event that elicits both awe and fear (Figure 18.13). Of course, hurricanes and tornadoes also attract a great deal of much-deserved attention. A single tornado outbreak or hurricane can cause billions of dollars in property damage as well as many deaths.

Thunderstorm Occurrence

Almost everyone has observed a small-scale phenomenon that is the result of the vertical movements of relatively warm, unstable air. Perhaps you have seen a dust devil over an open field on a hot day whirl its dusty load to great heights, or seen a bird glide effortlessly skyward on an invisible thermal of hot air. These examples illustrate the dynamic thermal instability that occurs during the development of a *thunderstorm.* A **thunderstorm** is simply a storm that generates lightning and thunder. It frequently produces gusty winds, heavy rain, and hail. A thunderstorm may be produced by a single cumulonimbus cloud and influence only a small area or it may be associated with clusters of cumulonimbus clouds covering a large area.

Figure 18.13 Cumulonimbus clouds can produce lightning, thunderstorms, and other forms of severe weather. This lightning display occurred over Tucson, Arizona. (Photo by T.A. Wiewandt/DRK Photo)

Thunderstorms form when warm, humid air rises in an unstable environment. Various mechanisms can trigger the upward air movement needed to create thunderstorm-producing cumulonimbus clouds. One mechanism, the unequal heating of Earth's surface, significantly contributes to the formation of *air-mass thunderstorms*. These storms are associated with the scattered puffy cumulonimbus clouds that commonly form *within* maritime tropical air masses and produce scattered thunderstorms on summer days. Such storms are usually short-lived and seldom produce strong winds or hail.

In contrast, thunderstorms in a second category not only benefit from uneven surface heating but are associated with the lifting of warm air, as occurs along a front or a mountain slope. Moreover, diverging winds aloft frequently contribute to the formation of these storms because they tend to draw air from lower levels upward beneath them. Some of the thunderstorms in this second category may produce high winds, damaging hail, flash floods, and tornadoes. Such storms are described as *severe*.

At any given time there are an estimated 2000 thunderstorms in progress on Earth. As we would expect, the greatest number occur in the tropics where warmth, plentiful moisture, and instability are always present. About 45,000 thunderstorms take place each day, and more than 16 million occur annually around the world. The lightning from these storms strikes Earth 100 times each second (see Box 18.2)!

Annually, the United States experiences about 100,000 thunderstorms and millions of lightning strikes. A glance at Figure 18.14 shows that thunderstorms are most frequent in Florida and the eastern Gulf Coast region, where such activity is recorded between 70 and 100 days each year. The region on the east side of the Rockies in Colorado and New Mexico is next, with thunderstorms occurring on 60 to 70 days each year. Most of the rest of the nation experiences thunderstorms on 30 to 50 days annually. Clearly, the western margin of the United States has little thunderstorm activity. The same is true for the northern tier of states and for Canada, where warm, moist, unstable mT air seldom penetrates.

Stages of Thunderstorm Development

All thunderstorms require warm, moist air, which, when lifted, will release sufficient latent heat to provide the buoyancy necessary to maintain its upward flight. This instability and associated buoyancy are triggered by a number of different processes, yet all thunderstorms have a similar life history.

Because instability and buoyancy are enhanced by high surface temperatures, thunderstorms are most common in the afternoon and early evening. Surface heating is generally not sufficient in itself to cause the growth of towering cumulonimbus clouds. A solitary cell of rising hot air produced by surface heating alone can, at best, produce a small cumulus cloud, which would evaporate within 10 to 15 minutes.

The development of 12-kilometer (or on rare occasions 18-kilometer) cumulonimbus towers requires a continuous supply of moist air. Each new surge of warm air rises higher than the last, adding to the height of the cloud (Figure 18.15). These updrafts must occasionally reach speeds over 100 kilometers (60 miles) per hour to accommodate the size of hailstones they are capable of carrying upward.

Usually within an hour the amount and size of precipitation that has accumulated is too much for the updrafts to support, and in one part of the cloud downdrafts develop, releasing heavy precipitation. This represents the most active stage of the thunderstorm. Gusty winds, lightning, heavy precipitation, and sometimes hail are experienced.

Eventually downdrafts dominate throughout the cloud. The cooling effect of falling precipitation, coupled with the influx of colder air aloft, marks the end of the thunderstorm activity. The life span of a single cumulonimbus cell within a thunderstorm complex is only an hour or two, but as the storm moves, fresh supplies of warm, water-laden air generate new cells to replace those that are dissipating.

Tornadoes

Tornadoes are local storms of short duration that must be ranked high among nature's most destructive forces. Their sporadic occurrence and violent winds cause many deaths each year. Tornadoes are violent windstorms that take the form of a rotating column of air or *vortex* that extends downward from a cumulonimbus cloud.

Pressures within some tornadoes have been estimated to be as much as 10 percent lower than immediately outside the storm. Drawn by the much lower pressure in the center of the vortex, air near the ground rushes into the tornado from all directions. As the air streams inward, it is spiraled upward around the core until it eventually merges with the airflow of the parent thunderstorm deep in the cumulonimbus tower. Because of the tremendous pressure gradient associated with a strong tornado, maximum winds can sometimes approach 480 kilometers (300 miles) per hour.

Some tornadoes consist of a single vortex, but within many stronger tornadoes are smaller whirls called *suction vortices* that rotate within the main vortex (Figure 18.16). Suction vortices have diameters of only about 10 meters (30 feet) and rotate very rapidly. This structure accounts for occasional observations of virtually total destruction of one building while another one, just 10 or 20 meters away, suffers little damage.

BOX 18.2

PEOPLE AND THE ENVIRONMENT:
Lightning Safety*

Lightning ranks second only to floods in the number of storm-related deaths each year. Although the number of reported lightning deaths in the United States annually averages about 85, many go unreported. About 100 people are estimated to be killed and more than 500 injured by lightning every year in the United States.

Warnings, statements, and forecasts are routinely issued for floods, tornadoes, and hurricanes but not for lightning. Why is this the case? The answer relates to the wide geographic occurrence and frequency of lightning.

The magnitude of the cloud-to-ground lightning hazard is understood better today than ever before. Lightning occurs in the United States every day in summer and nearly every day during the rest of the year. In an average year the National Lightning Detection Network identifies about 22,000,000 cloud-to-ground flashes. Because lightning is so widespread and strikes the ground with such great frequency it is not possible to warn each person of every flash. For this reason, lightning can be considered the most dangerous weather hazard that many people encounter each year.

Being aware of and following proven safety guidelines can greatly reduce the risk of injury or death. Individuals are ultimately responsible for their personal safety and should take appropriate action when threatened by lightning.

No place is absolutely safe from the threat of lightning, but some places are safer than others.

- Large enclosed structures (substantially, constructed buildings) tend to be much safer than smaller or open structures. The risk for lightning injury depends on whether the structure incorporates lightning protection, the types of construction materials used, and the size of the structure.

- In general, fully enclosed metal vehicles such as cars, trucks, buses, vans, and fully enclosed farm vehicles, etc., with the windows rolled up, provide good shelter from lightning. Avoid contact with metal or conducting surfaces outside or inside the vehicle.

- *Avoid* being in or near high places and open fields, isolated trees, unprotected gazebos, rain or picnic shelters, baseball dugouts, communications towers, flagpoles, light poles, bleachers (metal or wood), metal fences, convertibles, golf carts, and water (ocean, lakes, swimming pools, rivers, etc.).

- When inside a building, *avoid* use of the telephone, taking a shower, washing your hands, doing dishes, or any contact with conductive surfaces with exposure to the outside such as metal door or window frames, electrical wiring, telephone wiring, cable TV wiring, plumbing, and so forth.

Safety guidelines for individuals include the following:

- Generally speaking, if individuals can see lightning and/or hear thunder, they are already at risk. Louder or more frequent thunder indicates that lightning activity is approaching, thus increasing the risk for lightning injury or death. If the time delay between seeing the flash (lightning) and hearing the bang (thunder) is less than 30 seconds, the individual should be in, or seek, a safer location. Be aware that this method of ranging has severe limitations in part due to the difficulty of associating the proper thunder to the corresponding flash.

- High winds, rainfall, and cloud cover often act as precursors to actual cloud-to-ground strikes, and these should motivate individuals to take action. Many lightning casualties occur in the beginning, as the storm approaches, because people ignore these precursors. Also, many lightning casualties occur after the perceived threat has passed. Generally, the lightning threat diminishes with time after the last sound of thunder but may persist for more than 30 minutes. When thunderstorms are in the area but not overhead, the lightning threat can exist even when it is sunny, not raining, or when clear sky is visible.

- When available, pay attention to weather-warning devices such as NOAA (National Oceanic and Atmospheric Administration) weather radio and/or credible lightning detection systems; however, do not let this information override good common sense.

*Material in this box is based on "Updated Recommendations for Lightning Safety—1998, in *Bulletin of the American Meteorological Society*, Vol. 80, No.10, October 1999, pp. 2035–39.

Tornado Occurrence and Development

Tornadoes form in association with severe thunderstorms that produce high winds, heavy (sometimes torrential) rainfall, and often damaging hail. Fortunately, less than 1 percent of all thunderstorms produce tornadoes. Nevertheless, a much higher number must be monitored as potential tornado producers. Although meteorologists are still not sure what triggers tornado formation, it is apparent that they are the product of the interaction between strong updrafts in the thunderstorm and the winds in the troposphere.

Tornadoes can form in any situation that produces severe weather, including cold fronts and tropical cyclones (hurricanes). Usually the most intense tornadoes are those that form in association with huge thunderstorms called *supercells*. An important precondition linked to tornado formation in severe thunderstorms is the development of a mesocyclone. A *mesocyclone* is a vertical cylinder of rotating air, typically about 3 to 10 kilometers (2 to 6 miles) across, that develops in the updraft of a severe thunderstorm (Figure 18.17). The formation of this large vortex often precedes tornado formation by 30 minutes or so.

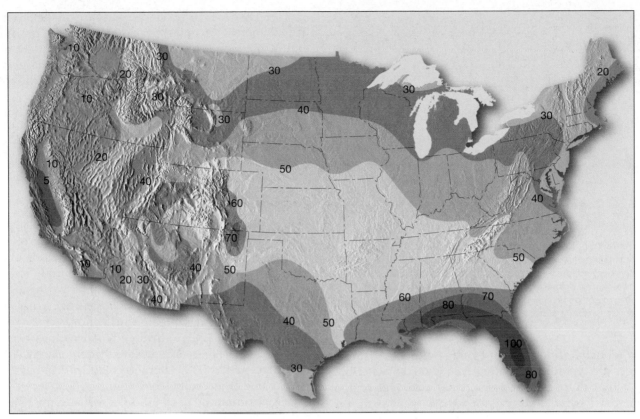

Figure 18.14   Average number of days per year with thunderstorms. Because of its close proximity to the source region for warm, humid, and unstable air masses, the Gulf Coast receives much of its precipitation from thunderstorms. (Source: Environmental Data Service, NOAA)

The formation of a mesocyclone does not necessarily mean that tornado formation will follow. Only about half of all mesocyclones produce tornadoes. Because this is the case, forecasters cannot determine in advance which mesocyclones will spawn tornadoes.

General Atmospheric Conditions.   Severe thunderstorms and hence tornadoes are most often spawned along the cold front of a middle-latitude cyclone. Throughout spring, air masses associated with middle-latitude cyclones are most likely to have greatly contrasting conditions. Continental polar air from Canada may still be very cold and dry, whereas maritime tropical air from the Gulf of Mexico is warm, humid, and unstable. The greater the contrast when these air masses meet, the more intense the storm.

These two contrasting air masses are most likely to meet in the central United States, because there is no significant natural barrier separating the center of the country from the Arctic or the Gulf of Mexico. Consequently, this region generates more tornadoes than any other area of the country or, in fact, the world. Figure 18.18, which depicts tornado incidence in the United States for a 27-year period, readily substantiates this fact.

Figure 18.15   Stages in the development of a thunderstorm. During the cumulus stage, strong updrafts act to build the storm. The mature stage is marked by heavy precipitation and cool downdrafts in part of the storm. When the warm updrafts disappear completely, precipitation becomes light, and the cloud begins to evaporate.

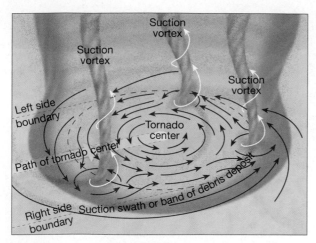

Figure 18.16 Some tornadoes have multiple suction vortices. These small and very intense vortices are roughly 10 meters (30 feet) across and move in a counterclockwise path around the tornado center. Because of this multiple vortex structure, one building might be heavily damaged and another one, just 10 meters away, might suffer little damage. (After Fujita)

Tornado Climatology. An average of about 770 tornadoes are reported annually in the United States. Still, the actual number that occur from one year to the next varies greatly. During a recent 45-year span, for example, yearly totals ranged from a low of 421 to a high of 1426.

Tornadoes occur during every month of the year. April through June is the period of greatest tornado frequency in the United States, and December and January are the months of lowest activity (Figure 18.18 graph inset). Of the nearly 35,000 confirmed tornadoes reported over the contiguous 48 states during the period, an average of almost five per day occurred during May. At the other extreme, a tornado was reported only every other day in January.

Profile of a Tornado. An average tornado has a diameter of between 150 and 600 meters (500 and 2000 feet), travels across the landscape at approximately 45 kilometers (30 miles) per hour, and cuts a path about 10 kilometers (6 miles) long.* Because tornadoes usually occur slightly ahead of a cold front, in the zone of southwest winds, most move toward the northeast. The Illinois example demonstrates this movement (Figure 18.19). Figure 18.19 also shows that many tornadoes do not fit the description of the "average" tornado.

Of the hundreds of tornadoes reported in the United States each year, over half are comparatively weak and short-lived. Most of these small tornadoes have lifetimes of three minutes or less and paths that seldom exceed 1 kilometer (0.6 mile) in length and 100 meters

*The 10-kilometer figure applies to documented tornadoes. Because many small tornadoes go undocumented, the real average path of all tornadoes is unknown but shorter than 10 kilometers.

(330 feet) wide. Typical wind speeds are on the order of 150 kilometers (96 miles) per hour or less. On the other end of the tornado spectrum are the infrequent and often long-lived violent tornadoes. Although large tornadoes constitute only a small percentage of the total reported, their effects are often devastating. Such tornadoes may exist for periods in excess of three hours and produce an essentially continuous damage path more than 150 kilometers (90 miles) long and perhaps a kilometer or more wide. Maximum winds range beyond 500 kilometers (310 miles) per hour.

Tornado Destruction

The potential for tornado destruction depends largely on the strength of the winds generated by the storm. Because tornadoes generate the strongest winds in nature, they have accomplished many seemingly impossible tasks, such as driving a piece of straw through a thick wooden plank and uprooting huge trees. Although it may seem impossible for winds to cause some of the fantastic damage attributed to tornadoes, tests in engineering facilities have repeatedly demonstrated that winds in excess of 320 kilometers (200 miles) per hour are capable of incredible feats (Figure 18.20A).

One commonly used guide to tornado intensity is the *Fujita intensity scale,* or simply the *F-scale* (Table 18.1). Because tornado winds cannot be measured directly, a rating on the F-scale is determined by assessing the worst damage produced by a storm.

Tornadoes take many lives each year, sometimes hundreds in a single day. When tornadoes struck an area stretching from Canada to Georgia on April 3, 1974, the death toll exceeded 300, the worst in half a century. Most tornadoes, however, do not result in a loss of life. In one statistical study that examined a 29-year period, there were 689 tornadoes that resulted in deaths. This figure represents slightly less than 4 percent of the total 19,312 reported storms.

Although the percentage of tornadoes that result in death is small, each tornado is potentially lethal. When tornado fatalities and storm intensities are compared, the results are quite interesting: The majority (63 percent) of tornadoes are weak (F0 and F1), and the number of storms decreases as tornado intensity increases. The dis-

Table 18.1 Fujita Intensity Scale.

| Scale | Wind Speed | | Damage |
	Km/Hr	Mi/Hr	
F0	<116	<72	Light damage
F1	116–180	72–112	Moderate damage
F2	181–253	113–157	Considerable damage
F3	254–332	158–206	Severe damage
F4	333–419	207–260	Devastating damage
F5	>419	>260	Incredible damage

Figure 18.17   The formation of a mesocyclone often precedes tornado formation. **A.** Winds are stronger aloft than at the surface (called *speed wind shear*), producing a rolling motion about a horizontal axis. **B.** Strong thunderstorm updrafts tilt the horizontally rotating air to a nearly vertical alignment. **C.** The mesocyclone, a vertical cylinder of rotating air, is established. **D.** If a tornado develops it will descend from a slowly rotating wall cloud in the lower portion of the mesocyclone. This supercell tornado hit the Texas Panhandle in May 1996. (Photo by Warren Faidley/Weatherstock)

tribution of tornado fatalities, however, is just the opposite. Although only 2 percent of tornadoes are classified as violent (F4 and F5), they account for nearly 70 percent of the deaths. If there is some question as to the causes of tornadoes, there certainly is no question about the destructive effects of these violent storms (Figure 18.20B)

Tornado Forecasting

Because severe thunderstorms and tornadoes are small and relatively short-lived phenomena, they are among the most difficult weather features to forecast precisely. Nevertheless, the prediction, the detection, and the monitoring of such storms are among the most important ser-

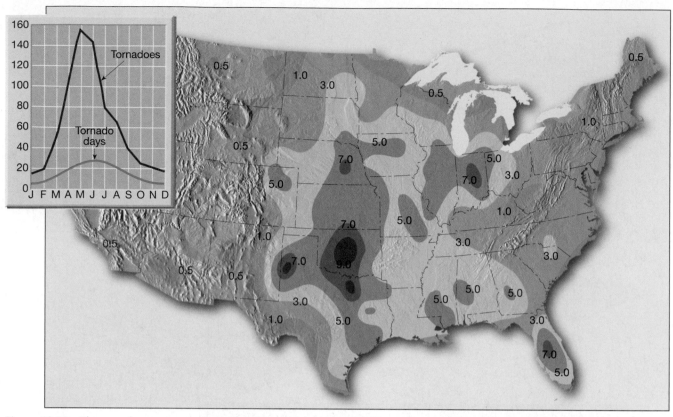

Figure 18.18   The map shows average annual tornado incidence per 10,000 square miles (26,000 square kilometers) for a 27-year period. The graph shows the average number of tornadoes and tornado days each month in the United States for the same period.

STUDENTS SOMETIMES ASK...
What is the most destructive tornado on record?

One tornado easily ranks above all others as the single most dangerous and destructive. Known as the Tri-state tornado, it occurred on March 18, 1925. Its path is labeled on Figure 18.19. Starting in southeastern Missouri, the tornado remained on the ground for 352 kilometers (219 miles), finally ending in Indiana. The losses included 695 dead and 2027 injured. Property losses were also great, with several small towns almost totally destroyed.

vices provided by professional meteorologists. The timely issuance and dissemination of watches and warnings are both critical to the protection of life and property.

The Storm Prediction Center (SPC) located in Norman, Oklahoma, is part of the National Weather Service (NWS) and the National Centers for Environmental Prediction (NCEP). Its mission is to provide timely and accurate forecasts and watches for severe thunderstorms and tornadoes.

Severe thunderstorm outlooks are issued several times daily. *Day 1* outlooks identify those areas likely to be affected by severe thunderstorms during the next 6 to 30 hours, and *day 2* outlooks extend the forecast

through the following day. Both outlooks describe the type, coverage, and intensity of the severe weather expected. Many local NWS field offices also issue severe weather outlooks that provide a more local description of the severe weather potential for the next 12 to 24 hours.

Tornado Watches and Warnings   **Tornado watches** alert the public to the possibility of tornadoes over a specified area for a particular time interval. Watches serve to fine-tune forecast areas already identified in severe weather outlooks. A typical watch covers an area of about 65,000 square kilometers (25,000 square miles) for a four- to six-hour period. A tornado watch is an important part of the tornado alert system because it sets in motion the procedures necessary to deal adequately with detection, tracking, warning, and response. Watches are generally reserved for organized severe weather events where the tornado threat will affect at least 26,000 square kilometers (10,000 square miles) and/or persist for at least three hours. Watches typically are not issued when the threat is thought to be isolated and/or short-lived.

Whereas a tornado watch is designed to alert people to the possibility of tornadoes, a **tornado warning** is issued by local offices of the National Weather Service when a tornado has actually been sighted in an area or is indicated by weather radar. It warns of a high prob-

Figure 18.19 Paths of Illinois tornadoes (1916–1969). Because most tornadoes occur slightly ahead of a cold front, in the zone of southwest winds, they tend to move toward the northeast. Tornadoes in Illinois verify this. Over 80 percent exhibited directions of movement toward the northeast through east. (After John W. Wilson and Stanley A. Changnon, Jr., *Illinois Tornadoes*, Illinois State Water Survey Circular 103, 1971, pp. 10, 24)

ability of imminent danger. Warnings are issued for much smaller areas than watches, usually covering portions of a county or counties. In addition, they are in effect for much shorter periods, typically 30 to 60 minutes. Because a tornado warning may be based on an actual sighting, warnings are occasionally issued after a tornado has already developed. However, most warnings are issued prior to tornado formation, sometimes by several tens of minutes, based on Doppler radar data and/or spotter reports of funnel clouds.

If the direction and the approximate speed of the storm are known, an estimate of its most probable path can be made. Because tornadoes often move erratically, the warning area is fan-shaped downwind from the point where the tornado has been spotted. Improved forecasts and advances in technology have contributed to a significant decline in tornado deaths over the last 50 years.

Doppler Radar Many of the difficulties that once limited the accuracy of tornado warnings have been reduced or eliminated by an advancement in radar technology called **Doppler radar.** Doppler radar not only performs the same tasks as conventional radar but also has the ability to detect motion directly. Doppler radar can detect the initial formation and subsequent development of a *mesocyclone.* Almost all mesocyclones produce damaging hail, severe winds, or tornadoes. Those that produce tornadoes (about 50 percent) can sometimes be distinguished by their stronger wind speeds and their sharper gradients of wind speeds.

It should also be pointed out that not all tornado-bearing storms have clear-cut radar signatures and that other storms can give false signatures. Detection, therefore, is sometimes a subjective process and a given display could be interpreted in several ways. Consequently, trained observers will continue to form an important part of the warning system in the foreseeable future.

Although some operational problems exist, the benefits of Doppler radar are many. As a research tool, it is not only providing data on the formation of tornadoes

A.

B.

Figure 18.20 **A.** The force of the wind during a tornado near Wichita, Kansas, in April 1991 was enough to drive this piece of metal into a utility pole. (Photo by John Sokich/NOAA) **B.** Damage from a tornado that struck Xenia, Ohio, on September 20, 2000. (Photo by Craig Holman/Reuters NewMedia/CORBIS).

but is also helping meteorologists gain new insights into thunderstorm development, the structure and dynamics of hurricanes, and air-turbulence hazards that plague aircraft. As a practical tool for tornado detection, Doppler radar has significantly improved our ability to track thunderstorms and issue warnings.

Hurricanes

Most of us view the weather in the tropics with favor. Places like Hawaii and the islands of the Caribbean are known for their lack of significant day-to-day variations. Warm breezes, steady temperatures, and rains that come as heavy but brief tropical showers are expected. It is ironic that these relatively tranquil regions sometimes produce the most violent storms on Earth.

The whirling tropical cyclones that on occasion have wind speeds attaining 300 kilometers (185 miles) per hour are known in the United States as **hurricanes** and are the greatest storms on Earth. Out at sea, they can generate 15-meter (50-foot) waves capable of inflicting destruction hundreds of kilometers from their source. Should a hurricane smash onto land, strong winds coupled with extensive flooding can impose billions of dollars in damage and great loss of life (Figure 18.21).

The vast majority of hurricane-related deaths and damage are caused by relatively infrequent yet powerful storms. Hurricane Floyd was one such storm. In Sep-

tember 1999 Floyd brought flooding rains, high winds, and rough seas to a large portion of the Atlantic Seaboard (Figure 18.22). More than 2.5 million people evacuated their homes from Florida, north to the Carolinas, and beyond. It was the largest peacetime evacuation in U.S. history. Torrential rains falling on already saturated ground created devastating inland flooding. Floyd was the deadliest hurricane to strike the U.S. mainland since Agnes in 1972. Most of the deaths were due to drowning from freshwater floods. Each hurricane season, storms like Floyd remind us of our vulnerability to these great forces of nature.

Hurricanes are becoming a growing threat because more and more people are living and working along and near coasts. At the close of the twentieth century, more than 50 percent of the U.S. population lived within 75 kilometers (45 miles) of a coast. This number is projected to increase substantially in the early decades of the twenty-first century. The concentration of such large numbers of people near the shoreline means that hurricanes and other large storms place millions at risk. Moreover, the potential costs of property damage are incredible.

Profile of a Hurricane

Most hurricanes form between the latitudes of 5 degrees and 20 degrees over all the tropical oceans except those of the South Atlantic and eastern South Pacific. They

Figure 18.21 Hurricane damage along the North Carolina coast caused by Hurricane Floyd in September, 1999. (Photo by Bruce Roberts/Photo Researchers, Inc.)

A.

B.

Figure 18.22 Hurricane Floyd formed from a tropical disturbance (easterly wave) that moved westward from the coast of Africa on September 2, 1999. The system became a tropical depression on September 7. The path on this map begins at that point. The hurricane ravaged portions of the Bahamas on September 13 and 14 and posed a serious threat to Florida. The satellite image shows the storm at this point. After weakening somewhat, Floyd eventually made landfall near Cape Fear, North Carolina, on September 16. Torrential rains were responsible for massive inland flooding. (AP/Wide World Photos)

515

are known by different names in various parts of the world. In the western Pacific, they are called *typhoons* and in the Indian Ocean, *cyclones.* The North Pacific has the greatest number of storms, averaging 20 per year. Fortunately for those living in the coastal regions of the southern and eastern United States, fewer than five hurricanes, on the average, develop annually in the warm sector of the North Atlantic.

Although many tropical disturbances develop each year, only a few reach hurricane status. By international agreement, a hurricane has wind speeds in excess of 119 kilometers (74 miles) per hour and a rotary circulation. Mature hurricanes average 600 kilometers (375 miles) in diameter and often extend 12,000 meters (40,000 feet) above the ocean surface. From the outer edge to the center, the barometric pressure has on occasion dropped 60 millibars, from 1010 millibars to 950 millibars. The lowest pressures ever recorded in the Western Hemisphere are associated with these storms. A steep pressure gradient generates the rapid, inward-spiraling winds of a hurricane.

As the inward rush of warm, moist surface air approaches the core of the storm, it turns upward and ascends in a ring of cumulonimbus towers (Figure 18.23). This doughnut-shaped wall of intense convective activity surrounding the center of the storm is called the **eye wall.** It is here that the greatest wind speeds and heaviest rainfall occur. Surrounding the eye wall are curved bands of clouds that trail away in a spiral fashion. Near the top of the hurricane the airflow is outward, carrying the rising air away from the storm center, thereby providing room for more inward flow at the surface.

At the very center of the storm is the **eye** of the hurricane (Figure 18.23). This well-known feature is a zone about 20 kilometers (12.5 miles) in diameter where precipitation ceases and winds subside. It offers a brief but deceptive break from the extreme weather in the enormous curving wall clouds that surround it. The air within the eye gradually descends and heats by compression, making it the warmest part of the storm. Although many people believe that the eye is characterized by clear blue skies, such is usually not the case because the subsidence in the eye is seldom strong enough to produce cloudless conditions. Although the sky appears much brighter in this region, scattered clouds at various levels are common.

Hurricane Formation and Decay

A hurricane is a heat engine that is fueled by the latent heat liberated when huge quantities of water vapor condense. The amount of energy produced by a typical hurricane in just a single day is truly immense—roughly equivalent to the entire electrical energy production of the United States in a year. The release of latent heat warms the air and provides buoyancy for its upward flight. The result is to reduce the pressure near the surface, which encourages a more rapid inward flow of air. To get this engine started, a large quantity of warm, moisture-laden air is required, and a continual supply is needed to keep it going.

Hurricanes develop most often in the late summer when water temperatures have reached 27°C (80°F) or higher and thus are able to provide the necessary heat and moisture to the air. This ocean-water temperature requirement accounts for the fact that hurricanes do not

Figure 18.23 Cross section of a hurricane. Note that the vertical dimension is greatly exaggerated. The eye, the zone of relative calm at the center of the storm, is a distinctive hurricane feature. Sinking air in the eye warms by compression. Surrounding the eye is the eye wall, the zone where winds and rain are most intense. Tropical moisture spiraling inward creates rain bands that pinwheel around the storm center. Outflow of air at the top of the hurricane is important because it prevents the convergent flow at lower levels from "filling in" the storm. (After NOAA)

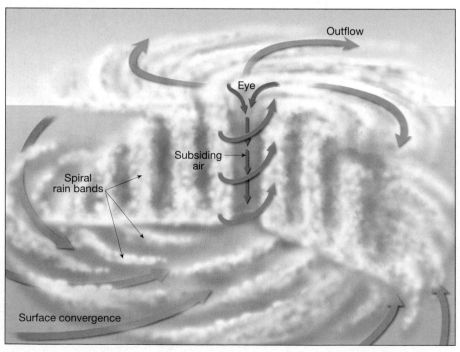

form over the relatively cool waters of the South Atlantic and the eastern South Pacific. For the same reason, few hurricanes form poleward of 20 degrees of latitude. Although water temperatures are sufficiently high, hurricanes do not form within 5 degrees of the equator, because the Coriolis effect is too weak to initiate the necessary rotary motion.

Many tropical storms begin as disorganized arrays of clouds and thunderstorms that develop weak pressure gradients but exhibit little or no rotation. Such areas of low-level convergence and lifting are called *tropical disturbances.* Most of the time these zones of convective activity die out. However, occasionally tropical disturbances grow larger and develop a strong cyclonic rotation.

Many tropical disturbances occur each year, but only a few develop into full-fledged hurricanes. By international agreement, lesser tropical cyclones are given different names based on the strength of their winds. When a cyclone's strongest winds do not exceed 61 kilometers (38 miles) per hour, it is called a **tropical depression.** When winds are between 61 and 119 kilometers (38 and 74 miles) per hour, the cyclone is termed a **tropical storm.** It is during this phase that a name is given (Andrew, Mitch, Opal, etc.). Should the tropical storm become a hurricane, the name remains the same. Each year between 80 and 100 tropical storms develop around the world. Of them, usually half or more eventually reach hurricane status.

Hurricanes diminish in intensity whenever they (1) move over ocean waters that cannot supply warm, moist tropical air, (2) move onto land; or (3) reach a location where the large-scale flow aloft is unfavorable. Whenever a hurricane moves onto land, it loses its punch rapidly. The most important reason for this rapid demise is the fact that the storm's source of warm, moist air is cut off. When an adequate supply of water vapor

? STUDENTS SOMETIMES ASK...
*Why are hurricanes given names,
and who picks the names?*

Actually, the names are given once the storms reach tropical-storm status (winds between 61–119 kilometers per hour). Tropical storms are named to provide ease of communication between forecasters and the general public regarding forecasts, watches, and warnings. Tropical storms and hurricanes can last a week or longer, and two or more storms can be occurring in the same region at the same time. Thus, names can reduce the confusion about what storm is being described.

The World Meteorological Organization (affiliated with the U.N.) creates the lists of names. The names for Atlantic storms are used again at the end of a six-year cycle unless a hurricane was particularly noteworthy. Such names are retired to prevent confusion when storms are discussed in future years.

does not exist, condensation and the release of latent heat must diminish. In addition, friction from the increased roughness of the land surface rapidly slows surface wind speeds. This factor causes the winds to move more directly into the center of the low, thus helping to eliminate the large pressure differences.

Hurricane Destruction

A location only a few hundred kilometers from a hurricane—just one day's striking distance away—may experience clear skies and virtually no wind. Prior to the age of weather satellites, such a situation made the task of warning people of impending storms very difficult.

Once a storm develops cyclonic flow and the spiraling bands of clouds characteristic of a hurricane, it receives continuous monitoring. When hurricanes form, satellites are able to identify and track the storms long before they make landfall (see Figure 18.22). In the United States, early warning systems have greatly reduced the number of deaths caused by hurricanes. At the same time, however, there has been an astronomical rise in the amount of property damage. The primary reason, of course, has been the extensive building of homes and businesses in coastal areas.

Although the amount of damage caused by a hurricane depends on several factors, including the size and population density of the area affected and the shape of the ocean bottom near the shore, certainly the most significant factor is the strength of the storm itself. Based on the study of past storms, the *Saffir-Simpson scale* was established to rank the relative intensities of hurricanes (Table 18.2). Predictions of hurricane severity and damage are usually expressed in terms of this scale. When a tropical storm becomes a hurricane, the National Weather Service assigns it a scale (category) number. As conditions change, the category of the storm is reevaluated so that public-safety officials can be kept informed. By using the Saffir-Simpson scale, the disaster potential of a hurricane can be monitored and appropriate precautions can be planned and implemented. A rating of 5 on the scale represents the worst storm possible, and a 1 is least severe. Storms that fall into category 5 are rare. Only three such storms hit the United States in the twentieth century. One was Hurricane Camille in 1969 (Figure 18.24).

Table 18.2 Saffir–Simpson hurricane scale.

Scale Number (category)	Central Pressure (millibars)	Winds (km/hr)	Storm Surge (meters)	Damage
1	≥980	119–153	1.2–1.5	Minimal
2	965–979	154–177	1.6–2.4	Moderate
3	945–964	178–209	2.5–3.6	Extensive
4	920–944	210–250	3.7–5.4	Extreme
5	<920	>250	>5.4	Catastrophic

Figure 18.24 In 1969, Hurricane Camille hit the coast of Mississippi. It was a rare category 5 storm. These classic photos document the devastating force of the hurricane's 25-foot storm surge at Pass Christian, Mississippi. **A.** The Richelieu Apartments before the hurricane. This substantial-looking three-story building was directly across the highway from the beach. **B.** The same apartments after the hurricane. (Photos by Chauncey T. Hinman)

A.

B.

Damage caused by hurricanes can be divided into three categories: (1) storm surge, (2) wind damage, and (3) inland flooding.

Storm Surge The most devastating damage in the coastal zone is caused by the storm surge. It not only accounts for a large share of coastal property losses but is also responsible for 90 percent of all hurricane-caused deaths. A **storm surge** is a dome of water 65 to 80 kilometers (40 to 50 miles) wide that sweeps across the coast near the point where the eye makes landfall. If all wave activity were smoothed out, the storm surge is the height of the water above normal tide level. Thus, a storm surge commonly adds 2 to 3 meters (6 to 10 feet) to normal tide heights—to say nothing of tremendous wave activity superimposed atop the surge.

In the delta region of Bangladesh, for example, the land is mostly less than 2 meters (6.5 feet) above sea level. When a storm surge superimposed upon normal high tide inundated that area on November 13, 1970, the official death toll was 200,000; unofficial estimates ran to 500,000. This was one of the worst disasters of modern times. In May 1991 a similar event again struck Bangladesh. This time the storm took the lives of at least 135,000 people and devastated coastal villages in its path.

Wind Damage Destruction caused by wind is perhaps the most obvious of the classes of hurricane damage. For some structures, the force of the wind is sufficient to cause total ruin. Mobile homes are particularly vulnerable. In addition, the strong winds can create a dangerous barrage of flying debris. In regions with good building codes, wind damage is usually not as catastrophic as storm-surge damage. However, hurricane-force winds affect a much larger area than storm surge and can cause huge economic losses.

For example, winds associated with Hurricane Andrew in 1992 produced more than $20 billion of damage in southern Florida and Louisiana. It was the costliest natural disaster in U.S. history. Tornadoes are spawned by many hurricanes that strike the United States and represent another aspect of wind damage. Although they can be very destructive locally, tornadoes generally account for only a small percentage of the total storm damage.

Inland Flooding. The torrential rains that accompany most hurricanes represent a third significant threat—flooding. Whereas the effects of storm surge

? STUDENTS SOMETIMES ASK...
*What is the deadliest hurricane
to strike the United States?*

That distinction goes to a hurricane that struck an unprepared Galveston, Texas, on September 8, 1900. In fact, it was the deadliest natural disaster *of any kind* in U.S. history. The strength of the storm, Galveston's vulnerable coastal setting, and the lack of adequate warning were a deadly combination. The storm took the lives of 6000 people in the city and at least 2000 more elsewhere. Fortunately, due to an effective warning system, hurricanes are no longer the unheralded killers they once were.

timeters (6 to 12 inches) or more of rain as they move inland. A good example of such destruction was Hurricane Floyd, mentioned earlier. This storm dumped more than 48 centimeters (19 inches) of rain on Wilmington, North Carolina, 33.98 centimeters (13.38 inches) of it in a single 24-hour span.

To summarize, extensive damage and loss of life in the coastal zone can result from storm surge, torrential rains, and strong winds. When loss of life occurs, it is commonly caused by the storm surge, which can devastate entire barrier islands and low-lying land along the coast. Although wind damage is usually not as catastrophic as the storm surge, it affects a much larger area. Where building codes are inadequate, economic losses can be especially severe. Because hurricanes weaken as they move inland, most wind damage occurs within 200 kilometers of the coast. Far from the coast a weakening storm can produce extensive flooding long after the winds have diminished below hurricane levels. Sometimes the damage from inland flooding exceeds storm-surge destruction.

and strong winds are concentrated in coastal areas, heavy rains may affect places hundreds of kilometers from the coast for several days after the storm has lost its hurricane-force winds.

Hurricanes weaken rapidly as they move inland, yet the remnants of the storm can still yield 15 to 30 cen-

Chapter Summary

• An *air mass* is a large body of air, usually 1600 kilometers (1000 miles) or more across, which is characterized by a *sameness of temperature and moisture* at any given altitude. When this air moves out of its region of origin, called the *source region*, it will carry these temperatures and moisture conditions elsewhere, perhaps eventually affecting a large portion of a continent.

• Air masses are classified according to (1) the nature of the surface in the source region and (2) the latitude of the source region. *Continental (c)* designates an air mass of land origin, with the air likely to be dry; whereas a *maritime (m)* air mass originates over water, and therefore will be humid. *Polar (P)* air masses originate in high latitudes and are cold. *Tropical (T)* air masses form in low latitudes and are warm. According to this classification scheme, the *four basic types of air masses* are *continental polar (cP)*, *continental tropical (cT)*, *maritime polar (mP)*, and *maritime tropical (mT)*. Continental polar (cP) and maritime tropical (mT) air masses influence the weather of North America most, especially east of the Rocky Mountains. Maritime tropical air is the source of much, if not most, of the precipitation received in the eastern two-thirds of the United States.

• *Fronts* are boundaries that separate air masses of different densities, one warmer and often higher in moisture content than the other. A *warm front* occurs when the surface position of the front moves so that warm air occupies territory formerly covered by cooler air. Along a warm front, a warm air mass overrides a retreating mass of cooler air. As the warm air ascends, it cools adiabatically to produce clouds and, frequently, light-to-moderate precipitation over a large area. A *cold front* forms where cold air is actively advancing into a region occupied by warmer air. Cold fronts are about twice as

steep and move more rapidly than do warm fronts. Because of these two differences, precipitation along a cold front is usually more intense and of shorter duration than is precipitation associated with a warm front.

• The primary weather producers in the middle latitudes are *large centers of low pressure* that generally travel from *west to east*, called *middle-latitude cyclones*. These *bearers of stormy weather*, which last from a few days to a week, have a *counterclockwise circulation* pattern in the Northern Hemisphere, with an *inward flow of air* toward their centers. Most middle-latitude cyclones have a *cold front and frequently a warm front* extending from the central area of low pressure. *Convergence and forceful lifting along the fronts* initiate cloud development and frequently cause precipitation. As a middle-latitude cyclone with its associated fronts passes over a region, it often brings with it abrupt changes in the weather. The particular weather experienced by an area depends on the path of the cyclone.

• *Thunderstorms* are caused by the upward movement of warm, moist, unstable air, triggered by a number of different processes. They are associated with cumulonimbus clouds that generate heavy rainfall, thunder, lightning, and occasionally hail and tornadoes.

• *Tornadoes*, destructive, local storms of short duration, are violent windstorms associated with severe thunderstorms that take the form of a rotating column of air that extends downward from a cumulonimbus cloud. Tornadoes are most often spawned along the cold front of a middle-latitude cyclone, most frequently during the spring months.

• *Hurricanes*, the greatest storms on Earth, are tropical cyclones with wind speeds in excess of 119 kilometers (74 miles) per hour. These complex tropical disturbances develop over tropical ocean waters and are fueled by the latent heat liberated when huge quantities of water vapor condense. Hurricanes form most often in late summer when ocean-surface temperatures reach 27°C (80°F) or higher and thus are able to provide the necessary heat and moisture to the air. Hurricanes diminish in intensity whenever they (1) move over cool ocean water that cannot supply adequate heat and moisture, (2) move onto land, or (3) reach a location where large-scale flow aloft is unfavorable.

Key Terms

air mass (p. 494)
air-mass weather (p. 495)
cold front (p. 499)
continental (c) air mass (p. 495)
Doppler radar (p. 513)
eye (p. 516)
eye wall (p. 516)
front (p. 496)
hurricane (p. 514)
lake-effect snow (p. 495)

maritime (m) air mass (p. 495)
middle-latitude or midlatitude cyclone (p. 500)
occluded front (p. 500)
occlusion (p. 504)
overrunning (p. 497)
polar (P) air mass (p. 495)
source region (p. 495)
stationary front (p. 499)
storm surge (p. 517)

thunderstorm (p. 506)
tornado (p. 507)
tornado warning (p. 512)
tornado watch (p. 512)
tropical (T) air mass (p. 495)
tropical depression (p. 517)
tropical storm (p. 517)
warm front (p. 498)

Review Questions

1. Describe the weather associated with a continental polar air mass in the winter and in the summer. When would this air mass be most welcome in the United States?

2. What are the characteristics of a maritime tropical air mass? Where are the source regions for the maritime tropical air masses that affect North America? Where are the source regions for the maritime polar air masses?

3. Describe the weather along a cold front where very warm, moist air is being displaced.

4. The formation of an occluded front marks the beginning of the end of a middle-latitude cyclone. Why is this true?

5. For each of the weather elements that follow, describe the changes that an observer experiences when a middle-latitude cyclone passes with its center north of the observer; wind direction, pressure tendency, cloud type, cloud cover, precipitation, temperature.

6. Describe the weather conditions an observer would experience if the center of a middle-latitude cyclone passed to the south.

7. Briefly explain how the flow aloft aids the formation of cyclones at the surface.

8. Compare the wind speeds and sizes of middle-latitude cyclones, tornadoes, and hurricanes. How are thunderstorms related to each?

9. What is the primary requirement for the formation of thunderstorms?

10. Based on your answer to Question 9, where would you expect thunderstorms to be most common on Earth? In the United States?

11. Why do tornadoes have such high wind speeds?

12. What general atmospheric conditions are most conducive to the formation of tornadoes?

13. When is the tornado season? That is, during what months is tornado activity most pronounced?

14. Distinguish between a tornado watch and a tornado warning.

15. Which has stronger winds, a tropical storm or a tropical depression?

16. Why does the intensity of a hurricane diminish rapidly when it moves onto land?

17. Hurricane damage can be divided into three broad categories. Name them. Which category is responsible for the highest percentage of hurricane-related deaths?

18. Great damage and significant loss of life can take place a day or more after a hurricane has moved ashore and weakened. When this occurs, what is the likely cause?

19. A hurricane has slower wind speeds than a tornado, yet it inflicts more total damage. How might this be explained?

Examining the Earth System

1. Which parts of the Earth system interact in the Great Lakes region of North America to produce the high snowfalls on the downwind shores of the lakes? What term is applied to these heavy snows? Describe what causes this effect. (A good short review of this process can be found at **http://www.usatoday.com/weather/lakeeffect.htm**

2. Hurricanes are among the most severe storms experienced on Earth. When a hurricane makes landfall in the southeastern United States, what impact might it have on coastal lands, drainage networks (hydrosphere), and natural vegetation (biosphere)? (You may find some helpful information at the National Hurricane Center's site at **http://www.nhc.noaa.gov/**)

Web Resources

 The *Earth Science* Website uses the resources and flexibility of the Internet to aid in your study of the topics in this chapter. Written and developed by Earth science instructors, this site will help improve your understanding of Earth science. Visit **http://www.prenhall.com/tarbuck** and click on the cover of *Earth Science* 10e to find:

- **Online review quizzes.**
- **Web-based critical thinking and writing exercises.**
- **Links to chapter-specific Web resources.**
- **Internet-wide key term searches.**

http://www.prenhall.com/tarbuck

Autumn in Alaska's Denali National Park. This subpolar area shows a mix of coniferous forest and tundra vegetation.
(Photo by Yva Momatiuk and John Eastcott/Photo Researchers, Inc.)

19

CLIMATE

Anyone who has the opportunity to travel around the world will find such an incredible variety of climates that it is hard to believe they could all occur on the same planet. The broad diversity of climates around the globe is an important focus of this chapter. Climate strongly influences the nature of plant and animal life, the soil, and many external geological processes. Climate influences people as well.

Although climate has a significant impact on people, we are learning that people also have a strong influence on climate. The latter portion of this chapter examines the ways in which humans may be changing global climate.

The Climate System

The focus of this chapter is *climate*. In Chapter 15, where this term was introduced, climate was characterized as being an aggregate of weather. It consists not only of average atmospheric values but also involves the variability of elements and information on the occurrence of extreme events.

To understand and appreciate climate, it is important to realize that climate involves more than just the atmosphere. Indeed, we must recognize that there is a **climate system.** In addition to the atmosphere, it consists of the hydrosphere, solid Earth, biosphere, and cryosphere (Figure 19.1). The *cryosphere* refers to the ice and snow that exist at Earth's surface. Powered by the energy from the Sun, the climate system involves the exchanges of energy and moisture that occur among the five parts. These exchanges link the atmosphere to the other parts of the system to produce an integrated and extremely complex interactive unit.

> The atmosphere is the central component of the complex, connected, and interactive global environmental system upon which all life depends. Climate may be broadly defined as the long-term behavior of this environmental system. To understand fully and to predict changes in the atmospheric component of the climate system, one must understand the Sun, oceans, ice sheets, solid Earth, and all forms of life.*

The climate system provides a framework for the study of climate. The interactions and exchanges among the parts of the climate system create a complex network that links the five spheres. Changes to the system do not occur in isolation. Rather, when one part of this interactive unit changes, the other components also react. This well-established relationship is demonstrated often when we study climate and climate change.

---

*The American Meterological Society and the University Corporation for Atmospheric Research. Weather and the Nation's Well-Being. *Bulletin of the American Meteorological Society*, 73, no. 12 (December 1991): 2038.

World Climates

Previous chapters have already presented the spatial and seasonal variations of the major elements of weather and climate. Chapter 15 examined the controls of temperature and the world distribution of temperature. In Chapter 17, you studied the general circulation of the atmosphere and the global distribution of precipitation. You are now ready to investigate the *combined* effects of these variations in different parts of the world. The varied nature of Earth's surface and the many interactions that occur among atmospheric processes give every location on our planet a distinctive, even unique, climate. Our intention, however, is not to describe the unique climatic character of countless different locales. Instead, the purpose is to introduce the major climatic regions of the world. The discussion examines large areas and uses particular places only to illustrate the characteristics of these major climatic regions.

Temperature and precipitation are the most important elements in a climatic description because they have the greatest influence on people and their activities and also have an important impact on the distribution of such phenomena as vegetation and soils. Nevertheless, other factors are also important for a complete climatic description. When possible, some of these factors are introduced into our discussion of world climates.

Climate Classification

The distribution of the major atmospheric elements is, to say the least, complex. Because of the many differences from place to place as well as from time to time at a particular locale, it is unlikely that any two sites on Earth experience exactly the same weather conditions. The fact that the number of places on Earth is virtually infinite makes it readily apparent that the number of different climates must be extremely large.

Of course, having a great diversity of information to investigate is not unique to the study of the atmosphere; it is a problem that is basic to all science. To cope with such variety, it is not only desirable but essential to devise some means of classifying the vast array of data to be studied. By establishing groups consisting of items that have certain important characteristics in common, order and simplicity are introduced. Bringing order to large quantities of information not only aids comprehension and understanding, but it also facilitates analysis and explanation.

Over the past 100 years, many climate-classification systems have been devised. It should be remembered that the classification of climates (or of anything else) is not a natural phenomenon but the product of human ingenuity. The value of any particular classification is determined largely by its intended use. A system de-

Figure 19.1 The climate system involves the complex interactions that occur among the atmosphere, hydrosphere, solid earth, cryosphere, and biosphere. All of the components are represented in this scene in Colorado's Maroon Bells. (Photo by Pete Saloutos/The Stock Market)

signed for one purpose is not necessarily applicable to another.

In this chapter, we use a classification devised by Russian-born German climatologist Wladimir Köppen (1846–1940). As a tool for presenting the general world pattern of climates, the **Köppen classification** has been the best-known and most used system for decades. It is widely accepted for many reasons. For one, it uses only easily obtained data: mean monthly and annual values of temperature and precipitation. Furthermore, the criteria are unambiguous, relatively simple to apply, and divide the world into climatic regions in a realistic way.

Köppen believed that the distribution of natural vegetation was an excellent expression of the totality of climate. Consequently, the boundaries he chose were largely based on the limits of certain plant associations. Five principal groups were recognized; each group was designated by a capital letter as follows:

A. *Humid tropical.* Winterless climates; all months having a mean temperature above 18°C (64°F).

B. *Dry.* Climates where evaporation exceeds precipitation; there is a constant water deficiency.

C. *Humid middle-latitude.* Mild winters; the average temperature of the coldest month is below 18°C (64°F) but above −3°C (27°F).

D. *Humid middle-latitude.* Severe winters; the average temperature of the coldest month is below −3°C (27°F), and the warmest monthly mean exceeds 10°C (50°F).

E. *Polar.* Summerless climates; the average temperature of the warmest month is below 10°C (50°F).

Notice that four of the major groups (A, C, D, E) are defined on the basis of temperature characteristics, and the fifth, the B group, has precipitation as its primary criterion. Each of the five groups is further subdivided by using the criteria and symbols presented in Table 19.1. In addition, the world distribution of climates according to the Köppen classification is shown in Figure 19.2. You will be referred to this figure several times as Earth's climates are discussed in the following pages.

Humid Tropical (A) Climates

Within the A group of climates, two main types are recognized—wet tropical climates (Af and Am) and tropical wet and dry (AW).

The Wet Tropics

The constantly high temperatures and year-round rainfall in the wet tropics combine to produce the most luxuriant vegetation found in any climatic realm: the **tropical rain forest** (Figure 19.3).

The environment of the wet tropics characterizes almost 10 percent of Earth's land area. An examination of Figure 19.2 shows that Af and Am climates form a discontinuous belt astride the equator that typically extends 5° to 10° into each hemisphere. The poleward margins are most often marked by diminishing rainfall, but occasionally decreasing temperatures mark the boundary. Because of the general decrease in temperature with height in the troposphere, this climate region is restricted to elevations below 1000 meters. Consequently, the major interruptions near the equator are principally cooler highland areas.

Figure 19.2 Climates of the world based on the Köppen classification.

Data for some representative stations in the wet tropics are shown in Figure 19.4A,B. A brief examination reveals the most obvious features that characterize the climate in these areas.

1. Temperatures usually average 25°C (77°F) or more each month. Consequently, not only is the annual mean temperature high, but the annual temperature range is also very small.

2. The total precipitation for the year is high, often exceeding 200 centimeters (80 inches).

3. Although rainfall is not evenly distributed throughout the year, tropical rain forest stations

Table 19.1 Köppen system of climatic classification.

1st	2nd	3rd	
A			Average temperature of the coldest month is 18°C or higher.
	f		Every month has 6 cm of precipitation or more.
	m		Short, dry season; precipitation in driest month less than 6 cm but equal to or greater than 10–R/25 (R is annual rainfall in cm).
	w		Well-defined winter dry season; precipitation in driest month less than 10–R/25.
	s		Well-defined summer dry season (rare).
B			Potential evaporation exceeds precipitation. The dry-humid boundary is defined by the following formulas: (Note: R is the average annual precipitation in cm and T is the average annual temperature in °C.) R < 2T + 28 when 70% or more of rain falls in warmer 6 months. R < 2T when 70% or more of rain falls in cooler 6 months. R < 2T + 14 when neither half year has 70% or more of rain.
	S	Steppe	The BS-BW boundary is 1/2 the dry-humid boundary.
	W	Desert	
		h	Average annual temperature is 18°C or greater.
		k	Average annual temperature is less than 18°C.
C			Average temperature of the coldest month is under 18°C and above −3°C.
	w		At least 10 times as much precipitation in a summer month as in the driest winter months.
	s		At least three times as much precipitation in a winter month as in the driest summer month; precipitation in driest summer month less than 4 cm.
	f		Criteria for w and s cannot be met.
		a	Warmest month is over 22°C; at least 4 months over 10°C.
		b	No month above 22°C; at least 4 months over 10°C.
		c	One to 3 months above 10°C.
D			Average temperature of coldest month is −3°C or below; average temperature of warmest month is greater than 10°C.
	s		Same as under C.
	w		Same as under C.
	f		Same as under C.
		a	Same as under C.
		b	Same as under C.
		c	Same as under C.
		d	Average temperature of the coldest month is −38°C or below.
E			Average temperature of the warmest month is below 10°C.
	T		Average temperature of the warmest month is greater than 0°C and less than 10°C.
	F		Average temperature of the warmest month is 0°C or below.

*When classifying climatic data using Table 19.1, you should first determine whether the data meet the criteria for the E climates. If the station is not a polar climate, proceed to the criteria for B climates. If your data do not fit into either the E or B groups, check the data against the criteria for A, C, and D climates, in that order.

Figure 19.3 Unexcelled in luxuriance and characterized by hundreds of different species per square kilometer, the tropical rain forest is a broadleaf evergreen forest that dominates the wet tropics. Tutch River region, Sarawak, Malaysia. (Photo by Art Wolfe/Photo Researchers, Inc.)

Figure 19.4 By comparing these three climatic diagrams, the primary differences among the A climates can be seen. **A.** Iquitos, the Af station, is wet throughout the year. **B.** Monrovia, the Am station, has a short, dry season. **C.** As is true for all Aw stations, Normanton has an extended dry season and a higher annual temperature range than the others.

are generally wet in all months. If a dry season exists, it is very short.

Because places with an Af or Am designation lie near the equator, the reason for the uniform temperature rhythm experienced in such locales is clear: The intensity of solar radiation is consistently high. The vertical rays of the Sun are always relatively close, and changes in the length of daylight throughout the year are slight; therefore, seasonal temperature variations are minimal.

The region is strongly influenced by the equatorial low. Its converging trade winds and the accompanying ascent of warm, humid, unstable air produce conditions that are ideal for the formation of precipitation.

Tropical Wet and Dry

In the latitude zone poleward of the wet tropics and equatorward of the subtropical deserts lies a transitional climatic region called **tropical wet and dry.** Here the rain forest gives way to the *savanna,* a tropical grassland with scattered drought-tolerant trees (Figure 19.5). Because temperature characteristics among all A climates are quite similar, the primary factor that distinguishes the Aw climate from Af and Am is precipitation. Although the overall amount of precipitation in the tropical wet and dry realm is often considerably less than in the wet tropics, the most distinctive feature of this climate is not the annual rainfall total but the markedly seasonal character of the rainfall. The climatic diagram for Normanton, Australia (Figure 19.4C), clearly illustrates this trait. As the equatorial low advances poleward in summer, the rainy season commences and features weather patterns typical of the wet tropics. Later, with the retreat of the equatorial low, the subtropical high advances into the region and brings with it pronounced dryness. In some Aw regions such as India, Southeast Asia, and portions of Australia, the alternating periods of rainfall and dryness are associated with a well-established monsoon circulation (see Chapter 17).

Dry (B) Climates

It is important to realize that the concept of dryness is a relative one and refers to any situation in which a water deficiency exists. Climatologists define a dry climate as one in which the yearly precipitation is not as great as the potential loss of water by evaporation. Thus, dryness is not only related to annual rainfall totals but is also a function of evaporation, which in turn is closely dependent upon temperature.

To establish the boundary between dry and humid climates, the Köppen classification uses formulas that involve three variables: average annual precipitation, average annual temperature, and seasonal distribution of precipitation. The use of average annual temperature reflects its importance as an index of evaporation. The amount of rainfall defining the humid–dry boundary

Figure 19.5 Tropical savanna grassland in Tanzania's Serengeti National Park. This tropical savanna, with its stunted, drought-resistant trees, probably resulted from seasonal burnings carried out by native human populations. (Photo by Stan Osolinski/Dembinsky/Photo Associates)

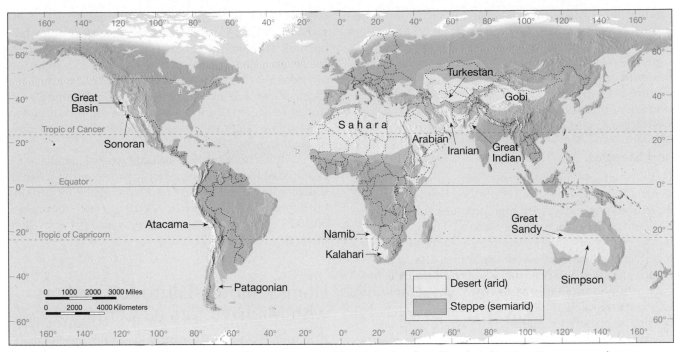

Figure 19.6   Arid and semiarid climates cover about 30 percent of Earth's land surface. No other climatic group covers so large an area.

increases as the annual mean temperature increases. The use of seasonal precipitation as a variable is also related to this idea. If rain is concentrated in the warmest months, loss to evaporation is greater than if the precipitation were concentrated in the cooler months.

Within the regions defined by a general water deficiency are two climatic types: **arid** or **desert** (BW) and **semiarid** or **steppe** (BS) (Figure 19.6). These two groups have many features in common; their differences are primarily a matter of degree. The semiarid is a marginal and more humid variant of the arid and represents a transition zone that surrounds the desert and separates it from the bordering humid climates.

STUDENTS SOMETIMES ASK...
*Do the terms "arid" and "drought"
mean about the same thing?*

The concept of drought differs from that of aridity. Drought is a *temporary* happening that is considered an atmospheric hazard. By contrast, aridity describes regions where low rainfall is a *permanent* feature of climate. Although natural disasters such as floods and hurricanes generate more attention, droughts can be just as devastating and carry a bigger price tag. On the average, droughts cost the United States $6 to $8 billion annually compared to $2.4 billion for floods and $1.2 to $4.8 billion for hurricanes.

Low-Latitude Deserts and Steppes

The heart of low-latitude dry climates lies in the vicinities of the Tropics of Cancer and Capricorn. A glance at Figure 19.6 shows a virtually unbroken desert environ-

ment stretching for more than 9300 kilometers (nearly 6000 miles) from the Atlantic coast of North Africa to the dry lands of northwestern India. In addition to this single great expanse, the Northern Hemisphere contains another, much smaller area of subtropical desert and steppe in northern Mexico and the southwestern United States. In the Southern Hemisphere, dry climates dominate Australia. Almost 40 percent of the continent is desert, and much of the remainder is steppe. In addition, arid and semiarid areas are found in southern Africa and make a limited appearance in coastal Chile and Peru.

The existence of this dry subtropical realm is primarily the result of the prevailing global distribution of air pressure and winds. Earth's low-latitude deserts and steppes coincide with the subtropical high-pressure belts (see Figure 17.12). Here, air is subsiding. When air

STUDENTS SOMETIMES ASK...
Are deserts always hot?

Actually, this is a misconception. Deserts can certainly be hot places. In fact, the record high temperature for the United States, 57°C (134°F), was set at Death Valley, California. Despite this and other remarkably high figures, deserts also experience cold temperatures. For example, the average daily minimum in January at Phoenix, Arizona, is 1.7°C (35°F), just barely above freezing. At Ulan Bator in Mongolia's Gobi Desert, the average *high* temperature in January is only −19°C (−2°F)! Although subtropical deserts lack a cold season, mid-latitude deserts do experience seasonal temperature changes.

sinks, it is compressed and warmed. Such conditions are just opposite of what is needed for cloud formation and precipitation. Therefore, clear skies, a maximum of sunshine, and drought are to be expected. The climate diagrams for Cairo, Egypt, and Monterrey, Mexico (Figure 19.7A,B), illustrate the characteristics of low-latitude dry climates.

Middle-Latitude Deserts and Steppes

Unlike their low-latitude counterparts, middle-latitude deserts and steppes are not controlled by the subsiding air masses associated with high pressure. Instead, these dry lands exist principally because of their positions in the deep interiors of large landmasses far removed from the oceans, which are the ultimate source of moisture for cloud formation and precipitation. In addition, the presence of high mountains across the paths of prevailing winds further acts to separate these areas from water-bearing, maritime air masses.

Windward sides of mountains are often wet. As prevailing winds meet mountain barriers, the air is forced to ascend, producing clouds and precipitation. By contrast, the leeward sides of mountains are usually much dryer and are often arid enough to be referred to as **rain shadow deserts.** Because many middle-latitude deserts occupy sites on the leeward sides of the mountains, they can also be classified as rainshadow deserts

(Figure 19.7C). In North America, the Coast Ranges, Sierra Nevada, and Cascades are the foremost mountain barriers. In Asia, the great Himalayan chain prevents the summertime monsoon flow of moist Indian Ocean air from reaching the interior. Because the Southern Hemisphere lacks extensive land areas in the middle latitudes, only a small area of desert and steppe is found in this latitude range, existing primarily in the rain shadow of the towering Andes.

In the case of middle-latitude deserts, we have an example of the impact of tectonic processes on climate. Rainshadow deserts exist by virtue of the mountains produced when plates collide. Without such mountain-building episodes, wetter climates would prevail where many dry regions exist today.

Humid Middle-Latitude Climates with Mild Winters (C Climates)

Although the term *subtropical* is often used for the C climates, it can be misleading. Although many areas with C climates do indeed possess some near-tropical characteristics, other regions do not. For example, we would be stretching the use of the term *subtropical* to describe the climates of coastal Alaska and Norway, which belong to the C group. Within the C group of climates, several subgroups are recognized.

Figure 19.7   Climatic diagrams for representative arid and semiarid stations. Stations **A.** and **B.** are in the subtropics, whereas **C.** is in the middle-latitudes. Cairo and Lovelock are classified as deserts; Monterey is a steppe. Lovelock, Nevada, may also be called a rain shadow desert.

Humid Subtropics

Located on the eastern sides of the continents, in the 25- to 40-degree latitude range, the **humid subtropical climate** dominates the southeastern United States, as well as other similarly situated areas around the world (Figures 19.2 and 19.8A). In the summer, the humid subtropics experience hot, sultry weather of the type one expects to find in the rainy tropics. Daytime temperatures are generally high, and because both mixing ratio and relative humidity are high, the night brings little relief. An afternoon or evening thunderstorm is also possible, for these areas experience such storms on an average of 40 to 100 days each year, the majority during the summer months.

As summer turns to autumn, the humid subtropics lose their similarity to the rainy tropics. Although winters are mild, frosts are common in the higher-latitude Cfa areas and occasionally plague the tropical margins as well. The winter precipitation is also different in character from the summer. Some is in the form of snow, and most is generated along fronts of the frequent middle-latitude cyclones that sweep over these regions.

Marine West Coast

Situated on the western (windward) side of continents, from about 40 to 65 degrees north and south latitude, is a climatic region dominated by the onshore flow of oceanic air. In North America, the **marine west coast climate** extends from near the U.S.–Canadian border northward as a narrow belt into southern Alaska (Figure 19.8B). The largest area of Cfb climate is found in Europe, for here there is no mountain barrier blocking the movement of cool maritime air from the North Atlantic.

The prevalence of maritime air masses means that mild winters and cool summers are the rule, as is an ample amount of rainfall throughout the year. Although there is no pronounced dry period, there is a drop in monthly precipitation totals during the summer. The reason for the reduced summer rainfall is the poleward migration of the oceanic subtropical highs. Although the areas of marine west coast climate are situated too far poleward to be dominated by these dry anticyclones, their influence is sufficient to cause a decrease in warm season rainfall.

Dry-Summer Subtropics

The **dry-summer subtropical climate** is typically located along the west sides of continents between latitudes 30 and 45 degrees. Situated between the marine west coast climate on the poleward side and the subtropical steppes on the equatorward side, this climatic region is best described as transitional in character. It

Figure 19.8 Each of these climatic diagrams represents one of the three main types of C climates: **A.** humid subtropical, **B.** marine west coast, and **C.** dry-summer subtropical.

is unique because it is the only humid climate that has a strong winter rainfall maximum, a feature that reflects its intermediate position (Figure 19.8C). In summer, the region is dominated by stable conditions associated with the oceanic subtropical highs. In winter, as the wind and pressure systems follow the Sun equatorward, it is within range of the cyclonic storms of the polar front. Thus, during the course of a year, these areas alternate between becoming a part of the dry tropics and an extension of the humid middle latitudes. Although middle-latitude changeability characterizes the winter, tropical constancy describes the summer.

As was the case for the marine west coast climate, mountain ranges limit the dry-summer subtropics to a relatively narrow coastal zone in both North and South America. Because Australia and southern Africa barely reach to the latitudes where dry-summer climates exist, the development of this climatic type is limited on these continents as well. Consequently, because of the arrangement of the continents, and of their mountain ranges, inland development occurs only in the Mediterranean basin. Here the zone of subsidence extends far to the east in summer; in winter, the sea is a major route of cyclonic disturbances. Because the dry-summer climate is particularly extensive in this region, the name *Mediterranean climate* is often used as a synonym.

Humid Middle-Latitude Climates with Severe Winters (D Climates)

The C climates that were just described characteristically have mild winters. By contrast, D climates experience severe winters. Two types of D climates are recognized, the humid continental and the subarctic. Climatic diagrams of representative locations are shown in Figure 19.9. The D climates are land-controlled climates, the result of broad continents in the middle latitudes. Because continentality is a basic feature, D climates are absent in the Southern Hemisphere where the middle-latitude zone is dominated by the oceans.

Humid Continental

The **humid continental climate** is confined to the central and eastern portions of North America and Eurasia in the latitude range between approximately 40 and 50 degrees north latitude. It may at first seem unusual that a continental climate should extend eastward to the margins of the ocean. However, because the prevailing atmospheric circulation is from the west, deep and persistent incursions of maritime air from the east are not likely to occur.

Both winter and summer temperatures in the humid continental climate can be characterized as

Figure 19.9 *D* climates are associated with the interiors of large landmasses in the mid-to-high latitudes of the Northern Hemisphere. Although winters can be harsh in Chicago's humid continental (Dfa) climate, the subarctic environment (Dfc) of Moose Factory is more extreme.

A.

B.

relatively severe. Consequently, annual temperature ranges are high throughout the climate.

Precipitation is generally greater in summer than in winter. Precipitation totals generally decrease toward the interior of the continents as well as from south to north, primarily because of increasing distance from the sources of mT air. Furthermore, the more northerly stations are also influenced for a greater part of the year by drier polar air masses.

Wintertime precipitation in humid continental climates is chiefly associated with the passage of fronts connected with traveling middle-latitude cyclones. Part of this precipitation is in the form of snow, the proportion increasing with latitude. Although precipitation is often considerably less during the cold season, it is usually more conspicuous than the greater amounts that fall during summer. An obvious reason is that snow remains on the ground, often for extended periods, and rain, of course, does not.

Subarctic

Situated north of the humid continental climate and south of the polar tundra is an extensive **subarctic climate** region covering broad, uninterrupted expanses from western Alaska to Newfoundland in North America and from Norway to the Pacific coast of Russia in Eurasia. It is often referred to as the *taiga* climate, for its extent closely corresponds to the northern coniferous forest region of the same name (Figure 19.10). Although scrawny, the spruce, fir, larch, and birch trees in the taiga represent the largest stretch of continuous forest on the surface of Earth.

Here in the source regions of continental polar air masses, the outstanding feature is certainly the domi-

nance of winter. Not only is it long but temperatures are also bitterly cold. Winter minimum temperatures are among the lowest ever recorded outside the ice sheets of Greenland and Antarctica. In fact, for many years, the world's coldest temperature was attributed to Verkhoyansk in east central Siberia, where the temperature dropped to −68°C (−90°F) on February 5 and 7, 1892. Over a 23-year period, this same station had an average monthly minimum of −62°C (−80°F) during January. Although exceptional temperatures, they illustrate the extreme cold that envelops the taiga in winter.

By contrast, summers in the subarctic are remarkably warm, despite their short duration. However, when compared with regions farther south, this short season must be characterized as cool. The extremely cold winters and relatively warm summers combine to produce the highest annual temperature ranges on Earth. Because these far northerly continental interiors are the source regions for cP air masses, there is very

STUDENTS SOMETIMES ASK...
How hot can it get in Alaska?

You might be surprised to learn that Alaska's record high temperature is 38°C (100°F)! The record was set June 27, 1915, at Fort Yukon, a town along the Arctic Circle in the interior of the state. It is also interesting to note that when statistics for the other states are examined, we find that all but one have record high temperatures in *excess* of 38°C. The one state that ties Alaska with the "lowest high" is Hawaii. Panaloa, on the south coast of the big island, recorded 38°C on April 27, 1931.

Figure 19.10   The northern coniferous forest is also called the taiga. Denali National Park, Alaska. (Photo by Carr Clifton)

limited moisture available throughout the year. Precipitation totals are therefore small, with a maximum occurring during the warmer summer months.

Polar (E) Climates

Polar climates are those in which the mean temperature of the warmest month is below 10°C (50°F). Thus, just as the tropics are defined by their year-round warmth, the polar realm is known for its enduring cold. As winters are periods of perpetual night, or nearly so, temperatures at most polar locations are understandably bitter. During the summer months temperatures remain cool despite the long days, because the Sun is so low in the sky that its oblique rays are not effective in bringing about a genuine warming. Although polar climates are classified as humid, precipitation is generally meager. Evaporation, of course, is also limited. The scanty precipitation totals are easily understood in view of the temperature characteristics of the region. The amount of water vapor in the air is always small because low mixing ratios must accompany low temperatures. Usually precipitation is most abundant during the warmer summer months when the moisture content of the air is highest.

Two types of polar climates are recognized. The **tundra climate** (ET) is a treeless climate found almost exclusively in the Northern Hemisphere (Figure 19.11). Because of the combination of high latitude and continentality, winters are severe, summers are cool, and annual temperature ranges are high (Figure 19.12A). Further, yearly precipitation is small, with a modest summer maximum.

The **ice cap climate** (EF) does not have a single monthly mean above 0°C (32°F) (Figure 19.12B). Consequently, because the average temperature for all months is below freezing, the growth of vegetation is prohibited and the landscape is one of permanent ice and snow. This climate of perpetual frost covers a surprisingly large area more than 15.5 million square kilometers (6 million square miles), or about 9 percent of Earth's land area. Aside from scattered occurrences in high mountain areas, it is confined to the ice sheets of Greenland and Antarctica (see Figure 5.2, p. 132).

Highland Climates

It is a well-known fact that mountains have climate conditions that are distinctively different from those found in adjacent lowlands. Compared to nearby places at lower elevations, sites with **highland climates** are cooler and usually wetter. Unlike the world climate types already discussed, which consist of large, relatively homogeneous regions, the outstanding characteristic of highland climates is the great diversity of climatic conditions that occur.

The best-known climatic effect of increased altitude is lower temperatures. In addition, an increase in precipitation due to orographic lifting usually occurs at higher elevations. Despite the fact that mountain stations are colder and often wetter than locations at lower elevations, highland climates are often very similar to those in adjacent lowlands in terms of seasonal temperature cycles and precipitation distribution. Figure 19.13 illustrates this relationship.

Phoenix, at an elevation of 338 meters (1109 feet), lies in the desert lowlands of southern Arizona. By contrast, Flagstaff is located at an altitude of 2100 meters (7000 feet) on the Colorado Plateau in northern Arizona (Figure 19.14). When summer averages climb to 34°C (93°F) in Phoenix, Flagstaff is experiencing a pleasant 19°C (66°F), which is a full 15°C (27°F) cooler. Although the temperatures at each city are quite different, the pattern of monthly temperature changes for

Figure 19.11 The tundra in bloom north of Nome, Alaska. It is a region almost completely devoid of trees. Bogs and marshes are common, and plant life frequently consists of mosses, low shrubs, and flowering herbs. (Photo by Fred Bruemmer/DRK Photo)

Figure 19.12 These climatic diagrams represent the two basic types of polar climates. **A.** Barrow, Alaska, exhibits a tundra (ET) climate. **B.** Eismitte, Greenland, a station located on a massive ice sheet, is classified as an ice cap (EF) climate.

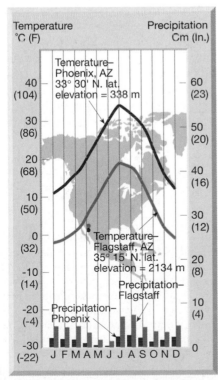

Figure 19.13 Climate diagrams for two stations in Arizona illustrate the general influence of elevation on climate. Flagstaff is cooler and wetter because of its position on the Colorado Plateau, nearly 1800 meters (6000 feet) higher than Phoenix.

each place is similar. Both experience their minimum and maximum monthly means in the same months. When precipitation data are examined, both places have a similar seasonal pattern, but the amounts at Flagstaff are higher in every month. In addition, owing to its higher altitude, much of Flagstaff's winter precipitation is in the form of snow. By contrast, all of the precipitation at Phoenix is rain.

Perhaps the terms *variety* and *changeability* best describe mountain climates. Because atmospheric conditions fluctuate rapidly with changes in altitude and exposure, a nearly limitless variety of local climates occur in mountainous regions. The climate in a protected valley is very different from that of an exposed peak. Conditions on windward slopes contrast sharply with those on the leeward sides, whereas slopes facing the Sun are unlike those that lie mainly in the shadows.

Human Impact on Global Climate

The proposals to explain global climatic change are many and varied. In Chapter 5, we examined some possible causes for ice-age climates. These hypotheses, which included the movement of lithosphere plates and variations in Earth's orbit, involved natural forcing mechanisms. Another natural forcing mechanism, discussed in Chapter 8, is the possible role of explosive volcanic eruptions in modifying the atmosphere. It is important to remember that these mechanisms, as well as others, not only have contributed to

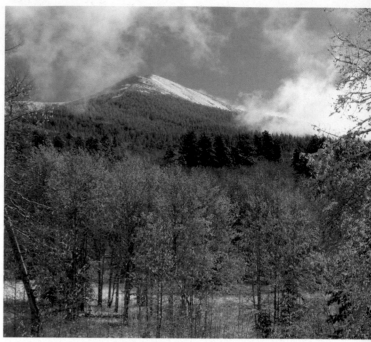

A.

B.

Figure 19.14 **A.** Only scanty drought-tolerant natural vegetation can survive in the hot, dry climate of southern Arizona, near Phoenix. (Photo by Charlie Ott Photography/Photo Researchers, Inc.) **B.** The natural vegetation associated with the cooler, wetter highlands near Flagstaff, Arizona, is much different from the desert lowlands. (Photo by Larry Ulrich/DRK Photo)

climatic changes in the geologic past but will also be responsible for future shifts in climate. However, when relatively recent and future changes in our climate are considered, we must also examine the possible impact of human beings. In this section, we shall examine the major way in which humans might be contributing to global climatic change. This impact results from the addition of carbon dioxide and other gases to the atmosphere.

Human influence of regional and global climate probably did not just begin with the onset of the modern industrial period. There is good evidence that humans have been modifying the environment over extensive areas for thousands of years. The use of fire as well as the overgrazing of marginal lands by domesticated animals have negatively affected the abundance and distribution of vegetation. By altering ground cover, such important climatological factors as surface albedo, evaporation rates, and surface winds have been, and continue to be, modified. Commenting on this aspect of human-induced climatic modification, the authors of one study observed:

> In contrast to the prevailing view that only modern humans are able to alter climate, we believe it is more likely that the human species has made a substantial and continuing impact on climate since the invention of fire.*

---

*Carl Sagan et al., Anthropogenic Albedo Changes and the Earth's Climate, *Science* 206(4425):1367.

It should be pointed out that when any hypothesis of climatic change is examined, whether it depends on natural or human-induced causes, a degree of caution must be exercised. It is safe to say that most, if not all, hypotheses are to some degree controversial and speculative. This is to be expected if we consider the fact that at present all of our models of Earth's climate are far from complete. Because atmospheric processes are so large and complex, they cannot be physically reproduced in laboratory experiments. Instead, the climate must be simulated mathematically with the aid of computers. Although such models are sophisticated enough

STUDENTS SOMETIMES ASK...
Do cities influence climate?

Yes, in lots of ways, including more fog and rain, as well as less sunshine and slower winds. The most studied and well-documented effect is the *urban heat island,* which refers to the fact that city temperatures are generally higher than surrounding rural areas. Several factors contribute to the heat island. The city's concrete and asphalt surfaces absorb and store more solar energy than the natural rural landscape. At night these stonelike materials help keep the city warmer. Urban temperatures are also higher because cities generate a great deal of waste heat (home heating, factories, cars, etc.). The "blanket" of pollutants also contributes by "trapping" heat that would otherwise escape.

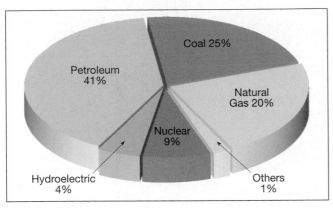

Figure 19.15 Paralleling the rapid growth of industrialization, which began in the nineteenth century, has been the combustion of fossil fuels, which has added great quantities of carbon dioxide to the atmosphere. The graph shows energy consumption in the United States, 2000. Fossil fuels (petroleum, coal, and natural gas) represent about 86 percent of the total. (Data from U.S. Department of Energy)

Carbon Dioxide, Trace Gases, and Global Warming

In Chapter 15, we learned that although carbon dioxide (CO_2) represents only about 0.036 percent of the gases that make up clean, dry air, it is nevertheless a meteorologically significant component. The importance of carbon dioxide lies in the fact that it is transparent to incoming short-wavelength solar radiation, but it is not transparent to some of the longer-wavelength outgoing terrestrial radiation. A portion of the energy leaving the ground is absorbed by carbon dioxide and subsequently reemitted, part of it toward the surface, thereby keeping the air near the ground warmer than it would be without carbon dioxide.

Thus, along with water vapor, carbon dioxide is largely responsible for the *greenhouse effect* of the atmosphere. Carbon dioxide is an important heat absorber, and it follows logically that any change in the air's carbon dioxide content could alter temperatures in the lower atmosphere.

CO_2 Levels Are Rising

Earth's tremendous industrialization of the past two centuries has been fueled—and still is fueled—by burning fossil fuels: coal, natural gas, and petroleum (Figure 19.15). Combustion of these fuels has added great quantities of carbon dioxide to the atmosphere.

to serve as primary tools for climate research, they cannot yet approach the actual complexity of the atmosphere. Computer models are powerful and essential aids, but climate forecasts based on such simulations still contain many uncertainties.

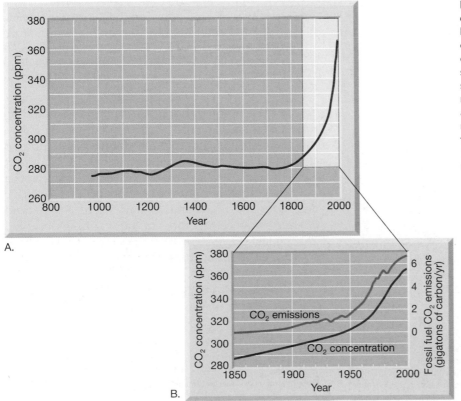

A.

B.

Figure 19.16 **A.** Carbon dioxide (CO_2) concentrations over the past 1000 years. Most of the record is based on data obtained from Antarctic ice cores. Bubbles of air trapped in the glacial ice provide samples of past atmospheres. The record since 1958 comes from direct measurements of atmospheric CO_2 taken at Mauna Loa Observatory, Hawaii. **B.** The rapid increase in CO_2 concentration since the onset of industrialization in the late 1700s is clear and has followed closely the rise in CO_2 emissions from fossil fuels.

Figure 19.17   Annual average global temperature variations for the period 1860–2001. The basis for comparison is the average for the 1961–90 period (the 0.0 line on the graph). Each narrow bar on the graph represents the departure of the global mean temperature from the 1961–1990 average for one year. For example, the global mean temperature for 1862 was more than 0.5°C (1°F) *below* the 1961–90 average, whereas the global mean for 1998 was more than 0.5°C above. (Specifically, 1998 was 0.56°C warmer.) The bar graph clearly indicates that there can be *significant variations from year to year.* But the graph also shows a trend. Estimated global mean temperatures have been above the 1961–90 average every year since 1978. Globally the 1990s was the warmest decade, and the years 1998 and 2001 the warmest years, since 1861. (Modified and updated after *G. Bell, et al. "Climate Assessment for 1998," Bulletin of the American Meteorological Society, Vol. 80, No. 5, May 1999, p. 54.*)

The use of coal and other fuels is the most prominent means by which humans add CO_2 to the atmosphere, but it is not the only way. The clearing of forests also contributes substantially because CO_2 is released as vegetation, is burned, or decays. Deforestation is particularly pronounced in the tropics, where vast tracts are cleared for ranching and agriculture or subjected to inefficient commercial logging operations.

Although some of the excess CO_2 is taken up by plants or is dissolved in the ocean, it is estimated that 45 to 50 percent remains in the atmosphere. Figure 19.16 shows CO_2 concentrations over the past thousand years based on ice-core records and (since 1958) measurements taken at Mauna Loa Observatory, Hawaii. The rapid increase in CO_2 concentration since the onset of industrialization is obvious and has closely followed the increase in CO_2 emissions from burning fossil fuels. Figure 19.16B shows this parallel growth.

The Atmosphere's Response

Given the increase in the atmosphere's carbon dioxide content, have global temperatures actually increased? The answer is yes. A report by the Intergovernmental Panel on Climate Change (IPCC)* indicates the following:

- During the twentieth century, the global average surface temperature increased by about 0.6°C (1°F).
- Globally it is very likely that the 1990s was the warmest decade and 1998 the warmest year since 1861 (Figure 19.17).
- New analyses of data for the Northern Hemisphere indicate that the increase in temperature in the twentieth century is likely to have been the largest in any century during the past 1000 years.

Are these temperature trends caused by human activities, or would they have occurred anyway? Scientists are cautious but seem convinced that human activities have played a significant role. An IPCC report in 1996 stated that "the balance of evidence suggests a discernible human influence on global climate."** Five years later the IPCC stated that "there is new and stronger evidence that most of the warming observed over the last 50 years is attributable to human activities."† What about the future? By the year 2100, models project atmospheric CO_2 concentrations of 540 to 970 ppm. With such an increase, how will global tempera-

*Intergovernmental Panel on Climate Change, *Climate Change 2001: The Scientific Basis.* Cambridge, UK: Cambridge University Press, 2001, p. 2. The Intergovernmental Panel on Climate Change is an authoritative group that was jointly established by the World Meteorological Organization and the United Nations Environmental Program to evaluate and make recommendations regarding the many aspects of global climate change.

**Intergovernmental Panel on Climate Change, *Climate Change* 1995: The Science of Climate Change, New York: Cambridge University Press, 1996, p. 4.
†IPCC. *Climate Change 2001: The Scientific Basis*, p. 10.

Figure 19.18 Methane is produced by *anaerobic* bacteria in wet places where oxygen is scarce (anaerobic means "without air," specifically oxygen). Such places include swamps, bogs, wetlands, and the guts of termites and grazing animals like cattle and sheep. Methane is also generated in flooded paddy fields ("artificial swamps") used for growing rice. These paddies are in India's Ganges lowlands. Mining of coal and drilling for oil and natural gas are other sources because methane is a product of their formation. (Photo by George Holton/Photo Researchers, Inc.)

tures change? Here is some of what the 2001 IPCC report has to say:[††]

- The globally averaged surface temperature is projected to increase by 1.4 to 5.8°C by the year 2100.
- The projected rate of warming is much larger than the observed changes during the twentieth century and is very likely to be without precedent during at least the last 10,000 years.
- It is very likely that nearly all land areas will warm more rapidly than the global average, particularly those at northern high latitudes in the cold season.

The Role of Trace Gases

Carbon dioxide is not the only gas contributing to a possible global increase in temperature. In recent years, atmospheric scientists have come to realize that the industrial and agricultural activities of people are causing a buildup of several trace gases that may also play a significant role. The substances are called *trace gases* because their concentrations are so much smaller than that of carbon dioxide. The trace gases that appear to be most important are methane (CH_4) (Figure 19.18), nitrous oxide (N_2O), and certain types of chlorofluorocarbons (CFCs). These gases absorb wavelengths of outgoing Earth radiation that would otherwise escape into space. Although individually their impact is modest, taken together the effects of these trace gases may be nearly as great as CO_2 in warming Earth.

Sophisticated computer models of the atmosphere show that the warming of the lower atmosphere caused by CO_2 and trace gases will not be the same everywhere (see Box 19.1). Rather, the temperature response in polar regions could be two to three times greater than the global average. One reason for such a response is an expected reduction in sea ice. This topic is explored more fully in the following section.

Climate-Feedback Mechanisms

Climate is a very complex interactive physical system. Thus, when any component of the climate system is altered, scientists must consider many possible outcomes. These possible outcomes are called **climate-feedback mechanisms.** They complicate climate-modeling efforts and add greater uncertainty to climate predictions.

What climate-feedback mechanisms are related to carbon dioxide and other greenhouse gases? The most important mechanism is that warmer surface temperatures increase evaporation rates. This in turn increases the water vapor in the atmosphere. Remember that water vapor is an even more powerful absorber of radiation emitted by Earth than is carbon dioxide. Therefore, with more water vapor in the air, the temperature increase caused by carbon dioxide and the trace gases is reinforced.

Recall that the temperature increase at high latitudes may be two to three times greater than the global average. This assumption is based in part on the likelihood that the area covered by sea ice will decrease as surface temperatures rise. Because ice reflects a much larger percentage of incoming solar radiation than does open water, the melting of the sea ice would replace a highly reflecting surface with a relatively dark surface (Figure 19.19). The result would be a substantial increase in the solar energy absorbed at the surface. This in turn would feed back to the atmosphere and magnify the initial temperature increase created by higher levels of greenhouse gases.

So far the climate-feedback mechanisms discussed have magnified the temperature rise caused by the buildup of carbon dioxide. Because these effects reinforce the initial change, they are called *positive-feedback mechanisms.* However, other effects must be classified as *negative-feedback mechanisms* because they produce results that are just the opposite of the initial change and tend to offset it.

[††]IPCC. *Climate Change 2001. The Scientific Basis*, p. 13.

BOX 19.1

PEOPLE AND THE ENVIRONMENT:
Aerosols from the "Human Volcano"

Increasing the levels of carbon dioxide and other greenhouse gases in the atmosphere is the most direct human influence on global climate. But it is not the only impact. Global climate is also affected by human activities that contribute to the atmosphere's aerosol content. *Aerosols* are the tiny, often microscopic, liquid and solid particles that are suspended in the air. Atmospheric aerosols are composed of many different materials, including soil, smoke, sea salt, and sulfuric acid. Natural sources are numerous and include such phenomena as dust storms and volcanoes. In Chapter 8 you learned that some explosive volcanoes (such as Mount Pinatubo) emit large quantities of sulfur dioxide gas high into the atmosphere. This gas combines with water vapor to produce clouds of tiny sulfuric acid aerosols that can lower air temperatures near the surface by reflecting solar energy back to space. So it is with sulfuric acid aerosols produced by human activities.

Presently the human contribution of aerosols to the atmosphere *equals* the quantity emitted by natural sources. Most human-generated aerosols come from the sulfur dioxide emitted during the combustion of fossil fuels and as a consequence of burning vegetation to clear agricultural land (Figure 19.A). Chemical reactions in the atmosphere convert the sulfur dioxide into sulfate aerosols, the same material that produces acid precipitation.

The aerosols produced by human activity act directly by reflecting sunlight back to space and indirectly by making clouds "brighter" reflectors. The second effect relates to the fact that sulfuric acid aerosols attract water and thus are es-

Figure 19.A One source of human-generated aerosols is power plants.

pecially effective as cloud condensation nuclei (tiny particles upon which water vapor condenses). The large quantity of aerosols produced by human activities (especially industrial emissions) trigger an increase in the number of cloud droplets that form within a cloud. A greater number of small droplets increases the cloud's brightness—that is, more sunlight is reflected back to space.

By reducing the amount of solar energy available to the climate system, aerosols have a net cooling effect. Studies indicate that the cooling effect of human-generated aerosols could offset a portion of the global warming caused by the growing quantities of greenhouse gases in the atmosphere. Unfortunately, the magnitude and extent of the cooling effect of aerosols is highly uncertain. This uncertainty is a significant hurdle in advancing our understanding of how humans alter Earth's climate.

It is important to point out some significant differences between global

warming by greenhouse gases and aerosol cooling. After being emitted, greenhouse gases such as carbon dioxide remain in the atmosphere for many decades. By contrast, aerosols released into the lower atmosphere remain there for only a few days or, at most, a few weeks before they are "washed out" by precipitation. Because of their short lifetime in the atmosphere, human-generated aerosols are distributed unevenly over the globe. As expected, they are concentrated near the areas that produce them, namely industrialized regions that burn fossil fuels and land areas where vegetation is burned.

Because the lifetime of human-generated aerosols in the atmosphere is short, the effect of the "human volcano" on today's climate is determined by the quantity of material emitted during the preceding couple of weeks. By contrast, the carbon dioxide released into the atmosphere remains for much longer spans and thus influences climate for many decades.

One probable result of a global temperature rise would be an accompanying increase in cloud cover due to the higher moisture content of the atmosphere. Most clouds are good reflectors of solar radiation. At the same time, however, they are also good absorbers and emitters of radiation emitted by Earth. Consequently, clouds produce two opposite effects. They are a negative-feedback mechanism because they increase albedo and thus diminish the amount of solar energy available to heat the atmosphere. On the other hand, clouds act as

a positive-feedback mechanism by absorbing and emitting radiation that would otherwise be lost from the troposphere (see Figure 15.25, p. 427).

Which effect, if either, is stronger? Atmospheric modeling shows that the negative effect of a higher albedo is dominant. Therefore, the net result of an increase in cloudiness should be a decrease in air temperature. The magnitude of this negative feedback, however, is not believed to be as great as the positive feedback caused by added moisture and decreased sea

Figure 19.19 A reduction in sea ice would act as a positive feedback mechanism because surface albedo would decrease and the amount of solar energy absorbed at the surface would increase. This aerial view shows the springtime breakup of sea ice near Antarctica. (Photo by D. N. Thomas)

ice. Thus, although increases in cloud cover may partly offset a global temperature increase, climate models show that the ultimate effect of the projected increase in CO_2 and trace gases will still be a temperature increase.

The problem of global warming caused by human-induced changes in atmospheric composition continues to be one of the most studied aspects of climate change. Although no models yet incorporate the full range of potential factors and feedbacks, the scientific consensus is that the increasing levels of atmospheric carbon dioxide and trace gases will lead to a warmer planet with a different distribution of climate regimes.

Some Possible Consequences of Global Warming

What consequences can be expected if the carbon dioxide content of the atmosphere reaches a level that is twice what it was early in the twentieth century? Because the climate system is so complex, predicting the distribution of particular regional changes is still very speculative. It is not yet possible to pinpoint specifics, such as where or when it will become drier or wetter. Nevertheless, plausible scenarios can be given for larger scales of space and time. As computers become larger and data improve, scientists will gradually develop models that can provide more specific and reliable results.

As noted, the magnitude of the temperature increase will not be the same everywhere. The temperature rise will probably be smallest in the tropics and increase toward the poles. As for precipitation, the models indicate that some regions will experience significantly more precipitation and runoff. However, other regions will experience a decrease in runoff due to reduced precipitation or because of greater evaporation caused by higher temperatures.

Water Resources and Agriculture

Such changes could profoundly alter the distribution of the world's water resources and hence affect the productivity of agricultural regions that depend on rivers for irrigation water. For example, a 2°C warming and 10 percent precipitation decrease in the region drained by the Colorado River could diminish the river's flow by 50 percent or more. Because the present flow of the river barely meets current demand for irrigation agriculture, the negative effect would be serious. Many other rivers are the basis for extensive irrigated agriculture, and the projected reduction of their flow could have equally grave consequences. In contrast, large precipitation increases in other areas would increase the flow of some rivers and bring more frequent destructive floods.

Harder to estimate is the effect on nonirrigated crops that depend on direct rainfall for moisture. Some places will no doubt experience productivity loss due to a decrease in rainfall or increase in evaporation. Still, these losses may be offset by gains elsewhere. Warming in the high latitudes could lengthen the growing season, for instance. This in turn could allow expansion of agriculture into areas that are presently not suited to crop production.

Sea-Level Rise

Another impact of a human-induced global warming is a probable rise in sea level. How is a warmer atmosphere related to a global rise in sea level? The most obvious connection, the melting of glaciers, is important,

Figure 19.20 The slope of a shoreline is critical to determining the degree to which sea-level changes will affect it. **A.** When the slope is gentle, small changes in sea level cause a substantial shift. **B.** The same sea-level rise along a steep coast results in only a small shoreline shift. **C.** As sea level gradually rises, the shoreline retreats, and structures that were once thought to be safe from wave attack are exposed to the force of the sea. (Photo by Kenneth Hasson)

but *not* the only factor. Also significant is that a warmer atmosphere causes an increase in ocean volume due to thermal expansion. Higher air temperatures warm the adjacent upper layers of the ocean, which in turn causes the water to expand and sea level to rise.

Research indicates that sea level has risen between 10 and 25 centimeters (4 and 8 inches) over the past century and that the trend will continue at an accelerated rate. Some models indicate that the rise may approach or even exceed 50 centimeters (20 inches) by the end of the twenty-first century. Such a change may seem modest, but scientists realize that any rise in sea level along a *gently* sloping shoreline, such as the Atlantic and Gulf coasts of the United States, will lead to significant erosion and severe, permanent inland flooding (Figure 19.20). If this happens, many beaches and wetlands will be eliminated and coastal civilization would be severely disrupted.

Changing Weather Patterns

Atmospheric scientists also expect that weather patterns will change as a result of the projected global warming. Potential weather changes include the following:

- Although uncertain, a higher frequency and greater intensity of hurricanes because of warmer ocean temperatures.
- Shifts in the paths of large-scale cyclonic storms, which in turn would affect the distribution of precipitation and the occurrence of severe storms, including tornadoes.
- More intense heat waves and droughts in some regions and fewer such events in other places.

The impact on climate of an increase in atmospheric CO_2 and trace gases is obscured by many unknowns and uncertainties. The changes will probably be gradual environmental shifts, imperceptible from year to year. Nevertheless, the effects, accumulated over decades, will have powerful economic, social, and political consequences.

Policymakers are confronted with responding to the risks posed by emissions of greenhouse gases in the face of significant scientific uncertainties. However, they are also faced with the fact that climate-induced environmental changes cannot be reversed quickly, if at all, owing to the lengthy time scales associated with the climate system.

Chapter Summary

• *Climate* is the aggregate of weather conditions for a place or region over a long period of time. Earth's *climate system* involves the exchanges of energy and moisture that occur among the atmosphere, hydrosphere, solid Earth, biosphere, and *cryosphere* (the ice and snow that exist at Earth's surface).

• Climate classification brings order to large quantities of information, which aids comprehension and understanding, and facilitates analysis and explanation. *Temperature and precipitation are the most important elements in a climatic description.* Many climate classifications have been devised, with the value of each determined by its intended use. The *Köppen classification,* which uses mean monthly and annual values of temperature and precipitation, is a widely used system. The boundaries Köppen chose were largely based on the limits of certain plant associations. Five principal climate groups, each with subdivisions, were recognized. Each group is designated by a capital letter. Four of the climate groups (A, C, D, and E) are defined on the basis of temperature characteristics, and the fifth, the B group, has precipitation as its primary criterion.

• *Humid tropical (A) climates* are winterless, with all months having a mean temperature above 18°C. *Wet tropical climates* (Af and Am), which lie near the equator, have constantly high temperatures and enough rainfall to support the most luxuriant vegetation (tropical rain forest) found in any climatic realm. *Tropical wet and dry climates* (Aw) are found poleward of the wet tropics and equatorward of the subtropical deserts, where the rain forest gives way to the tropical grasslands and scattered drought-tolerant trees of the savanna. The most distinctive feature of this climate is the seasonal character of the rainfall.

• *Dry (B) climates,* in which the yearly precipitation is less than the potential loss of water by evaporation, are subdivided into two types: *arid* or *desert* (BW) and *semiarid* or *steppe* (BS). Their differences are primarily a matter of degree, with semiarid being a marginal and more humid variant of arid. Low-latitude deserts and steppes coincide with the clear skies caused by subsiding air beneath the subtropical high-pressure belts. Middle-latitude deserts and steppes exist principally because of their position in the deep interiors of large landmasses far removed from the ocean. Because many middle-latitude deserts occupy sites on the leeward sides of mountains, they can also be classified as *rain shadow deserts.*

• *Middle-latitude climates with mild winters* (C climates) occur where the average temperature of the coldest month is below 18°C but above −3°C. Several C climate subgroups exist. *Humid subtropical climates* (Cfa) are located on the eastern sides of the continents, in the 25- to 40-degree latitude range. Summer weather is hot and sultry, and winters are mild. In North America, the *marine west coast climate* (Cfb, Cfc) extends from near the United States–Canada border northward as a narrow belt into southern Alaska. The prevalence of maritime air masses means that mild winters and cool summers are the rule. *Dry-summer subtropical climates* (Csa, Csb) are typically located along the west sides of continents between latitudes 30 and 45 degrees. In summer, the regions are dominated by stable, dry conditions associated with the oceanic subtropical

highs. In winter they are within range of the cyclonic storms of the polar front.

• *Humid middle-latitude climates with severe winters* (D climates) are land-controlled climates that are absent in the Southern Hemisphere. The D climates have severe winters. The average temperature of the coldest month is −3°C or below, and the warmest monthly mean exceeds 10°C. *Humid continental climates* (Dfa, Dfb, Dwa, Dwb) are confined to the eastern portions of North America and Eurasia in the latitude range between approximately 40 and 50 degrees north latitude. Both winter and summer temperatures can be characterized as relatively severe. Precipitation is generally greater in summer than in winter. *Subarctic climates* (Dfc, Dfd, Dwc, Dwd) are situated north of the humid continental climates and south of the polar tundras. The outstanding feature of subarctic climates is the dominance of winter. By contrast, summers in the subarctic are remarkably warm, despite their short duration. The highest annual temperature ranges on Earth occur here.

• *Polar (E) climates* are summerless, with the average temperature of the warmest month below 10°C. Two types of polar climates are recognized. The *tundra climate* (ET) is a treeless climate found almost exclusively in the Northern Hemisphere. The *ice cap climate* (EF) does not have a single monthly mean above 0°C. As a consequence, the growth of vegetation is prohibited, and the landscape is one of permanent ice and snow.

• Compared to nearby places of lower elevation, *highland climates* are cooler and usually wetter. Because atmospheric conditions fluctuate rapidly with changes in altitude and exposure, these climates are best described by their variety and changeability.

• Humans have been modifying the environment for thousands of years. By altering ground cover with the use of fire and the overgrazing of land, people have modified such important climatological factors as surface albedo, evaporation rates, and surface winds.

• By adding carbon dioxide and other trace gases (methane, nitrous oxide, and chlorofluorocarbons) to the atmosphere, humans may be significantly contributing to global warming.

• When any component of the climate system is altered, scientists must consider the many possible outcomes, called *climate-feedback mechanisms.* Changes that reinforce the initial change are called *positive-feedback mechanisms.* On the other hand, *negative-feedback mechanisms* produce results that are the opposite of the initial change and tend to offset it.

• Because the climate system is very complex, predicting specific regional changes that may occur as the result of increased levels of carbon dioxide in the atmosphere is highly speculative. However, some potential consequences of global warming include: (1) altering the distribution of the world's water resources and therefore the productivity of agricultural regions that depend on rivers for irrigation, (2) a probable rise in sea level, and (3) a change in weather patterns, such as a higher frequency and greater intensity of hurricanes and shifts in the paths of large-scale cyclonic storms.

Key Terms

arid climate (p. 530)
climate feedback mechanisms (p. 541)
climate system (p. 524)
desert climate (p. 530)
dry-summer subtropical climate
    (p. 533)
highland climate (p. 536)

humid continental climate (p. 534)
humid subtropical climate (p. 533)
ice cap climate (p. 536)
Köppen classification (p. 525)
marine west coast climate (p. 533)
polar climate (p. 536)
rain shadow desert (p. 532)

semiarid climate (p. 530)
steppe climate (p. 530)
subarctic climate (p. 535)
tropical rain forest (p. 526)
tropical wet and dry climate (p. 530)
tundra climate (p. 536)

Review Questions

1. List the five parts of the climate system.

2. Why is classification often a helpful or even necessary task in science?

3. What climatic data are needed in order to classify a climate using the Köppen scheme?

4. (a) What primary factor distinguishes Aw climates from Af and Am?
   (b) How is this difference reflected in the vegetation?

5. Describe the influence of the equatorial low and the subtropical high on the precipitation regime in the Aw climate.

6. Why is the amount of precipitation that defines the humid–dry boundary variable?

7. What is the primary reason (control) for the existence of the dry subtropical realm (BWh and BSh)?

8. What is the primary cause for the existence of the middle-latitude deserts and steppes?

9. Describe and explain the differences between summertime and wintertime precipitation in the humid subtropics (Cfa).

10. Why is the marine west coast climate (Cfb and Cfc) represented by only slender strips of land in North and South America, and why is it very extensive in Western Europe?

11. In this chapter the dry-summer subtropics were described as transitional. Explain why this statement is true.

12. Why is the humid continental climate confined to the Northern Hemisphere?

13. Although generally characterized by small precipitation totals, subarctic and polar climates are considered humid. Explain.

14. Describe and explain the annual temperature range one should expect in the realm of the *taiga*.

15. Although polar regions experience extended periods of almost perpetual sunlight in the summer, temperatures remain cool. Explain.

16. The tundra climate is not confined solely to high latitudes. Under what circumstances might the ET climate be found in more equatorward locations?

17. Where are EF climates most extensively developed?

18. The Arizona cities of Flagstaff and Phoenix are relatively close to one another yet have contrasting climates (see Figure 19.13). Briefly explain why the differences occur.

19. Why has the atmosphere's carbon dioxide level been rising for more than 150 years?

20. How are temperatures in the lower atmosphere likely to change as carbon dioxide levels continue to increase? Why?

21. Is carbon dioxide the only gas that may contribute to a future global temperature change?

22. What are the main sources of human-generated aerosols? What effect do these aerosols have on temperatures in the troposphere? How long do aerosols remain in the lower atmosphere before they are removed? (See Box 19.1.)

23. List three possible consequences of a global warming.

Examining the Earth System

1. The Köppen climate classification is based on the fact that there is an excellent association between natural vegetation (biosphere) and climate (atmosphere). Briefly describe the climate conditions (temperature and precipitation) and natural vegetation associated with each of the following Köppen climates: Af, BWh, Dfc, ET.

2. How might the burning of fossil fuels, such as the gasoline to run your car, influence global temperature? If such a temperature change occurs, how might sea level be affected? How might the intensity of hurricanes change? How might these changes impact people who live on a beach or barrier island along the Atlantic or Gulf coasts?

3. What are climate-feedback mechanisms? Give an example of a positive-feedback mechanism and a negative-feedback mechanism.

Web Resources

The *Earth Science* Website uses the resources and flexibility of the Internet to aid in your study of the topics in this chapter. Written and developed by Earth science instructors, this site will help improve your understanding of Earth science. Visit **http://www.prenhall.com/tarbuck** and click on the cover of *Earth Science 10e* to find:

- **Online review quizzes.**
- **Web-based critical thinking and writing exercises.**
- **Links to chapter-specific Web resources.**
- **Internet-wide key term searches.**

http://www.prenhall.com/tarbuck

Moon over Mount Humphreys, in California's eastern Sierras. (Photo by Galen Rowell)

CHAPTER 20

ORIGIN OF MODERN ASTRONOMY

Earth is one of nine planets and numerous smaller bodies that orbit the Sun. The Sun is part of a much larger family of perhaps 100 billion stars that make up the Milky Way, which in turn is only one of billions of galaxies in an incomprehensibly large universe. This view of Earth's position in space is considerably different from that held only a few hundred years ago, when Earth was thought to occupy a privileged position as the center of the universe. This chapter unfolds with some events that led to modern astronomy. In addition, it examines Earth's place in time and space.

Long before recorded history, which began about 5000 years ago, people were aware of the close relationship between events on Earth and the positions of heavenly bodies, the Sun in particular. People noted that changes in the seasons and floods of great rivers like the Nile in Egypt occurred when the celestial bodies, including the Sun, Moon, planets, and stars, reached a particular place in the heavens. Early agrarian cultures, which were dependent on the weather, believed that if the heavenly objects could control the seasons, they must also strongly influence all Earthly events. This belief undoubtedly was the reason that early civilizations began keeping records of the positions of the celestial objects. The Chinese, Egyptians, and Babylonians in particular are noted for this.

These cultures recorded the locations of the Sun, Moon, and the five planets visible to the unaided eye as these objects moved slowly against the background of "fixed" stars. In addition, the Chinese kept quite accurate records of comets and "guest stars" (Figure 20.1). Today we know that a "guest star" is really a normal star, usually too faint to be visible, that increases its brightness as it explosively ejects gases from its surface, a phenomenon we call a *nova* (*novus* = new).

A study of Chinese archives shows that they recorded every appearance of the famous Halley's comet for at least 10 centuries. However, because this comet appears only once every 76 years, they were unable to link these appearances to establish that what they saw was the same object each time. Thus, like most ancients, the Chinese considered comets to be mystical. Comets were seen as bad omens and were blamed for a variety of disasters, from wars to plagues (Figure 20.2).

Ancient Astronomy

The "Golden Age" of early astronomy (600 B.C.–A.D. 150) was centered in Greece. The early Greeks have been criticized, and rightly so, for using philosophical arguments to explain natural phenomena. However, they did rely on observational data as well. The basics of geometry and trigonometry, which they had developed, were used to measure the sizes and distances of the largest-appearing bodies in the heavens—the Sun and the Moon.

Early Greeks

Many astronomical discoveries have been credited to the Greeks. They held the **geocentric** (*geo* = Earth, *centric* = centered) view, believing that Earth was a sphere that stayed motionless at the center of the universe. Orbiting Earth were the Moon, Sun, and the known planets—Mercury, Venus, Mars, and Jupiter. Beyond the planets was a transparent, hollow sphere (**celestial sphere**) on which the stars traveled daily around Earth (this is how it looks, but of course the effect is actually caused by Earth's rotation about its axis). Some early Greeks realized that the motion of the stars could be explained just as easily by a rotating Earth, but they rejected that idea, because Earth exhibited no sense of motion and seemed too large to be movable. In fact, proof of Earth's rotation was not demonstrated until 1851, a topic we consider in Box 20.1.

To the Greeks, all of the heavenly bodies, except seven, appeared to remain in the same relative position

Figure 20.1 Comet Hale-Bopp. The two tails are about 10 million to 15 million miles long. (A Peoria Astronomical Society, Inc., photograph by Eric Clifton and Greg Neaveill)

Figure 20.2 The Bayeux Tapestry that hangs in Bayeux, France, shows the apprehension caused by Halley's comet in A.D. 1066. This event preceded the defeat of King Harold by William the Conqueror. (Photo by Giraudon/Art Resource with special authorization of the City of Bayeux)

to one another. These seven wanderers (*planetai* in Greek) included the Sun, the Moon, Mercury, Venus, Mars, Jupiter, and Saturn. Each was thought to have a circular orbit around Earth. Although this system was incorrect, the Greeks refined it to the point that it explained the apparent movements of all celestial bodies.

As early as the fifth century B.C., the Greeks understood what causes the phases of the Moon. Anaxagoras reasoned that the Moon shines by reflected sunlight, and because it is a sphere, only half is illuminated at one time. As the Moon orbits Earth, that portion of the illuminated half that is visible from Earth is always

changing. Anaxagoras also realized that an eclipse of the Moon occurs when it moves into Earth's shadow.

The famous Greek philosopher Aristotle (384–322 B.C.) concluded that Earth is spherical because it always casts a curved shadow when it eclipses the moon. Although most of the teachings of Aristotle were passed along and were considered infallible by many, his belief in a spherical Earth was abandoned during the Middle Ages.

The first Greek to profess a Sun-centered, or **heliocentric,** (*helios* = Sun, *centric* = centered), universe was Aristarchus (312–230 B.C.). Aristarchus also used

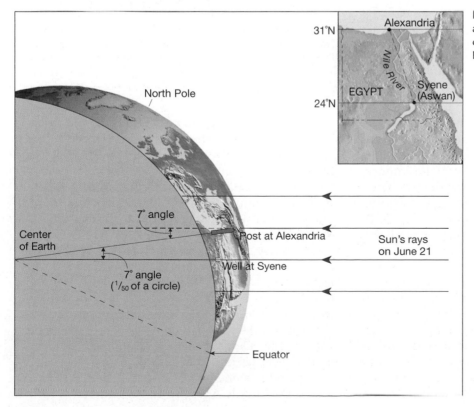

Figure 20.3 Orientation of the Sun's rays at Syene (Aswan) and Alexandria in Egypt on June 21 when Eratosthenes calculated Earth's circumference.

BOX 20.1

UNDERSTANDING EARTH:
Foucault's Experiment

Every schoolchild learns that Earth rotates on its axis once each day to produce periods of daylight and darkness. However, day and night and the apparent motions of the stars can be accounted for equally well by a Sun and celestial sphere that revolve around a stationary Earth. Copernicus realized that a rotating Earth greatly simplified the existing model of the universe and strongly advocated this as the correct view. He was unable, however, to *prove* that Earth rotates. The first substantial proof was presented 300 years after his death by the French physicist Jean Foucault.

In 1851 Foucault used a free-swinging pendulum to demonstrate that Earth does, in fact, turn on its axis. To envision Foucault's experiment, imagine a large pendulum swinging over the North Pole (Figure 20.A). Keep in mind that once a pendulum is put into motion, it continues swinging in the same plane unless acted upon by some outside force. Assume that a sharp stylus is attached to the bottom of this pendulum, marking the snow as it oscillates. When we observe the marks made by the stylus, we note that the pendulum is slowly but continually changing position. At the end of 24 hours it has returned to the starting position (Figure 20.A).

Figure 20.A   Apparent movement of a pendulum at the North Pole caused by Earth rotating beneath it.

Because no outside force acted on the pendulum to change its position, what we observed must have been Earth rotating beneath it. Foucault conducted a similar experiment when he suspended a long pendulum from the dome of the Pantheon in Paris. Today, Foucault pendulums can be found in a number of museums to recreate this famous scientific experiment (Figure 20.B).

Figure 20.B   Foucault pendulum housed in Museum of Science and Industry, Chicago. (Courtesy Museum of Science and Industry)

---

simple geometric relations to calculate the relative distances from Earth to the Sun and the Moon. He later used these data to calculate their sizes. As a result of observational errors beyond his control, he came up with measurements that were much too small. However, he did learn that the Sun was many times more distant than the Moon and many times larger than Earth. The latter fact may have prompted him to suggest a Sun-centered universe. Nevertheless, because of the strong influence of Aristotle, the Earth-centered view dominated Western thought for nearly 2000 years.

The first successful attempt to establish the size of Earth is credited to Eratosthenes (276–194 B.C.). Eratosthenes observed the angles of the noonday Sun in two Egyptian cities that were roughly north and south of each other—Syene (presently Aswan) and Alexandria (Figure 20.3). Finding that the angles differed by 7 degrees, or 1/50 of a complete circle, he concluded that

the circumference of Earth must be 50 times the distance between these two cities. The cities were 5000 *stadia* apart, giving him a measurement of 250,000 *stadia*. Many historians believe the *stadia* was 157.6 meters (517 feet), which would make Eratosthenes' calculation of Earth's circumference—39,400 kilometers (24,428 miles)—a measurement very close to the modern value of 40,075 kilometers (24,902 miles).

Probably the greatest of the early Greek astronomers was Hipparchus (second century B.C.), best known for his star catalogue. Hipparchus determined the location of almost 850 stars, which he divided into six groups according to their brightness. He measured the length of the year to within minutes of the modern value and developed a method for predicting the times of lunar eclipses to within a few hours.

Although many of the Greek discoveries were lost during the Middle Ages, the Earth-centered view that the Greeks proposed became established in Europe. Presented in its finest form by Cladius Ptolemy, this geocentric outlook became known as the **Ptolemaic system.**

The Ptolemaic System

Much of our knowledge of Greek astronomy comes from a 13-volume treatise, *Almagest* (the great work), which was compiled by Ptolemy in A.D. 141 and survived thanks to the work of Arab scholars. In this work, Ptolemy is credited with developing a model of the universe that accounted for the observable motions of the planets (Figure 20.4). The precision with which his model was able to predict planetary motion is attested to by the fact that it went virtually unchallenged, in principle if not in detail, for nearly 13 centuries.

In the Greek tradition, the Ptolemaic model had the planets moving in circular orbits around a motionless Earth. (The circle was considered the pure and perfect shape by the Greeks.) However, the motion of the planets, as seen against the background of stars, is not so simple. Each planet, if watched night after night, moves slightly eastward among the stars. Periodically, each planet appears to stop, reverse direction for a period of time, and then resume an eastward motion. The apparent westward drift is called **retrograde** (*retro* = to go back, *gradus* = walking) **motion.** This rather odd *apparent* motion results from the combination of the motion of Earth and the planet's own motion around the Sun.

The retrograde motions of Mars is shown in Figure 20.5. Earth has a faster orbital speed than Mars, so it overtakes its neighbor. While doing so, Mars *appears* to be moving backward, in retrograde motion. This is analogous to what a race-car driver sees out the side window when passing a slower car. The slower planet, like the slower car, appears to be going backward, although its actual motion is in the same direction as the faster-moving body.

A.

B.

Figure 20.4 The universe according to Ptolemy, second century A.D. **A.** Ptolemy believed that the star-studded celestial sphere made a daily trip around a motionless Earth. In addition he proposed that the Sun, Moon, and planets made trips of various lengths along individual orbits. **B.** Retrograde motion as explained by Ptolemy.

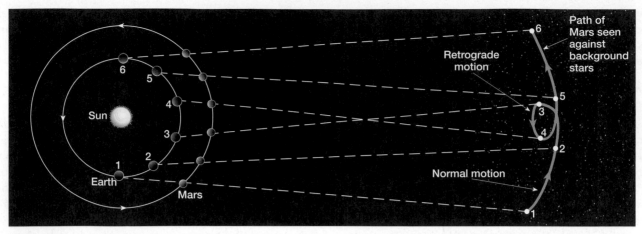

Figure 20.5 Retrograde (backward) motion of Mars as seen against the background of distant stars. When viewed from Earth, Mars moves eastward among the stars each day, then periodically appears to stop and reverse direction. This apparent westward drift is a result of the fact that Earth has a faster orbital speed than Mars and overtakes it. As this occurs, Mars appears to be moving backward; that is, it exhibits retrograde motion.

It is much more difficult to accurately represent retrograde motion using the incorrect Earth-centered model, but Ptolemy was able to do so (Figure 20.4B). Rather than using a simple circle for each planet's orbit, he showed that the planets orbited on small circles (*epicycles*), revolving along large circles (*deferents*). By trial and error, he found the right combinations of circles to produce the amount of retrograde motion observed for each planet. (An interesting note is that almost any closed curve can be produced by the combination of two circular motions, a fact that can be verified by anyone who has used the Spirograph™ design-drawing toy.)

It is a tribute to Ptolemy's genius that he was able to account for the planets' motions as well as he did, considering that he used an incorrect model. Some suggest that he did not mean his model to represent reality, but only to be used for calculating the positions of heavenly bodies. We probably will never know his intentions. However, the Roman Catholic Church, which dominated European thought for centuries, accepted Ptolemy's theory as the correct representation of the heavens, and this created problems for those who found fault with it.

? STUDENTS SOMETIMES ASK...

Why was the discovery of Jupiter's moons important?

According to the Ptolemaic (Earth-centered) model of the universe, all of the heavenly bodies revolved around Earth. When Galileo, using a crude telescope, saw four moons revolving around Jupiter, he knew that Earth was not the center of all motion. Consequently, at least one of the basic tenants of the Ptolemaic model had to be incorrect. Astronomers soon demonstrated that the other tenants of the Earth-centered model were also inconsistent with observations.

The Birth of Modern Astronomy

Modern astronomy was not born overnight. Its development involved a break from deeply entrenched philosophical and religious views that was brought about by the discovery of a new and greater universe governed by discernible laws. Let us now look at the work of five noted scientists involved in this transition: Nicolaus Copernicus, Tycho Brahe, Johannes Kepler, Galileo Galilei, and Sir Isaac Newton.

Nicolaus Copernicus

For almost 13 centuries after the time of Ptolemy, very few astronomical advances were made in Europe. The first great astronomer to emerge after the Middle Ages was Nicolaus Copernicus (1473–1543) from Poland (Figure 20.6). Copernicus became convinced that Earth is a planet, just like the other five then-known planets. The daily motions of the heavens, he reasoned, could be better explained by a rotating Earth. To counter the Ptolemaic objection that Earth would fly apart if it rotated, Copernicus suggested that the much larger celestial sphere would be even more likely to fly apart if it rotated!

Having concluded that Earth is a planet, Copernicus reconstructed the solar system with the Sun at the center and the planets Mercury, Venus, Earth, Mars, Jupiter, and Saturn orbiting around it. This was a major break from the ancient idea that a motionless Earth lies at the center of all movement. However, Copernicus retained a link to the past and used circles, which were considered to be the perfect geometric shape, to represent the orbits of the planets. Although these circular orbits were close to reality, they didn't quite match what people saw. Unable to get satisfactory agreement between predicted locations of the planets and their observed positions, Copernicus found it necessary to add

Figure 20.6   Polish astronomer Nicolaus Copernicus (1473–1543) believed that Earth was just another planet. (Yerkes Observatory Photograph/University of Chicago)

epicycles like those used by Ptolemy. The discovery that the planets have *elliptical* orbits would wait another century for the insights of Johannes Kepler.

Also, like his predecessors, Copernicus used philosophical justifications, such as the following, to support his point of view.

> . . . in the midst of all stands the Sun. For who could in this most beautiful temple place this lamp in another or better place than that from which it can at the same time illuminate the whole?

Copernicus's monumental work, *De Revolutionibus, Orbium Coelestium (On the Revolution of the Heavenly Spheres)*, which set forth his controversial ideas, was published as he lay on his deathbed. Hence, he never suffered the criticisms that fell on many of his followers.

The Copernican system challenged the primacy of Earth in the universe and was considered heretical by the Roman Catholic Church. Expounding the idea cost at least one person his life. Giordano Bruno was seized by the Inquisition, a Church tribunal, in 1600 and, refusing to denounce the Copernican theory, was burned at the stake.

Tycho Brahe

Tycho Brahe (1546–1601) was born of Danish nobility three years after the death of Copernicus. Reportedly, Tycho became interested in astronomy while viewing a solar eclipse that had been predicted by astronomers. He persuaded King Frederick II to establish an observatory, which he headed, near Copenhagen. There he designed and built pointers (the telescope would not be invented for a few more decades), which he used for 20 years to systematically measure the locations of the heavenly bodies (Figure 20.7). These observations, particularly of Mars, were far more precise than any made previously and are his legacy to astronomy.

Tycho did not believe in the Copernican (Sun-centered) system, because he was unable to observe an apparent shift in the position of stars that would be caused by Earth's motion. His argument went like this: If Earth does revolve along an orbit around the Sun, the position of a nearby star, when observed from extreme points in Earth's orbit six months apart, should shift with respect to the more distant stars. His *idea* was correct, and this apparent shift of the stars is called *stellar parallax* (see Figure 23.2, p. 625).

The principle of parallax is easy to visualize: Close one eye, and with your index finger vertical, use your eye to line up your finger with some distant object. Now, without moving your finger, view the object with your other eye and notice that the object's position appears to change. The farther away you hold your finger, the less the object's position seems to shift. Herein lay the flaw in Tycho's argument. He was right about parallax, but because the distance to even the nearest stars is enormous compared to the width of Earth's orbit, the shift that occurs is too small to be noticed by using the first primitive telescopes, let alone the unaided eye.

With the death of his patron, the King of Denmark, Tycho was forced to leave his observatory. It was probably his arrogant and extravagant nature that caused a conflict with the next ruler, so Tycho moved to Prague in the Czech Republic. Here, in the last year of his life, he acquired an able assistant, Johannes Kepler. Kepler retained most of the observations made by Tycho and put them to exceptional use. Ironically, the data Tycho collected to refute the Copernican view would later be used by Kepler to support it.

Johannes Kepler

If Copernicus ushered out the old astronomy, Johannes Kepler (1571–1630) ushered in the new (Figure 20.8). Armed with Tycho's data, a good mathematical mind, and, of greater importance, a strong faith in the accuracy of Tycho's work, Kepler derived three basic laws of planetary motion. The first two laws resulted from his inability to fit Tycho's observations of Mars to a circular orbit. Unwilling to concede that the discrepancies were a result of observational error, he searched for another solution. This endeavor led him to discover that the orbit of Mars is not a perfect circle but is elliptical (Figure 20.9). About the same time, he realized that the

Figure 20.7 Tycho Brahe (1546–1601) in his observatory, Uraniborg, on the Danish island of Hveen. Tycho (central figure) and the background are painted on the wall of the observatory within the arc of the sighting instrument called a quadrant. In the far right, Tycho can be seen "sighting" a celestial object through the "hole" in the wall. Tycho's accurate measurements of Mars enabled Johannes Kepler to formulate his three laws of planetary motion. (Courtesy of Thomas Clarke, McLaughlin Planetarium/ Royal Ontario Museum copyright ROM)

Figure 20.8 German astronomer Johannes Kepler (1571–1630) helped establish the era of modern astronomy by deriving three laws of planetary motion. (Smithsonian Institution Photo #56123)

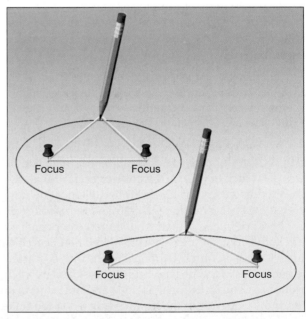

Figure 20.9 Drawing ellipses with various eccentricities. Using two straight pins for foci and a loop of string, trace out a curve while keeping the string taut, and you will have drawn an ellipse. The farther the pins (the foci) are moved apart, the more flattened (more eccentric) is the resulting ellipse.

orbital speed of Mars varies in a predictable way. As it approaches the Sun, it speeds up, and as it moves away from the Sun, it slows down.

In 1609, after almost a decade of work, Kepler proposed his first two laws of planetary motion:

1. The path of each planet around the Sun is an ellipse, with the Sun at one focus (Figure 20.9).

The other focus is symmetrically located at the opposite end of the ellipse.

2. Each planet revolves so that an imaginary line connecting it to the Sun sweeps over equal areas in equal intervals of time (Figure 20.10). This law of equal areas expresses geometrically the variations in orbital speeds of the planets.

Figure 20.10 illustrates the second law. Note that in order for a planet to sweep equal areas in the same amount of time, it must travel more rapidly when it is nearer the Sun and more slowly when it is farther from the Sun.

Kepler was very religious and believed that the Creator made an orderly universe. The uniformity he tried to find eluded him for nearly a decade. Then in 1619, Kepler published his third law in *The Harmony of the Worlds*.

3. The orbital periods of the planets and their distances to the Sun are proportional.

In its simplest form, the orbital period of revolution is measured in Earth years, and the planet's distance to the Sun is expressed in terms of Earth's mean distance to the Sun. The latter "yardstick" is called the **astronomical unit (AU)** and averages about 150 million kilometers (93 million miles).

Using these units, Kepler's third law states that the planet's orbital period squared is equal to its mean solar distance cubed ($p^2 = d^3$). Consequently, the solar distances of the planets can be calculated when their periods of revolution are known. For example, Mars has a period of 1.88 years, which squared equals 3.54. The cube root of 3.54 is 1.52, and that is the distance to Mars in astronomical units (Table 20.1).

Table 20.1 Period of revolution and solar distances of planets.

Planet	Solar Distance (AU)*	Period (years)
Mercury	0.39	0.24
Venus	0.72	0.62
Earth	1.00	1.00
Mars	1.52	1.88
Jupiter	5.20	11.86
Saturn	9.54	29.46
Uranus	19.18	84.01
Neptune	30.06	164.80
Pluto	39.44	247.70

*AU = astronomical unit.

Kepler's laws assert that the planets revolve around the Sun and therefore support the Copernican theory. Kepler, however, did fall short of determining the *forces* that act to produce the planetary motion he had so ably described. That task would remain for Galileo Galilei and Sir Isaac Newton.

Galileo Galilei

Galileo Galilei (1564–1642) was the greatest Italian scientist of the Renaissance (Figure 20.11). He was a contemporary of Kepler and, like Kepler, strongly supported the Copernican theory of a Sun-centered solar system. Galileo's greatest contributions to science were his descriptions of the behavior of moving objects. These he derived from experimentation. The method of using experiments to determine natural laws had essentially been lost since the time of the early Greeks.

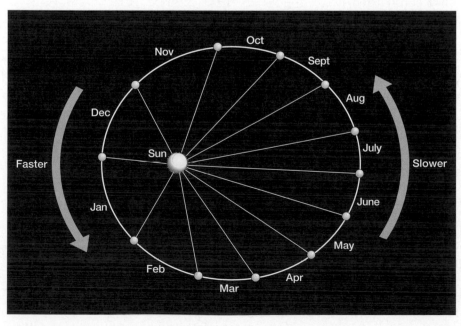

Figure 20.10   Kepler's law of equal areas. A line connecting a planet (Earth) to the Sun sweeps out an area in such a manner that equal areas are swept out in equal times. Thus, Earth revolves slower when it is farther from the Sun (aphelion) and faster when it is closest (perihelion). The eccentricity of Earth's orbit is greatly exaggerated in this diagram.

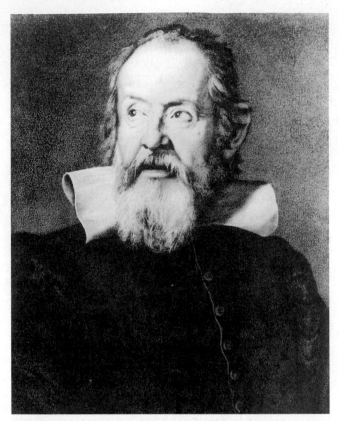

Figure 20.11 Italian scientist Galileo Galilei (1564–1642) used a new invention, the telescope, to observe the Sun, Moon, and planets in more detail than ever before. (Yerkes Observatory Photograph/University of Chicago)

All astronomical discoveries before Galileo's time were made without the aid of a telescope. In 1609, Galileo heard that a Dutch lens maker had devised a system of lenses that magnified objects. Apparently without ever seeing a telescope, Galileo constructed his own, which magnified distant objects to three times the size seen by the unaided eye. He immediately made others, the best having a magnification of about 30.

With the telescope, Galileo was able to view the universe in a new way. He made many important discoveries that supported the Copernican view of the universe, such as:

1. The discovery of four satellites, or moons, orbiting Jupiter. Galileo accurately determined their periods of revolution, which range from two to 17 days (Figure 20.12). This find dispelled the old idea that Earth was the only center of motion in the universe; for here, plainly visible, was another center of motion—Jupiter. It also countered the argument, frequently used by those opposed to the Sun-centered system, that the Moon would be left behind if Earth really revolved around the Sun.

Figure 20.12 Sketch by Galileo of how he saw Jupiter and its four largest satellites through his telescope. The positions of Jupiter's four largest Moons (drawn as stars) change nightly. You can observe these same changes with binoculars. (Yerkes Observatory Photograph/University of Chicago)

2. The discovery, through observation, that the planets are circular disks rather than just points of light, as was previously thought. This indicated that the planets must be Earth-like.

3. The discovery that Venus has phases just like the Moon, demonstrating that Venus orbits its source

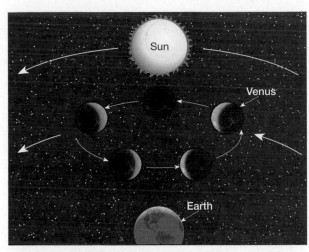

A. Phases of Venus as seen from Earth in the Earth-centered model.

D. Phases of Venus as seen from Earth in the Sun-centered model.

Figure 20.13   Using a telescope, Galileo discovered that Venus has phases just like the Moon. **A.** In the Ptolemaic (Earth-centered) system, the orbit of Venus lies between the Sun and Earth as shown in Figure 20.4A. Thus, in an Earth-centered solar system, only the crescent phase of Venus would be visible from Earth. **B.** In the Copernican (Sun-centered) system, Venus orbits the Sun and hence all of the phases of Venus should be visible from Earth. **C.** As Galileo observed, Venus goes through a series of Moonlike phases. Further, Venus appears smallest during the full phase when it is farthest from Earth and largest in the crescent phase when it is closest to Earth. This verified Galileo's belief that the Sun was the center of the solar system. (Photo courtesy of Lowell Observatory)

C.

of light—the Sun. Galileo saw that Venus appears smallest when it is in full phase and thus is farthest from Earth (Figure 20.13B,C). In the Ptolemaic system, as shown in Figure 20.13A, the orbit of Venus lies between Earth and the Sun, which means that only the crescent phase of Venus could be seen from Earth.

4. The discovery that the Moon's surface is not a smooth glass sphere, as the ancients had suspected and the Church had decreed. Rather, Galileo saw mountains, craters, and plains. He thought the plains might be bodies of water, and this idea was strongly promoted by others, as we can tell from the names given to these features (Sea of Tranquility, Sea of Storms, and so forth).

5. The discovery that the Sun (the viewing of which may have caused the eye damage that later blinded him) had sunspots (dark regions caused by slightly lower temperatures). He tracked the movement of these spots and estimated the rotational period of the Sun as just under a month. Hence, another heavenly body was found to have both "blemishes" and rotational motion.

In 1616 the Church condemned the Copernican theory as contrary to Scripture, and Galileo was told to abandon it. Unwilling to accept this verdict, Galileo began writing his most famous work, *Dialogue of the Great World Systems.* Despite poor health, he completed the project and in 1630 went to Rome, seeking permission from Pope Urban VIII to publish. Because the book was a dialogue that expounded both the Ptolemaic and Copernican systems, publication was allowed. However, Galileo's enemies were quick to realize that he was promoting the Copernican view at the expense of the Ptolemaic system. Sales of the book were quickly halted, and Galileo was called before the Inquisition. Tried and convicted of proclaiming doctrines contrary to religious doctrine, he was sentenced to permanent house arrest, under which he remained for the last 10 years of his life.

Despite this restriction, and his age, and his grief after the death of his eldest daughter, Galileo continued to work. In 1637 he became totally blind, yet during the next few years he completed his finest scientific work, a book on the study of motion. When Galileo died in 1642, the Grand Duke of Tuscany wanted to erect a monument in his honor, but fear that it might offend the Church prevailed, and it was never built. It was not until 1992 that Galileo was finally exonerated by the Roman Catholic Church.

? STUDENTS SOMETIMES ASK...

*Did Galileo drop balls of iron and
wood from the Leaning Tower of Pisa?*

Through experimentation, Galileo discovered that the acceleration of falling objects does not depend on their weight. According to some accounts, Galileo made this discovery by dropping balls of iron and wood from the Leaning Tower of Pisa to show that they would fall together and hit the ground at the same time. Despite the popularity of this legend, Galileo probably did not attempt this experiment. In fact, it would have been inconclusive because of the effect of air resistance. However, nearly four centuries later, this experiment was performed most dramatically on the airless Moon when David Scott, an *Apollo 15* astronaut, demonstrated that a feather and a hammer do, indeed, fall at the same rate.

Figure 20.14   English scientist Sir Isaac Newton (1642–1727) explained gravity as the force that holds planets in orbit around the Sun. (Yerkes Observatory Photograph/University of Chicago)

Sir Isaac Newton

Sir Isaac Newton (1642–1727) (Figure 20.14) was born in the year of Galileo's death. His many accomplishments in mathematics and physics led a successor to say that "Newton was the greatest genius that ever existed."

Although Kepler and those who followed attempted to explain the forces involved in planetary motion, their explanations were less than satisfactory. Kepler believed that some force pushed the planets along in their orbits. Galileo, however, correctly reasoned that no force is required to keep an object in motion. Galileo proposed that the natural tendency for a moving object (that is unaffected by an outside force) is to continue moving at a uniform speed and in a straight line. This concept, *inertia*, was later formalized by Newton as his first law of motion.

The problem, then, was not to explain the force that keeps the planets moving but rather to determine the force that *keeps them from going in a straight line out into space.* It was to this end that Newton conceptualized the force of *gravity*. At the early age of 23, he envisioned a force that extends from Earth into space and holds the Moon in orbit around Earth. Although others had theorized the existence of such a force, he was the first to formulate and test the *law of universal gravitation.* It states:

> Every body in the universe attracts every other body with a force that is directly proportional to their masses and inversely proportional to the square of the distance between them.

Thus, the gravitational force decreases with distance, so that two objects 3 kilometers apart have 3^2, or 9, times less gravitational attraction than if the same objects were 1 kilometer apart.

The law of gravitation also states that the greater the mass of the object, the greater its gravitational force. For example, the large mass of the Moon has a gravitational force strong enough to cause ocean tides on Earth, whereas the tiny mass of a communications satellite has no measurable effect on Earth. The mass of an object is a measure of the total amount of matter it contains. But more often mass is measured by determining the resistance an object exhibits in response to any effort made to change its state of motion.

Often we confuse the concept of *mass* with our notion of *weight.* Specifically, weight is the force of gravity acting upon an object. Therefore, weight varies when gravitational forces change. An object weighs less on the Moon than on Earth because the Moon is much less massive than Earth. However, unlike weight, the mass of an object does not change. For example, a person weighing 120 pounds on Earth weighs 1/6 as much, or 20 pounds, on the Moon, but the person's mass remains unchanged.

With his laws of motion, Newton proved that the force of gravity—combined with the tendency of a planet to remain in straight-line motion—results in the elliptical orbits discovered by Kepler. Earth, for example, moves forward in its orbit about 30 kilometers (18.5 miles) each second, and during the same second, the force of gravity pulls it toward the Sun about 0.5 centimeter (1/8 inch). Therefore, as Newton concluded, it is the combination of Earth's forward motion and its "falling" motion that defines its orbit (Figure 20.15). If gravity were somehow eliminated, Earth would move in a straight line out into space. Conversely, if Earth's forward motion suddenly stopped, gravity would pull it directly toward the Sun.

Up to this point, we have discussed Earth as if the only forces involved in its motion were caused by its gravitational relationship with the Sun. However, all bodies in the solar system have gravitational effects on Earth and on each other. For this reason, the orbit of Earth is not the perfect ellipse determined by Kepler. Any variance in the orbit of a body from its predicted path is called **perturbation** (*perturb* = disturb). For example, Jupiter's gravitational pull on Saturn reduces Saturn's orbital period by nearly one week from the predicted period. As we shall see, the application of this concept led to the discovery of the planet Neptune, because of Neptune's gravitational effect on the orbit of Uranus.

Newton used the law of universal gravitation to redefine Kepler's third law, which states the relationship between the orbital periods of the planets and their solar distances. When restated, Kepler's third law takes into account the masses of the bodies involved and thereby provides a method for determining the mass of a body when the orbit of one of its satellites is known. For example, the mass of the Sun is known from Earth's orbit, and Earth's mass has been determined from the orbit of the Moon. In fact, the mass of any body with a satellite can be determined. The masses of bodies that do not have satellites can be determined only if the bodies noticeably affect the orbit of a neighboring body or of a nearby artificial satellite.

Constellations

As early as 5000 years ago, people became fascinated with the star-studded skies and began to name the patterns they saw (see Box 20.2). These configurations, called **constellations** (*con* = with, *stella* = star), were named in honor of mythological characters or great heroes, such as Orion. It takes a good bit of imagination to make out the intended subjects, as most constellations were probably not thought of as likenesses in the first place. Although we inherited many of the constellations from Greek mythology, it is believed that Greek astronomers acquired most of theirs from the Babylonians, Egyptians, and Mesopotamians.

Although the stars that make up a constellation all appear to be the same distance from Earth, this is not the case. Some are many times farther away than others. Thus, the stars in a particular constellation are not associated with each other in any physical way. In addition, various cultural groups, including Native Americans, attached their own names, pictures, and stories to the constellations. For example, the constellation Orion, the Hunter, was known as the White Tiger to ancient Chinese astronomers.

Today, 88 constellations are recognized, and they are used to divide the sky into units, just as state boundaries divide the United States. Every star in the sky is in,

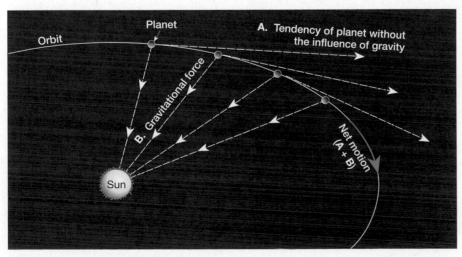

Figure 20.15 Orbital motion of Earth and other planets.

BOX 20.2

UNDERSTANDING EARTH:
Astrology—the Forerunner of Astronomy

Many people confuse astrology and astronomy to the point of believing these terms to be synonymous. Nothing can be further from the truth. *Astronomy* is a scientific probing of the universe to derive the properties of celestial objects and the laws under which the universe operates. *Astrology,* on the other hand, is based on ancient superstitions that hold that an individual's actions and personality are based on the positions of the planets and stars now, and at the person's birth. Scientists do not accept astrology, regarding it as a pseudoscience ("false science"). It is hoped that most people read horoscopes only as a pastime and do not let them influence daily living.

Apparently astrology had its origin more than 3000 years ago when the positions of the planets were plotted as they regularly migrated against the background of the "fixed" stars. Because the solar system is "flat," like a whirling Frisbee, the planets orbit the Sun along nearly the same plane. Therefore, the planets, Sun, and Moon all appear to move along a band around the sky known as the *zodiac*. Because Earth's Moon cycles through its phases about 12 times each year, the Babylonians divided the zodiac into 12 constellations (Figure 20.C). Thus, each

Figure 20.D   Stonehenge, an ancient observatory in England. On June 21–22 (summer solstice), the Sun can be observed rising above the heel stone. (Robin Scagell/Science Photo Library/Photo Researchers, Inc.)

successive full Moon can be seen against the backdrop of the next constellation.

The dozen constellations of the zodiac ("Zone of Animals," so named because some constellations represent animals) are Aries, Taurus, Gemini, Cancer, Leo, Virgo, Libra, Scorpio, Sagittarius, Capricorn, Aquarius, and Pisces. These names may be familiar to you as the astrological signs of the zodiac. When first established, the vernal equinox (first day of spring) occurred when the Sun was viewed against the constellation Aries. However, during each succeeding vernal equinox, the position of the Sun shifts very slightly against the background of stars. Now, over 2000 years later, the vernal equinox occurs when the Sun is in Pisces (Figure 20.C). In several years, it will occur when the Sun appears against Aquarius. (Hence, the "Age of Aquarius" is coming.)

Although astrology is not a science and has no basis in fact, it did contribute to the science of astronomy. The positions of the Moon, Sun, and planets at the time of a person's birth (sign of the zodiac) were considered to have great influence on that person's life.* Even the great astronomer Kepler was required to make horoscopes part of his duties. To make horoscopes for the future, astrologers attempted to predict the future positions of the celestial bodies. Consequently, some of the improvements in astronomical instruments were made because of the desire for more accurate predictions of events such as eclipses, which were considered highly significant in a person's life.

Even prehistoric people built observatories. The structure known as Stonehenge, in England, was undoubtedly an attempt at better solar predictions (Figure 20.D). At the time of midsummer in the Northern Hemisphere (June 21–22—the summer solstice), the rising Sun emerges directly above the heel stone of Stonehenge. Besides keeping this calendar, Stonehenge may also have provided a method of determining eclipses. The remnants of other early observatories exist elsewhere in the Americas, Europe, Asia, and Africa.

*It is interesting to note that 2000 years ago a person born on July 28 was considered a Leo because the Sun was in that constellation. During modern times the Sun appears in the constellation Cancer on this date, but individuals born during this time are still dubbed Leos.

Figure 20.C   The 12 constellations of the zodiac. Earth is shown in its autumn (September) position in orbit, from which the Sun is seen against the background of the constellation Virgo.

but is not necessarily part of, one of these constellations. Constellations therefore enable astronomers to roughly identify the area of the heavens they are observing. For the student, the constellations provide a good way to become familiar with the night sky.

Some of the brightest stars were given proper names, such as Sirius, Arcturus, and Betelgeuse. In addition, the brightest stars in a constellation are generally named in order of their brightness by the letters of the Greek alphabet—alpha (α), beta (β), and so on—followed by the name of the parent constellation. For example, Sirius, the brightest star in the constellation Canis Major (Larger Dog), is called Alpha (α) Canis Majoris.

Positions in the Sky

If you gaze away from the city lights on a clear night, you will get the distinct impression that the stars produce a spherical shell surrounding Earth. This impression seems so real that it is easy to understand why many early Greeks regarded the stars as being fixed to a crystalline celestial sphere. Although we realize that no such sphere exists, it is convenient to use this concept for locating stars.

One method for doing this, called the **equatorial system,** divides the celestial sphere into a coordinate system. It is very similar to the latitude–longitude system used for locations on Earth's surface (Figure 20.16). Because the celestial sphere appears to rotate around an imaginary line extending from Earth's axis, the north and south celestial poles are in line with the terrestrial North Pole and South Pole. The north celestial pole happens to be very near the bright star whose various names reflect its location: "pole star," Polaris, and North Star. To an observer in the Northern Hemisphere, the stars appear to circle Polaris, because it, like the North Pole, is in the center of motion (Figure 20.17). Figure 20.18 shows how to locate the North Star by using two stars in the very visible constellation called the Big Dipper.

Now, imagine a plane through Earth's equator, a plane that extends outward from Earth and intersects the celestial sphere. The intersection of this plane with the celestial sphere is the *celestial equator* (see Figure 20.16). In the equatorial system, the terms *declination,* which is analogous to latitude, and *right ascension,* which is analogous to longitude, are used (Figure 20.16). **Declination** (*delinare* = to turn away), like latitude, is the angular distance north or south of the celestial equator. **Right ascension** (*ascendere* = to climb up) is the angular distance measured eastward along the celestial equator from the position of the vernal equinox. (The *vernal equinox* is at the point in the sky where the Sun crosses the celestial equator, at the onset of spring.) While declination is expressed in degrees, right ascension is usually expressed in hours, where each hour is equivalent to 15 degrees. To visualize distances on the celestial sphere, it helps to remember that the Moon and Sun have an apparent width of about 0.5 degree.

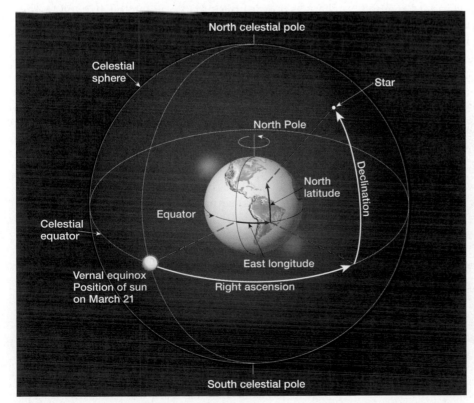

Figure 20.16   Astronomical coordinate system on the celestial sphere.

Figure 20.17 Star trails in the region of Polaris (north celestial pole) on a time exposure. (Courtesy of National Optical Astronomy Observatories)

Motions of Earth

The two primary motions of Earth are rotation and revolution. **Rotation** is the turning, or spinning, of a body on its axis. **Revolution** is the motion of a body, such as a planet or moon, along a path around some point in space. For example, Earth *revolves* around the Sun, and the Moon *revolves* around Earth. Earth also has another very slow motion known as **precession,** which is the slight movement, over a period of 26,000 years, of Earth's axis.

Rotation

The main consequences of Earth's rotation are day and night. Earth's rotation has become a standard method of measuring time because it is so dependable and easy to use. Each rotation equals about 24 hours. You may be surprised to learn that we can measure Earth's rotation

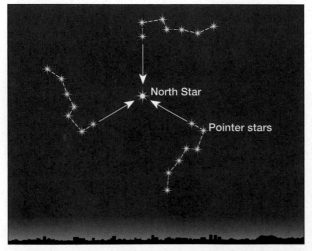

Figure 20.18 Locating the North Star (Polaris) from the pointer stars in the Big Dipper, which is part of the constellation Ursa Major. The Big Dipper is shown soon after sunset in December (lower figure), April (upper figure), and August (left).

in two ways, making two kinds of days. Most familiar is the **mean solar day,** the time interval from one noon to the next, which averages about 24 hours. Noon is when the Sun has reached its zenith (highest point in the sky).

The **sidereal** (*sider* = star, *at* = pertaining to) **day,** on the other hand, is the time it takes for Earth to make one complete rotation (360 degrees) with respect to a star other than our Sun. The sidereal day is measured by the time required for a star to reappear at the identical position in the sky where it was observed the day before. The sidereal day has a period of 23 hours, 56 minutes, and 4 seconds (measured in solar time), which is almost 4 minutes shorter than the mean solar day. This difference results because the direction to distant stars changes only infinitesimally, owing to Earth's slow revolution along its orbit, whereas the direction to the Sun changes by almost 1 degree each day. This difference is shown in Figure 20.19.

If it is not apparent why we use the mean solar day rather than the sidereal day as a measurement of our day, consider the fact that in sidereal time, "noon" occurs four minutes earlier each day. Therefore, after a span of six months, "noon" occurs at "midnight." Astronomers use sidereal time because the stars appear in the same position in the sky every 24 sidereal hours. Usually, an observatory will begin its sidereal day when the position of the vernal equinox is directly overhead—that is, over the meridian on which the observatory is located. Therefore, when the observatory's sidereal clock is the same as the star's right ascension, the star will be overhead, or at its highest point. For example, the brightest star in the heavens, Sirius, has a right ascension of 6 hours, 42 minutes, and 56 seconds and will be overhead when the clock at the observatory indicates that time.

Revolution

Earth revolves around the Sun in an elliptical orbit at an average speed of 107,000 kilometers (66,000 miles) per hour. Its average distance from the Sun is 150 million

Figure 20.19   The difference between a solar day and a sidereal day. Locations X and Y are directly opposite each other. It takes Earth 23 hours and 56 minutes to make one rotation with respect to the stars (sidereal day). However, notice that after Earth has rotated once with respect to the stars, point Y is not yet returned to the "noon position" with respect to the Sun. Earth has to rotate another 4 minutes to complete the solar day.

kilometers (93 million miles), but because its orbit is an ellipse, Earth's distance from the Sun varies. At **perihelion** (*peri* − near, *helios* = Sun), it is 147 million kilometers (91.5 million miles) distant, which occurs about January 3 each year. At **aphelion** (*apo* = away from, *helios* = Sun), Earth is 152 million kilometers (94.5 million miles) distant, which occurs about July 4.

Because of Earth's annual movement around the Sun, each day the Sun appears to be displaced among the constellations at a distance equal to about twice its width, or 1 degree. The apparent annual path of the Sun against the backdrop of the celestial sphere is called the **ecliptic** (Figure 20.20). Generally, the planets and the Moon travel in nearly the same plane as Earth. Hence,

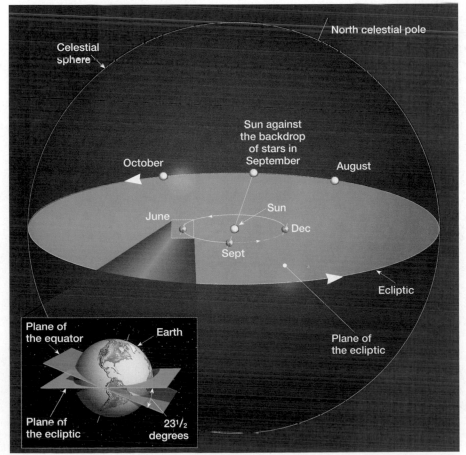

Figure 20.20   Earth's orbital motion causes the apparent position of the Sun to shift about 1 degree each day on the celestial sphere.

their paths on the celestial sphere lie near the ecliptic. (The most notable exception is Pluto, which has an orbit that is tilted 17 degrees to the plane of Earth's orbit.)

The imaginary plane that connects Earth's orbit with the celestial sphere is called the **plane of the ecliptic.** From the reference plane, Earth's axis of rotation is tilted about $23\frac{1}{2}$ degrees. Because of Earth's tilt, the apparent path of the Sun (ecliptic) and the celestial equator intersect each other at an angle of $23\frac{1}{2}$ degrees (Figure 20.20). This angle is very important to Earth's inhabitants. Because of the inclination of Earth's axis to the plane of the ecliptic, Earth exhibits its yearly cycle of seasons, a topic discussed in detail in Chapter 15.

When the *apparent* position of the Sun is plotted on the celestial sphere over a period of a year's time, its path intersects the celestial equator at two points (Figure 20.21). From a Northern Hemisphere point of view, these intersections are called the vernal (spring) equinox (March 20–21) and autumnal equinox (September 22–23). On June 21–22, the date of the summer solstice, the Sun appears $23\frac{1}{2}$ degrees north of the celestial equator, and six months later, on December 21–22, the date of the winter solstice, the Sun appears $23\frac{1}{2}$ degrees south of the celestial equator.

Precession

A third and very slow movement of Earth is called *precession*. Although Earth's axis maintains approximately the same angle of tilt, the direction in which the axis points continually changes. As a result, the axis traces a circle on the sky. This movement is very similar to the movement (wobble) of a spinning top (Figure 20.22A). At the present time, the axis points toward the bright

star Polaris. In A.D. 14,000, it will point toward the bright star Vega, which will then become the North Star for a few thousand years (Figure 20.22B). The period of precession is 26,000 years. By the year 28,000, Polaris will once again be the North Star.

Precession has only a minor effect on the seasons, because the angle of tilt changes only slightly. It does,

? STUDENTS SOMETIMES ASK...

Our book states that Earth is farthest from the Sun in July and closest to the Sun in January. What would the seasons be like if this situation were reversed?

The situation you proposed will, in fact, occur in about 13,000 years. (Recall that variations in Earth-Sun distance are not the primary cause of the seasons. Nevertheless, they do affect average seasonal temperatures.) Gradually, Earth's position will change so that in 13,000 years the Northern Hemisphere will experience winter when Earth is farthest from the Sun (aphelion), and summer will occur when our planet is closest to the Sun (perihelion). This will be, of course, just the opposite of the current situation. When perihelion occurs in July, average summer temperatures in the Northern Hemisphere will be warmer than they presently are. A summer in Montreal, Canada, might be more akin to a typical summer in Washington, D.C. today. Likewise, Southern Hemisphere winters will also be warmer than they are currently. When aphelion occurs in January, northern latitudes will experience more extreme winter temperatures whereas average summer temperatures south of the equator will be cooler.

Figure 20.21 The apparent position of the Sun plotted on the celestial sphere. The path of the Sun (ecliptic) crosses the celestial equator on two occasions each year, March 20–21 and September 22–23. These are known as the equinox positions because the lengths of daylight and darkness on Earth are equal.

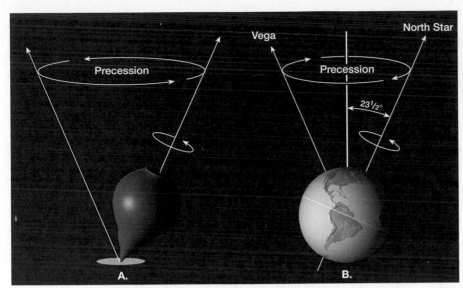

Figure 20.22   **A.** Precession illustrated by a spinning top. **B.** Precession of Earth causes the North Pole to point to different parts of the sky during a 26,000-year cycle. Today, the North Pole points to Polaris (North Star). In 13,000 years, Vega will be the North Star.

however, cause the positions of the seasons (equinox and solstice) to move slightly each year among the constellations.

In addition to its own movements, Earth shares numerous motions with the Sun. It accompanies the Sun as the entire solar system speeds in the direction of the bright star Vega at 20 kilometers (12 miles) per second. Also, the Sun, like other nearby stars, revolves around the galaxy, a trip that requires 230 million years to traverse at speeds approaching 250 kilometers (150 miles) per second. In addition, the galaxies themselves are in motion. We are presently approaching one of our nearest galactic neighbors, the Great Galaxy in Andromeda. In summary, the motions of Earth are many and complex, and its speed in space is very great.

Motions of the Earth–Moon System

Earth has one natural satellite, the Moon. In addition to accompanying Earth in its annual trek around the Sun, our Moon orbits Earth within a period of about one month. When viewed from above the North Pole, the direction of this motion is counterclockwise (eastward). Because the Moon's orbit is elliptical, its distance to Earth varies by about 6 percent, averaging 384,401 kilometers (238,329 miles).

The motions of the Earth–Moon system constantly change the relative positions of the Sun, Earth, and Moon. The results are some of the most obvious of astronomical phenomena, namely the *phases of the Moon* and the occasional *eclipses of the Sun and Moon.*

Phases of the Moon

The first astronomical phenomenon to be understood was the regular cycle of the **phases of the Moon.** On a monthly basis, we observe the phases as a systematic change in the amount of the Moon that appears illuminated (Figure 20.23). We will choose the "new-Moon" position in the cycle as a starting point. About two days after the new Moon, a thin sliver (*crescent phase*) appears low in the western sky just after sunset. During the following week, the illuminated portion of the Moon visible from Earth increases (*waxing*) to a half circle (*first-quarter phase*) and can be seen from about noon to midnight. In another week, the complete disk (*full-Moon phase*) can be seen rising in the east as the Sun is sinking in the west. During the next two weeks, the percentage of the Moon that can be seen steadily declines (*waning*), until the Moon disappears altogether (*new-Moon phase*). The cycle soon begins anew with the reappearance of the crescent Moon.

The lunar phases are a consequence of the motion of the Moon and the sunlight that is reflected from its surface (Figure 20.23B). Half of the Moon is illuminated at all times (note the inner group of Moon sketches in Figure 20.23A). But to an Earthbound observer, the percentage of the bright side that is visible depends on

Why do we sometimes see the Moon in the daytime?

During a full moon, the Moon and the Sun are on opposite sides of Earth, which causes the Moon to rise at sunset and set at sunrise. Thus, the full moon is visible only at night. However, during the other phases of the lunar cycle, the Moon and the Sun are not directly opposite each other, making it possible to see the Moon during daylight hours. For example, the waning-gibbous phase can be seen in the early morning hours and the waxing-gibbous Moon in the afternoon (see Figure 20.23). Although the crescent Moon is "out" shortly before sunset and after sunrise, you probably won't see it in the daytime. Why not?

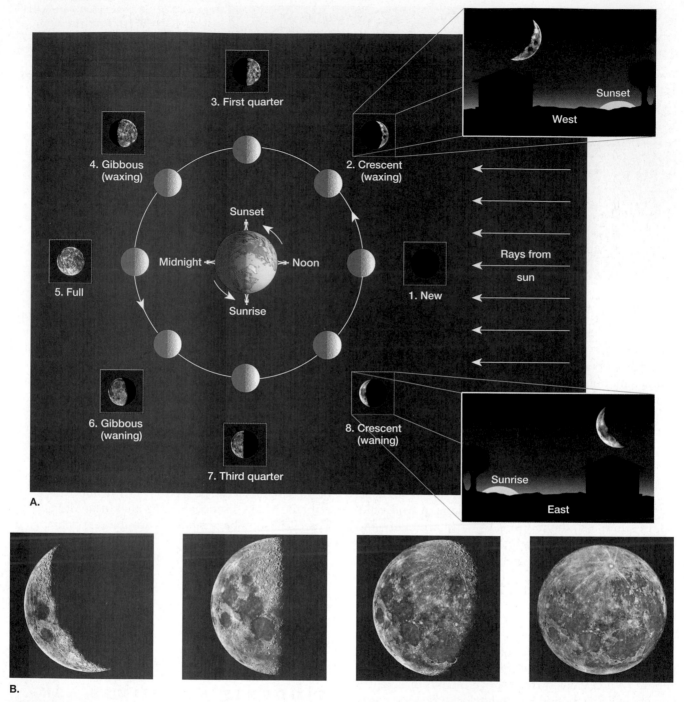

Figure 20.23 Phases of the Moon. **A.** The outer figures show the phases as seen from Earth. **B.** Compare these photographs with the diagram. (Courtesy of UCO/Lick Observatory Image)

the location of the Moon with respect to the Sun and Earth. When the Moon lies *between* the Sun and Earth, none of its bright side faces Earth, so we see the new-Moon ("no-Moon") phase. Conversely, when the Moon lies on the side of Earth opposite the Sun, all of its lighted side faces Earth, so we see the full Moon. At all positions between these extremes, an intermediate amount of the Moon's illuminated side is visible from Earth.

Lunar Motions

The cycle of the Moon through its phases requires $29\frac{1}{2}$ days, a time span called the **synodic month.** This cycle was the basis for the first Roman calendar. However, this is the *apparent period* of the Moon's revolution around Earth and not the true period, which takes only $27\frac{1}{3}$ days and is known as the **sidereal month.** The reason for the difference of nearly two days each cycle is shown in Figure 20.24. Note that as the Moon orbits

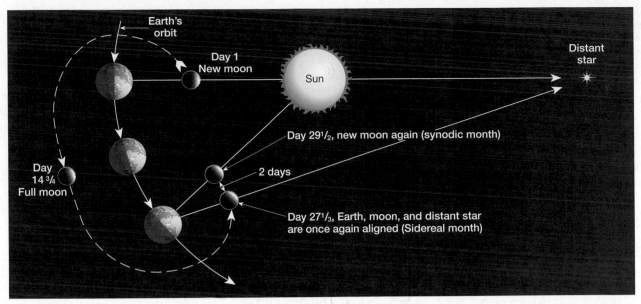

Figure 20.24   The difference between the sidereal month ($27\frac{1}{3}$ days) and the synodic month ($29\frac{1}{2}$ days). Distances and angles are not shown to scale.

Earth, the Earth–Moon system also moves in an orbit around the Sun. Consequently, even after the Moon has made a complete revolution around Earth, it has not yet reached its starting position, which was directly between the Sun and Earth (new-Moon phase). The additional motion to reach the starting point takes another two days.

An interesting fact concerning the motions of the Moon is that its period of rotation about its axis and its revolution around Earth are the same—$27\frac{1}{3}$ days. Because of this, the same lunar hemisphere always faces Earth. All of the manned *Apollo* missions were confined to the Earth-facing side. Only orbiting satellites and astronauts have seen the "back" side of the Moon.

Because the Moon rotates on its axis only once every $27\frac{1}{3}$ days, any location on its surface experiences periods of daylight and darkness lasting about two weeks. This, along with the absence of an atmosphere, accounts for the high surface temperature of 127°C

(261°F) on the day side of the Moon and the low surface temperature of -173°C (-280°F) on its night side.

Eclipses

Along with understanding the Moon's phases, the early Greeks also realized that eclipses are simply shadow effects. When the Moon moves in a line directly between Earth and the Sun, which can occur only during the new-Moon phase, it casts a dark shadow on Earth, producing a **solar eclipse** (*eclipsis* = to fail to appear) (Figure 20.25). Conversely, the Moon is eclipsed (**lunar eclipse**) when it moves within Earth's shadow, a situation that is possible only during the full-Moon phases (Figure 20.26).

Why then does a solar eclipse not occur with every new-Moon phase and a lunar eclipse with every full-Moon phase? They would, if the orbit of the Moon lay exactly along the plane of Earth's orbit (the plane of the

Figure 20.25   Solar eclipse. Observers in the zone of the umbral shadow see a total solar eclipse. Those in the penumbra see a partial eclipse. The path of the solar eclipse moves eastward across the globe.

Figure 20.26 Lunar eclipse. During a total lunar eclipse the Moon's orbit carries it into the dark shadow of Earth (umbra). During a partial eclipse only a portion of the Moon enters the umbra.

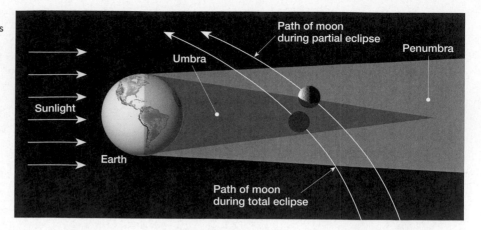

ecliptic). However, the Moon's orbit is inclined about 5 degrees to the plane that contains Earth and the Sun (Figure 20.27). Thus, during most new-Moon phases, the shadow of the Moon misses Earth (passes above or below); and during most full-Moon phases, the shadow of Earth misses the Moon. Only when a new- or full-Moon phase occurs where the Moon's orbit crosses the plane of the ecliptic can an eclipse take place.

Because these conditions are normally met only twice a year, the usual number of eclipses is four. These occur as a set of one solar and one lunar eclipse, followed six months later with another set (Figure 20.27). Occasionally the alignment is such that three eclipses can be squeezed into a one-month period. These occur as a solar eclipse flanked by two lunar eclipses, or vice versa. Furthermore, it occasionally happens that the first set of eclipses occurs at the very beginning of a year, and a third set occurs before the year ends, resulting in six eclipses in that year. More rarely, if one of these sets is a three-eclipse kind, the total number of eclipses in a year can reach seven, which is the maximum.

During a total lunar eclipse, Earth's circular shadow can be seen moving slowly across the disk of the full Moon. When totally eclipsed, the Moon is completely within Earth's shadow but still is visible as a coppery disk, because Earth's atmosphere bends and

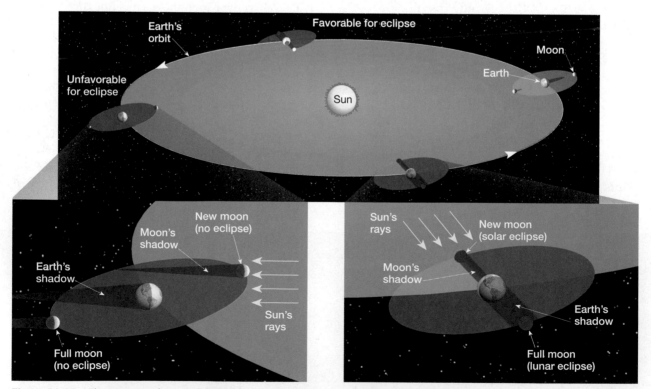

Figure 20.27 The Moon's orbit is inclined about 5 degrees to the plane that contains the Sun and Earth. Thus, during most new-Moon phases, the shadow of the Moon misses Earth (passes above or below), and during most full-Moon phases, the shadow of Earth misses the Moon. Only when a new- or full-Moon phase occurs where the Moon's orbit crosses the Earth-Sun plane can an eclipse take place. These conditions are met at roughly six-month intervals.

Figure 20.28 This sequence of photos starting from the upper left to the lower right shows the stages of a total solar eclipse. (From *Foundations of Astronomy*, Third Edition, by Michael Seeds, reprinted by permission of the author and Wadsworth Publishing Company)

transmits some long-wavelength light (red) into its shadow. A total eclipse of the Moon can last up to four hours and is visible to anyone on the side of Earth facing the Moon.

During a total solar eclipse, the Moon casts a circular shadow that is never wider than 275 kilometers (170 miles), about the size of South Carolina. Anyone observing in this region will see the Moon slowly block the Sun from view and the sky darken (Figure 20.28). Near totality, a sharp drop in temperature of a few degrees is experienced. The solar disk is completely blocked for at most seven minutes, because the Moon's shadow is so small, and then one edge reappears.

At totality, the dark Moon is seen covering the complete solar disk, and only the Sun's brilliant white outer atmosphere is visible (see Figure 22.21). Total solar eclipses are visible only to people in the dark part of the Moon's shadow (*umbra*), while a partial eclipse is seen by those in the light portion (*penumbra*) (see Figure 20.25).

Partial solar eclipses are more common in the polar regions, because it is this zone that the penumbra covers when the dark umbra of the Moon's shadow just misses Earth. A total solar eclipse is a rare event at any given location. The next one that will be visible from the contiguous United States will take place on August 21, 2017.

Chapter Summary

• Early Greeks held the *geocentric* (Earth-centered) view of the universe, believing that Earth was a sphere that stayed motionless at the center of the universe. Orbiting Earth were the seven wanderers (*planetai* in Greek), which included the Moon, Sun, and the known planets Mercury, Venus, Mars, Jupiter, and Saturn. To the early Greeks, the stars traveled daily around Earth on a transparent, hollow sphere called the *celestial sphere*. In A.D. 141, *Claudius Ptolemy* presented the geocentric outlook of the Greeks in its most sophisticated form in a model that became known as the *Ptolemaic system*. The Ptolemaic model had the planets moving in circular orbits around a motionless Earth. To explain the *retrograde motion* of planets (the apparent westward or opposite motion that planets exhibit for a period of time as Earth overtakes and passes them), Ptolemy proposed that the planets orbited in small circles (*epicycles*), revolving along large circles (*deferents*).

• In the fifth century B.C., the Greek *Anaxagoras* reasoned that the Moon shines by reflected sunlight, and because it is a sphere, only half is illuminated at one time. *Aristotle* (384–322 B.C.) concluded that Earth is spherical. The first Greek to profess a Sun-centered, or *heliocentric*, universe was *Aristarchus* (312–230 B.C.). The first successful attempt to establish the size of Earth is credited to *Eratosthenes* (276–194 B.C.). The greatest of the early Greek astronomers was *Hipparchus* (second century B.C.), best known for his star catalogue.

• Modern astronomy evolved through the work of many dedicated individuals during the sixteenth and seventeenth centuries. *Nicolaus Copernicus* (1473–1543) reconstructed the solar system with the Sun at the center and the planets orbiting around it but erroneously continued to use circles to represent the orbits of planets. *Tycho Brahe's* (1546–1601) observations were far more precise than any made previously and are his legacy to astronomy. *Johannes Kepler* (1571–1630) ushered in the new astronomy with his three laws of planetary motion. After constructing his own telescope, *Galileo Galilei* (1564–1642) made many important discoveries that supported the Copernican view of a Sun-centered solar system. *Sir Isaac Newton* (1642–1727) was the first to formulate and test the law of universal gravitation, develop the laws of motion, and prove that the force of *gravity*, combined with the tendency of an object to move in a straight line (*inertia*), results in the elliptical orbits discovered by Kepler.

• As early as 5000 years ago people began naming the configurations of stars, called *constellations,* in honor of mythological characters or great heroes. Today, 88 constellations are recognized that divide the sky into units, just as state boundaries divide the United States.

• One method for locating stars, called the *equatorial system,* divides the celestial sphere into a coordinate system similar to the latitude-longitude system used for locations on Earth's surface. *Declination,* like latitude, is the angular distance north or south of the *celestial equator. Right ascension* is the angular distance measured eastward from the position of the *vernal equinox* (the point in the sky where the Sun crosses the celestial equator at the onset of spring).

• The two primary motions of Earth are *rotation* (the turning, or spinning, of a body on its axis) and *revolution* (the motion of a body, such as a planet or moon, along a path around some point in space). Another very slow motion of Earth is *precession* (the slow motion of Earth's axis that traces out a cone over a period of 26,000 years). Earth's rotation can be measured in two ways, making two kinds of days. The *mean solar day* is the time interval from one noon to the next, which averages about 24 hours. In contrast, the *sidereal day* is the time it takes for Earth to make one complete rotation with respect to a star other than the Sun, a period of 23 hours, 56 minutes, and 4 seconds. Earth revolves around the Sun in an elliptical orbit at an average distance from the Sun of 150 million kilometers (93 million miles). At *perihelion* (closest to the Sun), which occurs in January, Earth is 147 million kilometers from the Sun. At *aphelion* (farthest from the Sun), which occurs in July, Earth is 152 million kilometers distant. The imaginary plane that connects Earth's orbit with the celestial sphere is called the *plane of the ecliptic.*

• One of the first astronomical phenomena to be understood was the regular cycle of the phases of the Moon. The cycle of the Moon through its phases requires $29\frac{1}{2}$ days, a time span called the *synodic month.* However, the true period of the Moon's revolution around Earth takes $27\frac{1}{3}$ days and is known as the *sidereal month.* The difference of nearly two days is due to the fact that as the Moon orbits Earth, the Earth–Moon system also moves in an orbit around the Sun.

• In addition to understanding the Moon's phases, the early Greeks also realized that eclipses are simply shadow effects. When the Moon moves in a line directly between Earth and the Sun, which can occur only during the new-Moon phase, it casts a dark shadow on Earth, producing a *solar eclipse.* A *lunar eclipse* takes place when the Moon moves within the shadow of Earth during the full-Moon phase. Because the Moon's orbit is inclined about 5 degrees to the plane that contains the Earth and Sun (the plane of the ecliptic), during most new- and full-Moon phases no eclipse occurs. Only if a new- or full-Moon phase occurs as the Moon crosses the plane of the ecliptic can an eclipse take place. The usual number of eclipses is four per year.

Key Terms

aphelion (p. 565)
astronomical unit (AU) (p. 557)
celestial sphere (p. 550)
constellations (p. 561)
declination (p. 563)
ecliptic (p. 565)
equatorial system (p. 563)
geocentric (p. 550)
heliocentric (p. 551)

lunar eclipse (p. 569)
mean solar day (p. 564)
perihelion (p. 565)
perturbation (p. 561)
phases of the Moon (p. 567)
plane of the ecliptic (p. 566)
precession (p. 564)
Ptolemaic system (p. 553)

retrograde motion (p. 553)
revolution (p. 564)
right ascension (p. 563)
rotation (p. 564)
sidereal day (p. 564)
sidereal month (p. 568)
solar eclipse (p. 569)
synodic month (p. 568)

Review Questions

1. Why did the ancients believe that celestial objects had some control over their lives?

2. Describe what produces the retrograde motion of Mars. What geometric arrangements did Ptolemy use to explain this motion?

3. What major change did Copernicus make in the Ptolemaic system? Why was this change philosophically significant?

4. What was Tycho Brahe's contribution to science?

5. Does Earth move faster in its orbit near perihelion (January) or near aphelion (July)? Keeping your answer to the previous question in mind, is the solar day longest in January or July?

6. Use Kepler's third law ($p^2 = d^3$) to determine the period of a planet whose solar distance is
   (a) 10 AU
   (b) 1 AU
   (c) 0.2 AU

7. Use Kepler's third law to determine the distance from the Sun of a planet whose period is
   (a) 5 years.
   (b) 10 years.
   (c) 10 days.

8. Did Galileo invent the telescope?

9. Explain how Galileo's discovery of a rotating Sun supported the Copernican view of a Sun-centered universe.

10. Using a diagram, explain why the fact that Venus appears full when it is smallest supports the Copernican view and is inconsistent with the Ptolemaic system.

11. Newton learned that the orbits of the planets are the result of two actions. Explain these actions.

12. Of what value are constellations to modern-day astronomers?

13. Explain the difference between the mean solar day and the sidereal day.

14. What is the approximate length of the cycle of the phases of the Moon?

15. What is different about the crescent phase that precedes the new-Moon phase and that which follows the new-Moon phase?

16. What phase of the Moon occurs approximately one week after the new Moon? Two weeks?

17. When you observe the crescent phase early in the evening, is the visible Moon waxing (growing) or waning (declining)? (See Figure 20.23.)

18. What phenomenon results from the fact that the Moon's period of rotation and revolution are the same?

19. The Moon rotates very slowly (once in $27\frac{1}{3}$ days) on its axis. How does this affect the lunar surface temperature?

20. Describe the locations of the Sun, Moon, and Earth during a solar eclipse and during a lunar eclipse.

21. How many eclipses normally occur each year?

22. Solar eclipses are slightly more common than lunar eclipses. Why, then, is it more likely that your region of the country will experience a lunar eclipse?

23. How long can a total eclipse of the Moon last? How about a total eclipse of the Sun?

Examining the Earth System

1. Currently, Earth is closest to the Sun (perihelion) in January (147 million kilometers/91.5 million miles) and farthest from the Sun in July (152 million kilometers/94.5 million miles). As the result of the precession of Earth's axis, 13,000 years from now perihelion (closest) will occur in July and aphelion (farthest) will take place in January. Assuming no other changes, how might this change *average* summer temperatures for your location? What about *average* winter temperatures? What might the impact be on the biosphere and hydrosphere? (To aid your understanding of the effect of Earth's orbital parameters on the seasons, you may want to review the section on "Variations in Earth's Orbit" in Chapter 5 and also investigate this National Oceanic and Atmospheric Administration (NOAA) Web site at **http://www.ngdc.noaa.gov/paleo/milankovitch.html**)

2. In what ways do the interactions between Earth and its Moon influence the Earth system? If Earth did not have a Moon, would the atmosphere, hydrosphere, solid Earth, and biosphere be any different? Explain.

Web Resources

The *Earth Science* Website uses the resources and flexibility of the Internet to aid in your study of the topics in this chapter. Written and developed by Earth science instructors, this site will help improve your understanding of Earth science. Visit **http://www.prenhall.com/tarbuck** and click on the cover of *Earth Science 10e* to find:

• **On-line review quizzes.**
• **Web-based critical thinking and writing exercises.**
• **Links to chapter-specific Web resources.**
• **Internet-wide key term searches.**

http://www.prenhall.com/tarbuck

Meteor Crater, near Winslow, Arizona. This cavity is about 1.2 kilometers (0.75 mile) across and 170 meters (560 feet) deep. The solar system is cluttered with meteoroids and other objects that can strike Earth with explosive force. (Photo by Michael Collier)

CHAPTER 21

TOURING OUR SOLAR SYSTEM

When people first recognized that the planets resembled Earth more than the stars, excitement grew. Could intelligent life exist on these other planets, or elsewhere in the universe? Space exploration has rekindled this interest. So far, no evidence of extraterrestrial life within our solar system has emerged. Nevertheless, we study the other planets to learn about Earth's formation and early history. Recent space explorations have been organized with this goal in mind. To date, Mercury, Venus, Mars, Jupiter, Saturn, Uranus, and Neptune have been explored by space probes.

The Sun is the hub of a huge rotating system of nine planets, their satellites, and numerous small asteroids, comets, and meteoroids. An estimated 99.85 percent of the mass of our solar system is contained within the Sun. The planets collectively make up most of the remaining 0.15 percent. The planets, traveling outward from the Sun, are Mercury, Venus, Earth, Mars, Jupiter, Saturn, Uranus, Neptune, and Pluto (Figure 21.1).

Under control of the Sun's gravitational force, each planet is tethered in an elliptical orbit, and all of them travel in the same direction. The nearest planet to the Sun—Mercury—has the fastest orbital motion, 48 kilo-meters per second, and the shortest period of revolution around the Sun, 88 Earth-days. By contrast, the most distant planet, Pluto, has an orbital speed of 5 kilometers per second and requires 248 Earth-years to complete one revolution.

Imagine a planet's orbit drawn on a flat sheet of paper. The paper represents the planet's *orbital plane*. The orbital planes of seven planets lie within 3 degrees of the plane of the Sun's equator. The other two, the innermost and outermost, Mercury and Pluto, are inclined 7 and 17 degrees, respectively.

The Planets: An Overview

Astronomy
↳ The Planets: An Overview

Careful examination of Table 21.1 shows that the planets fall quite nicely into two groups: the **terrestrial** (Earth-like) **planets** (Mercury, Venus, Earth, and Mars) and the **Jovian** (Jupiter-like) **planets** (Jupiter, Saturn, Uranus, and Neptune). Pluto is not included in either category (see Box 21.2, p. 594)

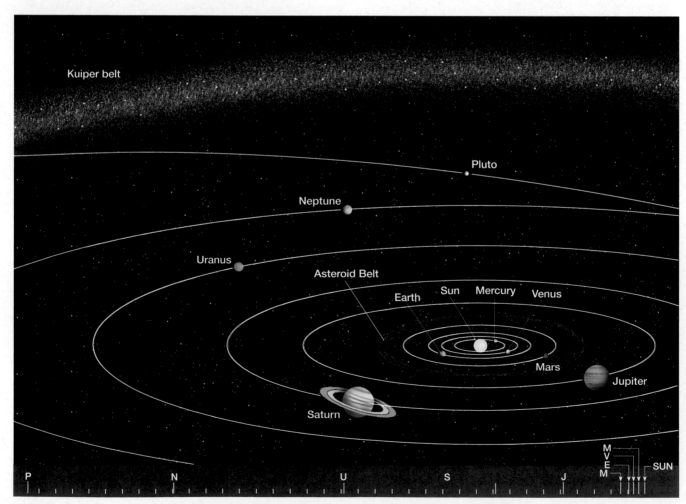

Figure 21.1 Orbits of the planets. Positions of the planets are shown to scale along bottom of diagram.

Table 21.1 Planetary data.

Planet	Symbol	AU*	Mean Distance from Sun Millions of Miles	Millions of Kilometers	Period of Revolution	Inclination of Orbit	Orbital Velocity mi/s	km/s
Mercury	☿	0.39	36	58	88^d	7°00′	29.5	47.5
Venus	♀	0.72	67	108	225^d	3°24′	21.8	35.0
Earth	⊕	1.00	93	150	365.25^d	0°00″	18.5	29.8
Mars	♂	1.52	142	228	687^d	1°51′	14.9	24.1
Jupiter	♃	5.20	483	778	12^{yr}	1°18′	8.1	13.1
Saturn	♄	9.54	886	1427	29.5^{yr}	2°29′	6.0	9.6
Uranus	♅	19.18	1783	2870	84^{yr}	0°46′	4.2	6.8
Neptune	♆	30.06	2794	4497	165^{yr}	1°46′	3.3	5.3
Pluto	♇	39.44	3666	5900	248^{yr}	17°12′	2.9	4.7

Planet	Period of Rotation	Diameter Miles	Kilometers	Relative Mass (Earth = 1)	Average Density (g/cm^3)	Polar Flattening (%)	Eccentricity	Number of Known Satellites**
Mercury	59^d	3015	4878	0.06	5.4	0.0	0.206	0
Venus	244^d	7526	12,104	0.82	5.2	0.0	0.007	0
Earth	23^h 56^m 04^s	7920	12,756	1.00	5.5	0.3	0.017	1
Mars	24^h 37^m 23^s	4216	6794	0.11	3.9	0.5	0.093	2
Jupiter	9^h 50^m	88,700	143,884	317.87	1.3	6.7	0.048	28
Saturn	10^h 14^m	75,000	120,536	95.14	0.7	10.4	0.056	30
Uranus	17^h 14^m	29,000	51,118	14.56	1.2	2.3	0.047	21
Neptune	16^h 03^m	28,900	50,530	17.21	1.7	1.8	0.009	8
Pluto	6.4^d	≈1425	≈2300	0.002	1.8	0.0	0.250	1

*AU = astronomical unit, Earth's mean distance from the Sun.
**Includes all satellites discovered as of April 2001.

The most obvious difference between the terrestrial and the Jovian planets is their size (Figure 21.2). The largest terrestrial planets (Earth and Venus) have diameters only one-quarter as great as the diameter of the smallest Jovian planet (Neptune). Also, their masses are only 1/17 as great as Neptune's. Hence, the Jovian planets are often called *giants*. Because of their relative locations, the four Jovian planets are referred to as the **outer planets**, whereas the terrestrial planets are called the **inner planets**. As we shall see, there appears to be a correlation between the positions of these planets and their sizes.

Other dimensions in which the two groups differ include density, chemical makeup, and rate of rotation. The densities of the terrestrial planets average about five times the density of water, whereas the Jovian planets have densities that average only 1.5 times that of water. One of the outer planets, Saturn, has a density only 0.7 times that of water, which means that Saturn would float if placed in a large enough water tank! Variations in the chemical compositions of the planets are largely responsible for these density differences.

The Interiors of the Planets

The substances that make up the planets are divided into three compositional groups: *gases, rocks,* and *ices,* based on their melting points.

1. The gases, hydrogen and helium, are those with melting points near absolute zero (−273°C or 0 kelvin).
2. The rocks are principally silicate minerals and metallic iron, which have melting points exceeding 700°C.
3. The ices include ammonia (NH_3), methane (CH_4), carbon dioxide (CO_2), and water (H_2O). They have intermediate melting points (for example, H_2O has a melting point of 0°C).

The terrestrial planets are dense, consisting mostly of rocky and metallic substances, with minor amounts of gases and ices. The Jovian planets, on the other hand, contain large amounts of gases (hydrogen and helium) and ices (mostly water, ammonia, and methane). This accounts for their low densities. The outer planets also

Mercury •

Venus

Earth

Mars

Sun

Jupiter

Saturn

Uranus

Neptune

Pluto •

Figure 21.2 The planets drawn to scale.

contain substantial amounts of rocky and metallic materials, which are concentrated in their cores.

The Atmospheres of the Planets

The Jovian planets have very thick atmospheres of hydrogen, helium, methane, and ammonia. By contrast, the terrestrial planets, including Earth, have meager atmospheres at best. The reason is that a planet's ability to retain an atmosphere depends on its mass and temperature.

Simply stated, a gas molecule can evaporate from a planet if it reaches a speed known as the **escape velocity.** For Earth, this velocity is 11 kilometers (7 miles) per second (about 25,000 miles per hour). Any material, including a rocket, must reach this speed before it can escape Earth's gravity and go into space.

The Jovian planets, because of their greater surface gravities, have escape velocities of 21 to 60 kilometers per second (45,000 to 134,000 miles per hour), much higher than the terrestrial planets. Consequently, it is more difficult for gases to evaporate from them. Also, because the molecular motion of a gas is temperature-dependent, at the low temperatures of the Jovian planets even the lightest gases are unlikely to acquire the speed needed to escape.

In contrast, a comparatively warm body with a small surface gravity, like our Moon, is unable to hold even heavy gases, like carbon dioxide and radon, and thus lacks an atmosphere. The slightly larger terrestrial planets of Earth, Venus, and Mars retain some heavy gases such as carbon dioxide, but even their atmospheres make up only an infinitesimally small portion of their total mass.

STUDENTS SOMETIMES ASK...
Why are the Jovian planets so much larger than the terrestrial planets?

According to the nebular hypothesis, the planets formed from a rotating disk of dust and gases that surrounded the Sun. The growth of planets began as solid bits of matter began to collide and clump together. In the inner solar system, the temperatures were so high that only metals and silicate minerals could form solid grains. It was too hot for ices of water, carbon dioxide, and methane to form. Thus, the innermost (terrestrial) planets grew mainly from the high melting point substances found in the solar nebula. By contrast, in the frigid outer reaches of the solar system, it was cold enough for ices of water and other substances to form. Consequently, the outer planets grew not only from accumulations of solid bits of metals and silicate minerals but also from large quantities of ices. Eventually, the outer planets became large enough to gravitationally capture even the lightest gases (hydrogen and helium), and thus grow to become "giant" planets.

In the remainder of this chapter, we will consider each planet briefly, plus the minor members of the solar system. First, however, let us visit Earth's companion in space, our own Moon.

Earth's Moon

Astronomy
↳ Earth's Moon

Earth now has hundreds of satellites, but only one natural satellite, the Moon, accompanies us on our annual journey around the Sun. Although other planets have moons, our planet-satellite system is unique in the solar system, because Earth's Moon is unusually large compared to its parent planet. The diameter of the Moon is 3475 kilometers (2150 miles), about one-fourth of Earth's 12,756 kilometers (7920 miles).

From calculation of the Moon's mass, its density is 3.3 times that of water. This density is comparable to that of *mantle* rocks on Earth but is considerably less than Earth's average density, which is 5.5 times that of water. Geologists have suggested that this difference can be accounted for if the Moon's iron core is small. The gravitational attraction at the lunar surface is one-sixth of that experienced on Earth's surface (a 150-pound person on Earth weighs only 25 pounds on the Moon). This difference allows an astronaut to carry a heavy life-support system with relative ease. If not burdened with such a load, an astronaut could jump six times higher than on Earth.

The Lunar Surface

When Galileo first pointed his telescope toward the Moon, he saw two different types of terrain—dark lowlands and bright, cratered highlands (Figure 21.3). Because the dark regions resembled seas on Earth, they were later named **maria** (*mar* = the sea). (singular **mare**). This name is unfortunate, because the Moon's surface is totally devoid of water. Figure 21.4 shows typical features of the lunar surface.

Today we know that the Moon has no atmosphere or water. Therefore, the weathering and erosion that continually modify Earth's surface are virtually lacking on the Moon. In addition, tectonic forces are not active on the Moon, so earthquakes and volcanic eruptions no longer occur. However, because the Moon is unprotected by an atmosphere, a different kind of erosion occurs: tiny particles from space (micrometeorites) continually bombard its surface and ever

Figure 21.3 Telescopic view from Earth of the lunar surface. The major features are the dark "seas" (maria) and the light highly cratered highlands. (UCO/Lick Observatory Image)

Figure 21.4 Block diagram illustrating major topographic features on the lunar surface.

so gradually smooth the landscape. Moon rocks become slightly rounded on top if they are long exposed at the lunar surface. Nevertheless, it is unlikely that the Moon has changed appreciably in the last 3 billion years, except for a few craters created by large meteorites.

Craters. The most obvious features of the lunar surface are craters. They are so profuse that craters-within-craters are the rule! The larger ones in the lower portion of Figure 21.3 are about 250 kilometers (150 miles) in diameter, roughly the width of Indiana. Most craters were produced by the impact of rapidly moving debris (meteoroids), a phenomenon that was considerably more common in the early history of the solar system than it is today.

By contrast, Earth has only about a dozen easily recognized impact craters. This difference can be attributed to Earth's atmosphere. Friction with the air burns up small debris before it reaches the ground. In addition, evidence for most of the craters that formed in Earth's history has been obliterated by erosion or tectonic processes.

The formation of an impact crater is illustrated in Figure 21.5. Upon impact, the high-speed meteoroid compresses the material it strikes, then almost instantaneously the compressed rock rebounds, ejecting material from the crater. This process is analogous to the splash that occurs when a rock is dropped into water, and it often results in the formation of a central peak, as seen in the large crater in Figure 21.6. Most of the ejected material (*ejecta*) lands near the crater, building

a rim around it. The heat generated by the impact is sufficient to melt some of the impacted rock. Astronauts have brought back samples of glass beads produced in this manner, as well as rock formed when angular fragments and dust were welded together by the impact.

A meteoroid only 3 meters (10 feet) in diameter can blast out a 150-meter- (500-foot-) wide crater. A few of the large craters, such as Kepler and Copernicus, shown in Figure 21.3, formed from the impact of bodies 1 kilometer or more in diameter. These two large craters are thought to be relatively young because of the bright *rays* (splash marks) that radiate outward for hundreds of kilometers.

Highlands. Densely pockmarked highland areas make up most of the lunar surface. In fact, all of the back side of the Moon is characterized by such topography. (Only astronauts have seen the back side, because the Moon rotates on its axis once with each revolution around Earth, always keeping the same side facing Earth.) Within the highland regions are mountain ranges. The highest lunar peaks reach elevations approaching 8 kilometers (5 miles), only 1 kilometer (0.6 mile) lower than Mount Everest.

Maria. The seas of basaltic lava originated when asteroids punctured the lunar surface, letting basaltic magma bleed out (Figure 21.7). Apparently the craters were flooded with layer upon layer of very fluid basaltic lava somewhat resembling the Columbia Plateau in the northwestern United States. The lava

Meteorite impact

Compressional
wave

Fused rock and meteorite

Compressional
wave

Rebound

Ejecta

Ejecta blanket

Fractured
bedrock

Figure 21.5   Formation of an impact crater. The energy of the rapidly moving meteoroid is transformed into heat energy and compressional waves. The rebound of the compressed rock causes debris to be ejected from the crater, and the heat melts some material, producing glass beads. Small secondary craters are formed by the material "splashed" from the impact crater. (After E. M. Shoemaker)

flows are often over 30 meters (100 feet) thick, and the total thickness of the material that fills the maria must approach thousands of meters.

Regolith.   All lunar terrains are mantled with a layer of gray, unconsolidated debris derived from a few billion years of meteoric bombardment (Figure 21.8). This soil-like layer, properly called **lunar regolith** (*rhegos* = blanket, *lithos* = stone), is composed of igneous rocks, breccia, glass beads, and fine *lunar dust.* In the maria that have been explored by *Apollo* astronauts, the lunar regolith is apparently just over 3 meters (10 feet) thick.

Lunar History

Although the Moon is our nearest planetary neighbor and astronauts have sampled its surface, much is still unknown about its origin. The most widely accepted model for the origin of the Moon is that during the formative period of the solar system, a Mars-size body impacted Earth. The impact would have liquefied Earth's surface and ejected huge quantities of crustal and mantle rock from an infant Earth. A portion of this ejected debris would have entered an orbit around Earth where it coalesced to form the Moon.

The giant-impact hypothesis is consistent with a number of facts known about the Moon. The ejected material would have been mostly iron-poor mantle and crustal rocks, which would account for the lack of a sizable iron core on the Moon. Further, the ejected material would have remained in orbit long enough to have lost the volatile materials (water) that the Moon lacks. Despite this supporting evidence, some questions remain unanswered.

Nevertheless, planetary geologists have worked out the basic details of the Moon's later history. One of their methods is to observe variations in crater density (quantity per unit area). The greater the crater density, the longer the topographic feature must have existed. From such evidence, scientists concluded that the Moon evolved in three phases: the original crust (highlands), maria basins, and rayed craters.

During its early history the Moon was continually impacted as it swept up debris. This continuous bombardment, and perhaps radioactive decay, generated enough heat to melt the Moon's outer shell and quite possibly the interior as well. Remnants of this original crust occupy the densely cratered highlands, which have been estimated to be as much as 4.5 billion years old—about the same age as Earth.

The second major event in the Moon's evolution was the formation of maria basins (see Figure 21.7). Radiometric dating of the maria basalts puts their age between 3.2 billion and 3.8 billion years, roughly a billion years younger than the initial crust. In places, the lava flows overlap the highlands, another testimonial to the younger age of the maria deposits.

Figure 21.6 The 20-kilometer-wide lunar crater Euler in the southwestern part of Mare Imbrium. Clearly visible are the bright rays, central peak, secondary craters, and the large accumulation of ejecta near the crater rim. (Courtesy of NASA)

Crater ray

Secondar cha

Central peak

Continuous ejec

Discontinuous ejecta

The last prominent features to form were the rayed craters, as exemplified by the Copernicus crater (see Figure 21.3). Material ejected from these young depressions is clearly seen blanketing the surface of the maria and many older rayless craters. Even a relatively young crater like Copernicus must be millions of years old. Had it formed on Earth, erosional forces would have long since obliterated it.

If photos of the Moon taken several hundreds of millions of years ago were available, they would reveal that the Moon changed little in the intervening years. By all measures, the Moon is a tectonically dead body wandering through space and time.

The Planets: A Brief Tour

Astronomy
↳ A Brief Tour of the Planets

Mercury: The Innermost Planet

Mercury, the innermost and second smallest planet, is hardly larger than Earth's Moon and is smaller than three other moons in the solar system. Like our own Moon, it absorbs most of the sunlight that strikes it, reflecting only 6 percent into space (Figure 21.9). This is

Ejected debris

Maria basalts

A.

B.

Figure 21.7 Formation of lunar maria. **A.** Impact of an asteroid-sized mass produced a huge crater hundreds of kilometers in diameter and disturbed the lunar crust far beyond the crater. **B.** Filling of the impact area with fluid basalts, perhaps derived from partial melting deep within the lunar mantle.

Figure 21.8 Astronaut Harrison Schmitt sampling the lunar surface. Notice the footprints (inset) in the lunar "soil." (Courtesy of NASA)

Figure 21.9 Photomosaic of Mercury. This view of Mercury is remarkably similar to the "far side" of Earth's Moon. (Courtesy of NASA)

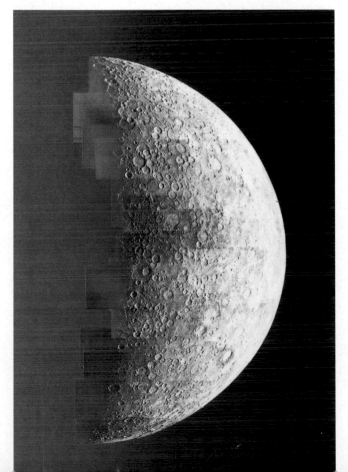

characteristic of terrestrial bodies that have no atmosphere. (Earth reflects about 30 percent of the light that strikes it, most of it from clouds.)

Mercury has cratered highlands, much like the Moon, and vast smooth terrains that resemble maria. However, unlike the Moon, Mercury is a very dense planet, which implies that it contains a very large iron core for its size. Also, Mercury has very long scarps that cut across the plains and craters alike. These scarps may have resulted from crustal shortening as the planet cooled and shrank.

Mercury revolves quickly but rotates slowly. One full day–night cycle on Earth takes 24 hours, but on Mercury it requires 179 Earth-days. Thus, a night on Mercury lasts for about three months and is followed by three

STUDENTS SOMETIMES ASK...
Do any nearby stars have planets?

Yes. Although this was long suspected, it was not until recently that the presence of extrasolar planets has been verified. Astronomers have found these bodies by measuring the telltale wobbles of nearby stars. The first apparent planet outside the solar system was discovered in 1995, orbiting the star 51 Pegasi, 42 light-years from Earth. Since that time, over two dozen Jupiter-size bodies have been identified, most of them surprisingly close to the stars they orbit.

583

Figure 21.10 This global view of the surface of Venus is computer generated from two years of Magellan Project radar mapping. The twisting bright features that cross the globe are highly fractured mountains and canyons of the eastern Aphrodite highland. (Courtesy of NASA/JPL)

tonic deformation must have been very active during the recent geologic past.

About 80 percent of the Venusian surface consists of subdued plains that are mantled by volcanic flows. Some lava channels extend hundreds of kilometers; one meanders 6800 kilometers across the planet. Thousands of volcanic structures have been identified, mostly small shield volcanoes, although over 1500 volcanoes greater than 20 kilometers (12 miles) across have been mapped (Figure 21.11). One is Sapas Mons, 400 kilometers (250 miles) across and 1.5 kilometers (0.9 mile) high. Many flows from this volcano erupted from its flanks rather than the summit, in the manner of Hawaiian shield volcanoes.

Only 8 percent of the Venusian surface consists of highlands that may be likened to continental areas on Earth. Tectonic activity on Venus seems to be driven by upwelling and downwelling of material in the planet's interior. Although mantle convection still operates on Venus, the processes of plate tectonics, which recycle rigid lithosphere, do not appear to have contributed to the present Venusian topography.

Before the advent of space vehicles, Venus was considered to be a potentially hospitable site for living organisms. However, evidence from space probes indicates otherwise. The surface of Venus reaches temperatures of 475°C (900°F), and the Venusian atmosphere is 97 percent carbon dioxide. Only scant water vapor and nitrogen have been detected. The Venusian atmosphere contains an opaque cloud layer about 25 kilometers (15 miles) thick, and has an atmospheric pressure that is 90 times that at Earth's surface. This hostile environment makes it unlikely that life as we know it exists on Venus.

Mars: The Red Planet

Mars has evoked greater interest than any other planet, for both scientists and nonscientists (see Box 21.1). When one imagines intelligent life on other worlds, little green Martians may come to mind. Interest in Mars stems mainly from this planet's accessibility to observation. All other planets within telescopic range have their surfaces hidden by clouds, except for Mercury, whose nearness to the Sun makes viewing difficult. Through the telescope, Mars appears as a reddish ball interrupted by some permanent dark regions that change intensity during the Martian year. The most prominent telescopic features of Mars are its brilliant white polar caps, resembling Earth's.

The Martian Atmosphere. The Martian atmosphere has only 1 percent the density of Earth's, and it is primarily carbon dioxide with tiny amounts of water vapor. Data from Mars probes confirm that the polar caps of Mars are made of water ice, covered by a thin layer of frozen carbon dioxide. As winter nears in either

months of daylight. Nighttime temperatures drop as low as −173°C (−280°F), and noontime temperatures exceed 427°C (800°F), hot enough to melt tin and lead. Mercury has the greatest extremes of any planet. The odds of life as we know it existing on Mercury are nil.

Venus: The Veiled Planet

Venus, second only to the Moon in brilliance in the night sky, is named for the goddess of love and beauty. It orbits the Sun in a nearly perfect circle once every 255 Earth-days. Venus is similar to Earth in size, density, mass, and location in the solar system. Thus, it has been referred to as "Earth's twin." Because of these similarities, it is hoped that a detailed study of Venus will provide geologists with a better understanding of Earth's evolutionary history.

Venus is shrouded in thick clouds impenetrable to visible light. Nevertheless, radar mapping by the unmanned *Magellan* spacecraft and by instruments on Earth has revealed a varied topography with features somewhat between those of Earth and Mars (Figure 21.10). Simply, radar pulses in the microwave range are sent toward the Venusian surface, and the heights of plateaus and mountains are measured by timing the return of the radar echo. These data have confirmed that basaltic volcanism and tectonic deformation are the dominant processes operating on Venus. Further, based on the low density of impact craters, volcanism and tec-

BOX 21.1

UNDERSTANDING EARTH:
Pathfinder—The First Geologist on Mars

On July 4, 1997, the *Mars Pathfinder* bounced onto the rock-littered surface of Mars and deployed its wheeled companion, *Sojourner*. For the next three months the lander sent back to Earth three gigabits of data, including 16,000 images and 20 chemical analyses. The landing site was a vast rolling landscape carved by ancient floods. The flood-deposit locale was selected in hope that a variety of rock types would be available for the rover *Sojourner* to examine.

Sojourner carried an alpha photon X-ray spectrometer (APXS) used to determine the chemical composition of rocks and Martian "soil" (regolith) at the landing site (Figure 21.A). In addition, the rover was able to take close-up images of the rocks. From these images, researchers concluded that the rocks were igneous. However, one hard, white, flat object named Scooby Doo was originally thought to be sedimentary rock, but the APXS data suggest its chemistry is like that of the soil found at the site. Thus, Scooby Doo is probably a well-cemented soil.

During its first week on Mars, *Sojourner's* APXS obtained data for a patch of windblown soil and a medium-sized rock, known affectionately as Barnacle Bill. Preliminary evaluation of the APXS data on Barnacle Bill

Figure 21.A   *Pathfinder's rover Sojourner (left) obtaining data on the chemical composition of a Martian rock known as Yogi. (Photo courtesy of NASA)*

shows that it contains over 60 percent silica. If these data are confirmed, it could indicate that Mars contains the volcanic rock andesite. Researchers had expected that most volcanic rocks on Mars would be basalt, which is lower in silica (less than 50 percent). On Earth, andesites are associated with tectonically active regions where oceanic crust is subducted into the mantle. Examples include the volcanoes of South America's Andes Mountains and the Cascades of North America.

Sojourner analyzed eight rocks and seven soils. The results thus far are only preliminary. Because these rocks are covered with a reddish dust that is high in sulfur, some controversy has arisen as to the exact composition of these Martian rocks. Some researchers believe they are all of the same composition. The differences in measurements, they claim, are the result of varying thickness of dust.

---

hemisphere, we see the equatorward growth of that hemisphere's ice cap as temperatures drop to −125°C (−193°F), and additional carbon dioxide is deposited.

Although the atmosphere of Mars is very thin, extensive dust storms occur and may cause the color changes observed from Earth-based telescopes. Hurricane-force winds up to 270 kilometers (170 miles) per hour can persist for weeks. Images from *Viking 1* and *Viking 2* revealed a Martian landscape remarkably similar to a rocky desert on Earth (Figure 21.12), with abundant sand dunes and impact craters partially filled with dust.

Mars' Dramatic Surface.   *Mariner 9*, the first spacecraft to orbit another planet, reached Mars in 1971 amid a raging dust storm. When the dust cleared, images of Mars' northern hemisphere revealed numerous large volcanoes. The biggest, Mons Olympus, is the size of Ohio and 23 kilometers (75,000 feet) high,

over two and a half times higher than Mount Everest. This gigantic volcano and others resemble Hawaiian shield volcanoes on Earth (Figure 21.13).

Most Martian surface features are old by Earth standards. The highly cratered southern hemisphere is probably similar in age to the lunar highlands (3.5 billion to 4.5 billion years old). Even the relatively fresh-appearing volcanic features of the northern hemisphere may be older than 1 billion years. This fact and the absence of Marsquake recordings by *Viking* seismographs point to a tectonically dead planet.

Another surprising find made by *Mariner 9* was the existence of several canyons that dwarf even Earth's Grand Canyon of the Colorado River. One of the largest, Valles Marineris, is thought to have formed by slippage of material along huge faults in the crustal layer. In this respect, it would be comparable to the rift valleys of Africa (Figure 21.14).

Figure 21.11 Computer-generated image of Venus. Near the horizon is Maat Mons, a large volcano. The bright feature below is the summit of Sapas Mons. (Courtesy David P. Anderson/SMU/NASA/Science Photo Library/Photo Researchers, Inc.)

Water on Mars? Not all Martian valleys have a tectonic origin. Some areas exhibit drainage patterns similar to those created by streams on Earth. In addition, *Viking* orbiter images have revealed unmistakable ancient islands in what is now a dry stream bed. When these streamlike channels were first discovered, some observers speculated that a thick water-laden atmosphere capable of generating torrential downpours once existed on Mars. If so, what happened to this water? The present Martian atmosphere contains only traces.

Many planetary geologists do not accept the premise that Mars once had an active water cycle similar to Earth's. Rather, they believe that many of the large streamlike valleys were created by the collapse of surface material caused by the slow melting of subsurface ice. If this is the case, these large valleys would be more akin to features formed by mass-wasting processes on Earth.

Images from the *Mars Global Surveyor* indicate that groundwater has recently migrated to the surface. These springlike seeps have created gullies where they emerge from valley and crater walls. Some of the escaping water may have initially frozen due to the average Martian temperatures that range between −70°C and −100°C. Eventually, however, it seeped out as a slurry of sediment, ice, and liquid that formed the gullies shown in Figure 21.15. Because water is an essential ingredient for life, astrobiologists are intrigued and enthusiastic about the potential for future understanding of this phenomenon.

The Martian Satellites. Tiny Phobos and Deimos, the two satellites of Mars, were not discovered until 1977 because they are only 24 and 15 kilometers, in

Figure 21.12 This picture of the Martian landscape by the *Viking 1* lander shows a dune field with features remarkably similar to many seen in the deserts of Earth. The dune crests indicate that recent wind storms were capable of moving sand over the dunes in the direction from lower right to upper left. The large boulder at the left is about 10 meters from the spacecraft and measures 1 by 3 meters. (Courtesy of NASA)

Figure 21.13   Image of Mons Olympus, an inactive shield volcano on Mars that covers an area about the size of the state of Ohio. (Courtesy of the U.S. Geological Survey)

diameter. Phobos is nearer to its parent than any other natural satellite in the solar system—only 5500 kilometers (3400 miles)—and requires just 7 hours and 39 minutes for one revolution. *Mariner 9* revealed that both satellites are irregularly shaped and have numerous impact craters.

It is likely that these moons are asteroids captured by Mars. A most interesting coincidence in astronomy and literature is the close resemblance between Phobos and Deimos and two fictional satellites of Mars described by Jonathan Swift in *Gulliver's Travels*, written about 150 years before these satellites were actually discovered.

? STUDENTS SOMETIMES ASK...
*Why are the volcanoes on Earth
so much smaller than those on Mars?*

The largest shield volcanoes form where plumes of hot rock rise from deep within a planet's interior. Earth is tectonically active, with moving plates that keep the crust in constant motion. For example, the Hawaiian Islands consist of a chain of shield volcanoes that formed as the Pacific plate moved over a relatively stationary mantle plume. On Mars, volcanoes such as Mons Olympus have grown to great size because the crust there remains stationary. Successive eruptions occur at the same location and add to the bulk of a single volcano rather than producing several smaller structures, as occurs on Earth.

Figure 21.14   This image shows the entire Valles Marineris canyon system, over 5000 kilometers long and up to 8 kilometers deep. The dark spots on the left edge of the image are huge volcanoes, each about 25 kilometers high. (Courtesy of U.S. Geological Survey)

Volcanoes

Valles
Marineris

Figure 21.15 Image from *Mars Global Surveyor* showing crater wall with large gullies that may have been cut by liquid water, mixed with soil, rocks, and ice. (NASA/JPL photo courtesy of National Geographic)

Jupiter: Lord of the Heavens

Jupiter, truly a giant among planets, has a mass two and a half times greater than the combined mass of all the remaining planets, satellites, and asteroids. In fact, had Jupiter been about 10 times larger, it would have evolved into a small star. Despite its great size, however, it is only 1/800 as massive as the Sun. Jupiter also rotates more rapidly than any other planet, completing one rotation in slightly less than 10 Earth-hours. The effect of this fast spin is to make the equatorial region bulge and to make the polar dimension flatten (see the Polar Flattening column in Table 21.1).

When viewed through a telescope or binoculars, Jupiter appears to be covered with alternating bands of multicolored clouds aligned parallel to its equator (Figure 21.16). The most striking feature is the *Great Red Spot* in the southern hemisphere (Figure 21.16). The Great Red Spot has been a prominent feature since is was first discovered more than three centuries ago. When *Voyager 2* swept by Jupiter in 1979, the Great Red Spot was the size of two Earth-size circles placed side by side. On occasion, it has grown even larger.

Images from *Pioneer 11* as it moved within 42,000 kilometers of Jupiter's cloud tops in 1974 indicated that the Great Red Spot is a counterclockwise-rotating (cyclonic) storm. It is caught between two jetstream-like bands of atmosphere flowing in opposite directions. This huge hurricane-like storm rotates once every 12 Earth-days. Although several smaller storms have been observed in other regions of Jupiter's atmosphere, none have survived for more than a few days.

Structure of Jupiter. Jupiter's hydrogen-helium atmosphere also has methane, ammonia, water, and sulfur compounds as minor constituents. The wind systems generate the light- and dark-colored bands that encircle this giant (Figure 21.17). Unlike the winds on Earth, which are driven by solar energy, Jupiter itself gives off nearly twice as much heat as it receives from the Sun. Thus, it is the *interior* heat from Jupiter that produces huge convection currents in the atmosphere.

Atmospheric pressure at the top of the clouds is equal to sea-level pressure on Earth. Because of Jupiter's

Figure 21.16 Artist's view of Jupiter with Great Red Spot visible in its southern hemisphere. Earth for scale.

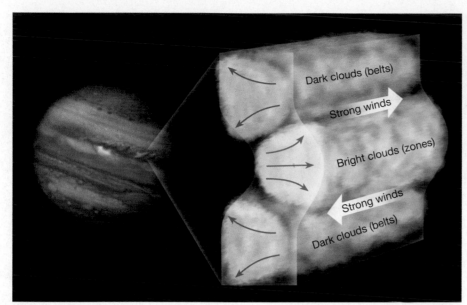

Figure 21.17 The structure of Jupiter's atmosphere. The areas of light clouds (zones) are regions where gases are ascending and cooling. Sinking dominates the flow in the darker cloud layers (belts). This convective circulation, along with the rapid rotation of the planet, generates the high-speed winds observed between the belts and zones.

immense gravity, the pressure increases rapidly toward its surface. At 1000 kilometers below the clouds, the pressure is great enough to compress hydrogen gas into a liquid. Consequently, Jupiter's surface is thought to be a gigantic ocean of liquid hydrogen. Less than halfway into Jupiter's interior, extreme pressures cause the liquid hydrogen to turn into *liquid metallic* hydrogen. Jupiter is also believed to contain as much rocky and metallic material as is found in the terrestrial planets, probably in a central core.

Jupiter's Moons.   Jupiter's satellite system, consisting of 28 moons discovered so far, resembles a miniature solar system. The four largest satellites, discovered by Galileo, travel in nearly circular orbits around the parent, with periods of from 2 to 17 Earth-days (Figure 21.18). The two largest Galilean satellites, Callisto and Ganymede, surpass Mercury in size, while the two smaller ones, Europa and Io, are about the size of Earth's Moon. The Galilean moons can be observed with binoculars or a small telescope and are interesting in their own right.

By contrast, Jupiter's four outermost satellites are very small (20 kilometers in diameter), revolve in a direction that is opposite (*retrograde motion*) the largest moons, and have orbits that are steeply inclined to the Jovian equator. These satellites appear to be asteroids that passed near enough to be captured gravitationally by Jupiter.

Images from *Voyagers 1* and *2* in 1979 revealed, to the surprise of almost everyone, that each of the four Galilean satellites is a unique geological world, as shown in Figure 21.18. The innermost of the Galilean moons, Io (Figure 21.18A), is one of three volcanically active bodies discovered in our solar system, the others being Earth and Neptune's moon Triton. To date, numerous active sulfurous volcanic centers have been

discovered. Umbrella-shaped plumes have been seen rising from the surface of Io to heights approaching 200 kilometers (120 miles) (Figure 21.19). The heat source for volcanic activity on Io is thought to be tidal energy generated by a relentless "tug of war" between Jupiter and the other Galilean satellites. Because Io is gravitationally locked to Jupiter, the same side always faces the giant planet, like Earth's Moon. The gravitational power of Jupiter and the other nearby satellites pulls and pushes on Io's tidal bulge as its slightly eccentric orbit takes it alternately closer to and farther from Jupiter. This gravitational flexing of Io is transformed into heat energy (similar to the back-and-forth bending

? STUDENTS SOMETIMES ASK...
Besides Earth, do any other bodies in the solar system have liquid water?

The planets closer to the Sun than Earth are considered too warm to contain liquid water, and those farther from the Sun are generally too cold to have water in the liquid form (although some features on Mars suggest that it may have had abundant liquid water at some point in its history). However, the best prospects of finding liquid water within our solar system lie beneath the icy surfaces of some of Jupiter's moons. For instance, Europa is suspected to have an ocean of liquid water hidden under its outer covering of ice. Detailed images sent back to Earth from the *Galileo* spacecraft have revealed that Europa's icy surface is quite young and exhibits cracks apparently filled with dark fluid from below. This suggests that under its icy shell, Europa must have a warm, mobile interior—and perhaps an ocean. Because the presence of water in the liquid form is a necessity for life as we know it, there has been much interest in sending an orbiter to Europa—and eventually a lander capable of launching a robotic submarine—to determine if it too may harbor life.

A. Io   B. Europa   C. Ganymede   D. Callisto

Figure 21.18   Jupiter's four largest moons (from left to right) called the Galilean moons because they were discovered by Galileo. **A.** The innermost moon, Io, is one of only three volcanically active bodies in the solar system. **B.** Europa, smallest of the Galilean moons, has an icy surface that is criss-crossed by many linear features. **C.** Ganymede, the largest Jovian satellite, contains cratered areas, smooth regions, and areas covered by numerous parallel grooves. **D.** Callisto, the outermost of the Galilean satellites, is densely cratered, much like Earth's moon. (Courtesy of NASA/NGS Image Collection)

of a paper clip) and results in Io's spectacular sulfurous volcanic eruptions.

One of the most unexpected discoveries made by *Voyager 1* was Jupiter's ring system. By analyzing how these rings scatter light, researchers concluded that the rings are composed of fine, dark particles, similar in size to smoke particles. Further, the faint nature of the rings indicates that these minute fragments are widely dispersed. The particles are thought to be fragments blasted by meteorite impacts from the surfaces of Metis and Adrastea, two small moons of Jupiter.

Saturn: The Elegant Planet

Requiring 29.46 Earth-years to make one revolution, Saturn is almost twice as far from the Sun as Jupiter, yet its atmosphere, composition, and internal structure are thought to be remarkably similar to Jupiter's. The most prominent feature of Saturn is its system of rings (Figure 21.20), discovered by Galileo in 1610. With his primitive telescope, the rings appeared as two small bodies adjacent to the planet. Their ring nature was discovered 50 years later by the Dutch astronomer Christian Huygens.

Saturn Close up.   In 1980 and 1981, fly-by missions of the nuclear-powered *Voyagers 1* and *2* space vehicles came within 100,000 kilometers of Saturn. More information was gained in a few days than had been acquired since Galileo first viewed this elegant planet telescopically.

1. Saturn's atmosphere is very dynamic, with winds roaring at up to 1500 kilometers (930 miles) per hour.

2. Large cyclonic "storms" similar to Jupiter's Great Red Spot, although smaller, occur in Saturn's atmosphere.

3. Eleven additional moons were discovered.

4. The rings of Saturn were discovered to be more complex than expected.

More recently, observations from ground-based telescopes and the Hubble Space Telescope have added to our knowledge of Saturn's ring system. In 1995 and 1996, when the positions of Earth and Saturn allowed the rings to be viewed edge-on, thereby reducing the glare from the main rings, Saturn's faintest rings and satellites became visible.

Planetary Ring Systems.   Until the recent discovery that Jupiter, Uranus, and Neptune also have ring systems, this phenomenon was thought to be unique to Saturn. Although the four known ring systems differ in detail, they share many attributes. They all consist of multiple concentric rings separated by gaps of various widths. In addition, each ring is composed of individual particles—"moonlets" of ice and rock—that circle the planet while regularly impacting one another.

Figure 21.19   A volcanic eruption on Io. This plume of volcanic gases and debris is rising over 100 kilometers (60 miles) above Io's surface. (Courtesy of NASA)

Most rings fall into one of two categories based on particle density. Saturn's main rings (designated A and B in Figure 21.20) and the bright rings of Uranus are tightly packed and contain "moonlets" that range in size from a few centimeters (pebble-size) to several meters (house-size). These particles are thought to collide

frequently as they orbit the parent planet. Despite the fact that Saturn's dense rings stretch across several hundred kilometers, they are very thin, perhaps less than 100 meters (300 feet) from top to bottom.

At the other extreme, the faintest rings, such as Jupiter's ring system and Saturn's outermost rings (designated E in Figure 21.20), are composed of very fine (smoke-size) particles that are widely dispersed. In addition to having very low particle densities, these rings tend to be thicker than Saturn's bright rings.

Recent studies have shown that the moons that coexist with the rings play a major role in determining their structure. In particular, the gravitational influence of these moons tends to shepard the ring particles by altering their orbits. The narrow rings appear to be the work of satellites located on either side that confine the ring by pushing back particles that try to escape.

More important, the ring particles are thought to be debris ejected from these moons. According to this view, material is continually being recycled between the rings and the ring moons. The moons gradually sweep up particles, which are subsequently ejected by collisions with large chunks of ring material, or perhaps by energetic collisions with other moons. It seems, then, that planetary rings are not the timeless features that we once thought; rather, they continually reinvent themselves.

The origin of planetary ring systems is still being debated. Did the rings form out of a flattened cloud of dust and gases that encircled the parent planet? In this scenario, the rings formed at the same time and of the same material as the planets and moons. Or, did the rings form later, when a moon or large asteroid was

Figure 21.20   A view of the dramatic ring system of Saturn.

Figure 21.21 Montage of the Saturnian satellite system. The moon Dione is in foreground; Tethys and Mimas are at lower right; Enceladus and Rhea are off ring's left; and Titan is upper right. (Photo courtesy of NASA)

gravitationally pulled apart after straying too close to a planet? Still another hypothesis suggests that a foreign body blasted apart one of the planet's moons. The fragments from this impact would tend to jostle one another and form a flat, thin ring. Researchers expect more light to be shed on the origin of planetary rings beginning in July 2004 when the Cassini spacecraft will begin a four-year tour of Saturn.

Saturn's Moons. The Saturnian satellite system consists of 30 named moons (Figure 21.21). (If you count the "moonlets" that comprise Saturn's rings, this planet has millions of satellites.) The largest, Titan, is bigger than Mercury and is the second-largest satellite in the solar system (after Jupiter's Ganymede). Titan and Neptune's Triton are the only satellites in the solar system known to have a substantial atmosphere. Because of its dense gaseous cover, the atmospheric pressure at Titan's surface is about 1.5 times that at Earth's surface. Another satellite, Phoebe, exhibits retrograde motion. It, like other moons with retrograde orbits, is most likely a captured asteroid or large planetesimal left over from the major episode of planetary formation.

Uranus and Neptune: The Twins

Earth and Venus have similar traits, but Uranus and Neptune are nearly twins. Only 1 percent different in di-

ameter, they are both bluish in appearance, which is attributable to the methane in their atmospheres (Figures 21.22 and 21.23). Their structure and composition are believed to be similar. Neptune, however, is colder, because it is half again as distant from the Sun's warmth as is Uranus.

Uranus: The Sideways Planet. A unique feature of Uranus is that it rotates "on its side." Its axis of rotation, instead of being generally perpendicular to the plane of its orbit, like the other planets, lies nearly parallel with the plane of its orbit. Its rotational motion, therefore, has the appearance of rolling, rather than the toplike spinning of the other planets. Because the axis of Uranus is inclined almost 90 degrees, the Sun is nearly overhead at one of its poles once each revolution, and then half a revolution later it is overhead at the other pole.

A surprise discovery in 1977 revealed that Uranus has a ring system. This find occurred as Uranus passed in front of a distant star and blocked its view, a process called *occultation* (*occult* = hidden). Observers saw the star "wink" briefly five times (meaning five rings) before the primary occultation and again five times afterward. Later studies indicate that Uranus has at least nine distinct belts of debris orbiting its equatorial region.

Figure 21.22 This image of Uranus was sent back to Earth by *Voyager 2* as it passed by this planet on January 24, 1986. Taken from a distance of nearly 1 million kilometers, little detail of its atmosphere is visible, except a few streaks (clouds) in the Northern Hemisphere (Courtesy of NASA)

Figure 21.23 This image of Neptune shows the Great Dark Spot (left center). Also visible are bright cirruslike clouds that travel at high speed around the planet. A second oval spot is at 54° south latitude on the east side of the planet. (Courtesy of the Jet Propulsion Laboratory)

Spectacular views from *Voyager 2* of the five largest moons of Uranus show quite varied terrains. Some have long, deep canyons and linear scars, whereas others possess large, smooth areas on otherwise crater-riddled surfaces. The Jet Propulsion Laboratory described Miranda, the innermost of the five largest moons, as having a greater variety of landforms than any body yet examined in the solar system.

Neptune: The Windy Planet Even when the most powerful telescope is focused on Neptune, it appears as a bluish fuzzy disk. Until the 1989 *Voyager 2* encounter, astronomers knew very little about this planet. However, the 12-year, nearly 3-billion-mile journey of *Voyager 2* provided investigators with so much new

STUDENTS SOMETIMES ASK...
Why does Uranus spin on its side?

The most likely explanation for the unusual sideways spin of Uranus is that it started out spinning the same way as the other planets, but then its spin was altered by a giant impact—which was probably very common when the planets were first formed. However, a giant impact would be very difficult to verify because it would not have left any crater on Uranus, which has no solid surface. Like many events that occurred early on in the formation of our solar system, the reason for Uranus's sideways spin may never be known for sure.

information about Neptune and its satellites that years will be needed to analyze it all.

Neptune has a dynamic atmosphere, much like those of Jupiter and Saturn (Figure 21.23). Winds exceeding 1000 kilometers per hour (600 miles per hour) encircle the planet, making it one of the windiest places in the solar system. It also has an Earth-size blemish called the *Great Dark Spot* that is reminiscent of Jupiter's Great Red Spot and is assumed to be a large rotating storm.

Perhaps most surprising are white, cirruslike clouds that occupy a layer about 50 kilometers above the main cloud deck, probably frozen methane. Six new satellites were discovered in the *Voyager* images, bringing Neptune's family to eight. All of the newly discovered moons orbit the planet in a direction opposite that of the two larger satellites. *Voyager* images also revealed a ring system around Neptune.

Triton, Neptune's largest moon, is a most interesting object. Its diameter is nearly that of Earth's Moon. Triton is the only large moon in the solar system that exhibits retrograde motion. This indicates that Triton formed independently of Neptune and was gravitationally captured.

Triton also has the lowest surface temperature yet measured on any body in the solar system, −200°C

(−391°F). Its atmosphere is mostly nitrogen with a little methane. Despite low surface temperatures, Triton displays volcanic-like activity. In 1989, *Voyager 2* detected active plumes that extended to an altitude of 8 kilometers and were blown downwind for more than 100 kilometers. Presumably, the surface layers of darker methane ice absorb solar energy more readily. Such surface warming vaporizes some of the underlying nitrogen ice. As subsurface pressures increase, explosive eruptions result.

Pluto: Planet X

Pluto lies on the fringe of the solar system, almost 40 times farther from the Sun than Earth. It is 10,000 times too dim to be visible to the unaided eye. Because of its great distance and slow orbital speed, it takes Pluto 248 Earth-years to orbit the Sun. Ever since its discovery in 1930, it has completed about one-fourth of a revolution. Pluto's orbit is noticeably elongated (highly eccentric), causing it to occasionally travel inside the orbit of Neptune, where it resided from 1979 through February 1999. There is no chance that Pluto and Neptune will ever collide, because their orbits are inclined to each other and do not actually cross (see Figure 21.1).

In 1978 the moon Charon was discovered orbiting Pluto. Because of its close proximity to the planet, the best ground-based images of Charon show it only as an elongated bulge. In 1990 the Hubble Space Telescope produced an image that clearly resolves the separation between these two icy worlds. Charon orbits Pluto once every 6.4 Earth-days at a distance 20 times closer to Pluto than our Moon is to Earth.

The discovery of Charon greatly altered earlier estimates of Pluto's size. Current data indicate that Pluto has a diameter of approximately 2,300 kilometers (1425 miles), about one-fifth the size of Earth, making it the smallest planet in the solar system (see Box 21.2). Charon is about 1300 kilometers across, exceptionally large in proportion to its parent.

The average temperature of Pluto is estimated at −210°C (−340°F), cold enough to solidify most gases that might be present. Thus, Pluto might best be described as a dirty iceball of frozen gases with lesser amounts of rocky substances.

BOX 21.2

UNDERSTANDING EARTH: Is Pluto Really a Planet?

Ever since Pluto's discovery in 1930, it has been a mystery on the edge of the solar system. At first, Pluto was thought to be about as large as Earth, but as better images were obtained, Pluto's diameter was estimated to be a little less than one half that of Earth. Then, in 1978, astronomers discovered that Pluto has a moon (Charon), whose brightness combined with its parent made Pluto appear much larger than it really was (Figure 21.B). Recent images obtained by the Hubble Space Telescope established the diameter of Pluto at only 2300 kilometers (1425 miles). This is about one-fifth that of Earth and less than half that of Mercury, long considered the runt of the solar system. In fact, seven moons, including Earth's moon, are larger than Pluto.

Even more attention was given to Pluto's status as a planet, when in 1992 astronomers discovered another icy body in orbit beyond Neptune. Soon hundreds of these objects were discovered forming a band similar to the asteroid belt between Mars and Jupiter. However, these orbiting bodies are made up of dust and ices, like comets, rather than metallic and rocky substances, like asteroids. Some astronomers believe that planetary objects even larger than Pluto may exist in this belt of icy worlds found in the outermost reaches of the solar system. In fact, one body larger than Pluto's moon, Charon, has already been discovered.

A growing number of astronomers assert that Pluto's small size and its location within a swarm of similar icy objects means that it should be reclassified as a minor planet, like asteroids and comets. Others insist that regardless of how Pluto's identity changes, demoting Pluto to a minor planet would dishonor astronomical history and confuse the public.

For now it seems that the International Astronomical Union, a group that has the power to vote on whether or not Pluto is a planet, is content with the status quo. Nevertheless, Pluto's planetary status will never be the same. It is now clear that Pluto is unique among the planets, being very different from the four rocky innermost planets, and unlike the four gaseous giants. Perhaps Pluto is best described as one of the largest members of a belt of millions of small, icy worlds (comets) that orbit in the outer reaches of our solar system.

Figure 21.B Pluto and its moon Charon. Earth is shown for scale.

Minor Members of the Solar System

Asteroids: Microplanets

Asteroids are smaller bodies that have been likened to "flying mountains." The largest, Ceres, is about 1000 kilometers (600 miles) in diameter, but most of the 50,000 that have been observed are only about 1 kilometer across. The smallest asteroids are assumed to be no larger than grains of sand. Most lie between the orbits of Mars and Jupiter and have periods of three to six years (Figure 21.24). Some have very eccentric orbits and travel very near the Sun, and a few larger ones regularly pass close to Earth and the Moon. Many of the most recent impact craters on the Moon and Earth were probably caused by collisions with asteroids. Inevitably, future Earth–asteroid collisions will occur (see Box 21.3).

Because many asteroids have irregular shapes, planetary geologists first speculated that they might be fragments of a broken planet that once orbited between Mars and Jupiter (Figure 21.25). However, the total mass of the asteroids is estimated to be only 1/1000 that of Earth, which itself is not a large planet. What happened to the remainder of the original planet? Others have hypothesized that several larger bodies once coexisted in close proximity and that their collisions produced numerous smaller ones. The existence of several families of asteroids has been used to support this explanation. However, no conclusive evidence has been found for either hypothesis.

In February 2001 an American spacecraft became the first visitor to an asteroid. Although it was not designed for landing, *NEAR Shoemaker* landed successfully and generated information that has planetary geologists intrigued and perplexed. Images obtained as the spacecraft drifted at the rate of 6 kilometers (4 miles) per hour toward the surface of Eros revealed a barren, rocky surface composed of particles ranging in size from fine dust to boulders up to 8 meters (26 feet) across. Researchers unexpectedly discovered that fine debris is concentrated in the low areas that form flat deposits that resemble ponds. Surrounding the low areas, the landscape is marked by an abundance of large boulders.

One of several hypotheses being considered as an explanation for the boulder-laden topography is seismic shaking, which would move the boulders upward. Analogous to what happens when a can of mixed nuts is shaken, the larger materials rise to the top while the smaller materials settle to the bottom.

Comets

Comets are among the most interesting and unpredictable bodies in the solar system. They have been compared to dirty snowballs, because they are made of frozen gases (water, ammonia, methane, carbon dioxide, and carbon monoxide) that hold together small pieces of rocky and metallic materials. Many comets travel in very elongated orbits that carry them far beyond Pluto. These comets take hundreds of thousands of years to complete a single orbit around the Sun. However, a few *short-period comets* (those having orbital periods of less than 200 years), such as Halley's comet, make regular encounters with the inner solar system.

When first observed, a comet appears very small, but as it approaches the Sun, solar energy begins to vaporize the frozen gases, producing a glowing head called the **coma** (Figure 21.26). The size of the coma varies greatly from one comet to another. Extremely rare ones exceed the size of the Sun, but most approximate the size of Jupiter. Within the coma, a small glowing nucleus with a diameter of only a few kilometers can sometimes be detected. As comets approach the Sun, some, but not all, develop a tail that extends for millions of kilometers. Despite the enormous size of their tails and comas, comets are relatively small members of the solar system.

The fact that the tail of a comet points away from the Sun in a slightly curved manner (Figure 21.26) led early astronomers to propose that the Sun has a repulsive force that pushes the particles of the coma away, thus forming the tail. Today, two solar forces are known

Figure 21.24 The orbits of most asteroids lie between Mars and Jupiter. Also shown are the orbits of a few known near-Earth asteroids. Perhaps a thousand or more asteroids have near-Earth orbits. Luckily, only a few dozen are thought to be larger than 1 kilometer in diameter.

BOX 21.3

EARTH AS A SYSTEM:
Is Earth on a Collision Course?

The solar system is cluttered with meteoroids, asteroids, active comets, and extinct comets. These fragments travel at great speeds and can strike Earth with the explosive force of a powerful nuclear weapon.

In the last few decades, it has become increasingly clear that comets and asteroids have collided with Earth far more frequently than was previously known. The evidence is giant impact structures. More than 100 have been identified (Figure 21.C). Many were once misunderstood to result from some volcanic process. Although most impact structures are so old that they no longer resemble impact craters, evidence of their intense impact remains (Figure 21.D). One notable exception is a very fresh-looking crater near Winslow, Arizona, known as Meteor Crater (see chapter-opening photo).

Evidence is mounting that about 65 million years ago a large asteroid about 10 kilometers (6 miles) in diameter col-lided with Earth. This impact may have caused the extinction of the dinosaurs, as well as nearly 50 percent of all plant and animal species (see Box 11.3).

More recently, a spectacular explosion has been attributed to the collision of our planet with a comet or asteroid. In 1908, in a remote region of Siberia, a "fireball" that appeared more brilliant than the Sun exploded with a violent force. The shock waves rattled windows and triggered reverberations heard up to 1000 kilometers away. The "Tunguska event," as it is called, scorched, de-limbed, and flattened trees to 30 kilometers from the epicenter. But expeditions to the area found no evidence of an impact crater nor any metallic fragments. Evidently, the explosion, which equaled at least a 10-megaton nuclear bomb, oc-curred a few kilometers above the surface. Most likely it was the demise of a comet or perhaps a stony asteroid. Why it exploded prior to impact is uncertain.

The dangers of living with these small but deadly objects from space

Figure 21.D Manicouagan, Quebec, is a 200-million-year-old eroded impact structure. The lake outlines the crater remnant, which is 70 kilometers (42 miles) across. Fractures related to this event extend outward for an additional 30 kilometers. (Courtesy of U.S. Geological Survey)

Figure 21.C World map of major impact structures. Others are being identified every year. (Data from Griffith Observatory)

again came to public attention in 1989 when an asteroid nearly 1 kilometer across shot past Earth. It was a near miss, about twice the distance to the Moon. Traveling at 70,000 kilometers (44,000 miles) per hour, it could have produced a crater 10 kilometers (6 miles) in diameter and perhaps 2 kilometers (1.2 miles) deep. As an observer noted, "Sooner or later it will be back." As it was, it crossed our orbit just six hours ahead of Earth. Statistics show that collisions of this tremendous magnitude should take place every few hundred million years and could have drastic consequences for life on Earth.

to contribute to this formation. One, *radiation pressure,* pushes dust particles away from the coma. The second, known as *solar wind,* is responsible for moving the ion-ized gases, particularly carbon monoxide. Sometimes a single tail composed of both dust and ionized gases is produced, but often two tails are observed (Figure 21.27).

As a comet moves away from the Sun, the gases forming the coma recondense, the tail disappears, and the comet returns to cold storage. Material that was blown from the coma to form the tail is lost from the comet forever. Consequently, it is believed that most comets cannot survive more than a few hundred close

orbits of the Sun. Once all the gases are expelled, the re-maining material—a swarm of tiny metallic and stony particles—continues the orbit without a coma or a tail.

Comets apparently originate in two regions of the outer solar system. Most short period comets are thought to orbit beyond Neptune in a region called the **Kuiper belt,** in honor of astronomer Gerald Kuiper, who had predicted their existence. (During the past decade over a hundred of these icy bodies have been discovered.) Like the asteroids in the inner solar sys-tem, most Kuiper belt comets move in nearly circular orbits that lie roughly in the same plane as the planets. A chance collision between two Kuiper belt comets, or

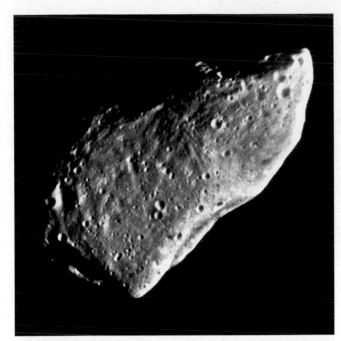

Figure 21.25   Image of asteroid 951 (Gaspra) obtained by the Jupiter-bound *Galileo* spacecraft. Like other asteroids, Gaspra is probably a collision-produced fragment of a larger body. (Courtesy of NASA)

the gravitational influence of one of the Jovian planets, may occasionally alter the orbit of a comet enough to send it to the inner solar system, and into our view.

Unlike Kuiper belt comets, long-period comets have orbits that are *not* confined to the plane of the solar system. These comets appear to be distributed in all directions from the Sun, forming a spherical shell around the solar system, called the **Oort cloud,** after the Dutch astronomer Jan Oort. Millions of comets are believed to orbit the Sun at distances greater than 10,000 times the Earth-Sun distance. The gravitational effect of a distant passing star is thought to send an occasional Oort cloud comet into a highly eccentric orbit that carries it toward the Sun. However, only a tiny portion of the Oort cloud comets have orbits that bring them into the inner solar system.

The most famous short-period comet is Halley's comet. Its orbital period averages 76 years, and every one of its 29 appearances since 240 B.C. has been recorded by Chinese astronomers. This record is a testimonial to their dedication as astronomical observers and to the endurance of their culture. When seen in 1910, Halley's comet had developed a tail nearly 1.6 million kilometers (1 million miles) long and was visible during the daylight hours.

In 1986 the unspectacular showing of Halley's comet was a disappointment to many people in the Northern Hemisphere. Yet it was during this most recent visit to the inner solar system that a great deal of new information was learned about this most famous of

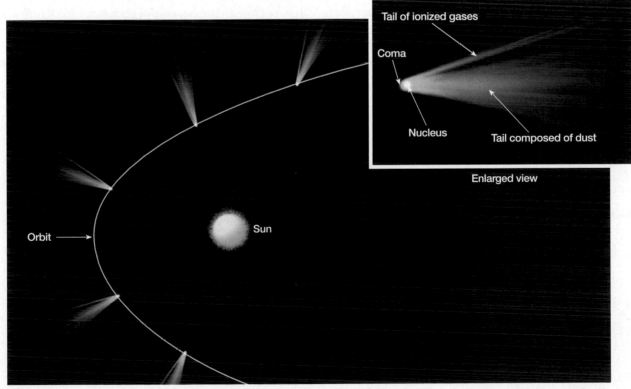

Figure 21.26   Orientation of a comet's tail as it orbits the Sun.

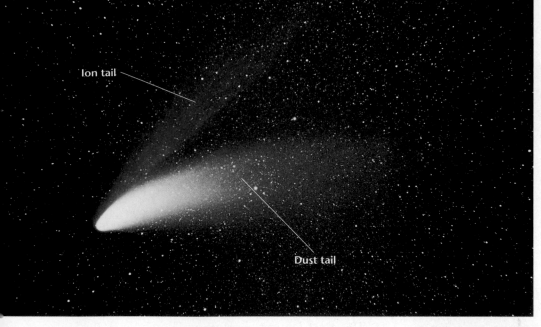

Figure 21.27 Comet Hale-Bopp. The two tails seen in the photograph are between 10 million and 15 million miles long. (Peoria Astronomical Society Photograph by Eric Clifton and Craig Neaveill)

comets. The new data were gathered by space probes sent to rendezvous with the comet. Most notably, the European probe *Giotto* approached to within 600 kilometers of the comet's nucleus and obtained the first images of this elusive structure.

We now know that the nucleus is potato-shaped, 16 kilometers by 8 kilometers. The surface is irregular and full of craterlike pits. Gases and dust that vaporize from the nucleus to form the coma and tail appear to gush from its surface as bright jets or streams. Only about 10 percent of the comet's total surface was emitting these jets at the time of the rendezvous. The remaining surface area of the comet appeared to be covered with a dark layer that may consist of organic material.

In 1997 the comet Hale-Bopp made for spectacular viewing around the globe. As comets go, the nucleus of Hale-Bopp was unusually large, about 40 kilometers (25 miles) in diameter. As shown in Figure 21.27, two tails nearly 15 million miles long extended from this comet. The bluish gas-tail is composed of positively charged ions, and it points almost directly away from the Sun. The yellowish tail is composed of dust and other rocky debris. Because the rocky material is more massive than the ionized gases, it is less affected by the solar wind and follows a different trajectory away from the comet.

Meteoroids

Nearly everyone has seen a **meteor,** popularly (but inaccurately) called a "shooting star." This streak of light lasts from an eyeblink to a few seconds and occurs when a small solid particle, a **meteoroid,** enters Earth's atmosphere from interplanetary space. The friction between the meteoroid and the air heats both and produces the light we see. Most meteoroids originate from any one of the following three sources: (1) interplanetary debris that was not gravitationally swept up by the planets during the formation of the solar system, (2) ma-

terial that is continually being displaced from the asteroid belt, or (3) the solid remains of comets that once traveled near Earth's orbit. A few meteoroids are believed to be fragments of the Moon, or possibly Mars, that were ejected when an asteroid impacted these bodies.

Although a rare meteoroid is as large as an asteroid, most are the size of sand grains and weigh less than 1/100 gram. Consequently, they vaporize before reaching Earth's surface. Some, called *micrometeorites,* are so tiny that their rate of fall becomes too slow to cause them to burn up, so they drift down as space dust. Each day, the number of meteoroids that enter Earth's atmosphere must reach into the thousands. After sunset on a clear night, a half dozen or more are bright enough to be seen with the naked eye each hour from anywhere on Earth.

Occasionally, meteor sightings increase dramatically to 60 or more per hour. These displays, called **meteor showers,** result when Earth encounters a swarm of meteoroids traveling in the same direction and at nearly the same speed as Earth. The close association of these swarms to the orbits of some short-term comets strongly suggests that they represent material lost by these comets (Table 21.2). Some swarms not associated with orbits of known comets are probably the remains of the nucleus of a long-defunct comet. The notable Perseid meteor shower that occurs each year around August 12 is believed to be the remains of the Comet 1862 III, which has a period of 110 years.

Meteoroids that are thought to be the remains of comets are small and only occasionally reach the ground. Most meteoroids large enough to survive the heated fall are thought to originate among the asteroids, where chance collisions modify their orbits and send them toward Earth. Earth's gravitational force does the rest.

The remains of meteoroids, when found on Earth, are referred to as **meteorites.** A few very large mete-

Table 21.2 Major meteor showers.

Shower	Approximate Dates	Associated Comet
Quadrantids	January 4–6	
Lyrids	April 20–23	Comet 1861 I
Eta Aquarids	May 3–5	Halley's comet
Delta Aquarids	July 30	
Perseids	August 12	Comet 1862 III
Draconids	October 7–10	Comet Giacobini-Zinner
Orionids	October 20	Halley's comet
Taurids	November 3–13	Comet Encke
Andromedids	November 14	Comet Biela
Leonids	November 18	Comet 1866 I
Geminids	December 4–16	

Figure 21.28 Iron meteorite protruding from desert sands. (Courtesy of *Aramco World Magazine*)

oroids have blasted out craters on Earth's surface that strongly resemble those on the lunar surface. The most famous is Meteor Crater in Arizona (see chapter-opening photo). This huge cavity is about 1.2 kilometers (0.75 miles) across, 170 meters (560 feet) deep, and has an upturned rim that rises 50 meters (165 feet) above the surrounding countryside. Over 30 tons of iron fragments have been found in the immediate area, but attempts to locate the main body have been unsuccessful. Based on erosion, the impact likely occurred within the last 20,000 years.

Prior to Moon rocks brought back by lunar explorers, meteorites were the only extraterrestrial materials that could be directly examined (Figure 21.28). Meteorites are classified by their composition: (1) **irons** mostly iron with 5 percent to 20 percent nickel, (2) **stony** silicate minerals with inclusions of other minerals, and (3) **stony-irons** mixtures. Although stony meteorites are probably more common, people find mostly irons. This is understandable, for irons withstand the impact better, weather more slowly, and are much easier for a layperson to distinguish from terrestrial rocks. Iron meteorites are probably fragments of once-molten cores of large asteroids or small planets.

One rare kind of meteorite, called a *carbonaceous chondrite*, was found to contain simple amino acids and other organic compounds, which are the basic building blocks of life. This discovery confirms similar findings in observational astronomy, which indicate that numerous organic compounds exist in the frigid realm of outer space.

If meteorites represent the makeup of Earth-like planets, as some planetary geologists suggest, then Earth must contain a much larger percentage of iron than is indicated by surface rocks. This is one reason why geologists suggest that Earth's core may be mostly iron and nickel. In addition, meteorite dating indicates that our solar system's age certainly exceeds 4.5 billion years. This "old age" has been confirmed by data from lunar samples.

Chapter Summary

• The planets can be arranged into two groups: the *terrestrial* (Earth-like) *planets* (Mercury, Venus, Earth, and Mars) and the *Jovian* (Jupiter-like) *planets* (Jupiter, Saturn, Uranus, and Neptune). Pluto is not included in either group. *When compared to the Jovian planets, the terrestrial planets are smaller, more dense, contain proportionally more rocky material, have slower rates of rotation, and lower escape velocities.*

• The lunar surface exhibits several types of features. Most *craters* were produced by the impact of rapidly moving interplanetary debris (*meteoroids*). Bright, densely cratered *highlands* make up most of the lunar surface. The dark, fairly smooth lowlands are called *maria* (singular, *mare*). Maria basins are enormous impact craters that have been flooded with layer upon layer of very fluid basaltic lava. All lunar terrains are mantled with a soil-like layer of gray unconsolidated debris, called *lunar regolith*, which has been derived from a few billion years of meteoric bombardment. Much is still unknown about the Moon's origin. One hypothesis suggests that a giant asteroid collided with Earth to produce the Moon. Scientists conclude that the *lunar surface evolved in three phases*: (1) *the original crust (highlands)*; (2) *maria basins*; and (3) *youthful rayed craters.*

• *Mercury* is a small, dense planet that has no atmosphere and exhibits the greatest temperature extremes of any planet. *Venus,* the brightest planet in the sky, has a thick, heavy atmosphere composed of 97 percent carbon dioxide, a surface of relatively subdued plains and inactive volcanic features, a surface atmospheric pressure 90 times that of Earth's, and surface temperatures of 475°C (900°F). *Mars*, the Red Planet, has a carbon dioxide atmosphere only 1 percent as dense as Earth's, extensive dust storms, numerous inactive volcanoes, many large canyons, and several valleys of debatable origin exhibiting drainage patterns similar to stream valleys on Earth. *Jupiter*, the largest planet, rotates rapidly, has a banded appearance caused by huge convection currents driven by the planet's interior heat, a *Great Red Spot* that varies in size, a thin ring system, and at least 16 moons (one of the moons, *Io*, is a volcanically active body). *Saturn* is best known for its system of rings. It also has a dynamic atmosphere with winds up to 930 miles per hour and storms similar to Jupiter's Great Red Spot. *Uranus* and *Neptune* are often called the twins because of similar structure and composition. A unique feature

of Uranus is the fact that it rotates on its side. Neptune has white, cirruslike clouds above its main cloud deck and an Earth-size *Great Dark Spot*, assumed to be a large rotating storm similar to Jupiter's Great Red Spot. *Pluto* is a small frozen world with one moon (Charon). Pluto's noticeably elongated orbit causes it to occasionally travel inside the orbit of Neptune, but with no chance of collision.

• The minor members of the solar system include the *asteroids, comets,* and *meteoroids.* Most asteroids lie between the orbits of Mars and Jupiter. No conclusive evidence has been found to explain their origin. Comets are made of frozen gases (water, ammonia, methane, carbon dioxide, and carbon monoxide) with small pieces of rocky and metallic material. Many travel in very elongated orbits that carry them beyond Pluto, and little is known about their origin. Meteoroids, small solid particles that travel through interplanetary space, become *meteors* when they enter Earth's atmosphere and vaporize with a flash of light. *Meteor showers* occur when Earth encounters a swarm of meteoroids, probably material lost by a comet. *Meteorites* are the remains of meteoroids found on Earth. The *three types of meteorites* (classified by their composition) are (1) *irons*, (2) *stony*, and (3) *stony-irons*. One rare kind of meteorite, called a *carbonaceous chondrite*, was found to contain amino acids and other organic compounds.

Key Terms

asteroid (p. 594)
coma (p. 595)
comet (p. 595)
escape velocity (p. 578)
inner planets (p. 577)
iron meteorite (p. 599)
Jovian planet (p. 576)

Kuiper belt (p. 596)
lunar regolith (p. 581)
maria (p. 580)
meteor (p. 598)
meteorite (p. 598)
meteoroid (p. 598)

meteor shower (p. 598)
Oort cloud (p. 597)
outer planets (p. 577)
stony-iron meteorite (p. 599)
stony meteorite (p. 599)
terrestrial planet (p. 576)

Review Questions

1. By what criteria are the planets placed into either the Jovian or terrestrial group?

2. What are the three types of materials thought to make up the planets? How are they different? How does their distribution account for the density differences between the terrestrial and Jovian planetary groups?

3. Explain why different planets have different atmospheres.

4. How is crater density used in the relative dating of features on the Moon?

5. Briefly outline the history of the Moon.

6. How are the maria of the Moon thought to be similar to the Columbia Plateau?

7. Why has Mars been the planet most studied telescopically?

8. What surface features does Mars have that are also common on Earth?

9. What evidence supports a water cycle on Mars? What evidence refutes the possibility of a wet Martian climate?

10. Why are astrobiologists intrigued about evidence that groundwater has seeped onto the surface of Mars?

11. The two "moons" of Mars were once suggested to be artificial. What characteristics do they have that would cause such speculation?

12. What is the nature of Jupiter's Great Red Spot?

13. Why are the Galilean satellites of Jupiter so named?

14. What is distinctive about Jupiter's satellite Io?

15. Why are the four *outer* satellites of Jupiter thought to have been captured rather than having been formed with the rest of the satellite system?

16. How are Jupiter and Saturn similar?

17. What two roles do ring moons play in the nature of planetary ring systems?

18. How are Saturn's satellite Titan and Neptune's Triton similar?

19. What three bodies in the solar system exhibit volcanic-like activity?

20. Where are most asteroids found?

21. What do you think would happen if Earth passed through the tail of a comet?

22. Where are most comets thought to reside? What eventually becomes of comets that orbit close to the Sun?

23. Compare meteoroid, meteor, and meteorite.

24. What are the three main sources of meteoroids?

25. Why are meteorite craters more common on the Moon than on Earth, even though the Moon is a much smaller target?

26. It has been estimated that Halley's comet has a mass of 100 billion tons. Further, this comet is estimated to lose 100 million tons of material during the few months that its orbit brings it close to the Sun. With an orbital period of 76 years, what is the maximum remaining life span of Halley's comet?

Examining the Earth System

1. On Earth the four major spheres (atmosphere, hydrosphere, solid Earth, and biosphere) interact as a system with occasional influence from our near-space neighbors. Which of these spheres are absent, or nearly absent, on the Moon? Because the Moon lacks these spheres, list at least five processes that operate on Earth but are absent on the Moon.

2. Among the planets in our solar system, Earth is unique because water exists in all three states (solid, liquid, and gas) on its surface. In what state(s) of matter is water found on Mercury, Venus, and Mars? How would Earth's hydrologic cycle be different if (a) its orbit was inside the orbit of Venus? (b) its orbit was outside the orbit of Mars?

3. If a large meteorite were to strike Earth in the near future, what effect might this event have on the atmosphere (in particular, average temperatures and climate)? Assuming the conditions persisted, how would the changes impact the biosphere?

Web Resources

The *Earth Science* Website uses the resources and flexibility of the Internet to aid in your study of the topics in this chapter. Written and developed by Earth science instructors, this site will help improve your understanding of Earth science. Visit **http://www.prenhall.com/tarbuck** and click on the cover of *Earth Science 10e* to find:

- On-line review quizzes.
- Web-based critical thinking and writing exercises.
- Links to chapter-specific Web resources.
- Internet-wide key term searches.

http://www.prenhall.com/tarbuck

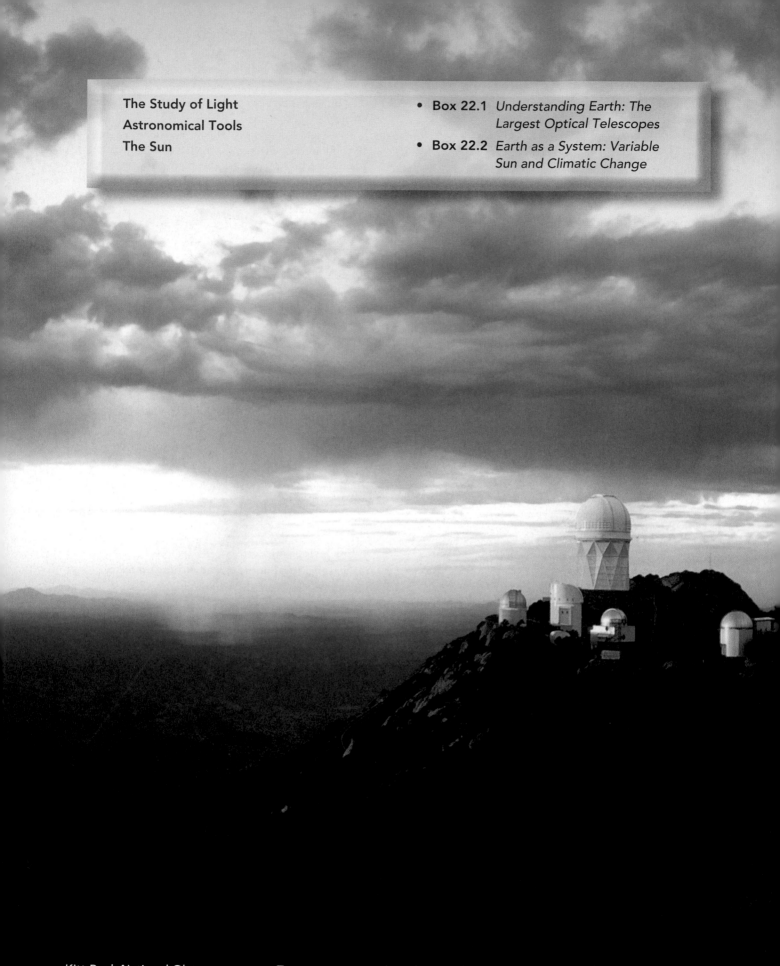

Kitt Peak National Observatory, near Tucson, Arizona. (Photo by David Parker/Photo Researchers, Inc.)

CHAPTER 22

LIGHT, ASTRONOMICAL OBSERVATIONS, AND THE SUN

Astronomers are in the business of gathering and studying light. Almost everything that is known about the universe beyond Earth comes by analyzing light from distant sources. Consequently, an understanding of the nature of light is basic to modern astronomy. This chapter deals with the study of light and the tools used by astronomers to gather light in order to probe the universe. In addition, we will examine the nearest source of light, our Sun. By understanding the processes that operate on the Sun, astronomers can better grasp the nature of more distant celestial objects.

The Study of Light

The vast majority of our information about the universe is obtained from the study of the light emitted from celestial bodies (Figure 22.1). Although visible light is most familiar to us, it constitutes only a small part of an array of energy generally referred to as **electromagnetic radiation.** Included in this array are gamma rays, X-rays, ultraviolet light, visible light, infrared radiation, and radio waves (see Figure 15.15, p. 418). All radiant energy travels through the vacuum of space in a straight line at the rate of 300,000 kilometers (186,000 miles) per second.* Over a 24-hour day, this equals a staggering 26 billion kilometers.

Nature of Light

Experiments have demonstrated that light can be described in two ways. In some instances light behaves like waves, and in others like discrete particles. In the wave sense, light is analogous to swells in the ocean. This motion is characterized by the wavelength, which is the distance from one wave crest to the next. Wavelengths vary from several kilometers for radio waves to less than a billionth of a centimeter for gamma rays. Most of these waves are either too long or too short for our eyes to detect.

The narrow band of electromagnetic radiation we can see is sometimes referred to as *white light*. However, white light consists of an array of waves having various wavelengths, a fact easily demonstrated with a prism (Figure 22.2). As white light passes through a prism, the color with the shortest wavelength, violet, is bent more than blue, which is bent more than green, and so forth (Table 22.1). Thus, white light can be separated into its component colors in the order of their wavelengths, producing the familiar rainbow of colors (Figure 22.2).

Wave theory, however, cannot explain some effects of light. In some cases, light acts like a stream of particles, analogous to infinitesimally small bullets fired

from a machine gun. These particles, called **photons,** can exert a pressure (push) on matter, which is called **radiation pressure.** Photons from the Sun are responsible for pushing material away from a comet to produce its tail. Each photon has a specific amount of energy, which is related to its wavelength in a simple way: *Shorter wavelengths* correspond to *more energetic* photons. Thus, blue light has more energetic photons than does red light.

Which theory of light—the wave theory or the particle theory—is correct? Both, because each will predict the behavior of light for certain phenomena. As George Abell, a prominent astronomer, stated about all scientific laws, "The mistake is only to apply them to situations that are outside their range of validity."

Spectroscopy

When Sir Isaac Newton used a prism to disperse white light into its component colors, he unknowingly initiated the field of **spectroscopy,** which is the study of properties of light that depend on wavelength. The rainbow of colors Newton produced is called a "continuous spectrum," because all wavelengths of light are included. It was later learned that two other types of spectra exist and that all three are generated under somewhat different conditions (Figure 22.3).

1. A **continuous spectrum** is produced by an incandescent solid, liquid, or gas under high pressure. It consists of an uninterrupted band of color (Figure 22.3A). One example would be light generated by a common lightbulb. (*Incandescent* means "to emit light when hot.")

2. A **dark-line spectrum (absorption spectrum)** is produced when white light is passed through a comparatively cool gas under low pressure. The gas absorbs selected wavelengths of light, so the spectrum that is produced appears as a continuous spectrum, but with a series of dark lines running through it (Figure 22.3B).

3. A **bright-line spectrum (emission spectrum)** is produced by a hot (incandescent) gas under low pressure. It is a series of bright lines of particular wavelengths, depending on the gas that produces them (Figure 22.3C). These bright lines appear in the exact location as the dark lines that are produced by this gas in a dark-line (absorption) spectrum.

The spectra of most stars are of the dark-line type. The importance of these spectra is that each element or compound that is in gaseous form (in a star, material is usually in the gaseous form) produces a unique set of spectral lines. When the spectrum of a star is studied, the spectral lines act as "fingerprints," which identify the elements present.

---

*Light rays are "bent" slightly when they pass nearby a very massive object such as the Sun.

Figure 22.1 The Trifid Nebula, in the constellation Sagittarius. This colorful nebula is a cloud of dust, plus hydrogen and helium gases. These gases are excited by the radiation of the hot, young stars within and produce a pink glow. (Courtesy of National Optical Astronomy Observatories)

In an admittedly oversimplified manner, we can imagine how the Sun and other stars create a dark-line spectrum. A continuous spectrum is produced in the interior of the Sun, where the gases are under very high pressure. When this light passes through the less dense gases of the solar atmosphere, they absorb selected wavelengths, which appear as dark lines in the spectrum.

When Newton studied solar light, he obtained a continuous spectrum. However, when a prism is used in conjunction with lenses, the solar spectrum can be dispersed even further. An instrument that does this is called a **spectroscope** (*spectro* = to look, *scope* = instrument). The spectrum of the Sun contains thousands of dark lines. Over 60 elements have been identified by matching these lines with those of elements known on Earth.

Two other facts concerning a radiating body are important. First, if the temperature of a radiating surface is increased, the total amount of energy emitted increases. The rate of increase is stated by the Stefan-Boltzmann law: The energy radiated by a body is

Figure 22.2 A spectrum is produced when sunlight (white light) is passed through a prism, which bends each wavelength at different angles.

605

Table 22.1 Colors and corresponding wavelengths.	
Color	**Wavelength (nanometers*)**
Violet	380–440
Blue	440–500
Green	500–560
Yellow	560–590
Orange	590–640
Red	640–750

*One nanometer is 10^{-9} meter.

directly proportional to the fourth power of its absolute temperature. For example, if the temperature of a star is doubled, the total radiation emitted increases by $2^4 (2 \times 2 \times 2 \times 2)$, or 16 times. Second, as the temperature of an object increases, a larger proportion of its energy is radiated at shorter wavelengths. To illustrate this, imagine a metal rod that is heated slowly. The rod first appears dull red (longer wavelengths) and later bluish white (shorter wavelengths). From this, it follows that blue stars are hotter than yellow stars, which are hotter than red stars.

The Doppler Effect

You may have heard the change in pitch of a car horn or ambulance siren as it passes by. When it is approaching, the sound seems to have a higher-than-normal pitch, and when it is moving away, the pitch sounds lower than normal. This effect, which occurs for both sound and light waves, was first explained by Christian Doppler in 1842 and is called the **Doppler effect.** The reason for the difference in pitch is that it takes time for the wave to be emitted. If the source is moving away, the beginning of the wave is emitted nearer to you than the end, which stretches the wave—that is, gives it a longer wavelength (Figure 22.4). The opposite is true for an approaching source.

In the case of light, when a source is moving away, its light appears redder than it actually is because its waves appear lengthened. Objects approaching have their light waves shifted toward the blue (shorter wavelength). Thus, if a source of red light approached you at a very high speed (near the speed of light), it would actually appear blue. The same effect would be produced if you moved and the light were stationary.

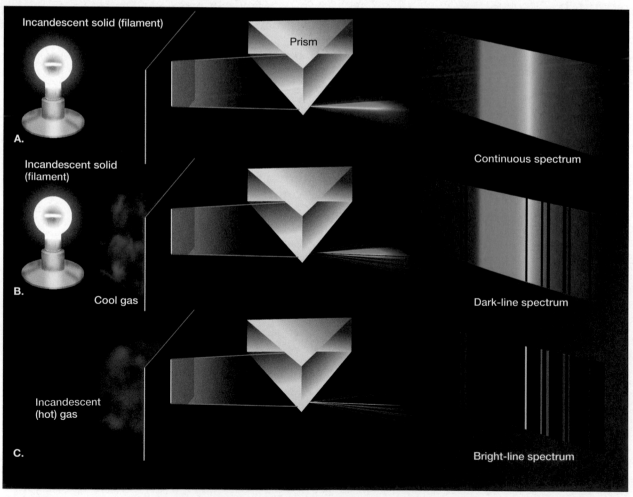

Figure 22.3   Formation of the three types of spectra. **A.** Continuous spectrum. **B.** Dark-line spectrum. **C.** Bright-line spectrum.

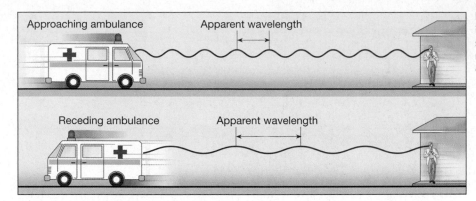

Figure 22.4   The Doppler effect, illustrating the apparent lengthening and shortening of wavelengths caused by the relative motion between a source and an observer.

Therefore, the Doppler effect reveals whether Earth is approaching or receding from a star or another celestial body. In addition, the amount of shift allows us to recalculate the rate at which the relative movement is occurring. Larger Doppler shifts indicate higher velocities; smaller Doppler shifts indicate slower velocities. Doppler shifts are generally measured from the dark lines in the spectra of stars by comparing them with a standard spectrum produced in the laboratory (Figure 22.5).

Astronomical Tools

Having examined the nature of light, we will now turn our attention to the tools astronomers use to intercept and study the energy emitted by distant objects in the universe (Figure 22.6). Because the basic principles of detecting radiation were originally developed through visual observations, we will consider optical telescopes first.

Refracting Telescopes

Galileo is considered to be the first person to use telescopes for astronomical observations. Having learned about the newly invented instrument, Galileo built one of his own that was capable of magnifying objects 30 times. Because this early instrument, as well as its modern counterparts, used a lens to bend or refract light, it is known as a **refracting telescope.**

The most important lens in a refracting telescope, the **objective lens,** produces an image by bending light from a distant object in such a way that the light converges at an area called the **focus** (*focus* = central point) (Figure 22.7). For an object such as a star, the image appears as a point of light, but for nearby objects it appears as an inverted replica of the original.

You can easily demonstrate the latter case by holding a lens in one hand and, with the other hand, placing a white card behind the lens. Now vary the distance between them until an image appears on the card. The distance between the focus (where the image appears) and the lens is called the **focal length** of the lens.

Astronomers usually study an image from a telescope by first photographing the image. However, if a telescope is used to examine an image directly, a second lens, called an **eyepiece,** is required (Figure 22.7). The eyepiece magnifies the image produced by the objective lens. In this respect, it is similar to a magnifying glass. Thus, the objective lens produces a very small, bright image of an object, and the eyepiece enlarges the image so that details can be seen.

Although used extensively in the nineteenth century, refracting telescopes suffer a major optical defect. As light passes through any lens, the shorter wavelengths of light are bent more than the longer wavelengths. (Recall the effect of a prism in separating the colors of the spectrum.) Consequently, when a refracting telescope is in focus for red light, blue and violet light are out of focus. The troublesome effect, known as **chromatic** (*chroma* = color) **aberration** (*aberrare* = to go astray), weakens the image and produces a halo of color around it. When blue light is in focus, a reddish halo appears, and vice versa. Although this effect cannot be eliminated completely, it is reduced by using a second lens made of a different type of glass.

A. Standard sodium lines

B. Red-shifted sodium lines

C. Large red-shifted sodium lines

D. Blue-shifted sodium lines

Figure 22.5   **A.** An illustration of standard sodium lines produced in a laboratory compared to sodium lines as they would appear when the source is receding (red shift, **B.** and **C.**) or approaching (blue shift, **D.**).

Figure 22.6 The 10-meter Keck Telescope in located at the summit of Hawaii's Mauna Kea volcano. (Photo by Roger Ressmeyer/CORBIS)

Reflecting Telescopes

Newton was so bothered by chromatic aberration that he built and used telescopes that reflected light from a shiny surface (mirror). Because reflected light is not dispersed into its component colors, the problem is avoided. **Reflecting telescopes** use a concave mirror that focuses the light in front of the objective (the mirror), rather than behind it, like a lens (Figure 22.8). The mirror is generally made of glass that is finely ground. In the case of the 5-meter (200-inch) Hale Telescope, the grinding is accurate to about 1 millionth of a centimeter. The surface is then coated with a highly reflective material, usually an aluminum compound.

To focus parallel incoming rays onto one spot, the mirror is ground to a special curved surface called a *paraboloid*. This is the same shape as that used for the reflector in the headlights of automobiles. In the case of the auto bulb, however, the light source is at the focus, and the light goes out in parallel rays rather than coming in.

Because the focus of a reflecting telescope is in front of the mirror, provisions have to be made to view the image without blocking too much of the incoming light. Figure 22.9 illustrates the most common arrangements. Most large telescopes employ more than one type. When using a very large reflecting telescope, the observer enters a viewing cage positioned at the focus to make the observations. The viewing cage blocks only about 10 percent of the total incoming light, and this is more than compensated for by the large objective (mirror) that is used.

In the good old days, an astronomer would spend numerous long nights outdoors in the cold mountain air, perched in a viewing cage. But advances in photographic materials and computer-enhancing technologies now allow indoor work and reduce the time required to obtain an image.

Nearly all large optical telescopes built today are reflectors. Among the reasons is the monumental task of producing a large piece of high-quality, bubble-free glass for refracting telescopes. Because light does not pass through a mirror, the glass for a reflecting telescope does not have to be of optical quality, nor does the instrument suffer from chromatic aberration. In addition, a lens can be supported only around the edge, so it sags. Mirrors, on the other hand, can be supported fully from behind.

Large reflecting telescopes of about 4 meters (150 inches) are located at Kitt Peak, Arizona (Figure 22.10); Mauna Kea, Hawaii; Cerro Tololo, Chile; and Siding Spring, Australia (see Box 22.1). By comparison, the largest refractor in the world is the 1-meter (40-inch)

Figure 22.7 Simple refracting telescope.

Figure 22.8   **A.** Diagram illustrating how concave mirrors, like those used in reflector telescopes, gather light. **B.** Preparation of the 2.4-meter mirror for the Hubble Space Telescope. (Courtesy of Space Telescope Science Institute)

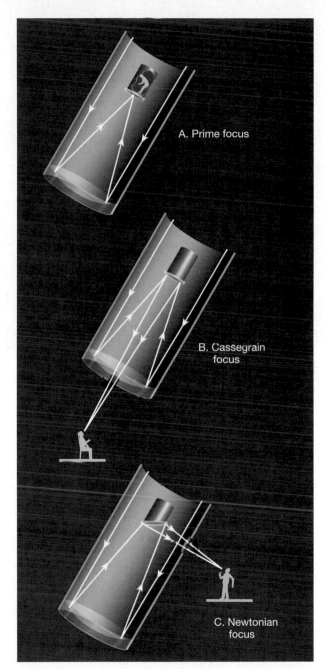

Figure 22.9   Viewing methods used with reflecting telescopes. **A.** Prime focus method (only used in very large telescopes). **B.** Cassegrain method (most common). **C.** Newtonian method.

telescope at Yerkes Observatory in Williams Bay, Wisconsin. This instrument was built before the turn of the twentieth century.

Properties of Optical Telescopes

Telescopes have three properties that aid astronomers in their work. They provide the observer with light-gathering power, resolving power, and magnifying power.

Light-Gathering Power. Because most celestial objects are very faint sources of light, astronomers are *most* interested in improving the *light-gathering* power of their instruments. As seen in Figure 22.11, a telescope with a large lens (or mirror) intercepts more light from distant objects and thereby produces brighter images. As very distant stars appear very dim as well, a great deal of light must be collected before the image is bright enough to be seen. Consequently, telescopes with large objectives "see" farther into space than do those with small objectives.

Figure 22.10 The 3.9-meter Anglo-Australian telescope. This modern reflecting telescope is one of the world's most advanced. (Royal Observatory Edinburgh/Photo Researchers, Inc.)

Resolving Power. Another advantage of telescopes with large-diameter objectives is their greater *resolving power* (Figure 22.12), which allows for sharper images and finer detail. For example, with the unaided eye, the Milky Way appears as a vague band of light in the night sky, but even a small telescope is capable of resolving (separating it into) individual stars. Even so, the condition of Earth's atmosphere (which is known as *seeing*) greatly limits the resolving power of telescopes on Earth. On a night when the stars twinkle, the seeing is poor because the air is moving rapidly. This causes the image to move about, blurring it. Conversely, when the stars shine steadily, the seeing is described as good. Even under ideal conditions, however, some blurring occurs, eliminating the fine details. Thus, even the largest telescopes cannot photograph lunar features less than 0.5 kilometer (0.3 mile) in size.

To eliminate the problems of Earthbound viewing, the United States built the Hubble Space Telescope, which was put into Earth orbit in April 1990 (Figure 22.13). This 2.4-meter (94-inch) space telescope has 10 billion times more light-gathering power than the human eye. As the technical problems that hampered the early operation of this instrument have been corrected, many spectacular images have been received. For example, the Hubble Space Telescope has

provided images that clearly resolve the separation between Pluto and its moon, Charon.

Magnifying Power. When you think of the power of a telescope, you probably think of its *magnifying power*, the ability to make an object larger. Magnification is calculated by dividing the focal length of the objective by the focal length of the eyepiece. Thus, the magnification of a telescope can be changed by simply changing the eyepiece. However, increased magnification does not necessarily improve the clarity of the image. What can be viewed telescopically is limited by atmospheric conditions and the resolving power of the telescope. Any part of an image that is not clear at low magnification will appear only as a larger blur at higher magnification. Furthermore, increasing magnification spreads out the light and decreases the brightness of the object. Thus, astronomers describe telescopes not in terms of their magnification but by the diameter of the objective mirror or lens, because it is this factor that determines both the light-gathering power and the resolving power of a telescope.

Most modern telescopes have supplemental devices that enhance the image. A simple but important example is a photographic plate that can be exposed for long periods of time, thereby collecting enough light from a star to make an image that otherwise would be undetectable. One of the latest advances makes use of high-speed computers that adjust the optics to partially remove the distortion caused by the atmosphere. This process greatly enhances the image sharpness.

Detecting Invisible Radiation

As we said, sunlight is made up of more than just the radiation that is visible to our eyes. Gamma rays, X-rays, ultraviolet radiation, infrared radiation, and radio waves are also produced by stars. Photographic film that is sensitive to ultraviolet and infrared radiation has been developed, thereby extending the limits of our vi-

BOX 22.1

UNDERSTANDING EARTH:
The Largest Optical Telescopes

The main focus of a telescope is to collect as much light as possible. The larger the telescope's lens or mirror, the more light it collects, allowing for viewing of fainter objects. Because an important focus of astronomy involves observing very distant and hence very faint cosmic sources, very large telescopes are essential.

Until recently, the largest telescopes were limited to mirrors about 5 meters (200 inches) in diameter because the task of casting, cooling, and polishing large glass mirrors to very fine tolerances (less than the thickness of a human hair) was enormously time consuming and expensive. For example, the construction of the 5-meter mirror for the Hale Telescope on Mount Palomar, California, began 1934 and was not completed until 1948. However, during the last decade, with the aid of high-tech manufacturing techniques, several large-diameter telescopes have been built, and several more are being planned.

Recently, two twin 8.1-meter telescopes went into operation. Gemini North is located in the Northern Hemisphere, on Hawaii's Mauna Kea. Located at an altitude of nearly 4,200 meters (13,800 feet), the Mauna Kea Observatory is the world's highest. Its twin, Gemini South, is located in Chile, on the western slopes of the Andes Mountains. Mauna Kea also houses the Japanese 8.3-meter Subaru telescope. (Subaru is the Japanese name for the Pleiades.)

A different innovative design is employed in the recently completed twin 10-meter Keck Telescope, also located on Mauna Kea (Figure 22.A). These instruments, operated by the California Institute of Technology and the University of California, use a mosaic of 36 six-sided 1.8-meter mirrors carefully positioned by the computer to give the optical effect of a 10-meter (400-inch) mirror.

To date, the largest optical telescope in terms of total light-gathering is the European Southern Observatory's Very Large Telescope (VLT), located at Cerro Paranal, Chile. It consists of four separate 8.2-meter instruments that work independently, or in conjunction with each other. When working in tandem, these telescopes have 10 times the light-gathering capacity of the 5-meter Hale Telescope and therefore can "see" cosmic objects that are 10 times dimmer.

Figure 22.A   Mirror of the 10-meter Keck Telescope. The mirror was constructed from 36 hexagonal segments. (Photo by Roger Ressmeyer/CORBIS)

Figure 22.11   Comparison of the light-gathering ability of two lenses.

Parallel rays from a distant star

A.

B.

Figure 22.12 Appearance of the galaxy in the constellation Andromeda using telescopes of different resolution. (Courtesy of AURA)

Figure 22.13 The deployment of the Hubble Space Telescope in Earth orbit, April 24, 1990, from Space Shuttle *Discovery*. (Courtesy of NASA)

sion. However, most of this radiation cannot penetrate our atmosphere, so balloons, rockets, and satellites must transport cameras "above" the atmosphere to record it.

Of great importance is a narrow band of radio radiation that does penetrate the atmosphere. One particular wavelength is the 21-centimeter (8-inch) line produced by neutral hydrogen (hydrogen atoms that lack an electrical charge). Measurement of this radiation has permitted us to map the galactic distribution of hydrogen, the material from which stars are made.

The detection of radio waves is accomplished by big dishes called **radio telescopes** (Figure 22.14A). In principle, the dish of one of these telescopes operates in the same manner as the mirror of an optical telescope. It is parabolic in shape and focuses the incoming radio waves on an antenna, which absorbs and transmits these waves to an amplifier, just like any radio antenna.

Because radio waves are about 100,000 times longer than visible radiation, the surface of the dish need not be as smooth as a mirror. In fact, except for the shortest radio waves, a wire mesh is a good reflector. On the other hand, because radio signals from celestial sources are very weak, large dishes are necessary to intercept an adequate signal. The largest radio telescope is a bowl-shaped antenna hung in a natural depression in Puerto Rico (Figure 22.15). It is 300 meters (1000 feet) in diameter and has some directional flexibility in its movable antenna. The largest steerable types have about 100-meter (330-foot) dishes like that currently being constructed at the National Radio Astronomy Observatory in Green Bank, West Virginia.

Radio telescopes also have rather poor resolution, making it difficult to pinpoint the radio source. Pairs or groups of telescopes are used to reduce this problem. When several radio telescopes are wired together, the resulting network is called a **radio interferometer** (Figure 22.14B).

Radio telescopes have some advantages over optical telescopes. They are much less affected by turbulence in the atmosphere, clouds, and the weather in general. No protective dome is required, which reduces the cost of construction, and "viewing" is possible 24 hours a day. More important, radio telescopes can "see" through interstellar dust clouds that obscure our view at visible wavelengths. Radio signals from distant points in the universe pass unhindered through the dust, giving us an unobstructed view. Furthermore, radio telescopes can detect clouds of gases too cool to emit visible light. These cold gas clouds are important because they are the sites of star formation.

Radio telescopes are, however, hindered by human-made radio interference. Thus, while optical telescopes are placed on remote mountaintops to reduce interference from city lights, radio telescopes are often hidden in valleys to block human-made radio interference.

Radio telescopes have revealed such spectacular events as the collision of two galaxies. Of even greater

A.   **B.**

Figure 22.14   A. The 43-meter (140-foot) steerable radio telescope at Green Bank, West Virginia. The dish acts like the mirror of a reflector-type optical telescope to focus radio waves onto the antenna. The antenna is the small round object supported on four "legs" above the dish. B. Twenty-seven identical antennas operate together to form the Very Large Array near Socorro, New Mexico. (Courtesy of National Radio Astronomy Observatory)

interest was the discovery of *quasars* (quasi-stellar radio sources). These perplexing objects are the most distant things known in the universe and will be examined further in Chapter 23.

The Sun

The Sun is one of the 200 billion stars that make up the Milky Way Galaxy. Although the Sun is of no significance to the universe as a whole, to us on Earth it is the primary source of energy (see Box 22.2). Everything from the fossil fuels we burn in our automobiles and power plants to the food we eat is ultimately derived from solar energy (Figure 22.16). The Sun is also important to astronomers, since it is the only star close enough to permit study of the surface. Even with the largest telescopes, other stars appear only as points of light.

Because of the Sun's brightness and its damaging radiation, it is not safe to observe it directly. However, a small telescope will project its image on a piece of cardboard held behind the telescope's eyepiece, and the Sun can be studied safely in this manner. This basic method is used in several telescopes around the world, which keep a constant vigil of the Sun. One of the finest is at the Kitt Peak National Observatory in southern Arizona (Figure 22.17). It consists of a 150-meter sloped enclosure that directs sunlight to a mirror situated below ground. From the mirror, an 85-centimeter (33-inch) image of the Sun is projected to an observing room, where it is studied.

Compared to other stars of the universe, many of which are larger, smaller, hotter, cooler, more red, or

Figure 22.15   The 300-meter (1000-foot) radio telescope at Arecibo, Puerto Rico. (Courtesy of National Astronomy and Ionosphere Center's Arecibo Observatory, operated by Cornell University under contract with the National Science Foundation)

BOX 22.2

EARTH AS A SYSTEM:
Variable Sun and Climatic Change

Among the most persistent hypotheses of climatic change have been those based on the idea that the Sun is a variable star and that its output of energy varies through time. The effect of such changes would seem direct and easily understood: Increases in solar output would cause the atmosphere to warm, and reductions would result in cooling. This notion is appealing because it can be used to explain climatic changes of any length or intensity. Still, there is at least one major drawback: No major long-term variations in the total intensity of solar radiation have yet been measured outside the atmosphere. Such measurements were not even possible until modern satellite technology became available. Now that it is possible, we will need many years of records before we begin to sense how variable (or invariable) energy from the Sun really is.

Several proposals for climatic change based on a variable Sun relate to *sunspot cycles.* The most conspicuous and best-known features on the surface of the Sun are the dark blemishes called *sunspots.* Although their origin is uncertain, it has been established that sunspots are huge magnetic storms that extend from the Sun's surface deep into the interior. Moreover, these spots are associated with the Sun's ejection of huge masses of particles that, on reaching Earth's atmosphere, interact with gases there to produce auroral displays.

Along with other solar activity, the number of sunspots seem to increase and decrease in a regular way, creating a cycle of about 11 years. The graph in Figure 22.B shows the annual number of sunspots, beginning in the early 1700s. However, this pattern is not always as regular as it appears.

There have been prolonged periods when sunspots have been absent or nearly so. Moreover, these events correspond closely with cold periods in Europe and North America. Conversely, periods characterized by plentiful sunspots have correlated well with warmer times in these regions.

Referring to these matches, some scientists have suggested that such correlations make it appear that changes on the Sun are an important cause of climate change. But other scientists seriously question this notion. Their hesitation stems in part from subsequent investigations using different climate records from around the world that failed to find a significant correlation between solar activity and climate. Even more troubling is that no testable physical mechanism exists to explain the purported effect.

Figure 22.B   Mean annual sunspot numbers.

Figure 22.16   The Sun is the source of more than 99 percent of all energy on Earth. (Photo by Thomas Dimock/The Stock Market)

Figure 22.17   The unique Robert J. McMath Solar Telescope at Kitt Peak, near Tucson, Arizona. Movable mirrors at the top follow the Sun, reflecting its light down the sloping tunnel. (Photo by Kent Wood/Photo Researchers, Inc.).

more blue, the Sun is an "average star." However, on the scale of our solar system, it is truly gigantic, having a diameter equal to 109 Earth diameters (1.35 million kilometers) and a volume of 1.25 million times as great as that of Earth. Yet, because of its gaseous nature, the density is only one quarter that of solid Earth, nearly the density of water.

Structure of the Sun

For convenience of discussion, we divide the Sun into four parts: the solar interior; the visible surface, or photosphere; and the two layers of its atmosphere, the chromosphere and the corona (Figure 22.18). Because the Sun is gaseous throughout, no sharp boundaries exist between these layers. The Sun's interior makes up all but a tiny fraction of the solar mass, and unlike the outer three layers, it is not accessible to direct observation. We shall discuss the visible layers first.

Photosphere. The **photosphere** (*photos* = light, *sphere* = a ball) is aptly named, for it radiates most of the sunlight we see and therefore appears as the bright disk of the Sun. Although it is considered to be the Sun's "surface," it is unlike most surfaces to which we are accustomed. The photosphere consists of a layer of incandescent gas less than 500 kilometers (300 miles) thick, having a pressure less than 1/100 of our atmosphere. Furthermore, it is neither smooth nor uniformly bright, as the ancients had imagined. It has numerous blemishes.

When viewed through a telescope under ideal conditions, the photosphere's grainy texture is apparent. This is the result of numerous relatively small, bright markings called **granules** (*granum* = small grain), which are surrounded by narrow, dark regions (Figure 22.19). Granules are typically the size of Texas, and they owe their brightness to hotter gases that are rising from below. As this gas spreads laterally, cooling causes it to darken and sink back into the interior. Each granule survives only 10 to 20 minutes, while the combined motion of new granules replacing old ones gives the photosphere the appearance of boiling. This up-and-down movement of gas is called *convection*. Besides producing the grainy appearance of the photosphere, convection is believed to be responsible for the transfer of energy in the uppermost part of the Sun's interior (Figure 22.18).

The composition of the photosphere is revealed by the dark lines of its absorption spectrum (see Figure 22.3). When these fingerprints are compared to the spectra of known elements, they indicate that most of the elements found on Earth also occur on the Sun. When the strengths of the absorption lines are analyzed, the relative abundance of the elements can be determined. These studies reveal that 90 percent of the Sun's surface atoms are hydrogen, almost 10 percent are helium, and

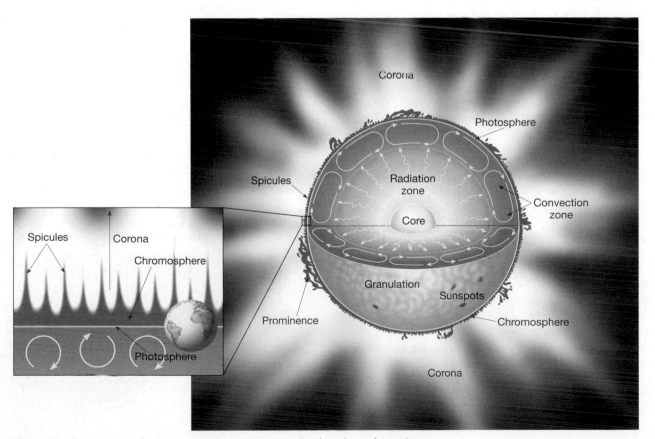

Figure 22.18   Diagram of solar structure in cutaway view. Earth is shown for scale.

Figure 22.19 Granules of the solar photosphere. The granules appear as yellowish-orange patches, while the dark areas shown are small sunspots. Each granule is about the size of Texas and lasts for only 10 to 20 minutes before being replaced by a new granule. (Courtesy of National Optical Astronomy Observatories)

only minor amounts of the other detectable elements are present. Other stars also indicate similar disproportionate percentages of these two lightest elements, a fact we shall consider later.

STUDENTS SOMETIMES ASK...

*If the Sun is an enormous ball
of hot gas, how can it have a surface?*

That's a good observation. Actually, the visible surface of the Sun, the photosphere, is a layer of gas about 500 kilometers (300 miles) thick. It is from this layer that we receive most sunlight. Although more light is emitted from the layer below the photosphere, that light is absorbed in the overlying layers of gas. Further, above the photosphere, the gas is less dense and thus unable to radiate much light. The photosphere is the layer that is dense enough to emit ample light yet has a density low enough to allow light to escape. Since the photosphere emits most of the light we see, it appears as the outermost surface of the Sun.

Chromosphere. Just above the photosphere lies the **chromosphere** (color sphere), a relatively thin layer of hot, incandescent gases a few thousand kilometers thick. The chromosphere is observable for a few moments during a total solar eclipse or by using a special instrument that blocks out the light from the photosphere. Under such conditions, it appears as a thin red rim around the Sun. Because the chromosphere consists of hot, incandescent gases under low pressure, it produces a bright-line spectrum that is nearly the reverse of the dark-line spectrum of the photosphere. One of the bright lines of hydrogen

contributes a good portion of its total light and accounts for this sphere's red color.

A study of the chromospheric spectrum conducted in 1868 revealed the existence of an unknown element. It was named helium, from *helios*, the Greek word for Sun. Originally, helium was thought to be an element unique to the stars, but 27 years later it was discovered in natural-gas wells on Earth.

The top of the chromosphere contains numerous **spicules** (*spica* = point), flamelike structures that extend upward as much as 10,000 kilometers into the lower corona, almost like trees that reach into our atmosphere (Figure 22.20). Spicules are produced by the turbulent motion of the granules below.

Corona. The outermost portion of the solar atmosphere, the **corona** (*corona* = crown), is very tenuous and, as with the chromosphere, is visible only when the brilliant photosphere is covered (Figure 22.21). This envelope of ionized gases normally extends a million kilometers from the Sun and produces a glow about half as bright as the full Moon.

At the outer fringe of the corona, the ionized gases have speeds great enough to escape the gravitational pull of the Sun. The streams of protons and electrons that boil from the corona constitute the **solar wind.** They travel outward through the solar system at very high speeds (250 to 800 kilometers a second) and eventually are lost to interstellar space. During its journey, the solar wind interacts with the bodies of the solar system, continually bombarding lunar rocks and altering their appearance. Although Earth's magnetic field prevents the solar winds from reaching our surface, these winds do affect our atmosphere, as we shall discuss later.

Studies of the energy emitted from the photosphere indicate that its temperature averages about 6000 K (10,000°F). Upward from the photosphere, the temperature unexpectedly increases, exceeding 1 million K at the top of the corona. It should be noted that although

Figure 22.20 Spicules of the chromosphere on the edge of the solar disk. (Courtesy of Sacramento Peak Observatory, Air Force Cambridge Research Laboratories)

Figure 22.21 Solar corona photographed during a total eclipse. (Photo by Lawrence Burr/Gamma Liaison)

the coronal temperature exceeds that of the photosphere by many times, it radiates much less energy overall because of its very low density.

Surprisingly, the high temperature of the corona is probably caused by sound waves generated by the convective motion of the photosphere. Just as boiling water makes noise, the energetic sound waves generated in the photosphere are believed to be absorbed by the gases of the corona and thereby raise their temperatures.

The Active Sun

Sunspots. The most conspicuous features on the surface of the Sun are the dark blemishes called **sunspots** (Figure 22.22A). Although large sunspots were occasionally observed before the advent of the telescope, they were generally regarded as opaque objects located somewhere between the Sun and Earth. In 1610 Galileo concluded that they were residents of the solar surface, and from their motion he deduced that the Sun rotates on its axis about once a month.

Observations made later indicated that not all parts of the Sun rotate at the same speed. The Sun's equator rotates once in 25 days, while a place located 70 degrees from the solar equator, whether north or south, requires 33 days for one rotation. If Earth rotated in a similar disjointed manner, imagine the consequences! The Sun's nonuniform rotation is a testimonial to its gaseous nature.

Sunspots begin as small dark pores about 1600 kilometers (1000 miles) in diameter. Although most pores last for only a few hours, some grow into blemishes many times larger than Earth and last for a month or more. The largest spots often occur in pairs surrounded by several smaller spots. An individual spot contains a black center, the *umbra*, (umbra = shadow), which is rimmed by a lighter region, the *penumbra* (paene = almost, umbra = shadow) (Figure 22.22B). Sunspots appear dark only by contrast with the brilliant photosphere, a fact accounted for by their temperature, which is about 1500 K less than that of the solar surface. If these

A.

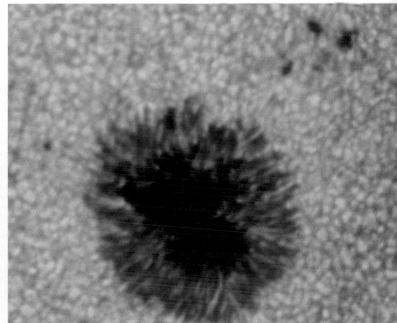

B.

Figure 22.22 A. Large sunspot group on the solar disk. (Celestron 8 photo courtesy of Celestron International) B. Sunspots having visible umbra (dark central area) and penumbra (lighter area surrounding umbra). (Courtesy of National Optical Astronomy Observatories)

dark spots could be observed away from the Sun, they would appear many times brighter than the full moon.

During the early nineteenth century, it was believed that a tiny planet named Vulcan orbited between Mercury and the Sun. In the search for Vulcan an accurate record of sunspot occurrences was kept. Although the planet was never found, the sunspot data collected did reveal that the number of sunspots observable on the solar disk varies in an 11-year cycle.

First, the number of sunspots increases to a maximum, with perhaps a hundred or more visible at a given time. Then, over a period of five to seven years, their numbers decline to a minimum, when only a few or even none are visible. At the beginning of each cycle, the first sunspots form about 30 degrees from the solar equator, but as the cycle progresses and their numbers increase, they form nearer the equator. During the period when sunspots are most abundant, the majority form about 15 degrees from the equator. They rarely occur more than 40 degrees away from the Sun's equator, or within 5 degrees of it.

Another interesting characteristic of sunspots was discovered by astronomer George Hale, for whom the Hale Telescope is named. Hale deduced that the large spots are strongly magnetized, and when they occur in pairs, they have opposite magnetic poles. For instance, if one member of the pair is a north magnetic pole, then the other member is a south magnetic pole, as with the north and south poles of Earth's magnetic field. Also, every pair located in the same hemisphere is magnetized in the same manner. However, all pairs in the *other* hemisphere are magnetized in the opposite manner. At the beginning of each sunspot cycle, the situation reverses, and the polarity of these sunspot pairs is opposite those of the previous cycle. The cause of this change

in polarity—in fact, the cause of sunspots themselves—is not fully explained. However, other solar activity varies in the same cyclic manner as sunspots do, indicating a common origin.

Prominences. Among the more spectacular features of the active Sun are **prominences** (*prominere* = to jut out). These huge cloudlike structures, consisting of concentrations of chromospheric gases, are best observed when they are on the edge, or limb, of the Sun, where they often appear as great arches that extend well into the corona (Figure 22.23). Many prominences have the appearance of a fine tapestry and seem to hang motionless for days at a time, but motion pictures reveal that the material within them is continually falling like luminescent rain. On the other hand, eruptive prominences rise almost explosively away from the Sun. These active prominences reach velocities up to 1000 kilometers (620 miles) per second and may leave the Sun entirely. Whether eruptive or quiescent, prominences are ionized chromospheric gases trapped by magnetic fields that extend from regions of intense solar activity.

Solar Flares. These are the most explosive events associated with sunspots. **Solar flares** are brief outbursts that normally last an hour or so and appear as a sudden brightening of the region above a sunspot cluster. During their existence, enormous quantities of energy are released, much of it in the form of ultraviolet, radio, and x-ray radiation. Simultaneously, fast-moving atomic particles are ejected, causing the solar wind to intensify noticeably. Although a major flare could conceivably endanger a manned space flight, they are relatively rare. About a day after a large outburst, the ejected particles reach Earth and

Figure 22.23 A huge solar prominence. (Sacramento Peak Observatory, Air Force Cambridge Research Laboratory)

Figure 22.24 Aurora borealis (Northern Lights) as seen from Alaska. The same phenomenon occurs toward the South Pole, where it is called the Aurora australis (Southern Lights). (Photo by Michio Hoshino/Minden Pictures)

disturb the ionosphere,* affecting long-distance radio communications.

The most spectacular effects of solar flares, however, are the **auroras,** also called the Northern and Southern Lights (Figure 22.24). Following a strong solar flare, Earth's upper atmosphere near its magnetic poles is set aglow for several nights. The auroras appear in a wide variety of forms. Sometimes the display consists of vertical streamers with considerable movement. At other times, the auroras appear as a series of luminous expanding arcs or as a quiet, almost foglike, glow. Auroral displays, like other solar activities, vary in intensity with the 11-year sunspot cycle.

The Solar Interior

The interior of the Sun cannot be observed directly. For that reason, all we know about it is based on information acquired from the energy it radiates and from theoretical studies. The source of the Sun's energy, **nuclear fusion** ($fusus$ = to melt), was not discovered until the late 1930s.

Deep in its interior, a nuclear reaction called the **proton-proton chain** converts four hydrogen nuclei (protons) into the nucleus of a helium atom. The energy released from the proton-proton reaction results because some of the matter involved is actually converted to energy. This can be illustrated by noting that four hydrogen atoms have a combined atomic mass of 4.032 (4×1.008), whereas the atomic mass of helium is 4.003, or 0.029 less than the combined mass of the hydrogen. The tiny missing mass is emitted as energy according to Einstein's formula $E = mc^2$, where E equals energy, m equals mass, and c equals the speed of light. Because the speed of light is very great, the amount of energy released from even a small amount of mass is enormous.

The conversion of just one pinhead's worth of hydrogen to helium generates more energy than burning thousands of tons of coal. Most of this energy is in the form of high-energy photons that work their way toward the solar surface, being absorbed and reemitted many times until they reach an opaque layer just below the photosphere. Here, convection currents serve to transport this energy to the solar surface, where it radiates through the transparent chromosphere and corona (see Figure 22.18).

Only a small percentage (0.7 percent) of the hydrogen in the proton-proton reaction is actually converted to energy. Nevertheless, the Sun is consuming an estimated 600 million tons of hydrogen each second, with about 4 million tons of it being converted to energy. As hydrogen is consumed, the product of this reaction, helium, forms the solar core, which continually grows in size.

Just how long can the Sun produce energy at its present rate before all of its fuel (hydrogen) is consumed? Even at the enormous rate of consumption, the Sun has enough fuel to last easily another 100 billion years. However, evidence from other stars indicates that the Sun will grow dramatically and engulf Earth long before all of its hydrogen is gone. It is thought that a star the size of the Sun can exist in its present stable state for 10 billion years. As the Sun is already 5 billion years old, it is "middle-aged."

To initiate the proton-proton reaction, the Sun's internal temperature must have reached several million degrees. What was the source of this heat? As previously noted, the solar system is believed to have formed from an enormous cloud of dust and gases (mostly hydrogen) that condensed gravitationally. The consequence of squeezing (compressing) a gas is to increase its temperature. Although all of the bodies in the solar system were compressed, the Sun was the only one, be-

---

*The ionosphere is a complex atmospheric zone of ionized gases extending between about 80 and 400 kilometers (50 and 250 miles) above Earth's surface.

cause of its size, that became hot enough to trigger the proton-proton reaction. Astronomers currently estimate its internal temperature at 15 million K.

The planet Jupiter is basically a hydrogen-rich gas ball; if it were about 10 times more massive, it too might have become a star. The idea of one star orbiting another may seem odd, but recent evidence indicates that about 50 percent of the stars in the universe probably occur in pairs or multiples!

Chapter Summary

• Visible light constitutes only a small part of an array of energy, generally referred to as *electromagnetic radiation*. Light, a type of electromagnetic radiation, can be described in two ways: (1) as waves and (2) as a stream of particles, called *photons*. The wavelengths of electromagnetic radiation vary from several kilometers for *radio waves* to less than a billionth of a centimeter for *gamma rays*. The shorter wavelengths correspond to more energetic photons.

• *Spectroscopy* is the study of the properties of light that depend on wavelength. When a prism is used to disperse visible light into its component parts (wavelengths), one of three possible types of *spectra* is produced (a *spectrum*, the singular form of *spectra*, is the light pattern produced by passing light through a prism). The three types of spectra are (1) *continuous spectrum*, (2) *dark-line (absorption) spectrum*, and (3) *bright-line (emission) spectrum*. The spectra of most stars are of the dark-line type. Spectroscopy can be used to determine (1) the state of matter of an object (solid, liquid, high- or low-pressure gas; (2) the composition of gaseous objects; (3) the temperature of a radiating body; and (4) the motion of an object. Motion (direction toward or away and velocity) is determined using the *Doppler effect*—the apparent change in the wavelength of radiation emitted by an object caused by the relative motions of the source and the observer.

• There are two types of optical telescopes: (1) the *refracting telescope*, which uses a *lens* as its *objective* to bend or refract light, so that it converges at an area called the *focus*; and (2) the *reflecting telescope*, which uses a *concave mirror* to focus (gather) the light. When examining an image directly, both types of telescopes require a second lens, called an *eyepiece*, which magnifies the image produced by the objective.

• Telescopes have three properties that aid astronomers: (1) *light-gathering power*, which is a function of the size of the objective—large objectives gather more light and therefore "see" farther into space; (2) *resolving power*, which allows for sharper images and finer details, is the ability of a telescope to separate objects that are close together, e.g., Pluto and its moon, Charon; and (3) *magnifying power*, the ability to make an object larger. Most modern telescopes have supplemental devices that enhance the image.

• Invisible radio-wave radiation is detected by "big dishes" called *radio telescopes*. A parabolic-shaped dish, often consisting of a wire mesh, operates in the same manner as the mirror of a reflecting telescope. Radio telescopes have poor resolution, making it difficult to pinpoint a radio source. To reduce this problem, several can be wired together into a network called a *radio interferometer*. The advantages of radio telescopes over optical telescopes are that radio telescopes are less affected by the weather, they are less expensive to construct, "viewing" is possible 24 hours a day, they can detect material in the universe too cool to emit visible radiation, and they can "see" through interstellar dust clouds.

• The *Sun* is one of the 200 billion stars that make up the Milky Way Galaxy. The Sun can be divided into four parts: (1) the *solar interior*, (2) the *photosphere* (visible surface), and the two layers of its atmosphere, (3) the *chromosphere* and (4) *corona*. The photosphere radiates most of the light we see. Unlike most surfaces, it consists of a layer of incandescent gas less than 500 kilometers (300 miles) thick, with a grainy texture consisting of numerous relatively small, bright markings called *granules*. Just above the photosphere lies the chromosphere, a relatively thin layer of hot, incandescent gases a few thousand kilometers thick. At the edge of the uppermost portion of the solar atmosphere, called the *corona*, ionized gases escape the gravitational pull of the Sun and stream toward Earth at high speeds, producing the *solar wind*.

• Numerous features have been identified on the active Sun. *Sunspots* are dark blemishes with a black center, the *umbra*, which is rimmed by a lighter region, the *penumbra*. The number of sunspots observable on the solar disk varies in an 11-year cycle. *Prominences*, huge cloudlike structures best observed when they are on the edge, or limb, of the Sun, are produced by ionized chromospheric gases trapped by magnetic fields that extend from regions of intense solar activity. The most explosive events associated with sunspots are *solar flares*. Flares are brief outbursts that release enormous quantities of energy that appear as a sudden brightening of the region above sunspot clusters. During the event, radiation and fast-moving atomic particles are ejected, causing the solar wind to intensify. When the ejected particles reach Earth and disturb the ionosphere, radio communication is disrupted and the *auroras*, also called the Northern and Southern Lights, occur.

• The source of the Sun's energy is *nuclear fusion*. Deep in the solar interior, at a temperature of 15 million K, a nuclear reaction called the *proton-proton chain* converts four hydrogen nuclei (protons) into the nucleus of a helium atom. During the reaction some of the matter is converted to the energy of the Sun. A star the size of the Sun can exist in its present stable state for 10 billion years. As the Sun is already 5 billion years old, it is a "middle-aged" star.

Key Terms

aurora (p. 619)
bright-line (emission) spectrum
   (p. 604)

chromatic aberration (p. 607)
chromosphere (p. 616)
continuous spectrum (p. 604)

corona (p. 616)
dark-line (absorption) spectrum
   (p. 604)

Review Questions

1. What term is used to describe the collection that includes gamma rays, X-rays, ultraviolet light, visible light, infrared radiation, and radio waves?

2. Which color has the longest wavelength? The shortest?

3. What is spectroscopy?

4. Describe a continuous spectrum. Give an example of a natural phenomenon that exhibits a continuous spectrum.

5. What produces emission lines (bright lines) in a spectrum?

6. What can be learned about a star (or other celestial objects) from a dark-line (absorption) spectrum?

7. How can astronomers determine whether a star is moving toward or away from Earth?

8. The primary objective of a refracting telescope is a mirror: true or false?

9. What three properties do telescopes have that aid astronomers?

10. Of the three properties listed in Question 9, which is the least important to astronomers?

11. What is the advantage of a space telescope over a similar instrument on Earth?

12. Why do astronomers seek to design telescopes with larger and larger objectives?

13. Why do all large optical telescopes use mirrors to collect light rather than lenses?

14. With reflecting telescopes, why are special viewing systems needed?

15. Explain the following statement: "Photography has extended the limits of our vision."

16. Why would the Moon make a good site for an observatory?

17. Why are radio telescopes much larger than optical telescopes?

18. What are some of the advantages of radio telescopes over optical telescopes?

19. Compare the diameter of the Sun to that of Earth.

20. Describe the photosphere, chromosphere, and corona.

21. List the features associated with the active Sun and describe each.

22. Explain how a sunspot can be very hot and yet appear dark.

23. What is the solar wind?

24. What "fuel" does the Sun consume?

25. What happens to the matter that is consumed in the proton-proton chain reaction?

Examining the Earth System

1. Of the two sources of energy that power the Earth system, the Sun is the main driver of Earth's external processes. If the Sun increased its energy output by 10 percent what would happen to global temperatures? What effect would this temperature change have on the percentage of water that exists as ice? What would be the impact on the position of the ocean shoreline? Speculate as to whether the change in temperature might produce an increase or decrease in the amount of surface vegetation? In turn, what impact might this change in vegetation have on the level of atmospheric carbon dioxide? How would such a change in the amount of carbon dioxide in the atmosphere affect global temperatures?

Web Resources

The *Earth Science* Website uses the resources and flexibility of the Internet to aid in your study of the topics in this chapter. Written and developed by Earth science instructors, this site will help improve your understanding of Earth science. Visit **http://www.prenhall.com/tarbuck** and click on the cover of *Earth Science 10e* to find:

- On-line review quizzes.
- Web-based critical thinking and writing exercises.
- Links to chapter-specific Web resources.
- Internet-wide key term searches.

http://www.prenhall.com/tarbuck

Stars embedded in clouds of dust and gases produce colorful emission nebulae. (Photo courtesy of Royal Observatory, Edinburgh/Photo Researchers, Inc.)

CHAPTER 23

BEYOND OUR SOLAR SYSTEM

The star Proxima Centauri is about 4.3 light-years away, roughly 100 million times farther than the Moon. Yet, other than our own Sun, it is the closest star to Earth. This fact suggests that the universe is incomprehensibly large. What is the nature of this vast cosmos beyond our solar system? Are the stars distributed randomly, or are they organized into distinct clusters? Do stars move, or are they permanently fixed features, like lights strung out against the black cloak of outer space? Does the universe extend infinitely in all directions, or is it bounded? To consider these questions, this chapter will examine the universe by taking a census of the stars—the most numerous objects in the night sky.

Properties of Stars

The Sun is the only star whose surface we can observe. Yet a great deal is known about the universe beyond our solar system. In fact, more is known about the stars than about our outermost planet, Pluto. This knowledge hinges on the fact that stars, and even gases in the "empty" space between stars, radiate energy in all directions into space (Figure 23.1). The key to understanding the universe is to collect this radiation and unravel the secrets it holds. Astronomers have devised many ingenious methods to do just that. We will begin by examining stellar distances and some intrinsic properties of stars, including *color, brightness, mass, temperature,* and *size.*

Measuring Distances to the Stars

Measuring the distance to a star is very difficult. Obviously, we cannot journey to the star. Nevertheless, astronomers have developed some *indirect* methods of measuring stellar distances. The most basic of these measurements is called stellar parallax.

Recall from Chapter 20 that **stellar parallax** is the extremely slight back-and-forth shifting in the apparent position of a nearby star due to the orbital motion of Earth. To review, the principle of parallax is easy to visualize. Close one eye, and with your index finger in a vertical position, use your eye to line up your finger with some distant object. Now, without moving your finger, view the object with your other eye and notice that its position appears to have changed. The farther away you hold your finger, the less its position seems to shift.

In practice, parallax is determined by photographing a nearby star against the background of distant stars. Then, six months later, when Earth has moved halfway around its orbit, a second photograph is taken. When these photographs are compared, the position of the nearby star appears to have shifted with respect to the background stars. Figure 23.2 illustrates this shift and the parallax angle determined from it. The nearest stars have the largest parallax angles, while those of distant stars are too slight to measure. Recall that the sixteenth-century astronomer Tycho Brahe was unable to detect stellar parallax, leading him to reject the idea that Earth orbits the Sun.

It should be emphasized that parallax angles are very small. The parallax angle to the nearest star, Proxima Centauri, is less than 1 second of arc, which equals 1/3600 of a degree. To put this in perspective, fully extend your arm and raise your little finger. Your finger is roughly 1 degree wide. Try doing this on a moonlit night, covering the Moon with your finger. The Moon is only about 1/2 degree wide. Now imagine detecting a movement that is only 1/3600 as wide as your finger. It

Figure 23.1   Lagoon Nebula. It is in glowing clouds like these that gases and dust particles become concentrated into stars. (Courtesy of National Optical Astronomy Observatories)

Original photo

Line of sight

Parallax angle

Earth's orbit

Sun

Apparent shift

Nearby star

Distant stars

Line of sight six months later

Photo taken 6 months later

Figure 23.2 Geometry of stellar parallax. The parallax angle shown here is enormously exaggerated to illustrate the principle. Because the distance to even the nearest stars is thousands of times greater than the Earth–Sun distance, the triangles that astronomers work with are quite long and narrow, making the angles that are measured very small.

should be apparent why Tycho Brahe, without the aid of a telescope, was unable to observe stellar parallax.

The distances to stars are so large that the conventional units, such as kilometers or astronomical units, are often too cumbersome to use. A better unit to express stellar distance is the **light-year,** which is the distance light travels in one Earth-year—about 9.5 trillion kilometers (5.8 trillion miles).

In principle, the method used to measure stellar distances is elementary and was known to the ancient Greeks. But in practice, measurements are greatly complicated because of the tiny angles involved and because the Sun, as well as the star being measured, also have actual motion through space. The first accurate stellar parallax was not determined until 1838. Even today, parallax angles for only a few thousand of the nearest stars are known with certainty.

Most stars have such small parallax shifts that accurate measurement is not possible. Fortunately, a few other methods have been derived for estimating distances to these stars. Also, the Hubble Space Telescope, which is not hindered by Earth's turbulent, light-distorting atmosphere, is expected to obtain accurate parallax distances for many additional remote stars.

Stellar Brightness

Three factors control the apparent brightness of a star as seen from Earth: *how big* it is, *how hot* it is, and *how far away* it is. The stars in the night sky are a grand assortment of sizes, temperatures, and distances, so their brightnesses vary widely.

Apparent Magnitude. Stars have been classified according to their brightness since the second century B.C., when Hipparchus placed about 1000 of them into six categories. The measure of a star's brightness is called its **magnitude.** Some stars may appear dimmer

than others only because they are farther away. Therefore, a star's brightness, *as it appears when viewed from Earth,* has been termed its **apparent magnitude.**

When numbers are employed to designate relative brightness, the larger the magnitude number, the dimmer the star. *Stars that appear the brightest are of the first magnitude, whereas the faintest stars visible to the unaided eye are of the sixth magnitude.* With the invention of the telescope, many stars fainter than the sixth magnitude were discovered.

In the mid-1800s, a method was developed for comparing the brilliance of stars using magnitude. Just as we can compare the brightness of a 50-watt lightbulb to that of a 100-watt bulb, we can compare the brightness of stars having different magnitudes. It was determined that a first-magnitude star was about 100 times brighter than a sixth-magnitude star. Therefore, on the scale that was devised, two stars that differ by 5 magnitudes have a ratio in brightness of 100 to 1. Hence, a seventh-magnitude star is 100 times brighter than a twelfth-magnitude star.

It follows, then, that the brightness ratio of two stars differing by only one magnitude is about 2.5.* Thus, a star of the first magnitude is about 2.5 times brighter than a star of the second magnitude. Table 23.1 shows some magnitude differences and the corresponding brightness ratios.

Because some stars are brighter than first-magnitude stars, zero and negative magnitudes were introduced. The brightest star in the night sky, Sirius, has an apparent magnitude of −1.4, about 10 times brighter than a first-magnitude star. On this scale, the Sun has an apparent magnitude of −26.7. At its brightest, Venus has a mag-

---

*Calculations: 2.512 × 2.512 × 2.512 × 2.512 × 2.512, or 2.512 raised to the fifth power, equals 100.

Table 23.1 Ratios of star brightness.	
Difference in Magnitude	**Brightness Ratio**
0.5	1.6:1
1	2.5:1
2	6.3:1
3	16:1
4	40:1
5	100:1
10	10,000:1
20	100,000,000:1

nitude of −4.3. At the other end of the scale, the 5-meter (200-inch) Hale Telescope can view stars with an apparent magnitude of 23, approximately 100 million times dimmer than stars that are visible to the unaided eye.

Absolute Magnitude. Astronomers are also interested in the "true" brightness of stars, called their **absolute magnitude**. Stars of the same luminosity, or brightness, usually do not have the same *apparent* magnitude, because their distances from us are not equal. To compare their true, or intrinsic, brightness, astronomers determine what magnitude the stars would have if they were at a standard distance of about 32.6 light-years. For example, the Sun, which has an apparent magnitude of −26.7, would, if located at a distance of 32.6 light-years, have an absolute magnitude of about 5. Thus, stars with absolute magnitudes greater than 5 (smaller numerical value) are intrinsically brighter than the Sun, but because of their distance, they appear much dimmer.

Table 23.2 lists the absolute and apparent magnitudes of some stars as well as their distances from Earth. Most stars have an absolute magnitude between −5 and 15, which puts the Sun near the middle of this range.

Stellar Color and Temperature

The next time you are outdoors on a clear night, take a good look at the stars and note their color (Figure 23.3). Some that are quite colorful can be found in the con-

Figure 23.3   Time-lapse photograph of stars in the constellation Orion. These star trails show some of the various star colors. It is important to note that the eye sees color somewhat differently than does photographic film. (Courtesy of National Optical Astronomy Observatories)

stellation Orion. Of the two brightest stars in Orion, Betelgeuse (α Orionis) is definitely red, whereas Rigel (β Orionis) appears blue.

Very hot stars with surface temperatures above 30,000 K emit most of their energy in the form of short-wavelength light and therefore appear blue. Red stars, on the other hand, are much cooler, generally less than 3000 K, and most of their energy is emitted as longer-wavelength red light. Stars with temperatures between 5000 and 6000 K appear yellow, like the Sun. Because color is primarily a manifestation of a star's temperature, this characteristic provides the astronomer with useful information about a star.

Binary Stars and Stellar Mass

One of the night sky's best-known constellations, the Big Dipper, appears at first glance to consist of seven stars. But anyone with good eyesight can resolve the second star in the handle of the Big Dipper (Mizar) as two stars. During the eighteenth century, astronomers used their new tool, the telescope, to discover numerous such star pairs. One of the stars in the pair is usually fainter than the other, and for this reason it was considered to be farther away. In other words, the stars were not considered true pairs but were thought only to lie in the same line of sight.

In the early nineteenth century, careful examination of numerous star pairs by William Herschel revealed that many stars actually orbit one another. The two stars are in fact united by their mutual gravitation. These pairs of stars, in which the members are far enough apart to be resolved telescopically, are called *visual binaries* (*binaries* = double). The idea of one star orbiting another may seem unusual, but evidence indicates that more than 50 percent of the stars in the universe occur in pairs or multiples.

Table 23.2 Distance, apparent magnitude, and absolute magnitude of some stars.			
Name	**Distance (Light-years)**	**Apparent Magnitude***	**Absolute Magnitude***
Sun	NA	−26.7	5.0
Alpha Centauri	4.27	0.0	4.4
Sirius	8.70	−1.4	1.5
Arcturus	36	−0.1	−0.3
Betelgeuse	520	0.8	−5.5
Deneb	1600	1.3	−6.9

*The more negative, the brighter; the more positive, the dimmer.

Binary stars are used to determine the star property most difficult to calculate—its mass. The mass of a body can be established if it is gravitationally attached to a partner, which is the case for any binary star system. Binary stars orbit each other around a common point called the *center of mass* (Figure 23.4). For stars of equal mass, the center of mass lies exactly halfway between them. If one star is more massive than its partner, their common center will be located closer to the more massive one. Thus, if the sizes of their orbits can be observed, a determination of their individual masses can be made. You can experience this relationship on a see-saw by trying to balance a person who has a much greater mass.

For illustration, when one star has an orbit half the size (radius) of its companion, it is twice as massive. If their combined masses are equal to three times the mass of the Sun, then the larger will be twice as massive as the Sun, and the smaller will have a mass equal to that of the Sun. Most stars have masses that range between 1/10 and 50 times the mass of the Sun.

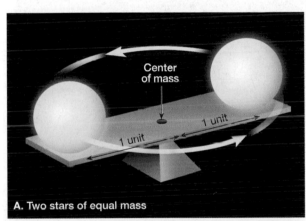

A. Two stars of equal mass

B. One star twice as massive as its companion

Figure 23.4 Binary stars orbit each other around their common center of mass. **A.** For stars of equal mass, the center of mass lies exactly halfway between them. **B.** If one star is twice as massive as its companion, it is twice as close to their common center of mass. Therefore, more massive stars have proportionately smaller orbits than do their less massive companions.

Hertzsprung-Russell Diagram

Early in the twentieth century, Einar Hertzsprung and Henry Russell independently studied the relation between the true brightness (absolute magnitude) and temperature of stars. From this research each developed a graph, now called a **Hertzsprung-Russell diagram, (H-R diagram),** that exhibits these intrinsic stellar properties. By studying H-R diagrams, we learn a great deal about the sizes, colors, and temperatures of stars.

To produce an H-R diagram, astronomers survey a portion of the sky and plot each star according to its luminosity (brightness) and temperature (Figure 23.5). Notice that the stars in Figure 23.5 are not uniformly distributed. Rather, about 90 percent of all stars fall along a band that runs from the upper-left corner to the lower-right corner of an H-R diagram. These "ordinary" stars are called **main-sequence stars.** As shown in Figure 23.5, *the hottest main-sequence stars are intrinsically the brightest, and vice versa.*

The luminosity of the main-sequence stars is also related to their mass. The hottest (blue) stars are about 50 times more massive than the Sun, whereas the coolest (red) stars are only 1/10 as massive. Therefore, on the H-R diagram, the main-sequence stars appear in a decreasing order, from *hotter, more massive* blue stars to *cooler, less massive* red stars.

Note the location of the Sun in Figure 23.5. The Sun is a yellow main-sequence star with an absolute magnitude of about 5. Because the magnitude of a vast majority of main-sequence stars lie between −5 and 15, and because the Sun falls midway in this range, the Sun is often considered an average star. However, more main-sequence stars are cooler and less massive than our Sun.

Just as all humans do not fall into the normal size range, some stars do not fit in with the main-sequence stars. Above and to the right of the main sequence in the H-R diagram in Figure 23.5 lies a group of very luminous stars called *giants,* or, on the basis of their color, **red giants.** The size of these giants can be estimated by comparing them with stars of known size that have the same surface temperature. We know that objects having equal surface temperatures radiate the same amount of energy per unit area. Therefore, any difference in the brightness of two stars having the same surface temperature is attributable to their relative sizes.

As an example, let us compare the Sun, which has been assigned a luminosity of 1, with another yellow star that has a luminosity of 100. Because both stars have the same surface temperature, they both radiate the same amount of energy per unit area. Therefore, for the more luminous star to be 100 times brighter than the Sun, it must have 100 times more surface area. It should be clear why stars whose plots fall in the upper-right position of an H-R diagram are called *giants.*

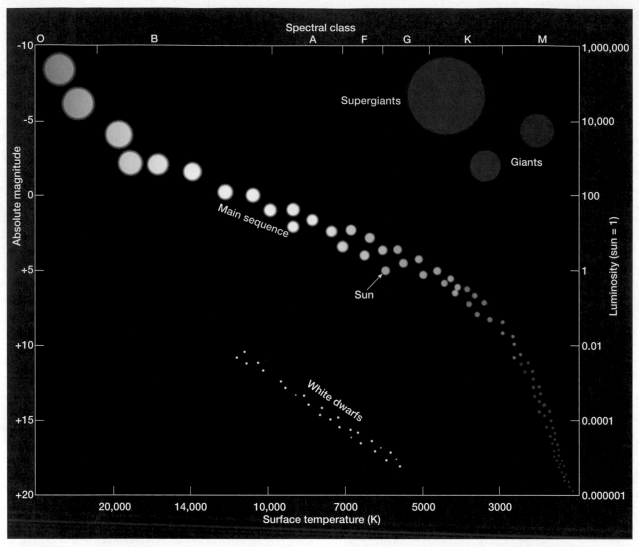

Figure 23.5 Idealized Hertzsprung-Russell diagram on which stars are plotted according to temperature and absolute magnitude.

Some stars are so large that they are called **super-giants.** Betelgeuse, a bright red supergiant in the constellation Orion, has a radius about 800 times that of the Sun. If this star were at the center of our solar system, it would extend beyond the orbit of Mars, and Earth would find itself inside the star! Other red giants that are easy to locate in our sky are Arcturus in the constellation Bootes and Antares in Scorpius.

In the lower-central portion of the H-R diagram, the opposite situation occurs. These stars are much fainter than main-sequence stars of the same temperature, and by using the same reasoning as before, they must be much smaller. Some probably approximate Earth in size. This group has come to be called **white dwarfs,** although not all are white.

Soon after the first H-R diagrams were developed, astronomers realized their importance in interpreting stellar evolution. Just as with living things, a star is born, ages, and dies. Owing to the fact that almost 90 percent

of the stars lie on the main sequence, we can be relatively certain that stars spend most of their active years as main-sequence stars. Only a few percent are giants, and perhaps 10 percent are white dwarfs.

After considering some variable stars and the nature of interstellar matter, we will return to the topic of stellar evolution.

Variable Stars

Stars that fluctuate in brightness are known as variables. Some, called **pulsating variables,** fluctuate regularly in brightness by expanding and contracting in size. The importance of one member of this group (cepheid variables) in determining stellar distances is discussed in Box 23.1. Other pulsating variables have irregular periods and are of no value for this purpose.

The most spectacular variables belong to a group known as **eruptive variables.** When one of these ex-

UNDERSTANDING EARTH:
Determining Distance from Magnitude

For a star too distant for parallax measurements, knowing its absolute and apparent brightness provides astronomers with a tool for approximating its distance. The apparent magnitude is measured with a photometer (light meter) attached to a telescope. If we also know a star's true brightness, we can determine just how far away that star would have to be for it to have the brightness we observe.

You use this same principle when you drive at night and estimate the distance to an oncoming car from the brightness of its headlights. You can do this because you know the true brightness of an automobile headlight. But how do astronomers determine the intrinsic brightness of a star? Fortunately, some stars have characteristics that provide the necessary data.

One important star group is called *cepheid variables*. These are pulsating stars that get brighter and fainter in a rhythmic fashion. The interval between two successive occurrences of maximum brightness of a pulsating variable is termed its *light period*. Most cepheid variables pulsate with periods of between two and 50 days. For example, the North Pole Star (Polaris) varies about 10 percent in brightness over a period of four days. In general, the longer the light period of a cepheid, the greater is its absolute magnitude (Figure 23.A). Thus, by determining the light period of a cepheid, its absolute magnitude can be calculated. When this absolute magnitude is compared to the apparent magnitude, a good approximation of its distance can be made.

Figure 23.A Relationship between the light period (two successive occurrences of maximum brightness) and absolute magnitude of pulsating stars (cepheid variables).

plosive events associated with these stars occurs, it appears as a sudden brightening of a star, called a **nova** (Figure 23.6). The term *nova*, meaning "new," was used by the ancients because these stars were unknown to them before their abrupt increase in luminosity.

During the outburst, the outer layer of the star is ejected at high speed. The "cloud" of ejected material occasionally can be captured photographically (Figure 23.7). A nova generally reaches maximum brightness in a couple of days, remains bright for only a few weeks, then slowly returns in a year or so to its original brightness. Because the star returns to its prenova brightness, we can assume that only a small amount of its mass is lost during the flareup. Some stars have experienced more than one such event. In fact, the process probably occurs repeatedly.

The modern explanation of novae proposes that they occur in binary systems consisting of an expanding red giant and a hot white dwarf that are in close proximity. Hydrogen-rich gas from the oversized giant encroaches near enough to the white dwarf to be gravitationally transferred. Eventually, enough of the hydrogen-rich gas is added to the dwarf to cause it to ignite explosively. Such a thermonuclear reaction rapidly heats and expands the outer gaseous envelope of the hot dwarf to produce a nova event. In a relatively short time, the white dwarf returns to its pre-nova state, where it remains inactive until the next buildup occurs.

Interstellar Matter

Between existing stars is "the vacuum of space." However, it is far from being a pure vacuum, for it is populated with accumulations of dust and gases. The name applied to these concentrations of interstellar matter is **nebula** (*nebula* = cloud). If this interstellar matter is close to a very hot (blue) star, it will glow and is called a **bright nebula.** The two main types of bright nebulae are known as *emission nebulae* and *reflection nebulae*.

Emission nebulae are gaseous masses that consist largely of hydrogen. They absorb *ultraviolet radiation* emitted by an embedded or nearby hot star. Because these gases are under very low pressure, they reradiate, or emit, this energy as *visible light*. This conversion of ultraviolet light to visible light is known as *fluorescence*, an effect you observe daily in fluorescent lights. A well-known emission nebula easily seen with binoculars is the sword of the hunter in the constellation Orion (Figure 23.8).

Reflection nebulae, as the name implies, merely reflect the light of nearby stars (Figure 23.9). Reflection nebulae are thought to be composed of rather dense clouds of large particles called **interstellar dust.** This

March 10, 1935

May 6, 1935

Figure 23.6   Photographs of Nova Herculis (a nova in the constellation Hercules), taken about 2 months apart, showing the decrease in brightness. (Courtesy of Lick Observatory)

view is supported by the fact that atomic gases with low densities could not reflect light sufficiently to produce the glow observed.

When a dense cloud of interstellar material is not close enough to a bright star to be illuminated, it is referred to as a **dark nebula.** Exemplified by the Horsehead Nebula in Orion, dark nebulae appear as opaque objects silhouetted against a bright background (Figure 23.10). Dark nebulae can also easily be seen as starless regions—holes in the heavens when viewing the Milky Way.

Although nebulae appear very dense, they actually consist of very thinly scattered matter. Because of their enormous size, however, the total mass of rarefied particles and molecules may be many times that of the Sun. Interstellar matter is of great interest to astronomers because it is from this material that stars and planets are formed.

Figure 23.7   Expanding "cloud" around Nova Persei. (Courtesy of Hale Observatories. Copyright by the California Institute of Technology and Carnegie Institution of Washington)

Stellar Evolution

The idea of describing how a star is born, ages, and then dies may seem a bit presumptuous, for many of these objects have life spans that surely exceed billions of years. However, by studying stars of different ages, astronomers have been able to piece together a plausible model for stellar evolution. The method that was used to create this model is analogous to what an alien being, upon reaching Earth, might do to determine the developmental stages of human life. By examining a large number of humans, this stranger would be able to observe the birth of human life, the activities of children and adults, and the death of the elderly. From this information, the alien would attempt to put the stages of human development into their proper sequence. Based on the relative abundance of humans in each stage of development, it would even be possible to conclude that humans spend more of their lives as adults than as toddlers. In a similar fashion, astronomers have pieced together the story of the stars.

Simply, stars exist because of gravity. The mutual gravitational attraction of particles in a thin, gaseous cloud causes the cloud to collapse. As the cloud is squeezed to unimaginable pressures, its temperature rises, igniting its nuclear furnace, and a star is born. A star is a ball of very hot gases, caught between the opposing forces of gravity trying to contract it and thermal nuclear energy trying to expand it. Eventually, all of a star's nuclear fuel will be exhausted and gravity takes over, collapsing the stellar remnant into a small, dense body.

Figure 23.8 The Orion Nebula is a well-known emission nebula. Bright enough to be seen by the naked eye, the Orion Nebula is located in the sword of the hunter in the constellation of the same name. (Courtesy of National Optical Astronomy Observatories)

Figure 23.9 A faint blue reflection nebula, in the Pleiades star cluster, is caused by the reflection of starlight from dust in the nebula. The Pleiades star cluster, just visible to the naked eye in the constellation Taurus, is spectacular when viewed through binoculars or a small telescope. (Palomar Observatories/California Institute of Technology [Caltech])

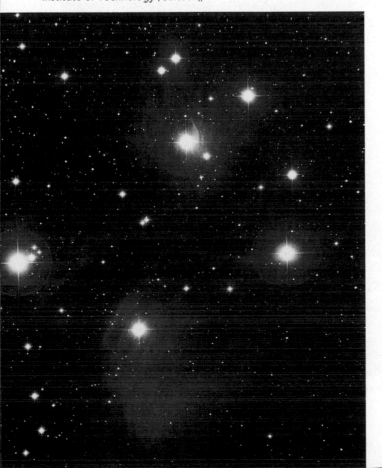

Star Birth

The birthplaces of stars are dark, cool interstellar clouds, which are comparatively rich in dust and gases (Figure 23.11). In the neighborhood of the Milky Way, these gaseous clouds consist of 92 percent hydrogen, 7 percent helium, and less than 1 percent of the remaining heavier elements. By some mechanism not yet fully understood, these thin gaseous clouds become concentrated enough to begin to contract gravitationally. One proposal to explain the triggering of stellar formation is a shock wave traveling from a catastrophic explosion (supernova) of a nearby star. But, regardless of the force that initiates the concentration of interstellar matter, once it is accomplished, mutual gravitational attraction of the particles squeeze the cloud, pulling every particle toward the center. As the cloud shrinks, gravitational energy (potential energy) is converted into energy of motion, or heat energy.

The initial contraction spans a million years or so. With the passage of time, the temperature of this gaseous body slowly rises, eventually reaching a temperature sufficiently high to cause it to radiate energy from its surface in the form of long-wavelength red light. Because this large red object is not hot enough to engage in nuclear fusion, it is not yet a star. The name **protostar** is applied to these bodies.

Protostar Stage

During the protostar stage, gravitational contraction continues, slowly at first, then much more rapidly (Figure 23.11). This collapse causes the core of the developing star to heat much more intensely than the outer envelope. When the core has reached at least 10 million

631

Figure 23.10 The Horsehead Nebula, a dark nebula in a region of glowing nebulosity in Orion. (© Anglo-Australian Observatory/Royal Observatory Edinburgh, photography by David Malin)

Figure 23.11 Diagram of stellar evolution of H-R diagram for a star about as massive as the Sun.

K, the pressure within is so great that groups of four hydrogen nuclei are fused together into single helium nuclei. Astronomers refer to this nuclear reaction as **hydrogen burning** because an enormous amount of energy is released. However, keep in mind that thermonuclear "burning" is not burning in the usual chemical sense.

Heat from hydrogen fusion causes the stellar gases to increase their motion. This in turn results in an increase in the outward gas pressure. At some point, this outward pressure exactly balances the inward force of gravity. When this balance is reached, the star becomes a stable main-sequence star (Figure 23.11). Stated another way, a stable main-sequence star is balanced between two forces: *gravity*, which is trying to squeeze it into a smaller sphere, and *gas pressure*, which is trying to expand it.

Main-Sequence Stage

From this point in the evolution of a main-sequence star until its death, the internal gas pressure struggles to offset the unrelenting force of gravity. Typically, hydrogen burning continues for a few billion years and provides the outward pressure required to support the star from gravitational collapse.

Different stars age at different rates. Hot, massive blue stars radiate energy at such an enormous rate that they substantially deplete their hydrogen fuel in only a few million years. By contrast, the very smallest (red) main-sequence stars may remain stable for hundreds of billions of years. A yellow star, such as the Sun, remains a main-sequence star for about 10 billion years. As the solar system is about 5 billion years old, it is comforting to know that the Sun will remain stable for another 5 billion years.

An average star spends 90 percent of its life as a hydrogen-burning main-sequence star. Once the hydrogen fuel in the star's core is depleted, it evolves rapidly and dies. However, with the exception of the least-massive (red) stars, a star can delay its death by burning other fuels and becoming a giant.

Red Giant Stage

The evolution to the red giant stage results because the zone of hydrogen burning continually migrates outward, leaving behind an inert helium core. Eventually, all of the hydrogen in the star's core is consumed. While hydrogen fusion is still progressing in the star's outer shell, no fusion is taking place in the core. Without a source of energy, the core no longer has enough pressure to support itself against the inward force of gravity. As a result, the core begins to contract.

Although the core cannot generate nuclear energy, it does grow hotter by converting gravitational energy into heat energy. Some of this energy is radiated outward, initiating a more vigorous level of hydrogen fu-

sion in the star's outer shell. This energy in turn heats and enormously expands the star's outer envelope, producing a giant body hundreds to thousands of times its main-sequence size (Figure 23.11).

As the star expands, its surface cools, which explains the star's reddish appearance. Eventually the star's gravitational force will stop this outward expansion. Once again, the two opposing forces, gravity and gas pressure, will be in balance, and this gaseous mass will be a stable but much larger star. Some red giants overshoot the equilibrium point and rebound like an overextended spring. Such stars continue to oscillate in size, becoming variable stars.

While the envelope of a red giant expands, the core continues to collapse and heat until it reaches 100 million K. At this incredible temperature, it is hot enough to initiate a nuclear reaction in which helium is converted to carbon. Thus, a red giant consumes both hydrogen and helium to produce energy. In stars more massive than the Sun, still other thermonuclear reactions occur that generate all the elements on the periodic table up to number 26, iron. Nuclear "burning" of elements heavier than iron requires an additional source of energy to keep the reaction progressing. Hence, these elements are not produced in ordinary stars.

Eventually, all the usable nuclear fuel in these giants will be consumed. The Sun, for example, will spend less than a billion years as a giant, and the more massive stars will pass through this stage even more rapidly. The force of gravity will again control the star's destiny as it squeezes the star into the smallest, most dense piece of matter possible.

Burnout and Death

Most of the events of stellar evolution discussed thus far are well documented. What happens to a star after the red giant phase is more speculative. We do know that a star, regardless of its size, must eventually exhaust all of its usable nuclear fuel and collapse in response to its immense gravitational force. With this in mind, we will now consider the final stage of stars in three different mass categories.

Death of Low-Mass Stars. Stars less than one half the mass of the Sun (0.5 solar mass) consume their fuel at a comparatively slow rate (Figure 23.12A). Consequently, these small, *cool red stars* may remain on the main sequence for up to 100 billion years. Because the interior of a low-mass star never attains sufficiently high temperatures and pressures to fuse helium, its only energy source is hydrogen fusion. Thus, low-mass stars never evolve to become bloated red giants. Rather, they remain as stable main-sequence stars until they consume their hydrogen fuel and collapse into a hot, dense *white dwarf*. As we shall see, white dwarfs are small, compact objects unable to support nuclear burning.

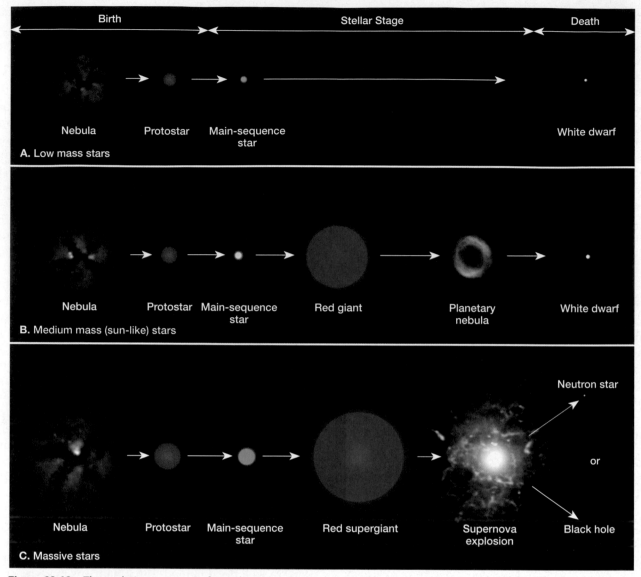

Figure 23.12   The evolutionary stages of stars having various masses.

Death of Medium-Mass (Sun-like) Stars. Main-sequence stars with masses ranging between half that of the Sun and three times that of the Sun evolve in essentially the same way (Figure 23.12B). During their giant phase, Sun-like stars fuse hydrogen and helium fuel at an accelerated rate. Once this fuel is exhausted, these stars (like low-mass stars) collapse into an Earth-size body of great density—a white dwarf. The density of a white dwarf is as great as physics will allow, short of destroying protons and electrons.

The gravitational energy supplied to a collapsing white dwarf is reflected in its high surface temperature. However, without a source of nuclear energy, a white dwarf becomes cooler and dimmer as it continually radiates its remaining thermal energy into space.

During their collapse from red giants to white dwarfs, medium-mass stars are thought to cast off their bloated outer atmosphere, creating an expanding spherical cloud of gas. The remaining hot, central white dwarf heats the gas cloud, causing it to glow. These often beautiful, gleaming spherical clouds are called **planetary nebulae.** A good example of a planetary nebula is the Helix Nebula in the constellation Aquarius (Figure 23.13). This nebula appears as a ring because our line of sight through the center traverses less gaseous material than at the nebula's edge. It is, nevertheless, spherical in shape.

Death of Massive Stars. In contrast to Sun-like stars, which expire gracefully, stars exceeding three solar masses have relatively short life spans and terminate in a brilliant explosion called a **supernova** (Figure 23.12C). During a supernova event, a star becomes millions of times brighter than its prenova stage (see Box 23.2). If one of the nearest stars to Earth produced such an outburst, its brilliance would surpass that of the Sun. Supernovae are rare; none have been observed in our galaxy since the advent of the telescope, although Tycho Brahe and Galileo each recorded one about 30 years apart. An even larger supernova was recorded in A.D. 1054 by the Chinese. Today, the remnant of this great outburst is the Crab nebula, shown in Figure 23.14.

A supernova event is thought to be triggered when a massive star consumes most of its nuclear fuel. Without a heat engine to generate the gas pressure required to balance its immense gravitational field, it collapses. This implosion is of cataclysmic proportion, resulting in a shock wave that moves out from the star's interior. This energetic shock wave destroys the star and blasts the outer shell into space, generating the supernova event.

Theoretical work predicts that during a supernova, the star's interior condenses into a very hot object, possibly no larger than 20 kilometers in diameter (see Box 23.3). These incomprehensibly dense bodies have been named *neutron stars*. Some supernovae events are thought to produce even smaller and most intriguing objects called *black holes*. We will consider the nature of neutron stars and black holes in the Stellar Remnants section.

H-R Diagrams and Stellar Evolution

The Hertzsprung-Russell diagrams have been very helpful in formulating and testing models of stellar evolution. They are also useful for illustrating the changes that take place in an individual star during its life span. Figure 23.11 shows on an H-R diagram the evolution of a star about the size of the Sun. Keep in mind that the star does not physically move along this path, but rather that its position on the H-R diagram represents the color

STUDENTS SOMETIMES ASK...
How will the Sun die?

In a few billion years, the Sun will exhaust the remaining hydrogen fuel in its core, an event that will trigger hydrogen fusion in the surrounding shell. As a result, the Sun's outer envelope will expand, producing a red giant that is hundreds of times larger and more luminous. The intense solar radiation will cause Earth's oceans to boil, and the solar winds will drive away Earth's atmosphere. In another billion years, the Sun will expel its outermost layer, producing a spectacular planetary nebula, while its interior will collapse to form a dense, small (planet size), white dwarf. Because of its small size, the Sun's energy output will be less than 1 percent of its current level. Gradually, the Sun will emit its remaining thermal energy, eventually becoming a cold, nonluminous body.

(temperature) and absolute magnitude (brightness) of the star at various stages in its evolution.

For example, on the H-R diagram, a protostar would be located to the right and above the main sequence (Figure 23.11). It is formed to the right because of its relatively cool surface temperature (red color), and above because it would be more luminous than a main-sequence star of the same color, a fact attributable to its large size. Careful examination of Figure 23.11 should help you visualize the evolutionary changes experienced by a star the size of our Sun. In addition, Table 23.3 provides a summary of the evolutionary history of stars having various masses.

UNDERSTANDING EARTH:
Supernova 1987A

The first naked-eye supernova in 383 years was discovered in the southern sky in February 1987 (Figure 23.B). This stellar explosion was officially named SN 1987A (SN stands for supernova, and 1987A indicates that it was the first supernova observed in 1987). Naked-eye supernovae are rare. Only a few have been recorded in historic times. Arab observers saw one in 1006, and the Chinese recorded one in 1054 at the present location of the Crab Nebula. In addition, the astronomer Tycho Brahe observed a supernova in 1572, and Kepler saw one shortly thereafter in 1604.

Prior to this event, researchers could only test their hypotheses on dim supernovae seen in distant galaxies. Thus, when SN 1987A occurred, astronomers quickly focused every available telescope in the Southern Hemisphere on this spectacular event. As one astronomer remarked, "This supernova is better studied by far than any supernova in history." More important, this event has allowed astronomers to use observational data to test their theoretical models of stellar evolution.

Supernova 1987A occurred about 170,000 light-years away in the Large Magellanic Cloud, a satellite galaxy to our own Milky Way. As expected, the supernova rapidly increased in brightness to a peak magnitude of 2.4, outshining all the other stars in the Large Magellanic Cloud. Also as predicted, within a few weeks it began to fade. However, SN 1987A did provide some surprises.

From old photographs taken of the area, researchers identified the exploded star as Sanduleak. Astronomers were surprised to find that the parent star was a hot blue star about 15 times the mass of our Sun. Recall that only cool red giants are thought to die as a supernova event. Further, the Hubble Space Telescope made another unexpected discovery. Its camera revealed a very large shell of gas that predates the supernova explosion by about 40,000 years.

Astronomers now think that Sanduleak was once a red supergiant that had blown away its outer shell, exposing a hot blue core. It is this ejected outer shell that appears in the image produced by the Hubble Space Telescope. Then, some 40,000 years later, the remaining hot core of the red supergiant collapsed, producing the supernova of 1987.

Despite these twists, the theory of stellar evolution has held up very well. Theory predicts that the expanding remnants to Supernova 1987A will be large enough to be observed by the turn of the twenty-first century. Thus, astronomers continue to monitor SN 1987A to unravel its secrets and to confirm or refute their ideas about the final stages of stellar evolution.

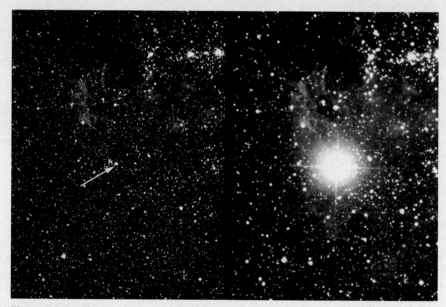

Figure 23.B The great Supernova 1987A. The photo on the left was made prior to the supernova and the one on the right was made following the event. (© Anglo-Australian Observatory, photography by David Malin)

Stellar Remnants

Eventually, all stars consume their nuclear fuel and collapse into one of three final states—white dwarf, neutron star, or black hole. Although different in some ways, these small, compact objects are all composed of incomprehensibly dense material and all have extreme surface gravity.

White Dwarfs

White dwarfs are extremely small stars with densities greater than any known terrestrial material. It is believed that white dwarfs once were low-mass or medium-mass stars whose internal heat was able to keep these gaseous bodies from collapsing under their own gravitational force.

Although some white dwarfs are no larger than Earth, the mass of such a dwarf can equal 1.4 times that of the Sun. Thus, their densities may be a million times greater than water. A spoonful of such matter would weigh several tons. Densities this great are possible only when electrons are displaced inward from their regular orbits, around an atom's nucleus, allowing the atoms to take up less than the "normal" amount of space. Material in this state is called **degenerate matter.**

In degenerate matter, the atoms have been squeezed together so tightly that the electrons are displaced much nearer to the nucleus. Degenerate matter

Figure 23.14 Crab Nebula in the constellation Taurus: the remains of the supernova of A.D. 1054. (UCO Lick Observatory Image)

uses electrical repulsion rather than molecular motion to support itself from total collapse. Although atomic particles in degenerate matter are much closer together than in normal Earth matter, they still are not packed as tightly as possible. Stars made of matter that has an even greater density are thought to exist.

As a star contracts into a white dwarf, its surface becomes very hot, sometimes exceeding 25,000 K. Even so, without a source of energy, it can only become cooler and dimmer. Although none have been observed, the terminal stage of a white dwarf must be a small, cold, nonluminous body called a *black dwarf.*

Neutron Stars

A study of white dwarfs produced what might at first appear to be a surprising conclusion. The smallest white dwarfs are the most massive, and the largest are the least massive. The explanation for this is that a more massive star, because of its greater gravitational force, is able to squeeze itself into a smaller, more densely packed object than can a less massive star. Thus, the smallest white dwarfs were produced from the collapse of larger, more massive stars than were the larger white dwarfs.

This conclusion led to the prediction that stars smaller and more massive than white dwarfs must exist.

Named **neutron stars,** these objects are thought to be the remnants of supernova events. In a white dwarf, the electrons are pushed close to the nucleus, whereas in a neutron star the electrons are forced to combine with protons to produce neutrons (hence the name). If Earth were to collapse to the density of a neutron star, it would have a diameter equivalent to the length of a football field. A pea-size sample of this matter would weigh 100 million tons. This is approximately the density of an atomic nucleus; thus, neutron stars can be thought of as large atomic nuclei.

During a supernova implosion, the envelope of the star is ejected (Figure 23.15), while the core collapses into a very hot neutron star about 20 kilometers (12.4 miles) in diameter. Although neutron stars have high surface temperatures, their small size would greatly limit their luminosity. Consequently, locating one visually would be extremely difficult.

However, theory predicts that a neutron star would have a very strong magnetic field. Further, as a star collapses, it will rotate faster, for the same reason ice skaters rotate faster as they pull in their arms. If the Sun were to collapse to the size of a neutron star, it would increase its rate of rotation from once every 25 days to nearly 1000 times per second. Radio waves generated by these rotating stars would be concentrated into two narrow zones that would align with the star's magnetic poles. Consequently, these stars would resemble a rapidly rotating beacon emitting strong radio waves. If Earth happened to be in the path of these beacons, the star would appear to blink on and off, or pulsate, as the waves swept past.

In the early 1970s, a source that radiates short pulses of radio energy, called a **pulsar** (pulsating radio source), was discovered in the Crab Nebula. Visual inspection of this radio source revealed it to be a small star centered in the nebula. The pulsar found in the Crab Nebula is undoubtedly the remains of the supernova of A.D. 1054 (see Figure 23.14). Thus, the first neutron star had been discovered.

Black Holes

Are neutron stars made of the most dense materials possible? No. During a supernova event, remnants of stars greater than three solar masses apparently collapse into objects even smaller and denser than neutron stars. Even though these objects would be very hot, their surface gravity would be so immense that even light could not escape the surface. Consequently, they would literally disappear from sight. These incredible bodies have appropriately been named **black holes.** Anything that moved too near a black hole would be swept in by its irresistible gravity and be devoured forever.

How can astronomers find an object whose gravitational field prevents the escape of all matter and energy? One strategy is to seek evidence of matter being

EARTH AS A SYSTEM:
From Stardust to You

During a supernova implosion, the internal temperature of a star may reach 1 billion K, a condition thought to produce very heavy elements such as gold and uranium. These heavy elements, plus the debris of novae and the planetary nebulae, are continually returned to interstellar space where they are available for the formation of other stars (Figure 23.C).

Astronomers believe that the earliest stars were made of nearly pure hydrogen. Fusion during the life and death of stars in turn produced heavier elements, some of which were returned to space. Because the Sun contains some heavy elements but has not yet reached the stage in its evolution where it could have produced them, it must be at least a second-generation star. Thus, our Sun, as well as the rest of the solar system, is believed to have formed from debris scattered from preexisting stars. If this is the case, the atoms in your body were produced billions of years ago inside a

Figure 23.C Eagle Nebula in the constellation Serpens. This gaseous nebula is the site of a recent star formation. (Courtesy of National Optical Astronomy Observatories)

star, and the gold in your jewelry was formed during a supernova event that occurred trillions of kilometers away. Without such events, the development of life on Earth would not have been possible.

---

rapidly swept into a region of apparent nothingness. Theory predicts that as matter is pulled into a black hole, it should become very hot and emit a flood of X-rays before being engulfed. Because isolated black holes would not have a source of matter to engulf, astronomers first looked at binary-star systems.

A likely candidate for a black hole is Cygnus X-1, a strong x-ray source in the constellation Cygnus. In this case, the x-ray source can be observed orbiting a supergiant companion with a period of 5.6 days. It appears that gases are pulled from this companion and spiral into the disc-shaped structure around the black hole (Figure 23.16). The result is a stream of X-rays. Because X-rays cannot penetrate our atmosphere efficiently, the existence of black holes was not confirmed until recently. The first x-ray sources were discovered in

Initial Mass of Interstellar Cloud (Sun = 1)	Main-Sequence Stage	Giant Phase	Evolution After Giant Phase	Terminal State (Final Mass*)
0.001	None (Planet)	No	None	Planet (0.001)
0.1	Red	No	None	White dwarf (0.1)
1–3	Yellow	Yes	Planetary nebula	White dwarf (< 1.4)
6	White	Yes	Supernova	Neutron star (1.4–3)
20	Blue	Yes (Supergiant)	Supernova	Black hole (> 3.0)

Table 23.3 Summary of evolution for stars of various masses.

*These mass numbers are estimates.

Figure 23.15 Veil Nebula in the constellation Cygnus is the remnant of an ancient supernova implosion. (Palomar Observatories/California Institute of Technology [Caltech])

1971 by detectors on satellites. Cygnus X-1 is such a source.

The Milky Way Galaxy

On a clear and moonless night away from city lights, you can see a truly marvelous sight—our own Milky Way Galaxy (Figure 23.17). With his telescope, Galileo discovered that this band of light was produced by countless individual stars that the unaided eye is unable to resolve. Today we realize that the Sun is actually a part of this vast system of stars, which number about 100 billion (Figure 23.18A). The milky appearance of our galaxy results because the solar system is located within the flat *galactic disk.* Thus, when it is viewed from the inside, a higher concentration of stars appears in the direction of the galactic plane than in any other direction. You can see this in the edge-on view in Figure 23.18B.

When astronomers began telescopically to survey the stars located along the plane of the Milky Way, it appeared that equal numbers lay in every direction. Could Earth actually be at the center of the galaxy? A better explanation was put forth. Imagine that the trees in an enormous forest represent the stars in the galaxy. After hiking into this forest a short distance, you look around. What you see is an equal number of trees in every direction. Are you really in the center of the forest? Not necessarily; anywhere in the forest, except at the very edge, you will seem to be in the middle.

Structure of the Milky Way Galaxy

Attempts to visually inspect the Milky Way Galaxy are hindered by the large quantities of interstellar matter that lie in our line of sight. Nevertheless, with the aid of radio telescopes, the gross structure of our galaxy has been determined. The Milky Way is a rather large spiral galaxy whose disk is about 100,000 light-years wide and about 10,000 light-years thick at the nucleus (Figure 23.18). As viewed from Earth, the center of the galaxy lies beyond the constellation Sagittarius.

Radio telescopes reveal the existence of at least three distinct *spiral arms,* with some showing splintering (Figure 23.19). The Sun is positioned in one of these arms about two-thirds of the way from the center, at a distance from the hub of about 30,000 light-years. The stars in the arms of the Milky Way rotate around the *galactic nucleus,* with the most outward ones moving the slowest, such that the ends of the arms appear to trail. The Sun and the arm it is in require about 200 million years for each orbit around the nucleus.

Figure 23.16 This illustration shows how astronomers believe a binary pair (red giant/black hole) might function.

Figure 23.17   Panorama of our galaxy, the Milky Way. Notice the dark band caused by the presence of interstellar dark nebulae. (Courtesy of Axel Mellinger)

Surrounding the galactic disk is a nearly spherical *halo* made of very tenuous gas and numerous globular clusters. These star clusters do not participate in the rotating motion of the arms but rather have their own orbits that carry them through the disk. Although some clusters are very dense, they pass among the stars of the arms with plenty of room to spare.

Galaxies

In the mid-1700s, German philosopher Immanuel Kant proposed that the telescopically visible fuzzy patches of light scattered among the stars were actually distant galaxies like the Milky Way. Kant described them as island universes. Each galaxy, he believed, contained billions of stars and, as such, was a universe in itself. The weight of opinion, however, favored the hypothesis that they were dust and gas clouds (nebulae) within our galaxy.

This matter was not resolved until the 1920s, when American astronomer Edwin Hubble was able to locate, within one of these fuzzy patches, some unique stars that are known to be intrinsically very bright. Because these very bright stars appeared only very faintly in the telescope, Hubble believed they must lie outside the Milky Way.

This fuzzy patch, which lies over a million light-years away, was named the Great Galaxy in Andromeda (Figure 23.20). Hubble had extended the universe far beyond the limits of our imagination, to include hundreds of billions of galaxies, each containing hundreds of billions of stars. It has been said that a million galaxies are found in that portion of the sky bounded by the cup of the Big Dipper. There are more stars in the heavens than grains of sand in all the beaches on Earth.

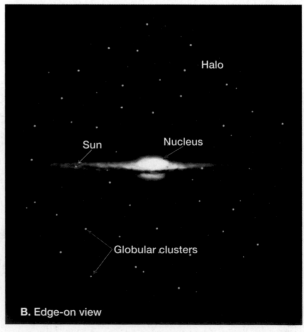

Figure 23.18   Structure of the visible portion of the Milky Way Galaxy.

Figure 23.19 If the Milky Way were photographed from a distance, it might appear like the spiral galaxy NGC 2997. (© Anglo-Australian Observatory, photography by David Malin)

Types of Galaxies

From the hundreds of billions of galaxies, several basic types have been identified: spiral, elliptical, and irregular.

The Milky Way and the Great Galaxy in Andromeda are examples of fairly large **spiral galaxies** (Figure 23.21). Andromeda can be seen with the unaided eye as a fuzzy fifth-magnitude object. Typically, spiral galaxies are disk-shaped, with a somewhat greater concentration of stars near their centers, but there are nu

Figure 23.20 Great Galaxy, a spiral galaxy, in the constellation Andromeda. The two bright spots to the left and right are dwarf elliptical galaxies. (Palomar Observatories/California Institute of Technology [Caltech])

merous variations. Viewed broadside, arms are often seen extending from the central nucleus and sweeping gracefully away. The outermost stars of these arms rotate most slowly, giving the galaxy the appearance of a fireworks pinwheel.

One type of spiral galaxy, however, has the stars arranged in the shape of a bar, which rotates as a rigid system. This requires that the outer stars move faster than the inner ones, a fact not easy for astronomers to reconcile with the laws of motion. Attached to each end of these bars are curved spiral arms. These have become known as **barred spiral galaxies** (Figure 23.22). Spiral galaxies are generally quite large, ranging from 20,000 to about 125,000 light-years in diameter. About 10 percent of all galaxies are thought to be barred spirals, and another 20 percent are regular spiral galaxies like the Milky Way.

The most abundant group, making up 60 percent of the total, is the **elliptical galaxies.** These are generally smaller than spiral galaxies. Some are so much smaller, in fact, that the term *dwarf* has been applied. Because these dwarf galaxies are not visible at great distances, a survey of the sky reveals more of the conspicuous large spiral galaxies. Although most elliptical galaxies are small, the very largest known galaxies (200,000 light-years in diameter) are also elliptical. As their name implies, elliptical galaxies have an ellipsoidal shape that ranges to nearly spherical, and they lack spiral arms. The two dwarf companions of Andromeda shown in Figure 23.20 are elliptical galaxies.

Only 10 percent of the known galaxies lack symmetry and are classified as **irregular galaxies.** The best-known irregular galaxies, the Large and Small Magellanic Clouds in the Southern Hemisphere, are easily visible with the unaided eye. Named after the ex-

A.

B.

Figure 23.21   Two views illustrating the idealized structure of spiral galaxies. (Courtesy of Hansen Planetarium/U.S. Naval Observatory)

plorer Ferdinand Magellan, who observed them when he circumnavigated Earth in 1520, they are our nearest galactic neighbors—only 150,000 light-years away.

One of the major differences among the galactic types is the age of the stars that make them up. The irregular galaxies are composed mostly of young stars, whereas the elliptical galaxies contain old stars. The Milky Way and other spiral galaxies consist of both young and old stars, with the youngest located in the arms.

Galactic Clusters

Once astronomers discovered that stars were associated in groups, they set out to determine whether galaxies also were grouped or just randomly distributed throughout the universe. They found that, like stars, galaxies are grouped in **galactic clusters** (Figure 23.23). Some abundant clusters contain thousands of galaxies. Our own, called the **Local Group,** contains at least 28

Figure 23.22   Barred spiral galaxy. (Palomar Observatories/California Institute of Technology [Caltech])

galaxies. Of these, 3 are spirals, 11 are irregulars, and 14 are ellipticals. Galactic clusters also reside in huge swarms called superclusters. From visual observations, it appears that superclusters may be the largest entities in the universe.

Red Shifts

You probably have noticed the change in pitch of a car horn or ambulance siren as it passes by. When it is approaching, the sound seems to have a higher-than-

Figure 23.23   A cluster of galaxies located about 1 million light-years from Earth. (Courtesy of NASA)

normal pitch, and when it is moving away, the pitch sounds lower than normal. This effect, which occurs for all wave motion, including sound and light waves, was first explained by Christian Doppler in 1842 and is called the *Doppler effect.* The reason for the difference in pitch is that it takes time for the wave to be emitted. If the source of the wave is moving away, the beginning of the wave is emitted nearer to you than the end of the wave, effectively "stretching" the wave. This gives it a longer wavelength (see Figure 22.4, p. 607). The opposite is true for an approaching source.

In the case of light, when a source is moving away, its light appears redder than it actually is, because its waves appear lengthened. Objects approaching have their light waves shifted toward the blue (shorter wavelength). Therefore, the Doppler effect reveals whether Earth and another celestial body are approaching or leaving one another. In addition, the amount of shift allows us to calculate the *rate* at which the relative movement is occurring. Large Doppler shifts indicate higher velocities; smaller Doppler shifts indicate lower velocities.

Expanding Universe

One of the most important discoveries of modern astronomy was made in 1929 by Edwin Hubble. Observations completed several years earlier revealed that most galaxies have Doppler shifts toward the red end of the spectrum. Recall that red shift occurs because the light waves are "stretched," indicating that Earth and the source are moving away from each other. Hubble set out to explain the predominance of red shift.

Hubble realized that dimmer galaxies were probably farther away than were brighter galaxies. Thus, he tried to determine whether a relation existed between the distances to galaxies and their red shifts. Using estimated distances based on relative brightness and the observed Doppler red shifts, Hubble discovered that galaxies that exhibit the greatest red shifts are the most distant.

A consequence of the universal red shift is that it predicts that most galaxies (except for a few nearby) are receding from us. Recall that the amount of Doppler red shift is dependent on the velocity at which the object is moving away. Greater red shifts indicate faster recessional velocities. Because more distant galaxies have greater red shifts, Hubble concluded that they must be retreating from us at greater velocities. This idea is currently termed **Hubble's law** and states that galaxies are receding from us at a speed that is proportional to their distance.

Hubble was surprised at this discovery because it implied that the most distant galaxies are moving away from us many times faster than those nearby. What type of cosmological theory can explain this fact? It was soon realized that an *expanding universe* can adequately account for the observed red shifts.

To help visualize the nature of this expanding universe, we will employ a popularly used analogy. Imagine a loaf of raisin bread dough that has been set out to rise for a few hours (Figure 23.24). As the dough doubles in size, so does the distance between all of the raisins. However, the raisins that were originally farther apart traveled a greater distance in the same time span than those located closer together. We therefore conclude that in an expanding universe, as in our analogy, those objects located farther apart move away from each other more rapidly.

Another feature of the expanding universe can be demonstrated using the raisin bread analogy. No matter which raisin you select, it will move away from all the other raisins. Likewise, no matter where one is located in the universe, every other galaxy (except those in the same cluster) will be receding. Edwin Hubble had indeed advanced our understanding of the universe. The Hubble Space Telescope is named in his honor.

The Big Bang

Did the universe have a beginning? Will it have an end? Cosmologists are trying to answer these questions, and that makes them a rare breed.

First and foremost, any viable theory regarding the origin of the universe must account for the fact that all galaxies (except for the very nearest) are moving away from us. Because all galaxies appear to be moving away from Earth, is our planet in the center of the universe? Probably not, because if we are not even in the center of our own solar system, and our solar system is not even

A. Raisin bread dough before it rises.

B. Raisin bread dough a few hours later.

6 cm   15 cm

2 cm   5 cm

Figure 23.24 Illustration of the raisin bread analogy of an expanding universe. As the dough rises, raisins originally farther apart travel a greater distance in the same time span as those located closer together. Thus, raisins (like galaxies in a uniform expanding universe) that are located farther apart move away from each other more rapidly than those located nearer to each other.

in the center of the galaxy, it seems unlikely that we could be in the center of the universe.

A more probable explanation exists: Imagine a balloon with paper-punch "dots" glued to its surface. When the balloon is inflated, each dot spreads apart from every other dot. Similarly, if the universe is expanding, every galaxy would be moving away from every other galaxy.

This concept of an expanding universe led to the widely accepted **big bang** theory. According to this theory, the entire universe was at one time confined to a dense, hot, supermassive ball. Then, about 15 billion years ago, a cataclysmic explosion occurred, hurling this material in all directions. The big bang marks the inception of the universe; all matter and space were created at that instant. The ejected masses of gas cooled and condensed, forming the stars that compose the galactic systems we now observe fleeing from their birthplace.

If the universe began with a big bang, how will it end? One view is that the universe will last forever. In this scenario, the stars will slowly burn out, being replaced by invisible degenerate matter and black holes that will travel outward through an endless, dark cold universe. The other possibility is that the outward flight of the galaxies will slow and eventually stop. Gravitational contraction would follow, causing the galaxies to collide and coalesce into the high-energy, high-density mass from which the universe began. This fiery death of the universe, the big bang operating in reverse, has been called the "big crunch."

Whether or not the universe will expand forever, or eventually collapse upon itself, depends on its average density. If the average density of the universe is more than an amount known as its *critical density* (about one atom for every cubic meter), the gravitational field is sufficient to stop the outward expansion and cause the universe to contract. On the other hand, if the density of the universe is less than the critical value, it will expand forever. Current estimates of the density of the universe place it below the critical density, which predicts an ever expanding, or *open universe*. Additional support for an open universe comes from studies that indicate the universe is expanding faster now than in the past. Hence, the view currently favored by most cosmologists is an expanding universe with no ending point.

It should be noted, however, that the methods used to determine the ultimate fate of the universe have substantial uncertainties. It is possible that previously undetected matter exists in great quantities in the universe. If this is so, the galaxies could, in fact, collapse in the "big crunch."

> Absence of evidence is not evidence of absence.
>
> —Anonymous

Chapter Summary

• One method for determining the distance to a star is to use a measurement called *stellar parallax*, the extremely slight back-and-forth shifting in a nearby star's position due to the orbital motion of Earth. *The farther away a star is, the less its parallax.* A unit used to express stellar distance is the *light-year*, which is the distance light travels in a year about 9.5 trillion kilometers (5.8 trillion miles).

• The intrinsic properties of stars include *brightness, color, temperature, mass,* and *size*. Three factors control the brightness of a star as seen from Earth: how big it is, how hot it is, and how far away it is. *Magnitude* is the measure of a star's brightness. *Apparent magnitude* is how bright a star appears when viewed from Earth. *Absolute magnitude* is the "true" brightness of a star if it were at a standard distance of about 32.6 light-years. The difference between the two magnitudes is directly related to a star's distance. Color is a manifestation of a star's temperature. Very hot stars (surface temperatures above 30,000 K) appear blue; red stars are much cooler (surface temperatures generally less than 3000 K). Stars with surface temperatures between 5000 and 6000 K appear yellow, like our Sun. The center of mass of orbiting *binary stars* (two stars revolving around a common center of mass under their mutual gravitational attraction) is used to determine the mass of the individual stars in a binary system.

• A *Hertzsprung-Russell diagram* is constructed by plotting the absolute magnitudes and temperatures of stars on a graph. A great deal about the sizes of stars can be learned from H-R diagrams. Stars located in the upper-right position of an H-R diagram are called *giants,* luminous stars of large radius. *Supergiants* are very large. Very small *white dwarf* stars are located in the lower-central portion of an H-R diagram. Ninety percent of all stars, called *main-sequence stars,* are in a band that runs from the upper-left corner to the lower-right corner of an H-R diagram.

• *Variable stars* fluctuate in brightness. Some, called *pulsating variables,* fluctuate regularly in brightness by expanding and contracting in size. When a star explosively brightens, it is called a *nova.* During the outburst, the outer layer of the star is ejected at high speed. After reaching maximum brightness in a few days, the nova slowly returns in a year or so to its original brightness.

• New stars are born out of enormous accumulations of dust and gases, called *nebula,* that are scattered between existing stars. A *bright nebula* glows because the matter is close to a very hot (blue) star. The two main types of bright nebulae are *emission nebulae* (which derive their visible light from the fluorescence of the ultraviolet light from a star in or near the nebula) and *reflection nebulae* (relatively dense dust clouds in interstellar space that are illuminated by reflecting the light of nearby stars). When a nebula is not close enough to a bright star to be illuminated, it is referred to as a *dark nebula.*

• Stars are born when their nuclear furnaces are ignited by the unimaginable pressures and temperatures in collapsing nebulae. New stars not yet hot enough for nuclear fusion are called *protostars.* When collapse causes the core of a protostar to reach a temperature of at least 10 million K, the fusion of hydrogen nuclei into helium nuclei begins in a process called *hydrogen burning.* The opposing forces acting on a star are *gravity* trying to contract it and *gas pressure (thermal nuclear energy)* trying to expand it. When the two forces are balanced, the star becomes a stable *main-sequence star.* When the hydrogen in a star's core is consumed, its outer envelope expands enormously and a *red giant* star, hundreds to thousands of times larger than its main-sequence size, forms. When all the usable nuclear fuel in these giants is exhausted and gravity takes over, the stellar remnant collapses into a small dense body.

• The *final fate of a star is determined by its mass.* Stars with less than one half the mass of the Sun collapse into hot, dense *white dwarf* stars. Medium-mass stars (between 0.5 and 3.0 times the mass of the Sun) become red giants, collapse, and end up as white dwarf stars, often surrounded by expanding spherical clouds of glowing gas called *planetary nebulae.* Stars more than three times the mass of the Sun terminate in a brilliant explosion called a *supernova.* Supernovae events can produce small, extremely dense *neutron stars,* composed entirely of subatomic particles called neutrons; or even smaller and more dense *black holes,* objects that have such immense gravity that light cannot escape their surface.

• The *Milky Way Galaxy* is a large, disk-shaped *spiral galaxy* about 100,000 light-years wide and about 10,000 light years thick at the center. There are three distinct *spiral arms* of stars, with some showing splintering. The Sun is positioned in one of these arms about two-thirds of the way from the galactic center, at a distance of about 30,000 light-years. Surrounding the galactic disk is a nearly spherical halo made of very tenuous gas and numerous *globular clusters* (nearly spherically shaped groups of densely packed stars).

• The various types of galaxies include (1) *irregular galaxies,* which lack symmetry and account for only 10 percent of the known galaxies; (2) *spiral galaxies,* which are typically disk-shaped with a somewhat greater concentration of stars near their centers, often containing arms of stars extending from their central nucleus; and (3) *elliptical galaxies,* the most abundant type, which have an ellipsoidal shape that ranges to nearly spherical and that lack spiral arms.

• Galaxies are not randomly distributed throughout the universe. They are grouped in *galactic clusters,* some containing thousands of galaxies. Our own, called the *Local Group,* contains at least 28 galaxies.

• By applying the *Doppler effect* (the apparent change in wavelength of radiation caused by the motions of the source and the observer) to the light of galaxies, galactic motion can be determined. Most galaxies have Doppler shifts toward the red end of the spectrum, indicating increasing distance. The amount of Doppler shift is dependent on the velocity at which the object is moving. Because the most distant galaxies have the greatest red shifts, Edwin Hubble concluded in the early 1900s that they were retreating from us with greater recessional velocities than were more nearby galaxies. It was soon realized that an *expanding universe* can adequately account for the observed red shifts.

• The belief in the expanding universe led to the widely accepted *Big Bang Theory.* According to this theory, the entire universe was at one time confined in a dense, hot, supermassive concentration. About 15 billion years ago, a cataclysmic explosion hurled this material in all directions, creating all matter and space. Eventually the ejected masses of gas cooled and condensed, forming the stellar systems we now observe fleeing from their place of origin.

Key Terms

absolute magnitude (p. 626)
apparent magnitude (p. 625)
barred spiral galaxy (p. 641)
big bang (p. 643)
black hole (p. 637)
bright nebula (p. 629)
dark nebula (p. 630)

degenerate matter (p. 636)
elliptical galaxy (p. 641)
emission nebula (p. 629)
eruptive variables (p. 628)
galactic cluster (p. 642)
Hertzsprung-Russell (H-R) diagram
    (p. 627)

Hubble's law (p. 643)
hydrogen burning (p. 633)
interstellar dust (p. 629)
irregular galaxy (p. 641)
light-year (p. 625)
Local Group (p. 642)
magnitude (p. 625)

main-sequence stars (p. 627)	protostar (p. 631)	spiral galaxy (p. 641)
nebula (p. 629)	pulsar (p. 637)	stellar parallax (p. 624)
neutron star (p. 637)	pulsating variables (p. 628)	supergiant (p. 628)
nova (p. 629)	red giant (p. 627)	supernova (p. 634)
planetary nebula (p. 634)	reflection nebula (p. 629)	white dwarf (p. 628)

Review Questions

1. How far away in light-years is our nearest stellar neighbor, Proxima Centauri? Convert your answer to kilometers.

2. What is the most basic method of determining stellar distances?

3. Explain the difference between a star's apparent and absolute magnitudes. Which one is an intrinsic property of a star?

4. What is the ratio of brightness between a twelfth-magnitude and fifteenth-magnitude star?

5. What information about a star can be determined from its color?

6. What color are the hottest stars? Medium-temperature stars? Coolest stars?

7. Which property of a star can be determined from binary-star systems?

8. Make a generalization relating the mass and luminosity of main-sequence stars.

9. The disk of a star cannot be resolved telescopically. Explain the method that astronomers have used to estimate the size of stars.

10. Where on an H-R diagram does a star spend most of its lifetime?

11. How does the Sun compare in size and brightness to other main-sequence stars?

12. Why is interstellar matter important to stellar evolution?

13. Compare a bright nebula and a dark nebula.

14. What element is the fuel for main-sequence stars? Red giants?

15. What causes a star to become a giant?

16. Why are less massive stars thought to age more slowly than more massive stars, even though they have much less "fuel"?

17. Enumerate the steps thought to be involved in the evolution of Sun-like stars.

18. What is the final state of a low-mass (red) main-sequence star?

19. What is the final state of a medium-mass (Sun-like) star?

20. How do the "lives" of the most massive stars end? What are the two possible products of this event?

21. Describe the general structure of the Milky Way Galaxy.

22. Compare the three general types of galaxies.

23. Explain why astronomers consider elliptical galaxies more abundant than spiral galaxies, even though more spiral galaxies have been sighted.

24. How did Edwin Hubble determine that the Great Galaxy in Andromeda is located beyond our galaxy?

25. What evidence supports the Big Bang Theory?

Examining the Earth System

1. Briefly describe how the atmosphere, hydrosphere, solid Earth, and biosphere are each related to the death of stars that occurred billions of years ago.

2. If a supernova explosion were to occur within the immediate vicinity of our solar system, what might be some possible consequences of the intense x-ray and gamma radiation that would reach Earth?

3. Scientists are continuously searching the Milky Way Galaxy for other stars that may have planets. What types of stars would most likely have a planet or planets suit-able for life as we know it? If you would like to investigate extra-solar planets on-line, you might find these two Websites helpful: *NOVA Online* at **http://www.pbs.org/wgbh/nova/worlds/** and the *Electronic Universe Project* at **http://zebu.uoregon.edu/galaxy.html**

4. Based upon your knowledge of the Earth system, the planets in our solar system, and the cosmos in general, speculate about the likelihood that extra-solar planets exist with atmospheres, hydrospheres, lithospheres, and biospheres, similar to Earth's. Explain your speculation.

Web Resources

 The *Earth Science* Website uses the resources and flexibility of the Internet to aid in your study of the topics in this chapter. Written and developed by Earth science instructors, this site will help improve your understanding of Earth science. Visit **http://www.prenhall.com/tarbuck** and click on the cover of *Earth Science 10e* to find:

- **On-line review quizzes.**
- **Web-based critical thinking and writing exercises.**
- **Links to chapter-specific Web resources.**
- **Internet-wide key term searches.**

http://www.prenhall.com/tarbuck

APPENDIX A
Metric and English Units Compared

Units

1 kilometer (km)	= 1000 meters (m)
1 meter (m)	= 100 centimeters
1 centimeter (cm)	= 0.39 inch (in.)
1 mile (mi)	= 5280 feet (ft)
1 foot (ft)	= 12 inches (in.)
1 inch (in.)	= 2.54 centimeters (cm)
1 square mile(mi^2)	= 640 acres (a)
1 kilogram (kg)	= 1000 grams (g)
1 pound (lb)	= 16 ounces (oz)
1 fathom	= 6 feet (ft)

Conversions

Length

When you want to convert: **multiply by:** **to find:**

to convert:	multiply by:	to find:
inches	2.54	centimeters
centimeters	0.39	inches
feet	0.30	meters
meters	3.28	feet
yards	0.91	meters
meters	1.09	yards
miles	1.61	kilometers
kilometers	0.62	miles

Area

When you want to convert: **multiply by:** **to find:**

to convert:	multiply by:	to find:
square inches	6.45	square centimeters
square centimeters	0.15	square inches
square feet	0.09	square meters
square meters	10.76	square feet
square miles	2.59	square kilometers
square kilometers	0.39	square miles

Volume

When you want to convert: **multiply by:** **to find:**

to convert:	multiply by:	to find:
cubic inches	16.38	cubic centimeters
cubic centimeters	0.06	cubic inches
cubic feet	0.028	cubic meters
cubic meters	35.3	cubic feet
cubic miles	4.17	cubic kilometers
cubic kilometers	0.24	cubic miles
liters	1.06	quarts
liters	0.26	gallons
gallons	3.78	liters

Masses and Weights

When you want to convert: **multiply by:** **to find:**

to convert:	multiply by:	to find:
ounces	20.33	grams
grams	0.035	ounces
pounds	0.45	kilograms
kilograms	2.205	pounds

Temperature

When you want to convert degrees Fahrenheit (°F) to degrees Celsius (°C), subtract 32 degrees and divide by 1.8.

When you want to convert degrees Celsius (°C) to degrees Fahrenheit (°F), multiply by 1.8 and add 32 degrees.

When you want to convert degrees Celsius (°C) to kelvins (K), delete the degree symbol and add 273. When you want to convert kelvins (K) to degrees Celsius (°C), add the degree symbol and subtract 273.

Figure A.1 Temperature scales.

Figure B.1 Earth's grid system.

A glance at any globe reveals a series of north-south and east-west lines that together make up Earth's grid system, a universally used scheme for locating points on Earth's surface. The north-south lines of the grid are called **meridians** and extend from pole to pole (Figure B.1). All are halves of great circles. A **great circle** is the largest possible circle that may be drawn on a globe; if a globe were sliced along one of these circles, it would be divided into two equal parts called **hemispheres**. By viewing a globe or Figure B.1, you can see that meridians are spaced farthest apart at the equator and converge toward the poles. The east-west lines (circles) of the grid are called **parallels**. As their name implies, these circles are parallel to one another (Figure B.1). Whereas all meridians are parts of great circles, all parallels are not. In fact, only one parallel, the equator, is a great circle.

Latitude and Longitude

Latitude may be defined as distance, measured in degrees, *north* and *south* of the equator. Parallels are used to show latitude. Because all points that lie along the same parallel are an identical distance from the equator, they all have the same latitude designation. The latitude of the equator is 0 degrees, whereas the north and south poles lie 90 degrees N and 90 degrees S, respectively.

Longitude is defined as distance, measured in degrees, *east* and *west* of the zero or *prime meridian*. Because all meridians are identical, the choice of a zero line is obviously arbitrary. However, the meridian that passes through the Royal Observatory at Greenwich, England, is universally accepted as the reference meridian. Thus, the longitude for anyplace on the globe is measured east or west from this line. Longitude can vary from 0 degrees along the prime meridian to 180 degrees, halfway around the globe.

It is important to remember that when a location is specified, directions must be given—that is, north or south latitude and east or west longitude (Figure B.2). If this is not done, more than one point on the globe is being designated. The only exceptions, of course, are places that lie along the equator, the prime meridian, or the 180-degree meridian. It should also be noted that while it is not incorrect to use fractions, a degree of latitude or longitude is usually divided into minutes and seconds. A minute (') is $1/60^{th}$ of a degree, and a second (") is $1/60^{th}$ of a minute. When locating a place on a map, the degree of exactness will depend on the scale of the map. When using a small-scale world map or globe, it may be difficult to estimate latitude and longitude to the nearest whole degree or two. On the other hand, when a large-scale map of an area is used, it is often possible to estimate latitude and longitude to the nearest minute or second.

Distance Measurement

The length of a degree of longitude depends on where the measurement is taken. Along the equator, which is a great circle, a degree of east-west distance is equal to approximately 111 kilometers (69 miles). This figure is found by dividing Earth's circumference—40,075 kilometers (24,900 miles)—by 360. However, with an increase in latitude, the parallels become smaller, and the length of a degree of longitude diminishes (see Table B.1). Thus, at about latitude 60 degrees N and S, a degree of longitude has a value equal to about half of what it was at the equator.

As all meridians are halves of great circles, a degree of latitude is equal to about 111 kilometers (69 miles), just as a degree of longitude along the equator is. However, Earth is not a perfect sphere but is slightly flattened at the poles and bulges slightly at the equator. Because of this, there are small differences in the length of a degree of latitude.

Determining the shortest distance between two points on a globe can be done easily and fairly accurately using the globe-and-string method. It should be noted here that the arc of a great circle is the shortest distance between two points on a sphere. To determine the great circle distance (as well as observe the great circle route) between two places, stretch a piece of string between the locations in question. Then, measure the length of the string along the equator (since it is a great circle with degrees marked on it) to determine the number of degrees between the two points. To calculate the distance in kilometers or miles, simply multiply the number of degrees by 111 or 69, respectively.

Figure B.2 Locating places using the grid system. For both diagrams: Point *A* is latitude 30 degrees N, longitude 60 degrees E; Point *B* is latitude 30 degrees S, longitude 30 degrees W; Point *C* is latitude 0 degrees, longitude 90 degrees W; Point *D* is latitude 45 degrees N, longitude 75 degrees W; Point *E* is approximately latitude 10 degrees N, longitude 25 degrees W.

Table B.1 Longitude as distance.

°Lat.	Length of 1° Long. km	Length of 1° Long. miles	°Lat.	Length of 1° Long. km	Length of 1° Long. miles	°Lat.	Length of 1° Long. km	Length of 1° Long. miles
0	111.367	69.172	30	96.528	59.955	60	55.825	34.674
1	111.349	69.161	31	95.545	59.345	61	54.131	33.622
2	111.298	69.129	32	94.533	58.716	62	52.422	62.560
3	111.214	69.077	33	93.493	58.070	63	50.696	31.488
4	111.096	69.004	34	92.425	57.407	64	48.954	30.406
5	110.945	68.910	35	91.327	56.725	65	47.196	29.314
6	110.760	68.795	36	90.203	56.027	66	45.426	28.215
7	110.543	68.660	37	89.051	55.311	67	43.639	27.105
8	110.290	68.503	38	87.871	54.578	68	41.841	25.988
9	110.003	68.325	39	86.665	53.829	69	40.028	24.862
10	109.686	68.128	40	85.431	53.063	70	38.204	23.729
11	109.333	67.909	41	84.171	52.280	71	36.368	22.589
12	108.949	67.670	42	82.886	51.482	72	34.520	21.441
13	108.530	67.410	43	81.575	50.668	73	32.662	20.287
14	108.079	67.130	44	80.241	49.839	74	30.793	19.126
15	107.596	66.830	45	78.880	48.994	75	28.914	17.959
16	107.079	66.509	46	77.497	48.135	76	27.029	16.788
17	106.530	66.168	47	76.089	47.260	77	25.134	15.611
18	105.949	65.807	48	74.659	46.372	78	23.229	14.428
19	105.337	65.427	49	73.203	45.468	79	21.320	13.242
20	104.692	65.026	50	71.727	44.551	80	19.402	12.051
21	104.014	64.605	51	70.228	43.620	81	17.480	10.857
22	103.306	64.165	52	68.708	42.676	82	15.551	9.659
23	102.565	63.705	53	67.168	41.719	83	13.617	8.458
24	101.795	63.227	54	65.604	40.748	84	11.681	7.255
25	100.994	62.729	55	64.022	39.765	85	9.739	6.049
26	100.160	62.211	56	62.420	38.770	86	7.796	4.842
27	99.297	61.675	57	60.798	37.763	87	5.849	3.633
28	98.405	61.121	58	59.159	36.745	88	3.899	2.422
29	97.481	60.547	59	57.501	35.715	89	1.950	1.211
30	96.528	59.955	60	55.825	34.674	90	0.000	0.000

APPENDIX C
World Soils

Figure C.1 shows the generalized pattern of global soil orders according to the *Comprehensive Soil Classification System* (CSCS). It should be examined in conjunction with Table C.1, which briefly describes each of the soil orders depicted on the map. To avoid subjective decisions as to classification (a problem that plagued earlier systems), the CSCS defined its classes strictly in terms of soil characterisitics. That is, it is based on features that can be observed or inferred.

The CSCS uses a hierarchy of six categories, or levels. The system recognizes 10 major global *orders* that can be further subdivided into *suborders*, *great groups*, *subgroups*, *families*, and *series*. Note, however, that on the scale of a world map such as Figure C.1, only the largest units (soil orders) can be shown and then only in an extremely generalized way. Although the distribution pattern of major soil orders is more complex than can be shown in Figure C.1, the major distinguishing regional properties of world soils are depicted.

Table C.1 World soil orders.

Entisols	Youngest soils on the Earth. Just beginning to develop in response to the weathering phenomena in the environment. Do not display natural horizons. Found in all climates. They weather slowly over thousands of years; consequently, volcanic ash deposits or sand deposits form the basis for entisols.
Vertisols	Soils containing large amounts of clay, which shrink upon drying and swell with the addition of water. Found in subhumid to arid climates, provided that adequate supplies of water are available to saturate the soil after periods of drought. Soil expansion and contraction exert stresses on human structures.
Inceptisols	Young soils that reveal developmental characteristics (horizons) in response to climate and vegetation. Exist from the Arctic to the tropics on young land surfaces. Common in alpine areas, on river floodplains, in stables and dune areas, and in areas once glaciated.
Aridsols	Soils that develop in dry places, such as the desert, where water—precipitation and groundwater—is insufficient to remove soluble minerals. Frequently irrigated for intensive agricultural production, although salt accumulation poses a problem.
Mollisols	Dark, soft soils that have developed under grass vegetation, generally found in prairie areas. Soil fertility is excellent because potential evaporation generally exceeds precipitation. Also found in hardwood forests with significant earthworm activity. Climatic range is boreal or alpine to tropical. Dry seasons are normal.
Spodosols	Soils found only in humid regions on sandy material. Range from the boreal coniferous forests into tropical forests. Beneath the dark upper horizon of weathered organic material lies a light-colored horizon of leached material, the distinctive property of this soil.
Alfisols	Mineral soils that form under boreal forests or broadleaf deciduous forests, rich in iron and aluminum. Clay particles accumulate in a subsurface layer in response to leaching in moist environments. Fertile, productive soils, because they are neither too wet nor too dry.
Ultisols	Soils that represent the products of long periods of weathering. Water percolating through the soil concentrates clay particles in the lower horizons (argillic horizons). Restricted to humid climates in the temperate regions and the tropics where the growing season is long. Abundant water and a long frost-free period contribute to extensive leaching, hence poorer soil quality.
Oxisols	Soils that occur on old land surfaces unless parent materials were strongly weathered before they were deposited. Generally found in the tropics and subtropical regions. Rich in iron and aluminum oxides, oxisols are heavily leached; hence are poor soils for agricultural activity. Few, if any, exist in the United States.
Histosols	Organic soils with little or no climatic implications. Can be found in any climate where organic debris can accumulate to form a bog soil. Dark, partially decomposed organic material commonly referred to as *peat*.

Source: Robert E. Norris et al., *Geography: An Introductory Perspective*, Columbus, Ohio: Merrill, 1982.

Figure C.1 Soil distribution. The pattern of global soil orders is remarkably similar to the pattern of major climates. Terminology employed is from the *Comprehensive Soil Classification System*.

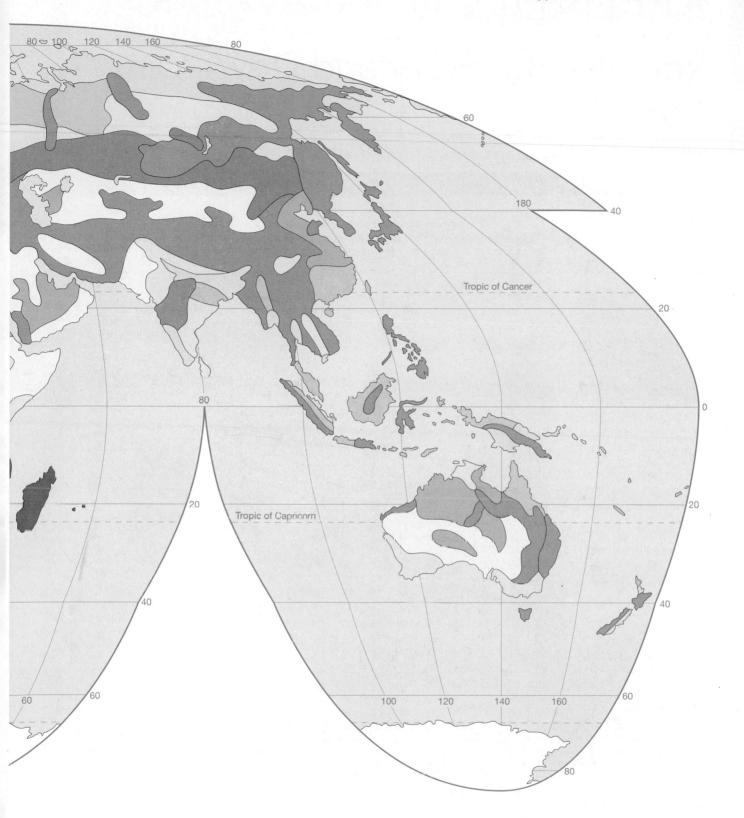

Tropic of Cancer

Tropic of Capricorn

APPENDIX D
Relative Humidity and Dew-Point Tables

Table D.1 Relative humidity (percent)

Dry bulb (°C)	Depression of Wet-Bulb Temperature (Dry-Bulb Temperature Minus Wet Bulb Temperature = Depression of the Wet Bulb)																					
	1	2	3	4	5	6	7	8	9	10	11	12	13	14	15	16	17	18	19	20	21	22
−20	28																					
−18	40																					
−16	48	0																				
−14	55	11																				
−12	61	23																				
−10	66	33	0																			
−8	71	41	13																			
−6	73	48	20	0																		
−4	77	54	32	11																		
−2	79	58	37	20	1																	
0	81	63	45	28	11																	
2	83	67	51	36	20	6																
4	85	70	56	42	27	14																
6	86	72	59	46	35	22	10	0														
8	87	74	62	51	39	28	17	6														
10	88	76	65	54	43	38	24	13	4													
12	88	78	67	57	48	38	28	19	10	2												
14	89	79	69	60	50	41	33	25	16	8	1											
16	90	80	77	62	54	45	37	29	21	74	7	1										
18	91	81	72	64	56	48	40	33	26	19	12	6	0									
20	91	82	74	66	58	51	44	36	30	23	17	11	5									
22	92	83	75	68	60	53	46	40	33	27	21	15	10	4	0							
24	92	84	76	69	62	55	49	42	36	30	25	20	14	9	4	0						
26	92	85	77	70	64	57	51	45	39	34	28	23	18	13	9	5						
28	93	86	78	71	65	59	53	45	42	36	31	26	21	17	12	8	4					
30	93	86	79	72	66	61	55	49	44	39	34	29	25	20	16	12	8	4				
32	93	86	80	73	68	62	56	51	46	41	36	32	27	22	19	14	11	8	4			
34	93	86	81	74	69	63	58	52	48	43	38	34	30	26	22	18	14	11	8	5		
36	94	87	81	75	69	64	59	54	50	44	40	36	32	28	24	21	17	13	10	7	4	
38	94	87	82	76	70	66	60	55	51	46	42	38	34	30	26	23	20	16	13	10	7	5
40	94	89	82	76	71	67	61	57	52	48	44	40	36	33	29	25	22	19	16	13	10	7

Relative Humidity Values

*To determine the relative humidity, find the air (dry-bulb) temperature on the vertical axis (far left) and the depression of the wet bulb on the horizontal axis (top). Where the two meet, the relative humidity is found. For example, when the dry-bulb temperature is 20°C and a wet-bulb temperature is 14°C, then the depression of the wet-bulb is 6°C (20°C − 14°C). From Table D-1, the relative humidity is 51 percent and from Table D-2, the dew point is 10°C.